Springer Tracts in Civil Engineering (STCE) publishes the latest developments in Civil Engineering—quickly, informally and in top quality. The series scope includes monographs, professional books, graduate textbooks and edited volumes, as well as outstanding PhD theses. Its goal is to cover all the main branches of civil engineering, both theoretical and applied, including:

- Construction and Structural Mechanics
- Building Materials
- Concrete, Steel and Timber Structures
- Geotechnical Engineering
- Earthquake Engineering
- Coastal Engineering; Ocean and Offshore Engineering
- Hydraulics, Hydrology and Water Resources Engineering
- Environmental Engineering and Sustainability
- Structural Health and Monitoring
- Surveying and Geographical Information Systems
- Heating, Ventilation and Air Conditioning (HVAC)
- Transportation and Traffic
- Risk Analysis
- Safety and Security

Indexed by Scopus

To submit a proposal or request further information, please contact: Pierpaolo Riva at Pierpaolo.Riva@springer.com, or Li Shen at Li.Shen@springer.com

More information about this series at http://www.springer.com/series/15088

Springer Tracts in Civil Engineering

Aiqun Li

Vibration Control for Building Structures

Theory and Applications

Springer

Aiqun Li
Beijing University of Civil Engineering
and Architecture
Beijing, China

ISSN 2366-259X ISSN 2366-2603 (electronic)
Springer Tracts in Civil Engineering
ISBN 978-3-030-40789-6 ISBN 978-3-030-40790-2 (eBook)
https://doi.org/10.1007/978-3-030-40790-2

This Springer imprint is published by the registered company Springer Nature Switzerland AG
The registered company address is: Gewerbestrasse 11, 6330 Cham, Switzerland

Preface

The author graduated from the Department of Civil Engineering, Southeast University of China, in December 1992, majored in structural engineering, and obtained the doctor's degree. Then, the author taught in Southeast University from 1993 to 2015, and now works in Beijing University of Civil Engineering and Architecture from 2015.

Since 1990, the author has paid attention to and entered the research field of structural vibration control, which has lasted for 30 years. In retrospect, the initial research only focused on the subject. With the deepening of the research, the questions that often linger in the author's mind include: how to scientifically recognize and describe the strong earthquakes and hurricanes; how to face the randomness and destructiveness of strong earthquakes and hurricanes; how to ensure the performance-based designs of building structure system and its resistances to earthquake and wind; and how to study appropriate high-performance vibration reduction and isolation technologies to ensure the building structure has higher and better disaster prevention ability.

According to the fortification goal of "no damage under small earthquakes and no collapse under large earthquakes," the houses under a strong earthquake are already "standing ruins." How to ensure that the houses on which people live are safe under the large earthquake and strong wind should be the common expectation of people in modern society.

The disaster investigation and experience of previous large earthquakes and gales show that earthquakes and gales are random and destructive. By improving the anti-seismic and anti-wind abilities of buildings, or expressed as, as long as buildings have the ability to resist large earthquakes and gales, buildings will certainly become a real "safe and secure beautiful home."

With the rapid development of the urbanization process of human society, the building structure has been developing toward the direction of higher height, larger span, and more complex structure. However, once a strong earthquake or hurricane occurs, whether these important buildings have the proper anti-vibration ability will test our managers, designers, and construction engineers. For the buildings located in high-intensity areas, buildings pursuing high-performance structures, hospital buildings, school buildings, lifeline buildings, no matter whether they are new or existing, the structural vibration reduction and isolation technology is presumably an important technical choice to make them have better seismic capacity.

This book is the part of the author's periodical academic achievements (1990–2019) in the research of structural vibration reduction control, including four parts: the basic principle of structural vibration reduction control, structural vibration reduction device, structural vibration reduction design method and structural vibration reduction engineering practice.

The theory, method, technology, and application in this book can also be used as reference for other engineering structure vibration reduction research and practice.

The research work of the author has been greatly supported by the National Natural Science Foundation of China (59238160, 50038010, 59408012, 59978009, and 51438002), the Key Projects in the National Science and Technology Pillar Program (2006BAJ03A04), and the National Key Research and Development Program of China (2017YFC0703602).

Thanks to Dr. Chen Xin, Dr. Zhou Guangpan, and Dr. Deng Yang in the author's team for their important contributions to the publication of this book.

Thanks to Dr. Jia Junbo, alumnus of Southeast University and academician of Norwegian Academy of Engineering, for his important contribution to the publication of this book.

I would like to dedicate this book to my two respected teachers who have passed away: Prof. Ding Dajun of Southeast University, a famous expert of civil engineering, and Prof. Cheng Wenrang of Southeast University, a famous expert of high-rise building structure.

In the process of research and writing, the author has learned and referred to a large number of works at home and abroad. I would like to extend my sincere thanks and respect to the original author!

Beijing, China
December 2019

Aiqun Li

Contents

Chapter 1
Summary

Abstract While the construction of civil engineering is developing towards the direction of higher height, larger span and more complex structure, structural vibration control technology has gradually become one of the most effective technical means to resist strong earthquakes and hurricanes due to its safety, economy and effective characteristics. This summary includes the following three parts: 1. Concept and Principle of Structural Vibration Control. Based on the basic principles of structural dynamics and earthquake engineering, the control mechanism of vibration reduction and base isolated structure is explained. 2. Classification and Basic Performance of Structural Vibration Control. The vibration control technology of building structures is divided into energy dissipation and vibration reduction technology, frequency modulation technology and isolation technology. The basic performance of these technologies is analyzed. 3. Development and Current Situation of Structural Vibration Control. Based on the analysis of current situation, the research of new high-performance damping device, the improvement of current design theory and method of vibration control are considered as the development trends in near future.

Global natural disasters occur frequently, resulting in the rapid growth of human life and property losses. In the global scope, during the decade of 2006–2015, natural disasters affect an average of 224 million people every year, making nearly 70,000 people lose their lives, while causing more than $135 billion of economic losses. Earthquake, wind and other dynamic disasters are one of the most destructive disasters. In recent years, strong earthquakes, hurricanes and other disasters have caused huge casualties and property losses. For example, Hurricane Katrina swept through New Orleans, USA in August 2005, causing economic losses of at least $75 billion, at least 1836 people lost their lives. In 2010, Haiti's 7.3-magnitude earthquake basically destroyed its capital, about 300,000 people died.

China has vast territory and rich resources, and is one of the countries suffering from the most natural disasters. According to the official statistics, there are 6.8 typhoons landing in China every year from the Western Pacific Ocean and the South China Sea. In the past 10 years, there have been many large-scale earthquakes in China, each time causing huge losses to people's lives and property, such as the 2008

A. Li, *Vibration Control for Building Structures*, Springer Tracts in Civil Engineering, https://doi.org/10.1007/978-3-030-40790-2_1

Wenchuan earthquake in Sichuan Province, with a magnitude of 8.0, 69,000 people were killed and 370,000 people were injured; 2010 Yushu earthquake in Qinghai Province, with a magnitude of 7.1, killed 2698 people. Further statistics show that in 2018 alone, the earthquake disaster caused more than 5000 houses to collapse, 31,000 houses to be seriously damaged, 108,000 houses to be generally damaged, with a direct economic loss of 2.9 billion yuan; the typhoon disaster caused 32.066 million people to be affected, 80 people died, 3 people missing, 24,000 houses to collapse, 43,000 houses to be severely damaged, and 162,000 houses to be generally damaged. The direct economic loss is 69.73 billion yuan. Every strong earthquake or hurricane will cause a lot of damage and collapse of civil engineering structures, and directly threaten the safety of people's lives and property, which brings a huge challenge to our scholars in the field of civil engineering.

With the continuous rise of national comprehensive strength, the construction of civil engineering is developing towards the direction of higher height, larger span and more complex structure, which is not only the development opportunity of civil engineering, but also a huge challenge to our research. At the same time, the rapid development of science and technology has brought many new perspectives to the disaster prevention and mitigation of civil engineering structures. Among them, structural vibration control technology has gradually become one of the most effective technical means to resist strong earthquakes and hurricanes due to its safety, economy and effective characteristics [1–3]. The concept of structural vibration control was first proposed by J. T. P. Yao, a Chinese American scholar [4]. After more than 40 years of development, it has gradually formed a new frontier research field including passive control, active control, semi-active control, hybrid control and isolation, involving dynamics, cybernetics, computer science, advanced materials and other disciplines [5–8].

1.1 Concept and Principle of Structural Vibration Control

In the early stage, in order to improve the ability of single engineering structure to resist earthquake, wind and other dynamic disasters, people determined the dynamic load of the building according to the site conditions of the building, the characteristics of the building itself and its functional requirements, and then carried out structural analysis, strength design and deformation calculation to ensure that the structure can meet the requirements of bearing capacity, deformation capacity and stability under the action of dynamic disasters. Especially when the earthquake intensity is large, the structure will enter the elastic–plastic state, so the structure is required to have enough elastic–plastic deformation capacity and ductility, in order to consume the earthquake energy, reduce the earthquake response, and make the structure "crack but not fall", which is the traditional seismic design method. The traditional design method of structural disaster resistance is to improve the disaster resistance performance of the structure itself by increasing the cross-section of the component, improving the bearing capacity and deformation capacity of the structure or the component. When the dynamic disaster occurs, the vibration energy absorbed is converted, stored and consumed by the movement of the structure and the deformation and damage of the structural component. Although this method has played a great role in improving the ability of structure to resist dynamic disasters, there are also a series of problems, such as difficult to ensure the safety of structure, limited adaptability, poor economy and difficult to repair after earthquake. Therefore, based on the traditional design theory and technology of structural disaster resistance, people put forward a more reasonable, safe and economic new theory and technology of structural disaster resistance, in order to ensure the safety of all kinds of structures under earthquake and strong wind.

Structural vibration control is such a new technology. Its concept was first used in military industry, aerospace industry and other fields. Since the 1970s, people gradually began to use it in the vibration control of civil engineering structures [7, 9, 10]. The structural vibration reduction control can be expressed as: the vibration response of the structure can be effectively controlled by reasonably setting the isolation or damping device in the structure, so that the response values of the structure under the action of earthquake, gale or other dynamic interference can be controlled within the allowable range [11, 12].

1.1.1 Structure Damping Principle

The idea of structural vibration reduction is to passively or actively provide the structure with a control force opposite to its motion direction, so as to reduce its motion response. Its basic principle can be simply explained by the following structural dynamic equation [11]:

$$[M]\{\ddot{x}(t)\} + [C]\{\dot{x}(t)\} + [K]\{x(t)\} + \{u(t)\} = F(t) \tag{1.1}$$

Among them, $\{F(t)\}$ is the external load vector, $\{x(t)\}$, $\{\dot{x}(t)\}$ and $\{\ddot{x}(t)\}$ are respectively the displacement, velocity and acceleration vectors of the particles, $[C]$, $[M]$ and $[K]$ are respectively the mass, damping and stiffness matrices of equivalent multiple degrees of freedom, $\{u(t)\}$ is the control force vectors, which can be expressed as the passive, active and semi-active control forces and the control forces provided by the frequency modulation damping device.

Structural vibration reduction is to change or adjust the dynamic characteristics or dynamic action of the structure by adjusting the natural frequency ω or natural period T of the structure (by changing $[K]$ and $[M]$) or increasing the damping $[C]$, or applying the control force, so as to reduce the response of the structure under the dynamic loads such as earthquake and wind. $\{\ddot{x}(t)\}_{\max}$, $\{\dot{x}(t)\}_{\max}$ and $\{x(t)\}_{\max}$ are the allowable structural acceleration, velocity and displacement response values to ensure the safety of the structure and the people, equipment and decoration facilities in the structure and in normal use, the dynamic response of the structure shall meet the following requirements:

$$\{\ddot{x}(t)\} \le \{\ddot{x}\}_{\max},\ \{\dot{x}(t)\} \le \{\dot{x}\}_{\max},\ \{x(t)\} \le \{x\}_{\max} \tag{1.2}$$

1.1.2 Structure Isolation Principle

Different from structural vibration reduction, the idea of structural isolation comes from the idea of separating the structure from the ground, so that even if the ground vibrates, the building will not be affected. Therefore, the period of the structure can be extended by introducing the isolation device, to avoid the frequency band of relatively concentrated seismic energy, to change the dynamic characteristics of the structure, and to restrain the displacement of the structure using the energy dissipation device, so as to achieve the goal of overall reducing the dynamic response of the structure. Although there are buildings built according to this idea in ancient times, the realization of modern isolation technology is only about 40 years old [13].

Based on the basic principles of structural dynamics and earthquake engineering, the mechanism of vibration reduction of base isolated structure can be explained. The predominant period of the typical ground motion is about 0.1–1.0 s, so the middle and low-rise structures with natural vibration period of 0.1–1.0 s are prone to resonance and damage during the earthquake. By reducing the stiffness of the structure through the isolation system, increasing the natural vibration period of the structure, thus avoiding the predominant period of ground motion, avoiding resonance and approaching resonance, the seismic action of the upper structure can be reduced to a large extent, so as to achieve the purpose of isolation. As shown in Fig. 1.1, generally, the middle and low-rise buildings have large rigidity and short period, and the seismic response of the structure is at point *A* of the seismic response spectrum [1, 14]. At point *A*, the acceleration response spectrum value of the structure is large, and the earthquake force on the structure is large. However, due to the large stiffness of the structure itself and the small displacement response spectrum value, the structure is in an elastic–plastic working state during a large earthquake, and the structure mainly consumes the input energy through the destruction of its own components. Compared with the aseismic structure, the natural vibration period of the isolation structure is greatly increased through setting the flexible isolation pad in the isolation layer, and the seismic response of the structure is located at the point *B* of the seismic response spectrum. At point *B*, the acceleration response spectrum value of the structure is relatively small, and the seismic force on the structure is relatively small. However, due to the existence of flexible isolation layer, the overall displacement of the structure is very large and mainly concentrated in the isolation layer, and the interlayer deformation of the upper structure is very small, which is in the overall translational state. For the isolated structure at point *B*, the seismic force

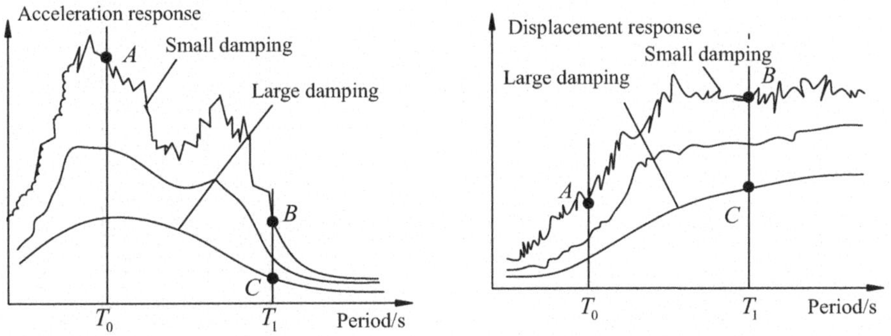

(a) Acceleration response spectrum (b) Displacement response spectrum

Fig. 1.1 Seismic response spectrum of structure

is greatly reduced, but the isolation layer may exceed the allowable displacement. In order to control the displacement, many kinds of dampers can be set in the isolation layer. This kind of isolation structure is located at the point *C* of the seismic response spectrum. At point *C*, not only the seismic force of the isolation structure is further reduced, but also the overall displacement of the isolation structure is greatly reduced, and the displacement of the isolation layer is effectively controlled.

1.2 Classification and Basic Performance of Structural Vibration Control Technology

According to the technical methods, the vibration control of building structure technology can be generally divided into isolation technology, energy dissipation technology, mass frequency modulation (FM) technology, active control technology, semi-active control technology, hybrid control technology and intelligent control technology; according to whether there is external energy input, it can be divided into passive control, active control and semi-active control [1, 11, 15–17]. This book comprehensively considers the characteristics of vibration reduction technology, installation position and dynamic principle, and divides the vibration reduction control technology of building structure into three categories: energy dissipation and vibration reduction technology, frequency modulation technology and isolation technology (as shown in Fig. 1.2). The basic performance of these technologies will be elaborated in the Sect. 1.2 of this book.

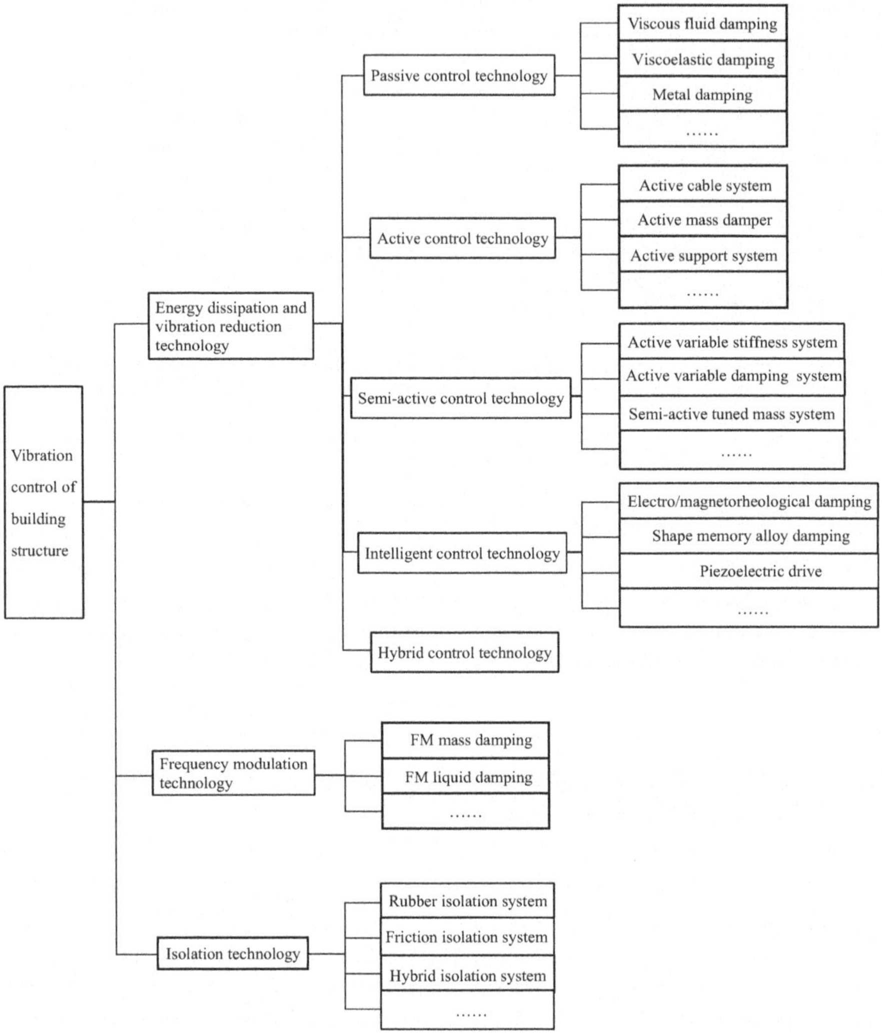

Fig. 1.2 Classification of vibration control technology for building structures

1.3 Development and Current Situation of Structural Vibration Control

Structural vibration control technology has been widely used in aviation, aerospace, navigation and machinery industry since 1920s [6]. With the development of anti-seismic and anti-wind design theory of civil engineering structure, traditional design method is difficult to meet the needs of anti-seismic and anti-wind design. According to the damage mechanism of structure under the action of earthquake and strong wind, and referring to the theory, technology and application results of structural vibration reduction control in mechanical, aerospace and other fields, the research and application of structural vibration reduction control technology of civil engineering are carried out.

In the early 1970s, Kelly, an American scholar, put forward that the metal mild steel yield energy dissipator, which is energy dissipation element and non-structural member, including torsion beam, bending beam and U-shaped steel device, should be installed in the structure to share and dissipate the vibration energy originally dissipated by the structural members [11]. This is the original passive energy dissipation and vibration reduction method of the structure. This method is a supplement and development to the seismic ductility design of the structure, which not only protects the safety of structural members under large earthquake, but also maintains the replaceability of energy consuming elements and the repairability of structure after earthquake. In addition to the mild steel yield dampers, various types of displacement dependent dampers such as metal yield dampers and friction dampers, velocity dependent dampers such as viscous dampers and viscoelastic dampers, frequency modulated vibration absorption damper such as frequency modulated mass dampers and frequency modulated liquid dampers have been successfully developed. These dampers reduce the dynamic response of the structure through their own passive dissipation and absorption of vibration energy. At present, these passive energy dissipation devices have formed different forms and standard models of commercial products. The corresponding analysis theory and design method of the passive damping structure have been relatively perfect, and gradually entered the relevant design specifications, regulations and guidelines, forming a more mature structure damping control technology. Passive energy dissipation technology has been widely used in buildings, structures, bridges and other civil engineering structures, and to some extent, it has withstood the test of earthquake and strong wind.

In the 1950s, Japanese scholar Kobori et al. put forward the seismic response control idea of active variable stiffness. In 1972, Professor J. T. P. Yao applied modern control theory to structural vibration control, put forward the concept of structural control of civil engineering, and started the theoretical and experimental research of active control of civil engineering structure; active control of structural vibration was developed on the basis of automatic control theory, which was based on the measured structural dynamic response or environmental interference. The optimal active control force is obtained using the active control algorithm, and the external energy drives the active control device to exert in the direction opposite to the direction of structural vibration to achieve structural control. Active mass damper or active mass driver (AMD), active tuned mass damper (tuned mass damper driven by active control actuator), active tendon and active brace are the main active control devices. Since active control is to restrain the vibration response of the structure by directly transforming the energy provided by the outside world into the structural control force, but for large civil engineering structures, it needs a lot of energy to transform into the control force, and there are still some difficulties in the engineering application. Kobori et al. applied the mechanical adjustable semi-active control devices such as active variable stiffness and variable damping to the structural vibration control in the 1980s. Based on the theory of structural active control, the idea of structural semi-active control is put forward, so as to find a practical method for the application of active control in practical engineering. The semi-active control of structure is mainly based on parameter control, and only a small amount of energy adjustment is needed to apply control force to the structure through the semi-active control device. Semi active control devices mainly include semi-active variable stiffness device, semi-active variable damping device, semi-active TMD/TLD, variable friction damping device, controllable fluid damper, etc. At present, structural active and semi-active control has been applied in nearly 70 high-rise buildings, TV towers and bridges in the world, greatly improving the seismic and wind resistance capacity of these structures.

In recent years, the development of intelligent materials and structures has become a new development direction of structural damping control. People have developed electric (magnetic) or temperature regulation passive damping devices, driving device of active control and variable damping devices of semi-active control, by using intelligent driving materials such as electro/magnetorheological fluids, shape memory materials, piezoelectric materials and electro/magnetostrictive materials. On this basis, the intelligent control theory and method of intelligent vibration damping structure and structure vibration are put forward. The principle of structure intelligent control is basically the same as that of active control, except that the actuator that applies control force to the structure is the intelligent actuator or intelligent damper developed by intelligent materials, and the effect of structure active control can be achieved with little external energy regulation. Compared with the vibration damping device developed by hydraulic or electric actuator and common materials, intelligent actuator or damper has the advantages of large output control force, small energy loss, fast response speed, etc., which will become a new generation of high-performance drive device or variable damping device for structural vibration control. At present, the intelligent actuator or intelligent damper developed mainly includes electric/magnetorheological damper, piezoelectric actuator, electric/magnetostrictive actuator and shape memory alloy dampers. Among them, the most typical one in the application of vibration control of civil engineering structures is the magnetorheological fluid (MR) damper developed by Lord Company in the United States, which has an energy consumption of 22 W and a maximum output control force of 200 kN. It can be used for passive energy dissipation and vibration reduction of structures through fixed magnetic field strength or as a variable damping device by adjusting the magnetic field strength to realize the semi active control of structural vibration. At present, MR damper has been used in the seismic response control of many buildings and the wind and rain vibration control of cable-stayed bridge, and has achieved good damping effect. Intelligent control of structure involves the intersection and synthesis of civil engineering, control theory, computer science, artificial intelligence, intelligent materials and structure, which has become a hot issue in the research of structural vibration control technology.

Compared with energy dissipation and frequency modulation, isolation is the earliest structural vibration control technology researched and applied at home and abroad. As early as 1881, the Japanese Kawai Hirohide put forward a method: first lay several layers of round wood horizontally and vertically staggered on the foundation, make concrete foundation on the round wood, and then build a house on the concrete foundation. In 1890, the German Bech-told proposed to use the rolling ball as the isolation base, and applied for the American patent. In 1909, British doctor Calantarients proposed the isolation method of separating the house from the foundation with talcum powder layer, which was patented in the United States. In the 1960s, the application of laminated rubber pad made the base isolation technology into a practical era: in 1969, the first isolation building using natural rubber pad was built in Yugoslavia, i.e. Bostanrauch primary school in Koubi City; in 1981, the first lead rubber isolation building in the world, William Clayton Government Office Building was built; in 1985, the first building using high damping rubber pad in the world was built in California, USA. Since then, many experimental researches, practical engineering construction and many strong earthquake tests have promoted the isolation technology to be more widely used. At present, the structure isolation technology has been more perfect, and the corresponding standard system has been improved, which makes it applied in more and more building structures.

At present, the basic theoretical research work of building structure vibration reduction technology has been improved day by day. In recent years, with the continuous emergence of innovative achievements of new disciplines such as computer and information science, new material science, etc., new theories, new methods and new materials such as self-reset concept, nonlinear vibration absorber, high energy consumption intelligent materials are gradually introduced into the research of building structure vibration reduction control, and the development of vibration reduction technology shows a diversified trend. How to develop practical and efficient new high-performance damping device, improve and perfect the practical design theory and method of vibration reduction starting from the practical problems of the project will be the future development trend of structural damping research. At the same time, the multi-dimensional damping device, the damping device considering the spatial characteristics, self-regulation, multi-functional, practical damping device and its scientific and technical problems will still be the urgent problems in the field of building structure.

References

1. Zhou, F.L. 1997. *Structural aseismic control*. Beijing: Seismological Press.
2. Ou, J.P. 2003. *Structural vibration control-active, semi-active and intelligent control*. Beijing: Science Press.
3. Soong, T.T., and F.G. Dargush. 1996. *Passive energy dissipation systems in structural engineering*. New York: Wiley.
4. Yao James, T.P. 1972. Concept of structural control. *Journal of the Structural Division* 98 (7): 1567–1574.
5. Housner, G.W., T.K. Caughey, A.G. Chassiakos, et al. 1997. Structural control: past, present, and future. *Journal of Engineering Mechanics* 123 (9): 897–971.
6. Spencer, B.F., and S. Nagarajaiah. 2003. State of the art of structural control. *Journal of Structural Engineering-ASCE* 129 (7): 845–856.

7. Basu, Biswajit, S. Bursi Oreste, Fabio Casciati, et al. 2014. A European Association for the control of structures joint perspective. Recent studies in civil structural control across Europe. *Structural Control and Health Monitoring* 21 (12): 1414–1436.
8. Ikeda, Yoshiki, Masashi Yamamoto, Takeshi Furuhashi, et al. 2019. Recent research and development of structural control in Japan. *Japan Architectural Review* 2 (3): 219–225.
9. Symans, M.D., F.A. Charney, A.S. Whittaker, et al. 2008. Energy dissipation systems for seismic applications: Current practice and recent developments. *Journal of Structural Engineering-ASCE* 134 (1): 3–21.
10. Ikeda, Y. 2009. Active and semi-active vibration control of buildings in Japan—Practical applications and verification. *Structural Control and Health Monitoring* 16 (7–8): 703–723.
11. Li, Aiqun. 2007. *Vibration control of engineering structure*. Beijing: China Machine Press.
12. Soong, T.T., and B.F. Spencer. 2002. Supplemental energy dissipation: State-of-the-art and state-of-the practice. *Engineering Structures* 24 (3): 243–259.
13. Japanese Architectural Society. 2006. *Isolation structure design*. Beijing: Earthquake Press.
14. Skinner. 1996. *Introduction to engineering isolation*. Beijing: Earthquake Press.
15. Teng, Jun. 2009. *Technology and method of structural vibration control*. Beijing: Science Press.
16. Li, Hongnan, and Linsheng Huo. 2007. *Multi-dimensional vibration control of structure*. Beijing: Science Press.
17. Zhou, Yun. 2006. *Reinforcement technology and design method of energy dissipation and shock absorption*. Beijing: Science Press.

Part I
Basic Principle of Structural Vibration Control

The structural vibration control is to reasonably set energy dissipation devices in the structure, so as to effectively control the structural vibration response under earthquake, strong wind or other dynamic disturbances within the allowable range. Since the 1920s, the vibration control technology has been widely used in the fields of aviation, aerospace, navigation and mechanical industry. In recent decades, with the increases in the height, span and complexity of building structures, the traditional seismic-resistant and wind-resistant design methods cannot meet the design requirements of structural performance. The structural vibration control technology has become an important technical means to solve the construction demand of high-intensity area, the technical challenge of complex structure design and the realistic need of people for high-performance structure after about 40 years' development, and is considered as a significant technological breakthrough with milestone significance.

According to the principle of structural dynamics, the structural vibration control technology can be divided into three categories: energy dissipation, frequency modulation and isolation. The basic principles of these three kinds of structural vibration control technology were introduced in this part from the perspective of structural dynamics theory, so as to clarify the basic concepts and internal mechanism of structural vibration control.

Chapter 2
Basic Principles of Energy Dissipation and Vibration Control

Abstract The basic principles of passive control, active control, semi-active control, intelligent control and hybrid control are introduced respectively in this chapter. For the passive control, motion equation of SDOF system, commonly used passive energy dissipation dampers and motion equation of passive vibration absorbing structural system are focally introduced. For the active and semi-active control, commonly used active and semi-active control strategies, motion equations of active and semi-active vibration absorbing system, structural state equation, structural active control algorithm and semi-active control algorithm are introduced systematically.

The energy dissipation and vibration reduction control is a kind of structural vibration control technology, in which the energy dissipation and vibration reduction devices are reasonably set up in the structure to reduce or control the dynamic response of the structure. Generally, it can be divided into passive control, active control, semi-active control, hybrid control and intelligent control.

2.1 Passive Control

The passive control means that the damper installed in the main structure generates the damper force passively along with the force and deformation of the structure, thus achieving the goal of structural vibration control [1]. The passive control is the most used control method, which does not need external energy.

2.1.1 Motion Equation of SDOF System

The typical single-particle building structure is shown in Fig. 2.1a, which can be further simplified to the typical system of single degree of freedom (SDOF) as shown in Fig. 2.1b. According to the principle of dynamic equilibrium, the motion equation of the system can be obtained as follows:

A. Li, *Vibration Control for Building Structures*, Springer Tracts in Civil Engineering, https://doi.org/10.1007/978-3-030-40790-2_2

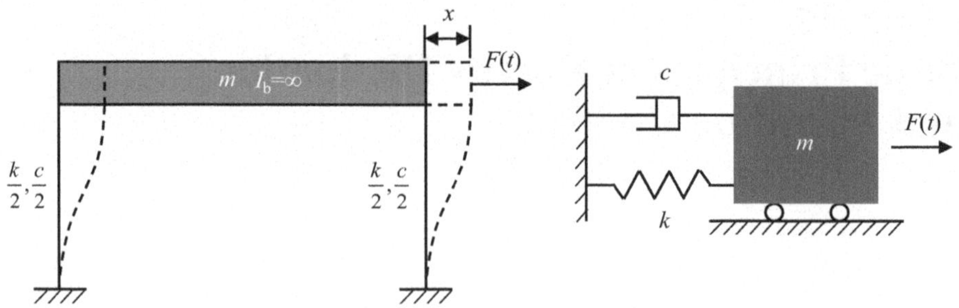

(a) Structural prototype (b) Computational model

Fig. 2.1 SDOF system

$$m\ddot{x} + c\dot{x} + kx = F(t) \tag{2.1}$$

where, m, c and k are the mass, damping and stiffness of the structure respectively; $F(t)$ is the external load applied to the structure system, and $F(t) = -m\ddot{x}_g$ when considering the earthquake action. $\ddot{x}_g$ is the acceleration of ground motion. The structure is in the free vibration state when $F(t) = 0$. Equation (2.1) can be simplified to:

$$m\ddot{x} + c\dot{x} + kx = 0 \tag{2.2}$$

where, ω_0 and ζ are the natural frequency and damping ratio of the structural system, respectively. The two sides of Eq. (2.2) are divided by the mass m at the same time, and $\frac{c}{m} = 2\zeta\omega_0$, $\frac{k}{m} = \omega_0^2$, so that Eq. (2.2) can be expressed as follows:

$$\ddot{x} + 2\zeta\omega_0\dot{x} + \omega_0^2 x = 0 \tag{2.3}$$

Plug $x = He^{\lambda t}$ into Eq. (2.2):

$$He^{\lambda t}\left(\lambda^2 + 2\zeta\omega_0\lambda + \omega_0^2\right) = 0 \tag{2.4}$$

and $\left(\lambda^2 + 2\zeta\omega_0\lambda + \omega_0^2\right) = 0$:

$$\begin{aligned}\lambda_{1,2} &= \left(-\zeta \pm \sqrt{\zeta^2 - 1}\right)\omega_0 \\ &= \left(-\zeta \pm i\sqrt{1 - \zeta^2}\right)\omega_0\end{aligned} \tag{2.5}$$

Plug λ and obtain the expression of x:

$$x = H_1e^{\lambda_1 t} + H_2e^{\lambda_2 t}$$

$$= H_1 e^{\left(-\zeta + i\sqrt{1-\zeta^2}\right)\omega_0 t} + H_2 e^{\left(-\zeta - i\sqrt{1-\zeta^2}\right)\omega_0 t}$$

Define the damping frequency as $\omega_c = \omega_0\sqrt{1-\zeta^2}$:

$$x = e^{-\zeta\omega t}\left(H_1 e^{i\omega_c t} + H_2 e^{-i\omega_c t}\right)$$

where, $e^{\pm i\omega_c t} = \cos\omega_c t \pm i\,\sin\omega_c t$, and $A = H_1 + H_2,\ B = i(H_1 - H_2)$:

$$x = e^{-\zeta\omega_0 t}(A\,\cos\omega_c t + B\,\sin\omega_c t) \tag{2.6}$$

Assuming that the initial condition is $t = 0,\ x(t) = x_0 = x(0)$ and $\dot{x}(t) = \dot{x}_0 = \dot{x}(0)$:

$$A = x_0,\ B = \frac{\dot{x}_0 + \zeta\omega_0 x_0}{\omega_c}$$

which are plugged into Eq. (2.6):

$$x = e^{-\zeta\omega t}\left[x_0\cos\omega t + \frac{(\dot{x}_0 + \zeta\omega_0 x_0)}{\omega}\sin\omega t\right] \tag{2.7}$$

The synthetic vector is $R = \sqrt{A^2 + B^2} = \sqrt{x_0^2 + \left(\frac{\dot{x}_0 + \zeta\omega_0 x_0}{\omega_c}\right)^2}$, and the phase angle is $\Phi = \tan^{-1}\frac{(\dot{x}_0 + \zeta\omega_0 x_0)/\omega_c}{x_0}$, Eq. (2.7) can be expressed as:

$$x = e^{-\zeta\omega t} R\,\cos(\omega_0 t - \Phi) \tag{2.8}$$

Equation (2.8) is the expression of free vibration response of the SDOF structure. It can be seen that when the damping ratio increases, the vibration response of the structure decreases exponentially. Increasing the structural damping can improve the attenuation velocity of structure vibration to a certain extent.

Considering the forced vibration, assuming that $F(t) = F_0\sin\omega t$, the two sides of Eq. (2.1) was divided by mass m:

$$\ddot{x} + 2\zeta\omega_0\dot{x} + \omega_0^2 x = \frac{F_0\sin\omega t}{m} \tag{2.9}$$

Plug $x = A\,\sin\omega t + B\,\cos\omega t$ into Eq. (2.9):

$$A = \frac{(F_0/m)\left(\omega_0^2 - \omega^2\right)}{\left(\omega_0^2 - \omega^2\right)^2 + (2\zeta\omega\omega_0)^2} \tag{2.10}$$

$$B = \frac{-(F_0/m)(2\zeta\omega\omega_0)}{\left(\omega_0^2 - \omega^2\right)^2 + (2\zeta\omega\omega_0)^2} \tag{2.11}$$

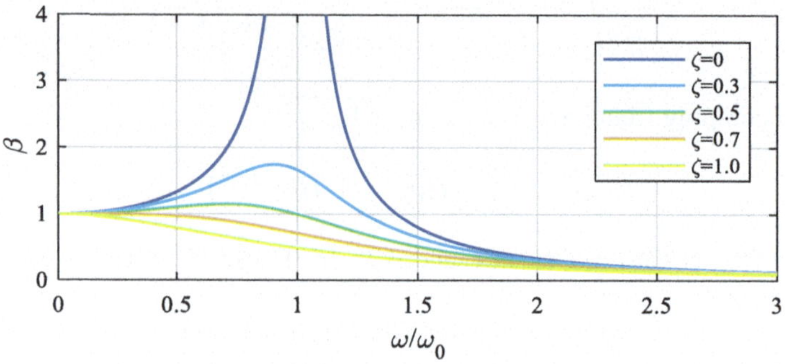

Fig. 2.2 Dynamic amplification coefficient of forced vibration for SDOF system

The synthetic vector is $R = \sqrt{A^2 + B^2} = \frac{F_0/k}{\sqrt{\left[1-(\omega/\omega_0)^2\right]^2+(2\zeta\omega/\omega_0)^2}}$, and the phase angle is $\Phi = \arctan\frac{2\zeta\omega/\omega_0}{1-(\omega/\omega_0)^2}$.

The displacement solution of the SDOF system can be expressed as:

$$\begin{aligned} x &= R\sin(\omega t - \Phi) \\ &= \frac{F_0/k}{\sqrt{\left[1-(\omega/\omega_0)^2\right]^2+(2\zeta\omega/\omega_0)^2}}\sin(\omega t - \Phi) \\ &= \beta(F_0/k)\sin(\omega t - \Phi) \end{aligned} \tag{2.12}$$

where, $\beta = \frac{1}{\sqrt{\left[1-(\omega/\omega_0)^2\right]^2+(2\zeta\omega/\omega_0)^2}}$, which is the amplification factor of structural dynamic action. Equation (2.12) can be understood as that the vibration response x of the particle is equal to the static response multiplied by the amplification factor of structural dynamic action. If $\beta > 1$, the forced vibration of the structure is amplification effect, and if $\beta < 1$, the forced vibration of the structure is characterized by attenuation effect. The demarcation line is $\beta = 1$, and $\zeta = 1/\sqrt{2} \approx 0.7$ can be obtained at this time. The relationship between the amplification factor of structural dynamic action and the frequency ratio and damping ratio is shown in Fig. 2.2.

1. Within the range of $(\omega/\omega_0) \leq 1.5$, if the damping ratio ζ of the structure system is smaller, the β value is larger, and the vibration of the structure is characterized by amplification effect. If the damping ratio ζ of the structure system is larger, the β value is smaller, and the vibration of the structure is characterized by attenuation effect. Therefore, if the energy dissipation devices are installed in the structural system to provide greater damping, the structural vibration response can be effectively reduced.
2. When the damping ratio of the structural system is $\zeta \geq 0.7$, the amplification factor of structural dynamic action is always less than 1 in all numerical ranges of (ω/ω_0).

2.1.2 Commonly Used Passive Energy Dissipation Dampers

The passive energy dissipation refers to that the passive energy dissipation damper installed in the structure dissipates the vibration energy with the structural deformation, thereby reduces the vibration response of the structure. The passive energy dissipation damper has been widely used in all kinds of structures because of the advantages of safety, economy and durability.

The passive damper can be divided into two types including velocity-dependent damper and displacement-dependent damper according to the characteristics of mechanical model [2]. The velocity-dependent damper dissipates energy mainly by throttling the resistance caused by viscous fluid or shear deformation of viscoelastic materials, and the viscous fluid damper and viscoelastic damper are commonly used. The displacement-dependent damper dissipates energy mainly by the plastic hysteresis or interface friction of metal materials, and the metal damper and friction damper are commonly used [1, 3–7]. In addition, there are vibration reduction devices using more than two kinds of energy dissipation principle or mechanism, which can be called composite damper. Four commonly used dampers and their general characteristics are shown in Table 2.1.

2.1.3 Motion Equation of Passive Vibration Absorbing Structural System

The typical single-particle building structure with damper is shown in Fig. 2.3a, which can be further simplified to the ideal system as shown in Fig. 2.3b. According to the principle of dynamic equilibrium, the motion equation of the system can be obtained as follows:

$$m\ddot{x} + c\dot{x} + kx + f_d = F(t) \tag{2.13}$$

where, f_d is the control force of vibration absorber, which can be approximately equal to the output force of damper when the stiffness of connecting support is large enough. The mechanical models of various types of dampers are shown in Table 2.1. More detailed mechanical characteristics will be introduced in the next part. Even if the main structure remains elastic, the motion equation changes from linear to non-linear after the damper is installed. In addition to the SDOF system with the linear viscous fluid damper or viscoelastic damper, the solution method in Sect. 2.1.1 is no longer applicable to the solution of Eq. (2.13). There are usually two ways to solve the above equations. One is to linearize the non-linear terms, and then solve them in frequency domain or time domain according to the method of solving SDOF linear equations. The other one is to solve them directly in time domain by numerical analysis methods, such as Newmark-β method and Runge-Kutta method. The relevant methods are introduced in detail in the corresponding literatures.

Table 2.1 Commonly used dampers

Type	Picture	Material and shape	Mechanical property
Viscous fluid damper		Material: polymer compound; Mechanisms: throttle/shear impedance type; Shape: cylindrical, planar	Mechanical model: $F_d = Cv^{\alpha}$ 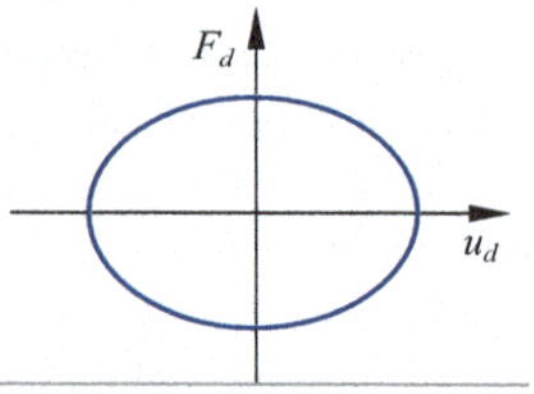
Viscoelastic damper		Material: propylene, diene and other compounds; Mechanism: shear impedance type; Shape: cylindrical, planar	Mechanical model: $F_d = Ku + Cv$ 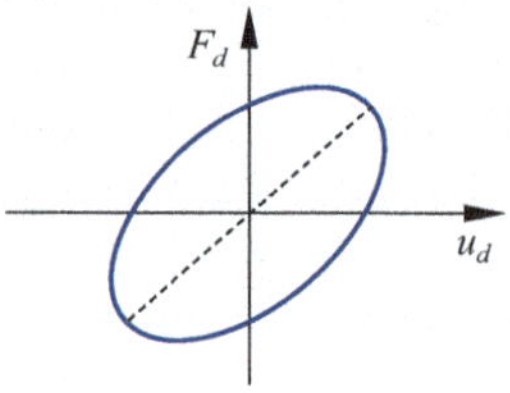
Metal damper		Material: steel, lead, etc.; Mechanism: plastic hysteresis impedance type; Shape: cylindrical, planar	Mechanical model: $F_d = Kf(u)$ 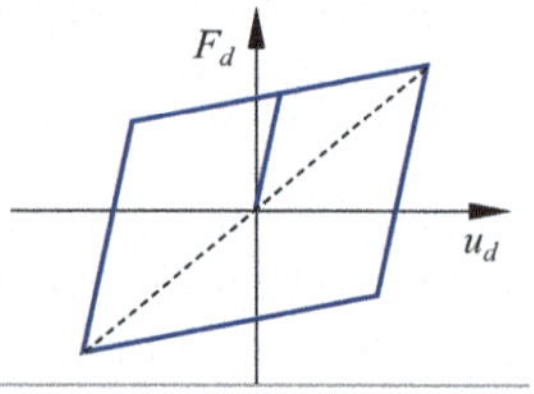
Friction damper		Material: friction material; Mechanism: friction hysteresis impedance; Shape: cylindrical, planar	Mechanical model: $F_d = Kf(u)$ 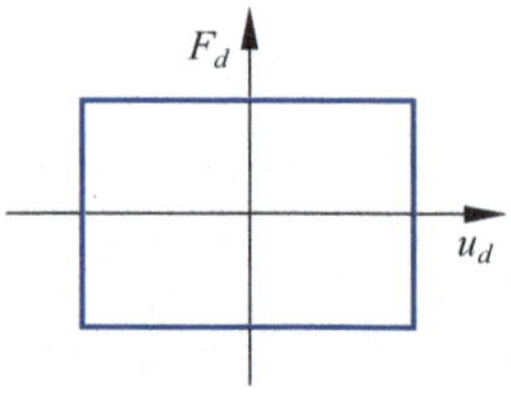

Although the general and exact solutions of Eq. (2.13) are difficult to obtain, the mechanism of passive structural control can still be analyzed from the perspective of energy. By integrating the relative displacement of the two sides of Eq. (2.1) with the integral interval of the whole duration of external load, the energy equation of the system is obtained as follows:

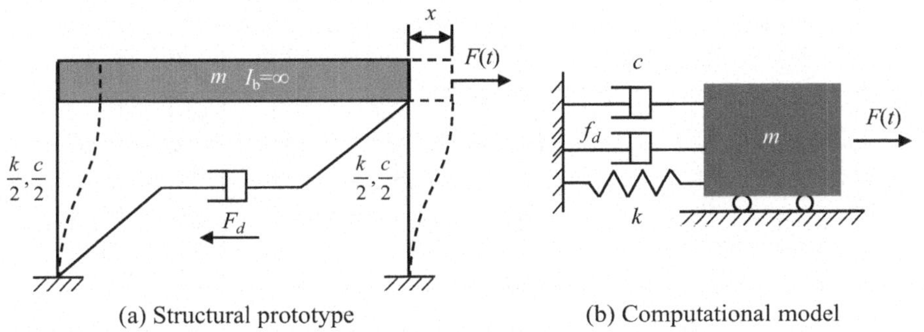

Fig. 2.3 Passive vibration absorbing structural system

$$\int_0^l dx \cdot m\ddot{x} + \int_0^l dx \cdot c\dot{x} + \int_0^l dx \cdot kx = \int_0^l dx \cdot F(t) \tag{2.14}$$

Plug $dx = \dot{x}dt$ into Eq. (2.14):

$$\int_0^l m\dot{x}\ddot{x}dt + \int_0^l c\dot{x}\dot{x}dt + \int_0^l kx\dot{x}dt = \int_0^l F(t)\dot{x}dt \tag{2.15}$$

The items on the left side of Eq. (2.15) are as follows:
Structural kinetic energy: $E_v = \int_0^l m\dot{x}\ddot{x}dt = \frac{1}{2}m\dot{x}^2$
Structural damping energy dissipation: $E_c = \int_0^l c\dot{x}\dot{x}dt = c\int_0^l \dot{x}^2 dt$
Elastic strain energy: $E_k = \int_0^l kx\dot{x}dt = \frac{1}{2}kx^2$
The right term of Eq. (2.15) is input energy: $E_i = \int_0^l F(t)\dot{x}dt$
Equation (2.15) can be expressed as:

$$E_v + E_c + E_k = E_i \tag{2.16}$$

Similarly, Eq. (2.13) can be expressed as follows;

$$\int_0^l m\dot{x}\ddot{x}dt + \int_0^l c\dot{x}\dot{x}dt + \int_0^l kx\dot{x}dt + \int_0^l f_d\dot{x}dt = \int_0^l F(t)\dot{x}dt \tag{2.17}$$

$$E_v + E_c + E_k + E_d = E_i \tag{2.18}$$

where, $E_d = \int_0^l F_d\dot{x}dt$ is the energy consumption of damper. Considering the plasticity of the structure, Eqs. (2.1) and (2.13) can be expressed as follows:

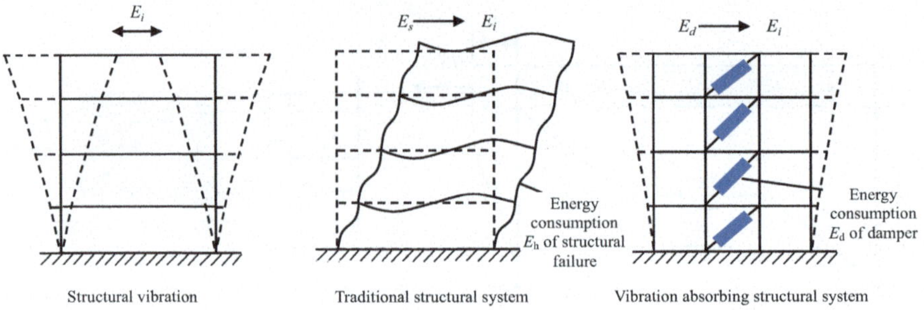

Fig. 2.4 Energy conversion approach of the structure

$$m\ddot{x} + c\dot{x} + f_s = F(t) \tag{2.19}$$

$$m\ddot{x} + c\dot{x} + f_s + f_d = F(t) \tag{2.20}$$

where, $f_s = f(x, \dot{x})$ is the hysteretic resilience of structural system. Similarly, the energy equation can be obtained as follows:

$$E_v + E_c + E_s = E_i \tag{2.21}$$

$$E_v + E_c + E_s + E_d = E_i \tag{2.22}$$

where, $E_s = E_k + E_h = \int_0^l f_s \dot{x} dt$ is the sum of elastic strain energy and structural hysteretic energy. Thus, the energy response equation defined by relative displacement is derived. The results show that the energy response equation defined by absolute displacement has little difference compared with the above equation when the period is within 0.3–5.0 s.

The comparisons between Eqs. (2.16) and (2.18), Eqs. (2.21) and (2.22) are shown in Fig. 2.4. (1) Compared with the traditional structural system, the energy dissipation term E_d of damper is added to the energy equation of the damping system; (2) For the traditional structural system (Fig. 2.4), the damping energy dissipation E_c is usually very small (about 5%) and can be neglected in order to terminate the structural vibration response ($E_v \to 0$), which will inevitably lead to serious damage or collapse ($E_s \to E_i$) of the main structure in order to consume the vibration energy of the input structure; (3) For the vibration-absorbing structure system (Fig. 2.4), the structural damping energy E_c is also usually very small (about 5%), which can be neglected. The damper firstly enters the working state of hysteretic energy consumption, and the energy dissipation function should be brought into full play, in order to rapidly attenuate the energy input for the structure ($E_d \to E_i$) and protect the main structure from damage ($E_s \to 0$). The structural vibration response is attenuated rapidly ($E_v \to 0$) to ensure the safety of structure in vibration.

2.2 Active and Semi-active Control

The active control is the control with external energy. The control force is exerted actively by the control device through external energy according to some control law. The semi-active control is generally a control with a small amount of external energy [8, 9]. Although the control force is passively generated by the movement of control device itself, the control device can actively adjust its parameters using the external energy during control process, thus playing a role of regulating the control force.

2.2.1 Commonly Used Active and Semi-active Control Strategies

The active control requires the real-time measurement of structural response or environmental disturbance. The active control algorithm based on modern control theory is used to calculate and solve the optimal control force based on accurate structural model. Finally, the optimal control force is achieved by the actuator under the external energy input. The active control actuator is usually hydraulic servo system or motor servo system, which usually needs large or even large energy drive.

At present, there are two main types of active control systems commonly used: The installation of the first one is similar to that of passive damper, as shown in Fig. 2.5a, b, which is usually installed between the structural frames as the Active Tendon System (ATS) or Active Brace System (ABS). The actuator actively exerts the forces opposite to structural deformation to control the structural response. The installation of the second type is similar to that of FM dampers, as shown in Fig. 2.5c, d. The Hybrid Mass Damper (HMD) or Active Mass Damper (AMD) are generally installed on the floor surface. The mass block is controlled by the actuator to exert reaction force on the structure in order to reduce the structural response.

The principle of semi-active structural control is basically the same as that of active structural control. Only a small amount of energy adjustment is needed by actuator to implement the control force so that they can actively utilize the reciprocating relative deformation or relative velocity of structural vibration to achieve the active optimal control force as far as possible. Therefore, the semi-active control actuator is usually a composite control system of passive stiffness or damper and mechanical active regulation. The commonly used installation methods of semi-active control system include the Tuned Mass Damper with Semi-active Continuous Variable Stiffness (SCVS-TMD), Active Variable Stiffness System (AVS), Active Variable Damping System (AVD), etc., which are installed in the similar way to Fig. 2.5. The principle of SCVS-TMD system is the same as that of tuned mass vibration reduction, but the SCVS-TMD system can continuously adjust the stiffness or frequency characteristics of the additional mass subsystem according to the real-time identified structural frequency. The control mode of AVS system is to control and change the stiffness of

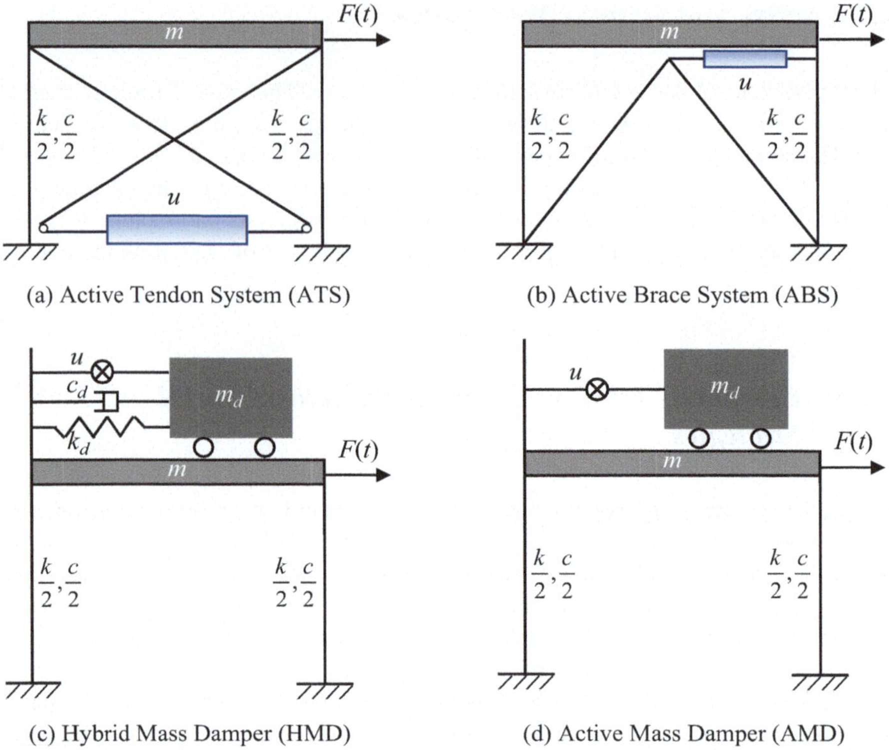

Fig. 2.5 Structural active control system

the system by a fast reaction locking device controlled by a computer, to avoid the influence of resonance and reduce structural response. The AVD system can adjust the damping force of the semi-active variable damping control device to make it equal to or close to the active optimal control force, to achieve the shock absorption effect close to active control. The semi-active variable damping control device can only achieve the speed-related control forces, but cannot achieve the displacement-related and speed-related control forces at the same time as the active control actuator.

Both active and semi-active control methods require the real-time measurement of structural response or environmental disturbance. Among them, the control based on structural response observation is called feedback control, while the active control based on structural environmental disturbance observation is called feed-forward control. The control principles are shown in Fig. 2.6.

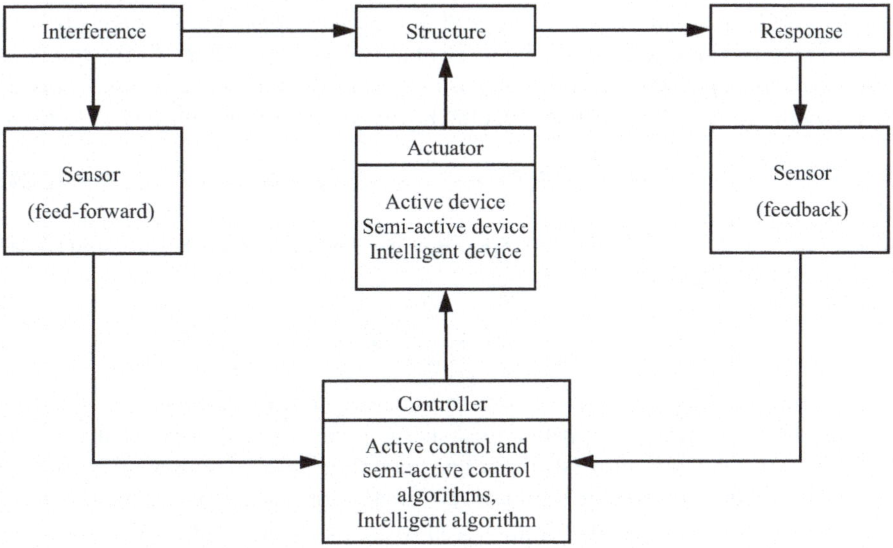

Fig. 2.6 Principles of structural active, semi-active and intelligent control

2.2.2 Motion Equations of Active and Semi-active Vibration Absorbing Systems

The computational models of typical active and semi-active control for the SDOF structure system is shown in Fig. 2.7.

Regardless of the control model, the motion equation of the vibration reduction system can be expressed as follows:

$$m\ddot{x} + c\dot{x} + kx = F(t) + u(t) \tag{2.23}$$

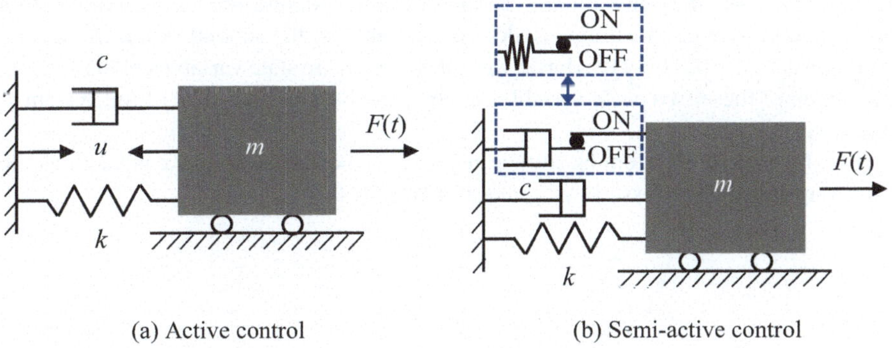

Fig. 2.7 Computational models of active and semi-active control

where, $u(t)$ is the control force of control device, the active control force $u(t)$ can be regarded as the linear combination of external excitation load $F(t)$, structural displacement response x, velocity response $\dot{x}$ and acceleration $\ddot{x}$ for the open-closed-loop control system. It is assumed that the linear feedback is established as follows:

$$u(t) = -G_0F(t) - G_1x - G_2\dot{x} - G_3\ddot{x} \tag{2.24}$$

where, G_i ($i = 0, 1, 2, 3$) is the corresponding feedback gain matrix. Equation (2.24) is plugged into Eq. (2.23):

$$(m + G_3)\ddot{x} + (c + G_2)\dot{x} + (k + G_1)x = (1 - G_0)F(t) \tag{2.25}$$

For the structure with active and semi-active control systems, the function of closed-loop control is to better feedback external excitation by changing the parameters of the structure (stiffness and damping). The function of open-loop control is to change (reduce or eliminate) external disturbance. If a reasonable control algorithm is selected and the optimal control force is determined, the structural response can be reduced and controlled.

2.2.3 *Structural State Equation*

The mathematical description of dynamic system is the basis of modern control theory and method of structural vibration control of civil engineering. Many control algorithms are derived from structural state equation. In the researches of active and semi-active control, the mathematical relationship between the input and output of structural system is generally established in the state space, which can be the state equation of continuous time or the difference equation of discrete time [1, 10]. For this reason, it is necessary to transform the motion equation introduced above into the state equation. As long as u and the two variables on the left side of Eqs. (2.1) and (2.25) are known, the other quantity $\ddot{x}$ can be uniquely determined, so that the motion law of the system can be uniquely determined, that is, the second-order differential equation actually has only two basic variables, which are state variables of the system. The value of these two basic variables at any time is called the state of the system at that time.

For the structure system as shown in Fig. 2.1, the sum of x and $\dot{x}$ is taken as the state variable and expressed by z_1 and z_2 as follows:

$$x = z_1, \dot{x} = z_2$$

and

$$\dot{z}_1 = z_2 \tag{2.26}$$

From Eq. (2.1),

$$\ddot{x} = -m^{-1}c\dot{x} - m^{-1}kx + m^{-1}F(t) \tag{2.27}$$

Plug Eq. (2.26) into Eq. (2.27):

$$\dot{z}_2 = -m^{-1}cz_2 - m^{-1}kz_1 + m^{-1}F(t) \tag{2.28}$$

Equations (2.26) and (2.28) are written in matrix form:

$$\begin{Bmatrix} \dot{z}_1 \\ \dot{z}_2 \end{Bmatrix} = \begin{bmatrix} 0 & 1 \\ -m^{-1}k & -m^{-1}c \end{bmatrix} \begin{Bmatrix} z_1 \\ z_2 \end{Bmatrix} + \begin{bmatrix} 0 \\ m^{-1} \end{bmatrix} F(t) \tag{2.29}$$

$A = \begin{bmatrix} 0 & 1 \\ -m^{-1}k & -m^{-1}c \end{bmatrix}$, $D = \begin{bmatrix} 0 \\ m^{-1} \end{bmatrix}$, $Z = \begin{Bmatrix} z_1 \\ z_2 \end{Bmatrix}$. Then, Eq. (2.29) can be expressed as follows:

$$\dot{Z} = AZ + DF(t) \tag{2.30}$$

Equation (2.30) is called the state equation of the system, and z is called the state vector. The relationship between structural displacement and state vector can be expressed as follows:

$$x = [1\ 0] \begin{Bmatrix} z_1 \\ z_2 \end{Bmatrix} = GZ \tag{2.31}$$

Equation (2.31) is called the output equation of the system, and x is the displacement response of the system, which is also called the output. Similarly, for the vibration reduction system shown in Fig. 2.7, Eq. (2.23) can be expressed as follows:

$$\ddot{x} = -m^{-1}c\dot{x} - m^{-1}kx + m^{-1}F(t) + m^{-1}u(t) \tag{2.32}$$

After transformation, the state equation expressed by matrix can be obtained as follows:

$$\begin{Bmatrix} \dot{z}_1 \\ \dot{z}_2 \end{Bmatrix} = \begin{bmatrix} 0 & 1 \\ -m^{-1}k & -m^{-1}c \end{bmatrix} \begin{Bmatrix} z_1 \\ z_2 \end{Bmatrix} + \begin{Bmatrix} 0 \\ m^{-1} \end{Bmatrix} u(t) + \begin{Bmatrix} 0 \\ m^{-1} \end{Bmatrix} F(t) \tag{2.33}$$

Assuming that $D = B = \begin{Bmatrix} 0 \\ m^{-1} \end{Bmatrix}$, Eq. (2.23) can be expressed as follows:

$$\dot{Z} = AZ + Bu(t) + DF(t) \tag{2.34}$$

where, B and D expressed by different symbols are generally unequal in the MDOF system. The optimal control system means that some performance indicators of the system have the best value when fulfilling the specific tasks required under certain specific conditions. According to the different uses of the system, different performance indicators can be proposed. The design of optimal control system is to select the optimal control law so that a certain performance index can reach the maximum (maximum or minimum).

1. **Calculating the optimal control using variational method**

The performance index function of the system is

$$e = \int_{t_0}^{t_1} H[\{z(t)\}, \{u(t)\}, t]dt \tag{2.35}$$

where, $\{z(t)\}$ is the n-dimensional state vector, $\{u(t)\}$ is the control force vector, $H[\{z(t)\}, \{u(t)\}, t]$ is the continuous differentiable pure quantity function of $\{z(t)\}$ and t, which can be selected according to the purpose of control system. In the integrable function of Eq. (2.35), the constraint relationship between $\{z(t)\}$ and $\{u(t)\}$ can be expressed by the state equation of the system described above. The state equation of the system is:

$$\{\dot{z}(t)\} = F[\{z(t)\}, \{u(t)\}, t], \qquad \{z(t_0)\} = \{z_0\} \tag{2.36}$$

The corresponding optimal control problem is to find an optimal control $\{u^*(t)\}$ among all the alternative controls $\{u(t)\}$ under satisfying Eq. (2.36), so that the system state $\{z(t)\}$ can reach the optimal trajectory $\{z^*(t)\}$ and the performance index e can reach the extreme value. This is a constrained extremum problem (the optimal control for linear systems is extremum control). The Lagrangian function is usually introduced:

$$\begin{aligned} &\tilde{H}[\{z(t)\}, \{u(t)\}, \{\dot{z}(t)\}, \lambda(t), t] \\ &\quad = H[\{z(t)\}, \{u(t)\}, t] + \lambda^{\mathrm{T}}(t)\{F[\{z(t), \{u(t)\}, t\}] - \{\dot{z}(t)\}\} \end{aligned} \tag{2.37}$$

where, $\lambda(t)$ is the Lagrange multiplier vector, also known as the common state vector. Equation (2.37) transforms the constrained extremum problems of Eqs. (2.35) and (2.36) into the unconstrained extremum problems. At this point, a new performance indicator can be defined as:

$$\varepsilon = \int_{t_0}^{t_1} \tilde{H}[\{z(t)\}, \{u(t)\}, \{\dot{z}(t)\}, \lambda(t), t]dt \tag{2.38}$$

Since the state equation of the system must be satisfied when the extreme value of ε is taken, the new performance index is actually the original performance index e. According to the variational principle, the necessary condition for the extreme value is Euler's formula:

$$\left.\begin{aligned} \frac{\partial \tilde{H}}{\partial\{u\}} - \frac{d}{dt}\frac{\partial \tilde{H}}{\partial\{\dot{u}\}} = 0 \\ \frac{\partial \tilde{H}}{\partial \lambda} - \frac{d}{dt}\frac{\partial \tilde{H}}{\partial \dot{\lambda}} = 0 \\ \frac{\partial \tilde{H}}{\partial\{z\}} - \frac{d}{dt}\frac{\partial \tilde{H}}{\partial\{\dot{z}\}} = 0 \end{aligned}\right\} \tag{2.39}$$

The $\tilde{H}$ of Eq. (2.39) is not explicit for $\dot{u}$, $\dot{\lambda}$,

$$\frac{\partial \tilde{H}}{\partial\{\dot{u}\}} = 0, \quad \frac{\partial \tilde{H}}{\partial \dot{\lambda}} = 0 \tag{2.40}$$

Equation (2.29) can be expressed as follows:

$$\left.\begin{aligned} \frac{\partial \tilde{H}}{\partial\{u\}} = 0 \\ \frac{\partial \tilde{H}}{\partial \lambda} = 0 \\ \frac{\partial \tilde{H}}{\partial\{z\}} - \frac{d}{dt}\frac{\partial \tilde{H}}{\partial\{\dot{z}\}} = 0 \end{aligned}\right\} \tag{2.41}$$

In addition, the extreme value of ε must satisfy the end condition:

$$\lambda(t_1) = 0 \tag{2.42}$$

2. **Riccati equation**

Next, the linear system with quadratic error measure is discussed using the extremum control principle. The state equation is set up as follows:

$$\{\dot{z}(t)\} = [A]\{z(t)\} + [B]\{u(t)\}, \ \{z(t_0)\} = \{z_0\} \tag{2.43}$$

The quadratic performance index is taken:

$$e = \int_{t_0}^{t_1} \left[\{z(t)\}^{\mathrm{T}}[Q]\{z(t)\} + \{u(t)\}^{\mathrm{T}}[R]\{u(t)\}\right]dt + \{z(t_1)\}^{\mathrm{T}}[S]\{z(t_1)\} \tag{2.44}$$

where, $[Q]$ and $[S]$ are semi-positive definite symmetric weight matrix, $[R]$ is the positive definite symmetric weight matrix, and Eq. (2.44) can also be written by the impulse function $\delta(t)$:

$$e = \int_{t_0}^{t_1} \left[\{z(t)\}^{\mathrm{T}}[Q]\{z(t)\} + \{z(t)\}^{\mathrm{T}}[S]\delta(t - t_1)\{z(t)\} + \{u(t)\}^{\mathrm{T}}[R]\{u(t)\} \right] dt \tag{2.45}$$

Similarly as Eq. (2.37), the Lagrange function is introduced firstly:

$$\begin{aligned} \tilde{H} = & \{z(t)\}^{\mathrm{T}}[Q]\{z(t)\} + \{z(t)\}^{\mathrm{T}}[S]\delta(t - t_1)\{z(t)\} + \{u(t)\}^{\mathrm{T}}[R]\{u(t)\} \\ & + \lambda^{\mathrm{T}}(t)([A]\{z(t)\} + [B]\{u(t)\} - \{\dot{z}(t)\}) \end{aligned} \tag{2.46}$$

According to the necessary conditions of Eqs. (2.41) and (2.42) controlled by the extreme value:

$$\left. \begin{aligned} & \frac{\partial \tilde{H}}{\partial \{u\}} = 2[R]\{u(t)\} + [B]^{\mathrm{T}}\lambda(t) = \{0\} \\ & \frac{\partial \tilde{H}}{\partial \lambda} = [A]\{z(t)\} + [B]\{u(t)\} - \{\dot{z}(t)\} = \{0\} \\ & \frac{\partial \tilde{H}}{\partial \{z\}} - \frac{d}{dt}\frac{\partial \tilde{H}}{\partial \{\dot{z}\}} = 2[Q]\{z(t)\} + 2[S]\{z(t)\}\delta(t - t_1) + [A]^{\mathrm{T}}\lambda(t) + \dot{\lambda}(t) = 0 \\ & \frac{d}{dt}\frac{\partial \tilde{H}}{\partial \{\dot{z}\}} = -\dot{\lambda}(t) \\ & \{z(t_0)\} = \{z_0\}, \ \lambda(t_1) = 0 \end{aligned} \right\} \tag{2.47}$$

and

$$\{u(t)\} = -\frac{1}{2}[R]^{-1}[B]^{\mathrm{T}}\lambda(t) \tag{2.48}$$

$$\{\dot{z}(t)\} = [A]\{z(t)\} + [B]\{u(t)\}, \ \{z(t_0)\} = \{z_0\} \tag{2.49}$$

$$\dot{\lambda}(t) = -2[Q]\{z(t)\} - [A]^{\mathrm{T}}\lambda(t) - 2[S]\{z(t)\}\delta(t - t_0), \ \lambda(t_1) = 0 \tag{2.50}$$

The equivalent form of Eq. (2.50) is

$$\dot{\lambda}(t) = -2[Q]\{z(t)\} - [A]^{\mathrm{T}}\lambda(t), \ \lambda(t_1) = 2[S]\{z(t_1)\} \tag{2.51}$$

Plug Eq. (2.48) into Eq. (2.49):

$$\{\dot{z}(t)\} = [A]\{z(t)\} - \tfrac{1}{2}[B][R]^{-1}[B]^{\mathrm{T}}\lambda(t), \ \{z(t_0)\} = \{z_0\} \tag{2.52}$$

Equations (2.51) and (2.52) show that there is a linear relationship between the state vectors $\{z(t)\}$ and $\lambda(t)$. The following linear transformation can be made when solving Eqs. (2.41) and (2.42):

$$\lambda(t) = 2[P]\{z(t)\} \tag{2.53}$$

Derivative Eq. (2.43),

$$\dot{\lambda}(t) = 2[\dot{P}]\{z(t)\} + 2[P]\{\dot{z}(t)\} \tag{2.54}$$

Plug Eqs. (2.51), (2.52), (2.53) into Eq. (2.54),

$$\begin{aligned} & -2[Q]\{z(t)\} - [A]^{\mathrm{T}}\lambda(t) \\ & = 2[\dot{P}]\{z(t)\} + 2[P]\left([A]\{z(t)\} - \frac{1}{2}[B][R]^{-1}[B]^{\mathrm{T}}\lambda(t)\right) \end{aligned} \tag{2.55}$$

After simplification, it can be obtained as:

$$\left([\dot{P}] + [P][A] + [A]^{\mathrm{T}}[P] + [Q] - [P][B][R]^{-1}[B]^{\mathrm{T}}[P]\right)\{z(t)\} = 0 \tag{2.56}$$

Because the linear transformation is valid for any $\{z(t)\}$. The coefficient matrix at the left of Eq. (2.56) is always equal to zero,

$$[\dot{P}] = -[P][A] - [A]^{\mathrm{T}}[P] - [Q] + [P][B][R]^{-1}[B]^{\mathrm{T}}[P] \tag{2.57}$$

Equation (2.57) is the Riccati matrix differential equation. $[P]$ can be proved to be symmetric matrices by Eqs. (2.51) and (2.53), and $[P(t_1)] = [S]$.

After the solution of [P] by Eq. (2.56), the optimal control can be obtained by Eqs. (2.48) and (2.53),

$$\{u^*(t)\} = -[R]^{-1}[B]^{\mathrm{T}}[P]\{z(t)\} \tag{2.58}$$

Or,

$$\{u^*(t)\} = [G]\{z(t)\},\ [G] = -[R]^{-1}[B]^{\mathrm{T}}[P] \tag{2.59}$$

where, $[G]$ is the feedback matrix of system. Plug Eq. (2.58) into Eq. (2.49),

$$\{\dot{z}(t)\} = \left([A] - [B][R]^{-1}[B]^{\mathrm{T}}[P]\right)\{z(t)\}, \quad \{z(t_0)\} = \{z_0\} \tag{2.60}$$

Thus, the optimal state $\{z^*(t)\}$ of the system can be obtained.

2.2.4 Structural Active Control Algorithm

In active control, the key problem is how to determine the control force vector $u(t)$, and the active control algorithm is the basis of determining the control force. The purpose is to make the active control system choose the appropriate gain matrix, find the best control parameters, achieve better performance index and the best control of the structure under satisfying its state equation and various constraints. Several commonly used structural active control algorithms are briefly introduced below.

2.2.4.1 Linear Optimal Control Algorithm

In this algorithm, the selection of control vectors $\{u(t)\}$ is to minimize the following objective function **J**:

$$\mathbf{J} = \int_0^{t_1} [\{z(t)\}^{\mathrm{T}}[Q]\{z(t)\} + \{u(t)\}^{\mathrm{T}}[R]\{u(t)\}]dt \tag{2.61}$$

where, t_1 is the duration of external excitation; $[Q]$ is the weight matrix of state vector of the structural system, which is a positive semi-definite matrix of $2n \times 2n$ order; $[R]$ is the weight matrix of control power vector of the structural system, which is a positive definite matrix of $r \times r$ order. The first term of Eq. (2.61) is the energy of structural vibration response, and the second term is the work done by control force, which is the energy input from the control system to the structural system. According to the optimal control solution method introduced in Sect. 2.2.3, the optimal control force vector $\{u(t)\}$ is obtained as follows:

$$\{u(t)\} = -\frac{1}{2}[R]^{-1}[B]^{\mathrm{T}}\{\lambda(t)\} \tag{2.62}$$

where, $\{\lambda(t)\}$ is the Lagrange time-dependent factor vector:

$$\left\{\dot{\lambda}(t)\right\} = -[A]^{\mathrm{T}}\{\lambda(t)\} - 2[Q]\{z(t)\},\ \{\lambda(t_1)\} = 0 \tag{2.63}$$

1. When closed-loop control is used:

$$\{\lambda(t)\} = 2[P]\{z(t)\} \tag{2.64}$$

The Riccati equation, which is the same as Eq. (2.57) of Sect. 2.2.3, can be obtained by plugging Eq. (2.64) into Eq. (2.63). The optimal control force vector $\{u(t)\}$ of active control is obtained by solving $[P]$:

$$\{u(t)\} = -[R]^{-1}[B]^{\mathrm{T}}[P]\{z(t)\} \tag{2.65}$$

The feedback gain matrix is $[G] = -[R]^{-1}[B]^{\mathrm{T}}[P]$.

2. When the open-closed loop control is adopted, and $\{\lambda(t)\} = [P]\{z(t)\} + [S]\{p(t)\}$, the corresponding Riccati equation can be set up as follows:

$$\left.\begin{array}{c}([\dot{P}]-[P][A]-[p(t)]+[A]^{\mathrm{T}}[P]+2[Q])\{z(t)\}+[S]\{\dot{p}(t)\} \\ +([\dot{S}]-\dfrac{1}{2}[P][B][R]^{-1}[B]^{\mathrm{T}}-[A][S]+[P][D])\{p(t)\}=0 \\ {[P(t_1)]=0,\,[S(t_1)]=0}\end{array}\right\} \tag{2.66}$$

However, the gain $[S]$ of open-loop control is usually not available in Eq. (2.66).

3. When the open-loop control is adopted:

$$\{\lambda(t)\} = 2[S]\{p(t)\} \tag{2.67}$$

The $[S]$ of open-loop control is also not available in Eq. (2.67). Thus, the open-closed loop control and open-loop control based on classical linear optimal control algorithm cannot be realized in active control.

2.2.4.2 Pole-Placement Method

The pole of the system is the eigenvalue of matrix $[A]$. The eigenvalues of a matrix can be either real or complex. When they are complex, they must appear in pairs. Considering the state equation of Eq. (2.34), the relationship between the eigenvalues λ_i of the system matrix $[A]$ and the modal frequency ω_i and damping ratio ξ_i of i order of the original structure system are as follows:

$$\begin{aligned}\lambda_i^{1,2} &= \beta \pm j\omega \\ &= -\xi_i\omega_i \pm j\omega_i\sqrt{1-\xi_i} \\ j &= \sqrt{-1}\end{aligned} \tag{2.68}$$

where, β and ω reflect the damping and frequency characteristics of the structural system, respectively. The eigenvalue of the system corresponds to a point on the complex plane (β, ω). If the control force vector $\{u(t)\}$ is the linear feedback of state vector $\{z(t)\}$:

$$\{u(t)\} = -[G]\{z(t)\} \tag{2.69}$$

Equation (2.69) is plugged into the state equation of the structure, and the equation of the closed-loop control system is obtained,

$$\{\dot{z}(t)\} = ([A] + [G][B])\{z(t)\} + [D]\{F(t)\} \tag{2.70}$$

At this time, the system matrix of the controlled structure has been changed into $([A] + [G][B])$, which produces new eigenvalue λ_i', new modal frequency ω_i' and modal damping ratio ξ_i'. It can be seen that the basic principle of pole placement method is to use state or output feedback to select the appropriate gain matrix $[G]$, so that the controlled structure has the desired stiffness matrix and damping matrix, leading to the pole change of the system, in order to obtain the desired system performance.

2.2.4.3 Instantaneous Optimal Control Algorithm

The basic idea of instantaneous optimum control (IOC) algorithm is to minimize the objective function **J** at any time t in the period of $0 \leq t \leq t_1$. Taking the instantaneous optimal objective function as:

$$\mathbf{J} = \{z(t)\}^{\mathrm{T}}[Q]\{z(t)\} + \{u(t)\}^{\mathrm{T}}[R]\{u(t)\} \tag{2.71}$$

Considering the state equation, the system matrix $[A]$ is assumed to have different eigenvalues. The eigenvalues of $[A]$ are used as column vectors to form $2n \times 2n$ order modal matrices $[\Phi]$, and the following transformations are made:

$$\{z(t)\} = [\Phi]\{\bar{z}(t)\} \tag{2.72}$$

Equation (2.72) is used to decouple the state equation:

$$\{\dot{z}(t)\} = [\Lambda]\{\bar{z}(t)\} + \{q(t)\}, \{\bar{z}(0)\} = 0 \tag{2.73}$$

where, $[\Lambda] = [\Phi]^{-1}[A][\Phi]$, $\{q(t)\} = [\Phi]^{-1}([B]\{u(t)\} + [D]\{F(t)\})$, $[\Lambda]$ is the diagonal matrix whose diagonal elements are the eigenvalues of $[A]$. For small time intervals Δt, the state vector modes $\{\bar{z}(t)\}$ can be expressed as:

$$\begin{aligned}\{\bar{z}(t)\} &= \int_0^{t-\Delta t} \mathrm{e}^{[\Lambda(t-\tau)]}\{q(\tau)\}d\tau + \int_{t-\Delta t}^{\mathrm{T}} \mathrm{e}^{[\Lambda(t-\tau)]}\{q(\tau)\}d\tau \\ &\approx \mathrm{e}^{[\Lambda]\Delta t}\{z(t-\Delta t)\} + \frac{1}{2}(\mathrm{e}^{[\Lambda]\Delta t}\{q(t-\Delta t)\} + \{q(t)\})\end{aligned} \tag{2.74}$$

Then the state vector $\{z(t)\}$ is solved as follows:

$$\{z(t)\} = [\Phi]\{d(t-\Delta t)\} + \frac{\Delta t}{2}([B]\{u(t)\} + [D]\{F(t)\}) \tag{2.75}$$

$$\{d(t-\Delta t)\} = \mathrm{e}^{[\Lambda]\Delta t}[\Phi]^{-1}\{z(t-\Delta t)\} + \frac{1}{2}([B]\{u(t-\Delta t)\} + [D]\{F(t-\Delta t)\}) \tag{2.76}$$

The instantaneous optimal control problem is transformed into the minimum value of **J** of Eq. (2.71) under the constraint condition Eq. (2.75). According to the variational method of Sect. 2.2.3 and extreme conditions, the optimal control problem can be solved by introducing the Hamiltonian function:

1. The instantaneous optimal control force vector $\{u(t)\}$ under closed-loop control is:

$$\{u(t)\} = -\frac{\Delta t}{2}[R]^{-1}[B]^{\mathrm{T}}Q]\{z(t)\} \tag{2.77}$$

2. The instantaneous optimal control force vector $\{u(t)\}$ under open-loop control is:

$$\{u(t)\} = -\frac{\Delta t}{2}\left([R] + \left(\frac{\Delta t}{2}\right)^2 [B]^{\mathrm{T}}[Q][B]\right)^{-1} [B]^{\mathrm{T}}[Q]([\Phi]\{d(t-\Delta t)\} + \frac{\Delta t}{2}[D]\{F(t)\}) \tag{2.78}$$

The instantaneous optimal control is only an asymptotic local optimal control, so the result is still not the optimal solution. However, this method does not need to solve Riccati equation, has less computation and does not depend on the characteristics and parameters of the controlled structure, so it can be applied to the active control of nonlinear time-dependent structural systems.

2.2.4.4 Independent Modal Space Control Algorithm

The independent modal space control algorithm is implemented in the state space. For a structural system with n degrees of freedom, its motion can be decomposed into n decoupled SDOF systems in modal coordinates. Assuming that the system has orthogonal damping, the following transformations can be defined:

$$\{z(t)\} = [\Phi]\{q(t)\} \tag{2.79}$$

where, $[\Phi]$ is the $2n \times 2n$ order modal matrix in the state space. The following decoupled modal equations are obtained by plugging Eq. (2.79) into the structural state equation:

$$m_j\ddot{q}(t) + c_j\dot{q}_j(t) + k_jq_j(t) = v_j(t) + F_j(t) \quad (j = 1, 2, 3, \ldots, n) \tag{2.80}$$

where, j represents the j order mode, $v_j(t)$ is the modal control force vector, and $\{v(t)\} = [v_1(t) \ldots v_n(t)]^{\mathrm{T}}$. The transformation relationship between $\{v(t)\}$ and the actual control force vector $\{u(t)\}$ is:

$$\{v(t)\} = [L]\{u(t)\} \tag{2.81}$$

$$[L] = [\Phi]^{\mathrm{T}}[H] \tag{2.82}$$

If set $\{v_j(t)\} = g_{1j}q_j(t) + g_{2j}\dot{q}_j(t)$, then the equations of Eq. (2.81) are independent of each other, and the control algorithm has been transformed into the control problem of "independent modal space". The modal control force $\{v(t)\}$ can be determined by appropriate control algorithm, and the actual control force vector $\{u(t)\}$ can be obtained by Eq. (2.82). The independent modal space control algorithm simplifies the system control problem greatly, and its effect is more obvious when only a few key modes are controlled. It must be noted that the number of controllers should not be less than the number of control modes when the algorithm is used to solve the modal control force.

2.2.4.5 Sliding Mode Control Algorithm

The sliding mode control (SMC) is an effective method for structural vibration control based on variable structure system theory. Its core idea is to determine a control law so that the response trajectory of the structure under control tends to sliding surface (switching surface), and the system motion on the sliding surface is stable. As the name implies, it is especially suitable for semi-active variable stiffness control and variable damping control with changing parameters of controlled system in the control process. The sliding mode control includes two aspects: the determination of sliding surface and the design of controller.

Firstly, the sliding surface should be determined. Considering the state equation and that $\{S(t)\}$ is zero, the equation of sliding surface is obtained:

$$\{S(t)\} = [P]\{z(t)\} = 0 \tag{2.83}$$

where, $[P]$ is an $m \times 2n$ order undetermined matrix, which can be solved using the pole placement method or linear quadratic regulator (LQR) method. Next, the Lyapunov direct method is used to design the sliding mode controller:

$$\begin{aligned} v(t) &= 0.5\{S(t)\}^{\mathrm{T}}\{S(t)\} \\ &= 0.5\{z(t)\}^{\mathrm{T}}[P]^{\mathrm{T}}[P]\{z(t)\} \end{aligned} \tag{2.84}$$

When the sliding mode is $t \to \infty$, the sufficient condition of $\{S(t)\} = 0$ is that the derivative of Eq. (2.84) is less than or equal to 0:

$$\dot{v}(t) = \{S(t)\}^{\mathrm{T}}\{\dot{S}(t)\} \le 0 \tag{2.85}$$

Plug the state equations of structural systems into Eqs. (2.84) and (2.85):

$$\begin{aligned}\dot{v}(t) &= [\lambda](\{u(t)\} - \{G(t)\}) \\ &= \sum_{i=1}^{m} \lambda_i (u_i(t) - G_i(t))\end{aligned} \tag{2.86}$$

where, $u_i(t)$ is the control force of the ith controller, and:

$$\left.\begin{aligned}[\lambda] &= \{S\}^{\mathrm{T}}[P][B] \\ \{G(t)\} &= -([P][B])^{-1}[P]([A]\{z(t)\} + [D]\{p(t)\})\end{aligned}\right\} \tag{2.87}$$

1. For the continuous controllers with no discontinuous points, the control force can be obtained as follows:

$$u_i(t) = G_i(t) - \delta_i \lambda_i \quad \text{or} \quad \{u(t)\} = \{G(t)\} - [\delta][\lambda]^{\mathrm{T}} \tag{2.88}$$

where, $\delta_i \geq 0$ is called sliding margin, $[\delta]$ is a diagonal matrix with diagonal elements δ_i. Therefore, $\dot{v}(t) = -[\lambda][\delta][\lambda]^{\mathrm{T}} \leq 0$ is always established in the whole control process. If the structure is stable in its uncontrolled state, the following control force is adopted:

$$u_i(t) = \alpha_i G_i(t) - \lambda_i \delta_i \tag{2.89}$$

The stable control can also be realized, and $0 \leq \alpha_i \leq 1$.

2. For discontinuous controllers, the control force is:

$$u_i(t) = \begin{cases} G_i(t) - \delta_i D(\|[\lambda]\| - \varepsilon_0) & \lambda_i > 0 \\ 0 & \lambda_i = 0 \\ G_i(t) + \delta_i D(\|[\lambda]\| - \varepsilon_0) & \lambda_i < 0 \end{cases} \tag{2.90}$$

where, $D(\|[\lambda]\| - \varepsilon_0)$ is a unit step function, then $D(\|[\lambda]\| - \varepsilon_0) = 0$ when $\|[\lambda]\| - \varepsilon_0 < 0$; and $D(\|[\lambda]\| - \varepsilon_0) = 1$ when $\|[\lambda]\| - \varepsilon_0 \geq 0$, $\|[\lambda]\|$ is the arbitrary norm of $[\lambda]$. ε_0 is the boundary thickness of slip surface, and the size of boundary layer can be designed through ε_0.

2.2.4.6 H_∞ State Feedback Control Algorithm

There are many kinds of algorithms in practical application of $\mathbf{H}_\infty$ control. The commonly used $\mathbf{H}_\infty$ state feedback control algorithm is introduced here. Considering that the system is represented by the equation of state, define the control output as:

$$\{\bar{z}(t)\} = \begin{Bmatrix} [\psi]\{z(t)\} \\ \{u(t)\} \end{Bmatrix} \tag{2.91}$$

where, $[\psi]$ is weight matrix. The control purpose is to ensure that the control output (assuming zero initial condition) satisfies the following condition:

$$\frac{\int_0^\infty \{\bar{z}(t)\}^{\mathrm{T}}\{\bar{z}(t)\}dt}{\int_0^\infty \{F(t)\}^{\mathrm{T}}\{F(t)\}dt} \le \gamma^2 \tag{2.92}$$

Equation (2.92) corresponds to that the infinite modulus of the transfer function from $\{F(t)\}$ to $\{\bar{z}(t)\}$ is less than γ:

$$\|[\mathbf{H}(s)]\|_\infty \left\| \begin{bmatrix} [\psi] \\ [\mathbf{G}] \end{bmatrix} (s[\mathbf{I}] - [\mathbf{A}']^{-1}[\mathbf{H}] \right\|_\infty \le \gamma \tag{2.93}$$

where, $[G]$ is the control gain matrix. It is assumed that all the states of the system can be obtained and the nominal system ($[A]$, $[B]$) can be controlled. For a given constant $\gamma > 0$, if there is a positive definite symmetric matrix $[Q]$, the algebraic Riccati equation has a positive definite symmetric solution $[P]$:

$$[A]^{\mathrm{T}}[P] + [P][A] + \gamma^{-2}[P][\mathbf{H}][\mathbf{H}]^{\mathrm{T}}[P] + [\psi]^{\mathrm{T}}[\psi] - [P][B][B]^{\mathrm{T}}[P] + [Q] = 0 \tag{2.94}$$

Then there is a state feedback control law:

$$\{u(t)\} = [G]\{z(t)\} = [B]^{\mathrm{T}}[P]\{z(t)\} \tag{2.95}$$

which satisfies:

1. $[A'] = [A] + [B][G]$ is stable matrix;
2. The infinite modulus of the transfer function matrix of the closed-loop (from $\{F(t)\}$ to $\{\bar{z}(t)\}$) is $\|[\mathbf{H}(s)]\|_\infty \le \gamma$, the controlled structural system is positive definite and has interference attenuation γ.

Therefore, the design steps of $\mathbf{H}_\infty$ state feedback control are as follows:

1. For a given constant γ, assuming that $[Q]$ is a small value, $[P]$ can be obtained by solving the Riccati equation Eq. (2.94);
2. If Eq. (2.94) has no positive definite solution, increase the value of γ until there is a positive definite solution;
3. If Eq. (2.94) has a positive definite solution, reduce the value of γ or adjust the weight matrix $[\psi]$;
4. For the solution of $[P]$, the control law of Eq. (2.95) is applied to obtain the control force.

2.2.4.7 Optimal Polynomial Control

Considering the state equation, the polynomial performance index **J** is defined as:

$$\mathbf{J} = \int_0^\infty [\{z(t)\}^{\mathrm{T}}[Q]\{z(t)\} + \{u(t)\}^{\mathrm{T}}[R]\{u(t)\}$$

$$+ \sum_{i=2}^{k} (\{z(t)\}^{\mathrm{T}}][M_i]\{z(t)\})^{i-1}(\{z(t)\}^{\mathrm{T}}][Q_i]\{z(t)\}) + \bar{h}(z(t))]dt \tag{2.96}$$

and

$$\bar{h}(z(t)) = \left(\sum_{i=2}^{k} (\{z(t)\}^{\mathrm{T}}][M_i]\{z(t)\})^{i-1}\{z(t)\}^{\mathrm{T}}[M_i]\right)$$

$$[B][R]^{-1}[B]^{\mathrm{T}}\left(\sum_{i=2}^{k} (\{z(t)\}^{\mathrm{T}}][M_i]\{z(t)\})^{i-1}[M_i]\{z(t)\}\right)$$

where, $[Q]$ and $[Q_i]$ are the semi-positive definite state weight matrices ($i = 2, 3 \ldots k$); $[R]$ is the positive definite control matrix;$[M_i]$ is the positive definite matrix ($i = 2, 3 \ldots k$); $[Q]$, $[Q_i]$ and $[R]$ can be arbitrarily selected,$[M_i]$ and $[Q_i]$ are implicit functions. By applying the solution of Hamilton-Jacobi-Bellman equation, the control force $\{u(t)\}$ that minimizes the performance index **J** is obtained:

$$\{u(t)\} = -[R]^{-1}[B][P]\{z(t)\} - [R]^{-1}[B]^{\mathrm{T}}(\sum_{i=2}^{k} (\{z(t)\}^{\mathrm{T}}][M_i]\{z(t)\})^{i-1}[M_i]\{z(t)\}) \tag{2.97}$$

where, $[P]$ and $[M_i]$ are solved by the following Riccati equation and Lyapunov equation, respectively:

$$[P][A] + [A]^{\mathrm{T}}[P] - [P][B][R]^{-1}[B]^{\mathrm{T}}[P] + [Q] = 0 \tag{2.98}$$

$$[M_i]([A] - [B][R]^{-1}[B]^{\mathrm{T}}[P]) + \left([A] - [B][R]^{-1}[B]^{\mathrm{T}}[P]\right)^{\mathrm{T}}[M_i] + [Q_i] = 0 \quad (i = 2, 3, \ldots, k) \tag{2.99}$$

2.2.4.8 Predictive Control Algorithm

In the optimal control algorithm, the Riccati equation must be solved to obtain the minimum value of objective function. The Model Predictive Control (MPC) algorithm, which is a computer control method developed in recent years, can avoid this and simplify the control problem. Assuming that the matrices of $[A]$ and $[B]$ in the state equation is known and remain constant throughout the process, the prediction model can be represented by the following discrete state equations:

$$\{\bar{x}(k+j|k)\} = [A]\{\bar{x}(k+j-1|k)\} + [B]\{\bar{u}(k+j-1|k)\} + [D]\{\bar{p}(k+j-1|k)\} \tag{2.100}$$

$$\{\bar{y}(k+j|k)\} = [L]\{\bar{x}(k+j|k)\} \tag{2.101}$$

where, $\{\bar{x}(k+j|k)\}$ and $\{\bar{y}(k+j|k)\}$ are the predictions of state vector and output at time $k+j$ by the time k, respectively. $\{\bar{u}(k+j|k)\}$ represents the actual control force, and $[L]$ is the output matrix. The control objective is to minimize the following objective function **J**:

$$J = \frac{1}{2}(\{\bar{y}(k+\lambda|k)\} - \{y_r(k+\lambda)\})^{\mathrm{T}}[Q](\{\bar{y}(k+\lambda|k)\} - \{y_r(k+\lambda)\}) \tag{2.102}$$

where, $\{y_r(k+\lambda)\}$ is the output vector at time $k+\lambda$; $[Q]$ is the weight matrix. At this point, the control force can be written in the following form:

$$\{u(k)\} = -[\beta]^{-1}[\alpha]\{x(k)\} + [\beta]^{-1}[\gamma]\{y_r(k+\lambda)\} \tag{2.103}$$

where,

$$[\alpha] = [w(\lambda)]^{\mathrm{T}}[L]^{\mathrm{T}}[Q][L][T(\lambda)],\ [\beta] = [w(\lambda)]^{\mathrm{T}}[L]^{\mathrm{T}}[Q][L][w(\lambda)],$$
$$[\gamma] = [w(\lambda)]^{\mathrm{T}}[L]^{\mathrm{T}}[Q],\ [w(\lambda)] = ([I] + [A]^1 + [A]^2 + \cdots + [A]^{\lambda-1})[B],$$
$$[T(\lambda)] = [A]^{\lambda}.$$

The predictive control algorithm is based on the continuous on-line rolling optimization, and the feedback correction is carried out continuously based on the error between the output results of measured and predictive models during optimization process. So it can overcome the influence of many uncertainties and complex factors in the structural system. In addition to the basic algorithms of active control mentioned above, there are other algorithms developed and derived from them, such as impulse control, compensation control, adaptive control, reasoning control, fuzzy control and μ theory control.

2.2.5 Structural Semi-active Control Algorithm

In semi-active control, the key problem is how to determine the parameter change of semi-active controller, which affects the control force vector $u(t)$. The semi-active control algorithm is the basis of determining the parameter change law of controller. The purpose is to find the change law of optimal control parameters along with the feedback results under the condition that the control system satisfies its state equation and various constraints. Then, the system can achieve better performance indicators and better structural control. The control algorithms of common semi-active control system are introduced below.

2.2.5.1 Active Variable Damping System

The control algorithm of structural active variable damping system generally includes two aspects: one is the determination of active (optimal) control force, i.e. command signal, which can be calculated and determined by any appropriate active control algorithm in Sect. 2.2.4; the other is using what rules to implement the damping force of active variable damping control device. The tracking command signal is determined according to the active control force combined with the performance of active variable damping control device. Among them, the determination algorithm of active control force has been introduced in Sect. 2.2.4. This section focuses on how to determine the active variable damping control force, that is, the active variable damping control algorithm of the structure.

The active variable damping control algorithm of structure mainly refers to the active control force $u(t)$ and considers the actual situation that the semi-active damping force may be realized, and sets the active variable damping force as close as possible to the active control force. The possible control forces of active variable damping devices have the following characteristics: (1) The direction of control force is limited, because the active variable damping control devices provide control forces in the form of damping forces, which need to passively depend on the speed of structural vibration (i.e. the relative speed between control devices). Therefore, it can only provide control force to prevent the movement of the structure. (2) Compared with the active control actuator, the maximum damping force of active variable damping device is not a special limitation. The maximum output forces of the both devices are limited under comparable conditions. According to the above characteristics, the active variable damping control algorithm expresses the control force that can be achieved by the active variable damping control device on the basis of the target requirements of active control force.

Assuming that the active control force vector obtained by active control algorithm is $\{u(t)\}$. The semi-active control force of the active variable damping control device corresponding to the active control force is $u_s(t)$. Considering that the control force is expressed at the right end of the motion or state equations. The relationships between the control force and viscous damping force $u_d(t)$ realized by the active variable damping control device are:

$$u_s(t) = -u_d(t) \tag{2.104}$$

$$u_d(t) = c_d(v, t)\dot{x} \tag{2.105}$$

Several main active variable damping control algorithms and their corresponding damping coefficients are given below.

1. Simple Bang-Bang control algorithm

Simple Bang-Bang control algorithm is also called simple ON/OFF algorithm, or two-stage fail-safe control algorithm. The corresponding active variable damping coefficient can be expressed as follows:

$$c_d(t) = \begin{cases} c_{d\max} & x\dot{x} > 0 \\ c_{d\min} & x\dot{x} \leq 0 \end{cases} \tag{2.106}$$

It can be seen that when the structure deviates from the balanced position, the active variable damping control device provides the maximum damping to the structure to prevent the vibration of structure; when the structure vibrates to the balanced position, the active variable damping control device provides the minimum damping to the structure, so that the structure can return to the balanced position as soon as possible. These algorithms often make the structure pass through the equilibrium position at a higher speed. Under the simple Bang-Bang control algorithm, the damping force of the active variable damping control device can be expressed as shown in Fig. 2.8.

2. Optimal Bang-Bang control algorithm

The optimal Bang-Bang variable damping control algorithm can be expressed as follows:

$$c_d(t) = \begin{cases} c_{d\max} & u\dot{x} > 0 \\ c_{d\min} & u\dot{x} \leq 0 \end{cases} \tag{2.107}$$

It can be seen that when the active control force and the vibration of the location of active variable damping control device, such as the inter-story location, are in the opposite direction, the maximum damping coefficient can be provided by the active

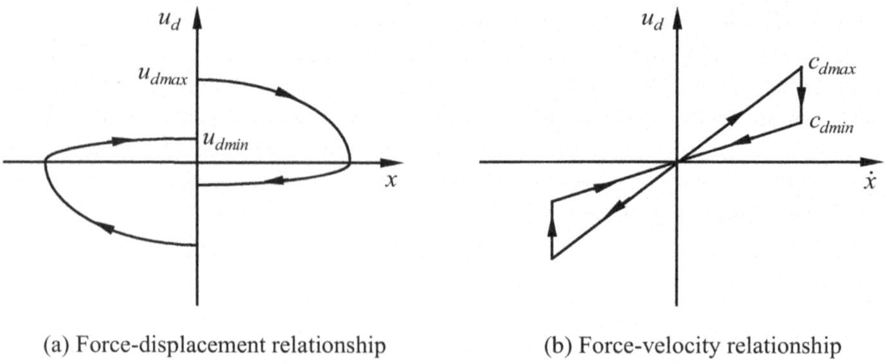

Fig. 2.8 Semi-active damping force determined by simple bang-bang control algorithm

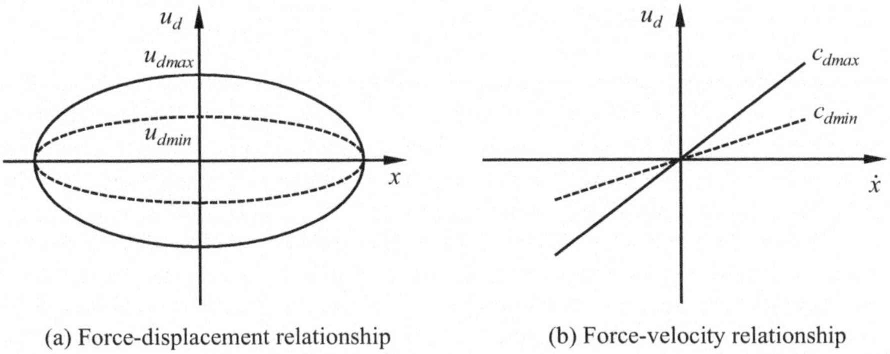

Fig. 2.9 Semi-active damping force determined by the control algorithms of Passive-off and Passive-on

variable damping control device, otherwise the minimum damping coefficient can be applied. This algorithm just reflects that the active variable damping control device can only exert the force to prevent the structure from moving, but cannot exert the force to push the structure to move. Two extreme cases of optimal Bang-Bang control algorithm are shown in Fig. 2.9. In fact, these two extreme cases are equivalent to the Passive-off control and Passive-on control, and that the active semi-damping force of the optimal Bang-Bang control algorithm varies between them.

3. Boundary Hrovat optimal control algorithm

In 1983, Hrovat proposed an active variable damping control method considering that the active variable damping control device is actually a variable parameter passive damper for structures. However, only the direction limitation of active variable damping force is considered in the Hrovat algorithm, and the limitation of active variable damping control force is not considered. The boundary Hrovat optimal control algorithm can be expressed as follows:

$$c_d(t) = \begin{cases} c_{d\max} & u\dot{x} < 0 \text{ and } |u\dot{x}| > c_{d\max} \\ c_d = |u/\dot{x}| & u\dot{x} < 0 \text{ and } |u\dot{x}| < c_{d\max} \\ c_{d\min} & u\dot{x} \geq 0 \end{cases} \tag{2.108}$$

Based on the optimal Bang-Bang control algorithm, the variable damping coefficient corresponding to Hrovat semi-active control force is added to the range of viscous damping coefficient, which can be realized by the active variable damping control device. That is, when $u\dot{x} < 0$ and $|u\dot{x}| < c_{d\max}$, the optimal control force u is achieved by adjusting the active variable damping control device.

2.2.5.2 Active Variable Stiffness System

When the semi-active control system shown in Fig. 2.7b represents the active variable stiffness system, the following control modes can be realized: (1) The stiffness controller is always in the locked state ("ON"), at which time the active variable stiffness control system always adds a stiffness element to the structure, and the mechanical model of the device is shown in Fig. 2.10a. (2) The stiffness controller is always in the open state ("OFF") when the active variable stiffness control does not add stiffness elements to the structure. When the active variable stiffness control device contains viscous damping characteristics, a damping element can be considered to be added to the structure. The mechanical model of the device is shown in Fig. 2.10b. (3) The stiffness controller determines the "ON" state or "OFF" state of the active variable stiffness device by some control algorithm. When the semi-active ON/OFF control algorithm is adopted and the damping characteristics of the device are neglected, the mechanical model of the device is shown in Fig. 2.10c; when the damping characteristics of the device are considered, the mechanical model is shown in Fig. 2.10d. Figure 2.10c shows that the active variable stiffness control system

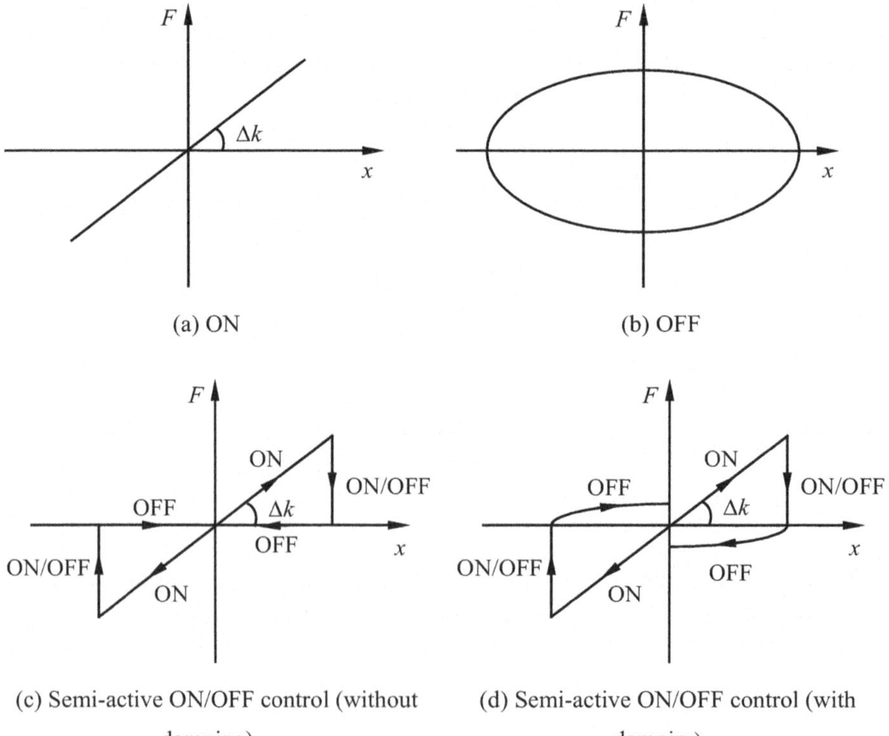

Fig. 2.10 Mechanical model of active variable stiffness control device

can consume part of the structural vibration energy by changing the stiffness of the control device, even without considering the damping characteristics of the device, thus increasing the structural damping.

The motion equations of the structural variable stiffness control system corresponding to the above three control states can be expressed as follows:
Passive-on:

$$m\ddot{x} + c\dot{x} + (k + \Delta k)x = F(t) \tag{2.109}$$

Passive-off:

$$m\ddot{x} + (c + \Delta c)\dot{x} + kx = F(t) \tag{2.110}$$

The semi-active control can be expressed by Eq. (2.23), where $u(t)$ is the control force. Since $u(t)$ is on the right side of the equation, which is exactly opposite to the symbol of control force of control device, that is, $u(t) = -\Delta kx$. When the Passive-on passive control is adopted, the control device only adds stiffness to the structure. This control method is very sensitive to the disturbance characteristics, and often increases the acceleration response of the structure. Sometimes, it is beneficial to reduce the displacement response of the flexible structure, just like the steel support in the steel structure. The use of Passive-off passive control is equivalent to installing passive viscous dampers with small damping coefficient on the structure. However, because the active variable stiffness control device usually only provides small damping, which generally cannot meet the requirements of vibration reduction. In the active variable stiffness control method, different control algorithms make the control force have different characteristics. If the ON/OFF control algorithm shown in Fig. 2.10c is adopted, the control force can be expressed as follows:

$$u(t) = \begin{cases} -\Delta kx & \dot{x}x > 0 \\ 0 & \dot{x}x \le 0 \end{cases} \tag{2.111}$$

The physical meaning of the control algorithm is that when the inter-layer relative velocity and displacement of the structure have identical sign, that is, when the structure vibrates in the direction away from the equilibrium point, the active variable stiffness control system adds stiffness to the structure; when the inter-layer vibration velocity and displacement of the structure have inverse sign, that is, when the structure vibrates from the position far from the equilibrium point to the equilibrium point, the active variable stiffness control system does not add stiffness to the structure.

2.3 Intelligent Control

The intelligent structural control usually includes two kinds, one is the intelligent control based on intelligent control algorithm, and the other is the intelligent control based on intelligent drive or intelligent damper.

The intelligent structural control is marked by the intelligent control algorithms such as fuzzy control, neural network control and evolutionary algorithm, which is different from the active control in that it does not need precise structural model and uses intelligent control algorithm to determine the relationship between the input or output feedback and control gain, while the control force is still implemented through the actuator under the input of external energy.

The intelligent structural control is marked by the intelligent driving materials and devices such as electro-magnetorheological fluids, piezoelectric materials, electro-magnetostrictive materials and shape memory materials. Its control principle is basically the same as that of active control, but the actuator implementing control force is an intelligent actuator or damper made of intelligent materials. The intelligent actuators usually need less external input energy than hydraulic or motor actuators, and can basically or completely achieve the optimal active control force. From this point of view, it has the same performance as active control actuators, but the way, speed and efficiency of the conversion from external energy into mechanical energy are different. Similar to the semi-active control devices, the smart dampers require only a small amount of energy adjustment in order to actively utilize the reciprocating relative deformation or velocity of structural vibration to achieve the optimal active control force as far as possible, but the adjustment mode, speed and effectiveness are different.

2.4 Hybrid Control

The so-called hybrid control is to apply both passive control and active control in structure, or applying more than one passive control device at the same time, to give full play to the strengths of each control form and each control device, overcome their weaknesses, and achieve better control effect.

References

1. Li, Aiqun. 2007. *Vibration control of engineering structure*. Beijing: China Machine Press.
2. Japan Isolation Structure Association. 2008. *Design and construction manual of passive damping structure*. Beijing: China Architecture & Building Press.
3. Li, Aiqun, Dajun Ding, and Zhengliang Cao. 1994. Analysis of mechanics and energy consumption of friction control device in building structure. *Mechanics in Engineering* 05: 44–46. (in Chinese).

4. Li, Aiqun. 1992. *Research on seismic control and control device of reinforced concrete shear wall structure*. Nanjing: Southeast University. (in Chinese).
5. Ye, Zhengqiang. 2003. *Research on theory, experiment and application of energy dissipation and vibration reduction technology with viscous fluid damper*. Nanjing: Southeast University. (in Chinese).
6. Chang, Yejun. 2003. *Research and application of viscoelastic damper and energy dissipation structure*. Nanjing: Southeast University. (in Chinese).
7. Zhang, Congjun. 2009. *Test of soft steel supported damper and research on performance-based design method of energy dissipation structure*. Nanjing: Southeast University. (in Chinese).
8. Ou, J.P. 2003. *Structural vibration control-active, semi-active and intelligent control*. Beijing: Science Press.
9. Ikeda, Y. 2009. Active and semi-active vibration control of buildings in Japan—Practical applications and verification. *Structural Control and Health Monitoring* 16 (7–8): 703–723.
10. Fei, Lu. 2002. *Study on some problems of active control project of wind vibration control for Nanjing TV tower*. Nanjing: Southeast University. (in Chinese).

Chapter 3
Basic Principle of Frequency Modulation Vibration Control

Abstract The commonly used frequency-modulation (FM) vibration control includes Tuned Mass Damper (TMD) and Tuned Liquid Damper (TLD). Basic principle of TMD and TLD, including motion equation, basic characteristics and basic construction of vibration control system are introduced respectively in this chapter.

The frequency-modulation (FM) vibration reduction refers to the conversion of the vibration energy of the structure itself to the passive frequency-modulated device when the natural frequency of vibration reduction device is modulated with the controlled frequency of the structure, so as to dissipate the vibration energy of the structure and control the structural dynamic response. The commonly used passive FM dampers include Tuned Mass Damper (TMD) and Tuned Liquid Damper (TLD).

3.1 FM Mass Vibration Control

The TMD system is composed of mass, spring and damper. By reasonably designing TMD, the frequency of TMD can be as close as possible to the basic frequency or excitation frequency of the main structure [1–3]. When the main structure is excited and vibrated, the TMD substructure will produce an inertial force opposite to the direction of structure vibration, which will reduce and control the vibration response of the main structure.

3.1.1 Motion Equation of FM Mass Vibration Control System

The typical single-particle structure with FM mass dampers is shown in Fig. 3.1a. The structure can be further simplified to an ideal system as shown in Fig. 3.1b. According to the principle of dynamic balance, the motion equation of the system can be obtained as follows:

A. Li, *Vibration Control for Building Structures*, Springer Tracts in Civil Engineering, https://doi.org/10.1007/978-3-030-40790-2_3

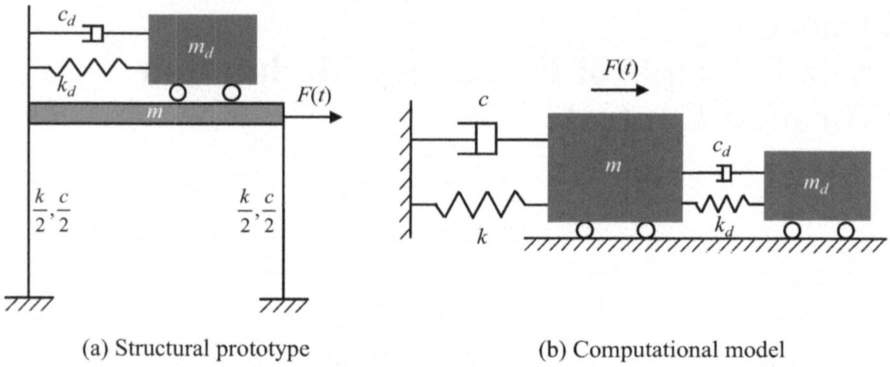

Fig. 3.1 FM mass vibration reduction system

$$m\ddot{x} + c\dot{x} + kx = F(t) + F_{TMD}(t) \tag{3.1}$$

$$\ddot{x}_d + 2\zeta_d\omega_d\dot{x}_d + \omega_d^2 x_d = -\ddot{x} \tag{3.2}$$

$$F_{TMD}(t) = c_d\dot{x}_d + k_d x_d \tag{3.3}$$

Plug Eq. (3.3) into Eqs. (3.1) and (3.2):

$$m\,\ddot{x} + (c + c_d)\dot{x} + (k + k_d)x - c_d\dot{x}_d - k_d x_d = F(t) \tag{3.4}$$

$$m_d\ddot{x}_d + c_d(\dot{x}_d - \dot{x}) + k_d(x_d - x) = 0 \tag{3.5}$$

where, m_d, c_d and k_d are the mass, damping coefficient and stiffness of FM mass damper, respectively; x_d, $\dot{x}_d$ and $\ddot{x}_d$ are the displacement, velocity and acceleration of FM mass damper, respectively. The damping coefficient ζ_d and circular frequency ω_d of FM mass dampers can be expressed as follows:

$$\zeta_d = \frac{c_d}{2\omega_d m_d} = \frac{c_d}{2\sqrt{k_d m_d}} \tag{3.6}$$

$$\omega_d = \frac{2\pi}{T_d} = \sqrt{\frac{k_d}{m_d}} \tag{3.7}$$

3.1.2 Basic Characteristics of FM Mass Vibration Control

In order to further analyze the basic characteristics of FM mass damper system, it is assumed that the FM mass damper system is periodically stimulated:

$$F(t) = F_0 \sin \omega_F t \tag{3.8}$$

where, ω_F is the circular frequency of the excitation, t is time. In order to facilitate discussion, according to the damping characteristics of the main structure, the basic characteristics of the FM mass vibration reduction system are discussed separately, which are divided into the undamped main structure and damped main structure [4].

1. **Undamped main structure**

Assuming that β is the dynamic magnification factor of displacement response of the main structure relative to the equivalent static displacement. The transfer function method is used to solve Eqs. (3.1) and (3.2) by introducing the parameters of λ and λ_d:

$$x = \beta \frac{F_0}{k} \tag{3.9}$$

$$\beta = \left[\frac{\left(\lambda^2 - \lambda_d^2\right)^2 + 4\zeta_d^2 \lambda^2 \lambda_d^2}{4\zeta_d^2 \lambda^2 \lambda_d^2 \left(\mu\lambda^2 + \lambda^2 - 1\right)^2 + \left[\mu\lambda^2\lambda_d^2 - \left(\lambda^2 - \lambda_d^2\right)\left(\lambda^2 - 1\right)\right]^2} \right]^{\frac{1}{2}} \tag{3.10}$$

where, $\frac{p_0}{k}$ is the maximum equivalent static displacement of the main structure under external harmonic excitation; λ is the ratio of external excitation frequency ω_F to the frequency ω of the main structure; λ_d is the ratio of frequency ω_d of substructure to the natural frequency ω of the main structure; μ is the ratio of mass m_d of substructure to the main structure mass m. The above parameters can be calculated according to the following equations:

$$\lambda = \frac{\omega_F}{\omega} \tag{3.11}$$

$$\lambda_d = \frac{\omega_d}{\omega} \tag{3.12}$$

$$\omega = \sqrt{\frac{k}{m}} \tag{3.13}$$

$$\mu = \frac{m_d}{m} \tag{3.14}$$

According to Eqs. (3.6) and (3.7), the application laws of TMD system in the vibration reduction design for building structures under external harmonic excitation are discussed separately.

1. Assuming that $\zeta_d = 0$ for the substructure, which is equivalent to that the TMD substructure without dampers is a pure undamped FM system, and Eq. (3.7) becomes:

$$\beta = \frac{|\lambda^2 - \lambda_d^2|}{|\mu\lambda^2\lambda_d^2 - \lambda^4 + \lambda^2 + \lambda^2\lambda_d^2 - \lambda_d^2|} \tag{3.15}$$

When the natural frequency ω_d of the substructure is equal to the external excitation frequency ω_F of the main structure, it can be obtained by Eqs. (3.7) and (3.9) that $\beta = 0$. Therefore, the main structure achieves the optimal FM control effect.

When $\omega_d = \omega$, $\lambda_d = 1$, if the main structure is excited by external resonance ($\omega_F = \omega$), then $\lambda = \lambda_d = 1$ is obtained by Eqs. (3.8) and (3.9). $\beta = 0$ is obtained by Eq. (3.12). Therefore, the main structure also achieves the optimal frequency modulation effect.

From the above analysis, for the undamped TMD substructures, when the natural frequency ω_d of substructure is equal to the external excitation frequency ω_F of the main structure, whether the main structure is subjected to external resonance excitation or not, the optimal frequency modulation effect can be achieved.

2. Assuming that $\zeta_d = 0$ for the substructure. When $\omega_d = \omega$, $\lambda_d = 1$, if the external excitation frequency ω_F deviates from the natural frequency ω of the main structure ($\omega_F \neq \omega$). According to Eq. (3.12), there may be a variety of situations: (a) the main structure may produce a large vibration response; (b) it may also approach the optimal frequency modulation effect; (c) it may also be between the two situations.

Assuming that $\mu = \frac{m_d}{m} = 0.05$, which in plugged into Eq. (3.12):

$$\beta = \frac{|\lambda^2 - 1|}{|\lambda^4 - 2.05\lambda^2 + 1|} \tag{3.16}$$

From Eq. (3.13), when $\lambda_d = 1$ and $\mu = 0.05$, it can be obtained that the variation curve of β_1 of the undamped TMD substructure system with respect to λ is shown in Fig. 3.2a. As can be seen from the figure, the curve is divided into three sections, two peaks appear near $\lambda = 0.894$ and 1.118, and $\beta \to +\infty$; when $\lambda = 1$ and $\beta = 0$, β rises very fast when λ leaves a very small frequency band with respect to 1; when $\lambda = 1$ and $\beta = 1$, it is equivalent to static load; when $\lambda > 2$ and $\beta \to 0$, it shows that the undamped TMD substructure is particularly suitable for long-span or high-flexible structures with low natural frequency ω.

Assuming that $\mu = \frac{m_d}{m} = 0.20$, it is derived from Eq. (3.12) that:

$$\beta = \frac{|\lambda^2 - 1|}{|\lambda^4 - 2.20\lambda^2 + 1|} \tag{3.17}$$

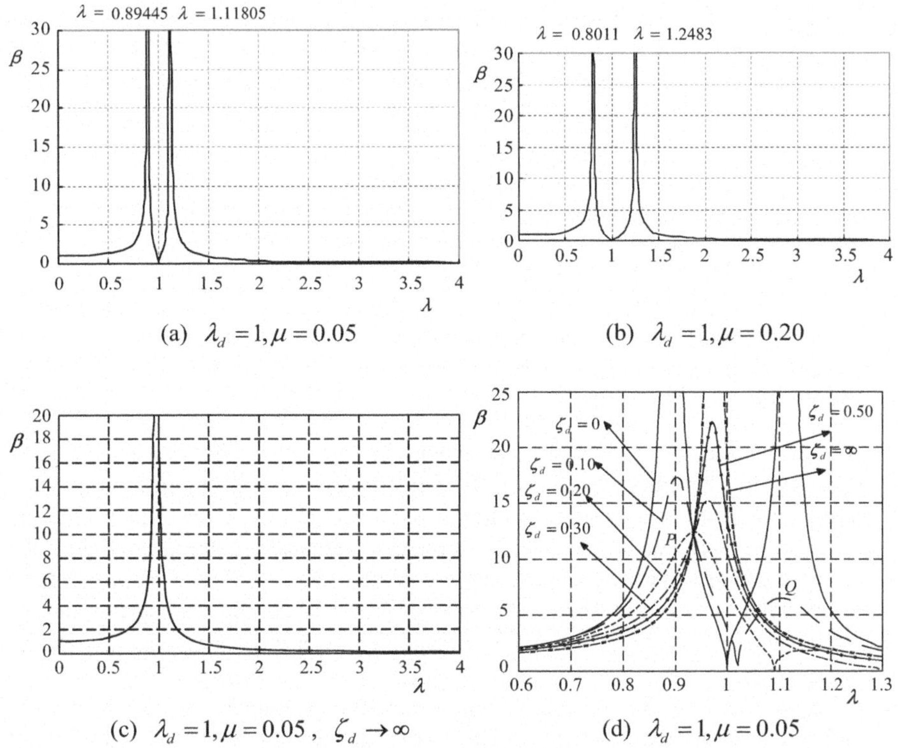

Fig. 3.2 Variation curve of β_1 of the undamped TMD substructure system with respect to λ

From Eq. (3.14), the variation curve of β of the undamped TMD substructure system with respect to λ, when $\lambda_d = 1$ and $\mu = 0.20$, is shown in Fig. 3.2b. The variation law of β with respect to λ reflected by Fig. 3.2b is similar to the results of Fig. 3.2a, except that the peak position of β changes. The β curve is still divided into three sections: when $\lambda = 1$ and $\beta = 0$, β rises very fast when λ leaves a very small frequency band with respect to 1 (i.e., slightly wider frequency band of Fig. 3.2a); $\beta \to +\infty$ near $\lambda = 0.8011$ and 1.2484; when $\lambda = 1$ and $\beta = 1$, it is equivalent to static load; when $\lambda > 2$, it still exists $\beta \to 0$.

From the above analysis, the following results can be obtained:

(a) When $\omega_d = \omega$, only when the external excitation frequency ω_p is equal to or very close to the resonance frequency ω, the optimal or better frequency modulation effect can be achieved. It shows that the undamped TMD substructure is only suitable for the external excitation in narrow band. If the external excitation frequency band is wider, the multiple TMD substructures with different frequencies can be installed in the main structure to control the external excitation in different frequency bands.

(b) If the natural frequency ω of the main structure is smaller than that of the external excitation frequency (i.e., the structure is very flexible, $\lambda > 2$, $\beta \to 0$), it may make the structure approach the optimal frequency modulation effect, which indicates that the undamped TMD substructure is especially suitable for the long-span or flexible structure with lower natural frequency ω, but not for the rigid structure with higher natural frequency.

(c) $\mu = \frac{m_d}{m}$, which is increased from 0.05 to 0.20, the shape of $\beta - \lambda$ curve has not changed. There are still two peaks of $\beta \to \infty$, but the peak value is slightly away from $\lambda = 1(\beta = 0)$, which makes the frequency band of effective control of TMD substructure increase slightly, but the increase is not obvious. The change trend and control effect have not been significantly improved, indicating that increasing the mass m_d of substructure cannot improve the control effect of TMD substructure system.

3. Assuming that $\mu = \frac{m_d}{m} = 0.05$, when $\zeta_d \to \infty$ for the substructure (the substructure damping is infinite), the substructure and the main structure are fixed (connected as a whole), and the actual frequency ω' of the whole structure is slightly lower than ω due to the influence of substructure damping, then if the main structure is excited by external resonance ($\omega_p = \omega'$), the main structure is equivalent to the traditional structure without damper, the main structure is in the resonance state ($\beta \to \infty$). The variation curve β with respect to λ is shown in Fig. 3.2c.
4. It is assumed that $\mu = \frac{m_d}{m} = 0.05$, when the natural frequency ω_d of substructure and the natural frequency ω of main structure are equal, if the damping of substructure are 0, 0.10, 0.20, 0.32, 0.50 and infinite, the variation curves of β of TMD substructure systems with respect to λ are shown in Fig. 3.2d. As can be seen from the figure:

   (a) If there is no substructure damping, the resonance response amplitude of the main structure is infinite.
   (b) If the damping ratio of the substructure is infinite, the substructure and the main structure are fixed (connected as a whole). The main structure is equivalent to the traditional structure without damper, and the resonance response amplitude of the main structure is infinite.
   (c) With the increase of substructure damping ratio, the maximum response amplitude of the main structure has a process from infinite to small and from small to infinite, so there must be a value ζ_d between the two extremes to minimize the peak value. For the two points (P and Q) of the $\beta - \lambda$ curve of Fig. 3.2d, β is independent with the damping ratio ζ_d, and the minimum peak amplitude can be obtained by selecting the two fixed points to reach the same height. The optimal frequency ratio λ_{dopt} in accordance with this process can be determined as follows:

$$\lambda_{dopt} = \frac{1}{1+\mu} \tag{3.18}$$

Table 3.1 Optimal parameter of TMD

Condition	Excitation		Optimized response		Optimized parameters of vibration absorber	
	Type	Position of action	Optimal parameter β	β_{opt}	λ_{dopt}	ζ_{dopt}
1	Force $p_0 e^{iwt}$	Structure	$\frac{kx}{p_0}$	$\left(1+\frac{2}{\mu}\right)^{\frac{1}{2}}$	$\frac{1}{1+\mu}$	$\sqrt{\frac{3\mu}{8(1+\mu)}}$
2	Force $p_0 e^{iwt}$	Structure	$\frac{m\ddot{x}}{p_0}$	$\left[\frac{2}{\mu(1+\mu)}\right]^{\frac{1}{2}}$	$\left(\frac{1}{1+\mu}\right)^{\frac{1}{2}}$	$\sqrt{\frac{3\mu}{8(1+\frac{\mu}{2})}}$
3	Acceleration $\ddot{x}_g e^{iwt}$	Foundation	$\frac{\omega^2 x}{\ddot{x}_g}$	$\left(\frac{2}{u}\right)^{\frac{1}{2}}(1+\mu)$	$\frac{(1-\frac{\mu}{2})^{\frac{1}{2}}}{1+\mu}$	$\sqrt{\frac{3\mu}{8(1+\mu)(1-\frac{\mu}{2})}}$
4	Acceleration $\ddot{x}_g e^{iwt}$	Foundation	$\frac{\ddot{x}_g+\ddot{x}}{\ddot{x}_g}$	$\left(1+\frac{2}{\mu}\right)^{\frac{1}{2}}$	$\frac{1}{1+\mu}$	$\sqrt{\frac{3\mu}{8(1+\mu)}}$

The amplitudes of the points of P and Q are as follows:

$$\beta = \sqrt{1 + \frac{2}{\mu}} \tag{3.19}$$

By adjusting ζ_{dopt}, the two fixed points P and Q of Fig. 3.2d are maximized:

$$\zeta_{dopt} = \sqrt{\frac{3\mu}{8(1+\mu)}} \tag{3.20}$$

Furthermore, the maximum dynamic amplification coefficients β of various excitation and response values and the corresponding optimal parameters of the absorbers can be summarized in Table 3.1.

2. **Damped main structure**

When the main structure is damped, the formula of β can also be established, but the invariant points P and Q no longer exist in the case of undamped structure, so the optimization values λ_d and ζ_d must be determined by numerical methods.

(1) Minimum displacement response optimization criterion

For harmonic excitation, the optimum criterion is to minimize the response amplitude β:

$$\beta_j = \sqrt{\frac{A_j^2 + B_j^2}{C^2 + D^2}} \tag{3.21}$$

Table 3.2 Response parameters A_j and B_j in various excitations

Condition	Excitation	Response parameters	Optimized response	A_j	B_j
1	$p_0 e^{iwt}$	x	$\frac{kx}{p_0}$	$\lambda_d^2 - \lambda^2$	$2\zeta_d \lambda_d \lambda$
2	$p_0 e^{iwt}$	$\dot{x}$	$\frac{kx}{p_0 \omega_s}$	$-2\zeta_d \lambda_d \lambda^2$	$\lambda(\lambda_d^2 - \lambda)^2$
3	$p_0 e^{iwt}$	$\ddot{x}$	$\frac{m\ddot{x}}{p_0}$	$-\lambda^2(\lambda_d^2 - \lambda)^2$	$-2\zeta_d \lambda_d \lambda^2$
4	$p_0 e^{iwt}$	Force of foundation	$\frac{F}{p_0}$	$\lambda_d^2 - \lambda^2 - 4\zeta_d \zeta \lambda_d \lambda^2$	$2\zeta_d \lambda_d \lambda + 2\zeta_d \lambda(\lambda_d^2 - \lambda^2)$
5	$\ddot{x}_g e^{iwt}$	x	$\frac{\omega_s^2 x}{\ddot{x}_g}$	$\lambda_d^2(1+\mu) - \lambda^2$	$2\zeta_d \lambda_d \lambda(1+\mu)$
6	$\ddot{x}_g e^{iwt}$	$\ddot{x} + \ddot{x}_g$	$\frac{\ddot{x}_g + \ddot{x}}{\ddot{x}_g}$	Same to condition 4	Same to condition 4
7	$\ddot{x}_g e^{iwt}$	$x + x_g$	$\frac{\omega_s^2(x_1 + x_g)}{\ddot{x}_g}$	$-\frac{A_4}{\lambda^2}$	$-\frac{B_4}{\lambda^2}$

where, j is the number of specific conditions given in Table 3.2, A_j and B_j are given in the same table, and:

$$C = (\lambda_d^2 - \lambda^2)(1 - \lambda^2) - \lambda_d^2 \lambda^2 \mu - 4\zeta_d \zeta \lambda_d \lambda^2 \tag{3.22}$$

$$D = 2\zeta_d \lambda_d \lambda(1 - \lambda^2 - \lambda^2 \mu) + 2\zeta \lambda(\lambda_d^2 - \lambda^2) \tag{3.23}$$

The optimal λ_d and ζ_d can be obtained by solving the equation combining $\frac{\partial \beta_j}{\partial \lambda} = 0$ or $\frac{\partial \beta}{\partial \zeta_d} = 0$, but the non-linear equation about λ_d and ζ_d must be solved.

(2) Minimum acceleration criterion

Because the large acceleration generated by excitation has adverse effects on the foundation shear force and comfort function of structural members, minimizing the acceleration is also a feasible optimization criterion. The empirical formulas for the optimizing stiffness and damping of TMD with minimum acceleration response are as follows:

$$\bar{\lambda}_{dopt} = \frac{1}{\sqrt{1+\mu}} + (0.096 + 0.88\mu - 1.8\mu^2)\zeta + (1.34 - 2.9\mu + 3\mu^2)\zeta \tag{3.24}$$

$$\bar{\zeta}_{dopt} = \sqrt{\frac{3\mu(1 + 0.49\mu - 0.2\mu^2)}{8(1+\mu)}} + (0.13 + 0.72\mu + 0.2\mu^2)\zeta + (0.19 + 1.6\mu - 4\mu^2)\zeta^2 \tag{3.25}$$

3.1.3 Construction of FM Mass Vibration Control

When the FM mass vibration reduction technology is applied to building structures, there are many ways to construct the mass, damping and stiffness elements to meet the needs of FM vibration reduction for different building structures. A variety of commonly used TMD structures are shown in Fig. 3.3, the most typical of which is shown in Fig. 2.3a. The mass element is placed on the rolling support, so that the mass element can slide freely in the horizontal direction; the spring and damper are set between the mass element and the adjacent vertical counterforce device, thus transferring the forces in the opposite direction to the structure. The representative projects are John Hancock Tower, Citicorp Center and Canadian National Tower. In addition, there are other TMD structures using support nodes to support mass elements. As shown in Fig. 2.3b, the laminated rubber bearings are used to provide stiffness, and the viscous fluid dampers are used to provide damping. As shown in Fig. 2.3c, the friction swing bearings are used, on the one hand, the swing mechanism formed by the radian of the bearings themselves is used to provide equivalent stiffness; on the other hand, the friction damping is provided by the internal friction of support. In terms of construction, the above three types of dampers in Fig. 2.3a–c are all supporting TMD, and the mass elements are supported on the floor.

Another construction type is suspension swing. The simplest suspension swing TMD is shown in Fig. 2.3d. The period of the substructure is determined by the length of suspension swing. The relative swing of mass element produces the horizontal force opposite to the floor, thus restraining the horizontal vibration of the structure. The representative project is Crystal Tower in Japan. However, when the TMD construction of simple suspension swing type is adopted in the building structure,

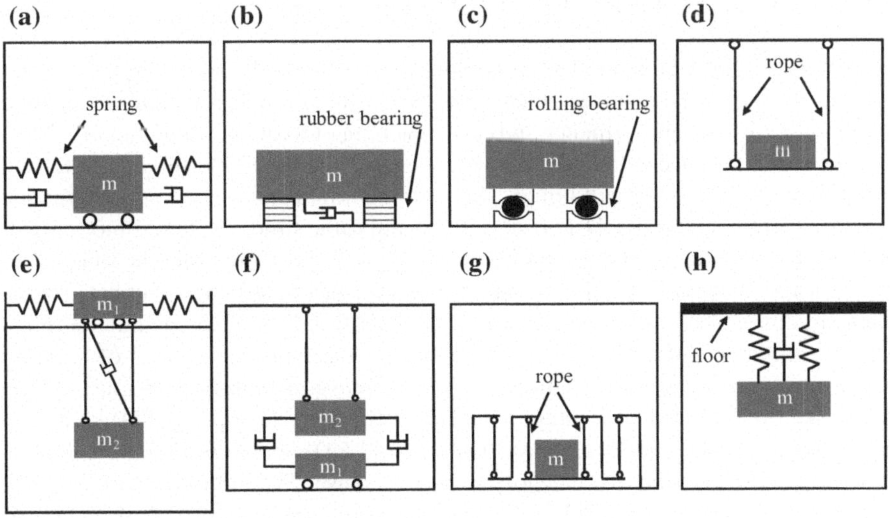

Fig. 3.3 Common construction forms of TMD

in order to match the optimal control frequency of the flexible structure, it usually needs a longer swing length, which leads to the higher requirement for building space. In order to obtain more practical installation space, it is necessary to improve the simple suspension swing TMD. One of the ideas is to improve the scheme that the swing period of substructure is determined only by the suspension length to the scheme that the swing period is determined simultaneously by the suspension length and mass. For example, the cantilever uses a rigid bar with counterweight (Fig. 3.3e) or two mass dampers, one sliding on the floor and the other acting as swing mass (Fig. 3.3f). Due to the use of multiple mass elements, the construction is much complex than the simple suspension swing, especially when the required substructure mass increases continuously. Another idea to save the installation space of TMD is the multi-stage suspension swing TMD. As shown in Fig. 3.3g, the multi-stage swing construction enlarges the effective swing length of the device by using internal rigid connecting rods, which is more practical in practical engineering. In recent years, with the increasing span and decreasing mass of building structures, the problem of pedestrian comfort caused by vertical vibration of long-span spatial structures has attracted more and more attention. The vertical TMD shown in Fig. 3.3h has been widely used in the large-span buildings and footbridges.

3.2 FM Liquid Vibration Control

3.2.1 Motion Equation of FM Liquid Vibration Control System

3.2.1.1 Sloshing Characteristics of FM Liquid Damper

The Tuned Liquid Damper (TLD) is composed of water tank and the liquid in water tank, which has the advantages of low cost, easy installation, less maintenance, good automatic activation performance and easy matching of frequency modulation. When the main structure is excited and vibrates, it causes the liquid sloshing in TLD, and forms waves on the surface of liquid. The sloshing liquid and wave produce dynamic pressure difference on the TLD box wall. At the same time, the liquid motion will also cause inertia force. The resultant force of dynamic pressure difference and liquid inertia force is opposite to the vibration direction of the structure and acts on the main structure. The vibration response of the structure is attenuated and controlled [2, 5]. Similar to TMD, TLD is also suitable for the vibration control of building structures, such as the wind-induced response control of high-rise structures. TLD is usually divided into two categories: Tuned Sloshing Damper (TSD) shown in Fig. 3.4a and Tuned Liquid Columns Damper (TLCD) shown in Fig. 3.4b. TLD is generally referred to the first category in narrow sense. The motion equation of FM liquid damper system is closely related to the liquid sloshing model of TLD, and these two types of TLD have different sloshing characteristics. The basic sloshing characteristics of rectangular tank and U-shaped tank are introduced.

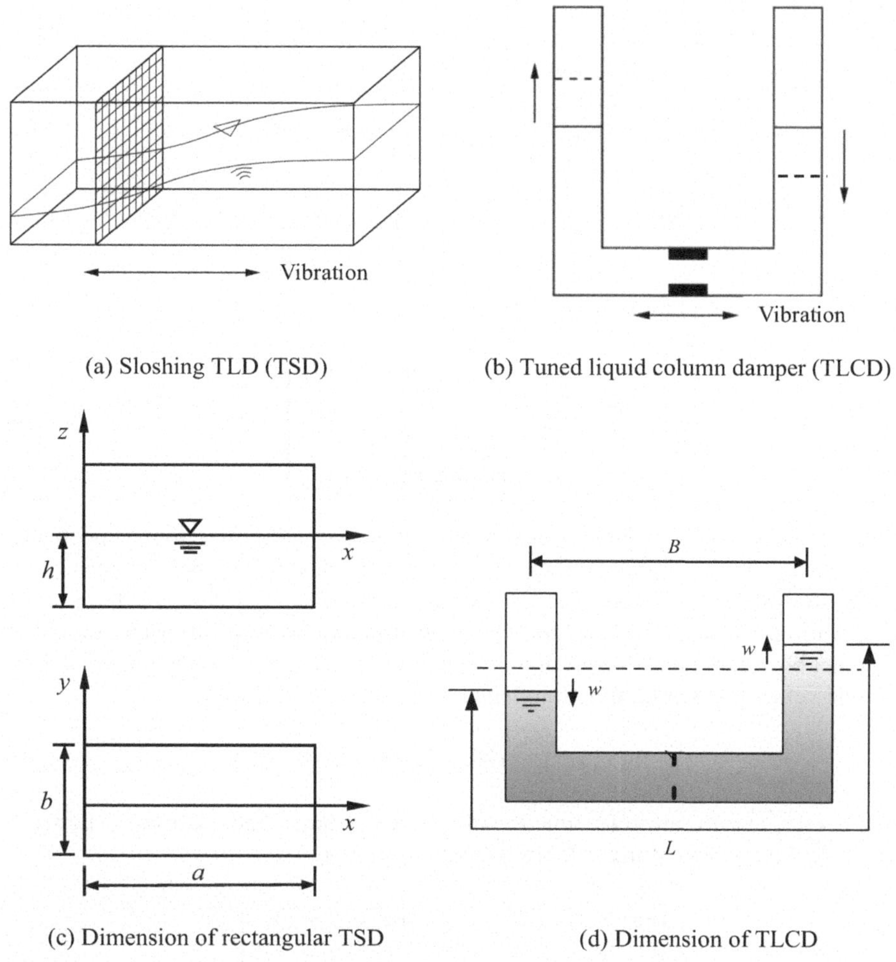

Fig. 3.4 Two kinds of commonly used TLD

1. Sloshing characteristic of TSD

Considering the shallow water wave theory, the motion of water in the shallow rectangular water tank shown in Fig. 3.4c is non-linear. At this time, the water in the tank can be divided into two layers, that is, the boundary layer near the bottom of tank and the water above the boundary. The thickness of boundary layer is very small, which is greatly affected by the friction between the boundary layer and bottom plate. The internal friction of water outside the boundary layer is very small, and its viscous damping is mainly caused by the viscosity of water surface. The motion of water in the rectangular shallow water tank can be described by the boundary layer equation Eq. (3.24) and the outer boundary layer equation Eq. (3.25), which are decomposed from the continuity equation Eq. (3.23) and the 2D Maniar-Stokes motion equation:

$$\frac{\partial u}{\partial x}+\frac{\partial w}{\partial z}=0 \tag{3.26}$$

$$\left.\begin{array}{l}\frac{\partial u}{\partial t}+u\frac{\partial u}{\partial x}+w\frac{\partial u}{\partial z}=-\frac{1}{\rho}\frac{\partial p}{\partial x}-\ddot{x}_s\\ \frac{\partial w}{\partial t}+u\frac{\partial w}{\partial x}+w\frac{\partial w}{\partial z}=-\frac{1}{\rho}\frac{\partial p}{\partial z}-g\end{array}\right\}(-(h-h_b)\leq z\leq\eta) \tag{3.27}$$

$$\left.\begin{array}{l}\frac{\partial u}{\partial t}+u\frac{\partial u}{\partial x}+w\frac{\partial u}{\partial z}=-\frac{1}{\rho}\frac{\partial p}{\partial x}+\nu\frac{\partial^2 u}{\partial z^2}-\ddot{x}_s\\ \frac{1}{\rho}\frac{\partial p}{\partial z}=-g\end{array}\right\}(-h\leq z<-(h-h_b)) \tag{3.28}$$

Their boundary conditions are as follows:

$$\left.\begin{array}{ll}u=0 & (x=\pm a)\\ w=0 & (z=-h)\\ w=\frac{\partial\eta}{\partial t}+u\frac{\partial\eta}{\partial x} & (z=\eta)\\ p=p_0=\text{constant} & (z=\eta)\end{array}\right\} \tag{3.29}$$

where, u and w are the velocities of water in x direction and z direction, respectively. η is the height of liquid surface, h_b is the thickness of boundary layer, p is the hydrodynamic pressure, ν is the dynamic viscosity of water, g is the gravity acceleration, and $\ddot{x}_s$ is the acceleration of the structure layer setting water tank. For the outer boundary layer, the motion of water can be considered as the potential flow, and the velocity potential function Φ can be assumed as follows:

$$\Phi(x,z,t)=G(x,t)\cosh[k(h+z)] \tag{3.30}$$

where, $G(x,t)$ is the arbitrary function of x, t. According to the relationship between water sloshing velocity and velocity potential function:

$$\left.\begin{array}{l}u(x,z,t)=\frac{\partial\Phi}{\partial x}=\frac{\partial G}{\partial x}\cosh[k(h+z)]\\ w(x,z,t)=\frac{\partial\Phi}{\partial z}=kG\ \sinh[k(h+z)]\end{array}\right\} \tag{3.31}$$

Plug Eq. (3.28) into Eqs. (3.23)–(3.25). After transformation, the governing equation of water movement in rectangular shallow water tank can be derived as follows:

$$\left.\begin{array}{l}\frac{\partial\eta}{\partial t}+h\sigma\frac{\partial[\phi u(\eta)]}{\partial x}=0\\ \frac{\partial}{\partial t}u(\eta)+(1-T_H^2)u(\eta)+g\frac{\partial\eta}{\partial x}+gh\sigma\phi\cdot\frac{\partial^2\eta}{\partial x^2}\cdot\frac{\partial\eta}{\partial x}=-\lambda u(\eta)-\ddot{x}_s\end{array}\right\} \tag{3.32}$$

$$\sigma=\frac{\tanh(kh)}{kh} \tag{3.33}$$

$$\phi=\frac{\tanh[k(h+\eta)]}{\tanh(kh)} \tag{3.34}$$

$$T_H=\tanh[k(h+\eta)] \tag{3.35}$$

where, λ is the dynamic viscous damping coefficient of water surface, $\lambda = \frac{1}{\eta+h} \cdot \frac{1}{\sqrt{2}}\sqrt{\omega_1 \nu}\left(1 + \frac{2h}{b} + s\right)$, $k = \frac{n\pi}{a}$ $(n = 0, 1, 2, \ldots)$; b is the width of water tank; and s is the viscous influencing factor of liquid surface, which is generally between 0 and 2. Considering the dimensionless and discretization of variables and parameters (dividing the water tank into n equal parts along the direction of vibration and discretizing the partial differential equations in each equal part), Eq. (3.30) can be transformed into the vector equations:

$$\left.\begin{aligned} \frac{\partial}{\partial t}\{\eta'\} &= \{f(t, \{\eta'\}, \{u'\})\} \\ \frac{\partial}{\partial t}\{u'\} &= \{g(t, \ddot{x}_s, \{\eta'\}, \{u'\})\} \end{aligned}\right\} \tag{3.36}$$

where, $\{\eta'\}^{\mathrm{T}} = [\eta'_0 \cdots \eta'_n]$ is the dimensionless wave height liquid level at the two ends of the vibration direction of water tank and the n equal points, and $\{u'\}^{\mathrm{T}} = [u'_0 \cdots u'_n]$ is the dimensionless horizontal velocity vector at the middle of the vibration direction of water tank. Considering that the feedback vibration control force of water tank to the structure is caused by the sloshing pressure of water on the side wall, the vibration reduction force $F_{TSD}(t)$ is as follows:

$$F_{TSD}(t) = b[\int_{-h}^{\eta_0} (P(t)|_{x=-a})dz - \int_{-h}^{\eta_0} (P(t)|_{x=a})dz] \tag{3.37}$$

The approximate expression of the vibration control force of water tank is:

$$F_{TSD} = \frac{1}{2}\rho g b(\eta_n^2 - \eta_0^2) \tag{3.38}$$

2. Sloshing characteristic of TLCD

As shown in Fig. 3.4c, the U-shaped pipe will drive the water tank to move together when the structure vibrates, and the movement of water tank will cause the water in the tubular water tank to sway. The horizontal inertia force produced by the water sway acts on the wall of water tank, which constitutes the vibration reduction force on the structure. The length L and width B of liquid in U-shaped tube and the mass of liquid in the tube are adjusted according to the need, so that the oscillation frequency of liquid in the tube is equal to or close to the natural frequency of the structure, so as to achieve the best vibration reduction effect. When the displacement of water from sloshing is w, the total inertia force of water in U-shaped tank is:

$$I_m = -[\rho A(L - B)\ddot{x}_k + \rho AB(\ddot{w} + \ddot{x}_k)] \tag{3.39}$$

where, I_m is the total inertia force of U-shaped tubular water tank; ρ is the density of water; A is the cross-sectional area of tubular water tank; B is the center distance of vertical pipe of the water tank; L is the total length of water center line in water tank; $\ddot{w}$ is the acceleration of the water sloshing away from the equilibrium position in water tank; and $\ddot{x}_k$ is the acceleration of the structural layer where the water tank is located. According to Darambel's principle, the motion equation of water sloshing in water tank can be obtained from equilibrium conditions as follows:

$$\rho AL\ddot{w} + \frac{\rho A}{2}\xi|\dot{w}|\dot{w}(t) + 2\rho Agw(t) = -\rho AB\ddot{x}_k \tag{3.40}$$

The above equation can be changed into:

$$\ddot{w} + 2\bar{\xi}_L\omega_L\dot{w} + \omega_L^2 w = -\frac{B}{L}\ddot{x}_k \tag{3.41}$$

where, ω_L is the natural frequency of water sloshing, and $\omega_L = \sqrt{2g/L}$; the non-linear damping ratio $\bar{\xi}_L$ of water tank is as follows:

$$\bar{\xi}_L = \sqrt{\frac{1}{2gL}}\frac{\xi|\dot{w}(t)|}{4} \tag{3.42}$$

where, ξ is the damping coefficient; considering the local damping caused by the baffle, along-the-way damping and local damping at the elbow are relatively small, so only the influence of baffle can be considered. The control force of water tank on the structure can be considered as the action of horizontal inertia force of water movement in U-shaped tubular water tank on the wall of water tank. In this way, the control force of U-shaped tubular water tank on the structure is as follows:

$$F_{TLCD}(t) = -(\rho AB\ddot{w}(t) + \rho AL\ddot{x}_k) \tag{3.43}$$

3.2.1.2 Motion Equation of Vibration Absorption System

The typical single-particle structure with FM liquid damper is shown in Fig. 3.5. According to the principle of dynamic equilibrium, the motion equation of the system is as follows:

$$m\ddot{x} + c\dot{x} + kx = F(t) - F_{TLD}(t) \tag{3.44}$$

$$\ddot{x}_d + 2\zeta_L\omega_L\dot{x}_d + \omega_L^2 x_d = -\ddot{x} \tag{3.45}$$

where, x_d, $\dot{x}_d$ and $\ddot{x}_d$ are the generalized displacement, velocity and acceleration of TLD sloshing, respectively. ζ_L and ω_L are the damping ratio and circular frequency of TLD, respectively. $F_{TLD}(t)$ is the control force of FM liquid damper, which is shown in Eq. (3.35). Considering TLCD, Eq. (3.42) can be changed to:

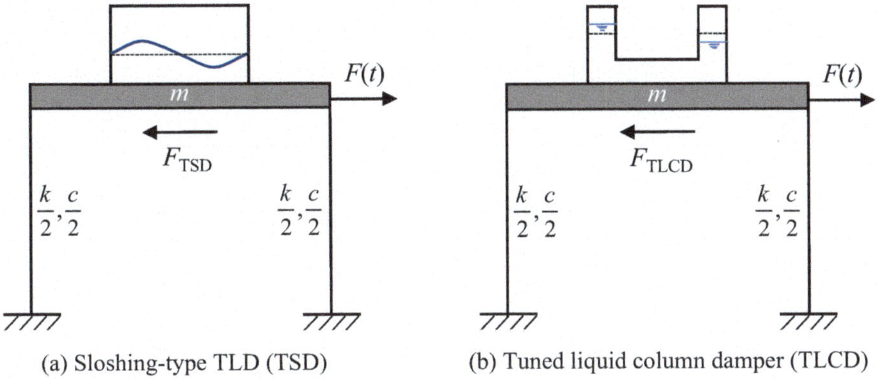

Fig. 3.5 FM liquid vibration absorption system

$$\ddot{x}_d + 2\zeta_L\omega_L\dot{x}_d + \omega_L^2 x_d = -(B/L)\cdot\ddot{x} \tag{3.46}$$

At this time, $F_{TLD}(t)$ in Eq. (3.41) can be expressed using Eq. (3.40).

3.2.2 Basic Characteristics of FM Liquid Vibration Control

Compared with Eqs. (3.1)–(3.2) and Eqs. (3.44)–(3.45), the motion equations of FM liquid vibration absorption system and FM mass vibration absorption system have similar forms, and the parameters are comparable. From the macro vibration-reduction effect, the two systems have similar vibration reduction control characteristics, so the basic characteristics of FM liquid vibration control can refer to the analysis results of FM mass vibration reduction control in Sect. 3.1.2.

References

1. Chen, Xin, Youliang Ding, Zhiqiang Zhang, et al. 2012. Investigations on serviceability control of long-span structures under human-induced excitation. *Earthquake Engineering and Engineering Vibration* 11 (1): 57–71.
2. Li, Aiqun. 2007. *Vibration control of engineering structure*. Beijing: China Machine Press.
3. Seto, K. 2010. *Structural vibration control*. Beijing: Mechanical Industry Press.
4. Ye, Zhengqiang. 2003. *Research on theory, experiment and application of energy dissipation and vibration reduction technology with viscous fluid damper*. Nanjing: Southeast University. (in Chinese).
5. Ren, Zhenghua. 1993. *Basic principle of TLD device and its application in wind vibration control of Nanjing TV tower*. Nanjing: Southeast University. (in Chinese).

Chapter 4
Basic Principle of Structural Isolation

Abstract The structural isolation is usually considered as the most effective technology to resist earthquake action. At first, motion equation of isolated structural system is derived. Then, in the section of basic characteristics of isolated structural system, response analysis and characteristics of isolated structural system are introduced. In the section of commonly used isolation devices for building structures, rubber isolation system, sliding isolation system and hybrid isolation system are introduced systematically.

The isolation is essentially the separation of structure and components from the ground motion or support motion that may cause damage. This separation or decoupling is achieved by increasing the flexibility of the system and providing appropriate damping. The basic idea of building structure isolation is to locate the superstructure on the isolation layer, absorb seismic energy through the deformation of isolation layer, control the seismic effect of superstructure and the deformation of isolation part, so as to reduce the seismic response of the structure and improve the seismic reliability of the building structure. Unlike the structural vibration control technology mentioned above, the structural isolation technology is usually only used for the structural response control under earthquake.

4.1 Motion Equation of Isolated Structural System

For base-isolated structures, the inter-story stiffness of the superstructure (such as brick-concrete structure, shear wall structure, frame-shear structure or frame structure) is far greater than the horizontal stiffness of isolation device. Therefore, the inter-story horizontal displacement angle of the superstructure is very small during earthquake, and the horizontal displacement of the structure system is concentrated in the isolation layer. The superstructure only moves horizontally when the earthquake occurs (Fig. 4.1a). If the swing rotation of superstructure is neglected, the structure can be simplified as the dynamic analysis model of single-particle isolation structure (Fig. 4.1b). The stiffness and damping of isolation device also approximately

A. Li, *Vibration Control for Building Structures*, Springer Tracts in Civil Engineering, https://doi.org/10.1007/978-3-030-40790-2_4

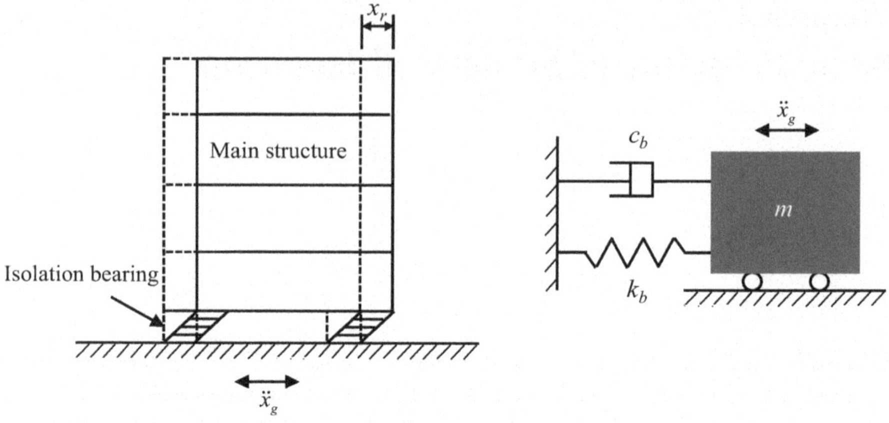

(a) Structural prototype (b) Computational model

Fig. 4.1 Isolated structural system

represent the stiffness and damping of the isolation structure system [1]. If the system is assumed to be a linear isolation system, the motion equation of the isolation structure system can be expressed as follows:

$$m\ddot{x}_r + c_b\dot{x}_r + k_b x_r = -m\ddot{x}_g(t) \tag{4.1}$$

$$m\ddot{x} + c_b\dot{x} + k_b x = c_b\dot{x}_g(t) + k_b x_g(t) \tag{4.2}$$

where, x_r, $\dot{x}_r$ and $\ddot{x}_r$ are the displacement, velocity and acceleration of the structure relative to the ground, respectively; x, $\dot{x}$ and $\ddot{x}$ are the displacement, velocity and acceleration of the structure, respectively; $x_g(t)$, $\dot{x}_g(t)$ and $\ddot{x}_g(t)$ are the displacement, velocity and acceleration of the ground, respectively. m is the total mass of superstructure; k_b is the restoring stiffness of isolation layer; and c_b is the damping coefficient of isolation layer. The other expressions of Eqs. (4.1) and (4.2) are:

$$\ddot{x}_r + 2\zeta_b\omega_b\dot{x}_r + \omega_b^2 x_r = -\ddot{x}_g(t) \tag{4.3}$$

$$\ddot{x} + 2\zeta_b\omega_b\dot{x} + \omega_b^2 x = 2\zeta_b\omega_b\dot{x}_g(t) + \omega_b^2 x_g(t) \tag{4.4}$$

where, $\zeta_b = \frac{c_b}{2\sqrt{mk_b}} = \frac{c_b}{2m\omega_b} = \frac{c_b T_b}{4\pi m}$, $T_b = 2\pi\sqrt{\frac{m}{k_b}}$, $\omega_b = \sqrt{\frac{k_b}{m}}$. For the SDOF isolation system mentioned above, given the values of ω_b and ζ_b of the structure, the response time history of the superstructure under earthquake can be obtained by Duhamel integral. Because the structure is usually designed according to the maximum response, the seismic response can be summarized by recording the maximum response of the single particle at only a set of ω_b and ζ_b. These maximum response values are seismic response spectra. The acceleration response spectrum $S_A(T_b, \zeta_b)$, the velocity response spectrum $S_V(T_b, \zeta_b)$ and displacement response spectrum $S_D(T_b, \zeta_b)$ are defined respectively as follows:

$$S_A(T, \zeta) = (\ddot{x}_g(t) + \ddot{x}(t))_{max} \tag{4.5}$$

$$S_V(T, \zeta) = \ddot{x}(t)_{max} \tag{4.6}$$

$$S_D(T, \zeta) = x(t)_{max} \tag{4.7}$$

The first mode of typical isolation structure is quite different from all other high-order modes. The first mode and the high-order modes can be separated in the analysis. The period, damping ratio and seismic response of isolation structure of the first mode are mainly determined by the characteristics of isolation system, basically independent of the period and damping of superstructure. In the first mode, the vertical distributions of horizontal displacement and acceleration are approximately rectangular, and the motions of all floors are approximately the same. Therefore, when evaluating the seismic response of the first mode, the isolation structure can be approximately regarded as a rigid mass. Except for special cases, the seismic response of linear isolation structure can be described by seismic response spectrum and the first mode. When the isolation system has strong nonlinearity, the seismic response of the main structure is still characterized by the first mode, but at this time the influence of higher mode should be paid attention to.

4.2 Basic Characteristics of Isolated Structural System

4.2.1 Response Analysis of Isolated Structural System

4.2.1.1 Displacement Response Analysis

In order to obtain the displacement x_r of the isolated structural system, the transformation function method is used to solve the relative displacement dynamic equation shown in Eq. (4.3). The dynamic response conversion function of the isolated structure system is assumed to be $G(\omega)$. The seismic acceleration of ground is $\ddot{x}_g = e^{i\omega g}$, the seismic displacement response of the isolated structure is $x_r = G(\omega)e^{i\omega g}$. The displacement response conversion function of the isolated structure system can be obtained by plugging $\ddot{x}_g$ and x_r into Eq. (4.3):

$$G(\omega) = \frac{x_r}{\ddot{x}_g} = \frac{1}{\omega_b^2}\sqrt{\frac{1}{[1-(\omega/\omega_b)^2]^2 + (2\zeta_b\omega/\omega_b)^2}} \tag{4.8}$$

The maximum absolute value of displacement response of superstructure can be obtained by arranging the terms of Eq. (4.8) and setting $\ddot{x}_g$ is the maximum absolute value of acceleration under earthquake,

$$|x_r| = \frac{|\ddot{x}_g|}{\omega^2} = \frac{1}{\omega_b^2} \frac{(\omega/\omega_b)^2}{\sqrt{[1-(\omega/\omega_b)^2]^2 + (2\zeta_b\omega/\omega_b)^2}} \tag{4.9}$$

Because the inter-story displacement of the superstructure is very small during earthquake, the horizontal displacement of the isolation structure system is basically characterized by the horizontal displacement of isolation device. Equation (4.9) is the basic formula for calculating the maximum absolute value $|x_r|$ of horizontal displacement of isolation device during earthquake. The value of $|x_r|$ is closely related to the ground acceleration $\ddot{x}_g$, the characteristic frequency ω of the site, the natural frequency ratio ω/ω_b (the ratio of characteristic frequency of the site to the natural frequency of isolation structure system), and the damping ratio ζ_b of isolation device. In general, the displacement of isolation can be reduced by adjusting the damping ratio ζ_b of isolation.

The magnification ratio of displacement response of isolation structure is defined as R_d. Since the horizontal displacement of isolation structure is mainly concentrated in the isolation device with smaller horizontal stiffness during earthquake, R_d can be understood as the ratio of displacement response $|x_r|$ (maximum absolute value) of isolation device to ground displacement $|x_g|$ (maximum absolute value), that is,

$$R_d = \frac{|x_r|}{|x_g|} = \frac{(\omega/\omega_b)^2}{[1-(\omega/\omega_b)^2]^2 + (2\zeta_b\omega/\omega_b)^2} \tag{4.10}$$

where, $|x_g| = \frac{|\ddot{x}_g|}{\omega^2}$, and R_d is the index to characterize the displacement amplification of isolation devices. When the site characteristics (ω) and the stiffness characteristics (k_b or ω_b) of the isolation device are fixed, the horizontal displacement of isolation device can be reduced by increasing the damping ratio ζ_b of isolation device.

4.2.1.2 Acceleration Response Analysis

The transformation function method is also used to obtain the acceleration response $\ddot{x}$ of the structural system. The dynamic response conversion function of the structure system is assumed to be $H(\omega)$, the site characteristic frequency of ground is ω; and the ground seismic acceleration is assumed to be $\ddot{x}_g = e^{i\omega g}$, the seismic acceleration response of the isolated structure is $\ddot{x} = H(\omega)e^{i\omega g}$. The dynamic response conversion function of the isolated structure system can be obtained by plugging $\ddot{x}_g$ and $\ddot{x}$ into Eq. (4.4):

$$H(\omega) = \frac{\ddot{x}}{\ddot{x}_g} = \sqrt{\frac{1 + (2\zeta_b\omega/\omega_b)^2}{[1 - (\omega/\omega_b)^2]^2 + (2\zeta_b\omega/\omega_b)^2}} \tag{4.11}$$

The physical meaning of the conversion function is the ratio of the seismic acceleration response of isolated structure to the ground seismic acceleration, which shows the attenuation effect of isolation structure on ground seismic acceleration. The acceleration response attenuation ratio of isolation structure is defined as R_a, which is the ratio of acceleration response of isolation structure to ground acceleration in earthquake:

$$R_a = \frac{\ddot{x}}{\ddot{x}_g} = \sqrt{\frac{1 + (2\zeta_b\omega/\omega_b)^2}{[1 - (\omega/\omega_b)^2]^2 + (2\zeta_b\omega/\omega_b)^2}} \tag{4.12}$$

Equation (4.12) represents that R_a is the important and basic formula for designing, calculating and controlling the isolation effect of isolated structure. If the characteristic frequency ω of building structure is known, the isolation device (the natural frequency ω_b, damping ratio ζ_b) can be reasonably selected to obtain the acceleration attenuation rate R_a of isolation structure, so as to ensure the safety of structure or the decoration, equipment and instruments of structure during earthquake. Equation (4.11) or Eq. (4.12) can be used to calculate the damping ratio ζ_b of isolated structure:

$$\zeta_b = \frac{1}{2(\omega/\omega_b)}\sqrt{\frac{1 - R_a^2[1 - (\omega/\omega_b)^2]^2}{R_a^2 - 1}} \tag{4.13}$$

Because the superstructure has little inter-story displacement and is basically in an elastic state during earthquake, its damping value is very small, so the damping ratio ζ_b can be approximately regarded as the damping ratio of isolation device. When the acceleration response attenuation ratio R_a of the isolated structure system is known, the damping ratio required by the isolation device (such as sandwich rubber cushion) can be obtained using Eq. (4.13).

4.2.2 Response Characteristics of Isolated Structural System

The seismic response of isolation structure mainly depends on the design of dynamic performance of isolation layer, that is, the hysteretic relationship of isolation layer and the design of parameters. At present, no matter what kind of isolation system is used, the main models simulating the actual shear-displacement hysteresis curve of isolation layer are linear and bilinear [2–4]. Figure 4.2a is a linear isolation system consisting of linear spring and viscous damper. The model is the same as that of Fig. 4.2 of Sect. 4.1. The force-displacement hysteresis curve has an effective slope

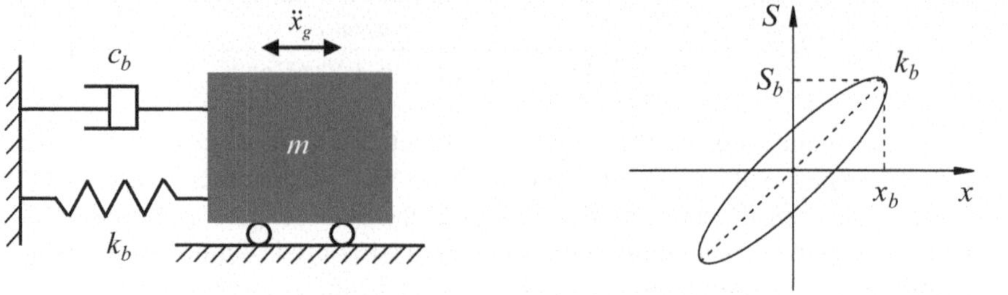

(a) Damped linear isolation system

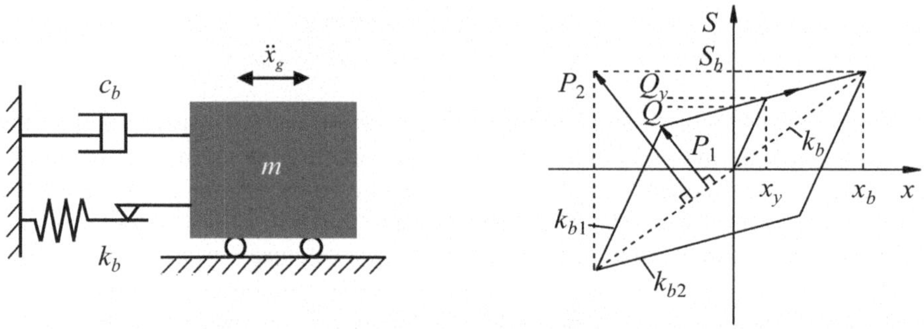

(b) Bilinear isolation system

Fig. 4.2 Main models for simulating various systems

(the dashed line), representing the stiffness of isolation system. Figure 4.2b is a bilinear isolation system. In this model, there are two linear springs, one of which is connected in series with a Coulomb damper. The hysteretic curve is bilinear and has two slopes, representing the initial stiffness and yield stiffness, respectively. Corresponding to the elastic and plastic deformations of isolation layer, when the initial stiffness is infinite and the yield stiffness is zero, the bilinear hysteresis curve is transformed into the Coulomb friction hysteresis curve. It can be seen that the Coulomb friction curve is a special kind of bilinear curve.

For linear isolation systems, the first mode of vibration is approximately rigid body motion, and the response dominates. The maximum displacement at any layer i of the structure is obtained,

$$x_i \approx S_D(T_b, \zeta_b) \tag{4.14}$$

The maximum inertial force on the mass m_i is as follows:

$$F_i \approx m_i S_A(T_b, \zeta_b) \tag{4.15}$$

The inertial forces of each layer are approximately in phase, so the shear force of each layer can be obtained through adding them together. The base shear can be given as follows:

$$S_b \approx mS_A(T_b, \zeta_b) \tag{4.16}$$

For the bilinear isolation system, the seismic response can still be obtained from the design seismic response spectrum, but the accuracy of the solution is worse than that of the linear isolation system. The seismic response analyses of a large number of structural systems with different isolators show that the first mode seismic response can be approximately obtained from the equivalent period T_B, equivalent damping ζ_B and seismic response spectrum of bilinear isolation system. From the parameters of isolator and bilinear spectral displacement $S_D(T_b, \zeta_b)$, the maximum base displacement x_B and the maximum base shear S_B can be derived as follows:

$$x_B \approx C_F S_D(T_B, \zeta_B) \tag{4.17}$$

$$S_B \approx Q_y + k_{b2}(x_b - x_y) \tag{4.18}$$

where, C_F is the correction coefficient derived from statistical data. For the North-South component of El-Centro wave in 1940, when the parameters T_B, ζ_B and Q_y/W of bilinear isolation system change in a large range, the correction coefficient C_F is approximately between 0.85 and 1.15. Because T_B and ζ_B are the functions of x_B and S_B, it needs to be calculated iteratively. When the bilinear isolation system is highly non-linear, there will usually be a larger acceleration response of higher-order mode, which leads to a significant increase in seismic force compared with the first mode response. However, because the shear force produced by higher modes is close to zero in the isolation layer, the first mode will still be the main mode of base shear.

4.3 Commonly Used Isolation Devices for Building Structures

In order to achieve obvious vibration absorption effect, the isolation device or isolation system must have the following four basic characteristics.

(1) Bearing characteristics: the isolation device has a large vertical bearing capacity, which effectively supports all the gravity and loads of the superstructure under the service condition of building structure. It has a large safety factor of vertical bearing capacity to ensure the absolute safety of building structure under the service condition and meet the using requirements.
(2) Seismic isolation characteristics: the isolation device has variable and enough horizontal stiffness in strong wind or small earthquakes, the horizontal displacement of the superstructure is very small and does not affect the using

requirements; in moderate intensity earthquakes, the horizontal stiffness is smaller, the superstructure slides horizontally, making the "rigid" seismic structure become a "flexible" isolation structure system, whose natural vibration period is greatly prolonged and far away from the natural vibration period of the superstructure and site characteristic period, thus effectively isolating the ground vibration and significantly reducing the seismic response of the superstructure.

(3) Reset characteristics: because of the horizontal elastic resilience of the isolation device, the isolation structure system has the function of instantaneous automatic reset in earthquake. After the earthquake, the superstructure returns to the initial state, which meets the requirements of normal use.

(4) Damping efficiency characteristics: the isolation device has enough damping c_b, that is, the load-displacement curve of the isolation device is full and has greater energy dissipation capacity. The displacement x_b of the superstructure can be significantly reduced by larger damping C.

According to the working principle and characteristics of isolation system, the commonly used isolation system can be divided into rubber isolation system, sliding isolation system and hybrid isolation system.

4.3.1 Rubber Isolation System

The rubber isolation system is supported by rubber isolation bearings in the isolation layer. The rubber isolation bearing, also known as sandwich rubber isolation cushion or laminated rubber isolation cushion, is composed of multi-layer rubber and multi-layer steel plate or other materials alternately overlapping. The rubber isolation system is a device with high vertical bearing capacity, small horizontal stiffness and large allowable value of horizontal displacement. The rubber isolation system can not only reduce the horizontal seismic action, but also withstand vertical earthquake. The rubber isolation system is suitable for the isolations of buildings, bridges, railways, equipment and other structures, which is the most widely used isolator in the world at present. The common rubber isolation systems include Laminated Rubber Bearing, Lead-Rubber Bearing (LRB) and High-damping Rubber Bearing (HDRB), etc. [5].

(1) The laminated rubber bearing (Fig. 4.3a) is composed of multi-layer rubber and steel plates, which makes the bearing cushion have the characteristics of low horizontal stiffness and high vertical stiffness. The main function of steel plate as stiffening layer is to realize the characteristics of high vertical bearing capacity and low horizontal stiffness of support through the restraint of steel plate on rubber layer. The hysteretic curve of this type of isolator is shown in Fig. 4.4a. The hysteresis loop has a small area, so the dampers are needed as energy absorbing elements in the isolation layer. The hysteretic characteristics of the isolation cushion are almost independent of the change of axial force and displacement history. The isolation cushion has stable elastic properties from small deformation to large deformation, and the calculation model of the structure is simple.

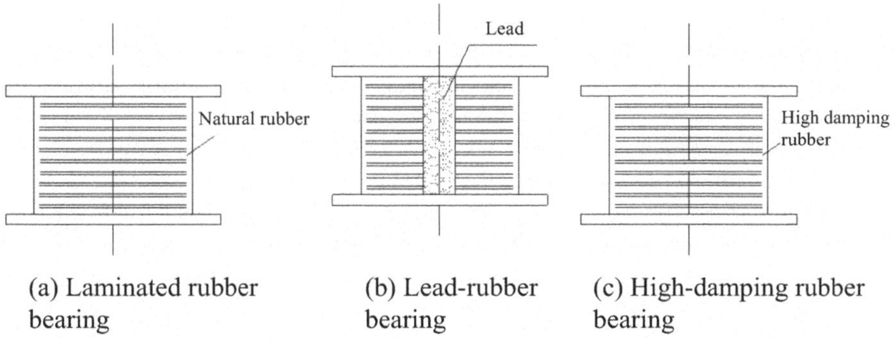

Fig. 4.3 Types of rubber isolation bearings

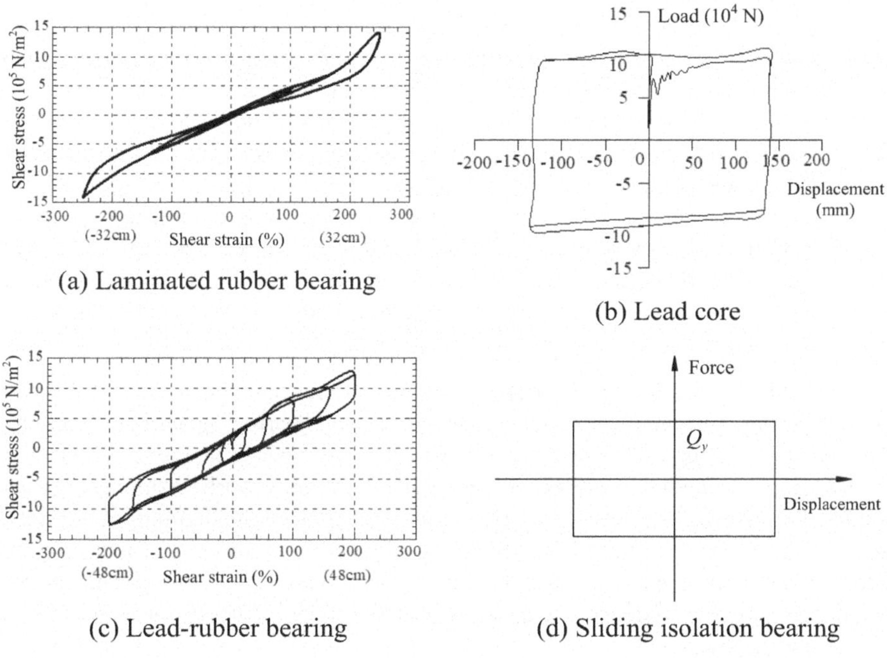

Fig. 4.4 Typical hysteretic curve of isolation bearing

(2) The lead rubber bearing (Fig. 4.3b) is the most used isolation device, which is composed of the rubber laminated cushion filled with lead core columns. Because the laminated rubber cushion may have excessive displacement under strong earthquakes and cannot resist environmental vibration, a high purity lead core is placed in the center of rubber cushion. Using the yielding behavior of lead itself, it can withstand environmental vibration in peacetime and play a role of energy consumption when earthquake comes. The rubber part provides lower lateral stiffness to prolong the structural period and reduce earthquake action.

Because the shear yield stress of lead is very small (about 10 MPa), it can produce hysteretic energy dissipation after yielding to reduce the displacement response of isolation structure, so the lead core is used as energy dissipation material. Compared with other metals, the lead can recrystallize rapidly at room temperature, which is not easy to produce strain hardening, so it can be reused for a long time repeatedly. The hysteresis curve of lead core is shown in Fig. 4.4b, which is approximately rectangular and has great energy dissipation capacity. The hysteretic behavior of lead rubber bearing under compression and shear is shown in Fig. 4.4c. The hysteretic characteristic is the combination of horizontal stiffness of laminated rubber bearing and the horizontal stiffness of lead core, and the hysteretic loop is double-linear. The hysteretic characteristics are related to shear deformation. When the hysteretic characteristics are equivalent to the linear damping relationship in the case of small deformation, the error of system analysis is small.

(3) The high damping rubber bearing (Fig. 4.3c) has the same isolation principle as lead rubber bearing. The main purpose is to prolong the structure period to reduce the seismic action, and use high damping materials to dissipate seismic energy and reduce the displacement of isolation structure. The high damping rubber support is a kind of synthetic rubber. The composition varies with different materials and components, and the mechanical properties are also quite different. The common characteristic is that it has the function of producing high damping. In addition, the shear modulus of high damping rubber bearing varies with temperature. If the temperature drops from 30 to 10 °C, the change of shear modulus may be as high as 50%, which may lead to a significant reduction of isolation effect.

Generally speaking, the rubber isolation bearings have the following characteristics: sufficient vertical stiffness and vertical bearing capacity; obvious and stable isolation effect, small horizontal stiffness, it can ensure that the basic period of the building is extended to 2–3 s or more; it can effectively absorb seismic energy and reduce the seismic response of superstructure with appropriate damping ratio; it can automatically reset in multiple earthquakes with stable elastic reset function; with simple construction, easy installation, detection and restoration; with repeated-load resistance, fatigue resistance and aging resistance; it can successfully withstand the test of real earthquakes.

4.3.2 Sliding Isolation System

The slip isolation system uses the sliding friction interface to isolate the transmission of seismic action, so that the seismic force on the isolation layer does not exceed the maximum friction force between the interfaces, thus greatly reducing the seismic response of superstructure. The Teflon-stainless steel plate is usually used as the friction surface, and the flexible graphite and its coating are also used as the friction surface.

According to the situation that the isolation layer has no restoring force, the sliding isolation system can be divided into two categories. One type is a non-resilient isolation structure, whose isolation layer components are mainly composed of pure friction sliding bearing or sand and other sliding friction materials. In order to improve its reliability, the safety lock structure should also be set up. The force-displacement hysteresis curve of the isolation layer of this kind of isolation system is shown in the Coulomb friction curve as shown in Fig. 4.4d. The other is the resilient sliding isolation structure, whose resilient components can be integrated with sliding bearings, such as R-FBI, EDF, SR-F and FPS, in order to reduce the residual displacement of the isolation layer after earthquake. The restoring force components and sliding bearings can also be set separately to facilitate the selection of design parameters and post-earthquake maintenance. The force-displacement hysteretic curve of the isolation layer of this kind of isolation system depends on the hysteretic characteristics of the specific restoring components. The actual hysteretic curve is generally a bilinear hysteretic curve obtained by adding the Coulomb friction curve and the hysteretic curve of the restoring components (Fig. 4.2b).

In addition, the sliding isolation system can be divided into ordinary sliding isolation structure and hybrid sliding isolation structure according to whether the components of the isolation layer bear vertical gravity load or not. In ordinary sliding isolation structures, only the sliding bearings bear gravity load; in hybrid isolation structures, the sliding bearings and restoring components all bear part of gravity load, and the isolation layers are generally composed of sliding bearings and rubber cushion bearings.

The sliding isolation bearings have the following characteristics: low cost; small net bearing height, easy coordination in architectural design; large vertical stiffness, strong bearing capacity; no fixed frequency of isolation system, insensitive to seismic waves and site types; good corrosion resistance and durability.

4.3.3 Hybrid Isolation System

The hybrid base isolation includes parallel base isolation with rubber bearings and sliding bearings, series base isolation and series-parallel base isolation, etc. The parallel base isolation and series base isolation are the basic forms of hybrid base isolation.

(1) The parallel isolation system refers to the parallel installation of rubber bearings and sliding bearings in the isolation layer, which respectively bear the gravity load of superstructure, as shown in Fig. 4.5a. The parallel base isolation can effectively improve the isolation effect by combining the two kinds of bearings and reasonable allocation. In parallel base isolation, the sliding bearing can provide greater damping, and has enough initial stiffness and bearing capacity to replace lead core and ensure the stability under normal conditions such as wind load and environmental vibration. The ordinary rubber cushion bearing

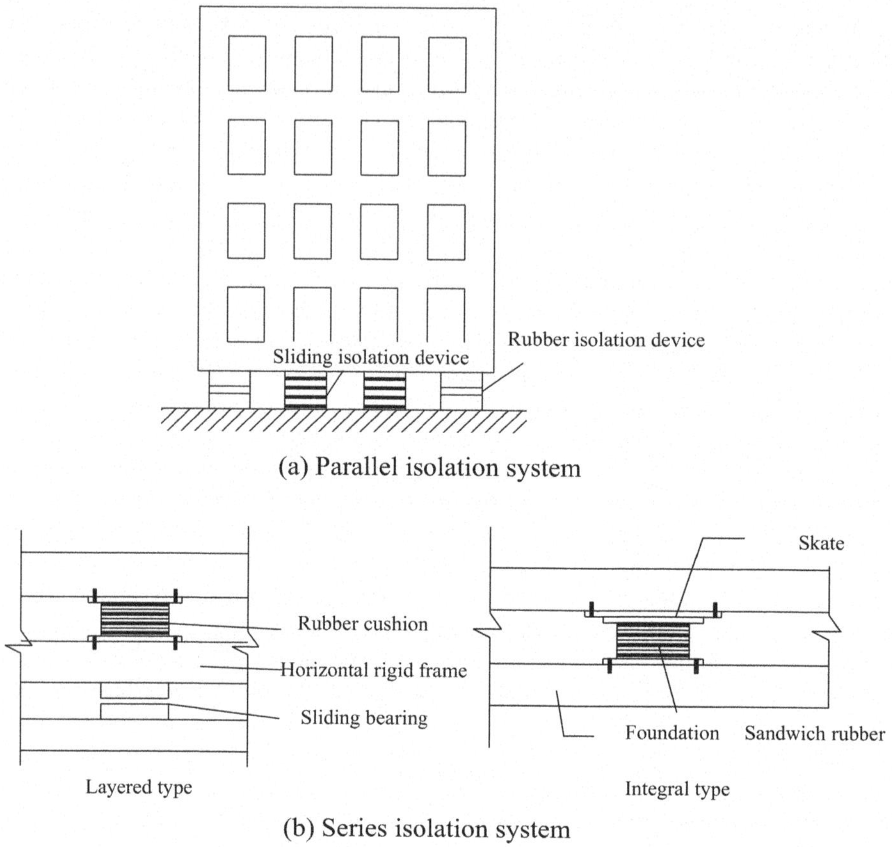

Fig. 4.5 Hybrid isolation system

can provide restoring force. Because the two kinds of isolation bearings bear part of the structural self-weight respectively, compared with the rubber cushion isolation, the size or quantity of rubber bearings can be reduced and the cost of isolation layer can be reduced. Compared with sliding isolation, the structural self-weight borne by sliding bearings decreases. If the same friction force is provided, the friction coefficient can be increased and the requirement of friction material is reduced, and the sensitivity of isolation effect to the change of friction coefficient is reduced, that is to say, the performance requirement and fluctuation range of friction material are relaxed. One of the problems in parallel base isolation is that with the increase of displacement, the ratio of vertical stiffness of the two isolation bearings will change, which will lead to different vertical deformations, resulting in the redistribution of self-weight among the bearings, especially the change of pressure and friction on the sliding device, which will affect the earthquake reaction of superstructure.

(2) The series isolation system refers to the series installation of building isolation rubber bearings and sliding bearings in the isolation layer to bear the gravity load of superstructure. The series base isolation is equivalent to two isolation layers. There are two kinds of structural forms, i.e. layered and integrated types, as shown in Fig. 4.5b. In order to coordinate the horizontal displacement between the upper and lower supports and keep the sliding bearings sliding uniformly, a planar integral frame is set up between the lower and upper supports in the layered type, which has a high cost. The integrated type is to remove the abovementioned planar frame and integrate the shear deformation function into a support. The horizontal lateral forces are transmitted directly through the sliding plate and rubber layer. The series isolation system uses more isolation bearings than other isolation systems mentioned above. The construction of isolation layer is complex and the cost is high. Therefore, the series isolation is not suitable for general buildings. For class A and B buildings or structures, when they require greater safety under strong earthquakes and strict isolation response of superstructure, such as nuclear power plants and dangerous goods storehouses, the series isolation systems can be considered under economic permission. The typical example of series base isolation is the EDF (France) isolation system in the mid-1980s, and the series base isolation was adopted in one of the nuclear power plants.

In the hybrid base isolation structure, the sliding bearings and rubber cushion bearings can be connected in series and then paralleled to form a more complex hybrid isolation system. The hysteretic relationship between the force and displacement of isolation layer in hybrid isolation system is complex. It is necessary to combine the hysteretic relationship of base isolation bearings properly according to the actual situation to form the practical hysteretic curve of force and displacement of isolation layer.

References

1. Li, Aiqun. 1991. Simplified solution of motion equations for base vibration isolator. *Journal of Southeast University (Natural Science Edition)* 21 (04): 130–135. (in Chinese).
2. Skinner. 1996. Introduction to engineering isolation. Beijing: Earthquake Press.
3. Zhang, Wenfang. 1999. *Test, theoretical analysis and application research on sliding isolation of multistory brick building foundation*. Nanjing: Southeast University. (in Chinese).
4. Mao, Lijun. 2004. *Research on building structures of sliding base-isolation*. Nanjing: Southeast University. (in Chinese).
5. Li, Aiqun. 2007. *Vibration control of engineering structure*. Beijing: China Machine Press.

Part II
Damping Devices of Building Structures

The damping devices of building structures are key devices for the successful application of vibration damping technology in building structures. Compared with the damping devices in the fields of aviation and machinery, the damping devices of civil structures must have the characteristics of large output force, fast response and good durability. In the three types of structural vibration reduction control technologies, including energy dissipation, frequency modulation and vibration isolation, the viscous fluid damper, viscoelastic damper, metal damper, frequency modulation damper, vibration isolation supports and other devices have been widely used in the practice of building structural engineering. In the research of more than 20 years, the author's team has carried out in-depth and systematic researches on the damping devices of building structures and has developed more than **30** kinds of various types of damping devices, and obtained more than **50** national invention patents. In addition, a series of products including energy dissipation, frequency modulation and vibration isolation have been formed with perfect system, independent property rights and excellent performance, through the cooperation platform of "production, study and research". Based on the first part of the vibration reduction principle, this part will introduce the research results of the author team in the field of high-performance damping materials and new vibration damping devices from the perspective of product development.

Chapter 5
Viscous Fluid Damper

Abstract The dissipating energy mechanism and characteristics, including types and characteristics of damping medium, energy dissipation mechanism and calculation models of viscous fluid damper are introduced. In the section of properties and improvement of viscous fluid materials, modification of viscous fluid damping materials, material property test of viscous fluid, experimental results and analysis are introduced. In the section of research and development of new viscous fluid damper, linear viscous fluid damper, nonlinear viscous fluid damper, variable damping viscous fluid damper and viscous fluid damping wall are introduced. About performance test of viscous fluid damper, including maximum damping force test, regularity test of damping force, test of loading frequency and temperature related performance of maximum damping force, pressure maintaining inspection, fatigue performance test are elaborated systematically.

5.1 Mechanism and Characteristics of Viscous Fluid Damper

The mechanism of vibration control by viscous fluid damper is to dissipate part of the energy of structural vibration through the viscous dissipation of viscous fluid materials in the damper, so as to reduce the structural vibration response.

5.1.1 Types and Characteristics of Damping Medium

The design reference period (50 years for general buildings) and the service life of building structures are relatively long, and the working environment is complex and diverse, which requires that the viscous fluid damper must have good stability and safety, so as to prevent the aging and failure of the devices in the service life. Therefore, the selection of damping medium is very important to the performance of viscous fluid damper. Its characteristics must meet the following requirements: (1) a certain viscosity; (2) low compressibility; (3) good chemical stability and high flash point; (4) poor sensitivity of viscosity to temperature and good cold resistance; (5) non-flammable, non-volatile, non-toxic and anti-aging performance.

A. Li, *Vibration Control for Building Structures*, Springer Tracts in Civil Engineering, https://doi.org/10.1007/978-3-030-40790-2_5

5.1.1.1 Fluid Viscosity

In fluids, when there is relative motion between two adjacent layers of fluid, the shear force parallel to the contact surface will be generated. The fast moving layer exerts dragging force on the slow moving layer, and the slow moving layer exerts retarding force on the fast moving layer. This pair of forces is relative in size and opposite in direction, and is an internal friction force. The property of fluid resisting the relative sliding or shear deformation of two layers of fluid is called fluid viscous. In 1686, British scientist Newton summed up the Newton's law of internal friction, also known as the Newton's law of viscosity, based on many experiments:

$$\tau = \mu \frac{du}{dy} \tag{5.1}$$

where τ is shear stress, $\frac{du}{dy}$ is shear rat, μ is dynamic viscosity or dynamic viscosity coefficient.

In fluid mechanics, the fluid that obeys Newtonian viscous law is usually called Newtonian fluid, while the fluid that has viscous but does not obey Newtonian viscous law is called non-Newtonian fluid. Common Newtonian fluids include air, water and various gases. Non-Newtonian fluids include grease, polymer solution and cement slurry. The flow curves of the two fluids are shown in Fig. 5.1a. According to whether the viscosity of non-Newtonian fluids is related to shear duration in simple shear flow, non-Newtonian fluids can be divided into non-time-varying non-Newtonian fluids and time-varying non-Newtonian fluids.

5.1.1.2 Time-Varying Non-Newtonian Fluids

The viscosity of time-varying non-Newtonian fluids is not only related to strain rate, but also to shear duration.

(1) **Thixotropic fluid and seismic condensation fluid**

The viscosity of thixotropic fluids decreases with the increase of time when the shear deformation rate remains unchanged. The viscosity of seismic condensation fluids increases with the increase of time. In addition, some fluids exhibit thixotropy at low shear rate, but their viscosity increases at high shear rate.

(2) **Viscoelastic fluid**

Viscoelastic fluid means that the deformation of non-Newtonian fluid has both recoverable elastic deformation and irrecoverable plastic deformation. It has elastic recovery or stress relaxation effect in the process of flow change. This kind of non-Newtonian fluid is called viscoelastic fluid.

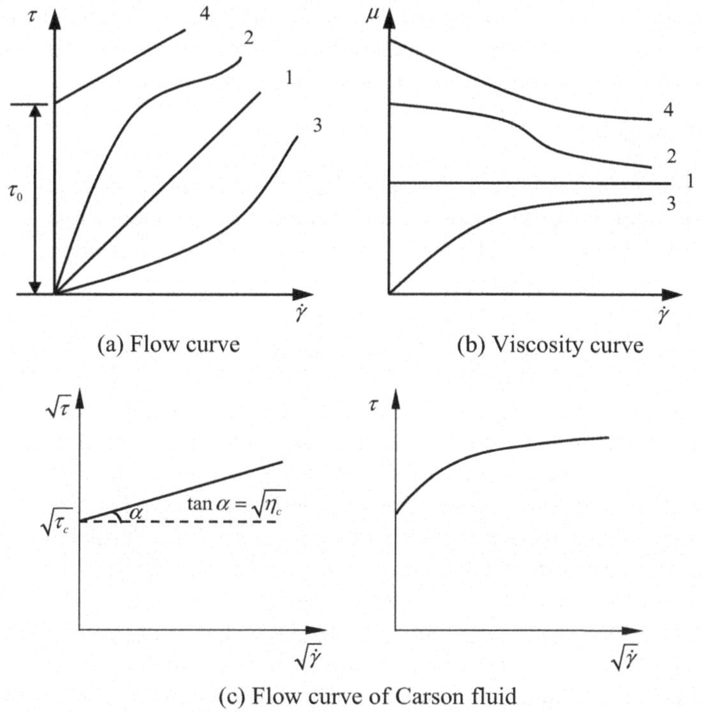

Fig. 5.1 Fluid types

5.1.1.3 Non-time-Varying Non-Newtonian Fluids

Non-time-varying non-Newtonian fluid is also known as generalized Newtonian fluid. Its shear force is only related to the speed of shear deformation, but has nothing to do with the time of shear action. The several types of viscosity curves of non-time-varying non-Newtonian fluids are shown in Fig. 2.1b.

(1) **Power law fluid**

The constitutive relation of power law fluid is as follows:

$$\tau = k\dot{\gamma}^n \tag{5.2}$$

where k is consistency coefficient, n is Non-Newtonian index, k and n are constants. When n < 1, it is called Pseudo-plastic fluid, or Shear-thinning fluid. When n > 1, the fluid whose viscosity decreases with the increase of shear rate is called Dilatant fluid, or shear thickening fluid. When n = 1, it's a Newtonian fluid, and then k is the viscosity. The shear thinning fluid is mainly composed of the polymer solution containing long-chain molecular structure and the suspension containing slender fibers or particles, such as hydraulic oil, silicone oil, and silica gel and so on. Shear thickening fluid is generally high concentration suspension containing irregular shape solid particles, such as starch paste, sesame paste, gum solution, etc.

(2) **Bingham fluid**

Bingham fluid means that when the shear stress is less than the yield stress, the fluid will not flow. Only when the shear stress is greater than the yield stress, it will begin to flow like Newtonian fluid, also known as plastic fluid. The constitutive relationship of Bingham fluid is as follows:

$$\tau - \tau_0 = \eta_\rho \dot{\gamma} \tag{5.3}$$

where τ_0 is yield stress, η_ρ is plastic viscosity, and the both are constant at a given temperature and pressure. The rheological properties of Bingham fluid are determined by the heterogeneity of its internal structure. In multiphase flow, the particles as dispersed phase are dispersed in continuous phase. Because of the strong interaction between dispersed particles, the yield stress is produced and the reticular structure is formed at rest. The flow can only be carried out when sufficient shear stress is applied to destroy the reticulated structure. Bingham fluid is a widely used material in engineering, such as asphalt, petroleum products, low temperature crude oil with high wax content, toothpaste and paint, and suspension of medium concentration. The Magnetorheological Fluid (MRF) commonly used in damper is also Bingham fluid. MRF is a mixed fluid consisting of non-magnetic carrier fluid, magnetic medium particles with high conductivity and low hysteresis, and surfactant.

(3) **Carson fluid**

The constitutive relationship of Carson fluid is as follows:

$$\sqrt{\dot{\gamma}} = \frac{\sqrt{\tau} - \sqrt{\tau_c}}{\sqrt{\eta_c}} \tag{5.4}$$

where η_c is the Carson viscosity, τ_c is the Carson yield value. The relationship between shear stress and shear strain velocity of Carson fluid is shown in Fig. 2.1c. Carson fluid generally includes some paints, plastic media, and blood of human and animal. At present, the viscous fluids used in viscous fluid energy dissipation dampers are mainly hydraulic oil, silicone oil, silicone-based gel and special suspension, and silicone oil is the most widely used.

5.1.1.4 Characteristics of Organosilicon Oil

Silicone oil is a kind of organosilicon polymer. It has elemental silicon in its molecular structure, and the main chain of the molecule is a skeleton composed of silicon atom and oxygen atom alternately. Silicone oil has the advantages of non-toxicity, odorless, non-corrosive and non-combustible. There are many kinds of silicone oil, among which dimethyl silicone oil is the most commonly used viscous medium of the viscous fluid energy dissipation damper, and the current research is also the most mature. Dimethyl silicone oil, or methyl silicone oil for short, is a colorless and transparent oily liquid with a density of 930–975 kg/m^3, which is insoluble in water, and has good hydrophobicity and good electrical insulation. It is the most basic and typical silicone oil, and it is also the most productive and widely used variety.

(1) **Viscosity characteristics**
The dynamic viscosity of methyl silicone oil generally varies from $10 \sim 3 \times 10^5$ cst. The viscosity of methyl silicone oil can be adjusted by the relative dosage of chain-stop agent, and can also be prepared with two different viscosities of silicone oil. Although the molecular weight distribution of organosilicon oil after blending is different between the two methods, they have no effect on its physical and chemical properties, which is very important for its application in engineering.

(2) **Viscosity-temperature property**
Methyl silicone oil has good resistance to high and low temperature, and can be used for a long time at high temperature. There is almost no oxidation below 150 °C. In addition, methyl silicone oil also has good cold resistance. Medium and low viscosity methyl silicone oil loses its fluidity at—65 to 50 °C. With the increase of the viscosity of methyl silicone oil, the cold resistance decreases gradually. In organic chemistry, the viscosity-temperature coefficient is the parameter that characterizes the change of the viscosity of methyl silicone oil with temperature. The viscosity-temperature coefficient of silicone oil with different viscosities is shown in Table 5.1. It can be seen from the table that the viscosity-temperature coefficient of methyl silicone oil is very small, that is, the viscosity changes little with temperature. This is because the interaction between the molecules of methyl silicone oil is very small, and the characteristics of molecular helix structure also have some influence.

$$\text{Viscosity-temperature coefficient} = 1 - \frac{\text{Viscosity at } 98.9\ ^\circ\text{C}\ \left(\text{mm}^2/\text{s}\right)}{\text{Viscosity at } 370.8\ ^\circ\text{C}\ \left(\text{mm}^2/\text{s}\right)}$$

(3) **Compressibility**
Methyl silicone oil has high compressibility, as shown in Table 5.2. When the viscous fluid energy dissipation damper is working, the viscous medium flows in the cylinder block. The compression of methyl silicone oil under pressure will form a certain elastic force, which is also one of the reasons for the dynamic stiffness of the viscous fluid energy dissipation damper, thus affecting the energy dissipation performance of the viscous fluid energy dissipation damper. However, relative to viscous, this effect is not very large and can be neglected when the frequency is small.

Table 5.1 Viscosity-temperature coefficients of various silicone oils

Viscosity	10	20	50	100	200	500	1000	12,500	30,000	60,000	100,000
Viscosity-temperature coefficient	0.56	0.59	0.59	0.60	0.60	0.60	0.61	0.61	0.61	0.61	0.61

Table 5.2 Compressibility of methyl silicone oil (%)

Pressure	Viscosity					
	0.65	1.0	2.0	12.8	100	12,500
50	6.3	5.4	4.8	4.4	4.5	4.5
100	10.0	8.8	8.2	7.3	7.3	7.3
150	12.6	11.4	10.7	9.5	9.5	9.3
200	14.6	13.4	12.7	11.3	11.2	11.0
300	17.8	16.5	15.8	14.2	16.5	13.8
500	Freeze at 401 MPa	20.7	20.1	18.1	20.7	17.7
1000		26.3	26.0	23.7	26.7	23.0
2000		31.7	31.5	29.1	31.7	28.1
4000		36.6	36.9	34.3	34.0	33.5

Note The unit of pressure is MPa and the unit of viscosity is mm^2/s

5.1.2 Energy Dissipation Mechanism of Viscous Fluid Damper

5.1.2.1 Basic Working Principle

Taking the most commonly used porous viscous fluid damper as an example, the basic working principle of viscous fluid damper is analyzed. This kind of damper is generally composed of cylinder block, piston, damper hole (or clearance, or both), viscous fluid damper material and guide rod. The piston moves reciprocating in the cylinder. There are some small holes on the piston to become damper hole (and the clearance between piston and cylinder), and the cylinder is filled with viscous fluid damper material. When the energy dissipation damper starts to work, the piston rod moves back and forth in the cylinder, and the damping medium flows through the damping hole. In this process, due to friction and other factors affecting the damping medium, some kinetic energy of the damping medium will be converted into irreversible heat energy, which just dissipates the energy input in the earthquake, thus protecting the structure and reducing the loss of life and property caused by disasters such as earthquakes. The conversion of kinetic energy to thermal energy of viscous fluid is mainly accomplished by friction energy dissipation and hole-shrinkage energy dissipation.

5.1.2.2 Friction Energy Dissipation

(1) **Loss along the way and local loss**

When the structure encounters the dynamic load such as earthquake action and wind load, the viscous fluid damper will be in working state, and the damping medium flows into the damping hole from the cylinder. Since the diameter of the damping hole is much smaller than the diameter of the cylinder. It is impossible to start a steady flow from entering the orifice, but to maintain a stable flow state after a certain distance. This change section is called the inlet start section, and the flow state after the start section is a fully developed flow state.

After the fluid flows into the pipeline at a uniform speed, due to the uniformity, a boundary layer is formed near the wall, and the boundary layer gradually expands toward the tube axis along the flow direction, and the speed of each section along the flow direction also changes continuously. After a certain distance l_1, the velocity distribution on the flow section can reach to a stable velocity profile, which is called the inlet start section.

When the fluid moves along the flow path, the friction occurs between the fluid micros or flow layers and between the fluid and solid wall, and the friction causes the fluid to lose a part of energy. This part of the lost energy is called the loss along the path, which is represented by E_l. The loss along the path is evenly distributed along the flow path, and its size is proportional to the length of the fluid flow.

When the fluid passes through local obstacles (such as valves, elbows, cross-section changes, etc.), the energy loss is caused by the violent momentum exchange between the fluid particles due to the change of the shape of flow channel, the disturbance of flow velocity, and the change of flow direction. This energy loss is called local loss, expressed in E_m. The size of the local loss is related to the type of obstruction, but it is usually concentrated on a shorter process.

In the viscous fluid damper, the energy loss generated by the movement of damping medium in the orifice mainly includes the loss along the path and the local loss due to the section changes at the inlet and outlet. The total energy loss is expressed by E_w, which shows:

$$E_w = E_l + E_m \tag{5.5}$$

(2) **Laminar flow and turbulence**

In fluid mechanics, when the fluid moves at a small speed, the particles do not move in the direction perpendicular to the mainstream, but in a clear hierarchy, do not mix with each other, and maintain a parallel steady beam state. This flow is called laminar flow. On the contrary, when the fluid moves at a faster speed, it does not maintain the regular steady state of the beam, but makes complex, irregular and random unsteady motion, which is called turbulence. The mean velocity of turbulence to laminar flow is smaller than that of turbulence to laminar flow, which is called the lower critical velocity. On the contrary, the critical velocity when laminar flow turns into turbulence is called the upper critical velocity. In addition, further experiments show that the critical flow velocity of the fluid is related to the diameter d, the dynamic viscosity μ of the fluid and the density ρ of fluid. The following dimensionless forms are written from these four physical quantities:

$$R_e = \frac{\rho v d}{\mu} \tag{5.6}$$

No matter how d, μ, ρ and ν change, the dimension is basically unchanged. This dimension is also called Reynolds number. The Reynolds number corresponding to the lower critical velocity is called the lower critical Reynolds number. Similarly, the Reynolds number corresponding to the upper critical velocity is called the upper critical Reynolds number. It can be seen that:

Lower critical Reynolds number:

$$R_e = \frac{\rho v d}{\mu} = \frac{v d}{\nu} = 2320 \tag{5.7}$$

Upper critical Reynolds number:

$$R_e = \frac{\rho v d}{\mu} = \frac{v d}{\nu} = 2320 \tag{5.8}$$

among them, ν is the kinematic viscosity of the fluid. In fluid mechanics, the lower critical Reynolds number is generally used as the criterion for fluid flow in a circular pipe. When $R_e = \frac{\rho \bar{v} d}{\mu} = \frac{\bar{v} d}{\nu} < 2320$, it is laminar flow; when $R_e = \frac{\rho \bar{v} d}{\mu} = \frac{\bar{v} d}{\nu} \geq 2320$, it is turbulent flow. Among them, $\bar{v}$ is the average velocity of fluid in the pipe. The lower critical Reynolds number is difficult to reach 2320, which is only about 2000, so the lower critical Reynolds number is taken as 2000.

(3) **Friction energy dissipation of Newtonian fluid**

When the fluid passes through the initial section of the inlet, it becomes a stable laminar flow. Because the diameter of the damping hole is small and the pressure inside the damping hole is relatively large, the gravity of the pressurized pipeline can be neglected. Damping holes (Fig. 5.2a) are considered as equal diameter circular tubes, and the damping medium is in laminar flow state. Taking the rectangular coordinate system as shown in Fig. 5.2b, the Navier-Stokes equation along the y axis (Navier-Stokes equation, abbreviated as N-S equation) is listed.

$$f_y - \frac{1}{\rho}\frac{\partial p}{\partial y} + \upsilon\left(\frac{\partial^2 v_y}{\partial x^2} + \frac{\partial^2 v_y}{\partial y^2} + \frac{\partial^2 v_y}{\partial z^2}\right) = \frac{\partial v_y}{\partial t} + \frac{\partial v_y}{\partial x} v_x + \frac{\partial v_y}{\partial y} v_y + \frac{\partial v_y}{\partial z} v_z \tag{5.9}$$

Because in steady laminar flow, the fluid particles do not flow transversely, but only flow axially, and:

$$v_x = v_z = 0,\ v_y = v \tag{5.10}$$

As mentioned above, the gravity of the pressurized pipeline can be neglected, so:

$$f_x = 0,\ \ f_y = 0,\ \ f_z = 0 \tag{5.11}$$

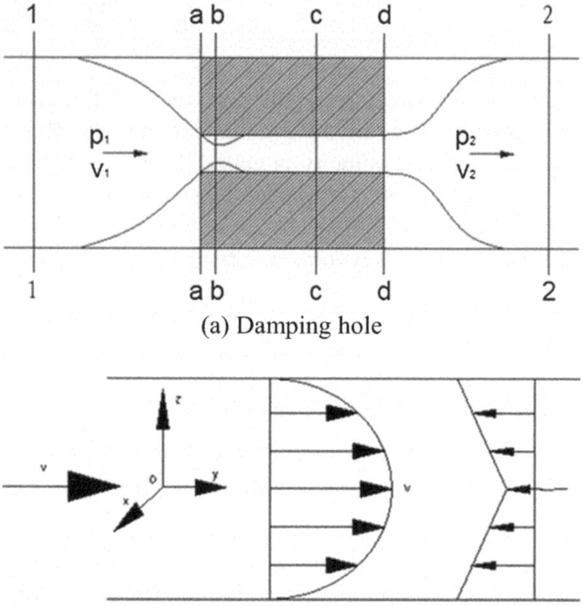

(a) Damping hole

(b) Laminar flow in circular tube

Fig. 5.2 Laminar flow in circular tube

And:

$$\frac{\partial v_x}{\partial t} = \frac{\partial v_y}{\partial t} = \frac{\partial v_z}{\partial t} = 0 \tag{5.12}$$

For incompressible fluids, the continuity differential equation is:

$$\frac{\partial v_x}{\partial x} + \frac{\partial v_y}{\partial y} + \frac{\partial v_z}{\partial z} = 0 \tag{5.13}$$

And:

$$\frac{\partial v_y}{\partial y} = 0 \tag{5.14}$$

By substituting the above conditions into the N-S equation, we can get that:

$$\frac{1}{\rho}\frac{\partial p}{\partial y} = \upsilon(\frac{\partial^2 v_y}{\partial x^2} + \frac{\partial^2 v_y}{\partial z^2}) \tag{5.15}$$

$$\frac{1}{\rho}\frac{\partial p}{\partial x} = 0 \tag{5.16}$$

$$\frac{1}{\rho}\frac{\partial p}{\partial z} = 0 \tag{5.17}$$

From the latter two equations, we can see that the pressure is independent of coordinates x and z, so there are:

$$\frac{\partial p}{\partial y} = \frac{dp}{dy} \tag{5.18}$$

Because of neglecting the influence of gravity, the fluid flows axisymmetrically in an equal diameter pipe, so the velocity is only a function of radius r and remains unchanged along the y axis. Therefore, in order to facilitate the integral of the equation, the cylindrical coordinates r, θ and y are chosen in the cylindrical coordinate system:

$$x^2 + z^2 = r^2, x = r\cos\theta, z = r\sin\theta \tag{5.19}$$

The first two formulas are derived separately:

$$2r\frac{\partial r}{\partial x} = 2x \tag{5.20}$$

$$\frac{\partial r}{\partial x} = \frac{x}{r} = \cos\theta \tag{5.21}$$

$$\frac{\partial x}{\partial x} = \frac{\partial}{\partial x}(r\cos\theta) = \cos\theta\frac{\partial r}{\partial x} - r\sin\theta\frac{\partial\theta}{\partial x} = \cos^2\theta - r\sin\theta\frac{\partial\theta}{\partial x} = 1 \tag{5.22}$$

Then:

$$\frac{\partial\theta}{\partial x} = -\frac{\sin\theta}{r} \tag{5.23}$$

and

$$\frac{\partial\theta}{\partial r} = \frac{\partial\theta}{\partial x}\frac{\partial x}{\partial r} = -\frac{\sin\theta}{r}\frac{1}{\cos\theta} \tag{5.24}$$

$$\begin{aligned}\frac{\partial^2 v}{\partial x^2} &= \frac{\partial}{\partial x}(\frac{\partial v}{\partial x}) = \frac{\partial}{\partial x}(\frac{\partial v}{\partial r}\frac{\partial r}{\partial x}) = \frac{\partial}{\partial r}(\frac{\partial v}{\partial r}\frac{\partial r}{\partial x})\frac{\partial r}{\partial x} = [\frac{\partial r}{\partial x}\frac{\partial^2 v}{\partial r^2} + \frac{\partial v}{\partial r}\frac{\partial}{\partial r}(\frac{\partial r}{\partial x})]\frac{\partial r}{\partial x}\\ &= (\frac{\partial r}{\partial x})^2\frac{\partial^2 v}{\partial r^2} + \frac{\partial v}{\partial r}\frac{\partial}{\partial r}(\cos\theta)\frac{\partial r}{\partial x} = (\frac{\partial r}{\partial x})^2\frac{\partial^2 v}{\partial r^2} + \frac{\partial v}{\partial r}(-\sin\theta)\frac{\partial\theta}{\partial r}\frac{\partial r}{\partial x}\\ &= \cos^2\theta\frac{\partial^2 v}{\partial r^2} + \frac{\partial v}{\partial r}(-\sin\theta)(-\frac{\sin\theta}{r}\frac{1}{\cos\theta})\cos\theta = \cos^2\theta\frac{\partial^2 v}{\partial r^2} + \frac{\sin^2\theta}{r}\frac{\partial v}{\partial r}\end{aligned} \tag{5.25}$$

Similarly:

$$\frac{\partial^2 v}{\partial z^2} = \sin^2\theta\frac{\partial^2 v}{\partial r^2} + \frac{\cos^2\theta}{r}\frac{\partial v}{\partial r} \tag{5.26}$$

Then:

$$\frac{\partial^2 v}{\partial x^2}+\frac{\partial^2 v}{\partial z^2}=\frac{\partial^2 v}{\partial r^2}+\frac{1}{r}\frac{\partial v}{\partial r} \tag{5.27}$$

By substituting the formulas above into the N-S equation:

$$\frac{\partial^2 v}{\partial r^2}+\frac{1}{r}\frac{\partial v}{\partial r}=\frac{1}{\mu}\frac{\partial p}{\partial y} \tag{5.28}$$

Since v is only a function of r, and p is only a function of y, the equation can be written as follows:

$$\frac{d^2 v}{dr^2}+\frac{1}{r}\frac{dv}{dr}=\frac{1}{\mu}\frac{dp}{dy} \tag{5.29}$$

If the pressure drop on the pipe length l is Δp,

$$\frac{dp}{dy}=-\frac{\Delta p}{l} \tag{5.30}$$

Where "–" indicates that the flow direction of the pressure increments dp in the pipe is negative:

$$\begin{aligned}\frac{d^2 v}{dr^2}+\frac{1}{r}\frac{dv}{dr}&=\frac{1}{\mu}\frac{\Delta p}{l}\\ \frac{1}{r}\frac{d}{dr}\left(r\frac{dv}{dr}\right)&=-\frac{1}{\mu}\frac{\Delta p}{l}\end{aligned} \tag{5.31}$$

After integration:

$$\begin{aligned}r\frac{dv}{dr}&=-\frac{1}{\mu}\frac{\Delta p r^2}{2l}+C_1\\ \frac{dv}{dr}&=-\frac{1}{\mu}\frac{\Delta p}{l}\frac{r}{2}+\frac{C_1}{r}\end{aligned} \tag{5.32}$$

After integration:

$$v=-\frac{\Delta p}{4\mu l}r^2+C_1\ln r+C_2 \tag{5.33}$$

The flow velocity v is finite at tube axis $r=0$, and $C_1=0$.
And $v=0$ at tube wall $r=r_0$, $C_2=\frac{\Delta p}{4\mu l}r_0$.
Then:

$$v = \frac{\Delta p}{4\mu l}(r_0^2 - r^2) \tag{5.34}$$

Equation (5.34) is the velocity distribution function of a fluid moving along an equal-diameter circular pipe as a constant laminar flow. It can be seen from the distribution function that the velocity is a parabola, the maximum velocity is at the middle line of the pipe, and the velocity near the wall of pipe is zero. In order to calculate the flow rate, the annular area of the element with width dr at the radius r of the cross section is taken. The flow through the area is as follows:

$$dq_V = v dA = 2\pi r v dr \tag{5.35}$$

Thus, the flow rate through the whole cross-section can be as follows:

$$Q = \int dq_V = \int_0^{r_0} 2\pi r v dr = \pi \frac{\Delta p}{2\mu l} \int_0^{r_0} [(r_0^2 - r^2) r] dr \tag{5.36}$$

After integration:

$$Q = \frac{\pi \Delta p}{8\mu l} r^4 \tag{5.37}$$

or

$$Q = \frac{\pi d^4}{128\mu l} \Delta p \tag{5.38}$$

The above formula is called Hagen-Poiseuille law. So the average velocity $\bar{V}$ is:

$$\bar{V} = \frac{Q}{A} = \frac{d^2}{32\mu l} \Delta p \tag{5.39}$$

The pressure drop Δp after the pipe length l is:

$$\Delta p = \frac{32\mu l}{d^2} \bar{V} \tag{5.40}$$

And $\Delta p = \frac{F_1}{A}$, $A = \frac{\pi(D^2 - D^2)}{4}$, then:

$$F_1 = \frac{\pi(D^2 - D'^2)}{4} \frac{32\mu l}{d^2} \bar{V} \tag{5.41}$$

If N is the number of damping holes on the piston, then the continuity equation of fluid flowing through the piston damping holes is as follows:

$$\frac{\pi(D^2 - D'^2)}{4}V = N\frac{\pi d^2}{4}\bar{V} \tag{5.42}$$

From the above two formulas:

$$F_1 = \frac{32\pi \mu l(D^2 - D'^2)^2}{4Nd^4}V = \frac{8\pi \mu l(D^2 - D'^2)^2}{Nd^4}V \tag{5.43}$$

where μ is consistency coefficient of fluid, l is the length of damping hole, D is the internal diameter of cylinder, $D^{'}$ is the diameter of piston rod, d is the diameter of damping hole, N is the number of damping holes, V is the motion velocity of fluid.

(4) **Friction energy dissipation of Power law fluids**

For power-law fluids, the constitutive relation is as follows:

$$\tau = k\dot{\gamma}^n = k(-\frac{du}{dr})^n \tag{5.44}$$

The velocity distribution law of non-Newtonian fluid in a circular tube is as follows:

$$u = -\int_0^r f(\tau)d\tau + C \tag{5.45}$$

where u is the velocity of fluid at r from the central axis of a circular pipe, R is the radius of damping hole, C is an integral constant. According to that the velocity of viscous fluid at the pipe wall is zero, $r = R, u = 0$, the integral constant C is calculated to be:

$$C = \int_0^R f(\tau)d\tau \tag{5.46}$$

Then:

$$u = \int_r^R f(\tau)d\tau \tag{5.47}$$

The distribution law of shear stress of viscous fluid along the radial direction is as follows:

$$\tau = \frac{\Delta pr}{2l} \tag{5.48}$$

among them, Δp is the pressure difference between the two ends of the damping hole, l is the length of damping hole. The shear stress τ_R at the wall of circular tube is as follows:

$$\tau_R = \frac{\Delta p R}{2l} \tag{5.49}$$

Then:

$$\tau = \tau_R \frac{r}{R} \tag{5.50}$$

Substitute the above equation into Eq. (5.47):

$$u = \frac{R}{\tau_R} \int_{\tau}^{\tau_R} f(\tau) d\tau \tag{5.51}$$

The flow rate of the pipe with length of l can be obtained by the integration of the above formula as follows:

$$Q = \int_0^R 2\pi r u dr \tag{5.52}$$

Through the partial integration of the above formula:

$$Q = \pi r^2 u /_0^R + \int_0^R \pi r^2 (-\frac{du}{dr}) dr \tag{5.53}$$

And $u = 0$ at $r = R$, $\pi r^2 u /_0^R = 0$, then:

$$Q = \frac{n\pi R^{\frac{3n+1}{n}}}{3n+1} (\frac{\Delta p}{2kl})^{\frac{1}{n}} \tag{5.54}$$

If the piston has N identical damping holes, the total flow rate is:

$$Q = N \frac{n\pi R^{\frac{3n+1}{n}}}{3n+1} (\frac{\Delta p}{2kl})^{\frac{1}{n}} \tag{5.55}$$

The pressure difference can be obtained by simplifying the above formula:

$$\Delta p = \frac{2kl(3n+1)^n Q^n}{(Nn\pi R^{\frac{3n+1}{n}})^n} \tag{5.56}$$

Substitute $\Delta p = \frac{F_1}{A}$, $Q = Av$, $A = \frac{\pi(D^2 - D'^2)}{4}$, $R = \frac{d}{2}$ into the above equation, the damping force formula is obtained as follows:

$$F_1 = \pi kl(D^2 - D'^2)[\frac{2(3n+1)(D^2 - D'^2)}{nNd^{(3n+1)/n}}]^n v^n \tag{5.57}$$

Considering that the energy dissipation damper is a closed space, according to the principle of fluid continuity, it is known that the piston's velocity is the same as that of the fluid, i.e. $V = v$, then the above formula can be written as follows:

$$F_1 = \pi kl(D^2 - D'^2)[\frac{2(3n+1)(D^2 - D'^2)}{nNd^{(3n+1)/n}}]^n V^n \tag{5.58}$$

$C_1 = \pi kl(D^2 - D'^2)[\frac{2(3n+1)(D^2 - D'^2)}{nNd^{(3n+1)/n}}]^n$, then the damping force produced by friction of power-law fluid is as follows:

$$F_1 = C_1 V^n \tag{5.59}$$

5.1.2.3 Energy Dissipation of Pore Shrinkage Effect

The local loss caused by the sudden change of fluid cross section is independent of the viscosity of fluid, so it can be assumed that the viscosity of fluid is zero, which is called the ideal fluid. The viscous frictional energy dissipation can be ignored. In addition, in order to simplify the derivation process, the fluid velocity is assumed to be non-negative. The flow pattern of the damped medium in the viscous fluid damper is shown in Fig. 5.2a. The damped medium flows through the damped hole from left to right. The 1–1 section is the section before the fluid flows into the damped hole, the 2–2 section is the section after the fluid flows out the damped hole, the a–a section is the import section, the b–b section is the section of the shrinkage of damped medium at the beginning segment of entrance, the c–c section is the section of the fully developed segment, and the d–d section is section at the outlet. The pressures before the inlet and after the outlet are p_1 and p_2, the velocities before the inlet and after the outlet are v_1 and v_2, respectively.

(1) **Local loss when the section suddenly expands**

According to the law of conservation of energy, the Bernoulli equations at section d–d and section 2–2 can be obtained:

$$\frac{p_d}{\rho g} + \frac{a_d v_d^2}{2g} = \frac{p_2}{\rho g} + \frac{a_2 v_2^2}{2g} + E_{\zeta 2} \tag{5.60}$$

Then:

$$E_{\zeta 2} = \frac{p_d - p_2}{\rho g} + \frac{v_d^2 - v_2^2}{2g} \tag{5.61}$$

In order to obtain the pressure difference between the two sections, the momentum correction coefficient of the two sections is taken as 1 for the momentum equation of the control volume between the two cross sections:

$$p_d A_d - p_2 A_2 + p(A_2 - A_d) = \rho q_v (v_2 - v_d) \tag{5.62}$$

where P_d is the pressure acting on the convex shoulder of the beam in the eddy region, it is proved by experiments that the formula can be written as follow:

$$(p_d - p_2)A_2 = \rho q_v (v_2 - v_d) \tag{5.63}$$

or

$$p_d - p_2 = \rho v_2 (v_2 - v_d) \tag{5.64}$$

Substitute the above equation into the expression of energy loss:

$$E_{\zeta 2} = \frac{\rho v_2 (v_2 - v_d)}{\rho g} + \frac{v_d^2 - v_2^2}{2g} \tag{5.65}$$

Lastly,

$$E_{\zeta 2} = \frac{(v_2 - v_d)^2}{2g} \tag{5.66}$$

The above formula is the formula for calculating the local loss of fluid passing through a sudden enlarged circular pipe. It is commonly called the Boda-Canow formula. v_d is the value of velocity reduction, which is called "lost velocity".

And

$$v_d \mathrm{A}_d = v_2 \mathrm{A}_2 \tag{5.67}$$

Then:

$$v_d = \frac{\mathrm{A}_2}{\mathrm{A}_d} v_2, \; v_2 = \frac{\mathrm{A}_d}{\mathrm{A}_2} v_d \tag{5.68}$$

Substitute the above equation into Eq. (5.66):

$$E_{\zeta 2} = \left(1 - \frac{A_d}{A_2}\right)^2 \frac{v_d^2}{2g} = \left(\frac{A_2}{A_d} - 1\right)^2 \frac{v_2^2}{2g} \tag{5.69}$$

And $\zeta_2 = \left(1 - \frac{A_d}{A_2}\right)^2$, $\zeta_2' = \left(\frac{A_2}{A_d} - 1\right)^2$. ζ_2 and ζ_2' are local loss factors of sudden enlarged section. In a viscous fluid damper, $A_2 \succ A_d$ and $\zeta_2 = 1$:

$$E_{\zeta 2} = \zeta_2 \frac{v_d^2}{2g} = \zeta_2' \frac{v_2^2}{2g} \tag{5.70}$$

(2) **Local loss when the section suddenly shrinks**

The formula for calculating the local loss when the section suddenly shrinks is as follows:

$$E_{\zeta 1} = \zeta_1 \frac{v_c^2}{2g} \tag{5.71}$$

$$\zeta_1 = 1 + \frac{1}{C_v^2 C_c^2} - \frac{2}{C_c} \tag{5.72}$$

Among them, C_c is the shrinkage coefficient and the ratio of section area A_b of shrinkage section to pipeline section area A_c. C_v is the velocity coefficient and the ratio of the actual average velocity V_b of shrinkage section to the ideal average velocity U_0. These coefficients are obtained experimentally from Weisbach as shown in Table 5.3. Because the diameter of damping hole is very small relative to the diameter of cylinder, it can be approximated to $\xi = 0.5$.

(3) **Damping force caused by local loss**

In summary, the total local loss of viscous fluid due to the shrinkage effect when flowing through damped holes can be obtained by adding Eqs. (5.70) and (5.71):

$$E_\zeta = E_{\zeta 1} + E_{\zeta 2} = \zeta_1 \frac{v_c^2}{2g} + \zeta_2 \frac{v_c^2}{2g} = (\zeta_1 + \zeta_2) \frac{v_c^2}{2g} \tag{5.73}$$

The Bernoulli equation at 1–1 and 2–2 sections is listed as follows:

$$\frac{p_1}{\rho g} + \frac{a_1 v_1^2}{2g} = \frac{p_2}{\rho g} + \frac{a_2 v_2^2}{2g} + E_\zeta \tag{5.74}$$

And $v_1 = v_2$, $a_1 = a_2$:

$$\frac{p_1 - p_2}{\rho g} = E_\zeta \tag{5.75}$$

$\Delta p = p_1 - p_2$, substitute Eq. (2.75) into the above equation:

$$\Delta p = E_\zeta \rho g = \rho(\zeta_1 + \zeta_2) \frac{v_c^2}{2} \tag{5.76}$$

Table 5.3 Local loss coefficient of abrupt reduction

A_c/A_D	0.01	0.10	0.20	0.30	0.40	0.50	0.60	0.70	0.80	0.90	1.0
C_c	0.618	0.624	0.632	0.643	0.659	0.681	0.715	0.755	0.813	0.892	1.00
C_v	0.98	0.982	0.984	0.986	0.988	0.990	0.992	0.994	0.993	0.998	1.00
ξ_1	0.49	0.458	0.421	0.377	0.324	0.264	0.195	0.126	0.065	0.02	0

Then:

$$v_c = \sqrt{\frac{2\Delta p}{\rho(\zeta_1 + \zeta_2)}} \tag{5.77}$$

And the rate of flow is as follows:

$$Q = v_c A_c = A_c \sqrt{\frac{2\Delta p}{\rho(\zeta_1 + \zeta_2)}} = \frac{\pi d^2}{4} \sqrt{\frac{2\Delta p}{\rho(\zeta_1 + \zeta_2)}} \tag{5.78}$$

If the number of damper holes on the piston is N, the total flow through the damper holes is:

$$Q_v = N \frac{\pi d^2}{4} \sqrt{\frac{2\Delta p}{\rho(\zeta_1 + \zeta_2)}} \tag{5.79}$$

Then:

$$\Delta p = \frac{8\rho(\zeta_1 + \zeta_2) Q_v^2}{N^2 \pi^2 d^4} \tag{5.80}$$

And $\Delta p = \frac{F_2}{A}$, $Q_v = Av$, $A = \frac{\pi(D^2 - D'^2)}{4}$, then:

$$F_2 = \frac{\pi \rho(\zeta_1 + \zeta_2)(D^2 - D'^2)^3}{8N^2 d^4} v^2 \tag{5.81}$$

And $C_2 = \frac{\pi\rho(\zeta_1+\zeta_2)(D^2-D'^2)^3}{8N^2 d^4}$. Finally, the damping force of viscous fluid damper caused by the hole shrinkage effect is obtained as follows:

$$F_2 = C_2 v^2 \tag{5.82}$$

where C_2 is called damping coefficient.

5.1.2.4 Theoretical Mechanics Model

In summary, the mechanical model of porous viscous fluid damper can be obtained:

$$F = F_1 + F_2 = C_1 V^n + C_2 V^2 \tag{5.83}$$

where, $C_1 = \pi k l (D^2 - D'^2)[\frac{2(3n+1)(D^2-D'^2)}{nNd^{(3n+1)/n}}]^n$, $C_2 = \frac{\pi\rho(\zeta_1+\zeta_2)(D^2-D'^2)^3}{8N^2 d^4}$. F_1 is the friction energy dissipation of viscous fluid, which is proportional to V^n. Once the flow parameters of viscous fluid are determined, the corresponding velocity V can also be determined, and then the friction energy consumption of this part can be known; F_2 is the energy consumption caused by the shrinkage effect of viscous fluid, where the values of ξ_1 and ξ_2 are all measured by experiments. Considering the damping hole characteristics of viscous fluid damper, taking $\xi_1 = 1.0$ and $\xi_2 = 0.5$ can truly reflect the actual situation. It can be seen from Eq. (5.83) that the magnitude of damping force is not only related to the speed, but also to the area of piston, the diameter and length of damping hole, etc. Because of the personality difference of viscous fluid damper, it is difficult to describe all viscous fluid dampers with one formula. However, if the formula is written as a polynomial of velocity V, considering that the damping force produced by the velocity multiples of viscous fluid damper is very small, α is generally less than 1, so the formula can be expressed as a classical formula, that is,

$$F = CV^{\alpha} \tag{5.84}$$

Among them, C is the coefficient of viscosity, and α is the velocity index considering the structure of energy dissipation damper. In addition, it should be noted that the above formula used many assumptions in the derivation process, and the basic theory of fluid mechanics itself is not very perfect, so whether the formula can accurately meet the actual situation, it needs to carry out experimental verification.

5.1.3 Calculation Model of Viscous Fluid Damper

In the last section, the energy dissipation mechanism and theoretical model of the classical pore energy dissipation viscous fluid damper are studied. However, at present, there are many kinds of viscous fluid damper and their structures are different, so the calculation models derived from them have certain differences. Here are several commonly used calculation models [1, 2]:

(1) The calculation model of viscous fluid damper with single rod is given by Shouyi, Takeda, Japan:

$$F_d = \frac{A^3}{(\alpha A_0)^2} \cdot \frac{\rho}{2} \cdot V^2 \tag{5.85}$$

Among them, F_d is the damping force of viscous fluid damper; A_0 is the area of damping hole; ρ is the density of fluid; A is the area of piston; α is the flow correction coefficient; V is the movement speed of piston. The formula is rough and can only be used in the segment with slower piston speed, and the effect of viscosity of fluid material is not considered, so it is difficult to determine the value of α.

(2) The calculation model of oil cylinder clearance viscous fluid damper is given by Jinping Ou of China:

$$F_d = c_d \cdot V^m \tag{5.86}$$

Among them, c_d is a constant related to the diameters of oil cylinder, piston, guide rod and fluid viscosity; m is an index between 0.79 and 0.87.

(3) The calculation model of viscous fluid damper given by Taylor Devices Inc. is as follows:

$$F_d = c_d \cdot V^\alpha \tag{5.87}$$

Among them, c_d is the damping constant; α is an index between 0.3 and 1.0.

(4) The relationship between the output force of G. W. Housner viscous fluid damper and the piston velocity can be expressed as follows:

$$F_d = c_d \cdot |V^\alpha|^n \tag{5.88}$$

Among them, c_d is the viscous damping coefficient, which is independent of the frequency of piston motion; n is an index between 0.3 and 0.75.

The expressions of the above formulas are different, and different numerical ranges are given for the relationship between the output force and the piston speed. However, all the formulas have one thing in common, that is to say, the viscous fluid damper is a velocity-dependent damper, and the main influencing factor of the damping force is the piston speed.

5.2 Properties and Improvement of Viscous Fluid Materials

It can be seen from the analysis in Sect. 5.1 that the main factors affecting the mechanical properties of viscous fluid damper are viscous fluid damper material and mechanical structure of damper. In order to improve the energy consumption capacity of damper, the author's team first carried out the research on material modification around commonly used viscous fluid damper material.

5.2.1 Modification of Viscous Fluid Damping Materials

Silicone oil is the most commonly used viscous fluid material in energy dissipation and vibration reduction devices of building structures. It behaves as Newtonian fluid when the shear rate is low. In order to improve its energy dissipation capacity, the addition of particle powder with very small particle diameter (micron level) into silicone oil is considered. Because of the existence of micro particles, the mutual movement between the media is hindered and the viscosity of damping material is increased. Due to the physical and chemical interaction between the damping medium and particles, a loose structure is formed. With the shear flow, the structure is gradually destroyed, and the apparent viscosity decreases with the increase of strain rate. In addition, when the medium is still, the fine particles are in a disordered and curled state in the silicone oil. As the flow proceeds, they are arranged along the flow direction. Obviously, the larger the strain rate is, the orderly the directional arrangement is, the smaller the flow damping is, and the smaller the apparent viscosity is. In this way, due to the existence of fine particles, the original silicon oil is changed from Newtonian fluid to non-Newtonian fluid, which increases the energy consumption capacity of damping materials, and at the same time, it can better adapt to the requirements of damper for the damping medium.

Therefore, the author's team selected five possible particle powder additives (Tab. 5.4) for research, including calcium carbonate, bentonite, spherical graphite, calcium sulfate whisker and silica powder [3]. The selected particles have the characteristics of viscous fluid damping medium, such as heat resistance and high temperature

Table 5.4 Filler and mix ratio

No.	Silicone oil	Filler type	Incorporation amount (%)	Order
SC-5	Methyl silicone oil 3000cP	Calcium carbonate	5	1
SC-10		Calcium carbonate	10	2
SC-15		Calcium carbonate	15	3
SP-5		Bentonite	5	4
SP-10		Bentonite	10	5
SP-15		Bentonite	15	6
SS-5		Spheroidal graphite	5	7
SS-10		Spheroidal graphite	10	8
SS-15		Spheroidal graphite	20	9
SJ-5		Calcium sulfate whisker	5	10
SJ-7		Calcium sulfate whisker	7	11
SJ-10		Calcium sulfate whisker	10	12
SJ-15		Calcium sulfate whisker	15	13
SG-2		Silica powder	2	14
SG-5		Silica powder	5	15

(a) Test instrument

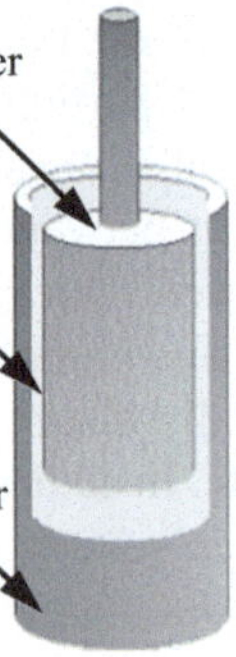

(b) Schematic diagram of material barrel

Fig. 5.3 Test equipment

resistance. In addition, after the filler and silicone oil are mixed and stirred evenly, a certain amount of loose structure can be formed in space, which increases the viscosity of silicone oil, and with the shear flow, the viscosity decreases continuously, so the nonlinearity of the silicone oil is stronger.

5.2.2 *Material Property Test of Viscous Fluid*

In order to study the effect of particle powder additives on the properties of silicone oil, the domestic dimethyl silicone oil was used in the test, two comparison groups were set with the stroke labeled as 3000# and 5000#, respectively. The label of silicone oil indicated the order of magnitude of dynamic viscosity (unit: CST, i.e. mm^2/s). Among them, the silicone oil of 3000# is mainly used to measure the effect of fine particles on silicone oil, while the silicone oil of 5000# is mainly used to measure the effect of temperature on high-grade silicone oil. The instrument used in this test is NXS-4C coal water slurry viscometer (Fig. 5.3), which is produced by Chengdu Instrument Factory. It is a coaxial cylinder up-rotation viscometer with microcomputer.

5.2.3 *Test Results and Analysis*

The viscosity and shear stress of each sample at different shear rates were determined according to the mix proportion shown in Table 5.4. The variation law of viscosity with shear rate, the variation law of shear stress with shear rate and variation law of viscosity with temperature were obtained.

5.2.3.1 Effects on Apparent Viscosity and Shear Stress

The apparent viscosity and shear stress of each group of samples at room temperature (25 °C) and shear rate of 25 s^{-1} are shown in Table 5.5. It can be seen from the table that the viscosity of the modified material increases significantly after the filler is added. In the test, the viscosity of the silicone oil of 3000# at room temperature is 3995 mPa s, after adding 10% calcium carbonate, the viscosity increases to 6680 mPa s, and after adding 5% silica powder, the viscosity increases to 7450 mPa s. The more fillers are added, the higher the viscosity is and the higher the shear stress is. The effect of filler on the viscosity of silicone oil is obvious. When the modified material is applied to the damper as damping medium, the damping force of the damper will increase correspondingly, and the lifting effect of different fillers is also different. It is feasible to improve the performance of silicone oil by adding fine particles. Figure 5.4a, b show that the apparent viscosity of the modified silicone oil is affected by the shear rate and temperature. It can be seen that the viscosity of the viscous fluid increases significantly with the adding of filler, but with the increase of the shear rate and temperature, the apparent viscosity gradually decreases, among which the pure silicone oil is the least affected, the calcium carbonate of 3000# + 10% is the most affected by the shear rate, silica powder of 3000# + 5% is the least affected by the temperature. When the damper is installed in the structure to play a role, the mechanical energy of the structure is finally converted into the internal energy of damping medium, which performed that the temperature of damping medium rises

Table 5.5 Apparent viscosity and shear stress of each sample under 25 s^{-1} at 25 °C

Sample	η (mPa s)	τ (Pa)
Silicone oil of 3000#, 25 °C	4020	100.5
Silicone oil of 5000#, 25 °C	5400	135
Calcium carbonate of 3000# + 5%	5280	132
Calcium carbonate of 3000# + 10%	6955	173.8
Calcium carbonate of 3000# + 15%	9376	234.5
Bentonite of 3000# + 5%	4995	124.8
Bentonite of 3000# + 10%	5180	129.5
Bentonite of 3000# + 15%	5845	146.1
Graphite of 3000# + 5%	4580	114.5
Graphite of 3000# + 10%	4720	118
Graphite of 3000# + 20%	5045	126.1
Whisker of 3000# + 5%	4688	117.2
Whisker of 3000# + 7%	5188	129.7
Whisker of 3000# + 10%	5885	147.1
Whisker of 3000# + 15%	6485	164.8
Silica powder of 3000# + 2%	5080	127
Silica powder of 3000# + 5%	8325	207

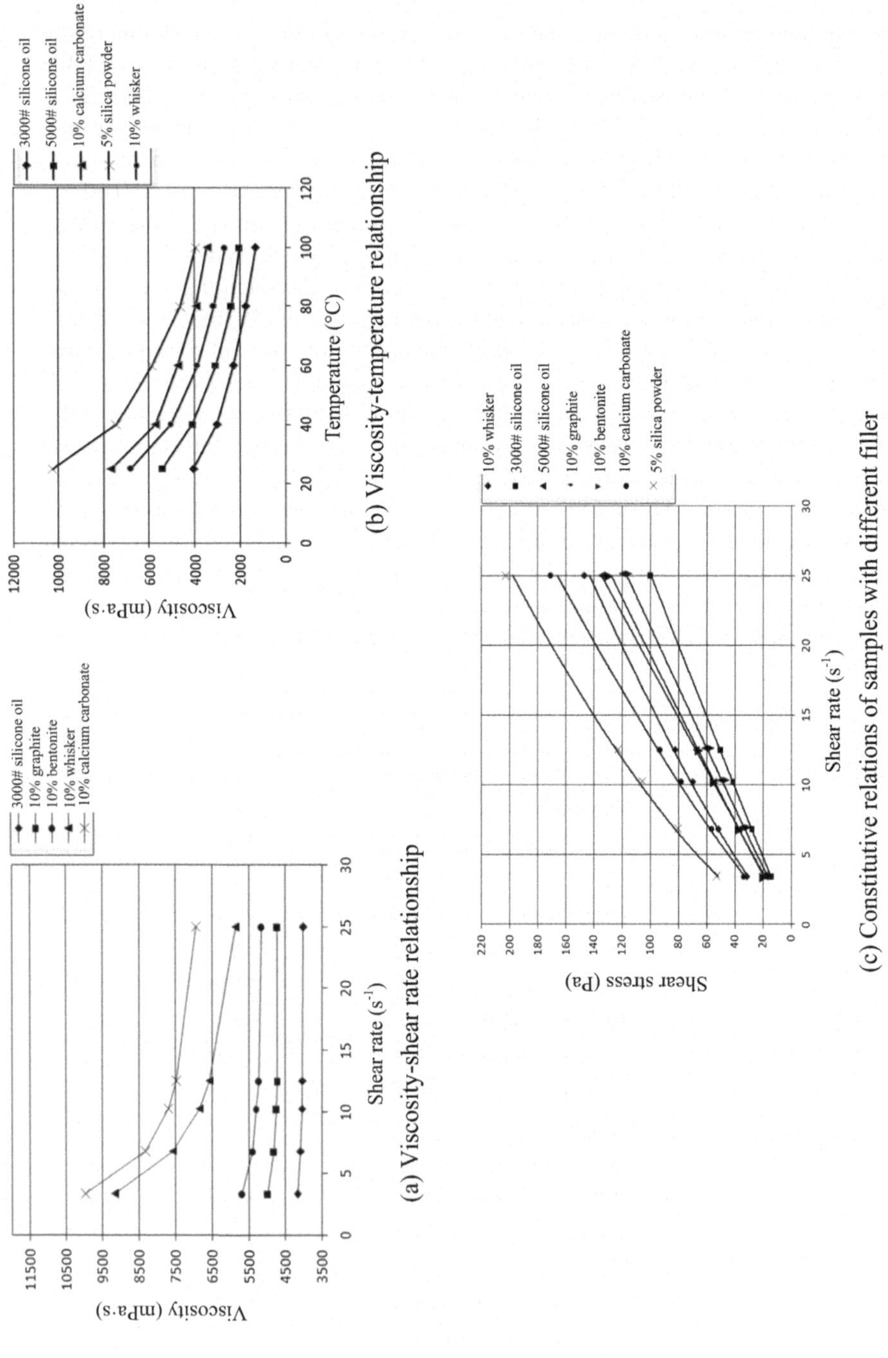

Fig. 5.4 Properties of modified silicone oil

obviously, and the energy is consumed. Therefore, the selection of silicone oil which is not affected by the temperature obviously as the damping medium is an important factor in making viscous fluid damper. Before making the damper, it is necessary to select the damping medium with good viscosity-temperature relationship and high temperature stability.

5.2.3.2 Effect on Mechanical Model

The results show that the constitutive relationship of shear thinning fluid is shown in Eq. (5.2). The consistency coefficient K of the modified material is directly increased by adding the fillers of different proportion into the silicone oil. Table 5.6 shows the constitutive relationship parameters obtained by fitting each sample, among which calcium carbonate powder, calcium sulfate whisker and silica powder are the most obvious. It can be seen from Fig. 5.4c that the nonlinearity of silicone oil is improved less by bentonite and graphite, and the nonlinearity of silicone oil is greatly enhanced by calcium carbonate and silica powder, and effect of calcium sulfate whisker on the nonlinearity improvement of silicone oil is the most obvious.

By summarizing the above data, we can get the factors that influence the modification effect: (1) the size of particle: the smaller the particle size is, the larger the volume of the particle with same mass is. After being added into the silicone oil, the larger the specific surface area contacting with the silicone oil is, the stronger the

Table 5.6 Constitutive parameters of each sample

Sample	K (Pa s^n)	n
Silicone oil of 3000#, 25 °C	4.257	0.983
Silicone oil of 5000#, 25 °C	5.594	0.978
Calcium carbonate of 3000# + 5%	6.426	0.937
Calcium carbonate of 3000# + 10%	12.10	0.816
Calcium carbonate of 3000# + 15%	22.88	0.734
Bentonite of 3000# + 5%	5.348	0.977
Bentonite of 3000# + 10%	5.984	0.951
Bentonite of 3000# + 15%	6.690	0.945
Graphite of 3000# + 5%	4.988	0.970
Graphite of 3000# + 10%	5.114	0.972
Graphite of 3000# + 20%	5.667	0.961
Whisker of 3000# + 5%	6.827	0.873
Whisker of 3000# + 7%	9.148	0.811
Whisker of 3000# + 10%	11.75	0.776
Whisker of 3000# + 15%	23.68	0.592
Silica powder of 3000# + 2%	6.699	0.912
Silica powder of 3000# + 5%	22.49	0.674

friction resistance is; (2) the structure of particles: the slender fiber is easier to form the spatial structure than the circular particle molecular structure, so the colloidal structure is formed after the powder is added with silicone oil and stirred evenly. When the shear rate is large, the structure is too late to recover, the viscosity begins to decline, and the shear stress decreases with it; (3) properties of the powder itself: some of the powder itself have thickening properties, such as bentonite, calcium sulfate whisker, etc. However, some powders have lubricity, such as spherical graphite.

It can be seen from the above studies that: (1) the method of adding particles can effectively improve the flow characteristics of silicone oil, but it has a great effect on the temperature. It is necessary to select viscous damping materials with high temperature stability or take other improvement measures to apply the materials to the actual products; (2) the more solid particles in the damping liquid, the more obvious the shear thinning of the mixture is, which can meet the requirements of non-linear characteristics in the damper design. However, it is found that the more solid powder is added, the more difficult the medium is to mix evenly, resulting in the powder in the mixture to form a block. Considering the small gap of the damper, it is bound to affect the flow stability of the medium. Therefore, it can be considered to improve using chemical additives such as dispersant and anti-settling agent; (3) the higher the powder content is, the smaller the flow index n is, but the viscosity of medium increases significantly, which brings many uncertain factors to the design of damper. At the same time, the derivation process of the constitutive relationship of damper shows that when the flow index of the fluid is too small, it affects the damping coefficient of the fluid damper and reduces the energy consumption capacity of the damper. Therefore, when making the damper of modified fluid material, it is one of the key points to select the appropriate fluid material after comprehensive consideration of various factors.

5.2.3.3 Effect on Mechanical Properties of Shock Absorber

In order to further compare the influence of viscous fluid modification on the mechanical properties of the damper, the dynamic performance test of the viscous fluid damper was carried out, with the maximum loading force of 1000 kN. The connection structure of the damper with the test bench is shown in Fig. 5.5. Sc-5 (5% of calcium carbonate), sc-10 (10% of calcium carbonate), sc-15 (15% of calcium carbonate), damping medium SG-2 (2% of silica powder), damping medium SG-5 (5% of silica powder) and damping medium SG-5 (10% of silica powder) were added to silicone oil 3000#, respectively. The loading condition was to apply a sinusoidal force of 0.25 and 0.75 Hz to the damper, control the input displacement amplitude of 5, 10, 15, 20 and 25 mm step by step from small to large, and carry out 5 cycles continuously. The specific loading condition is shown in Table 5.7.

Figure 5.6 shows the hysteretic curves of each sample under different working conditions. It can be seen that: (1) under the same working conditions, the maximum damping force of the damper increases significantly after adding the powder particles, and the larger the amount of powder particles added, the more the damping force

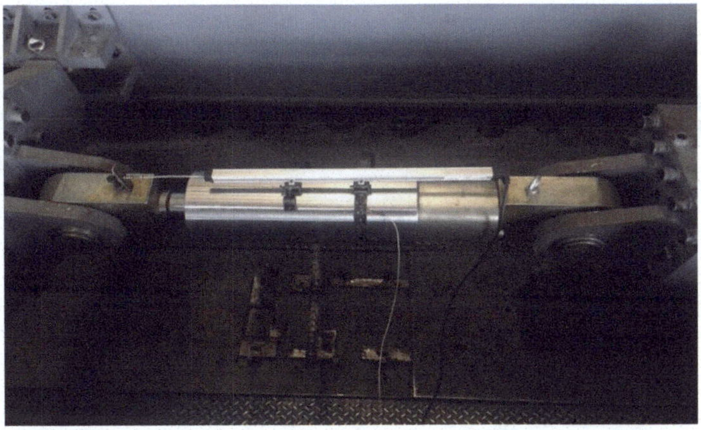

Fig. 5.5 Dynamic load test bench of damper

Table 5.7 Viscosity of samples at different temperatures under the shear rate of 10.2 s^{-1} (mPa s)

Working condition	Test frequency(Hz)	Horizontal displacement (mm)	Cycle times	Ambient temperature(°C)
1	0.25	±5	5	25
2	0.25	±10	5	25
3	0.25	±15	5	25
4	0.25	±20	5	25
5	0.25	±25	5	25
6	0.75	±5	5	25
7	0.75	±10	5	25
8	0.75	±15	5	25
9	0.75	±20	5	25
10	0.75	±25	5	25

increases; under the lower shear rate, the viscosity of the damping medium has a significant improvement on the damping force of damper; (2) at a higher shear rate, the maximum damping force of 3000# silicone oil is relatively large, which is because with the acceleration of piston motion, the fluid's thinning property begins to show, the viscosity gradually decreases, and the material with higher powder content drops more; the lifting effect of powder on silicone oil viscosity gradually decreases with the increase of piston motion speed, while the speed index of material decreases. The maximum damping force decreased compared with that before modification. In view of this phenomenon, the maximum damping force can be increased by properly reducing the damping aperture and increasing the viscosity of silicone oil; (3) in the case of the same filler content, the damping force of damper is much higher using

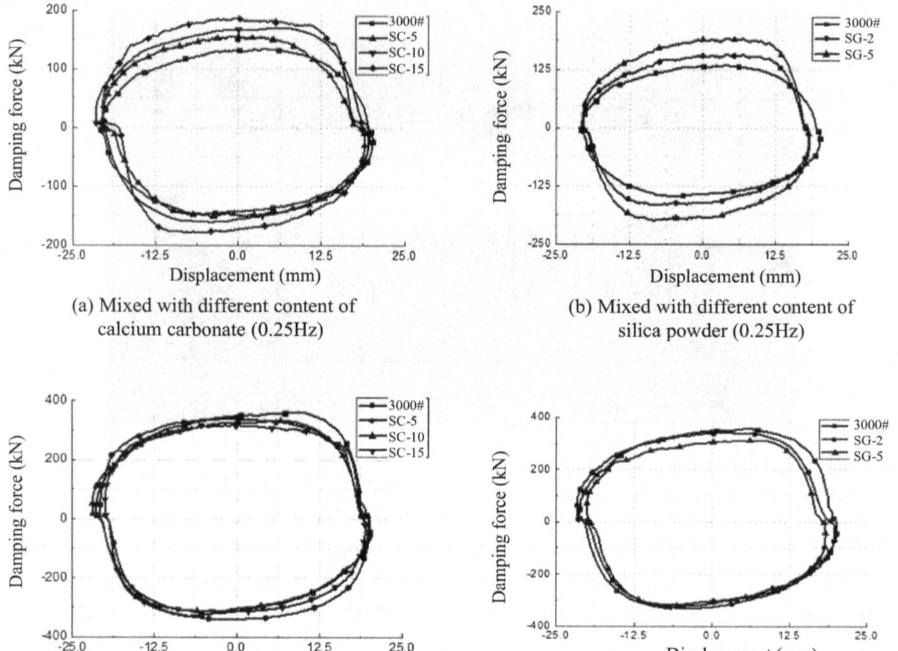

(a) Mixed with different content of calcium carbonate (0.25Hz)

(b) Mixed with different content of silica powder (0.25Hz)

(c) Mixed with different content of calcium carbonate (0.75Hz)

(d) Mixed with different content of silica powder (0.75Hz)

Fig. 5.6 Hysteresis curve of each sample under different working conditions

the mixture of silica powder and silicone oil than that using the mixture of calcium carbonate and silicone oil.

5.3 Research and Development of New Viscous Fluid Damper

5.3.1 Linear Viscous Fluid Damper

5.3.1.1 Viscous Fluid Damper with Double Exit Rods

The traditional viscous fluid damper of single-rod form has defects in structure, which is easy to cause the rapid change of the pressure in the cylinder, and the damping force produced by the damper is very unstable. The commonly used methods for this defect are additional adjusting cylinder and double-out-rod type. However, the method of additional adjusting cylinder makes the structure and processing of damper complex, and cannot provide large damping, which is not suitable for the control of large-scale structure, and its application prospect is limited. Based on this, the author team designed a kind of viscous fluid device of double-out-rod type with independent intellectual property rights [2]. The basic principle is shown in Fig. 5.7a. The main cylinder is filled with viscous fluid damping material, and there is no damping material in the auxiliary cylinder. When the piston moves to the left, part of the guide rod outside the cylinder originally enters the damper chamber, while the guide rod with the same volume on the back of piston is pushed out of the main cylinder and enters the auxiliary cylinder, and vice versa. In this way, the volume of the main cylinder is always kept constant, because the total volume of the cylinder will not change when the piston of the fluid damper of double-out-rod type moves, so that the pressure in the oil chamber will not change too much, thus avoiding the disadvantages of the single-out-rod type fluid damper.

The theoretical analysis and research show that the damping force of viscous fluid damper is mainly related to the effective area of piston (the cross-sectional area of piston or the cross-sectional area of main cylinder minus the cross-sectional area of guide rod), the size and length of damping hole, vibration frequency and displacement amplitude (the two essentially determine the moving speed of piston), temperature, damping material performance (viscosity, viscosity-temperature relationship, etc.).

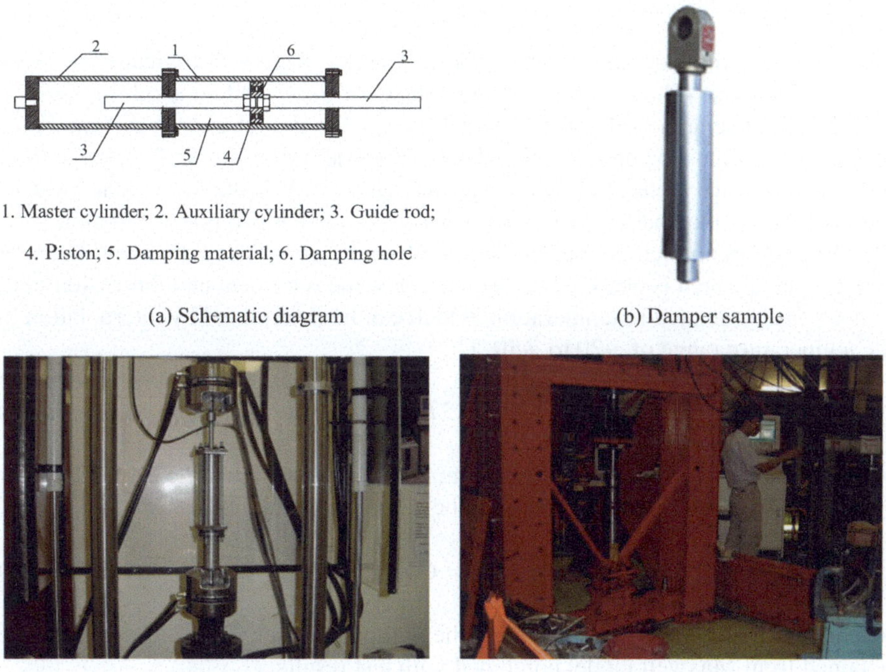

1. Master cylinder; 2. Auxiliary cylinder; 3. Guide rod;

4. Piston; 5. Damping material; 6. Damping hole

(a) Schematic diagram

(b) Damper sample

(c) Reduced scale model

(d) Full scale model

Fig. 5.7 Viscous fluid damper with double exit rods

That is, the larger the effective area of the piston is, the smaller the damping hole is, the faster the piston moves, the lower the ambient temperature is (within the working temperature range), the greater the viscosity of the damping material is, and the greater the damping force is. In order to obtain the mechanical model of damper by theoretical derivation and accumulate experience for product research and development, the author's team carried out three groups of damper samples at different frequencies, different displacement amplitude, different viscosity of silicone oil and dynamic tests at the same temperature in the Key Laboratory of Concrete and Prestressed Concrete Structures, Ministry of Education, Southeast University and the structural test laboratory of Department of Civil and Structural Engineering of Hong Kong University of Technology in 2000, respectively, including one set of reduced scale model tests (Hong Kong, Fig. 5.7c), two sets of full-scale model tests (Hong Kong and Nanjing, respectively, Fig. 5.7d), a total of 24 dampers with different parameter specifications.

According to the principle of hydrodynamics, regression analysis was carried out after processing the test data. Then the relationship between the damping force of the damper and the influencing factors was determined according to the following formula:

$$F = k' k_t k_\mu \rho \frac{A^3}{A_0^{1.5}} V \tag{5.89}$$

where F is the output damping force of the damper, in (N); μ is the dynamic viscosity of methyl silicone oil, in (Pa s); ρ is the density of methyl silicone oil, in (kg/m^3); A is the effective area of the piston, which is equal to the total cross-sectional area of the piston minus the cross-sectional area of guide rod, in (m^2); A_0 is the sum of the cross-sectional areas of all damping holes, in (m^2); V is the movement speed of the piston k_μ is the coefficient related to viscosity, which is a power function; k' is the correction coefficient (related to damper structure, processing accuracy, damping material, etc., which is obtained according to test measurement and dimensionless); k_t is a constant related to temperature, which can be determined by interpolation in the temperature range of −20 to +40 °C:

$$k_t = -0.2112T + 57.1712 \tag{5.90}$$

where T is the temperature of the damping medium, in °C. After the coefficients are combined, Eq. (2.1) can be written into the current general model form:

$$F = CV \tag{5.91}$$

where C is the damping coefficient of the damper. In fact, Eq. (5.91) can also be obtained from Maxwell model combined with test results.

According to the comparison between the calculation results of Eq. (5.89) and test results (see Fig. 5.8), the calculation results are in good agreement with the test results, which indicates that when the viscous fluid damper is used for vibration reduction

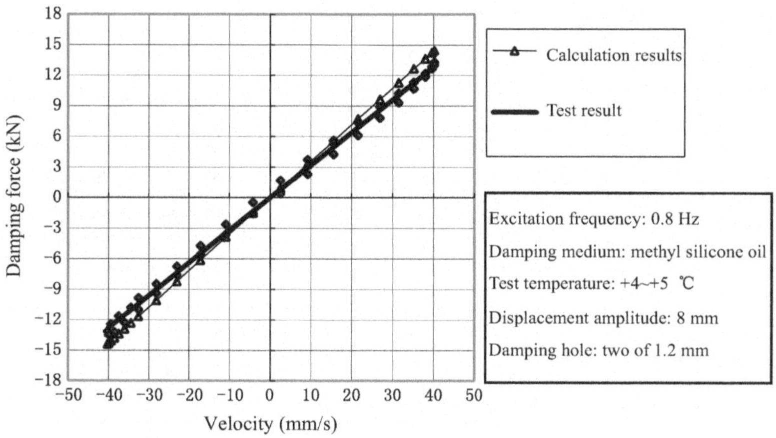

Fig. 5.8 Damping force-velocity relationship

design, the damper parameters, cylinder diameter, effective area of the piston, size and quantity of damper hole, viscosity of the damper material, etc. can be adjusted according to the design requirements to obtain the damper meeting different needs.

5.3.1.2 Control Valve Viscous Fluid Damper

It can be seen from Eq. (5.91) that the designed viscous fluid damper with double exit rods is a kind of linear viscous fluid damper, which is simple in structure, stable in performance and convenient in calculation, and has been widely used in practical engineering. However, the linear damper designed according to conventional small earthquake or wind vibration has too large output force in large earthquake, which brings difficulties to the design of structural joints and support systems. To solve this problem, there are two methods: one is to use nonlinear viscous fluid damper, which will be introduced in Sect. 5.3.2; the other is to limit the maximum output force of linear damper.

Based on the latter method, three kinds of control valve viscous fluid dampers were designed. The construction principle of the three kinds of control valves is shown in Fig. 5.9: when the moving speed of piston is small, the pressure in the cylinder cannot reach the opening pressure of the control valve, and the working performance of the damper is consistent with that of the conventional viscous damper; when the moving speed of piston is large, the pressure reaches or exceeds the opening pressure of the control valve, on the one hand, energy is consumed through damping holes, on the other hand, due to the overflow effect of the control valve, the output force of the damper is basically stable. The mechanical model is as follows:

$$F = \begin{cases} CV & (V < \frac{F_k}{C}) \\ C'V + F_k & (V \geq \frac{F_k}{C}) \end{cases} \tag{5.92}$$

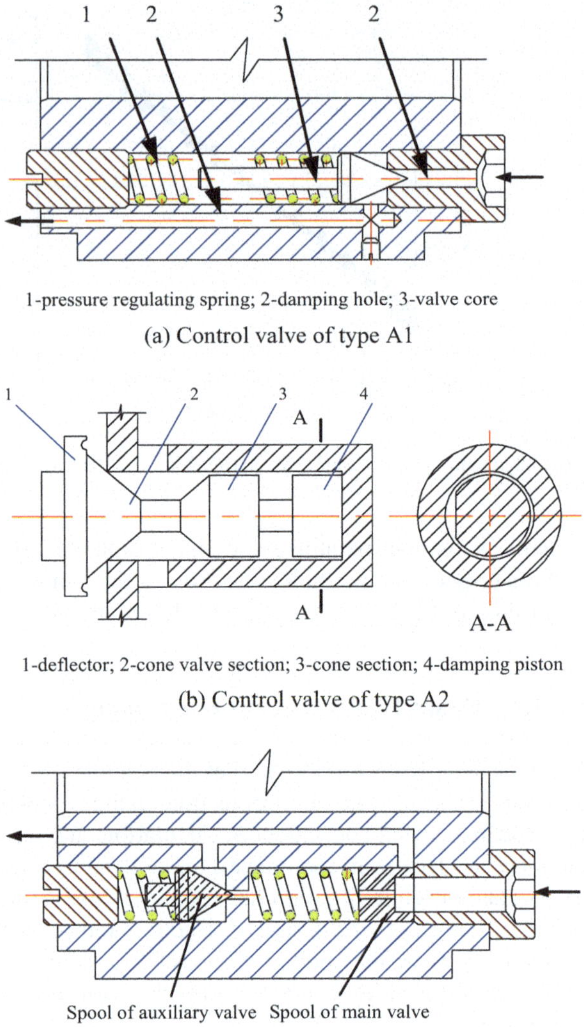

(a) Control valve of type A1

(b) Control valve of type A2

(c) Control valve of type B

Fig. 5.9 Schematic diagram of control valve

where, C and C' are the damping coefficients before and after opening, respectively, and F_k is the maximum output damping force of the damper when the control valve is opened.

In order to study the characteristics of the control valve viscous fluid damper and verify the mechanical model, three damper samples corresponding to three kinds of control valves were made (the other parameters and structures were the same), and the dynamic test was carried out in the MTS 1000 kN fatigue test machine in the structural laboratory of Southeast University. Among them, the comparison between the test and numerical simulation results of A2 type control valve viscous

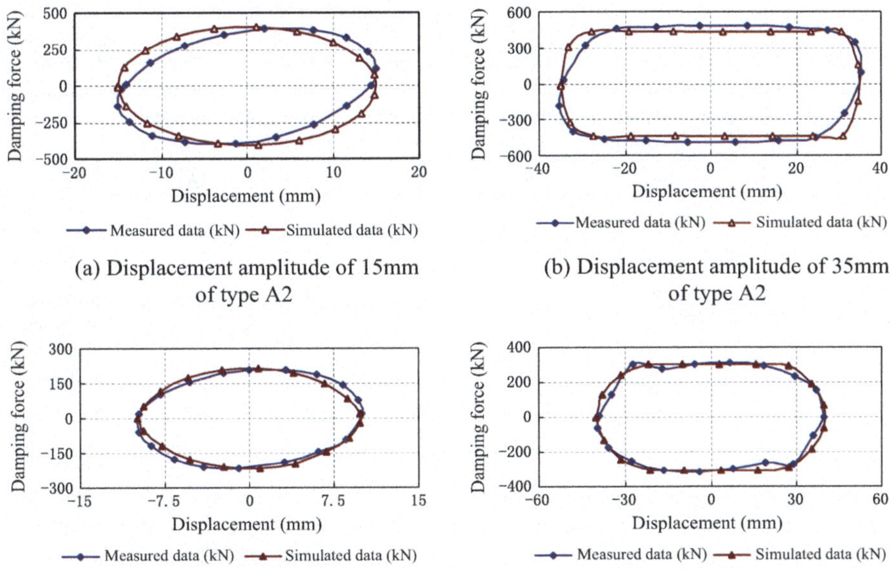

(a) Displacement amplitude of 15mm of type A2

(b) Displacement amplitude of 35mm of type A2

(c) Displacement amplitude of 10mm of type B

(d) Displacement amplitude of 40mm of type B

Fig. 5.10 Force-displacement curve of damper

fluid damper with loading frequency of 0.25 Hz are shown in Fig. 5.9. It can be seen from the figure that: (1) the designed control valve viscous fluid damper can meet the requirements of limiting the output force under large excitation; (2) the proposed mechanical model can better simulate the mechanical properties of the damper (Fig. 5.10 and Table 5.8).

5.3.2 Nonlinear Viscous Fluid Damper

Compared with the linear viscous fluid damper, the nonlinear viscous fluid damper has better energy dissipation capacity and lower output force at high speed, which is also widely used in practical engineering. In order to achieve the purpose of damper nonlinearity, there are usually two ideas: one is to improve the damping material (such as Sect. 5.2), the other is to design the damper structure. This section mainly introduces the research work of the author team in the latter aspect in recent years [4].

Table 5.8 Specification of control valve viscous damper for test

Number	Cylinder bore diameter (mm)	Guide rod diameter (mm)	Damping hole diameter (mm)	Damping hole length (mm)	Number of damping holes	Type of control valve	Damping medium
V1	180	90	1.5	10	2	A_1	Silicone oil based viscous material-1
V2	180	90	1.7	40	2	A_2	Silicone oil based viscous material-1
V3	180	90	1.7	30	2	B	Silicone oil based viscous material-1

5.3.2.1 Slender Hole Viscous Fluid Damper

The viscous fluid damper studied in this section has a long and thin damping hole on the piston (the structure of the damping hole is shown in Fig. 5.11a). When the damper works, with the reciprocating motion of the piston, the damping medium correspondingly flows from the high-pressure cavity on both sides of the piston to the low-pressure cavity through a long and thin damping channel. In the process of viscous fluid flowing through the damping channel repeatedly, the fluid dissipates the mechanical energy or work input from the outside due to overcoming the friction and other factors. When power-law fluid is used as damping medium, the mechanical model of damper can be expressed as follows:

$$F = CV^{\alpha} \tag{5.93}$$

where C is the damping coefficient and V is the piston velocity. In order to further study the designed damper, 16 samples with different specifications were made, and the dynamic test (as shown in Fig. 5.11b) was carried out on the MTS 1000 kN fatigue testing machine in the structural laboratory of Southeast University. The test includes the test conditions of different frequencies (0.1–1.5 Hz) and different amplitudes (15–50 mm). Figure 5.11c, d show the force-displacement curve of one of the samples. It can be seen that the viscous fluid damper with slender hole has obvious nonlinear characteristics, and the maximum damping force is more than 400 kN. In addition, the test results show that with the increase of the damping hole, keeping the input displacement amplitude, excitation frequency and damping medium unchanged, the maximum damping force decreases, the hysteresis loop tends to be flat, and the energy consumption capacity decreases. At the same time, with the help of the test

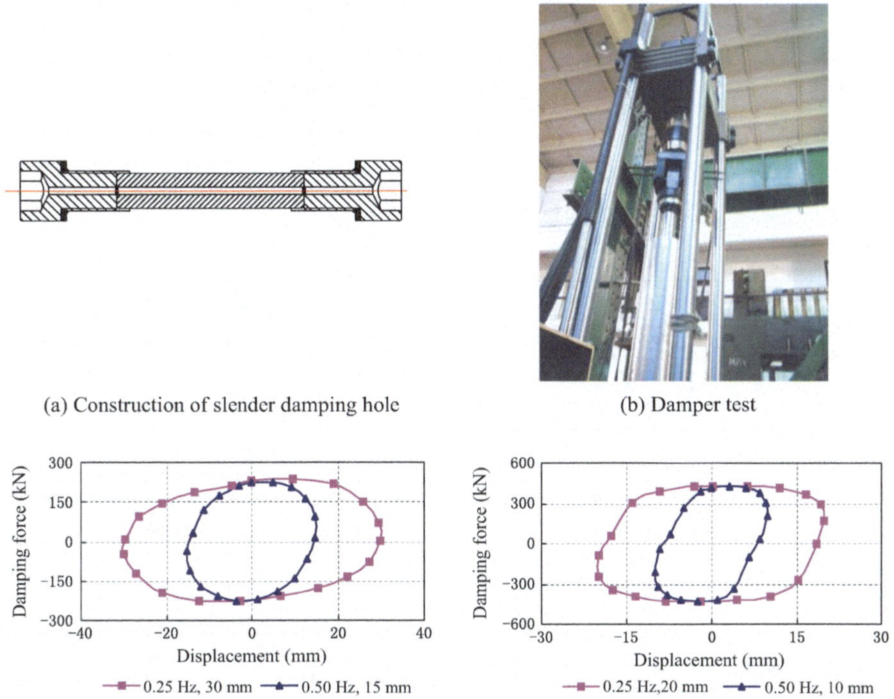

(a) Construction of slender damping hole

(b) Damper test

(c) Material damper of low apparent viscosity

(d) Material damper of high apparent viscosity

Fig. 5.11 Slender hole viscous fluid damper

results, the mass and instantaneous stiffness of the damping medium were discussed, and some useful experience was obtained.

5.3.2.2 Spiral Hole Viscous Fluid Damper

The viscous fluid damper studied in this section has a spiral type damping hole on the piston (see Fig. 5.12 for the structure of the damping hole). Spiral channel is a kind of curved pipe. In the field of fluid mechanics, the plane bend, twist pipe and spiral pipe are generally called curved pipe. According to the principle of hydrodynamics, the mechanical model of the damping medium using power-law fluid is derived by some simplification.

$$F = \frac{2\pi^2 RkL(D_1^2 - D_2^2)^{\alpha+1}}{Hd^{3\alpha+1}} \left[\frac{2(3\alpha + 1)}{n\alpha}\right]^{\alpha} V^{\alpha} \tag{5.94}$$

where, R is the radius of curvature of the spiral tube centerline; H is the pitch of the spiral tube; k is the consistency coefficient; α is the flow index; D_1 is the piston

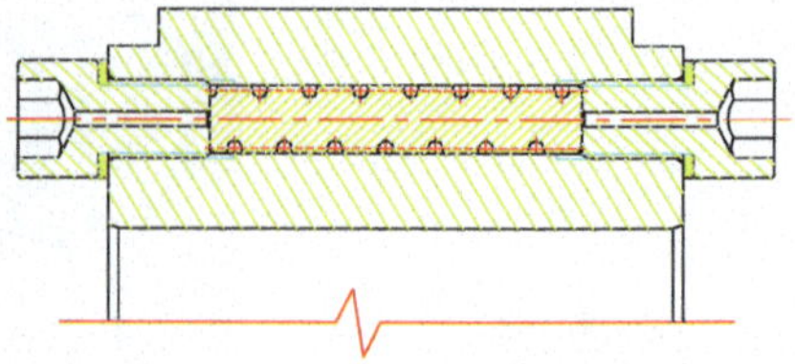

(a) Spiral hole structure

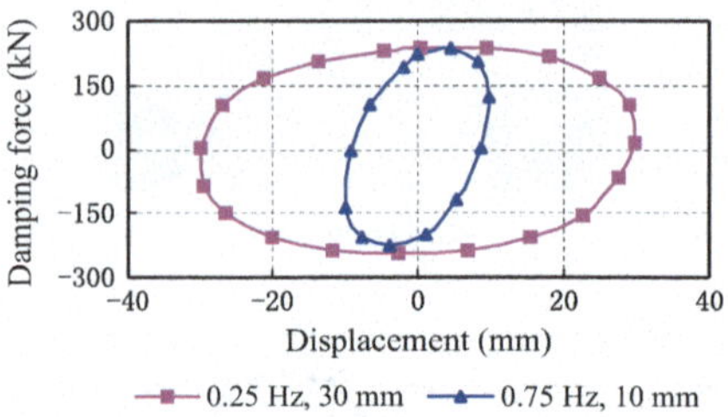

(b) Damper H1

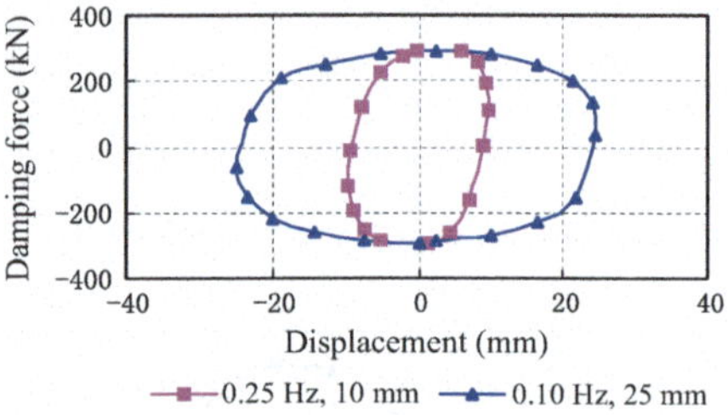

(c) Damper H2

Fig. 5.12 Spiral hole viscous fluid damper

Table 5.9 Specification of spiral hole viscous fluid damper for test

Number	Inner diameter of main cylinder	Diameter of piston guide rod	Damping hole diameter
H1	180	90	3.0
H2	180	90	4.0

outer diameter; D_2 is the guide rod diameter; d is the damping hole diameter; L is the axial length of the spiral channel. It can be seen that the form of the above formula is similar to Eq. (5.93), but the calculation formula of damping coefficient is different.

In this paper, two samples with different specifications (Table 5.9) are made, and the dynamic test was carried out on the MTS 1000 kN fatigue testing machine in the structural laboratory of Southeast University. The test method was the same as that in Fig. 5.11b, and the test condition was the same as that in Sect. 5.3.2.1. Figure 5.12b, c show the force-displacement curves of two test samples. It can be seen from the figure that the proposed spiral hole viscous fluid damper has obvious nonlinear characteristics, and the maximum damping force is about 300 kN. Through this test: (1) the influence of piston relative velocity, ambient temperature, damping hole structure and other factors on the mechanical properties of the damper were researched; (2) with the change of temperature, the size of the envelope area of the hysteresis loop changes, and with the increase of ambient temperature, the output damping force decreases, but the decrease is not obvious, and the fluctuation range of maximum output damping force is 15%; (3) compared with the slender hole viscous fluid damper, under the same condition, the maximum output damping force of the slender hole damper is smaller than that of the spiral hole damper, and the hysteretic curve of the latter is fuller and the energy consumption capacity is stronger. With the increase of the loading frequency, the instantaneous stiffness of the two kinds of dampers gradually appears. Under the same excitation, the stiffness of the latter is larger than the former; (4) the fatigue test shows that the force-displacement hysteretic curve is always very full. Compared with the 10,000th and 20,000th cycles after the 500th cycle, the shape and size of the hysteretic loop are basically unchanged. The fluctuation of maximum output damping force of tension and compression of the damper is within 10% and has high stability.

5.3.3 Other Viscous Fluid Damping Devices

With the deep understanding of viscous fluid damping technology and the need of practical engineering, in addition to the above linear viscous fluid damper and nonlinear viscous fluid damper, the research and development of other viscous damping devices are also carried out.

5.3.3.1 Variable Damping Viscous Fluid Damper

The variable damping viscous fluid damper is proposed as shown in Fig. 5.13a. Compared with Fig. 5.7a, the damper adds a specially designed damping rod based on the viscous fluid damper with double exit rods [5, 6]. In order to control the change of damping, make the damping smaller when the displacement is smaller, and larger when the displacement is larger, when working, the damper is controlled by the piston and damping rod together, and the damping groove in the piston stroke can be set on the damping rod (as shown in Fig. 5.13b as required). The mechanical model of the viscous fluid damper with variable damping and pluto fluid with two aperture segments can be expressed as follows:

$$F = C(x)V^{\alpha} \tag{5.95}$$

where $C(x)$ is the damping coefficient related to the piston movement distance x, which can be uniformly expressed as:

$$C(x) = H\int_{x-\frac{L}{2}}^{x+\frac{L}{2}} \frac{1}{d(x)^{3\alpha+1}}dx \tag{5.96}$$

where, L is the length of damping hole; dx is the trace of the length along the axial direction of the damping rod; $d(x)$ is the functional relationship between the damping hole diameter d and the axial position x. The damper samples were tested on MTS test

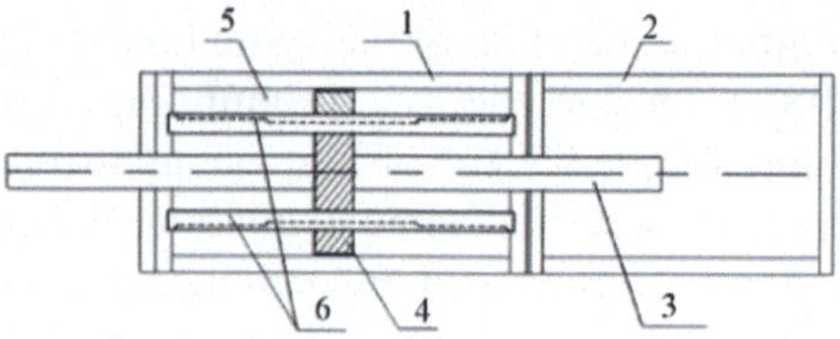

1-main cylinder; 2-auxiliary cylinder; 3-guide rod; 4-piston; 5-damping medium; 6-grooved damping rod

(a) Structural diagram of variable damping viscous fluid damper

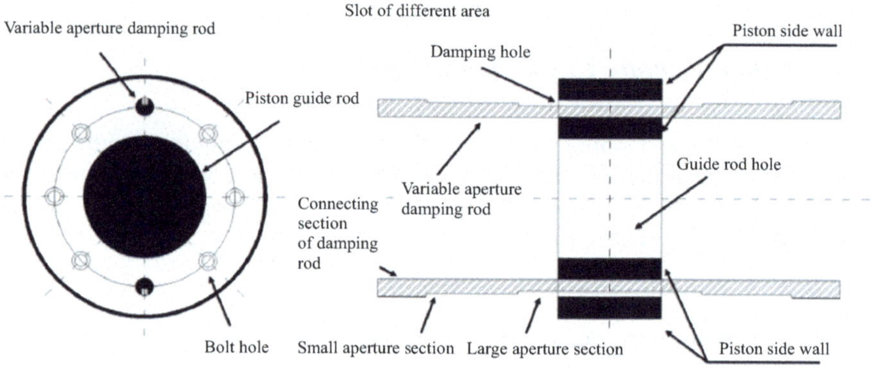

(b) Position diagram of piston and damper

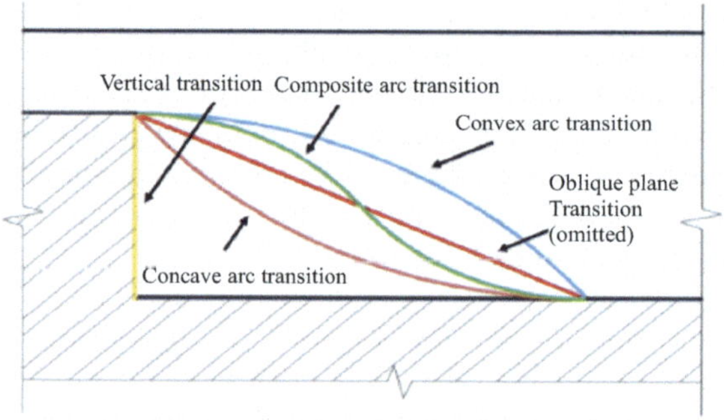

(c) Change mode of transition section

Fig. 5.13 Variable damping viscous fluid damper

machine in Structural Laboratory of Southeast University (as shown in Fig. 5.14a). Among them, the main parameters of the damper are as follows: the tonnage is 70 t, the outer diameter of the cylinder is 300 mm, the inner diameter of the cylinder is 180 mm, the diameter of the piston rod is 90 mm, the diameter of the damping rod is 20 mm, the stroke of the damper is 120 mm, and the number of the damper holes is 2, the viscosity of silicone oil is 30,000 cst. Different test pieces were designed through

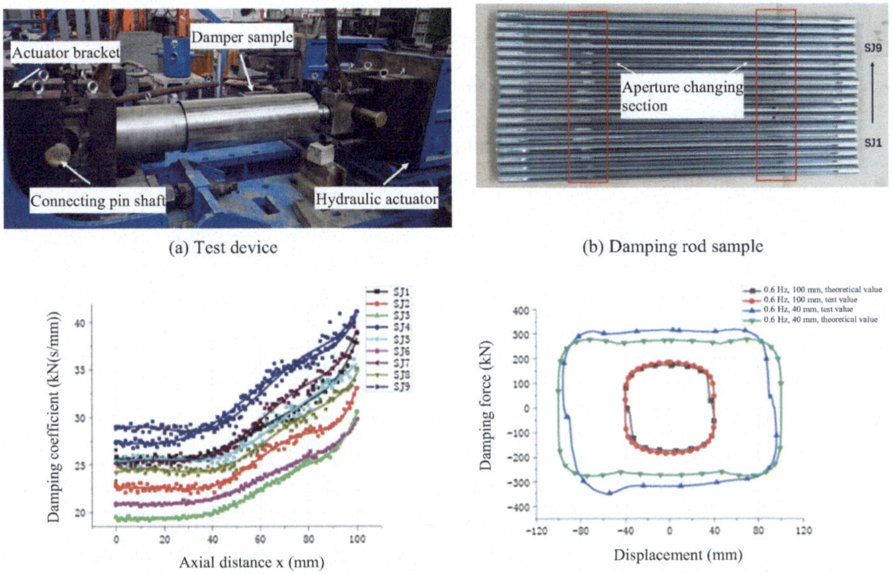

(a) Test device

(b) Damping rod sample

(c) Distribution of damping coefficient of each sample along the axial direction

(d) Comparison of theoretical and test values of SJ2

Fig. 5.14 Variable damping viscous fluid damper test

changing the change mode of aperture of transition section. The specific damping rod parameters are shown in Table 5.10, and the change mode of transition section is shown in Fig. 5.13c.

Figure 5.14c shows the distribution relationship of damping coefficient of each sample along the axial direction. According to the results of linear regression, except for SJ4, the slope of the linear regression results of damping coefficient of each sample in small displacement section is very small, and it can be considered that the damping coefficient C of each sample in the small displacement section is a constant number. The structure of SJ4 damping rod with variable aperture is vertical transition. Due to the drastic change of the aperture, on the one hand, because the damping medium in SJ4 cannot pass through the damping hole quickly, eddy current occurs before the vertical section of the aperture change. At the same time, in order to ensure the piston speed coordination, the damping medium compresses to a certain extent, the stiffness of the damper increases, and the damping force rises rapidly. The influence of stiffness is not considered, so the increment of damping force makes the calculated damping coefficient increase. On the other hand, the drastic change of the channel also increases the shrinkage effect when the damping medium passes through the damping hole when the piston approaches the change section of hole diameter, which also increases the damping energy consumption and the damping coefficient. Figure 5.14d shows the comparison between the theoretical calculation and test results. Due to the factors not considered in the theoretical calculation such

Table 5.10 Parameter of damping rod sample

Number	Curve of aperture transition section		Height-length ratio of transition section		Section area of damper hole	
	Change mode	Curve parameter	Height-length ratio	Length of transition section (mm)	Area ratio	Specific size (mm)
SJ1	Composite arc	R = 7.5 mm	1:3	9	1/1.6	5 × 5/5 × 8
SJ2	Composite arc	R = 12.5 mm	1:3	15	1/2.0	5 × 5/5 × 10
SJ3	Composite arc	R = 17.5 mm	1:3	21	1/2.4	5 × 5/5 × 12
SJ4	Radial straight line	–	1:0	0	1/2.0	5 × 5/5 × 10
SJ5	Composite arc	R = 32.5 mm	1:5	25	1/2.0	5 × 5/5 × 10
SJ6	Concave arc	R = 25 mm	1:3	15	1/2.0	5 × 5/5 × 10
SJ7	Convex arc	R = 25 mm	1:3	15	1/2.0	5 × 5/5 × 10
SJ8	Composite arc	R = 19.5 mm	1:3	9	1/2.0	8 × 3/8 × 6
SJ9	Composite arc	R = 19.5 mm	1:3	15	1/2.0	5 × 5/10 × 5

as stiffness and secondary flow in the actual test process, there are still some errors between the two, but the overall error is less than 15%.

5.3.3.2 Viscous Fluid Damping Wall

Viscous damping wall is a kind of wall-type velocity-dependent damper, which is mainly used in building structures. Since 1980s, it has been studied in foreign countries, but the domestic research started late, and the corresponding products and experimental research are less. The author's team designed a new type of viscous damping wall (as shown in Fig. 5.15a), which consists of an external steel box, a top plate, an internal steel plate and a viscous damping medium. The steel box is equipped with viscous damping medium. One or more transverse inner steel plates are placed between the steel box and the viscous damping material. In order to meet the needs of structural space and damping wall installation requirement, a bracket can be set at the lower part of the damping wall to connect the bottom of the bracket with the lower floor; generally, a reinforced concrete or fire-proof material should

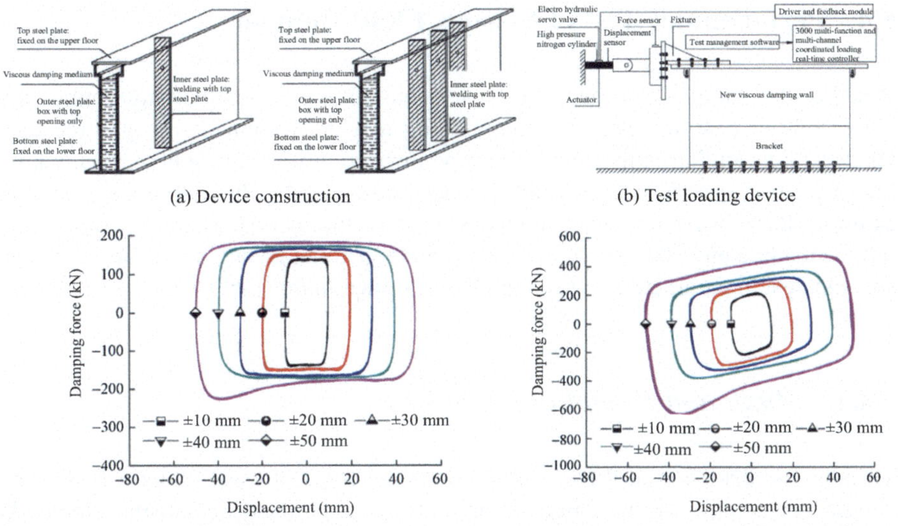

(a) Device construction (b) Test loading device

(c) Loading frequency 0.1Hz (d) Loading frequency 1.0Hz

Fig. 5.15 New viscous damping wall

be set outside the steel box to protect the wall from impact, corrosion, fire and other factors.

The samples were designed and manufactured and the dynamic test was carried out. The loading device is shown in Fig. 5.15b. The test was carried out at room temperature, which was 20 °C. Sine wave excitation was used to control the loading by input displacement. By applying displacement of different frequency and amplitude, the displacement, damping force and corresponding time of the new viscous damping wall were measured respectively, to obtain the dynamic characteristics of the damping force of the viscous damping wall with the change of loading frequency and displacement amplitude. Figure 5.15c, d show the force displacement curve of the viscous damping wall at loading frequencies of 0.1 and 1.0 Hz. The results show that the hysteretic curve of damping force displacement is full, and the shape of hysteretic curve is similar to round rectangle (between ellipse and rectangle) or ellipse, which is generally symmetrical about the origin, indicating that the energy dissipation effect of the new viscous damping wall is better. At the same time, the durability test of damping wall was also carried out. The results show that the damping force of damping wall changes slightly under the control of earthquake action, and the variation amplitude of recorded damping force every 10 cycles is less than 5%, which shows that the dynamic performance of new viscous damping wall is stable under the earthquake action; the damping force of damping wall of 1000 cycles decreases within 10% under the control of wind load.

5.4 Performance Test of Viscous Fluid Damper

Whether the damper can meet the basic requirements of engineering application still needs to pass the pressure maintaining test, maximum damping force test, temperature correlation test and other type inspection items. In this section, taking the inspection test of viscous fluid damper developed by the author's team as an example, these inspection items are introduced, including 60,000 times of high cycle fatigue loading and 30 times of low cycle and large tonnage fatigue loading test. The specific design parameters of the sample are shown in the Table 5.11.

5.4.1 Maximum Damping Force Test

Through the detection of the maximum damping force, it was verified whether the damper can achieve the maximum damping force under the design displacement. At the same time, the error between the actual output and the design value was compared. It was verified that the actual performance of the damper can be more close to the design goal by combining the above research results of damping hole construction parameters. According to the design damping coefficient $C = 450$ kN/(mm/s) and damping index $\alpha = 0.15$, combined with the damping force calculation formula $F = CV^{\alpha}$, the maximum speed required for the damper to reach the maximum damping force was $v = 205.2$ mm/s, so the test frequency was calculated as 0.71 Hz. The test working environment was 23 °C, and the loading cycle was 5. The test results are shown in Table 5.12. According to the analysis, the damper can meet the test requirements of the maximum damping force, which is within 1000 ± 15% of the design maximum damping force.

Table 5.11 Design parameter of damper sample for comparative test

Sample number	Maximum damping force (kN)	Design displacement (mm)	Limit displacement (mm)	Damping coefficient kN/(mm/s)	Damping index
GXN-VFD-NL-1000 × 46	1000	±46	100	450	0.15

Table 5.12 Test results of maximum damping force of damper

Sample number	Test displacement (mm)	Theoretical force (kN)	Measured force (circle 3)/(kN)	Error
GXN-VFD-NL-1000 × 46	±45.7	1000	1011.9	1.19%

Table 5.13 Comparison between the theoretical output force of damper and the measured results

Sample number	Condition	Test displacement (mm)	Theoretical force (kN)	Measured force (circle 3)/(kN)	Error (%)
GXN-VFD-NL-1000 × 46	1	±4.5	707.9	679.5	−4.01
	2	±9.1	785.5	757.9	−3.51
	3	±22.9	901.3	856.1	−5.01
	4	±32.1	948.0	895.2	−5.57
	5	±45.8	1000.1	1011.9	1.18
	6	±55.1	1027.9	1025.2	−0.2

5.4.2 Regularity Test of Damping Force

According to the test requirements, the displacement of $0.1A_0$, $0.2A_0$, $0.5A_0$, $0.7A_0$, $0.1A_0$, and $1.2A_0$ under the regular test condition of damping force were selected based on the set displacement A_0, and the loading frequency was 0.71 Hz. Table 5.13 shows the error between the theoretical and measured output forces. The theoretical damping coefficient of the damper was 450, and the damping index was 0.15. After the test results were fitted, the measured results were 412.48 and 0.163, respectively. The errors were 8.34 and 8.67%. The real performance of the damper products is close to the theoretical analysis results. On the one hand, the damper production technology level is high, on the other hand, the theoretical calculation formula has a certain prediction accuracy, which can be used to guide the production of viscous fluid dampers.

5.4.3 Test of Loading Frequency Related Performance of Maximum Damping Force

According to the test requirements, the frequency of the damping force regularity test was selected as $0.4f_0$, $0.7f_0$, $1.0f_0$, $1.3f_0$, $1.6f_0$ based on the damper basic frequency $f_0 = 0.71\text{Hz}$. At the same time, the displacement peak value of the load condition was modified to keep the maximum speed of each load condition as 205.2 mm/s, and the load frequency related performance of the maximum damping force of the damper was tested. The test results show that the deviation between the measured maximum damping force at each frequency and the measured maximum damping force at the fundamental frequency is not more than ±15%, and all hysteretic curves at the maximum speed are full, and the energy consumption effect is good (Table 5.14).

Table 5.14 Test results of loading frequency related performance of maximum damping force

Sample number	Condition	Test displacement (mm)	Theoretical force (kN)	Measured force (circle 3)/(kN)	Error (%)
GXN-VFD-NL-1000 × 46	1	±100.4	979.3	986.6	0.75
	2	±65.6	1004.2	1021.8	1.75
	3	±45.9	1004.2	1011.9	0.77
	4	±35.1	1004.2	994.6	−0.96
	5	±28.6	1004.2	1014.6	1.04

5.4.4 Test of Temperature Related Performance of Maximum Damping Force

The temperature related performance of the damper's maximum damping force was tested, the test temperature was −30 to 50 °C, and the interval of each working condition was 10 °C. The test results are shown in Table 5.15. Under the condition of frequency 0.71 Hz, displacement ±46 mm, test temperature −30 to 50 °C, the deviation of the measured maximum damping force was not exceed ±15%, and the change rate of the maximum damping force was not exceed ±10%. Moreover, the hysteresis curves under the used temperature are full, and the energy consumption effect is good.

Measured damping force variance of high performance sample: $D(x)_{GXN} = 53.49$. Change rate of maximum damping force of high performance: $r_{GXN} = \frac{1050.9-1027.3}{1027.3} \times 100\% = 2.30\%$.

Table 5.15 Test results of temperature related performance of maximum damping force

Sample number	Condition	Test displacement (mm)	Theoretical force (kN)	Measured force (circle 3)/(kN)	Error (%)
GXN-VFD-NL-1000 × 46	1	±45.9	1000	1027.3	2.73
	2	±45.6	1000	1028.8	2.88
	3	±45.8	1000	1033.1	3.31
	4	±45.7	1000	1035.8	3.58
	5	±45.7	1000	1038.5	3.85
	6	±45.6	1000	1039.6	3.96
	7	±45.7	1000	1043.7	4.37
	8	±45.8	1000	1045.2	4.52
	9	±45.9	1000	1050.9	5.09

5.4.5 *Pressure Maintaining Inspection*

In order to test the sealing performance of the damper, the pressure maintaining test of the damper was carried out. The damper was loaded with 1500 kN (1.5 times of the maximum damping force) for 3 min. In the test, the maximum damping force of the damper was 1507.1 kN, the minimum damping force was 1494.2 kN, and the damping force attenuation percentage was −0.86%. After inspection, the damper has no leakage and meets the requirement that the damping force attenuation value is not more than 5%.

5.4.6 *Fatigue Performance Test*

The fatigue performance of the high performance damper was tested by 60,000 high cycle fatigue loads and 30 low cycle large tonnage fatigue loads (Fig. 5.16a). The loading conditions are shown in Table 5.16, in which, the number of cycles of high cycle fatigue loading was divided into three times of concentrated loading according to the testing requirements, each time 20,000 cycles of concentrated loading were carried out, and the loading was carried out according to the working condition of 30 cycles for 0.5 Hz and ±12.5 mm. When the number of cycles reached 20,000, the loading was stopped, and the damper was removed from the loading platform to check and strengthen its appearance, connecting earrings and other positions. Due to the long duration of fatigue test, the ambient temperature was taken as the average room temperature in the loading cycle. During the loading process, the temperature sensor was used to measure the outer surface temperature of the damper cylinder in real time. If the outer surface temperature of the cylinder exceeds 60 °C, the loading will be stopped for a short time, and the loading will continue after the temperature basically recovered to room temperature.

It can be seen from the test results in Fig. 5.16b, c that the viscous fluid damper still has a relatively full hysteretic curve after 60,000 times of fatigue loading, and the damping force decreases with the increase of the number of loading cycles, but the decline is slow, which can still be regarded as normal operation of the damper. The damping force decreases 14.23% accumulatively when loaded to 60,000 cycles, which is within 15% and meets the requirements of engineering application. After loading, it was found in the disassembly inspection stage that the connection between the damper piston guide rod and the earring was loose due to the wear of the thread due to long-term loading. According to the hysteretic curve, the main reason for the decline of the peak value of damping force at 30,000-40,000 cycles was the loose connection of the damper, which caused the loss of the actual displacement of the piston, so the peak value of the damping force decreased. It can be seen from Fig. 5.16d, e that the hysteretic curve of damping force F-displacement d of the high performance damper obtained by the seismic fatigue loading test has a good coincidence, and the maximum damping force change is not more than ±15%;

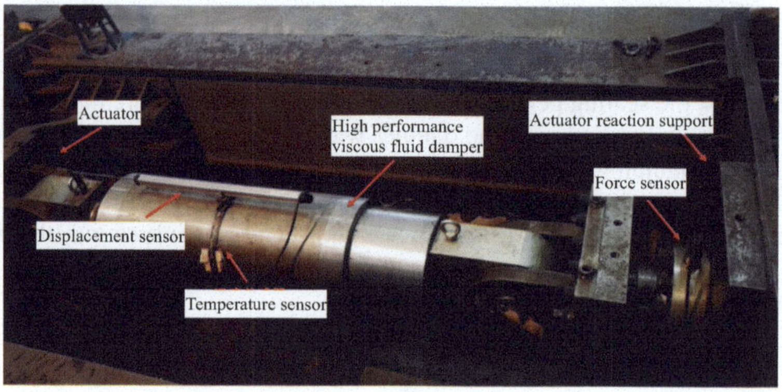

(a) Test specimen

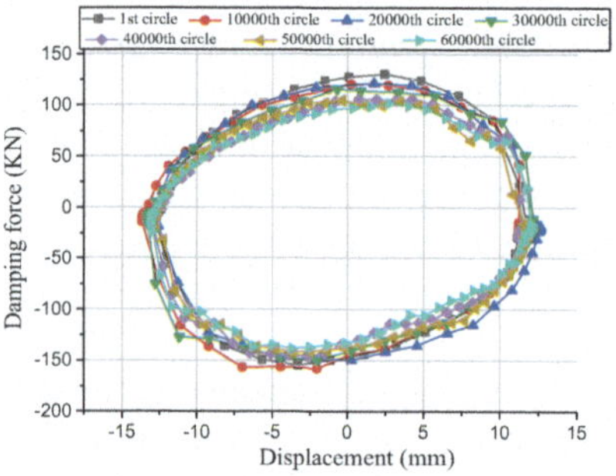

(b) High cycle fatigue hysteresis curve

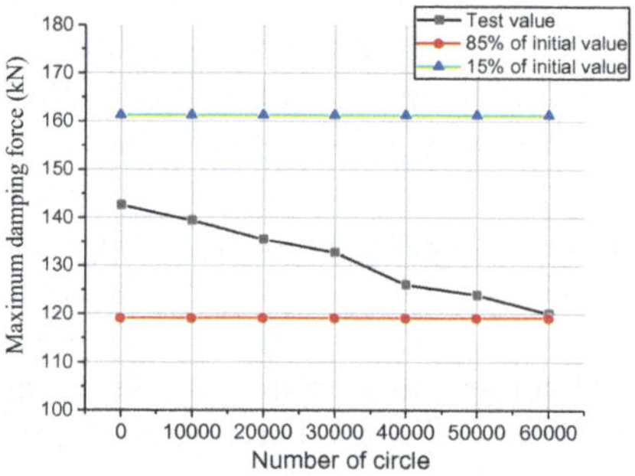

(c) Change of maximum damping force of high cycle fatigue

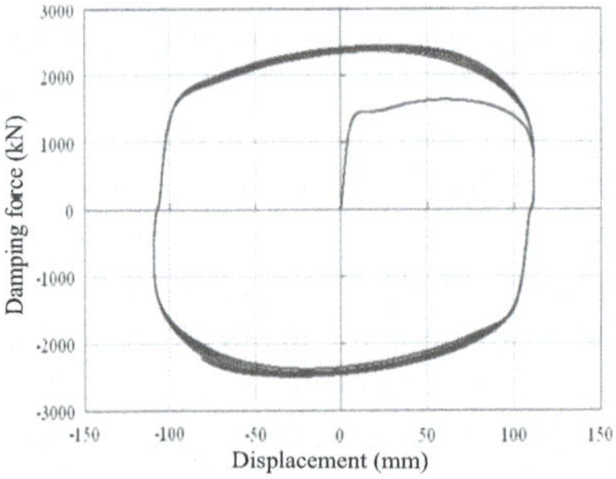

(d) Low cycle fatigue hysteresis curve

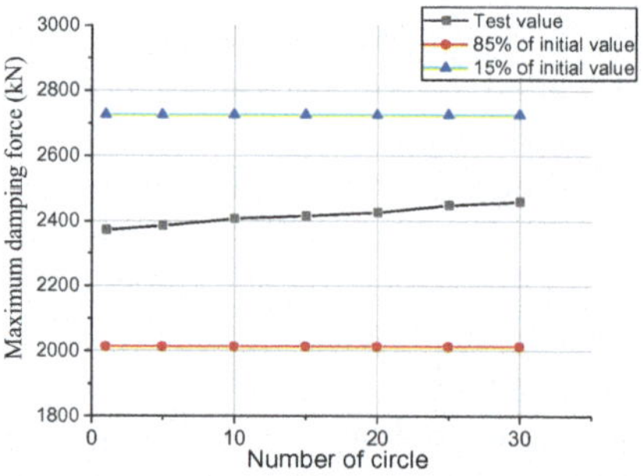

(e) Change of maximum damping force of low cycle fatigue

Fig. 5.16 Fatigue test of viscous fluid damper

Table 5.16 Loading condition of fatigue performance test

Inspection contents	Test frequency (Hz)	Horizontal displacement (mm)	Horizontal velocity (mm/s)	Number of cycles	Average ambient temperature (°C)
High cycle fatigue	0.5	±12.5	39.27	60,000	8
Low cycle fatigue	0.5	±110.0	345.40	30	8

after the disassembly test, it was confirmed that the damper had no oil leakage, and the piston rod and connecting piece had no damage. The high-performance damper shows excellent working stability in low cycle and large tonnage fatigue test.

References

1. Li, Aiqun. 2007. *Vibration control of engineering structure*. Beijing: China Machine Press.
2. Ye, Zhengqiang. 2003. *Research on theory, experiment and application of energy dissipation and vibration reduction technology with viscous fluid damper*. Nanjing: Southeast University. (in Chinese).
3. Liu, Bin. 2007. *Study on the modified viscous fluid machine damper*. Nanjing: Southeast University. (in Chinese).
4. Huang, Zhen. 2007. *Theoretical and experimental study on nonlinear viscous damper*. Nanjing: Southeast University. (in Chinese).
5. Liang, Shahe. 2010. *Experimental and analytical study of a new type controllable damping coefficient viscous damper*. Nanjing: Southeast University. (in Chinese).
6. Huang, Ruixin. 2011. *Research on TMD control of vibration response of high-rise structure with variable damping*. Nanjing: Southeast University. (in Chinese).

Chapter 6
Viscoelastic Damper

Abstract The energy dissipation mechanism of viscoelastic damper is through the hysteretic energy dissipation of viscoelastic material. Types, characteristics and calculation model of viscoelastic materials in viscoelastic damper are introduced. In the section of properties and improvement of viscoelastic materials, inorganic small molecule hybrid, blending of rubber and plastic, design and preparation of modified materials and long chain polymer blending method are introduced. In the section of research and development of new viscoelastic damper, laminated viscoelastic damper, cylindrical viscoelastic damper and viscoelastic damping wall are analyzed respectively.

6.1 Viscoelastic Damping Mechanism and Characteristics

The mechanism of viscoelastic damper is to dissipate part of the energy of structure vibration through the viscoelastic hysteretic energy dissipation of viscoelastic material in the damper, so as to reduce the vibration response of the structure.

6.1.1 Types and Characteristics of Viscoelastic Materials

Viscoelastic damping material is one of the most widely used damping materials, and the viscoelastic damping material for viscoelastic damper is usually polymer. In terms of micro conformation, the polymer consists of small and simple chemical units to form a long chain molecule, and then a three-dimensional dendrimer network is formed by winding and connecting the chemical or physical bonds (as shown in Fig. 6.1a). A long chain molecule is composed of more than 1000 atoms, with a molecular weight of 10^4–10^7 orders of magnitude, which is irregular and zigzag, and its stretching distance is far greater than the distances between the two terminals of the molecule under normal circumstances. Molecular motion is the link between molecular microstructure and molecular macro performance. There are two modes of

A. Li, *Vibration Control for Building Structures*, Springer Tracts in Civil Engineering, https://doi.org/10.1007/978-3-030-40790-2_6

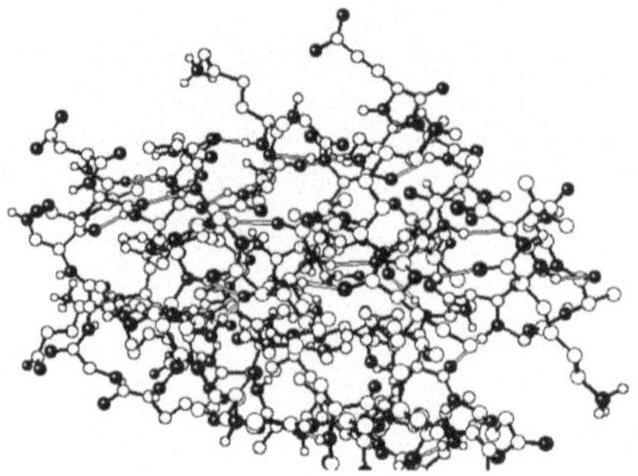

(a) Tree like three-dimensional molecular network

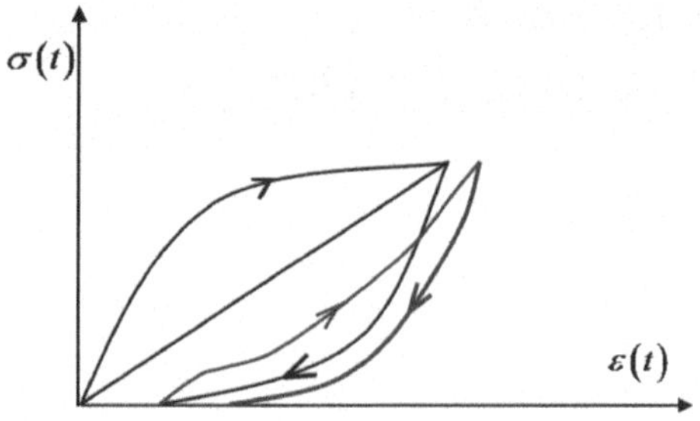

(b) Hysteretic curve of tension and pressure

Fig. 6.1 Properties of polymer

polymer motion under the external load, including molecular motion (including relative motion between molecules and free rotation of chemical units inside molecules) and molecular chain motion (including stretching and twisting of zigzag molecular chain and relative sliding and twisting of intermolecular chain segments). After the external force is removed, the motion of these two parts will produce partial recovery motion and the rest motion that cannot be recovered. Among them, the motion that can be recovered is the conversion of energy, which is manifested as the elasticity of viscoelastic materials, and the motion that cannot be recovered is the attenuation of energy, that is, the motion energy is transformed into heat energy to escape, which is manifested as the viscosity of viscoelastic materials.

The energy dissipation mechanism of general viscoelastic damping device is the conversion and attenuation of input dynamic load caused by the shear deformation of viscoelastic damping material between steel plates interlayer. This shear deformation cannot be recovered, which is called the phenomenon of dynamic displacement lagging behind the excitation force. The lag angle is δ ($\tan\delta$ is called loss factor), and the dynamic displacement and dynamic force change according to the law of harmony, which is described by the mathematical formula as follows:

$$F_g = F_0 \sin \omega t \quad \text{or} \quad F_g = F_0 e^{j\omega t} \tag{6.1}$$

$$\Delta = \Delta_0 \sin(\omega t - \delta) \quad \text{or} \quad \Delta = \Delta_0 e^{j(\omega t - \delta)} \tag{6.2}$$

where, F_g and Δ are the functions of the excitation force and dynamic displacement with time, F_0 and Δ_0 are the maximum excitation force and maximum dynamic displacement, respectively, $\Delta W = \oint \sigma(t) d\varepsilon(t) = \oint \sigma(t) \frac{d\varepsilon(t)}{dt} dt$. Figure 6.1b is the stress-strain curve of the polymer material in the tensile-shrinkage process, and the expression of work can be shown as $\Delta W = \oint F_g d\Delta = \pi F_0 \Delta_0 \sin\delta$, then:

$$\Delta W = \sigma_0 \varepsilon_0 \omega \int_0^{2\pi/\omega} \sin(\omega t)\cos(\omega t - \delta)dt \tag{6.3}$$

The integral of the above formula gives $\Delta W = \pi \sigma_0 \varepsilon_0 \sin \delta$, which shows that the energy consumption per unit volume is directly proportional to the sine of the lag angle in each cycle. Because the damping performance of polymer performs under the action of special force field, the energy consumption is related to the frequency of excitation force and temperature. When the temperature is specific, when the frequency of the excitation force is too fast or too slow, the chain segment motion cannot keep up with the external excitation or completely, and the internal friction is very small at this time. Only when the external excitation of the chain segment motion is in a half lag state of both up-to-date and backward, the internal friction is the largest. This period corresponds to the temperature transition zone of glass transition of the polymer. In this temperature range, the damping material has the most effective energy dissipation capacity to external excitation. How to increase the maximum value of internal friction in the glass transition region and how to expand the temperature region of high internal friction become the main direction to improve the efficiency of damping materials. Generally, the use of the damping materials with good application effect must meet the following conditions: (1) the peak value of loss factor should be high, and the corresponding peak value T_g should be consistent with the working temperature of the material; (2) the temperature range with loss factor greater than 0.7 should be wide, and $\Delta T_{0.7}$ should be suitable for the working environment of the material.

6.1.2 Calculation Model of Viscoelastic Damper

Although viscoelastic damper has many kinds and different structures, its performance is usually characterized by energy storage stiffness K_{d1}, energy dissipation stiffness K_{d2}, loss factor η and energy dissipation E_d per revolution.

$$K_{d1} = \frac{nG_1A}{t} \tag{6.4}$$

$$K_{d2} = \frac{nG_2A}{t} \tag{6.5}$$

$$\eta = \frac{G_2}{G_1} = \frac{K_{d2}}{K_{d1}} \tag{6.6}$$

The energy consumption E_d per cycle under sinusoidal load excitation is

$$E_d = \pi \gamma_0^2 G_2 V \tag{6.7}$$

where, K_{d1} is the energy storage rigidity of viscoelastic damper; K_{d2} is the loss rigidity of viscoelastic damper; G_1 is the energy storage shear modulus of viscoelastic material; G_2 is the loss shear modulus of viscoelastic material; η is the loss factor of viscoelastic material; A is the area of viscoelastic material layer; n is the number of viscoelastic layers; t is the thickness of viscoelastic material layer; γ_0 is the shear strain amplitude of viscoelastic damper; V is the volume of viscoelastic material layer. Because the main factors that affect the performance of viscoelastic materials are temperature, frequency and strain amplitude, the main factors that affect the performance of viscoelastic damper are also temperature, frequency and strain amplitude, especially temperature and frequency. Generally, G_1 decreases with the increase of temperature and increases with the increase of frequency; η has a maximum value with the change of temperature and a maximum value with the change of frequency. This shows that for the viscoelastic damper made of a specific viscoelastic material, its energy consumption performance is affected by temperature and frequency, and there is an optimal use temperature and frequency for the damper. In order to achieve a better balance between the simulation accuracy and efficiency, the macro phenomenological model is often used in engineering to describe the viscoelastic damper, and the commonly used description calculation models are mainly as follows:

(1) Kelvin model

Kelvin model is composed of elastic element and glue pot element in parallel, as shown in Fig. 6.2a, and its constitutive relation is

$$\tau = q_0\gamma + q_1\dot{\gamma} \tag{6.8}$$

among them, τ is shear stress, γ and $\dot{\gamma}$ are shear strain and shear strain rate, q_0 and q_1 are coefficients determined by the properties of viscoelastic materials. According to Eq. (6.4), the force displacement relationship of viscoelastic damper can be written as

$$F_d = K_d u_d + C_d \dot{u}_d \tag{6.9}$$

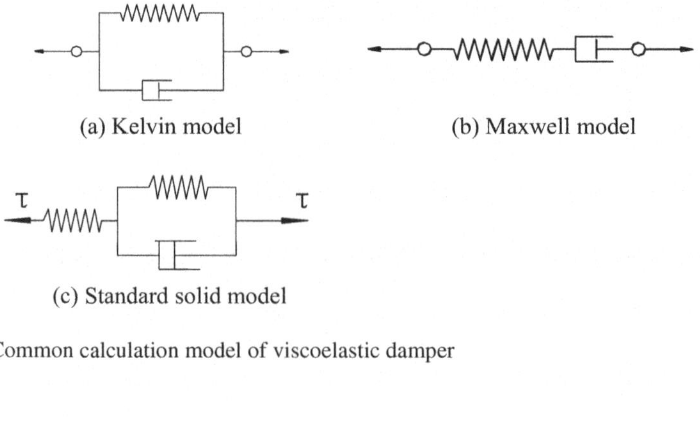

(a) Kelvin model

(b) Maxwell model

(c) Standard solid model

Fig. 6.2 Common calculation model of viscoelastic damper where, F_d is the output force of the viscoelastic damper, u_d and $\dot{u}_d$ are the relative displacement and speed of both ends of the viscoelastic damper, K_d and C_d are the equivalent stiffness and equivalent damping of the viscoelastic damper, respectively. For this smooth-shape hysteretic restoring force, the equivalent linear stiffness coefficient $K_d = K_{d1}$ is usually taken, namely:

$$K_d = \frac{nG_1A}{t} \tag{6.10}$$

The equivalent linear damping coefficient C_d is determined according to the principle of equal energy consumption by hysteresis and damping as follows:

$$C_d = \frac{E_d}{\pi\omega\gamma_0^2} \tag{6.11}$$

Substitute Eq. (6.7) into Eq. (6.11), we can get:

$$C_d = \frac{\pi\left(\frac{u_d}{t}\right)^2 \cdot G_2 \cdot n \cdot A \cdot t}{\pi\omega u_d^2} = \frac{nG_2A}{\omega t} \tag{6.12}$$

Substitute Eq. (6.6) into Eq. (6.12), we can get:

$$C_d = \frac{n\eta G_1A}{\omega t} \tag{6.13}$$

Substituting Eqs. (6.10) and (6.12) or Eq. (6.13) into Eq. (6.9), we can get

$$F_d = \frac{n \cdot G_1 \cdot A}{t} u_d + \frac{n \cdot G_2 \cdot A}{\omega t} \dot{u}_d \tag{6.14}$$

or

$$\begin{aligned} F_d &= \frac{nG_1A}{t} u_d + \frac{n\eta G_1A}{\omega t} \dot{u}_d \\ &= \frac{nG_1A}{t}\left(u_d + \frac{\eta}{\omega}\dot{u}_d\right) \end{aligned} \tag{6.15}$$

Kelvin model reflects the transient elastic response of viscoelastic damper, which can better reflect the creep and relaxation phenomenon of viscoelastic damper. It is a commonly used calculation and analysis model, but the model does not consider the influence of temperature, frequency and strain amplitude on the energy dissipation characteristics of viscoelastic damper.

(2) Maxwell model

Maxwell model simulates the viscoelastic damper as a series connection of an elastic element and a damping element, as shown in Fig. 6.2b. The total strain is

$$\gamma = \gamma' + \gamma'' \tag{6.16}$$

where, γ' is the strain of elastic element; γ'' is the strain of damping element. After the differential of Eq. (6.16), it can be derived

$$\tau + p_1\dot{\tau} = q_1\dot{\gamma} \tag{6.17}$$

$$p_1 = \frac{F}{G} \tag{6.18}$$

$$q_1 = F \tag{6.19}$$

where, p_1 and q_1 are coefficients determined by the properties of viscoelastic materials. Equation (6.17) is the constitutive equation of Maxwell model of viscoelastic damper. According to Eq. (6.17), the force displacement relationship of viscoelastic damper can be written as

$$F_d + P_1\dot{F}_d = Q_1\dot{u}_d \tag{6.20}$$

where, P_1 and Q_1 are the corresponding coefficients, which are generally determined by tests.

Maxwell model has clear concept and simple analysis, which can better reflect the relaxation phenomenon of viscoelastic damper and the change trend of storage modulus with frequency. However, within the scope of linear constitutive relation, the material shows typical fluid characteristics, that is, it can deform without limit under limited stress, cannot reflect the transient elastic response of viscoelastic damper, and cannot reflect the influence of environment temperature, excitation frequency and strain amplitude on the energy dissipation characteristics of viscoelastic dampers, which makes this model not suitable for the analysis of viscoelastic dampers.

(3) Standard linear solid model

This model simulates the viscoelastic damper as a series connection of elastic element and Kelvin element, as shown in Fig. 6.2c, and its constitutive relation is

$$\tau + p_1\dot{\tau} = q_0\gamma + q_1\dot{\gamma} \tag{6.21}$$

where, q_0, q_1 and p_1 are coefficients determined by the properties of viscoelastic materials respectively.

The standard linear solid model can not only reflect the relaxation and slight creep characteristics of viscoelastic damper, but also reflect the change trend of the performance of viscoelastic damper with frequency. However, it cannot reflect the influence of temperature on the performance of viscoelastic damper, and cannot accurately describe the influence of frequency on the performance of viscoelastic damper.

(4) Complex stiffness model

Under the excitation of sinusoidal load, the stress of viscoelastic damper leads phase angle α compared with strain, which has the following relationship

$$\gamma = \gamma_0 e^{i\omega t} \tag{6.22}$$

$$G^* = G_1 + iG_2 \tag{6.23}$$

$$\tau = G^* \gamma = (G_1 + iG_2)\gamma_0 e^{i\omega t} \tag{6.24}$$

among them, ω is the excitation frequency; γ_0 is the shear strain amplitude; G^* is the complex modulus of viscoelastic materials. From Eq. (6.24), the force displacement relationship of viscoelastic damper is

$$F_d = K^* u_d = (K_{d1} + iK_{d2}) u_{dm} e^{i\omega t} \tag{6.25}$$

where, u_{dm} is the displacement amplitude of shock absorber; K^* is the complex stiffness of shock absorber.

The concept of complex stiffness model is clear, easy to understand and widely used, but it is only suitable for the case of small strain. It does not reflect the influence of ambient temperature, excitation frequency and strain amplitude on the energy dissipation characteristics of the damper, and needs to solve the problem in the frequency domain.

(5) Four parameter model

In order to study the effect of frequency on the performance of viscoelastic dampers, Kasai proposed a four parameter model, and its stress-strain relationship is

$$\tau(t) + aD''[\tau(t)] = G\{\gamma(t) + bD''[\gamma(t)]\} \tag{6.26}$$

where, a and b are constants; G is elastic constant; α is index, $0 < \alpha < 1$; $D''[\tau(t)]$ and $D''[\gamma(t)]$ are defined as

$$\left.\begin{aligned} D''[\tau(t)] &= \frac{1}{\Gamma(1-\alpha)} \frac{d}{dt} \int_0^t \frac{\tau(t')}{(t-t')^\alpha} dt' \\ D''[\gamma(t)] &= \frac{1}{\Gamma(1-\alpha)} \frac{d}{dt} \int_0^t \frac{\gamma(t')}{(t-t')^\alpha} dt' \end{aligned}\right\} \tag{6.27}$$

The model can reflect the influence of frequency on the mechanical properties of viscoelastic damper, but it cannot reflect the influence of temperature and strain amplitude on the mechanical properties of viscoelastic damper. Moreover, the concept of the model is not clear, the calculation formula is complex, and it is less used in the actual structural analysis.

(6) Finite element model

In order to reflect the influences of ambient temperature, excitation frequency and strain amplitude on the energy dissipation characteristics of damper, Tsai established the finite element model of viscoelastic damper. R. l. Bagley gives the following relationship between stress and strain of viscoelastic damping materials.

$$\tau(t) = G_0\gamma(t) + G_1 D''[\gamma(t)] \tag{6.28}$$

where, $\tau(t)$ is the shear stress; $\gamma(t)$ is the shear strain; G_0 and G_1 are the basic model parameters; $D''[\gamma(t)]$ is defined as

$$D''[\gamma(t)] = \frac{1}{\Gamma(1-\alpha)}\frac{d}{dt}\int_0^t \frac{\gamma(t')}{(t-t')^{\alpha}}dt' \quad 0 < \alpha < 1 \tag{6.29}$$

where, $\Gamma(\bullet)$ is the Gama function. Since Eq. (6.28) does not reflect the effect of temperature on the properties of viscoelastic materials, to make up for this, G_0 and G_1 are calculated as follows:

$$G_0 = G_1 = A_0(1 + \mu \exp\{-\beta[\int \tau d\gamma + \theta(T - T_0)]\}) \tag{6.30}$$

where, T is the ambient temperature; T_0 is the reference temperature;α, A_0,β,$\mu$and$\theta$are the parameters to be determined by the test, which are all obtained under the temperature T_0.

Equation (6.30) shows that the basic model parameters G_0 and G_1 decrease with the increase of the total energy. The total energy includes the strain energy of viscoelastic materials and the energy obtained from the ambient temperature. The effect of ambient temperature is expressed in the form of initial strain energy stored in the material, and then affects the properties of viscoelastic material. In the process of earthquake, the temperature of viscoelastic materials increases due to the transformation of strain energy into heat energy, which is also reflected in Eq. (6.30). Because of the different strain rate and strain amplitude, the cumulative velocity of strain energy in Eq. (6.30) is different, so Eq. (6.30) also considers the influence of strain rate, strain amplitude and excitation frequency on viscoelastic materials.

Assuming that the strain varies linearly between time steps $(n-1)\Delta t$ and $n\Delta t$, the strain can be determined as follows:

$$\begin{aligned}\gamma(t') =& \left(n - \frac{t'}{\Delta t}\right)\gamma[(n-1)\Delta t] \\ &+ \left[\frac{t'}{\Delta t} - (n-1)\right]\gamma(n\Delta t) \quad (n-1)\Delta t \le t' \le n\Delta t\end{aligned} \tag{6.31}$$

By substituting Eq. (6.31) into Eqs. (6.28) and (6.29), the stress-strain relationship of viscoelastic damping material at any time $N\Delta t$ is obtained:

$$\tau(N\Delta t) = \left[G_0 + \frac{G_1(\Delta t)^{-\alpha}}{\Gamma(2-\alpha)}\right]\gamma(N\Delta t) + F(N\Delta t) \tag{6.32}$$

Among them, the pre aging $F(N\Delta t)$ of strain is

$$F(N\Delta t) = \frac{G_1(\Delta t)^{-\alpha}}{\Gamma(2-\alpha)}\{[(N-1)^{1-\alpha} + (-N+1-\alpha)N^{-\alpha}]\gamma(0) + \sum_{n=1}^{N-1}[-2(N-n)^{1-\alpha} + (N-n+1)^{1-\alpha} + (N-n-1)^{1-\alpha}]\gamma(n\Delta t)\} \tag{6.33}$$

The model considers the effects of temperature, frequency and strain amplitude on the performance of viscoelastic damper, which is more accurate, but the model is quite complex and difficult to apply.

(7) Equivalent standard solid model

The equivalent standard solid model is a model describing the change characteristics of viscoelastic damper with temperature and frequency established based on the temperature frequency equivalent principle and the standard linear solid model.

Although the shear modulus and loss factor of viscoelastic materials are both functions of temperature and frequency, their change relationships with temperature and frequency are not identical. When the temperature is in the glass transition temperature range of T_g to T_g+ 100 °C, there is an equivalent relationship between the temperature and frequency of most viscoelastic materials, that is, the effects of low temperature and high frequency are equivalent, and the effects of high temperature and low frequency are equivalent. If the effects of temperature and frequency on the properties of viscoelastic materials are considered, there will be

$$\left.\begin{aligned} G_1(\omega, T) &= G_1(\alpha_T\omega, T_0) \\ \eta(\omega, T) &= \eta(\alpha_T\omega, T_0) \end{aligned}\right\} \tag{6.34}$$

where T_0 is reference temperature; α_T is temperature conversion coefficient, which is a function of temperature T, which is determined by the following formula

$$\alpha_T = 10^{-12(T-T_0)/[525+(T-T_0)]} \tag{6.35}$$

The temperature and frequency ranges of viscoelastic damper used for building structure damping are $-30°\text{C} \leq T \leq 60°\text{C}$ and $0.1\text{Hz} \leq \omega \leq 10\text{Hz}$. In this range, in order to accurately describe the changing characteristics of the parameters G_1, G_2 and η of viscoelastic damper with temperature and frequency, the frequency in the standard linear solid model is changed to the reduced frequency $\alpha_T\omega$, which can be obtained as

$$\left.\begin{array}{c} G_1 = (q_0 + p_1 q_1 \alpha_T^c \omega^c)/(1 + p_1^2 \alpha_T^c \omega^c) \\ \eta = (q_1 - p_1 q_0)\alpha_T^d \omega^d / (q_0 + p_1 q_1 \alpha_T^{2d} \omega^{2d}) \end{array}\right\} \quad (6.36)$$

where, c and d are the indexes determined by the test; α_T is determined by Eq. (6.35).

Although the standard linear solid model can reflect the change trend of viscoelastic damper performance with frequency and the relaxation and creep characteristics of viscoelastic damper, it cannot reflect the influence of temperature on the performance of viscoelastic damper, nor can it accurately describe the influence rule of frequency on the performance of viscoelastic damper. The equivalent standard solid model combines the temperature frequency equivalent principle with the standard linear solid model, and improves it with the practical application range. It not only retains the advantages of the standard linear solid model, but also accurately describes the changing characteristics of performance of viscoelastic damper with temperature and frequency.

6.2 Properties and Improvement of Viscoelastic Materials

Most of the viscoelastic damper materials used in civil engineering come from the field of aerospace. However, there are some problems in the application of civil engineering structures, such as low loss factor of materials, high frequency of application and inconformity of using environment, which make it difficult for viscoelastic damper to play its proper damping effect in practical engineering. In order to solve this problem, two methods were used to study the material modification of Nitrile Butadiene Rubber (NBR), which is commonly used in viscoelastic damper: (1) inorganic small molecule hybrid, blending of rubber and plastic; (2) blending of long chain polymer [1].

6.2.1 Inorganic Small Molecule Hybrid, Blending of Rubber and Plastic

6.2.1.1 Design and Preparation of Modified Materials

Nitrile rubber, butyl rubber and their halides, chloroprene rubber and polyurethane are commonly used viscoelastic damping polymers. The damping materials popularized and applied to the seismic and shock absorption design of building structures usually adopt the nitrile rubber with good damping performance and excellent bonding performance with other materials. The nitrile butadiene rubber (NBR) is a kind of composite rubber made by the copolymerization of butadiene and acrylonitrile. Its molecular structure is shown in Fig. 6.3a, which has good properties such as large loss factor, oil resistance and aging resistance. NBR is a high damping material with

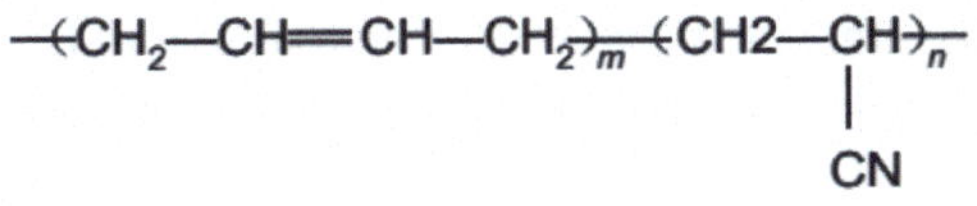

(a) Molecular structure diagram of NBR

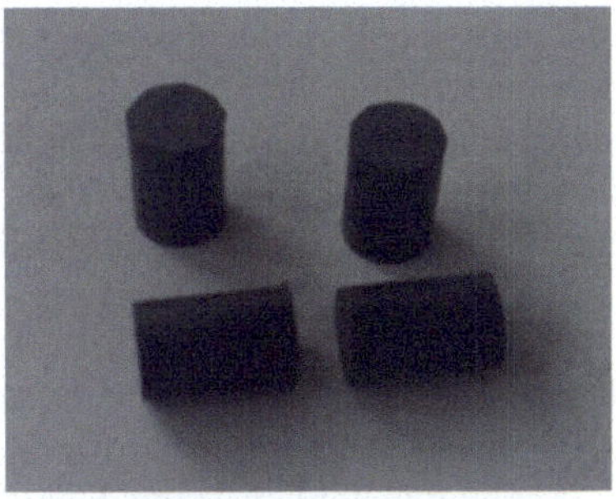

(b) Material test sample

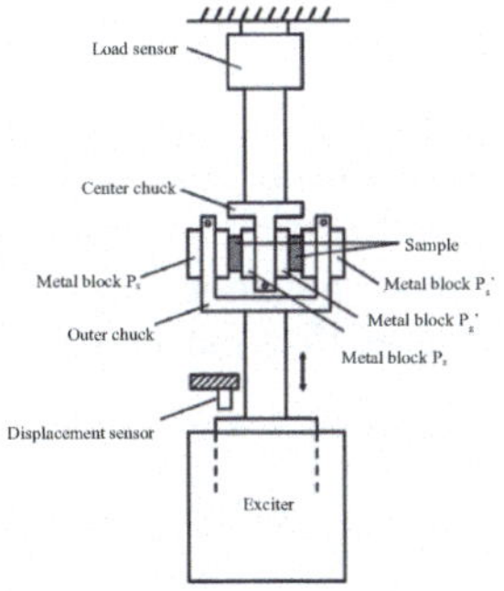

(c) Test loading diagram

(d) Test equipment

Fig. 6.3 Modification test of viscoelastic materials

high application prospects. The glass transition temperature range of homopolymer or random copolymer is very narrow, so in the design of viscoelastic damping materials, the methods of adding organic or inorganic small molecular materials, forming interpenetrating polymer network, copolymerization and blending are commonly used to improve the matrix materials. There are some problems in copolymerization, such as the low damping value between the two main transition regions; the materials obtained by interpenetrating polymer networks are generally used as coatings, but rarely used in the field of structure, and there are also some problems such as complex synthesis process, high cost and difficult industrialization. In this section, the matrix material used in the experiment is NBR with a large side chain. Therefore, the purpose of improving the damping property of the material is achieved by mixing rubber and plastic together and hybriding NBR with inorganic small molecules. Blending of rubber and plastic is the most common method to improve the damping property of polymer. The glass transition temperature of plastic and rubber is quite different, that is, at room temperature, when the plastic is in the glass state, the rubber is in the elastic state. Therefore, it is ideal to improve the damping property by rubber-plastic blending in theory. Chlorinated polypropylene with polarity was used in the test.

Inorganic small molecules can form weak connection and interface friction with polymers, which increases the friction resistance when the chain moves, so as to increase internal friction and obtain high damping performance. In the experiment, short carbon fiber, graphite powder, flake graphite, mica powder, high wear-resistant carbon black and fast-press out-carbon black were used as inorganic small molecules.

(1) **Test material**

The base material is NBR, and the material composition is: NBR 220, 2123 resin, zinc oxide, stearic acid, liquid oxidized paraffin, antimony trioxide, etc. The formula table is as shown in Table 6.1; the details and names of admixture are shown in Table 6.2; the above materials were provided by Changzhou LanJin Rubber & Plastic Co., Ltd. The loading sequence was determined according to the admixtures listed in Table 6.2, which is shown in Table 6.3.

(2) **Test instruments and preparation of blends**

Main test instruments and equipment: open type plasticizer, molding machine, electromagnetic flat vulcanizer, electronic universal testing machine, vulcanizer, shore hardness tester, etc. (the above is provided by Changzhou LanJin Rubber and Plastic Co., Ltd.); dynamic thermal analyzer (Rheogel-E4000 DMA tester, UBM company in Japan).

Table 6.1 Recipe table of NBR

Name of admixture	Consumption
NBR 220	600
2123 resin	300
Zinc oxide	30
Stearic acid	18
Sulfur	3
Accelerator DM	6
Antioxidant DFC-34	6

Table 6.2 Name list of admixture

Name of admixture	Abbreviation
Chlorinated polypropylene	CPP
Short carbon fiber	CRAB
200 graphite powder	GRAP
Flake graphite	FLAKE
Mica powder	MICA
High abrasion furnace black	HAF
Fast extrusion furnace black	FEF

Table 6.3 Test number table based on admixture combination

Number	Formula
1	NBR
2	NBR/CARB-10
3	NBR/CPP-10
4	NBR/GRAP-60
5	NBR/GRAP-60/FEF-20
6	NBR/GRAP-60/HAF-20
7	NBR/GRAP-80
8	NBR/GRAP-80/FEF-20
9	NBR/GRAP-80/HAF-20
10	NBR/GRAP-20/FEF-20/FLAKE-GRAP-20
11	NBR/GRAP-20/HAF-20/FLAKE-GRAP-20
12	NBR/GRAP-40/FEF-20/MICA-20
13	NBR/GRAP-40/HAF-20/MICA-20

Note the NBR is 100 g, and the number behind the admixture is the grams, such as NBR/CARB-10, i.e. 10 g short carbon fiber is added into 100 g NBR

Blending process:

(1) Preparation of matrix adhesive:

The matrix raw rubber is thinned several times on the plasticizer, then the roller spacing is increased, the roll is added, and the DM catalyst and antioxidant are added. The NBR masterbatch is prepared by making triangle bag and slicing.

(2) Preparation of blends:

According to the proportion given in Table 6.3, NBR and admixture are added to the plasticizer, blended, triangle bag is punched, after 24 h of storage, reverse refining is carried out, and the roll distance is adjusted, then the slice is obtained.

(3) Sample preparation:

Cut the sample piece according to the rolling direction, and then vulcanize it on the flat vulcanizer, place it for 24 h, cut the sample with the cutter, and make the cylinder with the test piece specification of 15 mm high and the diameter of 10 mm, as shown in Fig. 6.3b.

6.2.1.2 Test Conditions and Analysis Methods

(1) **Test condition**

The heating rate is 3 °C/min. The frequency is 3.5 Hz. Considering the working temperature range of the building, the temperature range of this test is −20 to 50 °C.

The purpose of the test is to improve the damping of materials within the temperature range under construction, and to expand the temperature range of the large damping. The consideration formula is:

Increase range of damping value is: $\varphi = \frac{\eta_i - \eta_0}{\eta_0} \times 100\%$
Large damping temperature range: $\Delta T_{\tan\delta>0.5} = T\mathrm{max}_{(\tan\delta>0.5)} - T\mathrm{min}_{(\tan\delta>0.5)}$

Among them, η_i is the peak value of damping value after adding additives; η_0 is the peak value of damping value of NBR; $T\mathrm{max}_{(\tan\delta>0.5)}$ is the highest temperature when the loss factor is greater than 0.5, and $T\mathrm{min}_{(\tan\delta>0.5)}$ is the lowest temperature when the loss factor is greater than 0.5. The difference between the two is the proposed large damping temperature range.

(2) **DMA thermal analysis method**

According to the different modes of load action, the research of polymer damping performance can be divided into two types: dynamic load and static load. In static load measurement, the creep and relaxation properties of materials are analyzed by applying a single direction of load to obtain some parameters, which takes a long time. This method is more suitable when the materials bear a long time of load. When the material is under dynamic load, the dynamic loading mode is suitable for the consideration of its dynamic mechanical properties, and can better reflect the performance under actual conditions. Considering the change of dynamic mechanical properties of viscoelastic materials in a certain temperature range is dynamic mechanical analysis (DMA for short, also called dynamic mechanical thermal analysis). It can not only provide the material elastic index but also obtain the viscosity index. Only a small sample can be tested continuously in the specified temperature domain/frequency domain, and the development curve of the material stiffness and damping with the change of temperature/frequency can be obtained quickly. Dynamic analysis has the following traditional techniques, such as differential thermal analysis (DTA) and differential scanning calorimetry (DSC), which are much less sensitive than dynamic mechanical thermal analysis in the determination of glass transition and secondary transition. Moreover, DMA method will not affect and destroy the structure of the material itself in the process of loading, and the viscoelastic response of the polymer material is very sensitive to the change of the morphological structure. This method is considered to have great advantages and important practical value in the study of the phase structure of multi-phase and multi-component polymer materials. With the development of dynamic mechanical thermal analyzer, it has been extended to measure the creep and stress relaxation of materials. Generally, the dynamic mechanical behavior of materials is the response under the action of alternating stress (or alternating strain), and the most commonly used alternating stress in dynamic mechanical analysis test is sinusoidal stress. According to the mechanical properties of rubber materials, the dynamic shear mode is adopted in this test. The loading diagram is shown in Fig. 6.3c. P_2 and P_2' are fixed on the rigid frame, and the excitation force is applied on P_1 and P_1' connected, to drive the specimen to conduct shear vibration. The form of sine alternating shear stress is $\tau(t) = \tau_0 \sin \omega t$.

According to the measured data and sample size, the shear storage modulus of the material is calculated as follows:

$$G' = \frac{F_A}{D_A} \times \frac{L}{A} \times \left(1 + \frac{L^2}{h^2} \times \frac{G'}{E'}\right) \times \frac{L_a + l}{L} \times \cos\delta \quad (6.37)$$

where, F_A is the load amplitude; D_A is the displacement amplitude; L is the specimen thickness; h is the specimen size of loading direction; δ is the phase difference; L_a is the apparent length of the specimen (the distance between the upper and lower chucks); $\frac{G'}{E'}$ is the bending correction term of the specimen, which is about 0.33 for rubber. Then the shear loss modulus of the material can be obtained from G', $G'' = G' \times \tan\delta$. The test was carried out in the laboratory of Institute of Polymer Alloy, School of Materials Science and Engineering, East China University of Science and Technology, using Rheogel-E4000 DNA tester (UBM Company, Japan) (Fig. 6.3d).

6.2.1.3 Test Results and Analysis

(1) **Basic mechanical properties**

Mechanical property: In accordance with GB/T528-1998, the tensile strength and elongation at break of the material were tested on a 500 N tensile testing machine (the instrument used is three force XLL-50, manufactured by Guangzhou Material Testing Machine Factory of China). The dumbbell shaped specimen was used, with a tensile rate of 500 mm/min, a test temperature of 17 °C and a humidity of 45%.
Shore A hardness: The Shore A hardness of materials is tested according to GB/T531-1999. The instrument used is Mitsubishi LX-A type shore rubber hardness tester (Shanghai made 01010007), the test temperature is 17 °C, and the humidity is 45% (Table 6.4).

(2) **Comparative analysis of results**

Figure 6.4a shows the change curve of loss factor with temperature under the three component ratios of NBR, NBR/CARP-10 and NBR/CPP-10. The curve compares the effect of adding the same amount of short carbon fiber (CARB) and chlorinated polypropylene (CPP) on the matrix material. It can be seen from the figure that the maximum loss factors of NBR after adding CARB-10 and CPP-10 are 1.00

Table 6.4 Basic mechanical properties of materials

	Performance	Value
Physical property	Tensile strength (MPa)	15
	Elongation (%)	400
	Permanent deformation (%)	18
	Hardness (shore A)	72
	Proportion (g/cm^3)	1.27

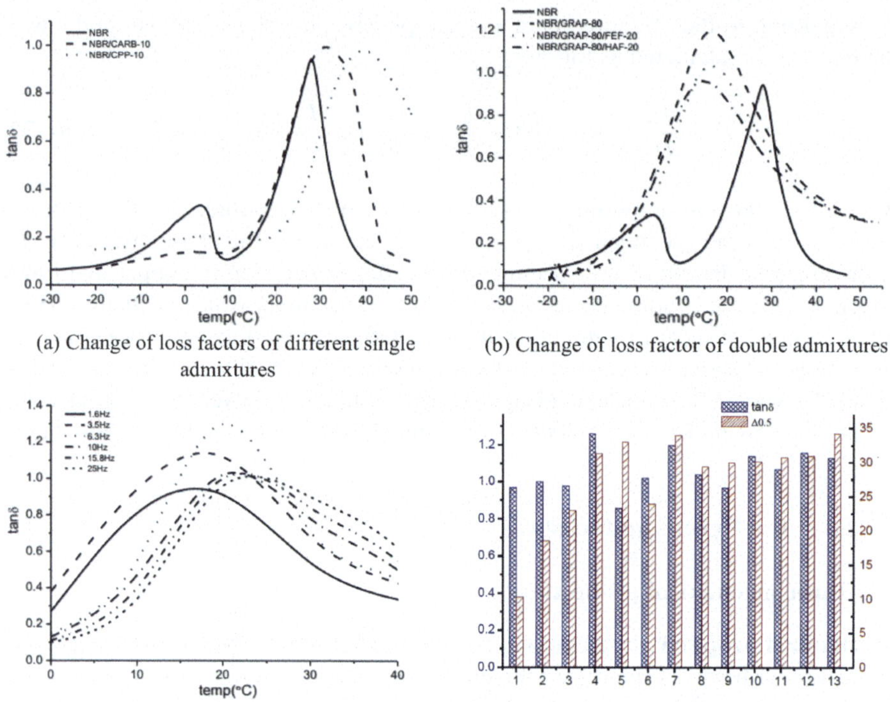

(a) Change of loss factors of different single admixtures

(b) Change of loss factor of double admixtures

(c) Variation of loss factor under different loading frequency

(d) Comparison of sample loss factor and Δ0.5

Fig. 6.4 Material test results

and 0.98, respectively, and the glass transition temperature are 30.8 °C and 39 °C, respectively. Compared with the matrix material, the loss factor increases by 3% and 1% respectively, the improvement effect is very little. It is possible that the incorporation of short carbon fiber and chlorinated polypropylene increases the NBR molecular structure and the activity of the molecular chain, so that the glass transition temperature is more inclined to the high temperature state. The temperature range with loss factor greater than 0.5 increased significantly, which were 18.5 °C (21–39.5 °C) and 23 °C (28–50 °C).

Figure 6.4b shows the change curve of loss factor with temperature under the four component ratios of NBR, NBR/GRAP-80, NBR/GRAP-80/FEF-20 and NBR/GRAP-80/HAF-20. It can be seen from the figure that after adding 80 parts of 200 mesh graphite powder, the temperature T_g is 16.5 °C, and the loss factor reaches 1.20, which is at a high level, and 24% higher than that of the matrix material; after adding 20 parts of FEF, the loss factor decreases to 1.04, which is 7% higher than that of the matrix material, and the temperature is 16 °C, which is not much different from that without FEF. The loss factor of blends without carbon black is higher The results show that the damping properties of the blends are not improved by adding high wear-resistant carbon black and soft quick pressing carbon black, and the improvement effect of the two is basically coincident, which may be related to the particle size of carbon black. The particle size of FEF is 26–30 nm, and the particle size of HAF is 40–48 nm, and the specific surface area and particle size are large, so the contact surface of carbon black and blends is small, and the surface friction force on NBR molecular structure is also small. It is suggested to try the one with the smallest particle size and the best reinforcement performance, such as N110 and N115. It can be seen that the addition of FEF is equivalent to that of HAF when 80 phr of 200 mesh graphite powder is added to NBR. The temperature domain when the loss factors of NBR/GRAP-80, NBR/GRAP-80/FEF-20 and NBR/GRAP-80/HAF-20 are larger than 0.5 are 34 (2.8 °C–36.8 °C), 29.4 (5.4 °C–34.8 °C) and 30 (3.9 °C–33.9 °C), respectively.

Figure 6.4c shows the change distribution of loss factor of NBR/GRAP-80 with temperature at each frequency value. 1.6–25 Hz basically covers the frequency range of viscoelastic damper under the vibration reduction condition. It can be seen from the figure that the improved material can work effectively in the working frequency range of viscoelastic damper, and its damping performance reaches the maximum when the external frequency is about 6.3 Hz.

Figure 6.4d shows the test results of loss factors and $\Delta 0.5$ of all sample. The study on the dynamic mechanical damping properties of blends shows that: (1) the damping property of original base material NBR is greatly improved by the admixture, among which the modification effect of 60 parts of 200 mesh graphite powder on the damping property of NBR is the most remarkable, the peak value of loss factor is increased to 1.26; the peak value of the material is inversely proportional to the amount of graphite powder in a certain temperature range. The main reason for the increase of energy consumption is that the internal friction of NBR chain increases with the addition of graphite powder. (2) the addition of admixtures will reduce the activity of NBR, so the peak value of $\tan\delta$ will move to the low temperature direction, and gradually move to the room temperature of building and structure. When 60 parts of 200 mesh graphite powder are added, the damping temperature range with $\tan\delta$ not less than 0.5 is widened to 31.3 °C, and the less the amount of graphite powder is, the more significant the effect on the expansion of damping temperature range is. (3) in the case of adding graphite powder, the influence of HAF and FEF carbon black on the damping performance is almost negligible, and the effects of FEF and HAF are basically the same. Carbon black with different particle size can be tried. Based on adding graphite powder, other admixtures can attenuate the peak value of loss factor and the large damping temperature range, and the effect of adding graphite powder alone is the best.

6.2.2 *Long Chain Polymer Blending Method*

6.2.2.1 Design and Preparation of Modified Materials

In 1955, ring opening metathesis polymerization (ROMP) polynorbornene was first obtained by Anderson et al. of DuPont Company with $TiCl_4/LiAlR_4$ catalyst system. Polynorbornene is a kind of polymer with different structure and properties, which is produced by the reaction of cyclopentadiene and ethylene via Diels and Alder reactions, synthesized by ring opening metathesis, addition polymerization, cationic or free radical polymerization. The properties of the three polymers are totally different, but the polynorbornene obtained by ring opening polymerization has a double bond main chain due to its large molecular weight, its excellent performance has been in-depth researched and widely used. The CIS and trans groups exist in the polynorbornene at the same time, so crystallization can be avoided. The glass transition temperature is 35–37 °C, and its outstanding performance is high damping. The loss factor can be as high as 3, or even higher, which can convert most of the vibration energy into heat energy. However, the problems of poor adhesion with metal and other base materials and brittleness of products limit its application in civil engineering. As a viscoelastic damping materials with important application value, NBR is playing an indispensable key role in the seismic response of building structures. The macromolecular chain of NBR has a large side group, and the nitrile side group of NBR has a strong polarity and strong interaction, so the damping performance is good. Although it is far less than PDLC, the adhesion between NBR and metal is excellent. It has been highly valued and popularized in the field of civil engineering disaster prevention and mitigation at home and abroad. From the above analysis, it can be seen that NBR and PDR can form a good complementary.

In addition, considering that adding plasticizers such as pine tar to the composite damping materials can improve the compatibility of the blends to a certain extent, and make the damping materials have better damping performance in a wide temperature range, based on this, the pine tar is added to NBR/NSX blends to adjust the properties of the blends, and the damping characteristics of NBR/NSX/A_2oil blends are investigated.

The main raw materials of the test: Nitrile Butadiene Rubber (NBR for short, made from the formula in Sect. 6.2.1), pine tar (A_2oil for short), norsorex (NSX for short), and relevant mechanical properties are shown in Table 6.5. The above products are provided by Changzhou LanJin Rubber and Plastic Co., Ltd. Secondly, NBR-pdr and NBR-PDR-PINE TAR blends were prepared: preparation → NBR plastic mixing → plastic compound parking → blending PDR and pine tar mixing → compound parking → vulcanization curve determination → tablet pressing, vulcanization and sample preparation. The loading sequence is determined according to the component ratio of NBR-PDR and NBR-PDR-TAR blends, as shown in Table 6.6.

Table 6.5 Basic mechanical properties of polynorbornene

	Performance	Value
Mechanical property	Tensile strength (MPa)	50
	Loss factor	3
	Resilience value (%)	18
	Hardness (Shore A)	63
	Wear resistance (mm^3)	35
	High friction (on polished steel plate)	$\mu = 1.20$

Table 6.6 Test number table according to component ratio

Number	Formula
1	NBR
2	NBR/NSX(80/20)
3	NBR/NSX(70/30)
4	NBR/NSX(50/50)
5	NBR/NSX(30/70)
6	NBR/NSX/A_2oil(80/20/10)
7	NBR/NSX/A_2oil(70/30/12)
8	NBR/NSX/A_2oil(30/70/30)
9	NBR/NSX/A_2oil(70/30/10)

6.2.2.2 Evaluation Method of Damping Performance

DMTA analysis method is adopted in this test. See Sect. 6.2.1 for test instrument content. The test conditions are as follows: the temperature rise rate is set as 3 °C/min; the material is used for seismic absorption of building structure, so the temperature range of this test is −20 to 50 °C according to the temperature range of building, and the frequency is set as 1.5 Hz.

There is no unified standard for the evaluation of damping materials. In many researches, the peak value of loss factor and the temperature range of large loss factor are regarded as one of the main standards to investigate the damping performance of polymer. For long-chain blends, a more scientific method is to take the TA value and LA value of polymer as the indexes of damping evaluation, that is, the area included under the temperature curve of loss factor ($\tan\delta - T$) and the area included under the temperature curve of loss modulus ($E'' - T$), and the calculation formula is as follows:

$$TA = \int_{T_0}^{T_R} \tan\delta dT \cong (\ln E'_G - \ln E'_R)\frac{R}{(E_a)_{avg}}\frac{\pi}{2}T_g^2 \tag{6.38}$$

$$LA = \int_{T_0}^{T_R} E''dT \cong (E'_G - E'_R)\frac{R}{(E_a)_{avg}}\frac{\pi}{2}T_g^{2\prime} \tag{6.39}$$

Among them, E''_G is the loss modulus of glassy state; E'_R is the loss modulus of rubber colloidal state; T_G is the minimum value of glass transition temperature; T_R is the maximum value of rubber colloidal transition temperature; R is the polymer gas constant; $(E_a)_{avg}$ is the average activation energy in the relaxation process.

6.2.2.3 Test Results and Analysis

Figure 6.5a shows the $\tan\delta - T$ curves of NBR and that of NBR after adding different component ratios of polynorbornene.$\tan\delta - T$ curve is a key factor to study the dynamic mechanical behavior of polymer. The larger the peak value of $\tan\delta$ is, the more intense the molecular motion is, and the better the energy dissipation and shock absorption function of polymer is. It can be seen from the curve group that

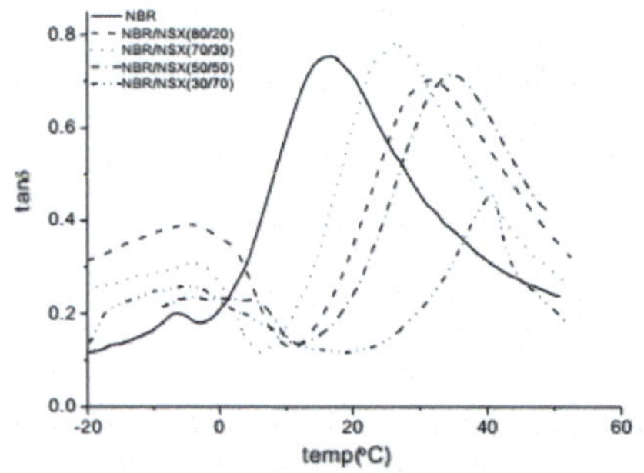

(a) $\tan\delta - T$curve of NBR/NSX blend

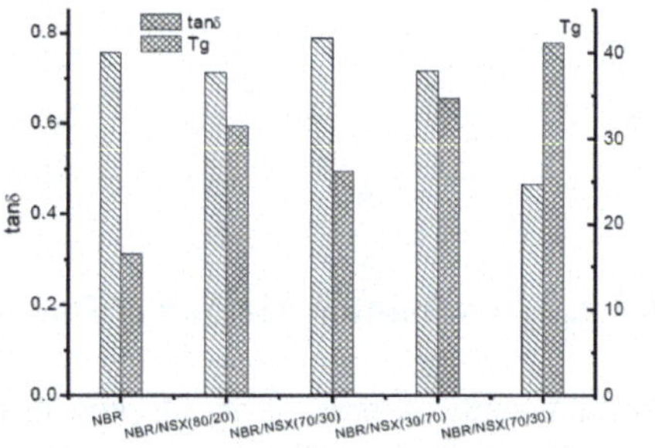

(b) Loss factor and transition temperature of NBR/NSX blend

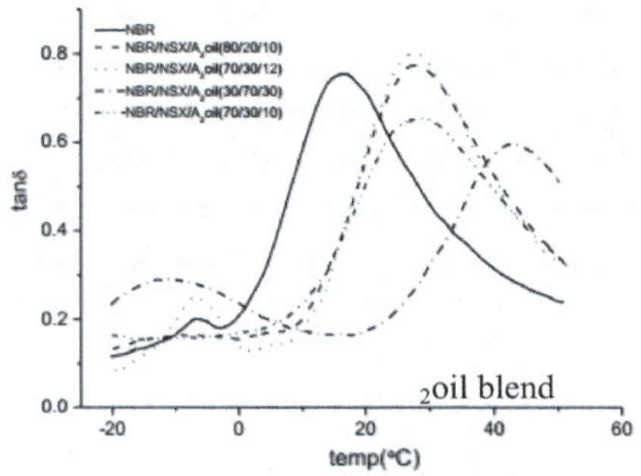

(c) $\tan\delta - T$ curve of NBR/NSX/ A_2oil blend

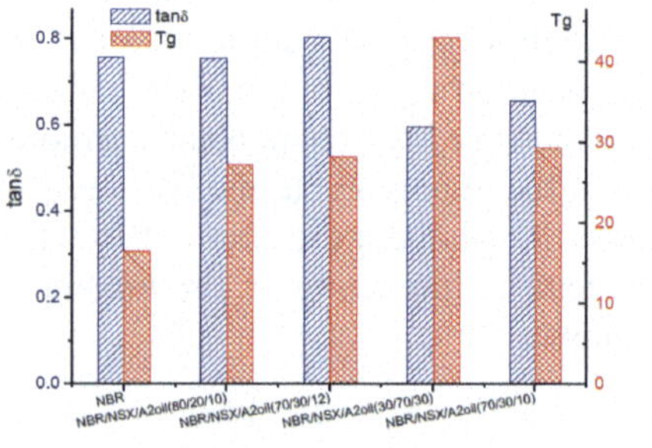

(d) Loss factor and transition temperature of NBR/NSX/A_2oil blend

Fig. 6.5 Material property test results of mixture

in the selected temperature range, NBR has two glass state temperature transition peaks, the first peak occurs at −7 °C, the value of tan δ is 0.20, the second occurs at 16.5 °C, the value of tan δ is 0.756, the tan δ of the first transition peak is slightly smaller, and T_g is lower than room temperature, the material's use temperature is in a bad situation. After adding PDL, the first peak value of the blend is increased. The NBR/NSX (70/30) increased the second peak of loss factor to 0.788. Although the expansion of $\Delta T_{\tan\delta \geq 0.5}$ of the base material (the temperature range with loss factor greater than 0.5) after adding NSX is not large, it has the effect of floating T_g of NBR to high temperature, which is good for the application of materials in seismic and shock absorption of building structures. The peak values of loss factors and transition temperatures for the five blends are summarized in Fig. 6.5b.

The tan $\delta - T$ curve of NBR/NSX/A_2oil blend is shown in Fig. 6.5c, d shows the peak value of loss factor and transition temperature of NBR/NSX/A2oil blend with different blending ratio. It can be seen from the figure that the loss factor of the first glass transition region of NBR is increased by adding NSX and A_2oil, but the damping effect is decreased by adding A_2oil. For the second glass transition region, when the ratio of the three components is NBR/NSX/A_2oil(80/20/10) and NBR/NSX/A_2oil(70/30/12), the second peak value of loss factor is 0.755 and 0.805 respectively, which is 6 and 13% higher than 0.712 of NBR matrix material, respectively. With the decrease of NBR content and the increase of NSX content in the blend, the loss factor of the blend decreases, which shows that increasing the amount of high damping polynorbornene is not the best, and it should be controlled at a certain amount. According to NBR/NSX/A_2oil(70/30/12) and NBR/NSX/A_2oil(70/30/10), under the same amount of NBR and NSX, the peak value of loss factor of blends is proportional to the amount of A_2oil.

6.3 Research and Development of New Viscoelastic Damper

6.3.1 Laminated Viscoelastic Damper

The Lanling brand viscoelastic damper developed by the research center of earthquake resistance and shock absorption of Southeast University and Jiangsu Changzhou Lanling rubber factory is an earlier product used in practical projects in China [2]. Most of the products are typical laminated plates, which are made by laminating and bonding the viscoelastic damping materials and constrained steel plates, and dissipate energy by the shear deformation of viscoelastic layer. The typical structural profile in which the damping material layer is made of double or multilayer 2301 viscoelastic material, and the performance index is shown in Table 6.6. The material has high storage shear modulus and loss factor within the range of ambient temperature and working frequency of general buildings, and can be applied to civil engineering structures. The shear adhesive strength between viscoelastic damping material and steel plate is 1.52 MPa, which adopts the hot press cementation process (Table 6.7).

Table 6.7 Main performance indexes of 2301 viscoelastic damping materials with local amplification

Project	Unit	Performance index
Tensile strength	MPa	17.5
Elongation	%	430
Permanent deformation	%	20
Maximum loss factor		1.4
Elastic	%	6
Storage shear modulus	MPa	2–100

In order to further study the mechanical properties of the damper, six specimens of viscoelastic damper with double damping materials, 460 mm in length, 100 mm in width and 64 mm in height, were designed, and the dynamic mechanical properties were tested on the pseudo dynamic testing machine of the structural and material testing center of Southeast University. The method of sinusoidal excitation was used to control the loading with shear displacement under different ambient temperature. By applying sinusoidal force of different frequency, the shear displacement and restoring force of viscoelastic damper under various shear strain amplitudes were measured respectively, so as to obtain the dynamic characteristics of damper with the change of temperature, vibration frequency and shear displacement. The test temperature range was 0–45 °C, frequency range was 0.15–4.0 Hz, and the shear deformation amplitude range was 1–30 mm.

(1) **Hysteretic curves under different working conditions**

Figure 6.6a shows the hysteretic curve of viscoelastic damper with the change of loading frequency under the same temperature and shear deformation amplitude. It can be seen that at the same temperature and shear deformation amplitude, with the increase of the excitation frequency, the slope of the long axis of the ellipse increases gradually, and the hysteresis loop tends to be full, which indicates that the stiffness of the viscoelastic damper increases with the increase of the excitation frequency, and the energy dissipation capacity increases with the increase of the excitation frequency.

Figure 6.6b shows the hysteretic curves of dampers at different ambient temperatures. At 0 °C, the slope of the long axis of the hysteresis loop increases obviously under various excitation frequencies. When the temperature increases to 8 and 12 °C, the hysteresis loop tends to be full, and the slope of the long axis decreases. From 12 to 17 °C and 20 °C, the degree of the fullness of the hysteresis loop decreases gradually, and the slope of the long axis of the ellipse becomes smaller and smaller. The fullness of hysteresis loop at 12 °C is the largest among various test temperatures. It can be seen that at 12 °C the energy dissipation capacity of viscoelastic damper is the strongest at °C.

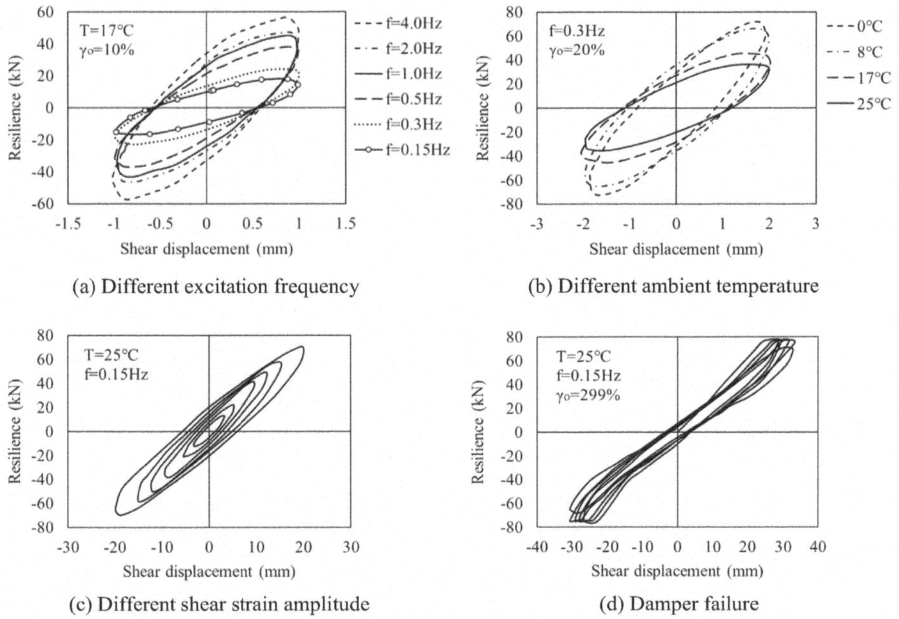

Fig. 6.6 Force displacement curves of viscoelastic damper under different working conditions

As shown in Fig. 6.6c, the input shear deformation amplitude increases step by step, and the shear deformation of damping material in the test piece increases with the increase of the input shear strain amplitude, but the slope of the long axis of the hysteresis loop and the fullness of the hysteresis loop have no obvious change.

At room temperature of 25 °C and loading frequency of 0.15 Hz, when the input shear deformation amplitude reaches 29.8 mm, that is, when the shear stress amplitude reaches 298%, the hot pressed cementitious layer of 2301 viscoelastic damping material and steel plate begins to tear in a small area. It is considered that the specimen is damaged, and the maximum recovery force is 76 kN at this time. After unloading, the residual deformation is still not obvious. Figure 6.7d is the force-shear displacement hysteresis curve at failure.

(2) Dynamic mechanical property index

Due to the existence of constrained steel plate layer and adhesive layer, the dynamic mechanical performance index of viscoelastic damper cannot be completely equal to the dynamic mechanical performance index of viscoelastic material. In fact, the loss factor of damper is often slightly smaller than that of viscoelastic material. Therefore, based on the above experimental data, the storage shear modulus and loss factor, which are important parameters of dynamic mechanical properties of viscoelastic dampers, are compared and analyzed. It can be seen from Fig. 6.7 that: (1) at the same frequency, the shear modulus of energy storage gradually decreases with the increase of temperature, with an order of magnitude difference between 0

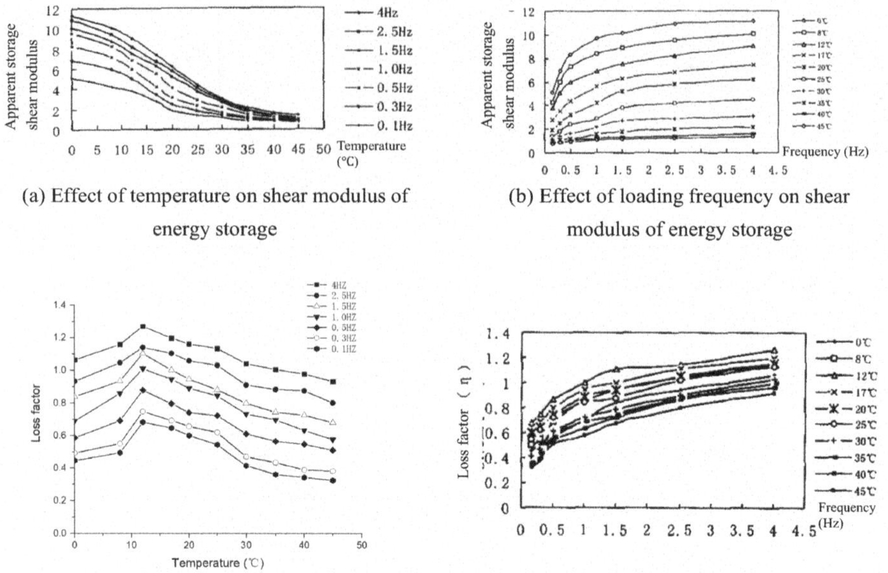

(a) Effect of temperature on shear modulus of energy storage

(b) Effect of loading frequency on shear modulus of energy storage

(c) Influence of temperature on loss factor

(d) Influence of loading frequency on loss factor

Fig. 6.7 Dynamic mechanical properties of viscoelastic damper

and 45 °C; (2) at the same temperature, the shear modulus of energy storage gradually increases with the increase of frequency; (3) at each temperature measured, the loss factor value at 12 °C is the largest, and gradually decreases with the increase or decrease of temperature; (4) at the same temperature, the loss factor increases with the increase of frequency.

(3) **Aging performance test**

Due to the long-term exposure of damping energy dissipation materials to the building environment, aging has become one of the important factors affecting the durability of viscoelastic dampers. For this reason, the aging performance test of viscoelastic damper was carried out. A total of 6 samples were designed for comparison test, including 3 aging samples and 3 non aging samples. The viscoelastic material used was 2301 viscoelastic damping material produced by Changzhou Lanling Rubber Factory.

It takes a long time to determine the aging performance of viscoelastic damper in natural state, so the method of high temperature accelerated aging is needed. The theory of high temperature accelerated aging is proposed by S. A. Arr-henius. The relationship between chemical reaction speed and temperature is as follows:

$$t = t_0 \times 10^{0.434E\left(\frac{1}{T}-\frac{1}{T_0}\right)/R} \tag{6.40}$$

Among them, t_0 and T_0 are the design service life (days) and the absolute temperature of the service environment (K); t and T are the time (days) and temperature (K) required for the test respectively; R is the gas constant of 8.31 J/mol K; E is the activation energy of rubber, with the value of 90.4 kJ/mol. According to this formula, the temperature and time of accelerated aging can be calculated relative to a certain ambient temperature and design service life.

The determination of the test temperature in the high temperature accelerated aging test must consider the need of the test speed and the consistency of the reaction process with the actual situation. Increasing the test temperature can shorten the test time, but increasing the test temperature unilaterally will increase the possibility of thermal decomposition and the migration volatility of the mixture, make the reaction process inconsistent with the actual situation, and affect the reliability of the test results. Viscoelastic materials belong to rubber, and their aging mechanism has not changed in the temperature range of 50–100 °C. Considering the variety of viscoelastic damping materials and the need of test speed, the test temperature was chosen as 80 °C, i.e. 353 K.

SC101 blast electric constant temperature drying oven is used in the test, and its constant temperature fluctuation is ±1 °C. The function of air blowing is to make the temperature in the box even, and to eliminate the volatiles produced in the aging process, and to supplement fresh air, so as to keep the air composition consistent and stable. Put the viscoelastic material samples to be aged and three viscoelastic damper samples to be aged together into SC101 blast electric thermostatic drying oven, the temperature was set to 80 °C, and take them out after 54 days. From Eq. (6.40), it can be seen that this is equivalent to the aging situation after 80 years when the ambient temperature is 20 °C. Then the dynamic mechanical property test was carried out for the aging sample, and the test equipment and test conditions were the same as those of the above-mentioned non aging sample.

Table 6.8 shows the performance comparison of viscoelastic damping material before and after aging. It can be seen that after aging, the tensile strength of the material increased, but the elongation and loss factor decreased in varying degrees. Figure 6.8 shows the change of storage shear modulus and loss factor of viscoelas-

Table 6.8 Comparison of aging properties of viscoelastic damping materials

Item	Unit	Non aging material	Aging materials
Tensile strength	MPa	17.5	19.3
Elongation	%	430	328
Permanent deformation	%	20	29
Maximum loss factor		1.4	1.1
Elastic	%	6	3.1
Storage shear modulus	MPa	2-100	5–110

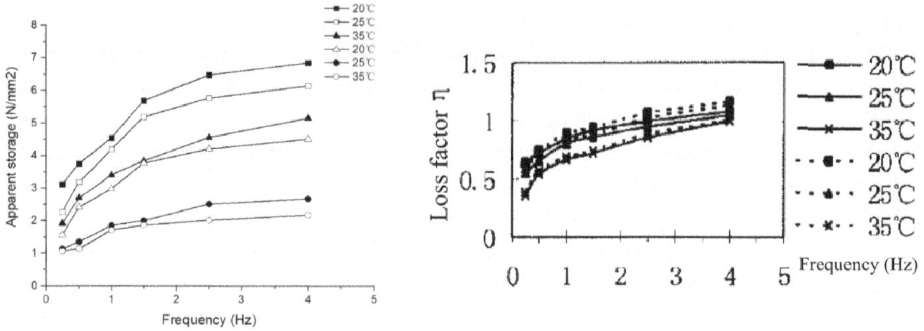

(a)Storage shear modulus (b) Loss factor

Fig. 6.8 Comparison of dynamic mechanical parameters of viscoelastic damper before and after aging

tic damper with environment temperature and loading frequency before and after aging. The results show that the storage shear modulus of aging viscoelastic damper increases and the loss factor decreases slightly, but the performance of the damper is still stable in general. Compared with Table 6.8 and Fig. 6.8, it can be seen that although the aging performance of viscoelastic damping material is poor, most of the damping material in the damper is wrapped by steel plate, only the surrounding is in contact with air, and the surrounding aged material plays a protective role on the internal material, which greatly reduces the aging speed. As the aging considered in the test is equivalent to 80 years at room temperature, which exceeds the design service life requirements of general buildings, so it meets the engineering requirements.

6.3.2 Cylindrical Viscoelastic Damper

In the large-span roof structure, viscoelastic damper is sometimes required to reduce the vibration response of the structure. At this time, the damper is usually installed on the roof structure in the form of support. In order to make the damper consistent with the steel roof structure in appearance, a cylindrical viscoelastic damper was designed, as shown in Fig. 6.9a [3, 4]. The damper is mainly composed of internal and external seamless steel pipes and viscoelastic materials of the overlapped part of two steel pipes. The material was 2301 viscoelastic damping material, and the hot pressing cementation process was still used between the material and the steel plate for bonding. The test was carried out in the same loading way as Sect. 6.3.1 (Fig. 6.9b). The test temperature range was 0–30 °C, the frequency range was 0.1–4.0 Hz, and the shear deformation amplitude range was 1–16.5 mm.

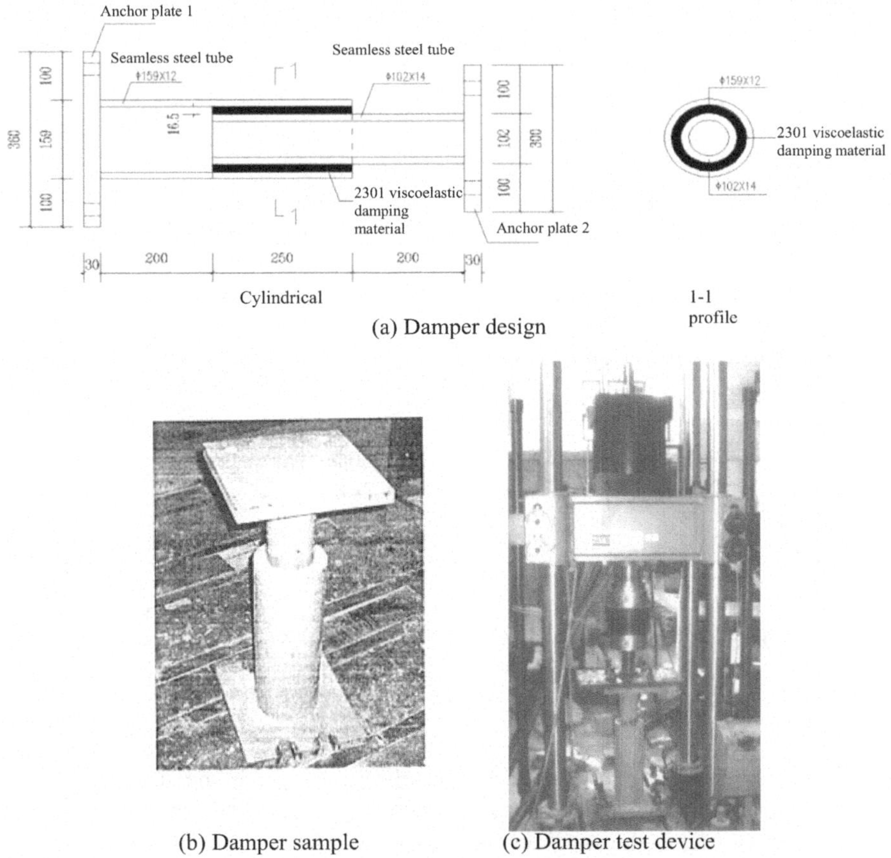

Fig. 6.9 Cylindrical viscoelastic damper

(1) Hysteretic curves under different working conditions

Figure 6.10a shows the hysteretic curves of viscoelastic dampers under different excitation frequencies. It can be seen that at the same temperature and shear deformation amplitude, with the increase of the excitation frequency, the slope of the long axis of the ellipse gradually increases, indicating that the stiffness of the viscoelastic damper increases with the increase of the excitation frequency. Figure 6.10b shows the hysteretic curve of viscoelastic damper under different ambient temperatures. When the temperature is increased to 10.2 and 11.8 °C, the hysteretic loop tends to be full and the slope of the long axis decreases. When the temperature increases from 11.8 to 14.5 °C, 17.5, 20 and 30 °C, the fullness of the hysteresis loop decreases gradually, and the slope of the long axis of the ellipse becomes smaller and smaller, which indicates that after 11.8 °C, with the increase of temperature, the initial stiffness of the tubular viscoelastic damper begins to decrease. The energy consumption capacity is also reduced to some extent. At 11.8 °C, the hysteresis loop is the largest,

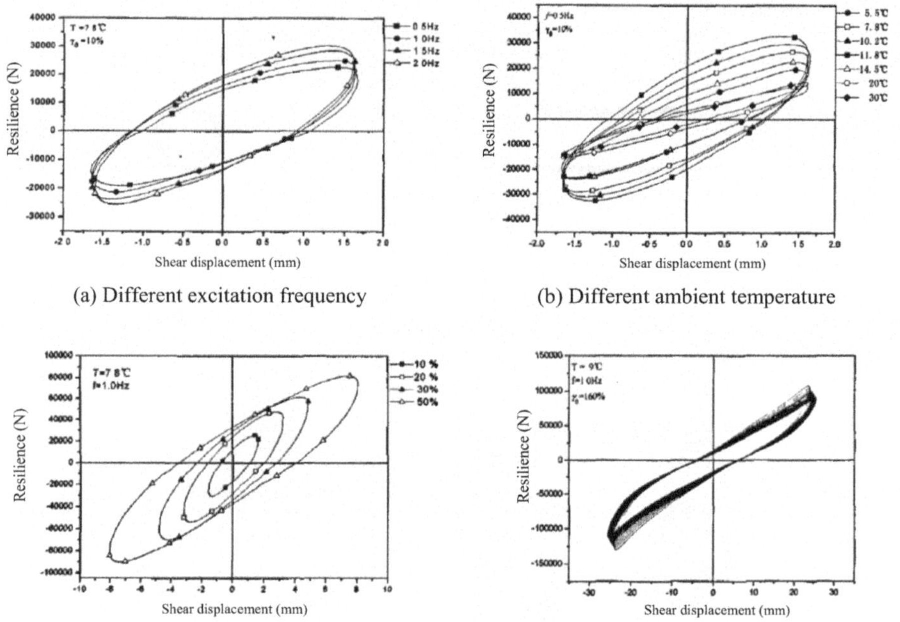

(a) Different excitation frequency (b) Different ambient temperature

(c) Different shear strain amplitude (d) Damper failure

Fig. 6.10 Force displacement curve of viscoelastic damper under different working conditions

and the energy dissipation capacity of viscoelastic damper is the strongest. However, at different shear displacement amplitude (Fig. 6.10c), the shear deformation of damping material in the specimen increases with the increase of the input shear strain amplitude, but the slope of the long axis of the hysteresis loop and the fullness of the hysteresis loop have no obvious change. When the temperature in the test room is 9 °C and the frequency is 1.0 Hz, when the input shear strain amplitude reaches 26.4 m, that is, when the shear strain amplitude reaches 160%, the hysteretic curve of the force and displacement of the tubular viscoelastic damper obtained from the test is in the reverse "s" shape (Fig. 6.10d), and there are obvious tear marks at the joint of the 2301 viscoelastic material and the steel pipe, which indicates that the tubular viscoelastic damper specimen has been damaged, and the maximum recovery force is only 75 kN. After unloading, there is obvious residual deformation.

(2) **Dynamic mechanical property index**

At the same time, the storage shear modulus and loss factor, which are important parameters of dynamic mechanical properties of viscoelastic dampers, are compared and analyzed. According to Fig. 6.11a, b, the rules similar to Sect. 6.3.1 can be obtained: (1) at the same frequency, the storage shear modulus gradually decreases with the increase of temperature; (2) at the same temperature, the storage shear modulus gradually increases with the increase of frequency; (3) at each measured temperature, the loss factor value at 11.8 °C is the largest, and gradually decreases

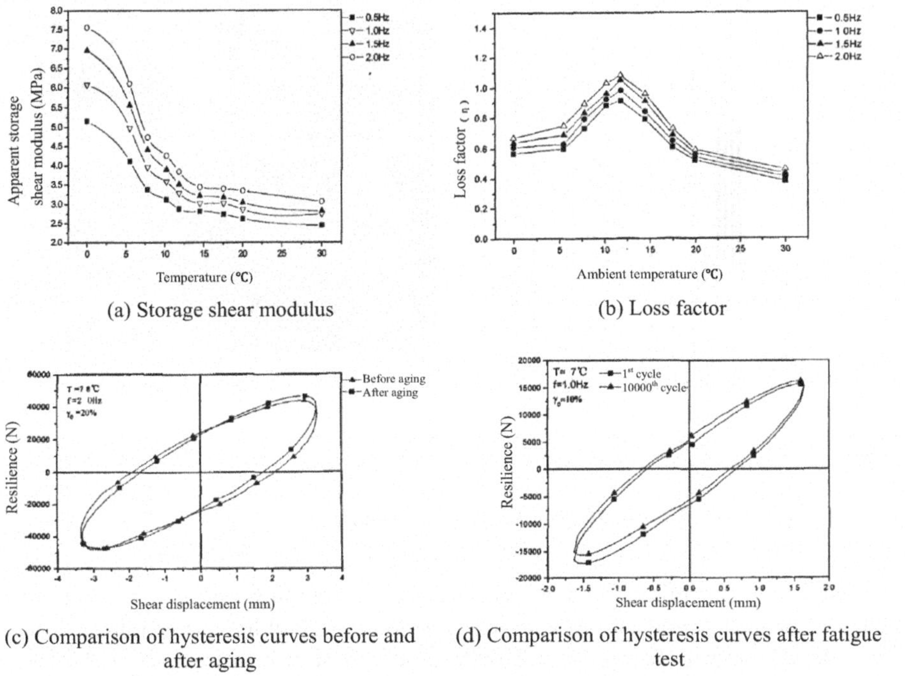

(a) Storage shear modulus

(b) Loss factor

(c) Comparison of hysteresis curves before and after aging

(d) Comparison of hysteresis curves after fatigue test

Fig. 6.11 Dynamic mechanical properties and durability of viscoelastic damper

with the increase or decrease of temperature; (4) at the same temperature, the loss factor increases with the increase of frequency.

(3) **Durability test**

According to the characteristics and application of viscoelastic damper, the durability test was carried out, mainly including aging performance test and fatigue test. Figure 6.11c shows the comparison of hysteretic curve of viscoelastic damper before and after aging: (1) the aged tubular viscoelastic damper was placed at 80 °C for 54 days, and compared with the non-aged tubular viscoelastic damper under the same conditions, although its energy consumption capacity was reduced to some extent, the aged tubular viscoelastic damper still had strong energy consumption capacity. (2) the hysteresis loop area before aging is only slightly larger than the hysteresis loop area after aging. This shows that after aging, the energy dissipation capacity of damper decreases little. which indicates that the cylindrical viscoelastic damper has a certain anti-aging property. Figure 6.11d shows the hysteretic curve of viscoelastic damper after 10,000 cycles of cyclic loading compared with that of the first cycle loading. The energy consumption capacity of viscoelastic damper decreases very little and has strong fatigue resistance. The analysis of the test data shows that the apparent storage modulus and loss factor of the damper decrease slightly after aging and fatigue, with a maximum reduction of 11.1% and 18.8%, respectively.

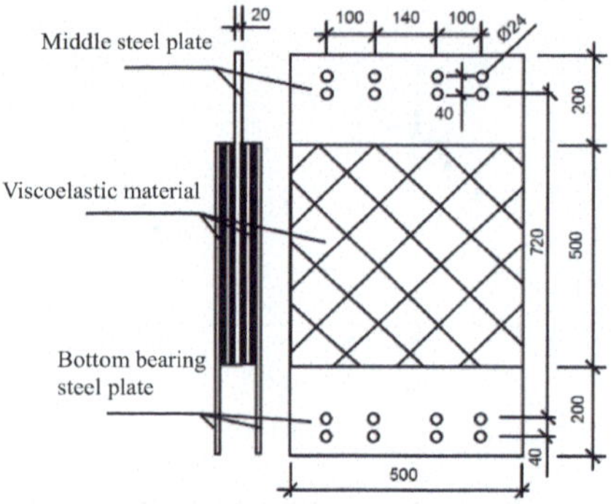

(a) Dampingwall design

(b) Damping wall test

Fig. 6.12 Design and test scheme of viscoelastic damping wall

6.3.3 "5 + 4" Viscoelastic Damping Wall

Similar to viscous fluid damping wall, viscoelastic damping wall is more and more suitable for the current vibration control of high-rise buildings. For this reason, a "5 + 4" viscoelastic damping wall is designed, as shown in Fig. 6.12a [1]. Each layer of four layers of viscoelastic material is 10 mm thick, the middle steel plate is 20 mm, and the four constraint steel plates on both sides are 12 mm. The effective size of the section of the viscoelastic damping wall is 500 mm × 500 mm, and the total area of the viscoelastic material layer is 10,000 mm^2. The middle steel plate and the outer steel plate are respectively extended by 200 mm up and down, and the total size is 900 mm × 500 mm. Based on this, a full-scale test model was made, and the multi working condition mechanical performance test was carried out on the electro-hydraulic servo test machine (Fig. 6.12b). Four frequencies of 0.5, 1.0, 1.5, 2.0 Hz and five displacement amplitudes of 22.5, 30.0, 37.5, 45.0 and 60.0 mm were selected for the test conditions.

Figure 6.13a shows the force displacement curve of damping wall under different displacement amplitudes when the loading frequency is 1.5 Hz. The hysteresis loop is not a smooth ellipse, but the edge angle appears. It shows that the viscoelastic material of the damping wall is mainly composed of the viscosity produced by friction damping elements between molecules and between molecules and fillers. Figure 6.13b–h shows the change trend of the mechanical performance indexes such as the maximum damping force, storage stiffness, loss stiffness, equivalent damping coefficient, shear storage modulus, shear storage modulus and loss factor. By comparison, it can be seen that: (1) the relationship between the equivalent damping coefficient and frequency is the reducing relationship, while the relationship between the maximum damping force, loss stiffness, loss factor and shear loss modulus and frequency is zero correlation; (2) the storage stiffness and shear storage modulus show a parabolic change in displacement correlation. When the displacement amplitude reaches 45.0 mm and the frequency is 0.5 Hz, the peak values of storage stiffness and shear storage modulus are 3.54 kN/mm and 2.7 MPa, respectively; (3) when the frequency is 2.0 Hz and the displacement amplitude is 22.5 mm, the peak value of loss factor is 0.77.

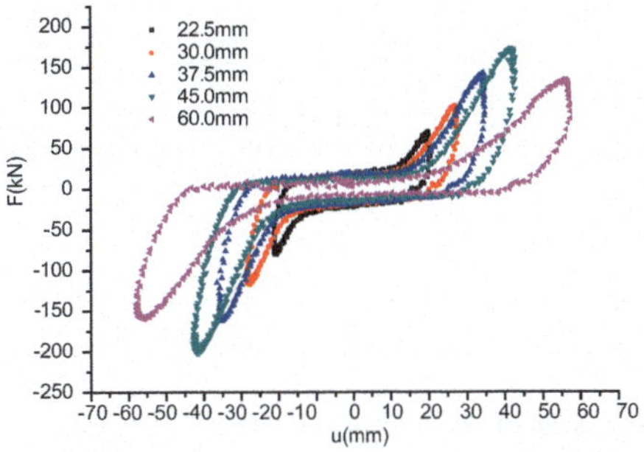

(a) Hysteresis curve at loading frequency of 1.5 Hz

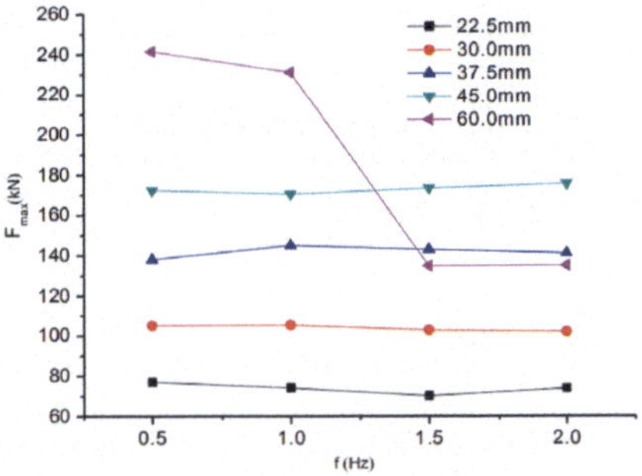

(b) Maximum damping force

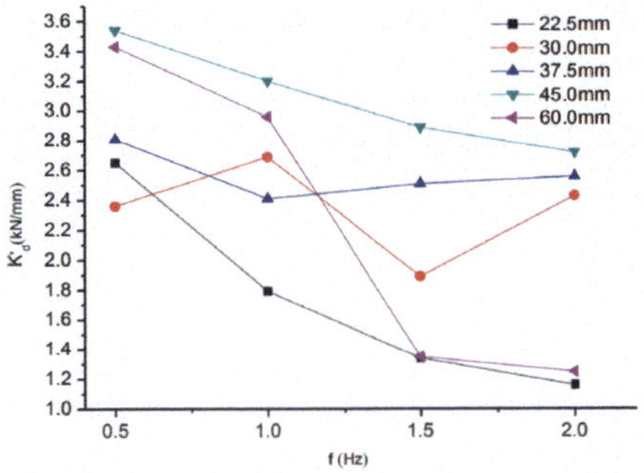

(c) Storage stiffness

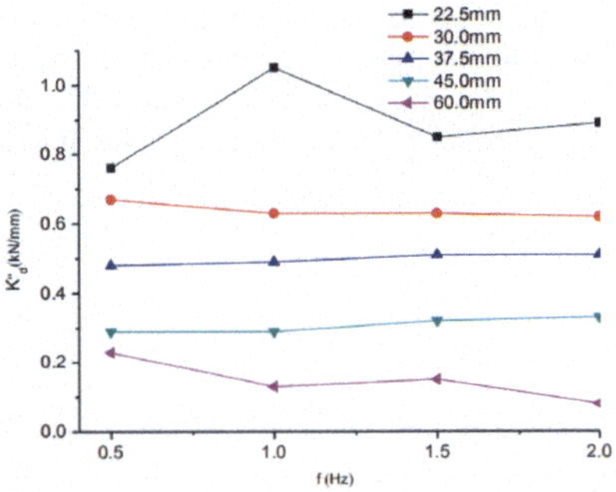

(d) Loss stiffness

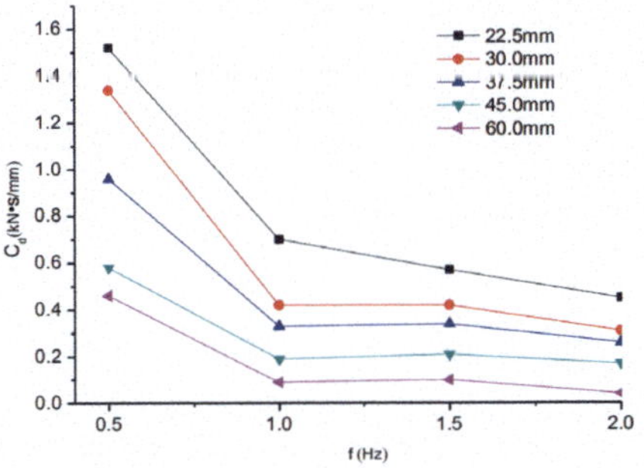

(e) Equivalent damping coefficient

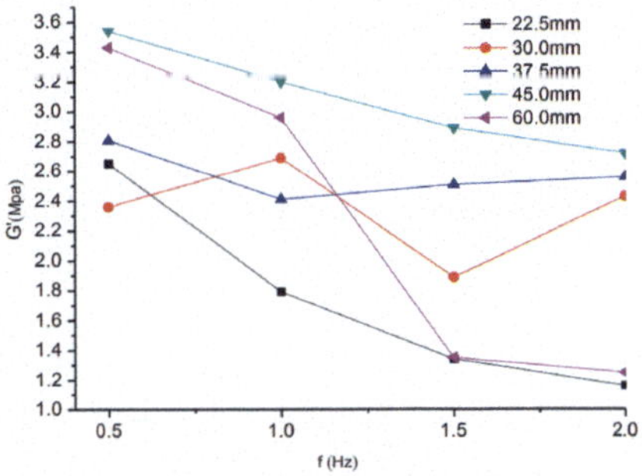

(f) Shear storage modulus

Fig. 6.13 Test results and analysis

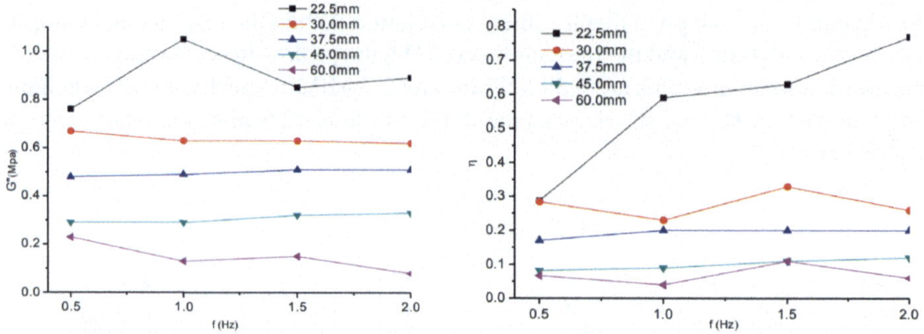

(g) Shear storage modulus (h) Loss factor

Fig. 6.13 (continued)

References

1. Junhong, Xu. 2015. *Experimental Study on the New Kind of Viscoelastic Damping Wall*. Nanjing: Southeast University. (in Chinese).
2. Sui, Y. 2002. *Experimental Study and Engineering Application of Viscoelastic Damper and Energy Dissipation Support*. Nanjing: Southeast University. (in Chinese).
3. Chang, Y. 2003. *Research and Application of Viscoelastic Damper and Energy Dissipation Structure*. Nanjing: Southeast University. (in Chinese).
4. Yi, S. 2006. *Study on Wind-Induced Vibration Control of Large Cantilevered Steel Space Truss With Tube Viscoelastic Damper*. Nanjing: Southeast University. (in Chinese).

Chapter 7
Metal Damper

Abstract Mechanism and characteristics of metal damping, including basic principle of metal damper, properties of steel with low yield point, and type and calculation performance of metal damper are introduced. Four calculation model, i.e. ideal elastoplastic model, bilinear model, Ramberg-Osgood model and Bouc-Wen model are analyzed and discussed. About tension-compression type metal damper, working mechanism of buckling proof brace and new buckling proof support are interpreted. About shear type metal damper, stress mechanism of unconstrained shear steel plate, buckling proof design of in-plane shear yield type energy dissipation steel plate, main performance parameters of buckling prevention shear energy dissipation plate and new shear metal damper are introduced. About bending metal damper, drum-shaped open hole soft steel damper and curved steel plate damper are elaborated.

7.1 Mechanism and Characteristics of Metal Damping

The mechanism of metal damper to control the vibration of structure is to dissipate part of the energy of structure vibration through the plastic energy dissipation of metal material in the damper, so as to reduce the vibration response of structure.

7.1.1 Basic Principle of Metal Damper

The commonly used materials of metal damper include mild steel (or low yield point steel), lead and shape memory alloy. In order to study the basic performance of metal damper, it is necessary to understand the mechanism of metal plastic deformation. Figure 7.1a shows the stress-strain curve of a typical metal under simple tension. At the beginning, the stress becomes proportional to the strain, and the proportion constant is the elastic modulus E. At this stage of the stress-strain curve, energy is not consumed during loading and unloading, and the relationship between the two can be expressed as follows:

A. Li, *Vibration Control for Building Structures*, Springer Tracts in Civil Engineering, https://doi.org/10.1007/978-3-030-40790-2_7

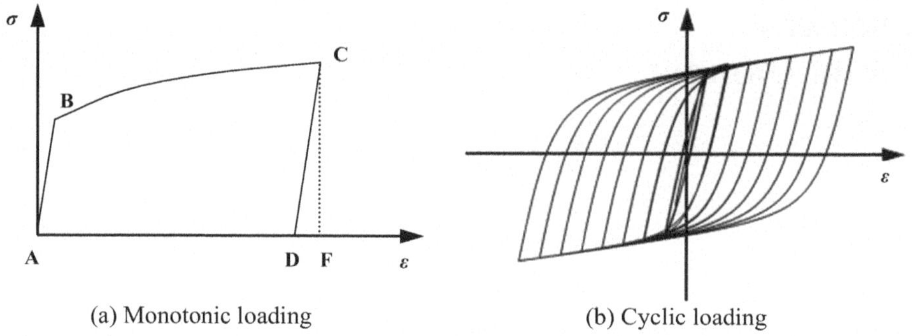

Fig. 7.1 Stress-strain curve of the typical metal

$$\sigma = E\varepsilon \tag{7.1}$$

Similarly, the relationship between shear stress τ and shear strain γ is:

$$\tau = G\gamma \tag{7.2}$$

where G is the shear modulus. If the strain continues to increase, it will reach a yield point (yield point B in Fig. 7.1a). When the stress increases further, the material will produce plastic deformation, the E_{di} curve will rise with a gentler slope, and the stress-strain curve of soft steel material in the plastic stage will rise with a medium slope. If it is released gradually during the stress strengthening stage, its curve is the CD segment in Fig. 7.1a. During unloading, the metal will not return along the original path, but will decline in a straight line parallel to the elastic stage (i.e. Young's modulus and shear modulus); and the metal material will no longer return to its initial state, resulting in residual deformation (i.e. increased plastic deformation).

In Fig. 7.1a, the area of ABCF represents the total input work, the area of CDF represents the elastic energy stored in the metal material at point C, and it is also the elastic energy released when unloading to point D. The area of the ABCD envelope is equal to the energy consumed by the metal material in the whole loading process.

In practice, the metal damper may be in the state of cyclic elastic-plastic deformation, so we need to consider the stress-strain curve of the metal specimen making elastic-plastic cycle in the range of different strain amplitude. Figure 7.1b shows the cyclic loading stress-strain curve for typical mild steel metallic materials. During the plastic cyclic loading, the stress will increase with the extension of strain range, and the hysteretic energy dissipation before fracture is often used to dissipate the vibration energy of the structure.

7.1.2 *Properties of Steel with Low Yield Point*

Generally, the steel used for structural seismic design not only has strength and ductility requirements, but also considers fatigue, welding and other properties. In the earthquake, the vibration damper enters into the yield state before the structural member, and uses the hysteretic deformation in the plastic stage to consume energy, so as to achieve the purpose of protecting the structure itself. Therefore, it is required that the steel used for the energy absorber has a lower yield point and a narrow yield range (Table 7.1), and has good processing and plastic deformation capacity.

7.1.2.1 Basic Mechanical Properties

1. **Tensile property**

The basic mechanical properties of low yield point steel refer to yield strength, tensile strength, elongation, etc. Wen Donghui of Baoshan Iron and Steel Co., Ltd. conducted experimental research on the mechanical properties of BLY160 steel produced by Baosteel, and the results are shown in Table 7.2.

The author's team carried out an experimental study on the tensile properties of LY160 steel produced by WISCO. In order to consider the influence of loading rate on the material performance, three different loading rates of 0.005, 0.03 and 1.0 mm/s are used for the test. See Table 7.3 for details.

Table 7.4 shows the mechanical properties test results of each test piece, in which the yield strength of low yield point steel is $f_y = \sigma_{0.2}$, $\sigma_{0.2}$ is the stress corresponding to the residual strain of 0.2%. It can be seen from the analysis of test results that there is no obvious yield platform for the tensile curve of low yield point steel, the elongation is about 1.5 times of that of ordinary steel, and it has good plastic

Table 7.1 Main properties of low yield point steel

Brand name	Yield strength/MPa	Tensile strength/MPa	Yield ratio/%	Elongation/%	Impact property	
					Temperature/°C	A_{kv}/J
LY100	80 120	200–300	≤60	>50	0	≥27
LY160	140–180	220–320	≤80	≥45	0	≥27
LY225	205–245	300– 400	≤80	≥40	0	≥27

Table 7.2 Tensile test of BLY160 steel

Thickness/mm	Yield strength/MPa	Tensile strength/MPa	Yield ratio/%	Elongation/%
Target value	140–180	220–320	≤80	≥45
12	171 158 165	270 265 270	61 61 62	63 60 61
16	159 162 166	275 270 270	56 57 55	58 60 62
20	162 174	285 285	70 71	57 60

Table 7.3 Number and size of material sample

Specimen number	Loading rate/mm/s	Thickness/mm	Width/mm	Remarks
R1	0.005	12.28	20.27	The same batch of low yield steel plates are cut by wire. 12 mm thick steel plate obtained by milling, grinding, etc.
R2	0.03	12.32	20.62	
R3	1	12.30	20.20	

Table 7.4 Property test results of mild steel

Specimen number	Loading rate/mm/s	Yield strength f_y/MPa	Tensile strength f_u/MPa	Yield ratio/%	Elongation/%
R1	0.005	143	280	51.1	52.7
R2	0.03	152	286	53.1	49.1
R3	1.0	183	300	61.0	45.2

deformation capacity, and its stress-strain curve is shown in Fig. 7.1. Figure 7.2 is the stress-strain curve of Q235 steel in tensile test at different rates. It can be seen from Figs. 7.1 and 7.2 that low yield point steel is similar to Q235 steel, which has gone through linear elastic stage, yield stage, stress strengthening stage and necking stage respectively, but the yield strength and ultimate strength of low yield point steel are relatively low, and there is no obvious yield platform.

2. **Fatigue property**

According to the requirements of the interpretative test in 《The method of axial constant amplitude low cycle fatigue test for metallic materials》 (GB15248-2008), the team designed the test piece as shown in Fig. 7.3a to carry out the low cycle fatigue test on LY160 steel produced by WISCO. The test piece is in the state of

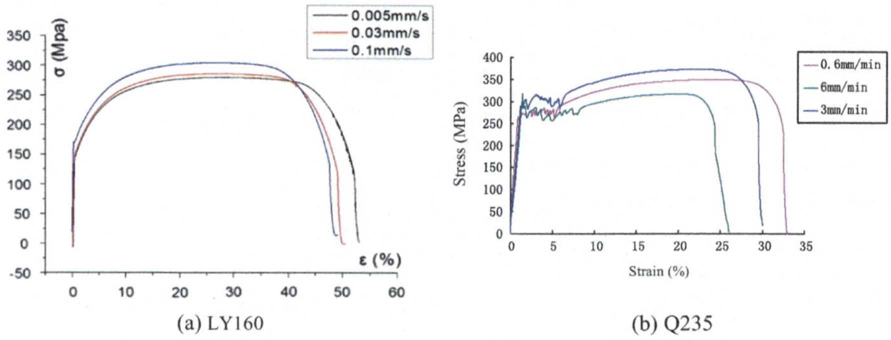

Fig. 7.2 Stress-strain curve of steel

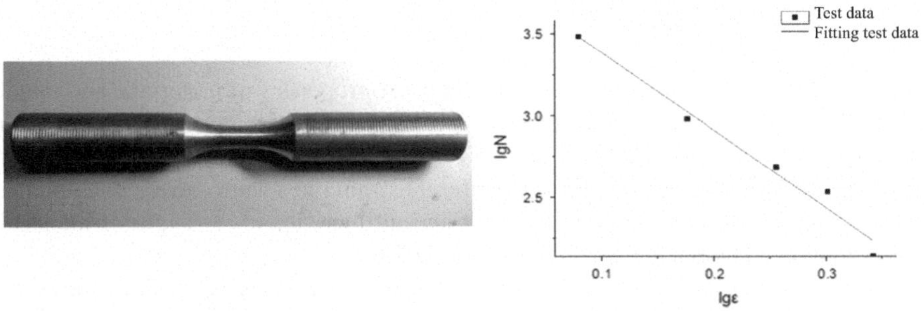

(a) Specimen (b) S-N curve

Fig. 7.3 LY160 steel fatigue test

Table 7.5 Constant amplitude low cycle test results (Loading frequency of 1 Hz)

Strain amplitude (%)	1	1.2	1.5	1.8	2.0	2.2
Cycle number	More than 10,000 times; manual stop	3062	960	488	346	139

large strain. The test results are shown in Table 7.6. According to the Mason-Coffin formula $\varepsilon_t = aN_f^c$, take the logarithm on both sides, the straight line $y = -0.3293x - 0.9153$, $R^2 = 0.9447$ was obtained after fitting, as shown in Fig. 7.3b. The parameters $a = 0.122$, $c = -0.3293$ can be obtained (Table 7.5).

Table 7.6 Effect of loading rate on steel properties

Serial number	Material	Strain rate (μ/s)	Yield strength (MPa)	Increase percentage of yield strength (%)	Tensile strength (MPa)	Increase percentage of tensile strength (%)
1-1	Q235	88	259.2	0	349.9	0
1-2		439	255.2	−1.5	316.9	−9.4
1-3		877	283.0	9.2	373.5	7.2
2-1	BLY160	100	168	0	265	0
2-2		1000	189	12.7	281	6.0
2-3		10,000	206	22.6	291	9.8
3-1	LY160	45	143	0	280	0
3-2		273	152	6.3	286	2.1
3-3		909	183	28.0	300	7.1

7.1.2.2 Rate Correlation

The loading rate will affect the strength and other properties of materials, and metal materials are no exception. The correlation relationships of material properties with rates of Q235 steel, BLY160 steel and LY160 steel were studied. The tensile tests of Q235, BLY160 and LY160 steels with different strain rates were carried out, and the influence of the rate on the yield strength and ultimate tensile strength of the steel was compared. The results are shown in Table 7.6.

For the influence of loading rate, the test results of various steels show the same trend, that is, with the increase of loading rate, the yield strength and ultimate tensile strength are improved, and the increase of yield strength is greater than that of ultimate tensile strength. The influence of rate on Q235 steel is less than 10%, which can be ignored in engineering application; the influence on the yield strength of low yield point steel is relatively large, when the rate is large, its improvement degree is obvious, which can reach more than 20%, and this rate exists in the actual earthquake (1–3 Hz), even higher for different types of energy dissipation devices. Therefore, in the design of damper the yield strength should be corrected, considering the influence of rate, and the specific correction method and coefficient need further experimental exploration; while the influence of rate on the ultimate tensile strength is relatively small, so the influence of rate on the ultimate tensile strength of steel can be ignored.

7.1.2.3 Time Effect Correlation

In production practice, it is known that the fatigue life will be affected if the fatigue test is interrupted and then continued. In recent decades, many scholars have also noticed and studied the influence of intermittent test on fatigue life. The research shows that the intermittence greatly improves the fatigue life of alloy steel, and the total intermittent time plays a major role, while the intermittent period, the number of intervals and the interval time of each interval are secondary. Compared with alloy steel, low yield point steel is more sensitive to aging, which can increase the service life of quenched mild steel by 331%. In addition, the effect of intermittence before crack development life N_i is significant, while the intermittence after N_i has little or no improvement on fatigue life. Due to the influence of environment, the intermittence after N_i will appear damage and reduce fatigue life.

The failure mode of building seismic energy absorber shows obvious low cycle fatigue characteristics. Therefore, it is necessary to study the aging correlation of large strain and low cycle fatigue life of low yield point steel. The effect of intermittence on the fatigue life of LY160 steel material is studied at 2% strain of interest in engineering application. See Table 7.7 for the test results. It can be seen that after 20 days of intermittent specimens after 30 cycles of loading, the fatigue life has a significant increase of more than 20%; after 20 days, the fatigue life of the specimens after 100 cycles of loading has no significant increase of less than 5%.

Similar to high cycle fatigue, intermittence can also significantly improve the low cycle fatigue life of low yield point steel. The principle is the same as that of high cycle fatigue. However, it is worth noting that the low cycle fatigue stress is large,

Table 7.7 Comparison of fatigue life with or without intermittence

	Without intermittence	Intermittence after 30 cycles	Intermittence after 100 cycles
Total life/time	303	30 + 334 = 364	100 + 217 = 317
Increase amplification (%)	–	20.13	4.62

and the environmental factors are easy to have adverse effects on the cracks. Some atoms or groups in the environment will penetrate the cracks and cause damage, thus weakening the beneficial effect of intermittence on the fatigue life. However, the adverse effect of this environment is real. The actual strong earthquake is often accompanied by rainfall, and the humid environment also increases the possibility of deepening of crack damage.

Low cycle fatigue is different from high cycle fatigue to some extent. High cycle fatigue pays attention to the time when intermittence occurs. When it occurs at a later time point (after the fatigue crack initiation life), the intermittence effect is small, while the effect is obvious at an earlier time point. This is because the high cycle fatigue life is long, and it may have been fully aged in the fatigue process, so the later intermittence has no effect. Generally, the time of earthquake action is about ten seconds to tens of seconds. Such a short time is not enough to give full play to the aging effect of materials. Therefore, for the low cycle fatigue life, intermittence will always improve the fatigue life.

7.1.3 Type and Calculation Performance of Metal Damper

7.1.3.1 Type of Metal Damper

Metal damper is one of the earliest and most widely used damper. Because of its stable mechanical properties and good plasticity, there are many kinds of metal damper. Overall, the damper materials can be divided into mild steel damper and lead damper; on the mechanical mechanism, it can also be divided into tension compression type, shear type, bending type, extrusion type and combination type.

As shown in Fig. 7.4a, X-type mild steel damper is one of the most commonly used metal damper. It is composed of several X-shaped steel plates. The vibration energy of the structure is dissipated by the lateral bending yield of the X-shaped steel plate to reduce the vibration response of the structure. The advantage of the damper is that every point at the same thickness of the steel plate will yield at the same time, which will give full play to the plastic properties of the steel plate material and greatly improve its energy consumption capacity. Based on this idea, researchers have developed various shapes or openings of steel plate dampers (such as triangle mild steel damper, Fig. 7.4b), becoming a large type of metal damper.

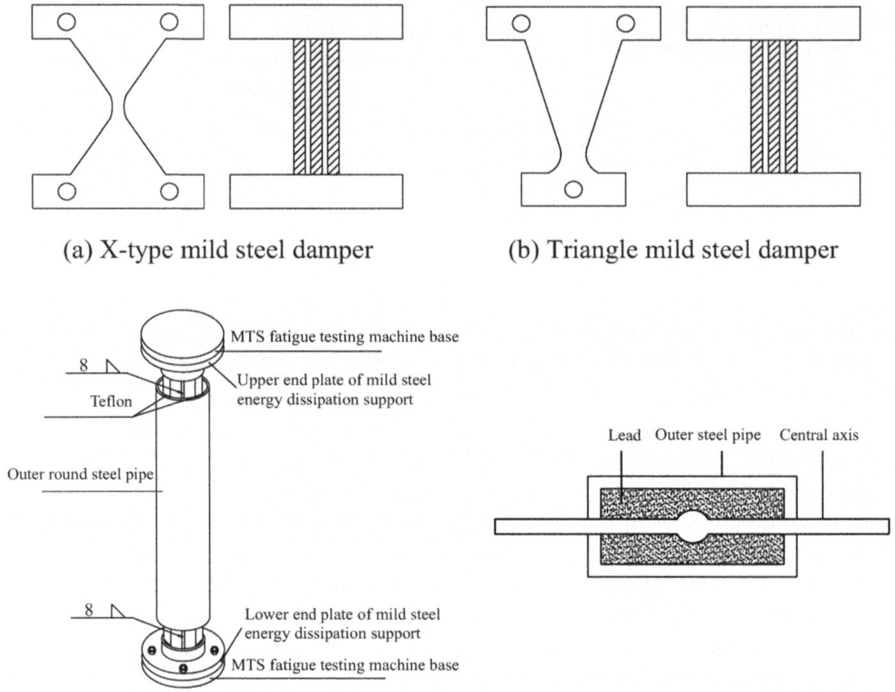

(a) X-type mild steel damper (b) Triangle mild steel damper

(c) Restrained buckling brace (d) Lead squeeze damper

Fig. 7.4 Metal damper

Buckling brace is a new type of metal damper developed in recent years, as shown in Fig. 7.4c. It uses the cross shaped core steel bar made of low yield steel such as Q235 as the energy dissipating member, and the outer circular steel tube provides the lateral restraint for it. An arc-shaped notch is arranged in the middle part of the four plate legs of the cross shaped core steel bar. The purpose of setting four arc-shaped defects is to adjust and control the moment when the energy dissipation support starts to work and to artificially set the position where the core cross steel plate has excessive plastic deformation failure. That is, the cross steel plate has plastic deformation. Because of the stress concentration at the arc-shaped defects, the failure position must start from the position of four arc-shaped notches in the middle. The gap between the cross core steel bar and the outer round steel tube can be filled and isolated by polytetrafluoroethylene strip. Its function is to isolate the outer round steel tube and the cross core steel bar, to ensure the free tension and compression deformation of the cross core steel bar, but at the same time, it can make the outer round steel tube transmit the lateral restraint to the cross core steel bar through the polytetrafluoroethylene strip, to prevent the cross core steel bar from buckling and breaking.

The lead extrusion damper is composed of outer steel tube, central axis and lead, as shown in Fig. 7.4d. When the central axis moves relative to the pipe wall, the lead in the pipe is extruded through the extrusion port to produce plastic deformation, which dissipates a lot of vibration energy in the process of plastic deformation, thus reducing the vibration response of the structure. The advantages of the damper are that the lead has recrystallization characteristics, the damper is not affected by work hardening or fatigue, and has good stability and durability.

7.1.3.2 Calculation Model of Metal Damper

The restoring force model of metal damper can be divided into two types: one is curve restoring force model with many parameters, the other is simplified linear restoring force model. The curvilinear model has higher precision, more practical, but the parameter determination and calculation are more complex; the second type is widely used in practice.

1. **Ideal elastoplastic model**

The ideal elastic-plastic model is the simplest mechanical model in the restoring force model of metal damper. The initial elastic stiffness is determined by yield load P_y and yield displacement d_y.

$$k_e = P_y/d_y \tag{7.3}$$

As shown in Fig. 7.5a, when the displacement value of the metal energy absorber exceeds d_y, the value of the force is equal to P_y. The energy consumed W_d in each cycle is equal to the area enveloped by the hysteresis curve between points (P_y, d_u) and $(-P_y, -d_u)$.

$$W_d = 4P_y(d_u - d_y), \text{ and } d_u \geq d_y \tag{7.4}$$

2. **Bilinear model**

In the bilinear model, the skeleton curves under forward and reverse loading are replaced by two broken lines. The unloading stiffness is not degraded. The inflexion of the reverse loading line (i.e. point C and point E in Fig. 7.5b) is determined according to the condition that the dissipative energy of the structure or component is equal (i.e. the area around the broken line is equal to the area around the test curve).

As can be seen from Fig. 7.5b, the slope of the first straight line OA represents the initial stiffness k_u (i.e. elastic stiffness) of the structure or component. Point A is yield point, and the corresponding load is called yield load P_y, and the corresponding displacement is called yield displacement d_y. The slope of AB segment decreases, that is, the rigidity decreases. If the rigidity after yielding is k_d, the decrease coefficient of rigidity is $P = k_d/k_u$. Point B in the figure is the corresponding point of ultimate load P_u and ultimate displacement d_u. Segment BC is in the working state of unloading

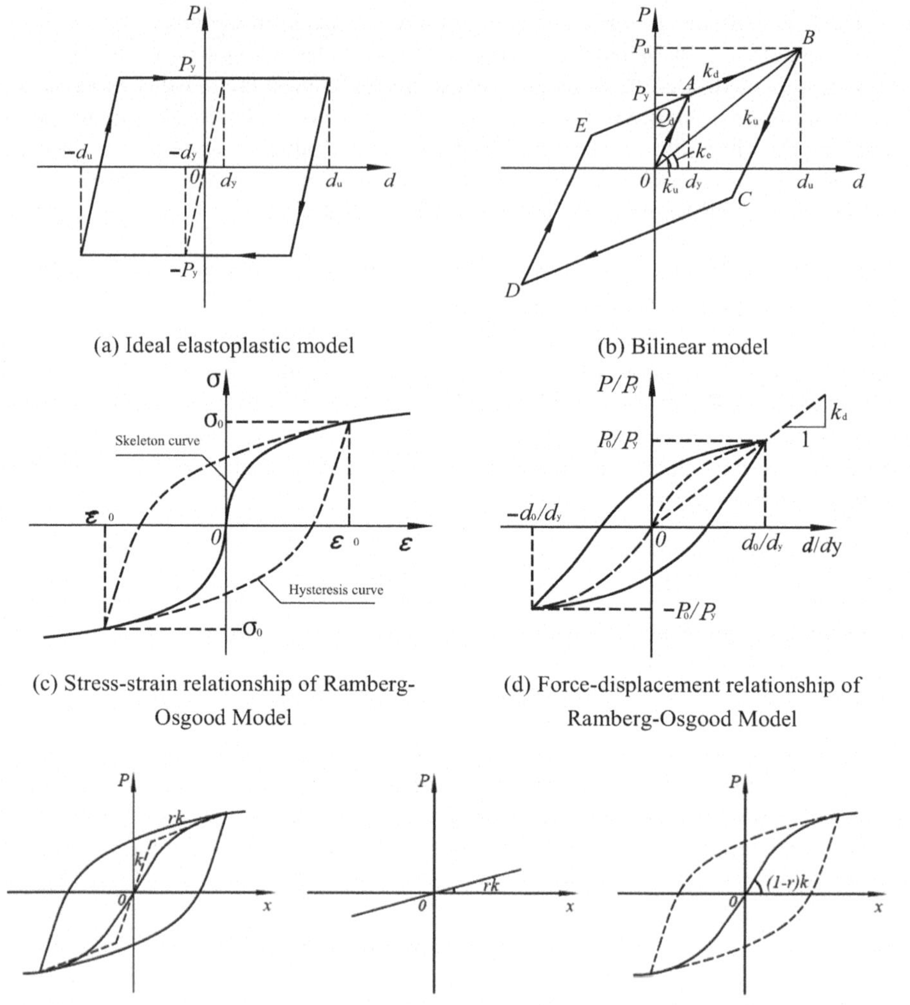

(a) Ideal elastoplastic model

(b) Bilinear model

(c) Stress-strain relationship of Ramberg-Osgood Model

(d) Force-displacement relationship of Ramberg-Osgood Model

(e) Breakdown diagram Bouc-Wen Model

Fig. 7.5 Calculation model of metal damper

and reverse loading after yielding, and maintains the elastic rigidity k_u. The segment CD is in reverse loading state, and the de segment DE indicates reloading after unloading. The whole curve is similar to the elastic-plastic stress-strain relationship of steel, so it is also called strain hardening model.

The slope of the line connecting the origin and the peak point of hysteresis curve is defined as the effective stiffness k_e:

$$k_e = k_d + \frac{Q_d}{d_u}, \quad \text{among them } d_u \geq d_y \tag{7.5}$$

Yield displacement:

$$d_y = \frac{Q_d}{k_u - k_d} \tag{7.6}$$

Hysteresis area (energy consumed per cycle):

$$W_d = 4Q_d(d_u - d_y) \tag{7.7}$$

Effective damping ratio:

$$\beta_e = \frac{4Q_d(d_u - d_y)}{2\pi k_e d_u^2} \tag{7.8}$$

Make $y = \frac{d_u}{d_y} \geq 1$, $a = \frac{Q_d}{k_d d_y} \geq 1$, then:

$$\beta_e = \frac{2a}{\pi}\frac{y - 1}{(y + a)y} \tag{7.9}$$

From Eq. (7.9), when $y = 1$, $\beta_e = 0$; when $y \to \infty$,$\beta_e \to 0$, $\to 0$; when $d\beta_e/dy = 0$, get the maximum value of β_e, at this time, $y = 1 + \sqrt{1 + a}$. Then:

$$\beta_e = \frac{2a}{\pi}\frac{1}{2(1 + a)^{1/2} + (2 + a)} \tag{7.10}$$

Substitute $a = \frac{Q_d}{k_d d_y}$ into Eq. (7.9):

$$a = \frac{k_u - k_d}{k_d} \tag{7.11}$$

It can be seen from the above formula that the maximum effective damping ratio k_u is related to the ratio with k_d. Therefore, k_u has little influence on the effective stiffness value k_e, but has great influence on the maximum damping ratio β_e.

3. **Ramberg-Osgood Model**

In 1943, Ramberg and Osgood firstly proposed the three parameter stress-strain curve of steel, which is the famous Ramberg-Osgood curve. Ramberg-Osgood Model (referred to as RO Model) is also called multi curve model, which is often used to describe the stiffness degradation model. RO Model consists of skeleton curve and hysteretic curve (as shown in Fig. 7.5c,). In 1996, Akazawa et al. proposed the expression of skeleton curve as follows:

$$\frac{\varepsilon}{\varepsilon_0} = \frac{\sigma}{\sigma_0}\left(1 + \alpha\left|\frac{\sigma}{\sigma_0}\right|^{\eta - 1}\right) \tag{7.12}$$

where, σ, σ_0, ε and ε_0 represent stress, yield force, strain and yield strain, respectively, and α and η are curve shape coefficients. Let the initial stiffness be K, and the relationship between σ_0 and ε_0 is:

$$\sigma_0 = K\varepsilon_0 \tag{7.13}$$

The expression of hysteretic curve is:

$$\frac{\varepsilon - \varepsilon_0}{2\varepsilon_0} = \frac{\sigma - \sigma_0}{2\sigma_0}\left(1 + \alpha\left|\frac{\sigma - \sigma_0}{2\sigma_0}\right|^{\eta-1}\right) \tag{7.14}$$

The relationship between the force and displacement of RO model is shown in Fig. 7.5d:

$$\frac{d}{d_y} = \frac{P}{P_y} + \alpha\left(\frac{P}{P_y}\right)^r \tag{7.15}$$

where, d is the displacement of the metal energy absorber; d_y is the displacement value of the feature point; P is the load acting on the metal energy absorber; P_y is the load of the feature point;α is a positive constant coefficient; r is a positive odd number greater than 1.

The size of the periodic curve is twice that of the contour curve, which depicts the relationship between force and displacement. The area enclosed by the hysteresis curve between point (P_0, d_0) and point $(-P_0, -d_0)$ is the energy consumed in a cycle.

$$W_d = 4d_y P_y[(r-1)/(r+1)](P/P_y)^{r+1} \tag{7.16}$$

4. **Bouc-Wen Model**

In the past research work, Caughcy uscd bilinear hysteretic model to study the random vibration of the system, and some people used piecewise linear hysteretic model to analyze the system response. However, in the piecewise linear model, the abrupt change of stiffness is difficult to truly reflect the yield characteristics of the system. Therefore, researchers began to find a method to analyze hysteretic system with smooth curve model. In 1967, Bouc first proposed a simple smooth hysteretic model controlled by differential equations. In 1976, Wen et al. improved Bouc's model, generalized this model, and proved that this model can produce a series of different hysteretic curves. As shown in Fig. 7.5e, the model can be regarded as the superposition of elastic force model and hysteretic force model. Then the hysteretic restoring force of Bouc-Wen Model is equal to the sum of elastic force and hysteretic force:

$$f = \gamma kx + (1-\gamma)F_y z \tag{7.17}$$

where, k is the initial elastic stiffness; r is the ratio of stiffness after yield to the elastic stiffness; F_y is the yield strength; x is the displacement of the energy dissipation and shock absorption device; z is the internal variable reflecting the hysteretic effect, defined by the following differential equation proposed by Wen (1976):

$$\dot{z} = \frac{k}{F_y}\left\{1 - |z| \exp\left[asign(\dot{x}\, z) + \beta\right]\right\}\dot{x}, \ \mathrm{sign}(x) = \begin{cases} 0(x = 0) \\ -1(x < 0) \\ 1(x > 0) \end{cases} \tag{7.18}$$

among them, α, β are the parameters that determine the shape of hysteretic curve; exp is the parameters that determine the size of the transition interval of yield position curve; and $\dot{x}$ is the displacement velocity of the energy dissipation and damping device.

Equations (7.17) and (7.18) can represent the general curvilinear hysteretic nonlinear model, which has strong adaptability and contains nonlinear damping and nonlinear stiffness. Therefore, they can approximately describe all kinds of smooth hysteretic curves. The shape of the hysteresis curve is determined by the parameters α and β, and the smoothness of the curve, that is, the size of the transition interval, is determined by the constant exp. By adjusting these coefficients, different hysteresis loops can be obtained. By comparison, it can be found that the hysteretic restoring force of the system has soft or hard characteristics with the different values of α and β. The so-called soft characteristic is that the restoring force of the system decreases with the increase of the absolute value of displacement; the so-called hard characteristic is that the restoring force of the system increases with the increase of the absolute value of displacement. When the ratio of α/β is large, the area surrounded by the hysteretic restoring force curve of the system is large, and the curve shape is relatively full, which indicates that the system consumes more energy in the process of vibration, otherwise, the energy consumption is less. Bouc-Wen Model can simulate the hysteretic curve of any shape by different parameters.

7.2 Tension-Compression Type Metal Damper

The commonly used tension-compression type metal damper is the bucking restrained brace (BRB), which is a kind of braced energy dissipating member. It is composed of core energy dissipating member, external restraint member and unbonded filler. It can yield and consume energy in tension and compression. The structure of buckling brace is simple and diverse, the principle is clear, the hysteretic energy dissipation capacity is stable, and it has been widely used at home and abroad [1].

7.2.1 Working Mechanism of Buckling Proof Brace

The basic design idea of buckling proof brace is to make the core members bear only axial force, the external restraint sleeve does not bear axial force, only plays the role of preventing the core members from buckling, and the core energy dissipation parts and the external restraint sleeve are separated by unbonded materials or spaces. Generally speaking, the selection and evaluation of the performance of BRB device mainly consider energy consumption capacity, stiffness and stability.

1. **Stiffness**

The stiffness of BRB includes elastic stiffness and plastic stiffness.

① Elastic stiffness, which is the stiffness of BRB when the material is in the linear elastic stage. At this time, it is considered that the core has not buckled. According to the knowledge of material mechanics:

$$K_u^c = \frac{1}{\frac{1}{K_1} + 2\frac{1}{K_2} + 2\frac{1}{K_3}} = \frac{EA_1A_2A_3}{L_1A_2A_3 + 2L_2A_1A_3 + 2L_3A_1A_2} \tag{7.19}$$

among them, E is the elastic modulus of steel; L_1, L_2 and L_3 are the length of working section, transition section and connection section respectively; A_1, A_2 and A_3 are the sectional area of working section, transition section and connection section respectively; K_1, K_2 and K_3 are the elastic rigidity of working section, transition section and connection section respectively, K_u^c is the overall elastic rigidity of buckling proof support.

② Plastic stiffness, which is the stiffness of BRB when the material enters the nonlinear stage. According to the constitutive relation of bilinear model, $E_t = \alpha E$, therefore:

$$K_d^c = \frac{1}{\frac{L_1}{E_tA_1} + \frac{2L_2}{EA_2} + \frac{2L_3}{EA_3}} = \frac{1}{\frac{L_1}{\alpha EA_1} + \frac{2L_2}{EA_2} + \frac{2L_3}{EA_3}} \tag{7.20}$$

2. **Stability**

BRB is a compression member with large slenderness ratio, so stability checking calculation must be carried out in design, which can be divided into overall stability checking calculation and core component stability checking calculation. In this paper, the stress and instability mode of BRB are discussed with the example of "-" shape core.

The global instability of the component is represented by the global buckling of the component as a whole, including the peripheral buckling restraint mechanism and core components, which occurs in the plane with less rigidity. The peripheral buckling restraint mechanism and the core member are separated by unbonded materials. In the derivation, it can be approximately considered that the buckling deformation of the buckling restraint mechanism and the core member is the same when they are compressed, and there is no friction between them. It is considered that there is only transverse distribution force $q(x)$, as shown in Fig. 7.6a.

The equilibrium equation of core components is as follows:

$$E_1 I_1 \frac{d^4 y}{dx^4} + P \frac{d^2 y}{dx^2} = -q(x) \tag{7.21}$$

The equilibrium equation of buckling restraint mechanism is as follows:

$$E_2 I_2 \frac{d^4 y}{dx^2} = q(x) \tag{7.22}$$

where, $E_1 I_1$ and $E_2 I_2$ are the flexural rigidity of the core member and peripheral restraint member respectively; eliminating $q(x)$ can obtain the critical load: $P_{cr,g} = \frac{\pi^2 (E_1 I_1 + E_2 I_2)}{(\mu l)^2}$, because $E_1 I_1$ is far less than $E_2 I_2$, the above formula can be simplified as $P_{cr} = \frac{\pi^2 E_2 I_2}{(\mu l)^2}$. In order to ensure that the member does not lose whole stability and realize the compression yield of the whole section of the supporting core member, the critical Eulerian load P_{cr} of the buckling restraint mechanism must be greater than the yield load F_{y} of the core member, that is to say, the restraint ratio ζ is defined as

$$\zeta = \frac{P_{cr}}{F_y} = \frac{\pi^2 E_2 I_2 / (\mu l)^2}{F_y} \geq 1 \tag{7.23}$$

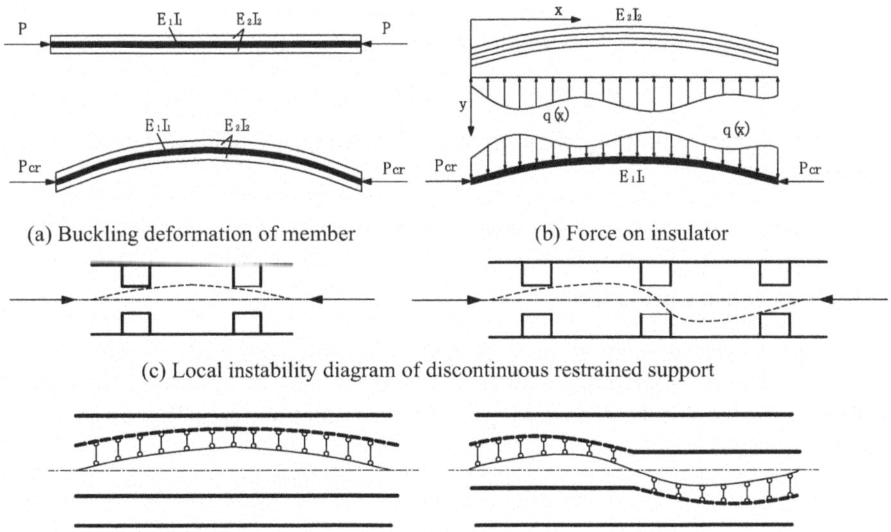

(a) Buckling deformation of member

(b) Force on insulator

(c) Local instability diagram of discontinuous restrained support

(d) Local instability diagram of continuous restrained support

Fig. 7.6 Stress and instability mode of buckling restrained brace

Considering the initial defect of the component that cannot be determined, it is suggested to take ζ as 1.5 in practical application. In addition to the overall instability, the core members of the buckling restrained braces may also have individual instability. For the support providing discontinuous constraint conditions, such as transverse stiffener, the section between the transverse stiffeners is the constraint weak area, which is prone to local instability, as shown in Fig. 7.6b. For the support with continuous constraint conditions provided by the longitudinal constraint type, when the lateral elastic support stiffness is insufficient or the clearance is too large, the instability of the core member due to the lack of effective constraint may occur, and the failure form is shown as single or multiple half wave buckling, as shown in Fig. 7.6c.

For discontinuous restrained supports, such as transverse stiffeners, the occurrence of local instability is related to the spacing of transverse stiffeners. The larger the spacing is, the lower the density of lateral support is, and the easier the local instability is. The lateral restraint spacing should be controlled reasonably to prevent the local instability of core components. The stability of the continuous constrained support can be calculated by simple mathematical method. The instability mode of longitudinal constraint is the instability of axial compression bar with two hinged ends under continuous elastic constraint, and its equilibrium equation is as follows:

$$E_1 I_1 \frac{d^4 y}{dx^4} + P\frac{d^2 y}{dx^2} + \beta y = 0 \tag{7.24}$$

where, β is the distributed elastic constant, and $\frac{P}{E_1 I_1 + E_2 I_2} = k^2$, the expression is $\beta = \frac{\pi^2 E_2 I_2 k^2}{l^2}$; solve the differential equation to get the buckling load of each order,

$$P_{cr,j} = n^2\pi^2 E_1 I_1/l^2 + \beta l^2/(n^2\pi^2) \tag{7.25}$$

Therefore, $P_{cr,j,\min} = 2\sqrt{\beta E_1 I_1}$, in order to avoid the instability of the core members and ensure that the yield is prior to the buckling, it should meet the following requirements:

$$P_{cr,j} \geq F_y \tag{7.26}$$

3. **Energy consumption index**

According to the regulation of FEMA356 on prototype test of damper, the performance indexes of damper mainly include secant stiffness K_{eff}, equivalent damping ratio ζ_{a} and hysteretic loop energy consumption area W_{c}, as shown in Fig. 7.7.

Secant stiffness K_{eff} and ζ_{a} are determined according to the following formula:

$$K_{eff} = \frac{|F^-| + |F^+|}{|D^-| + |D^+|} \tag{7.27}$$

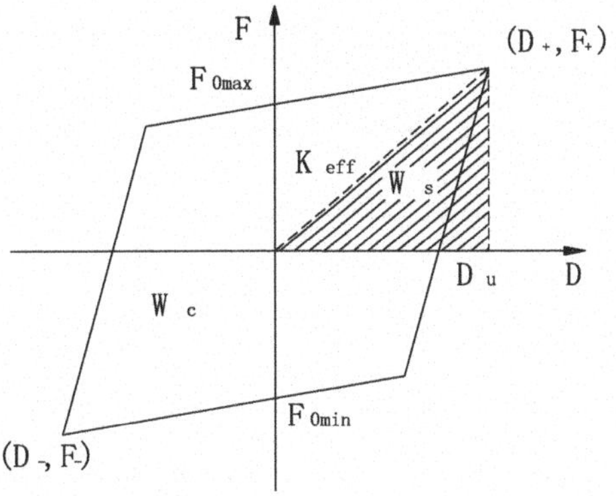

Fig. 7.7 Schematic diagram of energy consumption indexes F_{0max}, F_{0min}, W_c and K_{eff}

$$\zeta_a = \frac{W_c}{4\pi \cdot W_s} \tag{7.28}$$

W_c is the energy consumed by the energy dissipating member during one cycle of reciprocating under the expected displacement, which is equal to the area surrounded by each hysteretic curve, i.e.

$$W_c = 4F_y(1-\alpha)D_y(\mu - 1) \tag{7.29}$$

among them, F_y is the yield load of the energy dissipating member; D_y is the yield displacement of the energy dissipating member; displacement ratio μ is the ratio of the maximum displacement D_{max} to the yield displacement D_y of the energy dissipating member; stiffness ratio α is the ratio of the plastic stiffness to the elastic stiffness of the energy dissipating member. In practical use, the above performance parameters should be obtained through tests.

7.2.2 *Research and Development of New Buckling Proof Support*

7.2.2.1 Cross Core Restrained Buckling Support

The cross core restrained buckling brace designed by the author's team is shown in Fig. 7.8a [1, 2]. The cross core steel rod made of Q235 steel is used as the energy dissipation member, and the outer circular steel tube provides lateral restraint for it. Among them, the cross core steel rod is 1000 mm long, and the four plate legs are 40 mm × 40 mm long. In the middle of each plate limb, there is a segment of arc-shaped defect, with the arc length of 100 mm and the arc depth of 10 mm. There is no weld on the plate limb at the defect. The outer round steel pipe is 900 mm long, 90 mm inner diameter and 6 mm wall thickness. The gap between the cross core steel bar and the outer round steel pipe is filled and isolated by the polytetrafluoroethylene strip

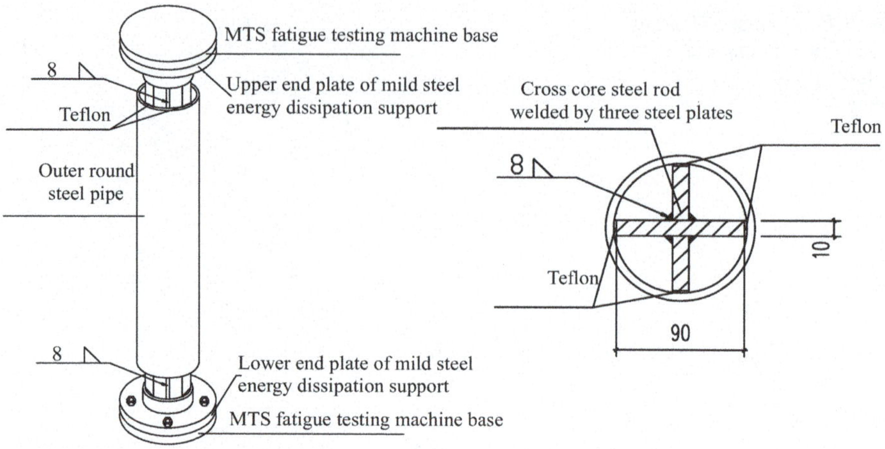

(a) Damper design

(b) Damper test

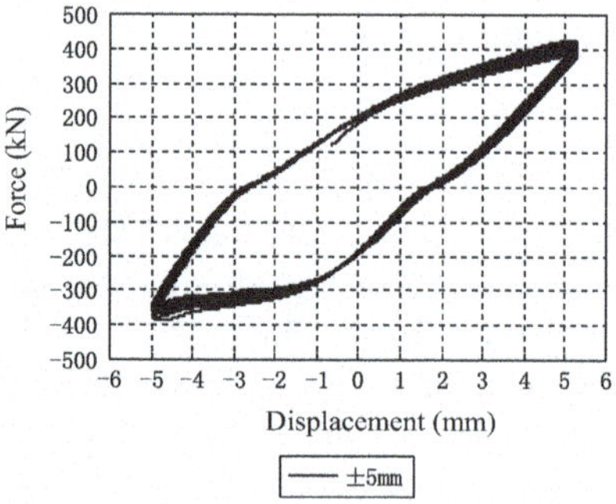

(c) Force displacement curve

Fig. 7.8 Cross core restrained buckling brace

with a thickness of 5 mm. After processing the specimens, the mechanical properties of restrained buckling brace were tested by the MTS fatigue testing machine of Southeast University laboratory (Fig. 7.8b). Figure 7.8c shows the force displacement curve of the restrained buckling support when the loading displacement is ± 5 mm. Combined with other test results, it can be seen that: (1) with the gradual increase of the load, the hysteretic curve is gradually full and has a stable hysteretic loop, but because the steel used is Q235 and the hysteretic curve is spindle shaped, the energy consumption capacity is relatively poor; (2) the mild steel energy dissipation support damper does not lose stability. The design of the outer circular steel tube can solve the stability problem of the damper under compression.

7.2.2.2 "—" Shaped Core Restrained Buckling Support

The "—" shaped cross rib restrained buckling support designed by the author's team is shown in Fig. 7.9a. Its core component is "—" shaped, which is the only difference compared with the cross core restrained buckling support, which is made of LYP160 low yield point steel [3]. The steel sleeve and stiffener tube are made of Q235 steel, and the unbonded material between the core component and the stiffener tube is made of 1 mm thick polytetrafluoroethylene. Among them, the longitudinal and transverse stiffeners are studied respectively, and the longitudinal stiffeners are selected to be used finally. Seven samples of different sizes were made, and the mechanical properties and fatigue performance tests were carried out (Fig. 7.9b). Figure 7.9c shows the force displacement curve of sample 6. Combined with the test results of other samples, it can be seen that: (1) the hysteretic curve of the damper designed is full, no obvious pinching phenomenon is found, the maximum compressive bearing capacity of sample 6 is 409.9 kN, the yield load is 286.5 kN, and the yield displacement is 1.40 mm, which is in good agreement with the theoretical results; (2) the sample experiences 30 cycles cyclic loading and 15 cycles of rapid addition of 16 mm. After loading, there is no obvious attenuation of energy consumption index under 24 mm loading amplitude, so it has better fatigue performance. The shape of the supporting core component after the test is shown in Fig. 7.9d, showing wave shape; the shape of the supporting core component after the failure is shown in Fig. 7.9e. Table 7.8 gives some performance parameters of the buckling proof brace designed by the author's research group for readers' understanding and design reference.

7.2.2.3 Double Steel Tube Restrained Buckling Support

In order to overcome the inherent defects of conventional buckling proof support, such as significant self-weight and easy to occur overall instability, a new type of double circular steel tube buckling proof energy absorber is proposed. The energy absorber is composed of outer constraint tube, core tube and inner constraint tube from outer layer to inner layer, as shown in Fig. 7.10a, b. The inner debonding layer is arranged between the core tube and the inner constraint tube, and the outer debonding layer is arranged between the core tube and the outer constraint tube; the core tube, the inner constraint tube and the outer constraint tube are connected together with pull bolts; the two ends of the outer constraint tube are respectively set with a compression sleeve, and the ends of the outer constraint tube and the compression sleeve are slotted, and the stiffening plate is located in the groove, and the connecting plates are fixed at both ends of the core tube and the stiffening plate. The connecting plate is connected with the main structure through bolts. Compared with the common "—" shaped and cross type buckling restrained brace, the main characteristics of the buckling restrained brace with double circular steel tube constraint are as follows: (1) it is a simple buckling restrained brace with double circular steel tube constraint. The buckling restrained brace with assembly type double circular steel tube constraint

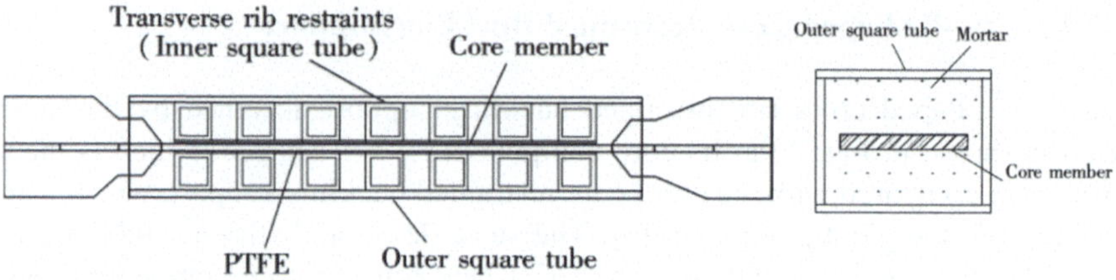

(a) Damper design

(b) Damper test

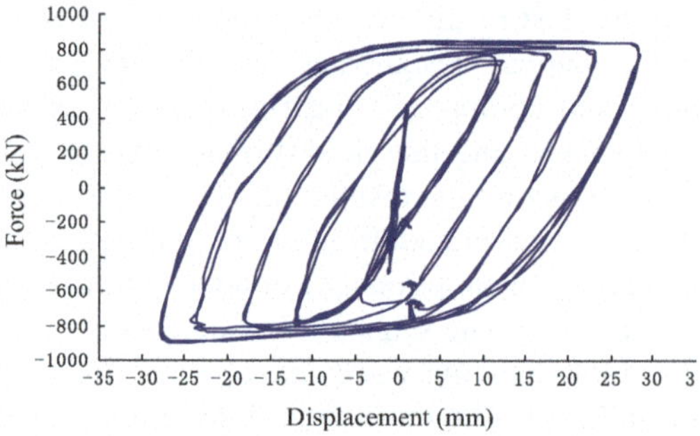

(c) Force displacement curve

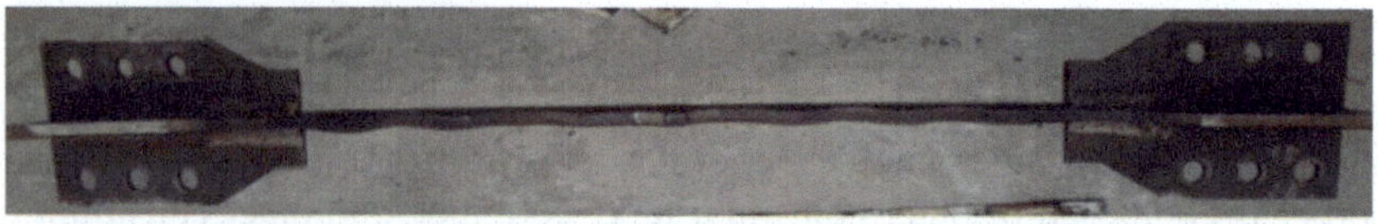

(d) Post loading shape of the core members of buckling proof brace

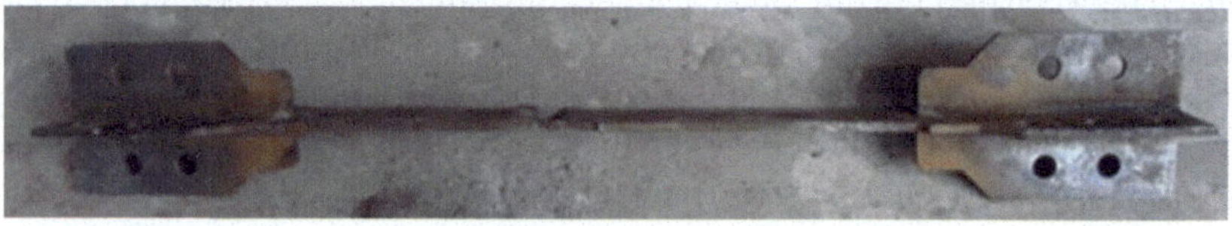

(e) Failure mode of local buckling of core members of buckling proof braces

Fig. 7.9 Restrained buckling brace with "—" shaped core transverse rib

Table 7.8 Performance parameters of buckling proof brace

Section size	Length/mm	Core material	Initial stiffness kN/mm	Yield displacement/mm	Yield load/kN
90 × 10	1200	Q235	160	≤1.2	220
100 × 12	1500		210	≤1.5	280
80 × 8	1150	BLY160	120	≤0.7	120
160 × 20	2700		210	≤1.3	500
220 × 20			430		700

is formed by the combination of the core steel tube and the constrained steel tube, which is easier to manufacture and solves the problem of long curing period brought by the filling of concrete mortar of traditional buckling restrained brace, and reduces the self-weight of the component. (2) circular steel pipe is used as the core part of the component, including the energy dissipation pipe and the restraint pipe, and the circular section will not have the overall buckling in the form of torsion, bending, bending and torsion. According to the current specifications, the local buckling conditions of the circular tube are easy to meet in the design. Due to the balanced performance of the section characteristics, the mechanical performance is more stable than the traditional "—" shaped and cross Sects. 7.3) the components are all steel components, which are made by machining. The clearance of the debonding layer between the core circular pipe and the internal and external constraint pipes can be controlled within the design range, and the machining accuracy is easier to control. At the same time, the discreteness of the steel is smaller, the performance is more stable and reliable, and the mechanical performance is more pure.

Based on this, three samples were designed and tested in the laboratory of university enterprise joint research center of Southeast University. The test load is shown in Fig. 7.10c. In the test, the quasi-static test loading method was adopted, and the 200t fatigue test machine was used for loading, and the loading was controlled by the digital controller and system software according to the preset loading history. In this test, displacement loading control was used to explore the yield displacement of the component. At the initial stage of loading, according to the theoretical calculation at the early stage, the yield displacement of the component under the ideal state was used to set a displacement value close to the yield displacement but smaller for one-way cycle, observed whether to yield, and gradually increased the displacement until the component yields. After finding out the yield displacement, the subsequent cyclic loading process was carried out by displacement control. According to the loading system of SEAOC-AISC, six cycles of one time of yield displacement, four cycles of two times of yield displacement, four cycles of four times of yield displacement, two cycles of six times of yield displacement and three cycles of four times of yield displacement were carried out successively. In order to ensure the uniformity and continuity of repeated loading, the loading and unloading rates should be consistent in the test. After the above test, if the component did not fail, it should be cycled

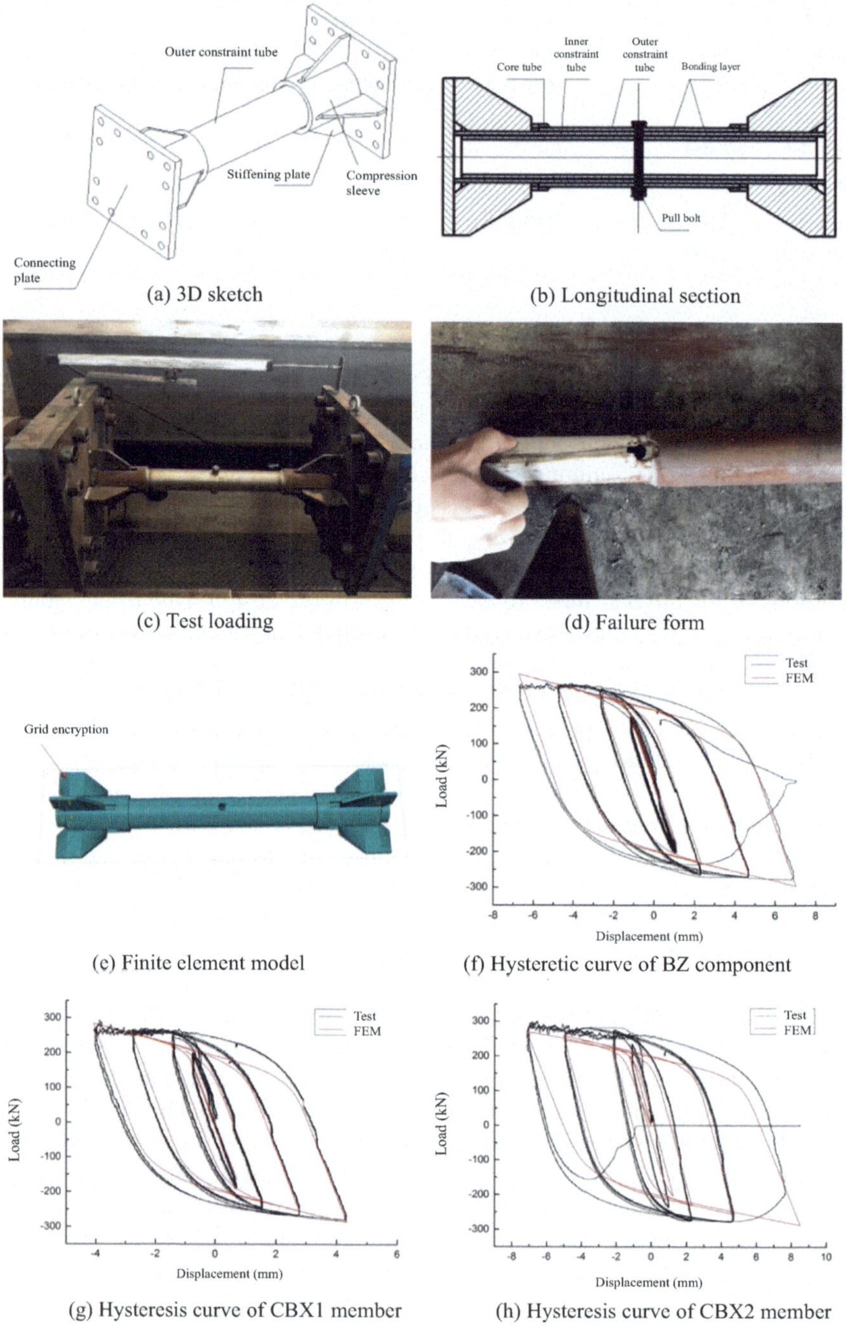

Fig. 7.10 Restrained buckling brace of double steel tube

Table 7.9 Quasi static test loading system

Load displacement amplitude	Cycle times	Loading rate (mm/s)
One time yield displacement of member	6	0.02
Twice yield displacement of member	4	0.02
Quadruple yield displacement of member	4	0.02
Six times yield displacement of member	2	0.02
Quadruple yield displacement of member	3	0.02
Quadruple yield displacement of member	Until failure	0.02

with four times of yield displacement until the component failed. That is to say, the whole loading process was divided into two parts: variable amplitude low cycle reciprocating loading part and constant amplitude low cycle reciprocating loading part. See Table 7.9 for the specific loading system, and Fig. 3.10 is the schematic diagram of variable amplitude low cycle reciprocating loading.

The three members are in tension state when they are damaged, and the core tube has no buckling deformation. In order to facilitate the observation of the failure mode of components, the wire cutting is used to separate the external constraints from the core pipe. See Fig. 7.10d for the failure mode obtained by cutting. It can be seen that the failure of the component is due to the tension of the core pipe, and the failure position is at the hole of the connecting bolt, which is basically consistent with the calculation result of the finite element model (Fig. 7.10e), indicating that it is effective to prevent the component from bending by using the internal and external constraints, that is, the gap between the constraint pipe and the core pipe is very small. At the same time, the internal and external constraint steel pipes provide enough restraint strength, which not only ensures that the member will not buckle as a whole, but also limits the amplitude of multi wave buckling of the core tube to not be too large, so that the core tube will not buckle. On the other hand, it is also reasonable to have tensile failure at the connection bolt, because the compressive strength is guaranteed after the core pipe is constrained, and the section at the connection bolt is weakened due to drilling, when the tensile force reaches a certain degree, the section will become a weak section to take the lead in tensile failure.

Figure 7.10f–h shows the force displacement curve of three test pieces obtained by the test and numerical simulation. It can be seen from the figure that: (1) the shapes of the hysteretic curve of three members of test are similar to that of the FEM, and the rigidities of loading and unloading of each cyclic are basically the same; (2) compared with the result of the finite element simulation, the yield strength of the test is larger, the hysteretic curve is fuller, and the actual energy consumption performance is better, and when the slenderness ratio is small, the matching degree of the two will be better with the increase of slenderness ratio, but the increase of slenderness ratio will make the matching degree worse after exceeding a certain limit value, which shows that there is an optimal slenderness ratio that makes the actual situation approximately consistent with the theoretical situation; (3) the six times yield displacement is the biggest difference between the test and the finite element simulation. In this displacement, two test members are damaged in tension. The finite element results show that the members still have stable bearing capacity and full hysteretic curve at this time, which also shows that the energy consumption performance of the members in the actual situation has a great relationship with the slenderness ratio.

7.3 Shear Type Metal Damper

7.3.1 Stress Mechanism of Unconstrained Shear Steel Plate

In order to effectively restrain the out of plane buckling of the shear plate, it is necessary to master the stress mechanism of the non-buckling restrained plate, the elastic buckling and ultimate shear strength of the web of the shear beam can be used for reference in the analysis of. According to the theory of shear plate, the failure of non-buckling restrained plate experiences elastic buckling stage and diagonal tension band stage. The stress in the direction of the main compressive stress remains unchanged after the plate buckling, and the increased external force load is mainly balanced by the formation of tension band. Therefore, the shear plate needs a strong restraint around to play a greater post buckling strength.

The upper and lower ends of the sheared steel plate are fixed on the end plate. When encountering the earthquake, the structure will have interlayer displacement, and the steel plate will also have deformation. The steel plate is similar to the frame column fixed at both ends from the stress form and deformation characteristics, bearing the joint action of shear and bending moment [4]. It can be seen from the structural mechanics that when the height width ratio h/w of the slab is large, the bending moment plays a controlling role, and the two ends of the slab first yield in tension and compression; when the height width ratio h/w of the slab is small, the shear force plays a controlling role, and the middle of the slab firstly yield in shear, and the stress diagram of the sheared slab is shown in Fig. 7.11.

The bending moment at the top of the steel plate is:

$$M = F\frac{h}{2} \tag{7.30}$$

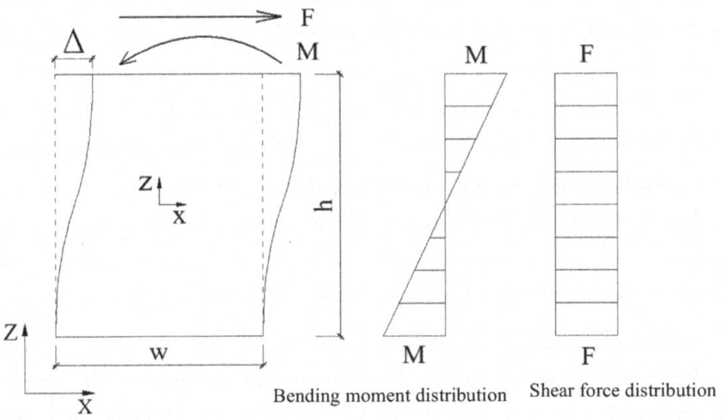

Fig. 7.11 Stress diagram of shear plate

Therefore, the maximum compressive (tensile) stress is:

$$\sigma = \frac{M}{W} = \frac{\frac{Fh}{2}}{\frac{tw^2}{6}} = \frac{3Fh}{tw^2} \tag{7.31}$$

When the compressive (tensile) stress reaches the yield strength, $\sigma = f_y$, namely:

$$\sigma = \frac{3Fh}{tw^2} = f_y \tag{7.32}$$

Then, the shear force corresponding to compression (tension) yield is:

$$F = \frac{tw^2 f_y}{3h} \tag{7.33}$$

The maximum shear stress of the shear plate is:

$$\tau = \frac{FS^*}{It} = \frac{Ft\frac{w^2}{8}}{\frac{t^2w^3}{12}} = \frac{3F}{2tw} \tag{7.34}$$

When the shear stress reaches the yield strength, $\tau = \frac{f_y}{\sqrt{3}}$, namely:

$$\tau = \frac{3F}{2tw} = \frac{f_y}{\sqrt{3}} \tag{7.35}$$

Therefore, the shear force at shear yield is:

$$F = \frac{2twf_y}{3\sqrt{3}} \tag{7.36}$$

When the compressive stress and shear stress reach yield at the same time:

$$F = \frac{tw^2 f_y}{3h} = \frac{2tw f_y}{3\sqrt{3}} \tag{7.37}$$

Therefore, the boundary height width ratio is:

$$\frac{h}{w} = \frac{\sqrt{3}}{2} \tag{7.38}$$

The deformation of steel plate consists of two parts: shear deformation and bending deformation. With the increase of the ratio of height to width, the proportion of shear deformation in the total deformation becomes smaller and smaller. The change of the ratio of height to width has an effect on the shear deformation of the steel plate.

The bending deformation of the sheared steel plate is as follows:

$$\begin{aligned}\Delta_M &= \int \frac{\overline{M}(y) M_P(y)}{EI(y)} dy = \frac{2}{EI}\left(\frac{1}{2} \times F \cdot \frac{h}{2} \cdot \frac{h}{2} \times \frac{2}{3} \times \frac{h}{2}\right) \\ &= \frac{Fh^3}{12EI} = \frac{Fh^3}{Ew^3 t}\end{aligned} \tag{7.39}$$

The shear deformation of the sheared steel plate is as follows:

$$\Delta_V = \int \overline{V}(y) d\eta = \int \frac{f_S \overline{V}(y) V_P(y)}{GA(y)} dy = \frac{f_S Fh}{GA} \tag{7.40}$$

where, f_S is the coefficient of non-uniformity of shear stress distribution, taking 1.2 for rectangular section. For steel, Poisson's ratio $\nu = 0.3$, $G = \frac{E}{2(1+\nu)} \approx 0.4E$. So shear deformation is:

$$\Delta_V = \frac{1.2Fh}{0.4Ewt} = \frac{3Fh}{Ewt} \tag{7.41}$$

The ratio of shear deformation to total deformation is:

$$\frac{\Delta_V}{\Delta} = \frac{\Delta_V}{\Delta_M + \Delta_V} = \frac{\frac{3Fh}{Ewt}}{\frac{Fh^3}{Ew^3 t} + \frac{3Fh}{Ewt}} = \frac{3}{\frac{h^2}{w^2} + 3} \tag{7.42}$$

When the ratio of height to width $\frac{h}{w} = 5$, $\frac{\Delta_V}{\Delta} = 10.71\%$, it can be seen that the proportion of shear deformation in the total deformation is very small, which can be ignored. According to the above analysis, the sheared steel plates can be classified according to the height width ratio: (1) when $h/w < \sqrt{3}/2$, the shear yield is prior to the tension compression yield, it is defined as "shear type steel plate";

(2) when $\sqrt{3}/2 \leq h/w \leq 5$, the tension compression yield is prior to the shear yield, but the shear deformation still accounts for a large proportion in the total deformation, it is still necessary to consider the influence of bending deformation and shear deformation simultaneously, which is defined as "bending shear type steel plate"; (3) when $h/w > 5$, the shear deformation has a small proportion in the total deformation, which can be ignored. Only considering the influence of bending deformation, it is defined as "bending steel plate".

From the above analysis, it can be seen that the mechanical properties of unrestricted shear steel plate are closely related to its height width ratio, height thickness ratio, width thickness ratio and other parameters.

1. **The effect of height width ratio on the mechanical properties of sheared steel plate**

With the increase of the ratio of height to width, the buckling shape of the steel plate changes from multiwave buckling to half wave buckling. The buckling shape of the shear steel plate is bi-directional bending at the edge and out of plane bulging at the middle. The buckling shape of the bending shear steel plate and the bending steel plate is two-way bending deformation of the edge.

With the increase of the ratio of height to width, the shear plates gradually change from elastic-plastic buckling to elastic buckling. At the same time, the effect of buckling on the bearing capacity is more and more significant. When the height width ratio $h/w = 0.5$, the bearing capacity curve shows an ideal elastic-plastic characteristic macroscopically, and the effect of buckling is negligible. With the increase of the ratio of height to width, the elastic stiffness and bearing capacity of the shear steel plate decrease gradually. When the ratio of height to width is large, the ultimate bearing capacity of the shear steel plate will generally decline after reaching the peak value.

2. **The effect of high thickness ratio on the mechanical properties of sheared steel plate**

When the height thickness ratio is small, the out of plane deformation of the sheared steel plate is very large; with the increase of the height thickness ratio, the out of plane deformation decreases gradually; when the height thickness ratio is large, there is almost no out of plane deformation of the sheared steel plate. According to the change rule of out of plane deformation with thickness, the sheared steel plates can be classified: (a) the sheared steel plates with great out of plane deformation are classified as "thin plates" with the increase of thickness, and the buckling occurs in the elastic-plastic stage; (b) the sheared steel plates with decreasing out of plane deformation with the increase of thickness are classified as "medium thick plates", and the buckling occurs in the elastic-plastic stage; (c) the sheared steel plates with small out of plane deformation with the increase of thickness are classified as "thick plates", and the shear buckling is not prior to the shear yield.

With the decrease of the ratio of height to thickness, the bearing capacity curve changes from elastic buckling to elastic-plastic buckling. At last, it shows ideal elastic-plastic characteristics. At the same time, the bearing capacity of steel plate is significantly improved, and the elastic stiffness is gradually increased, which shows that reducing the height thickness ratio is an effective way to improve the bearing capacity of steel plate.

3. **The influence of width thickness ratio on the mechanical properties of sheared steel plate**

With the increase of width thickness ratio, the bearing capacity curve changes from elastic-plastic buckling to elastic buckling. The buckling effect of steel plate is more and more obvious, and the bearing capacity increases gradually. When the width thickness ratio reaches about 12, the bearing capacity curve rises obviously, which means that the bearing capacity of the steel plate increases gradually with the displacement loading. When the width thickness ratio reaches 25, the bearing capacity curve shows the ideal elastic-plastic characteristics macroscopically. Then, with the increase of the width thickness ratio, the ultimate bearing capacity of the steel plate reaches the peak value, and there is generally a declining section.

4. **Comparison of the influences of high thickness ratio and width thickness ratio on the mechanical properties of sheared steel plat**

The ratio of height to thickness and the ratio of width to thickness are both important factors affecting the basic mechanical properties of steel plates. No matter bending shear steel plate or shear steel plate, the peak value of bearing capacity changes more with the width thickness ratio than with the height thickness ratio, which shows that the width thickness ratio has a more obvious impact on the bearing capacity of steel plate, so it is more effective to improve the bearing capacity of steel plate by changing the width thickness ratio in engineering practice.

The change range of out of plane deformation with the ratio of width to thickness is larger than that with the ratio of height to thickness, which shows that the ratio of width to thickness has a more obvious effect on out of plane deformation of bending shear steel. The change of out of plane deformation of shear steel plate with the ratio of height to thickness is larger than that with the ratio of width to thickness, which shows that the ratio of height to thickness has more obvious influence on out of plane deformation of shear steel plate.

5. **Influence of size on mechanical properties of sheared steel plate**

With the increase of the scale factor of the sheared plate size parameter, the influence of the plate buckling on the bearing capacity becomes less obvious. When the ratio coefficient is small, the curve has a significant decline after reaching the peak bearing capacity. When the ratio coefficient is large, the curve shows ideal elastic-plastic characteristics. At the same time, with the increase of steel plate size parameters, the elastic stiffness and bearing capacity of steel plate are significantly improved.

Whether it is shear steel plate or bending shear steel plate, the peak value of bearing capacity increases rapidly with the increase of size parameters, but the growth rate slows down gradually. So increasing the size of steel plate is another very effective way to improve its bearing capacity.

7.3.2 Buckling Proof Design of in-Plane Shear Yield Type Energy Dissipation Steel Plate

For the weakness of in-plane sheared yield type energy dissipation plate (sheared steel plate or sheared energy dissipation plate) prone to buckling, it is necessary to find a structural measure that can continuously provide buckling constraints for the shear plate to improve the mechanical properties of the shear plate. The restraint measures should be simple in construction, convenient in processing, clear in stress mechanism, small in influence on energy consumption performance of restrained steel plate, stable in restraint effect and material saving.

7.3.2.1 Buckling Proof Scheme Design

It is known that the buckling form of the bending shear steel plate is "half wave buckling, two-way bending deformation at the edge", and that of the shear steel plate is "multi wave buckling, two-way bending deformation at the edge, and out of plane bulging deformation at the middle". The traditional anti buckling measure is to weld stiffeners directly on the sheared steel plate, but the welding is easy to cause residual stress and residual deformation of the sheared steel plate, which will have an adverse impact on the energy consumption performance of the sheared steel plate (such as fatigue resistance, etc.), so a number of new anti buckling schemes are proposed, as shown in Fig. 7.12.

1. **Anti buckling scheme of frame**

At present, the flange part of in-plane shear yield type damper is usually made of ordinary steel, only considering the energy consumption of web part. Now soft steel material is used instead of ordinary steel as the flange of the sheared steel plate, and the "anti buckling scheme of frame" is proposed, which is mainly composed of one sheared steel plate, two end plates and two flange plates, and its structural form is shown in Fig. 7.12a. Because the flange is made of mild steel, on the one hand, it can restrain the deformation of the sheared steel plate and delay the buckling of steel plate; on the other hand, it can be used as an energy dissipation member to improve the energy dissipation capacity of the energy absorber. The flange is equivalent to the frame of the sheared steel plate, so that the boundary condition of the steel plate is fixed on four sides. After the buckling deformation of the steel plate, a tension band is formed along the diagonal direction, and the flange can provide a reliable anchoring effect, thus significantly improving the bearing capacity and energy dissipation capacity of the steel plate.

2. **Anti buckling scheme of splint**

When the size of the sheared steel plate is large, the restraint effect of the buckling prevention scheme of the frame is not ideal, such as the steel plate with the size of $w = 2000$ mm, $h = 2000$ mm, $t = 8$ mm. When sheared, the steel plate has a large out of

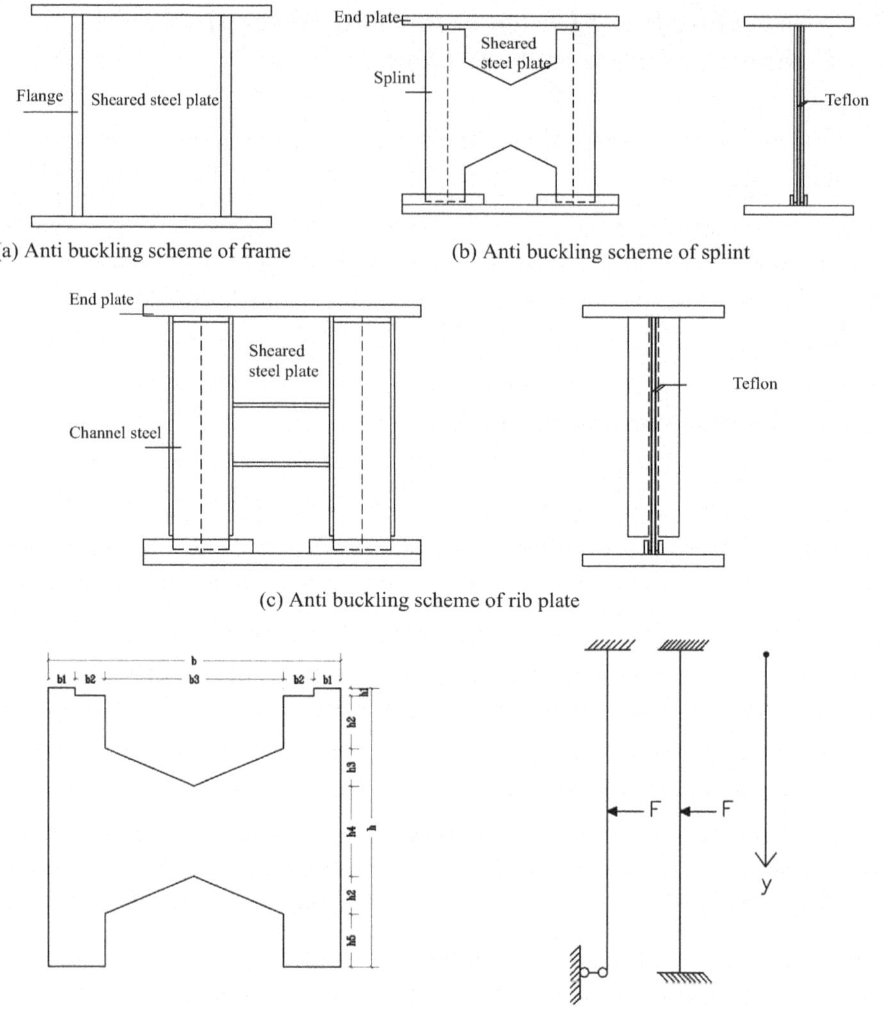

(a) Anti buckling scheme of frame (b) Anti buckling scheme of splint

(c) Anti buckling scheme of rib plate

(d) Dimension sketch of splint (e) Calculation diagram of constraint stiffness ratio

Fig. 7.12 Anti buckling scheme

plane deformation, the hysteresis curve has obvious pinching phenomenon, and the energy consumption capacity is poor. Now, from the idea of effectively restraining the out of plane deformation of the sheared steel plate, the "anti buckling scheme of splint" is proposed, which is mainly composed of one sheared steel plate, two end plates, two splints, four guide channel plates and polytetrafluoroethylene. The structural form is shown in Fig. 7.12b.

The upper end of the splint is welded with the end plate, the lower end is free, and the guide groove plate is used to limit its out of plane displacement, so that the influence of the splint on the in-plane stress of the sheared steel plate can be eliminated, and the out of plane deformation of the sheared steel plate can be restrained continuously and effectively. If the thickness of the splint is large and it has enough out of plane stiffness, there is no stiffener on the outside of the splint. If the thickness of the splint is small and the out of plane deformation of the steel plate cannot be restrained effectively, the stiffener can be set to increase the out of plane stiffness of the splint. The polytetrafluoroethylene isolation layer is set at the contact surface of the splint and the steel plate to reduce the influence of friction on the mechanical properties of the steel plate.

The anti-buckling scheme of splint can effectively restrain the out of plane deformation of the sheared steel plate, change the distribution of the main tensile stress and the main compressive stress of the sheared steel plate, weaken the role of the tension band, and change the energy dissipation mechanism of the out of plane buckling into that of the plane shearing, so as to improve the bearing capacity of the sheared steel plate and the mechanical properties of the sheared steel plate.

3. **Anti buckling scheme of rib plate**

When the size of the sheared steel plate is large, in order to have a good restraining effect, the splint must have enough thickness in the anti-buckling scheme of the splint, resulting in a large amount of steel, or the method of welding stiffeners to improve the out of plane stiffness of the restrained splint. At present, the channel steel with large out of plane stiffness is used as the restraint member instead of welding stiffener on the splint, and the "rib buckling prevention scheme" is proposed, which is mainly composed of one shear steel plate, two end plates, six channel steel, four guide channel plates and polytetrafluoroethylene, and its structure is shown in Fig. 7.12c.

On the one hand, the flange of channel steel is used as a stiffener to effectively restrain the out of plane deformation of the sheared steel plate, and on the other hand, the anti-torsional moment is formed to ensure the stability of the device. The upper end of channel steel is welded with the end plate, the lower end is free, and the guide channel plate is used to limit its out of plane displacement, which can not only eliminate the influence of channel steel participating in the forced state of sheared steel plate, but also effectively restrain the out of plane deformation of steel plate. The polytetrafluoroethylene isolation layer is set between the channel steel and the steel plate to reduce the influence of friction on the mechanical properties of the steel plate.

7.3.2.2 Restraint Stiffness Ratio of Buckling Prevention Scheme and Sheared Plate

In order to achieve the design principle of "out of plane buckling energy dissipation mechanism of steel plate changes to plane shear energy dissipation mechanism", the anti-buckling scheme must be able to provide sufficient lateral support, so as to effectively restrain the out of plane deformation of steel plate, improve the mechanical properties of steel plate, and ensure that the shear yield of the steel plate must be prior to the overall elastic buckling of the member. The ratio of the lateral stiffness of the buckling prevention scheme to the lateral stiffness of the shear plate is defined as the constraint stiffness ratio β. Taking the buckling prevention scheme of bending shear type splint as an example, the specific expression and recommended value range are derived.

For the same buckling shape of sheared steel plate, there is no significant difference between the restrained area of the anti-buckling plan of the rib and that of the anti-buckling plan of the splint, but the flexural rigidity of the rib is greater than that of the splint, so when the anti-buckling plan of the splint meets the requirements of the constraint rigidity ratio, the anti-buckling plan of the rib can also meet the requirements of constraint rigidity ratio.

1. **Restrained stiffness ratio of buckling prevention scheme of bending shear type splint**

The plane form of the splint in the anti-buckling scheme of the splint is determined according to the buckling form of the sheared steel plate, as shown in Fig. 7.12d, in which $b, b_1, b_2, b_3, h, h_1, h_2, h_3, h_4, h_5$ can be determined, the only parameter to be determined is the thickness t of the splint. It can be seen from the structural measures of the buckling prevention scheme that the upper and lower ends of the sheared steel plate are consolidated, the upper end of the splint is consolidated, and the lower end is hinged. The mechanical simplified model of the expression of its constraint stiffness ratio is shown in Fig. 7.12e.

The out of plane displacement of the sheared steel plate under the action of unit force is: $\Delta_1 = \frac{4}{EI}\left(\frac{1}{2} \times \frac{h}{4} \times \frac{Fh}{8} \times \frac{2}{3} \times \frac{h}{8}\right) = \frac{Fh^3}{192EI}$.

The lateral stiffness of the sheared steel plate is $I_1 = \frac{wt_1^3}{12}$, substitute it into the above formula to get: $\Delta_1 = \frac{Fh^3}{16Ewt_1^3}$.

Therefore, the out of plane lateral rigidity of the sheared steel plate is: $K_1 = \frac{16Ewt_1^3}{h^3}$.

The out of plane displacement of the splint under the action of unit force is: $\Delta_2 = \int \frac{\overline{M}(y)M_P(y)}{EI(y)}dy$.

Among them, the lateral stiffness of the splint is:

$$I_2 = \begin{cases} \frac{b_1t^3}{6} & 0 \le y \le h_1 \\ \frac{(b_1+b_2)t^3}{6} & h_1 < y \le h_1 + h_2, \\ \frac{b_3t^3}{12h_3}(y - h_1 - h_2 - h_3) + \frac{bt^3}{12} & h_1 + h_2 < y \le h_1 + h_2 + h_3 \\ \frac{bt^3}{12} & h_1 + h_2 + h_3 < y \le h_1 + h_2 + h_3 + h_4 \\ -\frac{b_3t^3}{12h_3}(y - h_1 - h_2 - h_3 - h_4) + \frac{bt^3}{12} & h_1 + h_2 + h_3 + h_4 < y \le h_1 + h_2 + 2h_3 + h_4 \end{cases} \tag{7.43}$$

Then,

$$K_2 = 1/\{\frac{27h^2h_1}{128Eb_1t^3} + \frac{3h_2(3h - 11h_1 - 11h_2)^2}{128E(b_1 + b_2)t^3} + \frac{121h_2^3}{128E(b_1 + b_2)t^3}$$

$$+\frac{125h^3}{5632Ebt^3}+\int\limits_{h_1+h_2}^{h_1+h_2+h_3}\frac{121y^2(3h-11h_1-11h_2)^2}{192Eh^2t^3[\frac{b_3}{h_3}(y-h_1-h_2-h_3)+b]}dy+\frac{25h_5^3}{128E(b_1+b_2)t^3}$$
$$+\frac{75h_4(h_3+h_5)^2}{64Ebt^3}+\frac{h_4(5h-10h_3-10h_5)^2}{8Ebt^3}$$
$$+\int_{h-h_3-h_5}^{h-h_5}\frac{75(h-y)^2}{64Et^3[-\frac{b_3}{h_3}(y-h_1-h_2-h_3-h_4)+b]}dy\} \tag{7.44}$$

The out of plane lateral stiffness of the splint is obtained:

$$K_2=1/\{\frac{27h^2h_1}{128Eb_1t^3}+\frac{3h_2(3h-11h_1-11h_2)^2}{128E(b_1+b_2)t^3}+\frac{121h_2^3}{128E(b_1+b_2)t^3}+\frac{125h^3}{5632Ebt^3}$$
$$+\int\limits_{h_1+h_2}^{h_1+h_2+h_3}\frac{121y^2(3h-11h_1-11h_2)^2}{192Eh^2t^3[\frac{b_3}{h_3}(y-h_1-h_2-h_3)+b]}dy+\frac{25h_5^3}{128E(b_1+b_2)t^3}+\frac{75h_4(h_3+h_5)^2}{64Ebt^3}$$
$$+\frac{h_4(5h-10h_3-10h_5)^2}{8Ebt^3}+\int\limits_{h-h_3-h_5}^{h-h_5}\frac{75(h-y)^2}{64Et^3[-\frac{b_3}{h_3}(y-h_1-h_2-h_3-h_4)+b]}dy\} \tag{7.45}$$

According to the definition, the restrained stiffness ratio of the buckling prevention scheme of bending shear type splint is:

$$\beta=\frac{16Ewt_1^3}{h^3}/\left\{\frac{27h^2h_1}{128Eb_1t^3}+\frac{3h_2(3h-11h_1-11h_2)^2}{128E(b_1+b_2)t^3}+\frac{121h_2^3}{128E(b_1+b_2)t^3}\right.$$
$$+\frac{125h^3}{5632Ebt^3}+\int\limits_{h_1+h_2}^{h_1+h_2+h_3}\frac{121y^2(3h-11h_1-11h_2)^2}{192Eh^2t^3[\frac{b_3}{h_3}(y-h_1-h_2-h_3)+b]}dy+\frac{25h_5^3}{128E(b_1+b_2)t^3}$$
$$+\frac{75h_4(h_3+h_5)^2}{64Ebt^3}+\frac{h_4(5h-10h_3-10h_5)^2}{8Ebt^3}$$
$$\left.+\int\limits_{h-h_3-h_5}^{h-h_5}\frac{75(h-y)^2}{64Et^3[-\frac{b_3}{h_3}(y-h_1-h_2-h_3-h_4)+b]}dy\right\} \tag{7.46}$$

where, t is the thickness of the splint, in mm, y is the distance to the upper end plate, in mm.

2. **Critical restraint stiffness ratio**

In order to ensure the effectiveness and economy of constraint structure, it is necessary to consider the reasonable range of constraint stiffness ratio, that is, critical constraint stiffness ratio. Take the above scheme as an example, calculate the deformation of six different splint thicknesses, and the maximum out of plane deformation is listed in Table 7.10. When the restraint stiffness is small, the splint is too weak to restrain the out of plane deformation of the steel plate. With the increase

Table 7.10 Out of plane deformation of steel plate under different restraint stiffness ratio

Serial number	Splint thickness/mm	Thickness of sheared steel plate/mm	Restraint stiffness ratio	Out of plane deformation/mm
1	2	8	0.00524	20.082
2	4	8	0.04192	10.534
3	6	8	0.14148	8.141
4	8	8	0.33536	4.252
5	10	8	0.655	2.463
6	12	8	1.13184	2.037

of constraint stiffness ratio, the out of plane deformation decreases gradually. When the restraint stiffness is large, the splint can provide enough lateral support and effectively restrain the development of the out of plane deformation of the sheared steel plate.

According to the stress distribution of six different thickness of splint, when the constraint stiffness is relatively small, the part of splint has entered into plasticity due to the development of out of plane deformation, and the stress development of shear plate is not sufficient due to the influence of buckling. When the constraint stiffness is large, the splint is completely elastic, and almost the whole section of the steel plate is in the plastic state. The stress distribution is uniform, and the mechanical properties are significantly improved. The analysis shows that if the restraint stiffness ratio is too small and the splint is too weak, the out of plane deformation of the sheared steel plate cannot be restrained at all, and at this time, the part of the splint itself has gone into plasticity, which will produce permanent residual deformation. If the restraint stiffness ratio is too large and the splint is too rigid, it will not improve the shear capacity of steel plate, but waste steel and reduce the economic index. Therefore, for the critical restraint stiffness ratio of the buckling prevention scheme of bending shear type splint, the recommended value is $0.4 < \beta < 1.0$.

7.3.3 Main Performance Parameters of Buckling Prevention Shear Energy Dissipation Plate

7.3.3.1 Shear Capacity of Buckling Proof Shear Energy Dissipation Plate

1. **Elastic buckling load of uniformly sheared plate**

There are three methods to solve the buckling load: balance method, energy method (Rayleigh-Ritz method, Galerkin method, numerical method) and dynamic method. The equilibrium equation of the plate belongs to the two-dimensional partial differential equation, which is difficult to be solved directly by the equilibrium method. The Galerkin method is now used to calculate the buckling load of simply supported plates on four sides.

For the uniformly sheared simply supported plate section with four sides as shown in Fig. 7.13a, where the surface force is $N_{xy} = N_{yx} = p_{xy} = p_{yx}$, the equilibrium

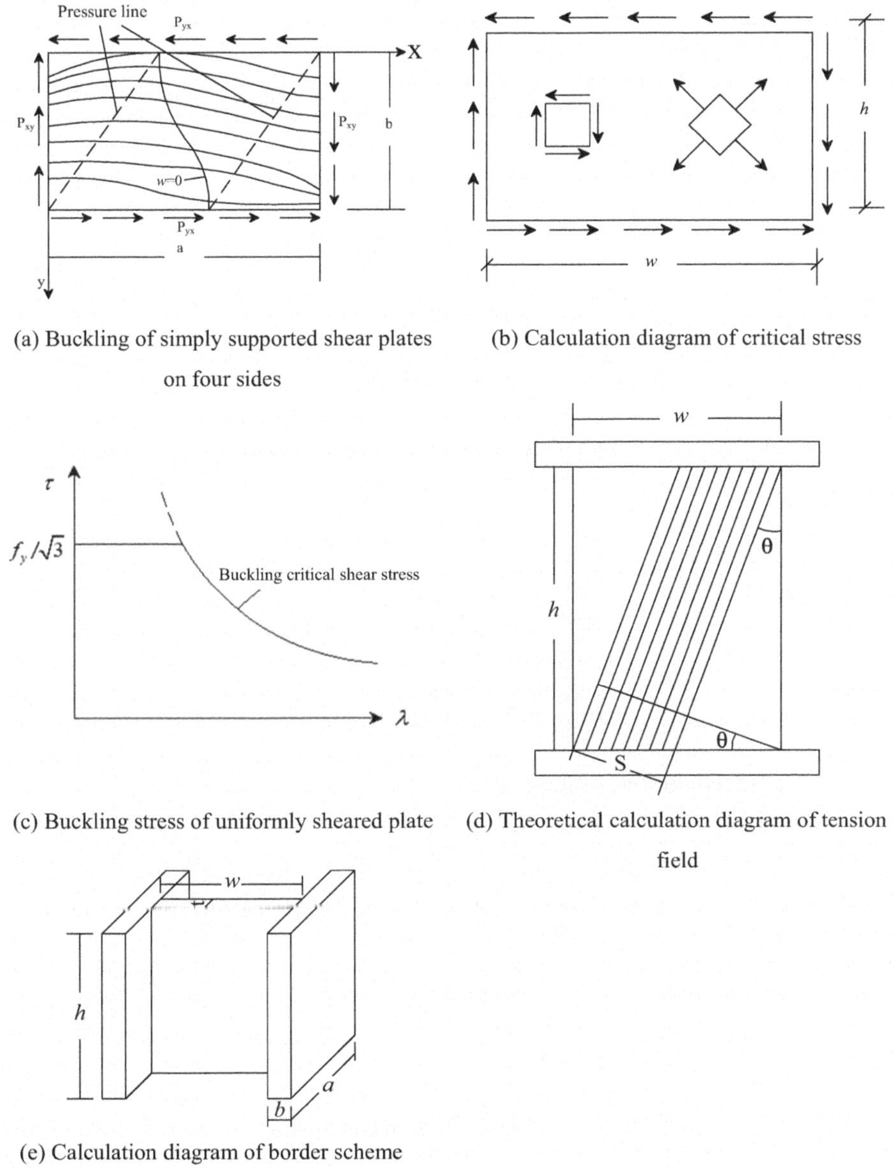

Fig. 7.13 Performance analysis of buckling resistant shear energy dissipation plate

partial differential equation of the plate is:

$$L(w) = \frac{\partial^4 w}{\partial x^4} + 2\frac{\partial^4 w}{\partial x^2 y^2} + \frac{\partial^4 w}{\partial y^4}) - \frac{2P_{xy}}{D}\frac{\partial^2 w}{\partial x \partial y} = 0 \tag{7.47}$$

Let the equation of deflection curve satisfying the geometric and natural boundary conditions be:

$$w = A_1 \sin\frac{\pi x}{a}\sin\frac{\pi y}{a} + A_2 \sin\frac{2\pi x}{a}\sin\frac{2\pi y}{a} + A_3 \sin\frac{3\pi x}{a}\sin\frac{3\pi y}{a} \tag{7.48}$$

The Galerkin equations are:

$$\begin{aligned} &\int_0^a\int_0^a L(w)\sin\tfrac{\pi x}{a}\sin\tfrac{\pi y}{a}dxdy = 0 \\ &\int_0^a\int_0^a L(w)\sin\tfrac{2\pi x}{a}\sin\tfrac{2\pi y}{a}dxdy = 0 \\ &\int_0^a\int_0^a L(w)\sin\tfrac{3\pi x}{a}\sin\tfrac{3\pi y}{a}dxdy = 0 \end{aligned} \tag{7.49}$$

By substituting the differential of Eqs. (7.47) into (7.49), we can get after integration:

$$\begin{aligned} &\frac{\pi^4}{a^2}A_1 - \frac{32P_{xy}}{9D}A_2 = 0 \\ &\quad -\frac{32P_{xy}}{9D}A_1 + \frac{16\pi^4}{a^2}A_2 - \frac{288P_{xy}}{25D}A_3 = 0 \\ &\quad -\frac{288P_{xy}}{25D}A_2 + \frac{81\pi^4}{a^2}A_3 = 0 \end{aligned} \tag{7.50}$$

The buckling conditions of the plate are:

$$\begin{vmatrix} \frac{\pi^4}{a^2} & -\frac{32P_{xy}}{9D} & 0 \\ -\frac{32P_{xy}}{9D} & \frac{16\pi^4}{a^2} & -\frac{288P_{xy}}{25D} \\ 0 & -\frac{288P_{xy}}{25D} & \frac{81\pi^4}{a^2} \end{vmatrix} = 0 \tag{7.51}$$

then

$$P_{cr} = 1.0585\frac{\pi^4 D}{a^2} = 10.447\frac{\pi^2 D}{a^2} \tag{7.52}$$

In general, energy method can only get approximate solution, the exact solution is $P_{cr} = 9.34\frac{\pi^2 D}{a^2}$: the difference is 11.85%. If the terms in Eq. (7.48) are added, the accuracy of the approximate solution can be improved. The expression of the exact solution of the buckling load of the shear plate is obtained by consulting the literature:

$$P_{cr} = k_s \frac{\pi^2 D}{l_{\min}^2} \tag{7.53}$$

where, k_s is the shear buckling coefficient.

For simply supported plates on four sides:

$$k_s = 5.34 + 4.0\left(\frac{l_{\min}}{l_{\max}}\right)^2 \tag{7.54}$$

For four side fixed support plate:

$$k_s = 8.98 + 5.6\left(\frac{l_{\min}}{l_{\max}}\right)^2 \tag{7.55}$$

where, $l_{\min}$ is the short side of the sheared plate and $l_{\max}$ is the long side of the sheared plate.

It can be seen from Eqs. (7.54) and (7.55) that there is no difference between the loaded side and the unloaded side under shear stress, only the difference between the long side and short side.

2. **Critical shear stress of uniformly sheared plate**

See Fig. 7.13b for the calculation diagram of simply supported shear plates on four sides. From Eq. (7.53):

$$P_{cr} = k_s \frac{\pi^2 E t^3}{12(1-\nu^2) l_{\min}^2} \tag{7.56}$$

So the critical shear stress of buckling is:

$$\tau_{cr} = k_s \frac{\pi^2 E}{12(1-\nu^2)} \left(\frac{t}{l_{\min}}\right)^2 \tag{7.57}$$

If the flexibility coefficient is defined as $\lambda = l_{\min}/t$, then the above formula is converted to:

$$\tau_{cr} = k_s \frac{\pi^2 E}{12(1-\nu^2)} \cdot \frac{1}{\lambda^2} \tag{7.58}$$

When the shape factor $m = \frac{l_{\max}}{l_{\min}}$ is fixed, the buckling shear stress is inversely proportional to λ^2, which is unified with the classical Euler formula $\sigma_{cr} = \frac{\pi^2 E}{\lambda^2}$. The relationship curve of buckling stress and flexibility coefficient of uniform shear plate is shown in Fig. 7.13c: (1) when the aspect ratio $h/w < 0.5$, the increase effect of τ_{cr} with the increase of h/w is not obvious, and only when $h/w \geq 0.5$, there is a significant increase of τ_{cr}. (2) when $h/w > 2.0$, the critical shear stress τ_{cr} is very high, and the steel plate will generally have strength failure at this time, so the buckling prevention measures lose practical significance.

3. **Shear capacity of buckling resistant shear energy dissipation plate**

Based on the theory of tension field, the expression of shear capacity of uniformly sheared plate is deduced. Its basic assumption is: the shear force of steel plate after buckling, part of which is borne by the shear force calculated by small deflection theory, part of which is borne by oblique tension field (film effect). The calculation diagram is shown in Fig. 7.13d. According to the theory of tension field, the shear bearing capacity V_u is equal to the sum of plate buckling shear force V_{cr} and tension field shear force V_t:

$$V_u = V_{cr} + V_t \tag{7.59}$$

The expression of plate buckling shear force V_{cr} is:

$$V_{cr} = \tau_{cr} \cdot w \cdot t \tag{7.60}$$

where, τ_{cr} is the buckling critical shear stress of the steel plate. The expression of shear force V_t in tension field is:

$$V_t = V \cdot \sin\theta = \sigma_t \cdot t \cdot s \cdot \sin\theta \tag{7.61}$$

among them, σ_t is the film tension of the tension field; s is the width of the oblique force transfer belt of the tension field; t is the thickness of steel plate; θ is the angle between the film tension and the vertical direction. The width s of the oblique force transfer belt in the tension field is:

$$s = w\cos\theta - h\sin\theta \tag{7.62}$$

After substituting Eq. (7.62) into Eq. (7.61), it can be simplified as follows:

$$V_t = \sigma_t \cdot t\left(\frac{w}{2}\sin 2\theta + \frac{h}{2}\cos 2\theta - \frac{h}{2}\right) \tag{7.63}$$

The optimum value of θ should make V_t has the maximum value, that is to say, it should meet the following requirements:

$$\frac{dV_t}{d\theta} = 0 \tag{7.64}$$

After derivation:

$$\tan 2\theta = \frac{w}{h} \tag{7.65}$$

After triangle transformation can be seen that: $\sin 2\theta = \frac{w}{\sqrt{h^2+w^2}}$, $\cos 2\theta = \frac{h}{\sqrt{h^2+w^2}}$; After substituting into Eq. (7.63), it can be simplified as follows:

$$V_t = \frac{\sigma_t wt}{2\left(\sqrt{1+\left(\frac{h}{w}\right)^2} + \frac{h}{w}\right)} \tag{7.66}$$

According to the theory of tension field, the tension of thin film is simplified, and the shear stress component provided by which is assumed to be $\sigma_t/\sqrt{3}$, when the shear yield occurs:

$$\frac{\sigma_t}{\sqrt{3}} + \tau_{cr} = f_{yv} \tag{7.67}$$

The above formula is deformed:

$$\sigma_t = \sqrt{3}(f_{yv} - \tau_{cr}) \tag{7.68}$$

After substituting into Eq. (7.63), it can be concluded that:

$$V_t = \frac{\sqrt{3}wt(f_{yv} - \tau_{cr})}{2\left(\sqrt{1+\left(\frac{h}{w}\right)^2} + \frac{h}{w}\right)} \tag{7.69}$$

The shear capacity V_u of steel plate is as follows:

$$V_u = V_{cr} + V_t = wt\left[\tau_{cr} + \frac{\sqrt{3}(f_{yv} - \tau_{cr})}{2\left(\sqrt{1+\left(\frac{h}{w}\right)^2} + \frac{h}{w}\right)}\right] \tag{7.70}$$

This is the result of strictly abiding by the theory of tension field. If two triangles outside the tension band are also considered to participate in the shear bearing, the above formula is modified as:

$$V_u = wt\left[\tau_{cr} + \frac{\sqrt{3}(f_{yv} - \tau_{cr})}{2\sqrt{1+\left(\frac{h}{w}\right)^2}}\right] \tag{7.71}$$

For the buckling proof schemes of rib plate and splint, as the boundary condition of the vertical edge is free edge, the tension field theory cannot be copied, and the correction coefficient $\sin 2\theta$ needs to be multiplied, so the shear bearing capacity expression of the shear steel plate at the core of the mild steel energy absorber is as follows:

Anti-buckling scheme of frame:

$$V_u = wt\left[\tau_{cr} + \frac{\sqrt{3}(f_{yv} - \tau_{cr})}{2\sqrt{1 + \left(\frac{h}{w}\right)^2}}\right] \tag{7.72}$$

For the buckling proof schemes of rib plate and splint:

$$V_u = wt\left[\tau_{cr} + \frac{\sqrt{3}(f_{yv} - \tau_{cr})}{2\sqrt{1 + \left(\frac{h}{w}\right)^2}}\right]\sin 2\theta \tag{7.73}$$

7.3.3.2 Other Performance Parameters of Bending Shear Type Energy Dissipation Plate

1. **Buckling proof schemes of rib plate and splint:**

Stiffness means the ability to resist deformation, equal to the ratio of force to displacement. The initial elastic rigidity of steel plate is equal to the ratio of loading force to corresponding deformation: $K = F/\Delta$. According to the above analysis, the deformation of steel plate is composed of bending deformation and shear deformation: $\Delta = \Delta_M + \Delta_V$, so the expression of initial elastic stiffness is: $K = \frac{F}{\Delta_M + \Delta_V}$. By substituting the expressions of Δ_M and Δ_V:

$$K = \frac{F}{\frac{Fh^3}{Ew^3t} + \frac{3Fh}{Ewt}} = \frac{1}{\frac{h^3}{Ew^3t} + \frac{3h}{Ewt}} = \frac{Ew^3t}{h^3 + 3hw^2} \tag{7.74}$$

The yield at the edge of bending shear steel plate is prior to the shear yield at the middle. The ultimate elastic load is as follows:

$$F_e = \frac{w^2 t f_y}{3h} \tag{7.75}$$

The elastic limit displacement is:

$$\Delta_e = \frac{F_e}{K} = \frac{\frac{tw^2 f_y}{3h}}{\frac{Ew^3t}{h^3+3hw^2}} = \frac{f_y(h^3 + 3hw^2)}{3Ehw} \tag{7.76}$$

Since the yield force F_y and the yield displacement Δ_y cannot be derived directly from the formula, now dimensionless parameter μ is introduced, so $F_y = \mu F_e$, $\Delta_y = \mu \Delta_e$. The μ value is obtained by finite element analysis or test data analysis. So the yield force is:

$$F_y = \frac{\mu w^2 t f_y}{3h} \tag{7.77}$$

The yield displacement is:

$$\Delta_y = \frac{\mu f_y (h^3 + 3hw^2)}{3Ehw} \tag{7.78}$$

2. **Anti buckling scheme of frame**

Figure 7.13e shows the calculation diagram of the buckling prevention scheme of frame. From the above derivation, it can be seen that $\Delta_M = \frac{Fh^3}{12EI}$, the moment of inertia I of the anti-buckling scheme of the frame is:

$$I = \frac{1}{12} t w^3 + 2 \times \left[\frac{1}{12} a b^3 + ab \left(\frac{b}{2} + \frac{w}{2} \right)^2 \right] \tag{7.79}$$

So the bending deformation is:

$$\begin{aligned} \Delta_M &= \frac{Fh^3}{12E\left\{\frac{1}{12}tw^3 + 2 \times \left[\frac{1}{12}ab^3 + ab\left(\frac{b}{2} + \frac{w}{2}\right)^2\right]\right\}} \\ &= \frac{Fh^3}{E[tw^3 + 2ab^3 + 6ab(b+w)^2]} \end{aligned} \tag{7.80}$$

It can be seen from the above formula: $\Delta_V = \frac{f_S Fh}{GA}$, where $f_S \approx \frac{A}{A_W}$, A is the area of the whole section, A_W is the section area of the web. For steel, Poisson's ratio $\nu = 0.3$, $G = \frac{E}{2(1+\nu)} \approx 0.4E$. Therefore, the shear deformation is:

$$\Delta_V = \frac{f_S Fh}{GA} \approx \frac{\frac{A}{A_W} Fh}{0.4EA} = \frac{5Fh}{2Ewt} \tag{7.81}$$

The exact solution of the buckling prevention scheme of the frame of bending and shearing steel plate is $f_s = 3.437$, the approximate solution is $f_s = 3.4$, the error is 1.08%; the exact solution of the buckling prevention scheme of shear steel plate frame is $f_s = 2.528$, the approximate solution is $f_s = 2.5$, the error is 1.10%. It can be seen that using the approximate formula $f_S \approx \frac{A}{A_W}$, not only the error is very small, meets the accuracy requirements, but also greatly simplifies the expression of core performance parameters. Therefore, the initial elastic stiffness is:

$$K = \frac{F}{\Delta_M + \Delta_V} = \frac{1}{\frac{h^3}{E[tw^3+2ab^3+6ab(b+w)^2]} + \frac{5h}{2Ewt}} \tag{7.82}$$

When the steel plate edge fiber just reaches yield:

$$\sigma = \frac{M\frac{w}{2}}{I} = \frac{F \cdot \frac{h}{2} \cdot \frac{w}{2}}{I} = f_y \tag{7.83}$$

The elastic limit load is obtained as follows:

$$F_e = \frac{f_y[tw^3 + 2ab^3 + 6ab(b+w)^2]}{3hw} \tag{7.84}$$

Elastic limit displacement:

$$\Delta_e = \frac{F_e}{K} = \frac{f_y[tw^3 + 2ab^3 + 6ab(b+w)^2]}{3hw}\left\{\frac{h^3}{E[tw^3 + 2ab^3 + 6ab(b+w)^2]} + \frac{5h}{2Ewt}\right\} \tag{7.85}$$

The yield force is:

$$F_y = \frac{\mu f_y[tw^3 + 2ab^3 + 6ab(b+w)^2]}{3hw} \tag{7.86}$$

The yield displacement is:

$$\Delta_y = \frac{\mu f_y[tw^3 + 2ab^3 + 6ab(b+w)^2]}{3hw}\left\{\frac{h^3}{E[tw^3 + 2ab^3 + 6ab(b+w)^2]} + \frac{5h}{2Ewt}\right\} \tag{7.87}$$

7.3.3.3 Other Performance Parameters of Shear Type Shear Energy Dissipation Plate

1. **Buckling proof schemes of rib plate and splint:**

The expression of the initial elastic stiffness of the shear steel plate is the same as that of the bending shear steel plate, $K = \frac{Ew^3t}{h^3+3hw^2}$.

The middle shear yield of the shear type steel plate is prior to the edge tension compression yield, so the elastic limit load is:

$$F_e = \frac{2wtf_y}{3\sqrt{3}} \tag{7.88}$$

The elastic limit displacement is:

$$\Delta_e = \frac{F_e}{K} = \frac{\frac{2twf_y}{3\sqrt{3}}}{\frac{Ew^3t}{h^3+3hw^2}} = \frac{2f_y(h^3+3hw^2)}{3\sqrt{3}Ew^2} \tag{7.89}$$

So the yield force is:

$$F_y = \frac{2\mu wtf_y}{3\sqrt{3}} \tag{7.90}$$

The yield displacement is:

$$\Delta_y = \frac{2\mu f_y(h^3+3hw^2)}{3\sqrt{3}Ew^2} \tag{7.91}$$

2. **Anti buckling scheme of frame**

The expression of the initial elastic rigidity of the shear steel plate is the same as that of the bending shear steel plate, namely:

$$K = \frac{F}{\Delta_M + \Delta_V} = \frac{1}{\frac{h^3}{E[tw^3+2ab^3+6ab(b+w)^2]} + \frac{5h}{2Ewt}} \tag{7.92}$$

When the middle fiber of steel plate begins to yield:

$$\tau = \frac{FS^*}{It} = f_y \tag{7.93}$$

where, S^* is the area moment of the rough section above the shear stress to the neutralization axis:

$$S^* = \frac{1}{4}ab(w+b) + \frac{1}{8}tw^2 \tag{7.94}$$

The elastic limit load is:

$$F_e = \frac{f_y[2t^2w^3 + 4ab^3t + 12abt(b+w)^2]}{6ab(w+b)+3tw^2} \tag{7.95}$$

So the elastic limit displacement is:

$$\Delta_e = \frac{f_y[2t^2w^3 + 4ab^3t + 12abt(b+w)^2]}{6ab(w+b)+3tw^2}\left\{\frac{h^3}{E[tw^3+2ab^3+6ab(b+w)^2]} + \frac{5h}{2Ewt}\right\} \tag{7.96}$$

The yield force is:

$$F_y = \frac{\mu f_y[2t^2w^3 + 4ab^3t + 12abt(b+w)^2]}{6ab(w+b) + 3tw^2} \tag{7.97}$$

The yield displacement is:

$$\Delta_y = \frac{\mu f_y[2t^2w^3 + 4ab^3t + 12abt(b+w)^2]}{6ab(w+b) + 3tw^2}\left\{\frac{h^3}{E[tw^3 + 2ab^3 + 6ab(b+w)^2]} + \frac{5h}{2Ewt}\right\} \tag{7.98}$$

7.3.4 Research and Development of New Shear Metal Damper

The research on the shear metal damper is mainly carried out from two aspects: one is to limit the out of plane buckling of the shear metal damper by using the buckling prevention measures proposed in Sect. 7.3.2 and improve the energy consumption mode of the shear steel plate; the other is to optimize the shape of the shear steel plate and improve the stress distribution mode when the shear steel plate yields.

7.3.4.1 Series of Buckling Proof Shear Mild Steel Dampers

Based on the research in Sects. 7.3.2 and 7.3.3, the test pieces of ordinary shear type soft steel damper, rib plate type buckling proof soft steel damper, splint type buckling proof soft steel damper and frame type buckling proof soft steel damper are designed, and their mechanical properties are tested. Figure 7.14a–d shows the steel plate deformation after the test of four types of mild steel dampers. It can be seen that the unconstrained shear energy dissipating plate has obvious out of plane bulging deformation after the destructive test, while the out of plane deformation of the shear energy dissipating plate adopting the buckling prevention design is relatively small, so as to ensure the stability of its performance under the repeated load.

Figure 7.14e shows two kinds of test hysteretic curves of soft steel damper. The comparison shows that: (1) under the action of low cycle repeated load, the hysteretic curve of unconstrained shear steel plate presents obviously pinched S-shape; after adopting the buckling prevention schemes of rib plate and splint, the hysteretic curve is full and the energy consumption capacity is improved; after adopting the buckling prevention scheme of frame, the hysteretic area is increased significantly. The energy consumption capacity has been greatly improved. (2) there is a certain degree of stiffness degradation after the new type of buckling prevention

(a) Unconstrained shear energy dissipation plate

(b) Shear energy dissipating plate with the buckling prevention scheme of splint

(c) Shear energy dissipating plate with the buckling prevention scheme of rib plate

(d) Shear energy dissipation plate of the buckling prevention scheme of frame

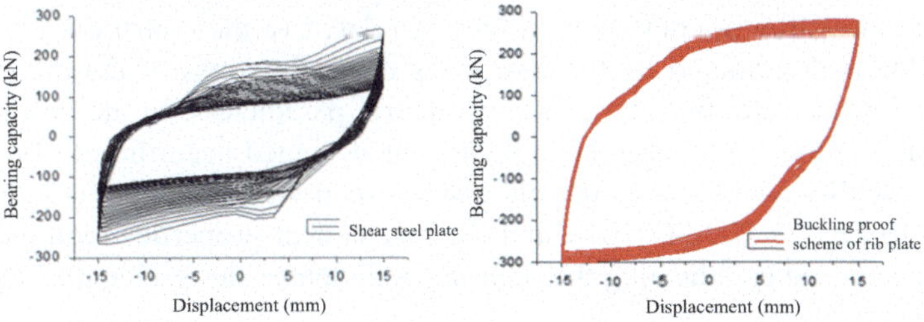

(e) Shear energy dissipation plate of the buckling prevention scheme of frame

Fig. 7.14 Test of shear type soft steel damper for buckling resistance shear energy dissipation plate enters into yield, but it is not obvious, and with the loading process, the bearing capacity can increase synchronously. (3) the constant amplitude displacement loading test shows that: the shear plates of the three buckling prevention schemes have no obvious damage, the hysteresis curve has no obvious deformation, the main design index error and attenuation are not more than 15%, and there is no obvious low cycle fatigue phenomenon, which meets the requirements of the current relevant codes and regulations in China. This kind of damper has stable energy consumption capacity and good ductility. It is an ideal damper.

7.3.4.2 Parabolic Shape Mild Steel Damper

In order to make the in-plane yielding soft steel damper not cause stress concentration at the hole and make the full-length section yield of the web more evenly, a parabola shaped soft steel damper is designed, as shown in Fig. 7.15a [5]. The idea is: if the yield occurs at the same time in the whole length of the member, the maximum normal stress of each section is equal and the yield stress is reached at the same time. In this case, $\frac{x}{b_x^2}$ should be a constant, that is:

$$b_x = a\sqrt{x} \tag{7.99}$$

where, a is the section coefficient, which can be any positive number. According to the theoretical derivation, the elastic stiffness, yield load and yield displacement of the parabola shaped mild steel damper with n-limb are:

$$K_u = cnK = cn\frac{0.5Eta^3}{(0.5l)^{3/2}} \tag{7.100}$$

$$D_y = \frac{(0.5l)^{3/2}\sigma_y}{0.5Eac} \tag{7.101}$$

$$P_y = nF_y = na^2\sigma_y t \tag{7.102}$$

where, c is the stiffness coefficient, which is related to the constraint conditions, and its value is determined by the test; σ_y is the yield stress of the material; see Fig. 7.15b for the definition of section geometric parameters. In the design of the damper, the section of the mild steel damper is designed according to Eq. (7.99), the elastic rigidity, yield load and yield displacement of the damper are calculated according to Eqs. (7.100)–(7.102), and the mechanical properties required by the design are achieved by adjusting the section coefficient a, design length l, thickness t and limb number n.

Four damper specimens were made of domestic LY225 low yield steel, and the mechanical property test was carried out on the 100 ton MTS equipment in the structural laboratory of Southeast University (Fig. 7.15c). Figure 7.15d shows the force displacement curve of the mild steel damper under standard loading. It can

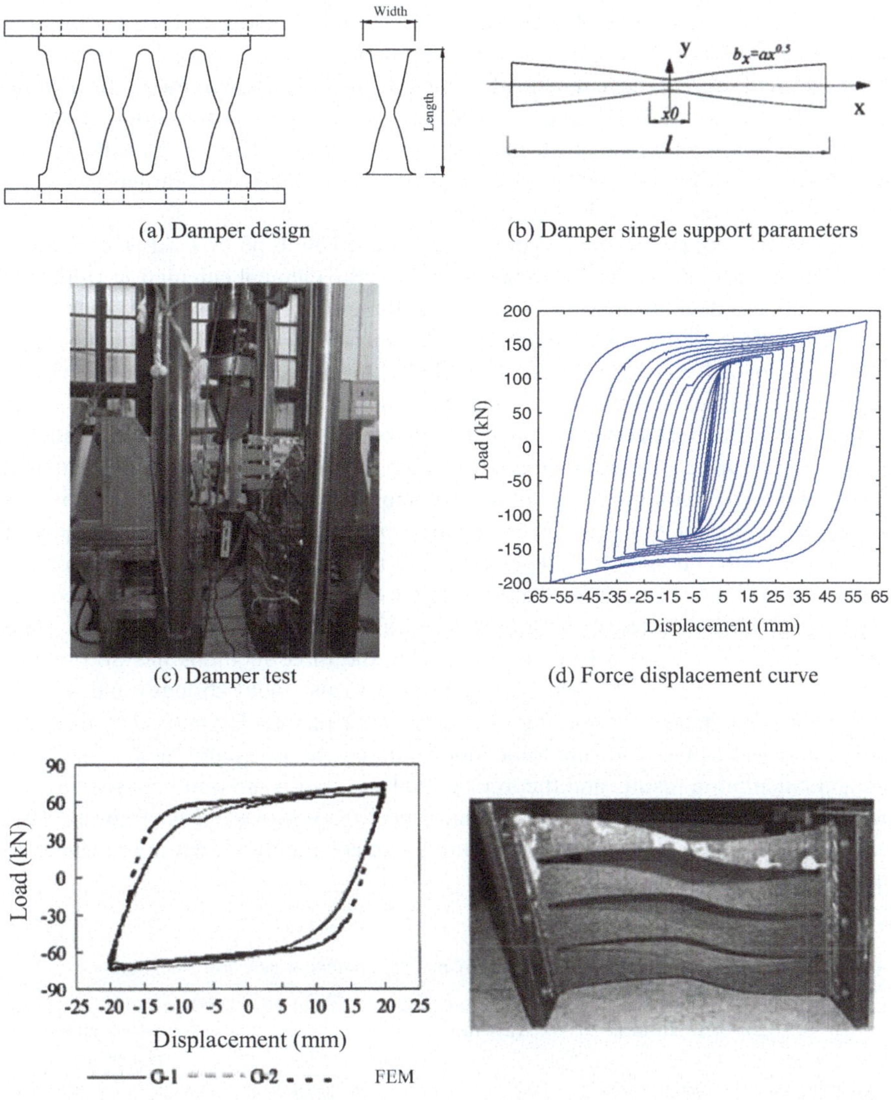

(a) Damper design

(b) Damper single support parameters

(c) Damper test

(d) Force displacement curve

(e) Hysteresis curve comparison

(f) Damper deformation after test

Fig. 7.15 Parabolic shaped mild steel damper

be seen that the hysteresis curve of the parabolic mild steel damper is full, without pinching, without obvious low cycle fatigue phenomenon, and has good energy consumption capacity. After the four specimens are subjected to standard loading and cyclic loading, relatively uniform plastic deformation is produced, and there is no crack on the surface or obvious damage. Figure 7.15f is the photo of SSD1 specimen after the test. Combined with other test results, it can be seen that: (1)

the test skeleton curve is analyzed, the skeleton curve is close to bilinear, and the mechanical model of the damper can be simplified as bilinear model; (2) when the deformation of the test piece reaches 15 times of the yield displacement, the bearing capacity is still not reduced, and the test piece shows good deformation capacity; (3) the ductility coefficient of all the test pieces is more than 15, and the energy consumption coefficient is also reached. 3. It shows that the specimen has good deformation and energy dissipation capacity.

The mechanical performance parameters of the new type of mild steel damper are calculated, and the comparison between the finite element calculation value and the theoretical calculation value is given in Table 7.11. In order to calculate the parameters of single test piece, the test measured value in the table is the test load data divided by 2. It can be seen from the table: (1) from the test data and finite element analysis, the stiffness coefficient c is taken as 0.84; the theoretical calculation value in the table is the data corrected after the stiffness coefficient c is determined. (2) the measured elastic stiffness is smaller than the theoretical value and the calculated value of finite element, but the yield displacement is larger. The main reason is that the elastic displacement measured by the test not only includes the displacement of the damper itself, but also the displacement of the connection section between the damper and MTS machine, as well as the relative displacement of the bolt connection, which leads to the decrease of the elastic stiffness and the increase of the yield displacement. (3) the yield force calculated by the three methods has no obvious change, because the yield force is only related to the yield strength and section coefficient of steel, and has nothing to do with other factors. It can be seen from the above analysis that the restoring force model parameters measured by the test, finite element calculation results and theoretical analysis results are quite consistent, and the hysteresis curve (Fig. 7.15e) is also relatively close. It shows that the finite element model is reasonable, and the design formula can accurately reflect the mechanical characteristics of the damper.

Table 7.11 Comparison of mechanical performance parameters of new soft steel damper

		Elastic stiffness (kN/mm)	Plastic stiffness (kN/mm)	Yield force (kN)	Yield displacement D_y(mm)
Theoretical calculation value		17.49	/	67.68	3.87
Calculated value of finite element		17.71	0.86	67.08	3.79
Test result	G-1	15.70	0.82	60.37	3.81
	G-2	16.64	0.79	64.58	3.88

7.4 Bending Metal Damper

In a wide variety of metal dampers, compared with the buckling proof energy dissipating brace and unbonded energy dissipating brace which utilize the energy dissipation of the axial yield deformation of metal, the accuracy requirements of manufacture and installation of energy dissipator which utilize the energy dissipation of out of plane bending yield deformation of metal are lower, the weight of energy dissipator itself is lighter, which is convenient for transportation and installation, and the connection form with structure is more flexible, which can minimize the obstacles to space permeability, which shows obvious advantages.

7.4.1 Research and Development of Drum-Shaped Open Hole Soft Steel Damper

7.4.1.1 Structural Design

According to the characteristics of stress distribution of in-plane yield soft steel damper, a kind of drum shaped open hole soft steel damper is designed. As shown in Fig. 7.16a, it is composed of several parallel core energy dissipation plates, top plates and bottom plates. The rectangular core energy dissipation plate with width of w, height of h and thickness of t is made of mild steel, and a drum shaped hole is opened in the middle. The hole is composed of two semicircles with radius of r on both sides and a rectangle with height of $2r$ and width of b in the middle. Several core energy dissipation plates are placed parallel to each other with a clear distance of d_0. The core energy dissipation plate is connected with the top plate and bottom plate by welding, the top plate and the bottom plate are provided with bolt holes, and the energy dissipator is connected with the support and the frame beam by high-strength bolts.

7.4.1.2 Main Performance Parameters

The core energy dissipation element of the drum shaped opening soft steel energy absorber is a soft steel plate with a drum shaped opening. It is defined that the elastic rigidity of the core energy dissipation plate is K, the elastic limit load is F_e, the elastic limit displacement is Δ_e, the yield load is F_y, and the yield displacement is Δ_y; the elastic rigidity of the energy dissipation device composed of n core energy dissipation plates is K_d, the elastic limit load is F_{de}, the elastic limit displacement is Δ_{de}, the yield load is F_{dy}, and the yield displacement is Δ_{dy}. The energy dissipation plates which are parallel to each other are connected in parallel:

$$K_d = nK,\ \ F_{de} = nF_e,\ \Delta_{de} = \Delta_e,\ F_{dy} = nF_y,\ \Delta_{dy} = \Delta_y$$

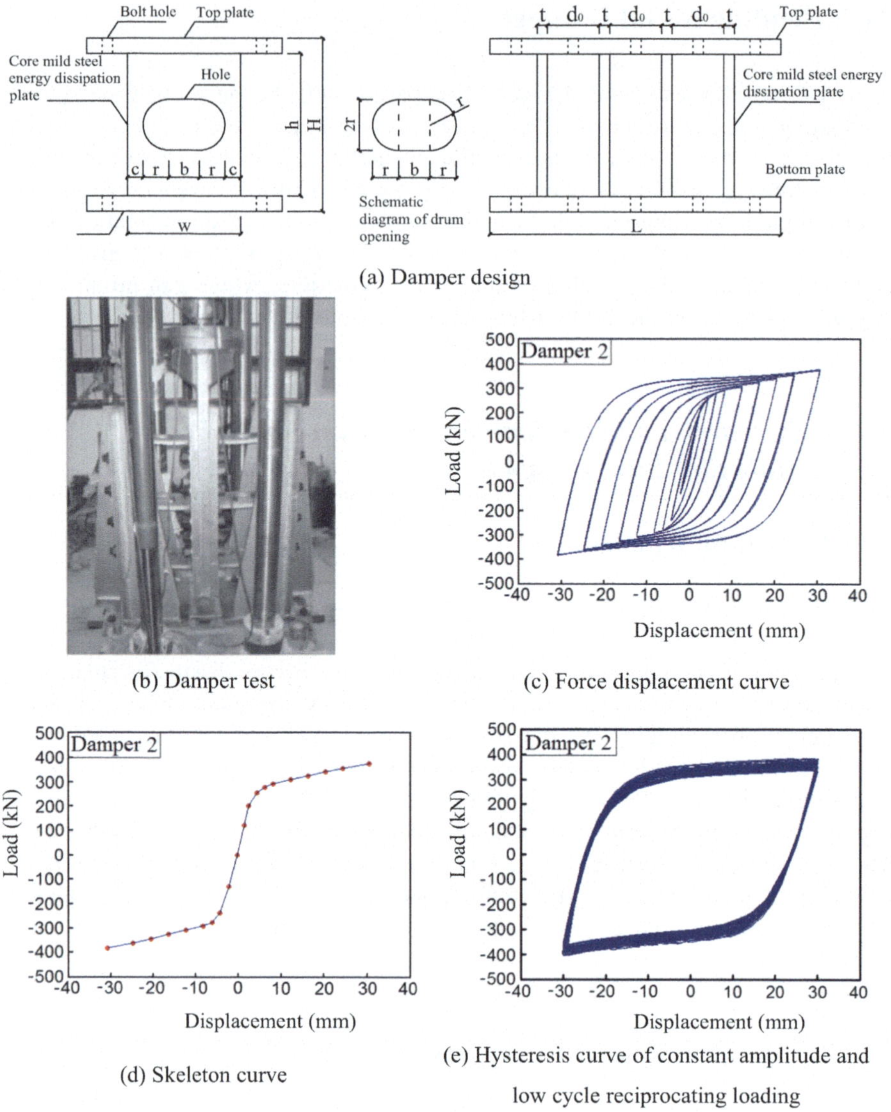

Fig. 7.16 Drum shaped opening mild steel damper

It is defined that the width direction of the energy dissipation plate is X direction, the height direction is Y direction and the thickness direction is Z direction. The main performance parameters of the damper can be deduced as follows:

1. **Elastic stiffness *K***

As the upper and lower ends of the core energy dissipation plate of the soft steel energy absorber are fixed on the top and bottom plates, it is equivalent to a beam fixed at both ends in terms of the stress mechanism. Considering the influence of its bending deformation and shear deformation, the deformation of the core energy dissipation plate can be obtained as follows:

$$\Delta = \Delta_{\mathrm{M}} + \Delta_{\mathrm{V}} \tag{7.103}$$

where, Δ_{M} and Δ_{V} are the deformation caused by bending moment and shear force of the fixed beam at both ends. For this example, Δ is the relative horizontal displacement of the upper and lower ends of the core energy dissipating plate, i.e. the lateral displacement of the energy dissipating plate, Δ_{M} and Δ_{V} are the lateral displacements caused by the bending deformation and shear deformation of the energy dissipating plate, respectively.

Taking the center of the energy dissipation plate as the origin, the top surface of the core plate is affected by the horizontal load F and the moment M.

$$\Delta_{\mathrm{M}} = \int \overline{M}(y)d\theta = \int \frac{\overline{M}(y)\, M_{\mathrm{P}}(y)}{EI(y)}dy \tag{7.104}$$

$$\Delta_{\mathrm{V}} = \int \overline{V}(y)d\eta = \int \frac{f_{\mathrm{s}}\,\overline{V}(y)\, V_{\mathrm{P}}(y)}{GA(y)}dy \tag{7.105}$$

where, $\overline{M}(y)$ and $\overline{V}(y)$ are the internal forces caused by the unit load, $d\theta = \frac{M_{\mathrm{P}}}{EI}ds$ and $d\eta = \frac{f_{\mathrm{s}} V_{\mathrm{P}}}{GA}ds$ are the deformations caused by the actual external cause, M_{P} and V_{P} are the bending moment and shear force caused by the load in the structure, E and G are the elastic modulus and shear modulus of the steel respectively, the Poisson's ratio ν of the steel is 0.3,$G = \frac{E}{2(1+\nu)} \approx 0.4E$, A is the sectional area, I is the moment of inertia of the section with respect to the neutral axis in the X direction, f_{s} is the coefficient of non-uniform distribution of the sectional shear stress. For rectangular section, $f_{\mathrm{s}} = 1.2$. Among them, $\overline{M}(y) = y$, $M_{\mathrm{P}}(y) = Fy$, $\overline{V}(y) = 1$, $V_{\mathrm{P}}(y) = F$. In order to facilitate integration, parabola is used to replace semi arc approximately:

$$I(y) = \begin{cases} \frac{1}{12}wt^3, & -\frac{h}{2} \le y < -r,\ r < y \le \frac{h}{2} \\ \frac{1}{12}\left[w - b - 2\left(r - \frac{y^2}{r}\right)\right]t^3, & -r \le y \le r \end{cases} \tag{7.106}$$

$$A(y) = \begin{cases} wt, & -\frac{h}{2} \le y < -r,\ r < y \le \frac{h}{2} \\ \left[w - b - 2\left(r - \frac{y^2}{r}\right)\right]t, & -r \le y \le r \end{cases} \tag{7.107}$$

Integrating and sorting out:

$$\Delta_{\mathrm{M}} = \frac{F}{Et^3}\left[\frac{h^3 - 8r^3}{w} + 12r\left(r - m \arctan\frac{r}{m}\right)\right] \tag{7.108}$$

$$\Delta_{\mathrm{V}} = \frac{3F}{Et}\left[\frac{h - 2r}{w} + r\left(r - m \arctan\frac{r}{m}\right)\right] \tag{7.109}$$

where $m = \sqrt{r(w - b - 2r)/2}$. Because t is relatively small, it can be seen from the comparison of Eqs. (7.108) and (7.109) that $\Delta_M \gg \Delta_V$. To sum up, the core energy dissipating plate is dominated by bending deformation, Δ_V can be ignored:

$$\Delta = \Delta_M = \frac{F}{Et^3}\left[\frac{h^3 - 8r^3}{w} + 12r\left(r - m \arctan\frac{r}{m}\right)\right] \tag{7.110}$$

The theoretical formula of the elastic stiffness of the core energy dissipation plate is

$$K = \frac{F}{\Delta} = \frac{F}{\Delta_M} = Et^3\left[\frac{h^3 - 8r^3}{w} + 12r\left(r - m \arctan\frac{r}{m}\right)\right]^{-1} \tag{7.111}$$

For the energy dissipator composed of n core energy dissipation plates, the theoretical formula of its elastic stiffness is

$$K_d = nK = nEt^3\left[\frac{h^3 - 8r^3}{w} + 12r\left(r - m \arctan\frac{r}{m}\right)\right]^{-1} \tag{7.112}$$

2. **Elastic limit load F_e and yield load F_y**

When the core plate reaches the elastic limit point, the surface fibers near the top and bottom of the core plate yield, that is, $\sigma_{max} = f_y$:

$$M_e = \frac{1}{6}\sigma\text{max}t^2 w = \frac{1}{6}f_y t^2 w \tag{7.113}$$

According to the equilibrium condition of the force, the elastic limit load of the core energy dissipation plate is

$$F_e = \frac{M_e}{h/2} = \frac{f_y t^2 w}{3h} \tag{7.114}$$

According to the hysteretic curve model of energy dissipator shown in Fig. 7.5, the parameter relationship between elastic limit point and yield point is set as

$$F_y = \alpha F_e;\ \Delta_y = \alpha \Delta_e \tag{7.115}$$

where, α is the conversion coefficient ($\alpha > 1$) between yield point 2 and elastic limit point 1, whose empirical value needs to be determined by finite element analysis and experimental research. When Eq. (7.114) is introduced into Eq. (7.115), the theoretical calculation formula of the yield force of the core energy dissipation plate can be obtained as follows:

$$F_y = \alpha F_e = \frac{\alpha f_y t^2 w}{3h} \tag{7.116}$$

For the energy dissipator composed of n core energy dissipation plates, the theoretical formulas of elastic limit load and yield load are as follows:

$$F_{de} = nF_e = \frac{n f_y t^2 w}{3h} \tag{7.117}$$

$$F_{dy} = nF_y = \frac{n\alpha f_y t^2 w}{3h} \tag{7.118}$$

3. **Elastic limit displacement Δ_e and yield displacement Δ_y**

The theoretical formulas of elastic stiffness, elastic limit load, elastic limit displacement, yield load and yield displacement derived above. According to $F_e = K\Delta_e$, $F_y = K\Delta_y$, $F_{de} = K_d\Delta_{de}$, $F_{dy} = K_d\Delta_{dy}$, the theoretical formulas of elastic limit displacement and yield displacement of core energy dissipation plate and energy dissipator can be obtained as follows:

$$\Delta_{de} = \Delta_e = \frac{f_y w}{3Eht}\left[\frac{h^3 - 8r^3}{w} + 12r\left(r - m \arctan\frac{r}{m}\right)\right] \tag{7.119}$$

$$\Delta_{dy} = \Delta_y = \frac{\alpha f_y w}{3Eht}\left[\frac{h^3 - 8r^3}{w} + 12r\left(r - m \arctan\frac{r}{m}\right)\right] \tag{7.120}$$

7.4.1.3 Mechanical Property Test

Three test specimens were made of low yield point steel with yield strength of 160 MPa and 225 MPa, and the mechanical properties were tested in the structural laboratory of Southeast University (Fig. 7.12b). Figure 7.12c–e shows the force displacement curve of one of the test pieces. Combined with other test results, it can be seen that: (1) the hysteretic curve of the soft steel damper is full, the pinch phenomenon is not obvious, and it has good ductility and energy dissipation capacity; (2) the opening can make the stress distribution of the energy dissipation plate more uniform after it enters into yield, which is conducive to improving the low cycle fatigue strength of the damper; (3) it can be seen that the attenuation of the performance parameter of this type of energy dissipation device is small under 30 cycles of reciprocating loading required by the code, and the energy dissipater breaks after 70–90 cycles of reciprocating cycle finally, so it has a large redundancy.

7.4.2 *Research and Development of Curved Steel Plate Damper*

7.4.2.1 Structural Design

The curved steel plate damper consists of three parts (as shown in Fig. 7.17a): left and right half arc steel plates, upper straight section steel plates and lower straight section steel plates [6]. The upper and lower flat steel plates can be connected with

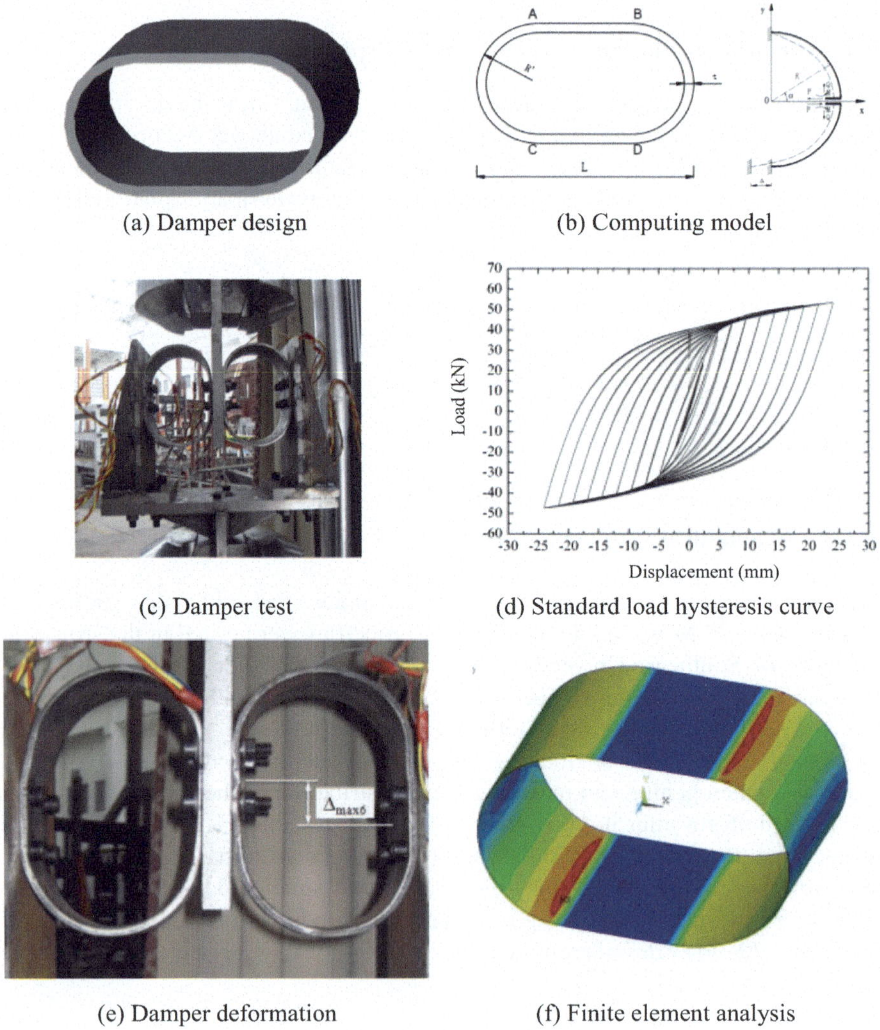

Fig. 7.17 Curved steel plate damper

beams and supports respectively by bolts, and the main energy consumption part is the left and right half arc steel plates. If the damper is set in the frame structure, the attached support is usually connected with the structure; if the damper is set in the shear wall structure, the damper is usually placed in the coupling beam or the shear wall with a joint.

In terms of energy dissipation mechanism, curved steel plate dissipates seismic energy through out of plane bending deformation. At present, one of the existing out of plane bending energy dissipation dampers is the stiffening energy dissipation device, which is composed of several parallel steel plates and fixed parts, such as X-shaped stiffening energy dissipation device, triangle stiffening energy dissipation device, diamond shaped opening stiffening energy dissipation device, etc. the yield displacement and yield force of a single steel plate of this energy dissipation device are very small, so it needs many pieces of combination to meet the requirements of dampers required by the structure. The other is the special-shaped steel plate energy dissipation device, that is, the energy dissipation of the plastic bending deformations of U-shaped and S-shaped steel plates are often used, which are used as the energy dissipation element in the isolation device and composite energy dissipation diagonal brace. The form of the curved steel plate damper proposed in this section is simpler than the first type of energy dissipating device, and the energy dissipation principle is more similar to that of the second type of energy dissipating device, and the curved steel plate damper itself forms a closed section, so that its overall energy dissipation performance and stability are better.

7.4.2.2 Main Performance Parameters

1. **Elastic stiffness**

Taking the semicircle arc steel plate of the energy dissipation part of the curved steel plate damper as the research object, it can be regarded as a curved beam fixed at both ends. Figure 7.17b shows the deformation diagram and internal force analysis diagram of the central axis of the semicircle arc steel plate when the displacement Δ at the support is unit 1. By the elastic center method, first of all, using the symmetry of the structure, M and Q are symmetric unknown forces, F is antisymmetric unknown forces, the force method equation can be simplified as:

$$\left.\begin{aligned} \delta_{11}M + \delta_{12}Q + \Delta_{1C} &= 0 \\ \delta_{21}M + \delta_{22}Q + \Delta_{2C} &= 0 \\ \delta_{33}F + \Delta_{3C} &= 0 \end{aligned}\right\} \tag{7.121}$$

Using the rigid arm, make δ_{12} and δ_{21} equal to zero, and further simplify the equation as follows:

$$\left.\begin{array}{l}\delta_{11}M+\Delta_{1C}=0\\ \delta_{22}Q+\Delta_{2C}=0\\ \delta_{33}F+\Delta_{3C}=0\end{array}\right\} \tag{7.122}$$

The bearing only has horizontal displacement, $\Delta_{2C}=\Delta_{3C}=0$, so the bending moment M and vertical force Q at the elastic center are both zero, and the equation is:

$$\delta_{33}F+\Delta_{3C}=0 \tag{7.123}$$

where

$$\begin{aligned}\delta_{33}&=\int\frac{\overline{M}_3^2}{EI}ds+\int\frac{\overline{F}_3^2}{EA}ds+\int\frac{\overline{Q}_3^2}{GA}ds\\ &=\int_{-\frac{\pi}{2}}^{\frac{\pi}{2}}\frac{R^2\sin^2\alpha}{EI}Rd\alpha+\int_{-\frac{\pi}{2}}^{\frac{\pi}{2}}\frac{\sin^2\alpha}{EA}Rd\alpha+\int_{-\frac{\pi}{2}}^{\frac{\pi}{2}}\frac{\cos^2\alpha}{GI}Rd\alpha\\ &=\frac{R\pi}{2}\left(\frac{R^2}{EI}+\frac{1}{EA}+\frac{1}{GA}\right)\end{aligned} \tag{7.124}$$

$$\Delta_{3C}=-1 \tag{7.125}$$

Equations (7.124) and (7.125) are introduced into Eq. (7.123):

$$F=-\frac{\Delta_{3C}}{\delta_{33}}=\frac{2}{R\pi\left(\frac{R^2}{EI}+\frac{1}{EA}+\frac{1}{GA}\right)} \tag{7.126}$$

F is the force at the end of the damper under unit displacement, i.e. the elastic stiffness of the semi-circular arc curved steel plate damper. Take the shear modulus of the steel as $G=0.4E$, and simplify it to obtain the elastic stiffness of the damper as follows:

$$K=\frac{Ebt^3}{R\pi(6R^2+7t^2/4)} \tag{7.127}$$

where, E is the elastic modulus of the material, b is the width of the curved steel plate damper, t is the thickness of the curved steel plate damper, and R is the radius of the curved steel plate damper.

2. **Yield strength**

According to the balance of force, the relationship between the elastic ultimate strength F_e and its bending moment M_e is as follows:

$$F_e\cdot 2R=2M_e \tag{7.128}$$

Then the elastic ultimate strength is:

$$F_e = \frac{M_e}{R} = \frac{f_y W_e}{R} = \frac{f_y bt^2}{6R} \tag{7.129}$$

where, W_e is the elastic resistance moment of the section. The value method of yield displacement is shown in Fig. 2.5. The transverse and longitudinal coordinates of intersection points of tangent OA and AB are yield displacement and yield strength respectively. Since the yield strength is greater than the elastic ultimate strength, the adjustment coefficient β is introduced, which is yield load $F_y = \beta F_e$. The adjustment factor β can be determined by relevant tests and numerical analysis.

$$F_y = \frac{\beta f_y bt^2}{6R} \tag{7.130}$$

The yield displacement can be expressed as follows:

$$\Delta_y = \frac{F_y}{K} = \frac{\beta f_y \pi (6R^2 + 7t^2/4)}{6Et} \tag{7.131}$$

According to Eqs. (7.127), (7.130) and (7.131), the elastic stiffness, yield strength and yield displacement of the curved steel plate damper can be calculated. By adjusting the relevant parameters, the damper that can meet the needs of the project can be designed.

7.4.2.3 Mechanical Property Test

According to the above design and model, the test pieces were designed and the test were carried out in the civil test center of Southeast University, and the loading equipment used was MTS50 ton fatigue test machine. The test was divided into four groups, eight test pieces, each group had two identical test pieces. The test device is shown in Fig. 7.17c. In order to connect with the fatigue testing machine, the fixture (see Fig. 7.17c) was designed. The inverted T-shaped clamp plate and T-shaped base were equipped with triangular stiffeners, and the components of the fixture were connected by bolts.

Figure 7.17a shows the hysteretic curve of test piece 3 under standard loading. It can be seen that: (1) the hysteretic curve of damper is full; the elastic rigidity is large, and the rigidity of damper decreases after yielding; (2) the maximum and minimum loads of damper are not the same, which may be caused by errors in installation; (3) four groups of test pieces were cycled for three times at each target displacement, and the three curves basically coincide, indicating that the performance of the damper is stable.

Four groups of test pieces experienced standard loading. When the displacement is small, the damper has no obvious change. With the increase of displacement, the steel oxide layer in the semicircle energy dissipation section of the damper rises and peels off, and the end of the semicircle arc peels off a lot. The damper produces plastic deformation in the process of reciprocating loading. When the maximum displacement is loaded, the deformation of the damper is shown in Fig. 7.17e. After 30 cycles of fatigue loading, the damper has obvious deformation, no cracks and no obvious damage with naked eyes, and the integrity is good.

According to the theoretical calculation formula of curved steel plate damper, combined with the test and finite element analysis data in this section (Fig. 7.17f), the comparison is listed in Table 7.12. The coefficient β in the calculation formula can be determined according to the test value and the finite element value. When $\beta = 1.78$ is calculated, the error is small. The maximum error of each mechanical property is listed in the table, and the other error values are small except that some errors of the elastic stiffness are greater than 10%. It shows that the finite element calculation and theoretical formula can reasonably reflect the performance of the damper.

Table 7.12 Comparison of mechanical parameters of curved steel plate damper

Specimen	Mechanical parameters	CSPD-1	CSPD-3	CSPD-4	CSPD-6
Test result	Yield displacement (mm)	4.37	2.66	1.86	4.52
	Yield bearing capacity (kN)	8.78	25.42	12.53	6.97
	Elastic rigidity (kN/mm)	1.92	9.56	6.74	1.54
Finite element simulation	Yield displacement (mm)	4.18	2.44	1.82	4.20
	Yield bearing capacity (kN)	9.34	27.11	13.85	7.56
	Elastic rigidity (kN/mm)	2.13	10.11	7.71	1.80
Theoretical calculation	Yield displacement (mm)	4.41	2.49	1.85	4.40
	Yield bearing capacity (kN)	8.90	25.58	13.81	7.21
	Elastic rigidity (kN/mm)	2.02	10.23	7.43	1.64
Maximum error	Yield displacement (mm)	5.50%	9.02%	2.20%	7.62%
	Yield bearing capacity (kN)	6.38%	6.65%	9.53%	7.80%
	Elastic rigidity (kN/mm)	9.86%	7.01%	12.58%	14.44%

References

1. Zhang, Congjun. 2009. *Test of Soft Steel Supported Damper and Research on Performance-Based Design Method of Energy Dissipation Structure*. Nanjing: Southeast University. (in Chinese).
2. Li, Aiqun. 2007. *Vibration Control of Engineering Structure*. Beijing: China Machine Press.
3. Zhao, Ming. 2007. *Experimental Study on Mechanical Behavior of Mild Steel Energy Dissipation Braced Damper*. Nanjing: Southeast University. (in Chinese).
4. Zheng, Jie, Aiqun Li, and Tong Guo. 2015. Analytical and experimental study on mild steel dampers with non-uniform vertical slits. *Earthquake Engineering and Engineering Vibration* 14 (1): 111–123.
5. Xu, Yanhong. 2013. *Theoretical and Experimental Research on the New Mild Steel Dampers Used in the Structure*. Nanjing: Southeast University. (in Chinese).
6. Zheng, Jie. 2015. *Theoretical and Experimental Research on a Curved Steel Damper Used in the Structures*. Nanjing: Southeast University. (in Chinese).

Chapter 8
Tuned Damping Device

Abstract Tuned Damping Device includes frequency modulation (FM) mass damper (TMD) and FM liquid damper (TLD). About TMD, the basic principle and performance characteristics of rubber supported TMD, suspended TMD, integrated ring tuned mass damper, adjustable stiffness vertical TMD and their calculation model are elaborated. About TLD, the basic principle and performance characteristics of rectangular liquid damper, circular liquid damper and ring liquid damper are introduced respectively.

When the natural frequency of the damping device is tuned to the controlled frequency of the structure, the vibration energy of the structure itself is transferred to the tuned damping device to dissipate energy and control the dynamic response of the structure. They are the oldest vibration control devices, which mainly include two categories: frequency modulation (FM) mass damper and frequency modulation liquid damper. In recent years, they have been widely used in the response control of various structures under narrow-band dynamic loads, such as wind, pedestrian and so on, and achieved good control effect and social and economic benefits [1, 2].

8.1 FM Mass Damper

As described in Sect. 3.1, in principle, TMD system is composed of mass, spring and damper. However, in practical engineering application, its construction form usually needs to be designed according to the characteristics of specific structure. In recent years, on the one hand, the author's team designed a unified TMD form for standardized structure to facilitate industrial production; on the other hand, designed targeted TMD structure combined with specific architectural features to achieve structural vibration control according to local conditions [3].

A. Li, *Vibration Control for Building Structures*, Springer Tracts in Civil Engineering, https://doi.org/10.1007/978-3-030-40790-2_8

8.1.1 *Rubber Supported TMD*

8.1.1.1 Ring TMD with Rubber Support

As mentioned above, for general buildings and structures, the structural design of TMD usually needs to consider the characteristics of the structure itself, and there is no fixed form. However, for the chimney, wind power tower, transmission tower and other self-standing high-rise structures, because of the relatively uniform structure form, the applicable TMD can be specifically designed. The author's team designed a rubber supported ring tuned mass damper (RSR-TMD) for this kind of structure, as shown in Fig. 8.1. The device is mainly composed of shell, mass block, rubber support, stiffening plate and damper. Its main principle is: the mass block provides the mass, the rubber bearing provides the horizontal stiffness of TMD, the damper provides the damping required by TMD, and the shell and stiffener transfer the reaction force to the main structure. The installation method is as follows: the mass

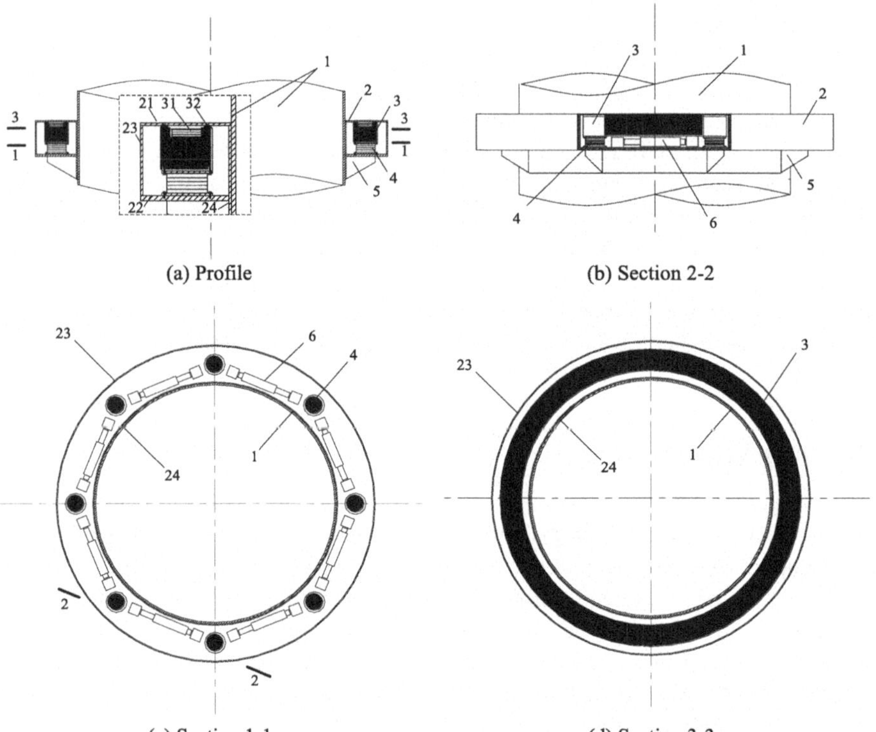

1. Main structure; 2. TMD shell; 3. Main mass block; 4. Rubber bearing; 5. Stiffening plate; 6. Damper; 21. Upper cover plate; 22. Lower cover plate; 23. Outer baffle; 24. Inner baffle; 31. Adjusting mass block; 32. Sliding pivot

Fig. 8.1 Ring TMD supported by rubber

block and the top plate of the rubber support are connected by bolts, and the sliding mass points form smooth sliding contact with the upper cover plate. The adjusting mass block is installed in the regulating groove of the main mass block and can be fixed in a proper way; the rubber support bottom plate is connected with the lower cover plate by bolts, the stiffening plate is welded with the lower cover plate and the inner cover plate, and the damper is hinged with the lower cover plate and mass block through the joint protruding from the mass block; the whole TMD can be connected with the main structure by bolt connection or welding according to the specific situation. Here, the rubber bearing adopts the general bearing without damping, and the damper adopts the series of viscous fluid damper developed by the research group, which is selected according to the specific design requirements; the lower cover plate and stiffening plate, as the main force transmission components, need to carry out the rigidity and strength checking calculation, and take the strengthening measures if necessary. It can be seen that the components providing mass, stiffness and damping of the device are separate, with small mutual influence and clear stress; through the sliding fulcrum, the excessive shaking of the mass block is avoided, and the influence of this factor on the damping effect is eliminated; the tuning frequency of the device can be adjusted after installation through the mass block according to the vibration characteristics of the actual structure to adjust the tuning frequency of the device and improve the real damping effect; the device structure is simple, which is easy to be made and has certain practical value.

It is assumed that the inner diameter of the mass block is $R_{m,1}$, the outer diameter is $R_{m,2}$, and the height is h_m; the rubber isolation bearings are uniformly and symmetrically arranged along the ring mass block, and the horizontal stiffness of each bearing is equal to k_{rub}; the dampers are installed considering the layout space and symmetry comprehensively, and the performance parameters of each damper are the same, the damping coefficient is C_d, and the damping index is α_d. The performance parameters of RSR-TMD are as follows:

$$m_{TMD} = \pi \rho_m h_m \left(R_{m,2}^2 - R_{m,1}^2\right) \tag{8.1}$$

$$k_{TMD} = n k_{rub} \tag{8.2}$$

$$C_{TMD} = C_d \sum_{j=1}^{m} \left(\cos \beta_j\right)^{1+\alpha_d} \tag{8.3}$$

$$\alpha = \alpha_d \tag{8.4}$$

where, ρ_m is the density of the mass block; n and m are the number of rubber bearings and dampers respectively; β_j is the angle between the j-th damper and the moving direction, belongs to the range of $(-\pi/2, \pi/2)$; m_{TMD} is the mass of the TMD mass block; k_{TMD} is the stiffness of TMD in the moving direction; C_{TMD} and α are the damping coefficient and damping index in the moving direction. It can be seen from the above formula that the mass of RSR-TMD depends on the geometric size and material density of the mass block; the rigidity depends on the number and rigidity of the rubber bearing; since the dampers are usually arranged symmetrically and evenly, the damping coefficient of TMD depends on the number, damping coefficient and damping index of viscous dampers, and the damping index of TMD is determined by the damping index of viscous dampers.

8.1.1.2 Variable Damping TMD of Rubber Bearing

In high-rise buildings, the fire water tank is an essential facility, and in the process of TMD vibration reduction design of such buildings, finding a reasonable space to set TMD is one of the keys to successful application. Therefore, the reasonable use of the existing space and quality of fire water tank to design appropriate TMD is gradually recognized by scholars and engineers. According to the actual needs of the project and the construction ideas in Fig. 3.3b, our team introduced the self-developed variable damping viscous fluid damper in Sect. 5.3.3.1, and designed a variable damping TMD with rubber support, as shown in Fig. 8.2. In the figure, the special-shaped water tank is used as the mass unit of TMD system. Four laminated rubber pads are set at the bottom of the water tank to completely separate the water tank from the ground, and a variable damping viscous fluid damper is installed between the laminated rubber pad and the water tank. The horizontal shear stiffness of the laminated rubber pad is used to provide the stiffness for the TMD system, so as to ensure that there is a definite natural vibration period between the inertial mass and the vibration mode resonance of the main structure; the viscous fluid damper is used to provide the damping for the TMD system. The design has been successfully applied to the wind-induced vibration control of Beijing Olympic multi-functional studio tower (Sect. 16.3).

8.1.2 Suspended TMD

8.1.2.1 Pendulum Ring Tuned Mass Damper

As described in Sect. 8.1.1.1, a Pendulum ring tuned mass damper (PR-TMD) is proposed for the standardized self-supporting high-rise structure. As shown in Fig. 8.3, the device is mainly composed of stiffening plate, shell, suspension sling, mass block and damper. The main principle is: the mass block provides the mass, the length of the sling controls the frequency of the TMD, the damper provides the required damping, and the shell and stiffener transfer the reaction force to the main structure. Its installation method is that: the mass block is hinged with the suspension cable, and the suspension cable is hinged with the upper cover plate; the whole TMD can be connected with the main structure by bolt connection or welding according to the specific situation; the damper is hinged with the upper cover plate and the mass block respectively, and the damper arrangement method

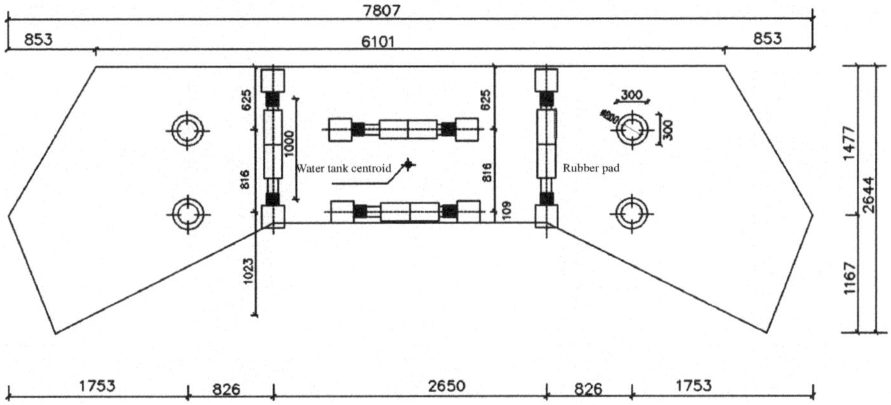

(a) Bottom layout

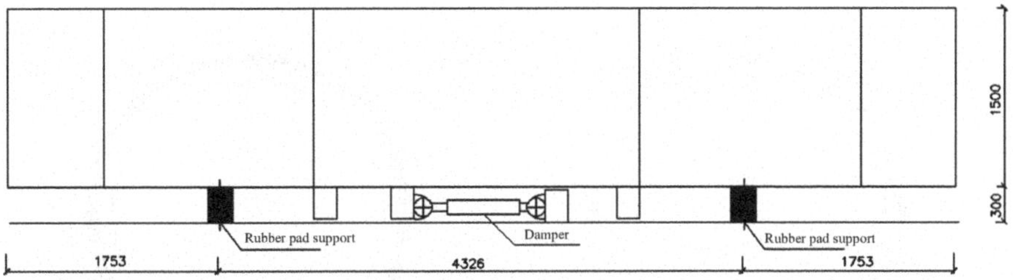

(b) Elevation layout

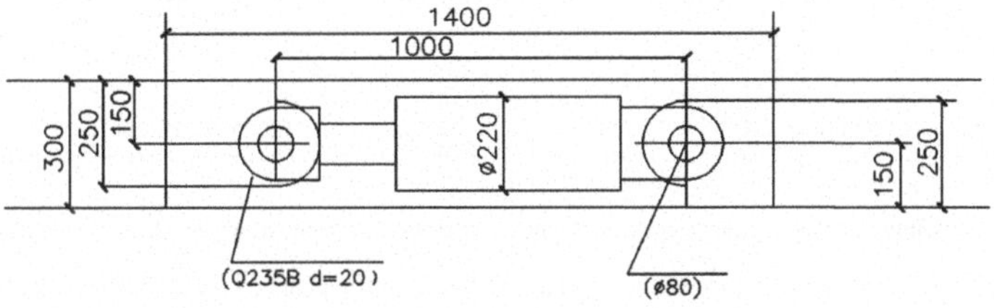

(c) Damper elevation

Fig. 8.2 Variable damping TMD of rubber bearing

of RSR-TMD (Fig. 8.1c) can also be used. The upper cover plate and stiffening plate, as the main force transmission components, need to be checked for rigidity and strength, and if necessary, the reinforcement measures should be taken; when using the damper arrangement of RSR-TMD, the lower cover plate also needs to be checked and strengthened. With the help of adjusting the effective length of the cable or increasing the small stiffness spring, the tuning frequency of the device can be adjusted according to the vibration characteristics of the actual structure to improve the real damping effect.

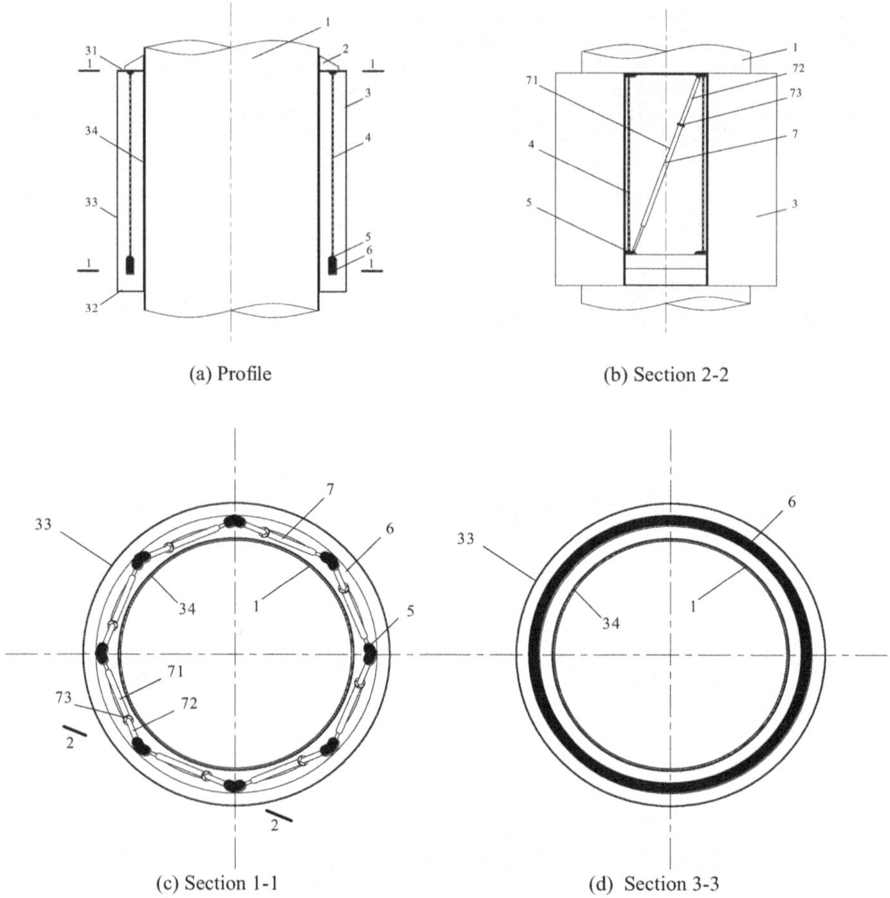

1. Chimney; 2. Stiffening plate; 3. TMD shell; 4. Suspension sling; 5. Suspension joint; 6. Mass block; 7. Damping support; 31. Upper cover plate; 32. Lower cover plate; 33. Outer baffle plate; 34. Inner baffle plate; 71. Damper; 72. Connecting rod; 73. Flange connection

Fig. 8.3 A suspension swing ring TMD

It is still assumed that the inner diameter of the mass block is $R_{m,1}$, outer diameter is $R_{m,2}$, and height is h_m; the distance from the cable node to the mass center is L_{TMD}; the damper is installed considering the layout space and symmetry comprehensively, and the performance parameters of each damper are the same, the damping coefficient is C_d, and the damping index is α_d. The performance parameters of PR-TMD are as follows:

$$m_{TMD} = \pi \rho_m h_m \left(R_{m,2}^2 - R_{m,1}^2\right) \tag{8.5}$$

$$k_{TMD} = \frac{m_{TMD} g}{L_{TMD}} \tag{8.6}$$

Using bottom plate damper to connect (Fig. 8.1c):

$$C_{TMD} = C_d \sum_{j=1}^{m} \left(\cos \beta_j\right)^{1+\alpha_d} \tag{8.7}$$

$$\alpha = \alpha_d \tag{8.8}$$

Using top plate damper to connect (Fig. 8.3c):

$$C_{TMD} = C_d \xi^{1+\alpha_d} \sum_{j=1}^{m} \left(\cos \beta_j\right)^{1+\alpha_d} \tag{8.9}$$

$$\alpha = \alpha_d \tag{8.10}$$

It can be seen from the above formula that the mass of PR-TMD depends on the geometric size and material density of the mass block; the rigidity depends on the mass and suspension height of mass block; since the dampers are usually arranged symmetrically and evenly, the damping coefficient of TMD depends on the number of viscous dampers, damping coefficient, damping index and the setting mode of damping support, and the damping index of TMD is still determined by the damping index of viscous dampers.

For the suspension ring TMD (PR-TMD, as shown in Fig. 8.4a, b) for vibration reduction test, the TMD mass block is 26.86 kg, and the spring stiffness is 1254 N/m. The inner diameter of the mass block is 0.4 m, the outer diameter is 0.5 m, the height is 0.048 m, the suspension height is 0.265 m, and the diameter of the suspension cable is 2 mm, totally 8 pieces. An acceleration sensor is arranged on the mass block, and then the mass block is moved and relaxed to make the mass block decay freely. Figure 8.4c shows the test results of TMD, PSD represents the power spectral density, and numbers 1 and 2 represent two orthogonal vertical directions respectively; Fig. 8.4d calculates the damping ratio of TMD, the mean value of four tests is 0.0264, and the root variance is 0.0063, which indicates that the friction between the cable itself and the connecting part has provided a certain viscous damping for TMD without damper.

8.1.2.2 Two Way Double Layer Compound Swing TMD

In the engineering practice of TMD vibration reduction design of high-rise structure, three requirements of structure should be met: long period, finite space and plane multi direction. According to the actual needs of the project and the construction ideas in Fig. 3.3g, our team introduced the independently developed viscous fluid

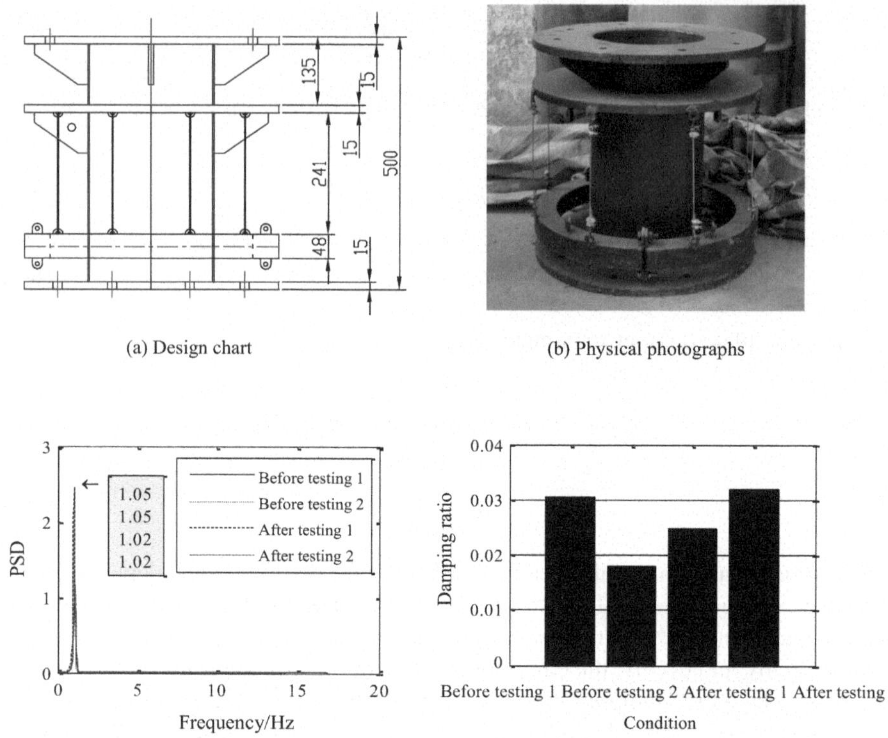

(a) Design chart

(b) Physical photographs

(c) Frequency

(d) Damping ratio

Fig. 8.4 Suspended ring TMD test

damper in Sect. 5.3, and designed the bidirectional double-layer composite swing TMD, as shown in Fig. 8.5. In this device, the water tank is hoisted by cable to form a TMD structure of suspension pendulum, which can realize the dynamic performance of TMD in many directions. At the same time, considering the contradiction between long period and limited space, the TMD is designed with the concept of double-layer suspension pendulum. The device is composed of cable, suspension frame, fire water tank, water tank bracket, viscous damper and other components. As a quality unit, the fire water tank is installed on the bracket of the water tank and suspended on the suspended outer frame through the bracket of the water tank. The outer frame is suspended on the main structure through the cable, forming a two-stage suspension system, which realizes the long-term suspension of TMD in a limited space. In order to provide reasonable damping for TMD, a viscous damper is set at the bottom of the water tank. One end of the damper is connected with the floor through the lower node buttress, and the other end is connected with the water tank through the upper node buttress. The damper and cable can be adjusted in a certain range according to the field test results to ensure better control effect.

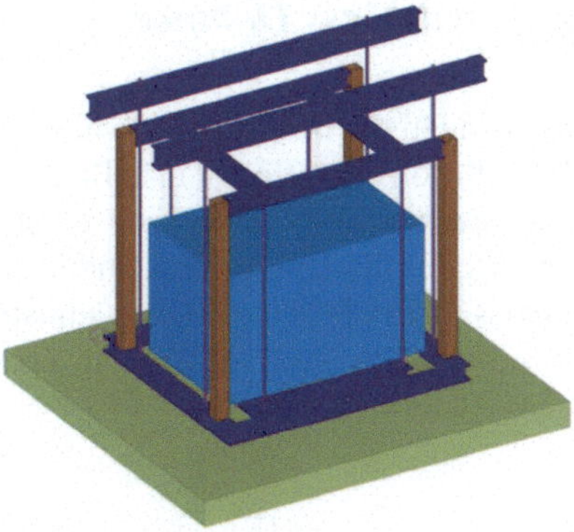

(a) 3D sketch

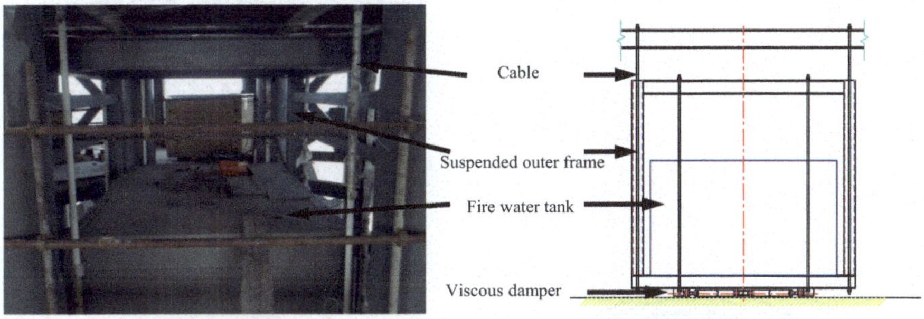

(b) Elevation layout

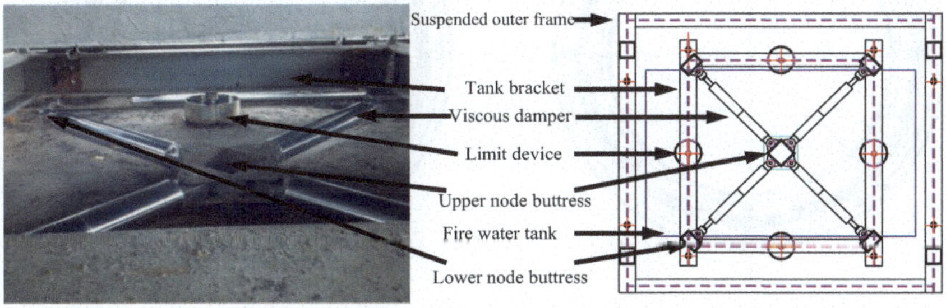

(c) Plan layout

Fig. 8.5 Double layer bidirectional compound swing TMD

8.1.3 Integrated Ring Tuned Mass Damper

For the standardized self-supporting high-rise structure, a new type of integrated ring tuned mass damper (IR-TMD) with consideration of installation space and durability. As shown in Fig. 8.6, the device is mainly composed of stiffening plate, shell, adjustable spring, mass piston, viscous fluid, partition, conduit and internal stiffening beam. Its main principle is: mass piston provides mass, adjustable spring provides rigidity, and viscous fluid flows through damping hole to provide damping when

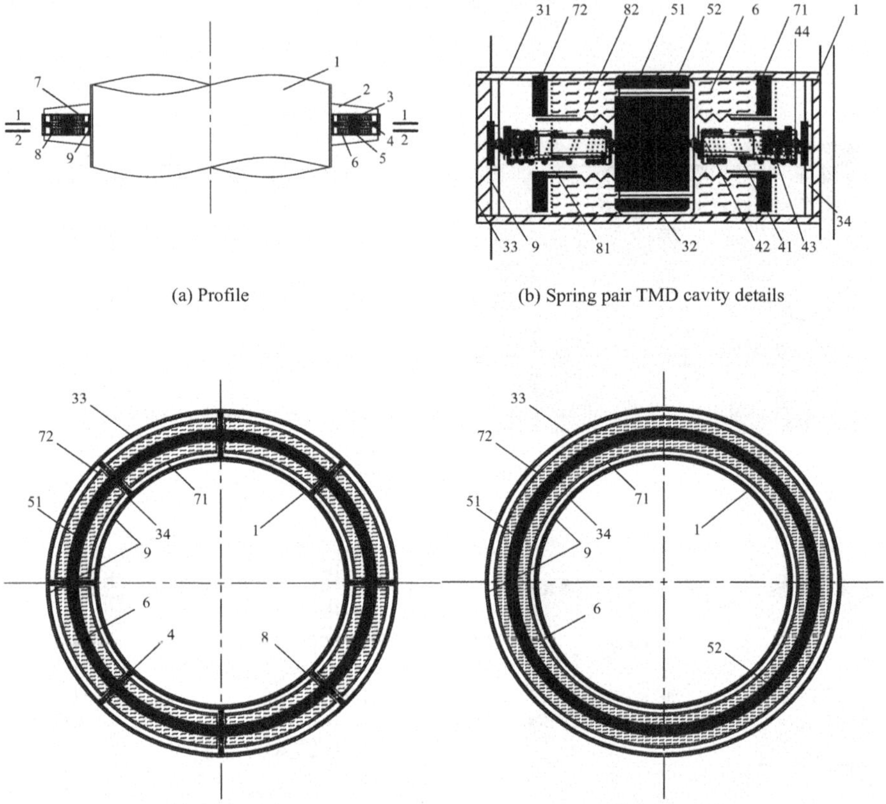

1. Chimney; 2. External stiffening plate; 3. TMD shell; 4. Adjustable spring; 5. Mass piston; 6. Viscous fluid; 7. Diaphragm; 8. Conduit; 9. Internal stiffening beam; 31. Upper cover plate; 32. Lower cover plate; 33. Outer baffle plate; 34. Inner baffle plate; 41. Spring ring; 42. Expansion guide rod; 43. Adjusting bolt; 44. Spring joint; 51. Piston body; 52. Damping hole; 71. Inner baffle plate; 72. Outer baffle plate; 81. Hard Conduit; 82. Flexible conduit.

Fig. 8.6 Integrated ring tuned mass damper

mass piston reciprocates. The installation method: the mass piston and adjustable spring are fully hinged in the horizontal plane, and the treatment process of viscous fluid damper piston and cylinder block is used to form the slide contact with the upper/lower cover plates; the adjustable spring and the inner/outer baffle plate are also fully hinged in the horizontal plane; the inner and outer baffle plates are welded with the upper/lower cover plates, and leak proof treatment is carried out; the connection between the conduit and the baffle plate is hard pipe connection, the connection between the conduit and the mass piston is flexible hose connection, and there is a certain gap with the adjustable spring. The specific length of the soft/hard pipe and gap width are determined by the TMD stroke; the viscous fluid is injected through the preset filling hole or before the installation of the top cover plate; the whole TMD can be connected with the main structure by bolt connection or welding according to the specific situation. In addition, TMD shell, stiffening plate and stiffening beam, as the main force transmission components, need to check the rigidity and strength, and take strengthening measures if necessary; the size of conduit needs to be determined according to the maximum displacement of TMD obtained by analysis; in order to reduce the decreasing effect on the damping effect caused by the self flow of fluid inside and outside the mass block, a soft connection partition can be set between the damping holes along the ring to limit the interaction flow between the fluids. With the help of the adjusting bolt (Fig. 8.6b), the tuning frequency of the device can be adjusted after installation according to the vibration characteristics of the actual structure to improve the real vibration reduction effect. Due to the reasonable integration of mass components, stiffness components and damping components in a small space, the required space of TMD is greatly reduced; and the performance of each direction is the same, which has a good damping effect for the vibration in different directions caused by wind load; because of the maturity of the current airtight leak proof and other related manufacturing processes, the TMD has a good reliability. When a smaller boundary dimension is needed, the spring pair can be reduced to only one spring adjacent to the main structure, and the other components are the same as the spring pair, so the TMD shape will be greatly reduced after reducing one external cavity.

It is still assumed that the inner diameter of mass block is $R_{m,1}$, outer diameter is $R_{m,2}$, and height is h_m; the radius of damping hole is R_d; the number of spring pairs is n; the total number of damping holes is m; the rigidity of each spring pair is equal to k_{stiff}, and the radius of conduit is R_{stiff}; Then the mass and stiffness of IR-TMD can be expressed as follows:

$$m_{TMD} = \rho_m \pi \left[h_m\left(R_{m,2}^2 - R_{m,1}^2\right) - mR_d^2\left(R_{m,2} - R_{m,1}\right)\right] \tag{8.11}$$

$$k_{TMD} = k_{stiff} \sum_{j=1}^{n} \left(\cos\beta_j\right)^2 \tag{8.12}$$

The damping coefficient and damping index should be analyzed by the principle of hydrodynamics. The viscous fluid in TMD is the product independently developed

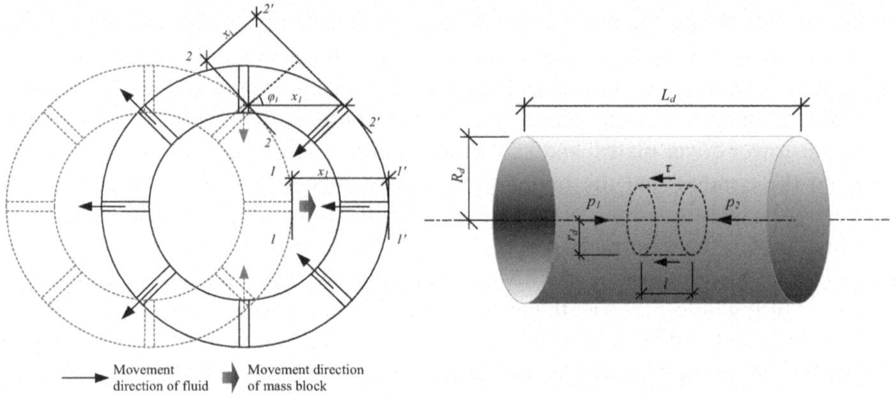

(a) Motion diagram (b) Fluid flow in the damping hole

Fig. 8.7 Analysis diagram of IR-TMD calculation model

by the research group. Because the viscosity of the fluid is large and the diameter of the damping hole is small, it can be known that the damping hole is generally laminar flow according to the criterion of flow pattern. For this kind of porous viscous fluid damping device, its damping is usually composed of friction damping caused by path loss affected by fluid viscosity when the fluid flows through the pores and the local loss damping caused by pore shrinkage.

As shown in Fig. 8.7a, with the movement of the mass block, the fluid flows in the opposite direction through the small hole. Since the small hole is very small relative to the mass block side surface, it is assumed that the fluid flows in the direction of the small hole, i.e. in the normal direction of the ring. In order to deduce the formula conveniently, it is assumed that the small hole perpendicular to the direction of motion also has the same fluid inflow. In fact, the resistance of the vertical motion direction of the small hole with relative nominal final expression is mutually offset, so it does not affect the final result. Firstly, the energy consumption of a single damping hole is studied. As shown in Fig. 8.7b, a damping hole is L_d in length and R_d in radius. Take a small cylinder whose axis coincides with the damping hole for analysis. The length of the cylinder is l and the radius is r_d. The cylinder flows in a constant and uniform laminar flow. On the two end faces of the cylinder, the pressures include $p_1$and p_2. On cylinder side surface, the pressure is perpendicular to the axis and the shear stress τ is parallel to the axis. Considering the symmetry, all the shear stresses are uniformly distributed on the sides. From the force balance along the axis direction, we can get:

$$(p_1 - p_2)\pi r_d^2 - T = 0 \tag{8.13}$$

where, $T = \tau S$ is the sum of the tangential stress around the cylinder; $S = 2\pi r_d l_d$ is the side area of the cylinder; so $\Delta p = p_1 - p_2$, the above formula can be expressed as:

$$\Delta p \pi r_d^2 = 2\pi r_d l_d \tau \tag{8.14}$$

Then:

$$\tau = \Delta p r_d / 2 l_d \tag{8.15}$$

Therefore, the shear stress on the hole wall is:

$$\tau_w = \Delta p R_d / 2 L_d \tag{8.16}$$

It is assumed that the constitutive equation of damping medium in TMD satisfies the power-law equation:

$$-(du_d/dr_d) = \dot{\gamma} = (\tau/k)^{1/\alpha'} \tag{8.17}$$

where k is the consistency coefficient and α' is the flow index. Substituting Eqs. (8.15) and (8.16) into the above formula, we can get:

$$-du_d = (\tau_w r_d/(R_d k))^{1/\alpha'} dr_d \tag{8.18}$$

It can be seen from the principle of hydrodynamics that, when $r_d = R_d$, $u_d = 0$, the following formula can be obtained by integrating the above formula with this as the boundary condition:

$$u_d(r_d) = \frac{R_d r_w^{1/\alpha'}}{(1/\alpha' + 1)k^{1/\alpha'}}\left[1 - \left(\frac{r_d}{R_d}\right)^{1/\alpha'+1}\right] \tag{8.19}$$

$$u_m = \frac{1}{\pi R_d^2}\int_0^{R_d} 2\pi r_d u_d(r_d) dr_d = \frac{R_d \tau_w^{1/\alpha'}}{(1/\alpha' + 3)k^{1/\alpha'}} \tag{8.20}$$

By integrating Eq. (8.19), the flow through a single damping hole can be obtained as follows:

$$\begin{aligned} Q_1 &= \int_0^{R_d} 2\pi r_d u_d(r_d) dr_d = \left(\frac{2\pi}{1/\alpha' + 1}\right)\left(\frac{\Delta p}{2kL_d}\right)^{1/\alpha'}\left(\frac{1/\alpha' + 1}{2(1/\alpha' + 3)}\right) R_d^{1/\alpha'+3} \\ &= \frac{\pi R_d^3 \tau_w^{1/\alpha'}}{(1/\alpha' + 3)k^{1/\alpha'}} \end{aligned} \tag{8.21}$$

The number of layers of damping holes is a. if there are B damping holes in each layer, there are: $m = ab$. As shown in Fig. 8.7a, the damping holes of the same vertical section are defined as a group, and the angle between the group i damping holes and the moving direction is φ_i, which belongs to the range of $(-\pi/2, \pi/2)$. When the horizontal displacement of the mass block is x_1, the distance change between the outside of the damping hole and the cylinder wall is:

$$x_i = x_1 \cos(\varphi_i) \tag{8.22}$$

The change speed between the outside of the small hole and the cylinder wall is:

$$v_i = \dot{x}_i = \dot{x}_1 \cos(\varphi_i) \tag{8.23}$$

The total amount of flow through a set of damping holes per unit time can be expressed as follows:

$$Q_i = a\left(\frac{2\pi}{1/\alpha'+1}\right)\left(\frac{\Delta p_i}{2kL_d}\right)^{1/\alpha'}\left(\frac{1/\alpha'+1}{2(1/\alpha'+3)}\right)R_d^{1/\alpha'+3} = a\frac{\pi R_d^3}{1/\alpha'+3}\left(\frac{\Delta p_i R_d}{2kL_d}\right)^{1/\alpha'} \tag{8.24}$$

$$Q_i = A_i v_i \tag{8.25}$$

where, A_i is the influence area of group i damping hole, approximately calculated by the mass block center line, $A_i = \pi\left(R_{m,1} + R_{m,2}\right)h_m/b - A_{stiff,i}$, $A_{stiff,i} = n_i \pi R_{stiff}^2$. According to Eq. (8.24), the pressure difference at both ends of a group of damping holes is:

$$\Delta p_i = \frac{2kL_d}{R_d}\left(\frac{3\alpha'+1}{a\alpha'\pi R_d^3}Q_i\right)^{\alpha'} \tag{8.26}$$

And the normal damping force produced by this group of damping holes is $F_i = \Delta p_i A_i$, so we can know:

$$\begin{aligned} F_i &= \pi\left(\left(R_{m,1} + R_{m,2}\right)h_m/b - n_i R_{stiff}^2\right) \\ &\quad \frac{2kL_d}{R_d}\left(\frac{3\alpha'+1}{a\alpha'\pi R_d^3}\pi\left(\left(R_{m,1} + R_{m,2}\right)h_m/b - n_i R_{stiff}^2\right)\dot{x}_1 \cos\varphi_i\right)^{\alpha'} \\ &= \frac{2\pi kL_d}{R_d}B_i\left(\frac{3\alpha'+1}{a\alpha' R_d^3}B_i \cos\varphi_i\right)^{\alpha'}\dot{x}_1^{\alpha'} \end{aligned} \tag{8.27}$$

Among them, $B_i = A_i/\pi$. Therefore, the damping force provided by the path loss along the mass motion direction is:

$$\begin{aligned} F &= \sum_{i=1}^{b} F_i \cos\varphi_i = \sum_{i=1}^{b} \frac{2\pi kL_d}{R_d}B_i\left(\frac{3\alpha+1}{a\alpha' R_d^3}B_i \cos\varphi_i\right)^{\alpha'}\dot{x}_1^{\alpha'} \cos\varphi_i \\ &= \left[\frac{2\pi kL_d}{R_d}\left(\frac{3\alpha'+1}{a\alpha' R_d^3}\right)^{\alpha'}\sum_{i=1}^{b} B_i(B_i \cos\varphi_i)^{\alpha'}\cos\varphi_i\right]\dot{x}_1^{\alpha'} \end{aligned} \tag{8.28}$$

Among them, the length of damping hole is $L_d = R_{m,2} - R_{m,1}$. When $\alpha' = 1$, the above formula is the expression of linear fluid. For the total local energy loss caused by shrinkage, it can be expressed as follows:

$$h = (\zeta_1 + \zeta_2)\frac{v_d^2}{2g} \tag{8.29}$$

Among them, ζ_1 and ζ_2 are the expansion loss coefficient and reduction loss coefficient respectively, v_d is the velocity in the middle of the small hole. Then Bernoulli equation is listed for the inlet section and the outlet section:

$$\frac{p_{in}}{\rho g} + \frac{a_{in}v_{in}^2}{2g} = \frac{p_{ou}}{\rho g} + \frac{a_{ou}v_{ou}^2}{2g} + h \tag{8.30}$$

where, p_{in} and p_{ou} are the pressures at the inlet and outlet respectively, v_{in} and v_{ou} are the flow velocities at the inlet and outlet respectively, $v_{in} = v_{ou}$, and $a_{in} = a_{ou} = 1$, $\Delta p = p_{in} - p_{ou}$. So the flow rate of a group of damping holes is:

$$Q_i = a\pi R_d^2 \sqrt{\frac{2\Delta p_i}{\rho(\zeta_1 + \zeta_2)}} \tag{8.31}$$

Then:

$$\Delta p_i = \frac{\rho(\zeta_1 + \zeta_2)}{2a^2\pi^2 R_d^4} Q_i^2 \tag{8.32}$$

$$\begin{aligned} F_i' &= \Delta p_i A_i \\ &= \frac{\rho(\zeta_1 + \zeta_2)}{2a^2\pi^2 R_d^4}\left[\pi\left(\left(R_{m,1} + R_{m,2}\right)h_m/b - n_i R_{stiff}^2\right)v_i\right]^2 \\ &\quad \pi\left(\left(R_{m,1} + R_{m,2}\right)h_m/b - n_i R_{stiff}^2\right) \\ &= \frac{\pi\rho(\zeta_1 + \zeta_2)}{2a^2 R_d^4} B_i^3(\cos\varphi_i)^2 \dot{x}_1^2 \end{aligned} \tag{8.33}$$

Therefore, the damping force provided by the shrinkage loss along the moving direction of the mass block can be obtained as follows:

$$\begin{aligned} F' &= \sum_{i=1}^{b} F_i' \cos\varphi_i = \sum_{i=1}^{b} \frac{\pi\rho(\zeta_1 + \zeta_2)}{2a^2 R_d^4} B_i^3(\cos\varphi_i)^2 \dot{x}_1^2 \cos\varphi_i \\ &= \left[\frac{\pi\rho(\zeta_1 + \zeta_2)}{2a^2 R_d^4} \sum_{i=1}^{b} B_i^3(\cos\varphi_i)^2 \cos\varphi_i\right]\dot{x}_1^2 \end{aligned} \tag{8.34}$$

Combining the above two damping forces, the damping force model of IR-TMD is obtained as follows:

$$F_{TMD,d} = F + F' = C_1 \dot{x}_{TMD}^{\alpha'} + C_2 \dot{x}_{TMD}^2 \quad (8.35)$$

where, $C_1 = \frac{2\pi k L_d}{R_d}\left(\frac{3\alpha'+1}{\alpha\alpha' R_d^3}\right)^{\alpha'} \sum_{i=1}^{b} B_i (B_i \cos\varphi_i)^{\alpha'} \cos\varphi_i$, $C_2 = \frac{\pi\rho(\zeta_1+\zeta_2)}{2a^2 R_d^4} \sum_{i=1}^{b} B_i^3 (\cos\varphi_i)^2 \cos\varphi_i$. In the preliminary calculation, ζ_1 and ζ_2 can be taken as 1 and 0.5, respectively. The above formula is based on a series of assumptions, and the real damper will be affected by many factors. Generally, it is difficult to realize the completely theoretical accurate calculation model of viscous fluid damper. Because the high-order energy consumption of viscous fluid damper accounts for a small proportion of damper energy consumption, it will be simplified as:

$$F = C_{TMD} \operatorname{sgn}(\dot{x}_{TMD}) |\dot{x}_{TMD}|^{\alpha} \quad (8.36)$$

where, C_{TMD} is viscous damping coefficient, which is related to C_1 and C_2; α is damping index, which is related to structure and damping material. They are expressed as $C_{TMD} = C_{\mathrm{m}} C_1, \alpha = \alpha_{\mathrm{m}} \alpha'$, where C_{m} and α_{m} are modified parameters by test data. It can be seen that the damping coefficient of IR-TMD is closely related to the construction size of mass block, the number of damping holes, the construction form and damping material; the damping index mainly depends on the flow index of damping material, and the construction also has a certain influence on it, but it is relatively small.

For the proposed IR-TMD, the influence of manufacturing parameters on its mechanical properties was studied. First of all, according to the analysis of the above calculation formula, the mass of TMD is related to the material and geometric dimension of the mass block, which is usually determined by the structural mass and selected according to the geometric dimension of the structure, with a small range of variation; when the influence of high-order energy consumption and correction parameters is ignored, the damping coefficient of TMD is directly proportional to the liquid viscosity coefficient K and the damping hole length L_d, in which the damping hole length is determined by the geometric size of mass lock. The damping index of TMD is usually the same as that of liquid and has little correlation with other parameters. Therefore, it is assumed that the number of springs is equal to the number of damping holes in each layer, and studies the influence of damping index, damping hole radius, damping hole number in each layer and spring pipe radius on TMD damping coefficient, as well as the influence of spring number on TMD stiffness. In order to facilitate the analysis, the mass block is made of iron, the outer radius is 1.35 m, the inner radius is 1.25 m, the height is 0.2 m, there are two layers of damping holes, the liquid viscosity coefficient is 1 N/(m/s)$^{\alpha}$, the stiffness of single spring is 1000 N/m, the influence of each parameter is analyzed: (1) the influence of liquid damping index is analyzed, the damping hole radius is 0.002 m, the total number of damping holes and springs in each layer are 4, the spring pipe radius is 0.05 m, the damping index changes from 0.2 to 1, and the analysis results are shown in Fig. 8.8a: with the increase of damping index, the damping coefficient increases, and the larger the damping index is, the greater the increase is. (2) make the damping index 0.6, the total number of damping holes and springs in each layer are 4, the radius of spring pipe is 0.05 m, and the radius of damping hole changes between 0.001 and 0.01 m. Analyze the influence of the radius of damping hole, as shown in Fig. 8.8b: with the increase of the radius of damping hole, the damping coefficient decreases, and the larger the damping radius is, the smaller the reduction is. (3) if the damping index is 0.6, the damping hole radius is 0.002 m, and the spring pipe radius is 0.05 m, and the total number of damping holes and springs in each layer vary from 4 to 16, the impact is shown in Fig. 8.8c: with the increase of the number, the damping coefficient decreases and the stiffness increases. (4) the damping index is 0.6, the damping hole radius is 0.002 m, the total number of damping holes and springs in each layer are 4, and the spring pipe radius is 0.01–0.05 m, and the influence is shown in Fig. 8.8d: with the increase of the pipe radius, the damping coefficient decreases

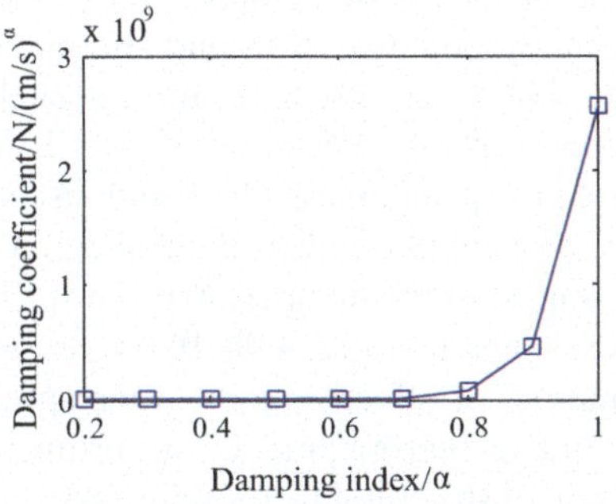

(a) Influence of damping index

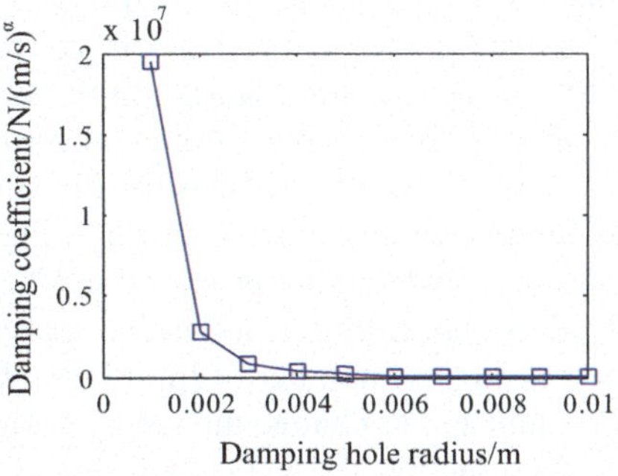

(b) Influence of damping hole radius

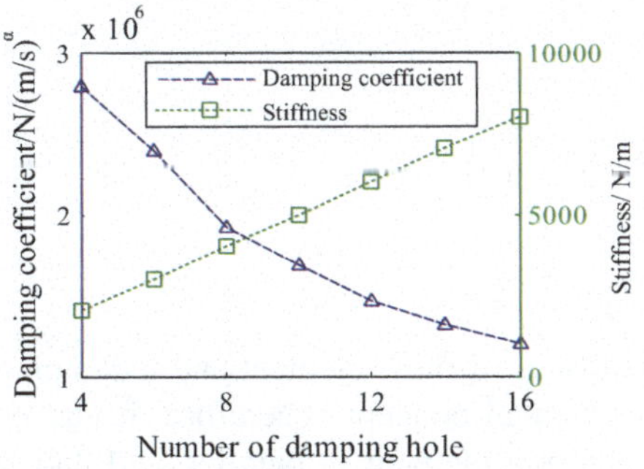

(c) The influence of the number of damping holes in each layer

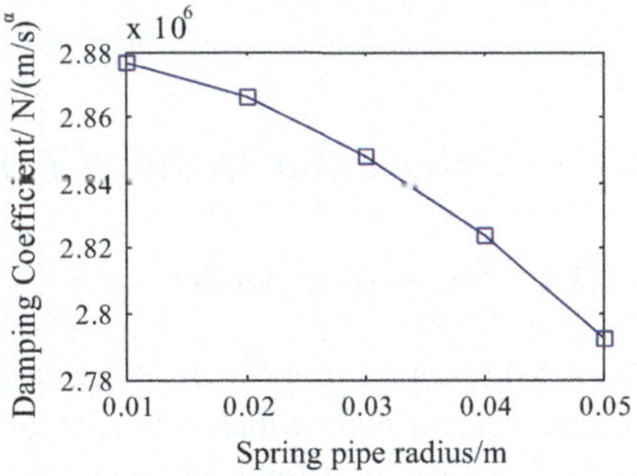

(d) Influence of spring pipe radius

Fig. 8.8 Influence of parameters on IR-TMD performance

gradually. To sum up, it can be seen from the change of specific value of damping coefficient along with the influence of each parameter that the pipeline radius has the least influence on it, the number of damping holes is the second, the radius of damping holes and damping index have a greater influence, and the parameters are reasonably selected according to the specific situation in the specific design.

8.1.4 Adjustable Stiffness Vertical TMD

The natural frequency of TMD is the key to reduce the dynamic response. However, there are some differences between the calculation model used in the actual design of TMD system and the real structure in normal use. The designed natural frequency of TMD must be different from the actual required natural frequency, which seriously affects the vibration reduction effect. Of course, it is neither economical nor practical to dismantle all the installed TMD and then conduct production and installation again. Considering that the natural frequency of TMD depends on the effective stiffness of spring, the effective stiffness of spring shock absorber can be realized by adjusting the diameter, pitch diameter, pitch, effective length, effective turns and single turn stiffness of spring wire. Based on the existing TMD, the research group uses the adjustable spring to replace common spring, a new type of adjustable stiffness TMD is developed. The new TMD is mainly composed of spring, mass block and viscous fluid damper. Its structure is shown in Fig. 8.9a. Among them, C is the damper, and the viscous fluid damper proposed in Chap. 5 is used as the damper; K is the spring with adjustable stiffness, and its basic principle is shown in Fig. 8.9b. By increasing or decreasing the number of limit bolts, the number of effective coils of the spring can be adjusted to change the spring stiffness. In engineering practice, according to the basic principle, the team has designed a variety of structural forms such as double side spring TMD (Fig. 8.9c), suspension mass TMD (Fig. 8.9d) and bottom bearing mass TMD (Fig. 8.9e).

8.1.5 Calculation Model of TMD

8.1.5.1 Mechanical Model

The TMD proposed above is all represented by a stiffness element and a damping element in parallel, which has a clear mechanical concept. Therefore, it can be represented by a stiffness element and a glue pot element in parallel, and then a mass element in series. As shown in Fig. 8.10, when the members connected with dampers meet the stiffness requirements, the elements between the mass block and the main structure are equivalent to a Kelvin model. In the figure, m_{TMD} is the mass of TMD mass block; k_{TMD} is the stiffness of TMD movement direction, generally the stiffness sum of spring in this direction and the equivalent stiffness considering the

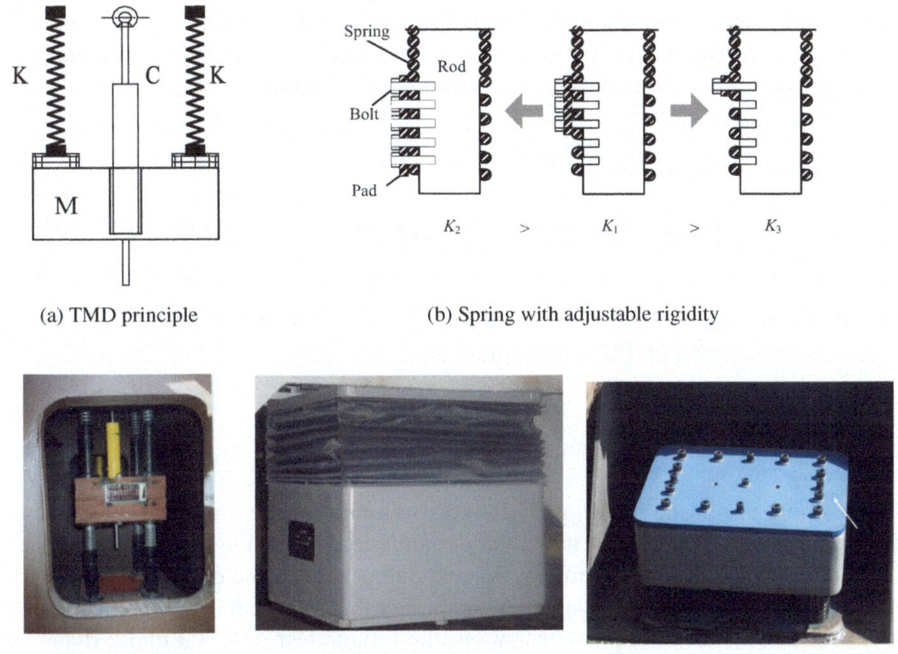

(a) TMD principle (b) Spring with adjustable rigidity

(c) Bilateral spring (d) Suspension mass (e) Bottom bearing mass

Fig. 8.9 Adjustable stiffness vertical TMD

Fig. 8.10 Computing model of TMD

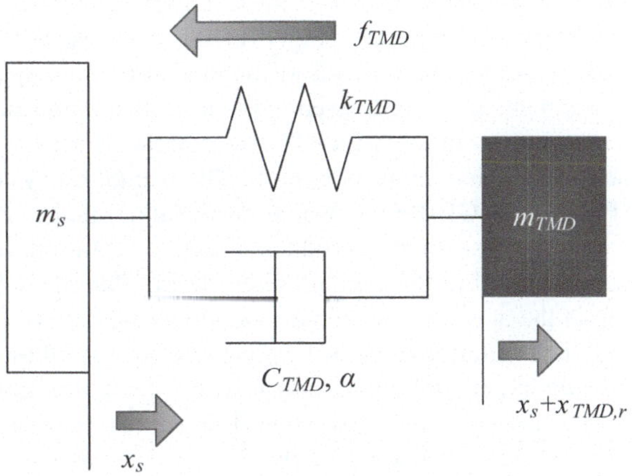

suspension length for suspension TMD; C_{TMD} and α are respectively the damping coefficient and damping index in the movement direction, and each parameter has been derived above. According to the principle of d'Alembert:

$$f_{TMD} = C_{TMD}\text{sgn}(\dot{x}_{TMD,r}) \cdot |\dot{x}_{TMD,r}|^{\alpha} + k_{TMD}x_{TMD,r} \tag{8.37}$$

$$m_{TMD}(\ddot{x}_s + \ddot{x}_{TMD,r}) + C_{TMD}\text{sgn}(\dot{x}_{TMD,r}) \cdot |\dot{x}_{TMD,r}|^{\alpha} + k_{TMD}x_{TMD,r} = 0 \tag{8.38}$$

Among them, f_{TMD} is the control force of TMD on the main structure, $x_{TMD,r}$ is the relative displacement of TMD mass block.

8.1.5.2 Fatigue Performance Test

In the whole life of the building, TMD keeps vibration state for a long time due to the action of wind, pedestrian and other dynamic loads. The long-term performance of its main component including spring and damper under the cyclic load is one of the focuses of research and design personnel. For this reason, the author's team took vertical TMD as an example to carry out the fatigue test of TMD under dynamic load.

Since spring is a common part of construction machinery, many researches have focused on its fatigue performance evaluation methods and tests, so this test mainly evaluates the performance degradation of viscous damper under long-term dynamic load. As shown in Fig. 8.11a, the test floor is a steel-concrete composite structure, and the TMD used in the test (Fig. 8.11b) is composed of eight springs, a viscous damper and a mass block, which is suspended at the mid span of the test floor. Two sets of instruments are mainly used in the test, including test system and excitation system. The test system includes 941-B acceleration sensor (Fig. 8.11c) and CRAS signal acquisition and analysis system. The excitation system includes JZQ-100 vibrator (Fig. 8.11d), GF-1000 power amplifier and signal source. In order to make the exciter excite the test floor to produce effective vibration, the exciter is fixed on the steel support fixed to the ground. Considering that the selection of excitation point will affect the accuracy of modal mode identification of the floor slab, in order to avoid that the excitation point is the node of modal amplitude is selected and the mode will be submerged and unrecognizable, the excitation point is set at point 7, that is, the position shown in the red circle (Fig. 8.11e). In order to make TMD vibrate greatly, the excitation frequency of the exciter is set as the floor fundamental frequency of 4.13 Hz. Eight channels are used for the test system, which are respectively connected with eight vibration pickups. 1–7 test points are arranged on the floor and 8 test points are arranged on the TMD. See Fig. 8.11f for the arrangement.

In order to obtain the change of TMD damping effect under the fatigue action, the acceleration of measuring points 1–7 under the action of exciter are tested before the fatigue test, and the acceleration of measuring points 1–7 are tested under the

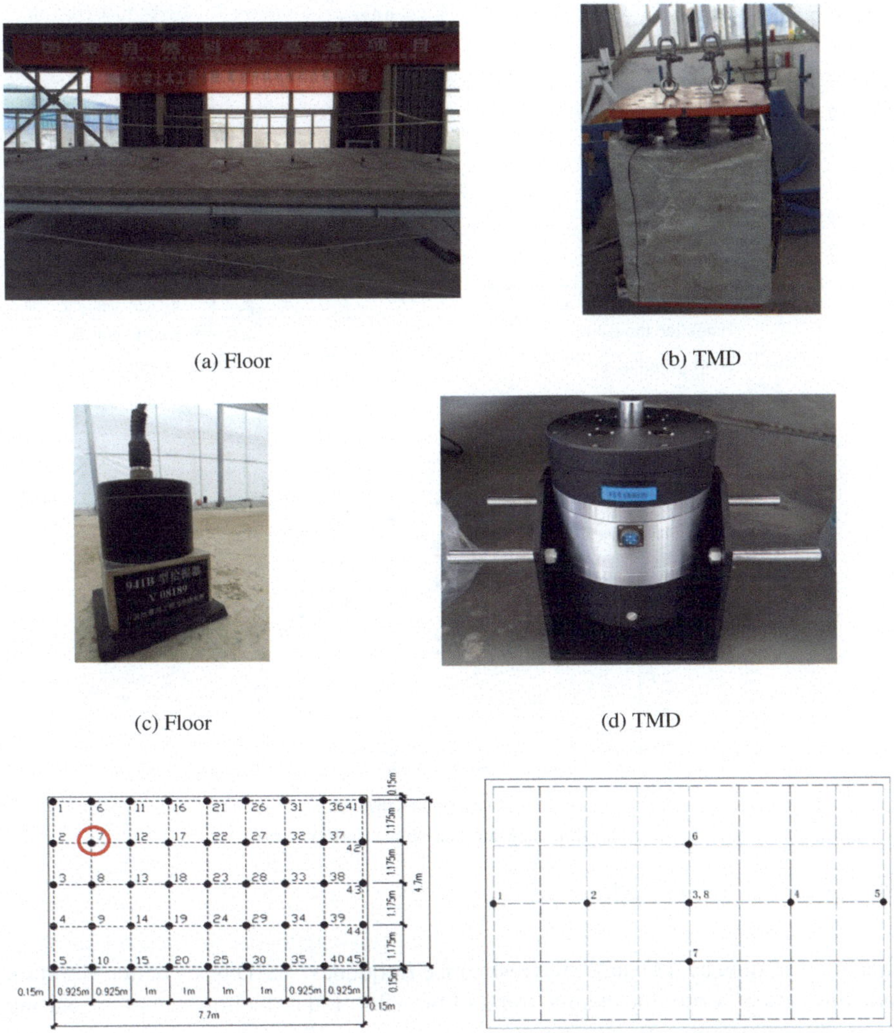

(a) Floor (b) TMD

(c) Floor (d) TMD

(e) Model node (f) Layout of measuring points

Fig. 8.11 Long term performance test of TMD

same vibration level of exciter after every 10,000 actions. The test results are shown in Fig. 8.12. It can be seen from the figure that: (1) the natural frequency of TMD shows a downward trend, and the change is obvious from 30,000 to 40,000 times. After 80,000 times of action, the frequency and stiffness of TMD decreased by 6.64% and 12.84% respectively; (2) the damping ratio of TMD fluctuated slightly. If the influence of error was ignored, it can be considered that the damping ratio ζ of TMD was almost unchanged, and the damping coefficient also decreased due to the

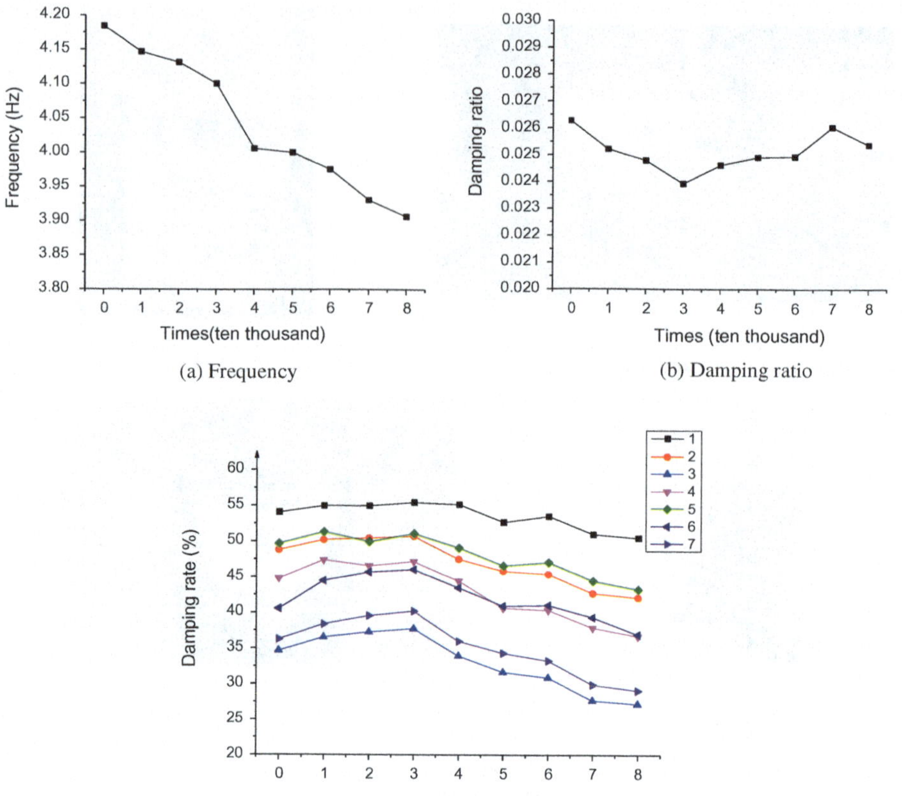

(a) Frequency

(b) Damping ratio

(c) Damping rate of each measuring point

Fig. 8.12 Analysis of test results

change of frequency. (3) under the resonance frequency, TMD has a better vibration reduction effect. From the change trend of the vibration reduction rate of measuring points 1–7, the vibration reduction rates of the first 30,000 times have an upward trend, and then slowly decline after 30,000 times, indicating that 4.10 Hz is the optimal vibration reduction frequency of TMD. (4) compared with the initial state, the vibration reduction rate of the test point 3 (TMD arrangement points) changes the most, and the vibration reduction rate of the test points far away from the test point 3 changes relatively little. (5) after 80,000 fatigue tests, the vibration reduction rate of each measuring point has decreased.

At the same time, the TMD fatigue performance test further verifies that the high performance viscous fluid damper proposed in Chap. 5 has excellent fatigue resistance and very stable working performance, but also exposes the following problems: on the one hand, in response to high frequency and persistent design conditions such as wind vibration and mechanical vibration, the superior stability of the high performance damper can meet the use requirements. However, due to the looseness of the connection between the damper and the structure, the energy dissipation effect of the damper is often reduced or even fails. On the other hand, although the improvement of damper assembly technology greatly reduces the installation error of damper, and reduces the deformation, displacement and wear of damper components caused by asymmetric installation, the cylinder rotation caused by extremely small asymmetric force under the high cycle working condition accumulated and increased continuously. During the disassembly inspection after loading, it is found that the cylinder barrel rotates a little around the shaft relative to the piston guide rod.

8.2 FM Liquid Damper

As mentioned in the Chap. 3 of this book, the basic principle of FM liquid damper is to absorb and dissipate the vibration energy of the main structure by using the vibration of liquid. However, how to reasonably design the form of FM liquid damper and the way of liquid damping according to the characteristics of the structure in practical engineering needs further exploration and research. The author's team has carried out theoretical and experimental research on different types of liquid damping method dampers around the FM liquid dampers of different shapes [3, 4].

8.2.1 Rectangular FM Liquid Damper

The basic mechanical properties of rectangular shallow water TLD have been introduced in Sect. 3.2.1. This section mainly introduces the mechanical properties of deep water TLD. In order to obtain the water sloshing law of deep water TLD, the calculation diagram is the same as Fig. 3.4c, and the following basic assumptions are made:

1. the water in the TLD is an ideal fluid, that is, the water in the tank is incompressible and non viscous, and only does non rotational motion driven by the structural reaction;
2. there is no friction and adhesion between water and tank wall in TLD;
3. the wave motion of water in TLD is micro amplitude and slow;
4. the water tank wall of TLD is rigid, and it will make the same horizontal movement with the structure layer;
5. the structural layer of TLD only moves horizontally in X direction, and its horizontal acceleration is $\ddot{x}_k(t)$.

The velocity potential function $\Phi(x, z, t)$ of water sloshing in rectangular deep water TLD can be expressed as the sum of two potential functions:

$$\Phi(x, z, t) = \Phi_1(x, z, t) + \Phi_2(x, z, t) \tag{8.39}$$

where, $\Phi(x, z, t)$ is the velocity potential function; x is the width of the rectangular tank; z is the depth of the rectangular tank; t is the time. They satisfy Laplace equation respectively:

$$\nabla^2\Phi_1 = 0 \tag{8.40}$$

$$\nabla^2\Phi_2 = 0 \tag{8.41}$$

The corresponding boundary conditions and initial conditions of the above two formulas are respectively:

$$\frac{\partial\Phi_1}{\partial z}|_{z=-h} = \frac{\partial\Phi_2}{\partial z}|_{z=-h} = 0 \tag{8.42}$$

$$\frac{\partial\Phi_2}{\partial t}|_{z=0} + g\eta_0^{(2)} = -\frac{\partial\Phi_1}{\partial t}|_{z=0} - g\eta_0^{(1)} \tag{8.43}$$

$$\eta_0^{(1)} = \int_0^t \frac{\partial\Phi_1}{\partial z}|_{z=0}dt,\ \eta_0^{(2)} = \int_0^t \frac{\partial\Phi_2}{\partial z}|_{z=0}dt \tag{8.44}$$

$$\frac{\partial\Phi_1}{\partial x}|_{z=0,a} = \dot{x}_k(t),\ \frac{\partial\Phi_2}{\partial x}|_{z=0,a} = 0 \tag{8.45}$$

According to Eqs. (8.39)–(8.45), the corresponding total velocity potential function of sloshing reaction of water in the rectangular TLD can be obtained:

$$\Phi(x, z, t) = -\sum_{i=1,3,\ldots}^{\infty}[b_i\omega_i(t)\frac{\cosh\frac{i\pi(z+h)}{a}}{\cosh\frac{i\pi h}{a}}\cos\frac{i\pi x}{a}] - (\frac{a}{2} - x)\dot{x}_k(t) \tag{8.46}$$

$$b_i - \frac{4a}{i^2\pi^2} \tag{8.47}$$

$$\omega_i^2(t) = \frac{i\pi g}{a}\tanh\frac{i\pi}{a}h \tag{8.48}$$

where, a is the width of the rectangular water tank; b is the height of the rectangular water tank; g is the acceleration of gravity; h is the depth of water; η_0 is the wave height of the water surface; $\omega_i(t)$ is the natural frequency of water sloshing; $\dot{x}_k(t)$ is the moving speed in the x direction of the water tank.

The control force of rectangular TLD on the structure is the resultant force of hydraulic pressure of water on the tank wall. According to the expression of sloshing velocity potential function Φ of water in rectangular TLD (Eq. 8.46), the hydrodynamic pressure at any point of water in rectangular TLD can be obtained as follows:

$$p_w(x, z, t) = -\rho \frac{\partial \Phi}{\partial t} \tag{8.49}$$

where, $p_w(x, z, t)$ is the hydraulic pressure at a point of the water in the rectangular TLD; ρ is the density of water. Therefore, the control force of the rectangular TLD on the structure, that is, the total force of the hydraulic pressure on the left and right water tanks of the TLD is:

$$\begin{aligned} F_{TLD}(t) &= [-\int_{-h}^{0} p_w(0, z, t)dz + \int_{-h}^{0} p_w(a, z, t)dz] \cdot b \\ &= -\rho abh[\ddot{x}_k(t) + \sum_{i=1,3,\cdots}^{\infty} \ddot{\omega}_i(t)\frac{2b_i}{i\pi h}\tanh\frac{i\pi h}{a}] \end{aligned} \tag{8.50}$$

The above formula can be further rewritten as:

$$F_{TLD}(t) = -M_T[\ddot{x}_k(t) + \sum_{n=1}^{\infty} d_n F_n \ddot{\omega}_n(t)] \tag{8.51}$$

$$M_T = \rho abh \tag{8.52}$$

$$F_i = \frac{a}{h\sigma_i}\tanh(\sigma_i \frac{h}{a}) \tag{8.53}$$

where, M_T is the mass of water; σ_i is the root of the first type Bessel function equation; $\ddot{x}_k(t)$ is the acceleration of x direction of water tank; $\sigma_i = (2i - 1)\pi$ and $d_i = \frac{8}{\sigma_i^2}$. The wave response of rectangular TLD water surface and its control force on the structure are changed from the combination of vibration shape generalized response of i equal to odd to vibration shape generalized response of $i = 1, 2 \ldots$. Therefore, the formula of natural frequency of water sloshing in rectangular TLD should also be changed to:

$$\omega_i(t) = \frac{g}{a}\sigma_i \tanh \sigma_i \frac{h}{a} \tag{8.54}$$

8.2.2 Circular FM Liquid Damper

8.2.2.1 Mechanical Model

The round shallow water TLD is shown in Fig. 8.13. Except for the boundary layer, the liquid in the container of TLD device can be considered as irrotational. In the irrotational flow field, Bernoulli Equation is established:

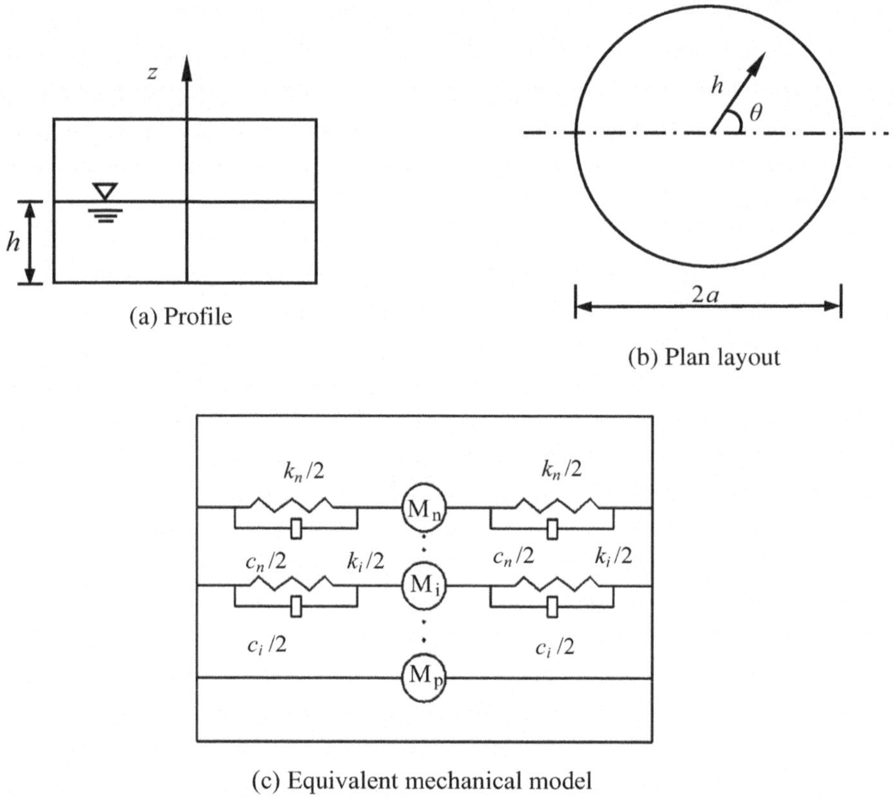

Fig. 8.13 Calculation diagram of circular TLD

$$\frac{p(x, z, t)}{\rho} + gz + \frac{1}{2}(u^2 + v^2 + w^2) = c \tag{8.55}$$

Among them, P is the internal pressure of the liquid; u, v, w are the projection of the liquid velocity on the x, y, z axis; c is constant; ρ is the density of liquid. For cylindrical vessels, the Bernoulli Equation shows that:

$$p(r, \theta, z, t) = g\rho(h - z) + \frac{\rho}{2}(v_r^2 + v_\theta^2 + w^2) \tag{8.56}$$

For shallow water fluctuations, there is

$$v_r = v_\theta \tag{8.57}$$

At the side wall of the vessel, it can be seen from the boundary:

$$v_r|_{r=a} = 0 \tag{8.58}$$

Substituting Eqs. (8.58) and (8.57) into Eq. (8.56) and omitting the influence of w:

$$p(r,\theta,z,t)|_{r=a} = g\rho(h-z)|_{r=a} \tag{8.59}$$

By omitting the viscous force of the boundary layer, the damping force provided by the liquid in the cylindrical container can be obtained:

$$p(r,\theta,z,t) = ag\rho \int_0^{2\pi}\int_0^{h_r} (h_r - z)dzd\theta \tag{8.60}$$

According to the analysis, the liquid in the container can be divided into two parts according to its movement characteristics, i.e. relatively static liquid and sloshing liquid. Among them, the sloshing liquid can be simulated by n mutually independent mechanical models of "mass-spring-damping", as shown in Fig. 8.13c. According to the principle that the dynamic effect of TLD liquid sloshing is equal to that of the mechanical model, it can be inferred that:

$$\begin{aligned} F_{TLD} &= -m_0\ddot{x}_0 - \sum_{n=1}^{\infty} m_n(\ddot{x}_0 + \ddot{x}_n) \\ &= -m\ddot{x}_0\left[1 + \sum_{n=2}^{\infty}\frac{m_n}{m}\frac{\ddot{x}_n}{\ddot{x}_0}\right] \end{aligned} \tag{8.61}$$

$$\frac{m_n}{m} = \frac{2\tanh\left(e_n\bar{h}\right)}{e_n\left(e_n^2-1\right)\bar{h}},\ \frac{m_0}{m} = 1 - \sum_{n=1}^{\infty}\frac{m_n}{m} \tag{8.62}$$

Among them, $m = \rho\pi a^2 h$ is the total mass of liquid, m_0 and m_n are non sloshing masses and n-order sloshing masses respectively. It can be seen from Eq. (8.61) that the total control force of TLD consists of two parts, one is the inertial force of non sloshing liquid moving together with the container, the other is the inertial force provided by each order sloshing liquid moving in waves.

$$\frac{F_n}{F_{TLD}} = \frac{-m_n(\ddot{x}_0 + \ddot{x}_n)}{-m_0\ddot{x}_0 - \sum\limits_{n=1}^{\infty} m_n(\ddot{x}_0 + \ddot{x}_n)} = \frac{a_n(1+b_n)}{1+\sum_{n=1}^{\infty} a_n b_n} \tag{8.63}$$

where, $a_n = \frac{m_n}{m}$, $b_n = \frac{\ddot{x}_n}{\ddot{x}_0}$, according to the base excitation theory, $b_n = \frac{\lambda^2}{\sqrt{\left(1-\lambda^2\right)^2 + (2\zeta_L\lambda)^2}}$, $\lambda = \frac{k_0}{k_n}$, ζ_L is the damping ratio of sloshing liquid.

8.2.2.2 Dynamic Test

In 1990s, the author carried out the dynamic test research of circular TLD. Three kinds of liquid, including purified water, glycerin solution (density 1.265 g/cm^3) and $ZnSO_4$ solution (density 1.13 g/cm^3), were selected, and three ways of increasing damping were designed, which are metal grating perpendicular to the direction of movement, radial partition boards and surface bubble plates. Six groups of dynamic tests were carried out. See Table 8.1 for specific test conditions and Fig. 8.14a for test process. The time history curve of sloshing of liquid level wave height of TLD was measured in the test, as shown in Fig. 8.14b. Based on this, the damping ratio of liquid sloshing of TLD is calculated (Table 8.2). It can be seen that the damping ratio of purified water is only 0.88% without any damping measures. After damping, the damping ratio of liquid sloshing is significantly increased.

Table 8.1 Dynamic test conditions of TLD

Test condition	1	2	3	4	5	6
Liquid type	Purified water	Purified water	Purified water	Purified water	Glycerin solution	$ZnSO_4$ solution
Damping mode	No	Grille	Divider plate	Foam board	No	No

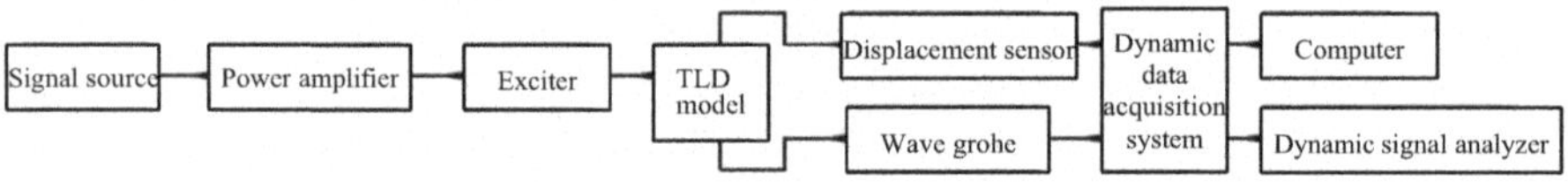

(a) Test plan and data acquisition

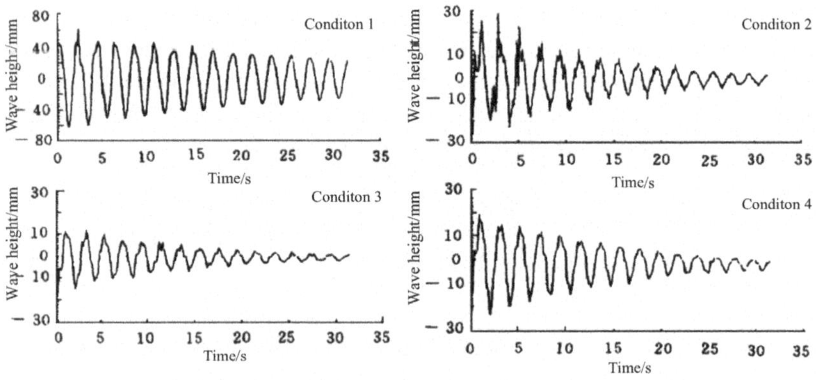

(b) Measured time history of wave height of TLD

Fig. 8.14 Dynamic test of circular TLD

Table 8.2 Measured damping ratio of TLD

Test condition	1	2	3	4
Damping ratio (%)	0.88	2.33	2.37	2.3

8.2.3 Ring FM Liquid Damper

8.2.3.1 Mechanical Model

In recent years, due to the limitation of structure function and shape, the demand of ring TLD is increasing. For its mechanical model, many scholars have carried out theoretical and experimental researches. It is assumed that the water motion in the ring TLD is small amplitude, and the water is an ideal incompressible potential flow, and the elastic deformation of the side wall of the vessel is not considered. By assuming that the velocity potential function of the water motion satisfies the Laplace equation in polar coordinates and the corresponding boundary conditions, the corresponding velocity potential function can be obtained by using the semi inverse method (Fig. 8.15):

$$\Phi(r,\theta,z,t) = a\sin\theta\left[\sum_{n=1}^{\infty}\dot{q}_n(t)\frac{d_n'\cosh\left(\sigma_n\frac{z+h}{a}\right)}{\cosh\left(\sigma_n\frac{h}{a}\right)}R_1^n\left(\sigma_n\frac{r}{a}\right)+\frac{r\dot{x}_0(t)}{a}\right] \tag{8.64}$$

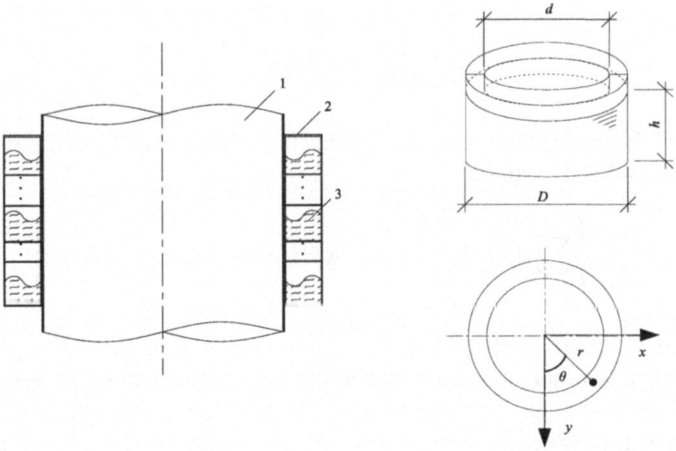

(a) Profile

(b) Three dimensional and plane diagrams

1. Structure; 2. Water tank; 3. Tuning liquid.

Fig. 8.15 Ring TLD

Among them, r, θ, z, t are the circumferential, radial, vertical and time coordinates respectively; a is the outer radius of the water tank $D/2$; h is the static height of water in the water tank; $q_n(t)$ is the generalized coordinate of water tank shaking; x_0 is the water tank displacement. The calculation expression of other parameters is as follows:

$$d'_n = \frac{2\left(1 - k' R_1^n\left(k'\sigma_n\right)\right)}{\left(\sigma_n^2 - 1\right) - \left(k'^2\sigma_n^2 - 1\right) R_1^{n2}(k'\sigma_n)}$$

$$R_1^n\left(\sigma_n \frac{r}{a}\right) = \alpha_n J_1\left(\sigma_n \frac{r}{a}\right) + \beta_n Y_1\left(\sigma_n \frac{r}{a}\right)$$

$$\alpha_n = \frac{1}{R_1(\sigma_n)} = \frac{1}{J_1(\sigma_n) + \gamma_n Y_1(\sigma_n)}$$

$$\beta_n = \gamma_n \alpha_n$$

$$\gamma_n = -\frac{J_1'(\sigma_n)}{Y_1'(\sigma_n)} = -\frac{J_1'(k'\sigma_n)}{Y_1'(k'\sigma_n)}$$

where, $J_1\left(\sigma_n \frac{r}{a}\right)$ and $Y_1\left(\sigma_n \frac{r}{a}\right)$ are the first order Bessel functions of Class I and class II, respectively; k' is the ratio of the internal and external diameter of the ring section; n is the modal order, $n = 1, 2, 3 \ldots$; σ_n is the solution of $J_1'(\sigma_n) Y_1'\left(k'\sigma_n\right) - J_1'\left(k'\sigma_n\right) Y_1'(\sigma_n) = 0$. The circular frequency of water shaking is expressed as:

$$\omega_{L_n} = \sqrt{\frac{g}{a}\sigma_n \tanh\left(\sigma_n \frac{h}{a}\right)} \tag{8.65}$$

The sum of dynamic pressure on the internal and external water tank is the control force of TLD on the structure:

$$F_{TLD} = m M_L \left\{ \sum_{n=1}^{\infty} \ddot{q}_n(t) F_n d'_n \left[1 + k' R_1^n\left(k'\sigma_n\right)\right] + \left(1 + k'^2\right)\ddot{x}_0(t) \right\} \tag{8.66}$$

where, m is the quantity of RS-TLD; $F_n = \frac{1}{h_0 \sigma_n} \tanh(\sigma_n h_0)$, $h_0 = h/a$; M_L is the total mass of liquid in the container. Generally, only the first mode of water tank is used in the vibration control of structure, so the calculation model of RS-TLD can be expressed as follows:

$$\omega_L = \sqrt{\frac{g}{a}\sigma_1 \tanh\left(\sigma_1 \frac{h}{a}\right)} \tag{8.67}$$

$$F_{TLD} = m M_L \left\{ \ddot{q}_1(t) F_1 d'_1 \left[1 + k' R_1^1\left(k'\sigma_1\right)\right] + \left(1 + k'^2\right)\ddot{x}_k(t) \right\} \tag{8.68}$$

$$\ddot{q}_1(t) + 2\zeta_L\omega_L\dot{q}_1(t) + \omega_L^2 q_1(t) = -\ddot{x}_k(t) \tag{8.69}$$

Among them, x_k is the displacement of the structural node where the water tank is installed; ζ_L is the viscous damping ratio of the water movement. Generally, the damping ratio of pure water tank is less than 1%, which is difficult to meet the requirements of the optimal damping ratio. It is found that the damping ratio of pure water decreases with the increase of water depth ratio, and its value varies from 0.5 to 0.8%; however, a larger damping ratio can be obtained by setting steel wire mesh. When the water depth ratio $h/a = 0.677$, the damping ratio of the structure can be expressed as follows:

$$\zeta_L = 0.064 + 0.005(A - 0.7) \tag{8.70}$$

where, A is the amplitude of water at $r = a$, $\theta = 0$, unit is cm, which belongs to the range of 0.7–4.3.

It can be seen from the analysis of the above expressions that the performance parameters of RS-TLD mainly depend on σ_1, and its value is only related to k', while σ_1 given in the literature is only the result of time. Although the intermediate value can be obtained by linear difference, k' is usually beyond this range for high-rise steel chimney, so this paper first compiles a program to solve its value. The comparison between the calculation results of the program and the corresponding results in the literature is given in Table 8.3. Since only the first mode is usually considered, and only the values of the intermediate parameters in the first mode are given, it can be seen that the calculation results of the program are almost the same as those in the literature. The program compiled in this paper can be used to solve the mechanical properties of RS-TLD.

Equation (8.67) shows that the sloshing frequency of RS-TLD is mainly related to the water tank section and water depth. Figure 8.16 shows the influence of each parameter on the sloshing frequency of the water tank: (1) in Fig. 8.16a, the inner diameter of the water tank is 1.24 m, k' varies from 0.5 to 0.9, and the water depth varies from 0.3 m to 1.5 m. It can be seen from the figure that the sloshing frequency of the water tank increases with the increases of k' and water depth; (2) in Fig. 8.16b, k' is 0.85. The inner diameter of water tank changes from 0.5 to 2.5 m, and the water depth changes from 0.3 to 1.5 m. It can be seen that the frequency decreases with

Table 8.3 Comparison of the intermediate calculation parameters in the RS-TLD model

Intermediate parameter		σ_n	γ_n	α_n	β_n	d'_n
$k' = 0.1$	Literature	1.80360	−0.02562	1.70300	−0.04362	0.82520
	Program calculation	1.80347	−0.02577	1.70275	−0.04389	0.82517
$k' = 0.2$	Literature	1.70530	−0.09038	1.65700	−0.14980	0.7990
	Program calculation	1.70511	−0.09055	1.65685	−0.15002	0.79888
$k' = 0.3$	Literature	1.58200	−0.16650	1.59300	−0.26520	0.77980
	Program calculation	1.58206	−0.16621	1.59304	−0.26478	0.78006

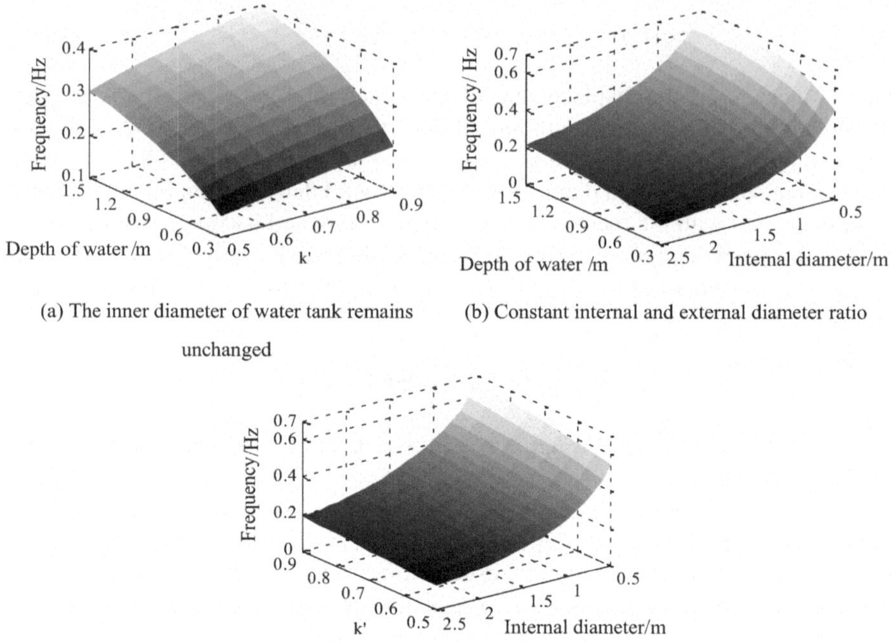

(a) The inner diameter of water tank remains unchanged

(b) Constant internal and external diameter ratio

(c) Water depth unchanged

Fig. 8.16 Influence of each parameters on RS-TLD sloshing frequency

the increase of inner diameter; (3) in Fig. 8.16c, the water depth is 1 m, k' changes from 0.5 to 0.9, and the inner diameter of water tank changes from 0.5 m to 2.5 m. To sum up, it can be seen that in the changing range of parameters of the RS-TLD applicable to high-rise steel chimneys, the change of inner diameter has the greatest impact on frequency, followed by water depth, with the smallest impact caused by k'.

8.2.3.2 Dynamic Test

In order to verify the calculation model of the ring TLD and further determine the liquid sloshing damping ratio of the ring TLD, the dynamic test of the ring TLD was carried out.

1. **Test piece and test device**

The ring TLD water tank is made of acrylic (plexiglass). The outer radius of the water tank is 225 mm, the inner radius is 140 mm, the height is 300 mm, and the wall thickness of plexiglass is 12 mm. Because the base plate is connected with the sensor by bolts, in order to ensure that there is no damage between the bolts and the glass water tank, the thickness of the inner diameter base plate is increased to 24 mm. After calculation, it can meet the strength requirements without damage. The ring TLD water tank is divided into eight parts on average, and a clamping groove (Fig. 8.17a) is added at the corresponding position, with a transverse length of 5 mm, a thickness of 5 mm, and a depth of 10.5 mm. Nine kinds of thin steel sheets of 3, 5, and 10 mm

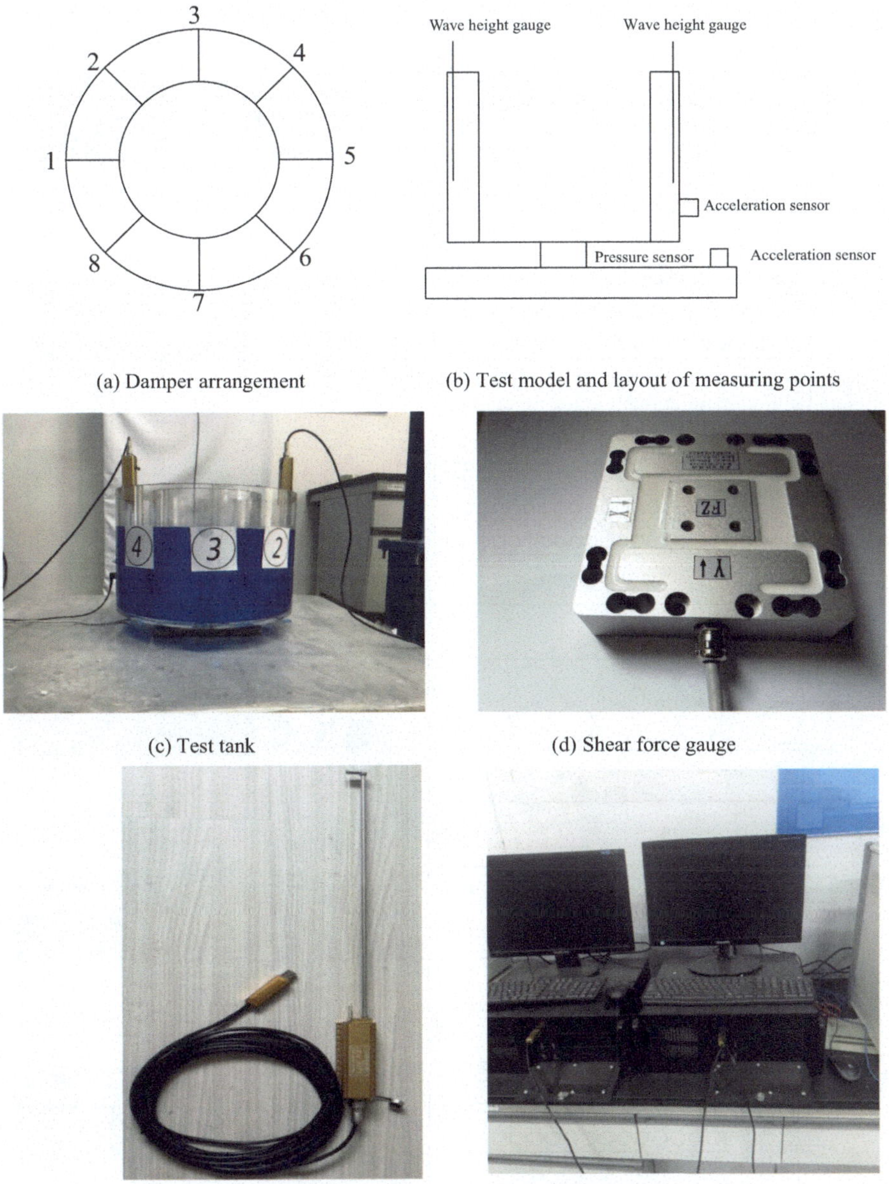

(a) Damper arrangement

(b) Test model and layout of measuring points

(c) Test tank

(d) Shear force gauge

(e) Wave height gauge

(f) Acquisition system

Fig. 8.17 Ring TLD test

are made respectively (Fig. 8.18). A circular hole is opened on the thin steel sheet, and the hole rate is calculated. According to the arrangement of different working conditions, it is put into the TLD water tank as required for test. The arrangement of the water tank and the connection with the vibration table are shown in Fig. 8.17b,

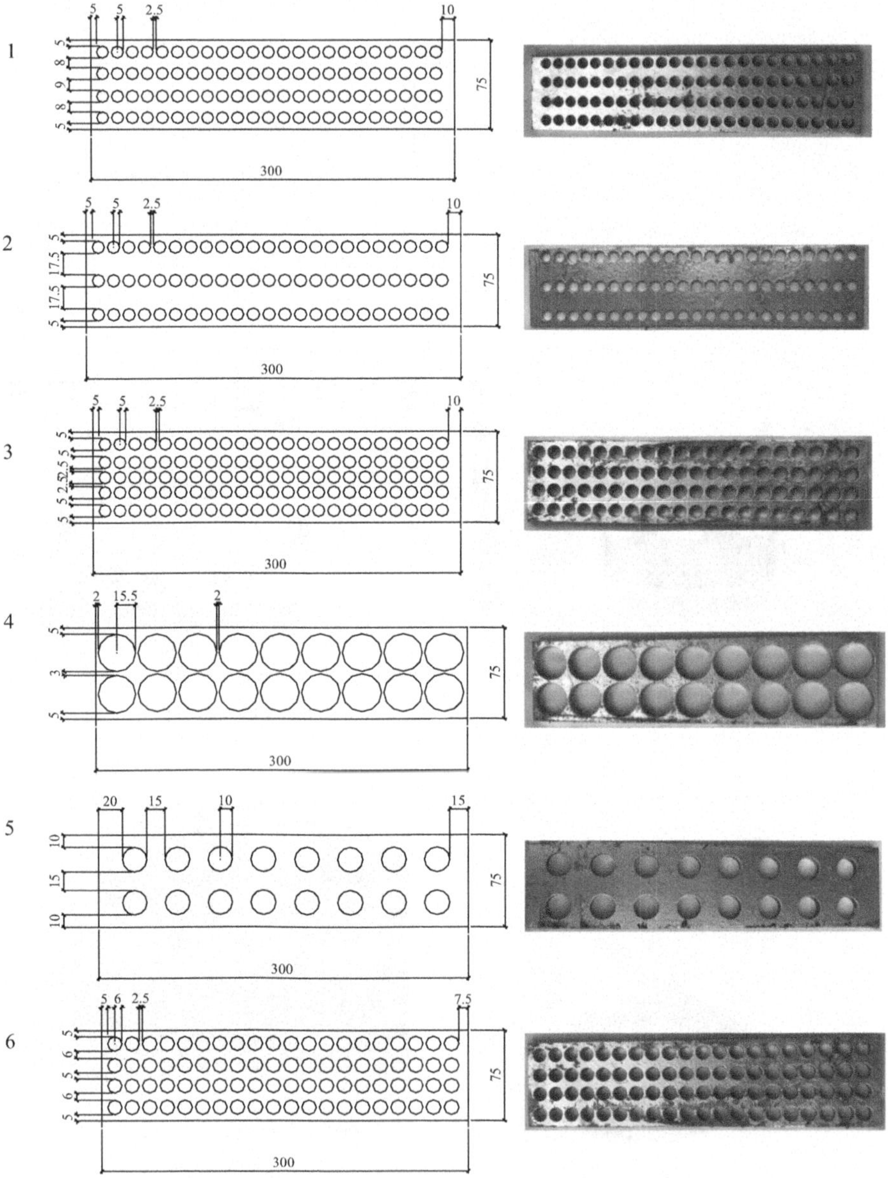

Fig. 8.18 Damping steel sheet for test

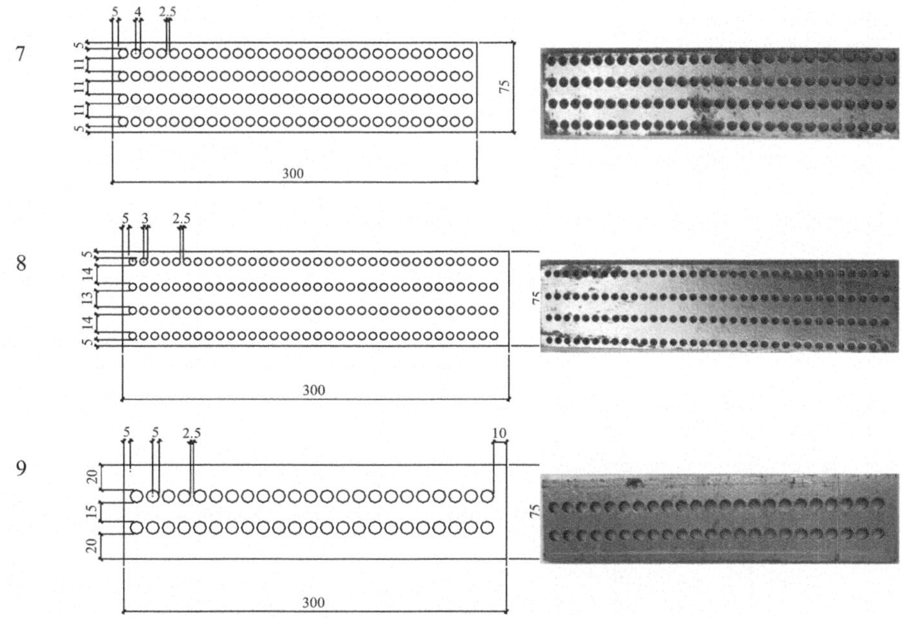

Fig. 8.18 (continued)

c. A three-way force sensor (Fig. 8.17d) is placed between the ring water tank and the vibration table to measure the shear force of TLD when the vibration table is shaking.

In the test, the liquid is pure water, and more than 500 groups of test conditions of the same liquid height, different liquid height and different frequency ratio with same liquid height were carried out respectively. For the convenience of comparison, the amplitude of each condition was 2 cm.

2. **Analysis of test results**

(1) Ring TLD frequency
Input sine wave in the vibration table, change the height of water in the water tank, record the height time history and shear force time history of water surface shaking with the water height of 80, 100, 120, 150, 180 and 200 mm; input the frequency of 0.9, 0.94, 0.98, 1.02, 1.06 and 1.1 times of the original frequency, record the height time history and shear force time history of water surface shaking with the water height of 80, 100, 120, 150, 180 and 200 mm. The relationship between frequency and water depth ratio is shown in Fig. 8.19 under the condition of the change of water surface height through the conversion of logarithm attenuation of water surface shaking amplitude or logarithm attenuation of shear force amplitude. The maximum error between the given formula and the test value is 1.3%. The first mode frequency of water tank sloshing increases with the increase of

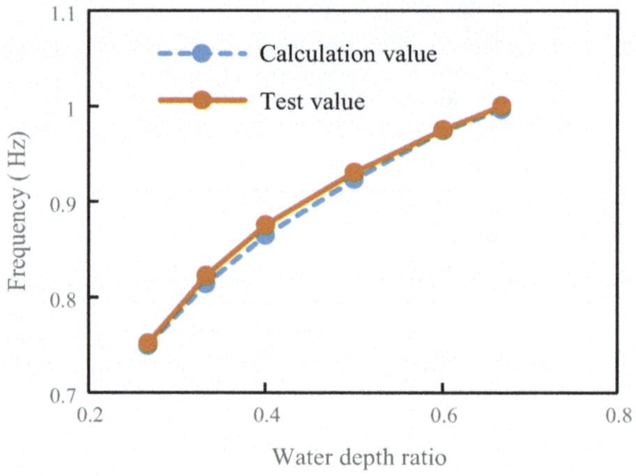

Fig. 8.19 Comparison of test and theoretical values of shaking frequency

water depth ratio, and the test value is slightly lower than the value calculated by the formula.

(2) Liquid damping ratio

The change of liquid damping ratio with amplitude and water depth ratio under pure water condition is shown in Fig. 8.20. It can be seen that: (1) when the height of water surface is fixed, the damping ratio of the first mode will increase with the increase of amplitude of water surface shaking under free vibration; (2) when the height of water surface changes, the damping ratio will gradually decrease with the increase of water depth ratio; (3) the damping ratio of the first mode of liquid under pure water condition is less than 1% in most cases and only slightly higher than 1% when the water depth is relatively small, which has a certain gap from the optimal liquid damping ratio required in the usual design.

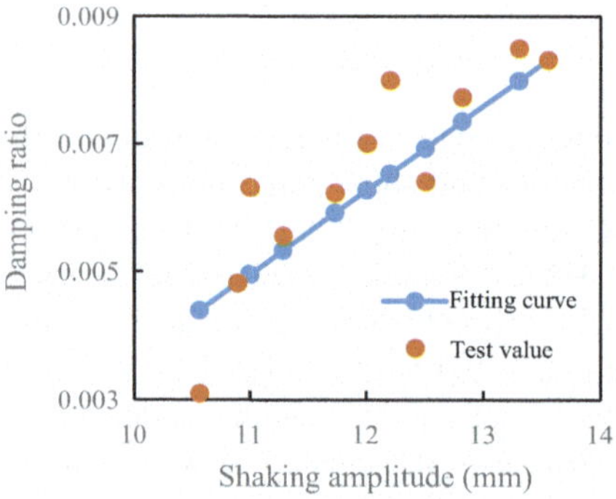

(a) Change with amplitude

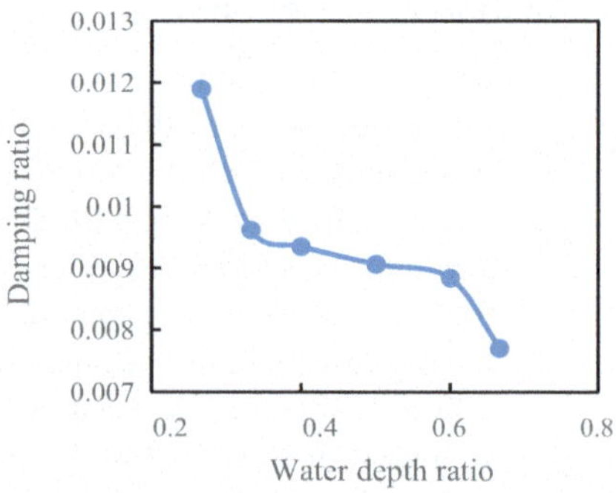

(b) Change with water depth ratio

Fig. 8.20 Damping ratio of pure water

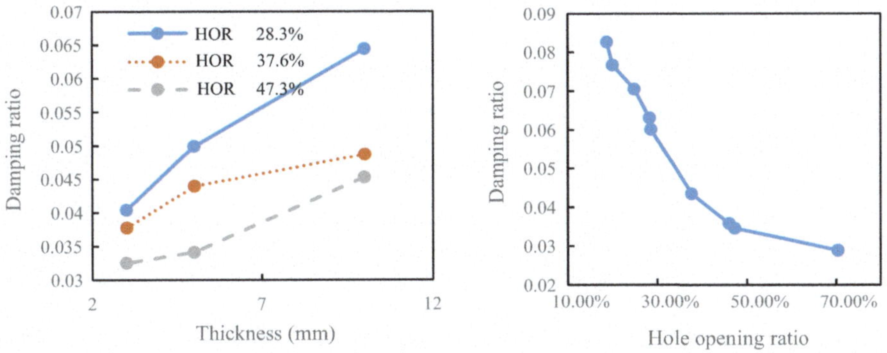

(a) Change with amplitude

(b) Change with water depth ratio

Fig. 8.21 Effect of damping steel sheet on liquid damping ratio

In order to improve the liquid damping ratio, insert a damping steel plate along the radial direction into the ring TLD, and the size is shown in Fig. 8.18. Carry out the dynamic test of ring type TLD under different working conditions, and the test results are shown in Fig. 8.21. It can be seen from the figure that the damping ratio decreases with the increases of sheet thickness, hole opening ratio (HOR), liquid amplitude and water depth ratio.

References

1. Chen, Xin, Youliang Ding, Zhiqiang Zhang, et al. 2012. Investigations on serviceability control of long-span structures under human-induced excitation. *Earthquake Engineering and Engineering Vibration* 11 (1): 57–71.
2. Zhang, Zhiqiang. 2003. *Study on vibration control of wind vibration and seismic response of Hefei TV tower*. Nanjing: Southeast University. (in Chinese).
3. Chen, Xin. 2012. *Theoretical and experimental study on vibration control of high-rise steel chimneys under wind load*. Nanjing: Southeast University. (in Chinese).
4. Ren, Zhenghua. 1993. *Basic principle of TLD device and its application in wind vibration control of Nanjing TV tower*. Nanjing: Southeast University. (in Chinese).

Chapter 9
Isolation Bearing of Building Structure

Abstract The mechanical properties of the isolation bearing play an important role in the isolation effect. About the high-performance rubber isolation bearing, its damping mechanism and characteristics, structure and classification, shape coefficient, vertical and horizontal performance of rubber bearing are introduced respectively. About the improved rubber isolation bearing with low shear modulus, the improvement of low shear modulus, vertical and horizontal mechanical property test, and numerical simulation of isolation bearing are discussed. The Honeycomb sandwich rubber isolation bearing, dish spring composite multi-dimensional isolation bearing and rubber composite sliding isolation bearing are also researched respectively.

The main purpose of building structure isolation is to set up isolation layer to reduce the earthquake action from the ground to the upper structure. The mechanical properties of the isolation bearing which makes up the isolation layer play an important role in the isolation effect. Through conduction the research work on isolation bearing, on the one hand, the mechanical properties of the isolation bearing materials are improved or new high-performance materials are selected. On the other hand, a new type of structure with better performance and more reasonable stress is proposed around the mechanical structure of the support.

9.1 High Performance Rubber Isolation Bearing

9.1.1 Damping Mechanism and Characteristics of Rubber Bearing

9.1.1.1 Structure and Classification of Rubber Isolation Bearing

Rubber isolation bearing refers to the isolation device composed of multilayer rubber and multilayer steel plate or other materials, as shown in Fig. 9.1a. Because sandwich steel plate is added in the rubber layer, and the rubber layer is closely bonded with the sandwich steel plate, when the rubber pad bears the vertical load, the transverse

A. Li, *Vibration Control for Building Structures*, Springer Tracts in Civil Engineering, https://doi.org/10.1007/978-3-030-40790-2_9

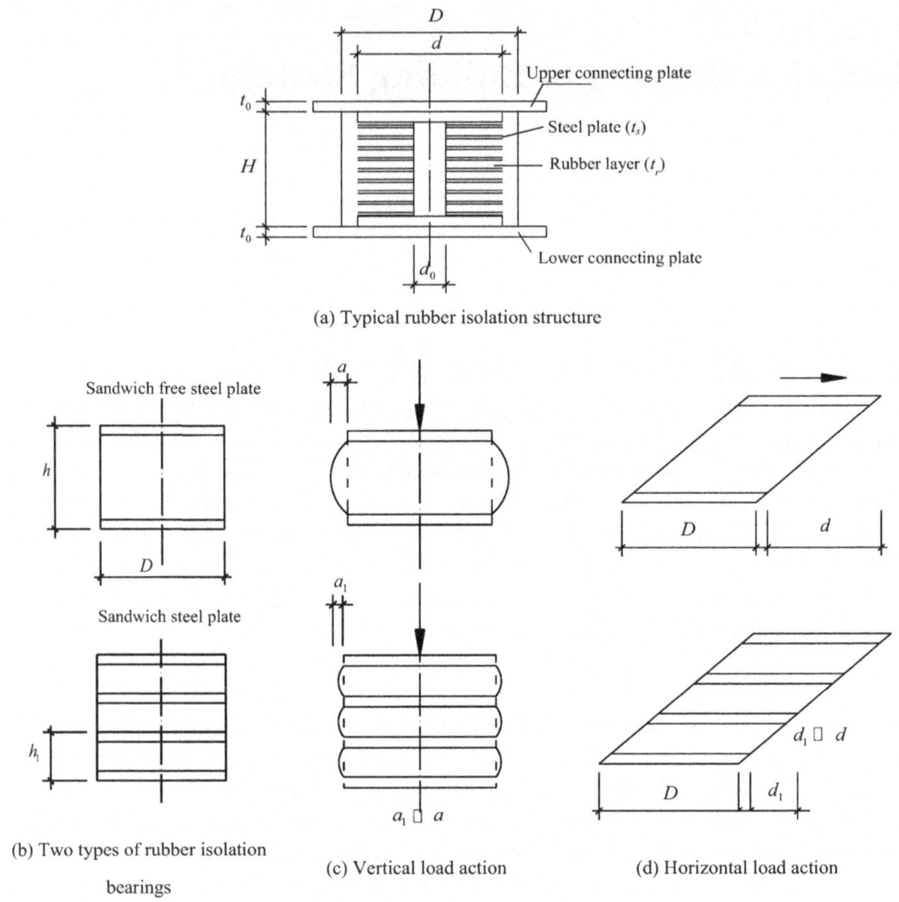

Fig. 9.1 Typical rubber isolation bearing and its stress and deformation

deformation of the rubber layer is constrained (Fig. 9.1b), that is, $a_1 \ll a$, The sandwich rubber pad has a large vertical bearing capacity and vertical stiffness. When the rubber pad bears the horizontal load (Fig. 9.1c), the relative lateral displacement of the rubber layer is greatly reduced, that is, $d_1 \ll d$. The rubber pad can achieve a large overall lateral displacement without instability, and maintain a small horizontal stiffness (only 1/500–1/1500 of the vertical stiffness). Due to the close bond between sandwich steel plate and rubber layer, the rubber layer can also bear certain tension under the vertical earthquake action, which makes the rubber pad become an ideal isolation device with great vertical bearing capacity, small horizontal stiffness, large horizontal lateral displacement allowance and able to bear the vertical earthquake action.

The main structural requirements of isolation rubber bearing are as follows: sandwich steel plate (with a thickness of t_s) and rubber pad (with a thickness of t_r) are closely bonded to ensure the deformation constraint of steel plate on rubber, so that rubber has high vertical compression bearing capacity and certain tensile capacity, large horizontal deformation capacity and anti complex load fatigue capacity; Under the multi-dimensional ground motion of multiple earthquakes (horizontal earthquake action, vertical earthquake action, torsion action, etc.), the isolation device can work reliably; a lateral protective layer is set to make the rubber pad have higher aging resistance (high and low temperature aging resistance, ozone aging resistance), water resistance, acid and alkali corrosion resistance, fire resistance, etc.; reliable upper and lower connecting plates are set to make the rubber pad reliably connect with the upper and lower structures (components); lead or viscous material core can be set or the rubber material with high damping can be used to make the isolation rubber bearing have enough damping ratio. According to whether there is inserted lead core in the middle hole, the building isolation rubber bearing can be divided into two types: ordinary type (lead-free core type) and lead core type [1]. Its section shape is generally circular or rectangular.

9.1.1.2 Shape Coefficient of Rubber Isolation Bearing

The shape coefficient of rubber isolation bearing is an important geometric parameter to ensure the bearing capacity and deformation capacity of rubber bearing.

1. **First shape factor S_1**

The first shape factor S_1 is defined as the ratio of the effective bearing area of each rubber layer in the rubber bearing to its free surface area, that is:

For circular rubber bearings:

$$S_1 = \frac{\pi(d^2 - d_0^2)/4}{\pi(d + d_0)t_r} = \frac{d - d_0}{4t_r} \tag{9.1}$$

For rectangular rubber bearing:

$$S_1 = \frac{ab}{2(a + b)t_r} \tag{9.2}$$

Among them, d is the diameter of the effective bearing surface of the rubber layer; d_0 is the diameter of the opening hole in the middle of the rubber bearing; t_r is the thickness of the single rubber layer; a is the long side size of the rectangular section rubber bearing; b is the short side size of the rectangular section rubber bearing.

S_1 indicates the restraint degree of the steel plate in the rubber bearing to the deformation of the rubber layer. Therefore, the larger S_1 value is, the greater the compression bearing capacity is and the greater the vertical stiffness of the rubber bearing is. The value of S_1 is generally taken according to the research results and application experience at home and abroad:

$$S_1 \geq 15 \tag{9.3}$$

When the above formula is satisfied, the ultimate compressive strength of the rubber bearing can reach 100–120 MPa. If the design pressure is 15 MPa, the safety factor of its compressive bearing capacity can reach 6.7–8.0, so that the isolation building structure has enough safety reserves.

2. **Second shape factor S_2**

The second shape factor S_2 is defined as the ratio of the diameter of the effective bearing body of the rubber bearing to the total thickness of the rubber, i.e.

$$S_2 = \frac{d}{nt_r} \tag{9.4}$$

where n is the total number of rubber layers. S_2 represents the ratio of width to height of compression body of rubber pad, that is, the stability of rubber pad under compression. The larger the S_2 value is, the coarser the rubber bearing is, the better the compression stability is, and the greater the critical load of compression instability is. However, the larger S_2 is, the greater the horizontal stiffness of the rubber bearing is, and the smaller the horizontal ultimate deformation capacity is. Therefore, S_2 cannot be too small or too large. The value of S_2 should not be less than 5 generally according to the research results and application experience at home and abroad. If the horizontal deformation capacity of rubber bearing is required to be large, S_2 is taken as low value, and the design bearing capacity is also taken as low value; otherwise, S_2 is taken as high value, and the design bearing capacity is also taken as high value.

9.1.1.3 Vertical Performance of Rubber Isolation Bearing

The vertical performance of rubber bearing includes vertical stiffness, vertical deformation, vertical ultimate compressive stress and vertical ultimate tensile stress.

1. **Vertical stiffness**

The vertical stiffness of rubber bearing refers to the vertical force exerted under unit vertical displacement of rubber bearing under vertical pressure, which is recorded as k_v. The main factors that affect the vertical stiffness of the sandwich rubber pad are the mechanical properties, shape coefficient, vertical axial compressive stress and horizontal shear deformation of the rubber material. The reasonable value of the vertical stiffness can make the upper structure of the isolation structure system not appear too large vertical deformation under normal load, and reasonably determine the vertical natural vibration period of the isolation structure, which can avoid the resonance effect in the earthquake (or other vibration).

The rubber bearing used in actual engineering must be tested in full scale, and the measured vertical stiffness value is taken as the design basis. It is required to apply the load to the design load on the rubber bearing in the test, take the vertical load corresponding to the axial compression stress of $(1 \pm 30\%)\sigma_{\mathrm{d}}$, load three times repeatedly, and draw the relationship curve between the vertical load and the vertical displacement. Take the results of the third reciprocating loading, and calculate the vertical stiffness as follows:

$$K_{\mathrm{v}} = \frac{P_1 - P_2}{\delta_1 - \delta_2} \tag{9.5}$$

Among them, σ_{d} is the design axial compression stress of rubber bearing; P_1 is the vertical load with an average stress of $1.3\sigma_{\mathrm{d}}$; P_2 is the vertical load with an average stress of $0.7\sigma_{\mathrm{d}}$; δ_1 is the vertical displacement with a vertical load of p_1; δ_2 is the vertical displacement with a vertical load of p_2. The measured vertical stiffness value shall be within ±20% of design value of the product and within ±10% of average value; the average value shall be within ±10% of the design value of the product.

2. **Vertical deformation performance**

When the rubber bearing is compressed vertically, the load displacement curve shows the characteristics of hardening spring, and it is approximately linear near the design axial force. Take the vertical load corresponding to the axial compressive stress of $(1 \pm 30\%)\sigma_{\mathrm{d}}$, the load displacement curve should be relatively stable in the process of 5 times of reciprocating loading, and there should be no obvious abnormality.

3. **Vertical ultimate compressive stress**
   The vertical ultimate compressive stress of rubber bearing refers to the bearing compressive stress when the tested bearing is damaged under the axial pressure when the rubber bearing is at a certain horizontal displacement.

   (1) **Axial ultimate compressive stress**

When the horizontal displacement of rubber bearing is zero, the measured ultimate compressive stress is axial ultimate compressive stress. It is not only an important index to ensure the normal use of rubber bearings at ordinary times, but also an important index to directly affect other mechanical properties of bearings in earthquake.

Under the axial pressure, the lateral bulge of rubber layer is restrained by sandwich steel plate, which makes the rubber bearing have great bearing capacity. Under the vertical pressure, the rubber bearing loses its bearing capacity because of the penetrating fracture of the steel plate under the radial tensile stress.

The factors affecting the axial ultimate compressive stress of rubber bearing are: the ultimate tensile yield strength σ_{y} of sandwich steel plate, the ratio $t_{\mathrm{s}}/t_{\mathrm{r}}$ of thickness of steel plate and rubber layer in rubber bearing and the first shape factor S_1. In order to ensure that the rubber layer of the rubber bearing is tightly bonded with the steel plate, and under the conditions of $S_1 \geq 15$ and $S_2 \geq 3$, σ_{vmax} can generally reach 95–120 MPa. Considering the safety factor of axial bearing capacity, in practical engineering application, the value of design axial stress of rubber bearing is:

$$\text{General works: } \sigma_v = 15\,\text{MPa} \tag{9.6}$$

$$\text{Important works: } \sigma_v = 10\,\text{MPa} \tag{9.7}$$

(2) **Ultimate shear compression stress**

The ultimate shear compression stress of rubber isolation bearing refers to the vertical ultimate compressive stress of bearing under horizontal shear deformation. When the earthquake occurs, the isolation of engineering structure is realized by the horizontal deformation of rubber bearing. Therefore, the horizontal shear deformation capacity and shear pressure bearing capacity of rubber bearing are important indexes to ensure the normal operation of the bearing during earthquake.

Rubber bearing has great axial compression bearing capacity under axial compression (vertical) load. When the lateral displacement of support occurs, the effective area of bearing decreases, the stress of the compression part of the core increases sharply, and the tensile stress may appear in the local area. The ultimate bearing capacity of the core compression part is greatly improved due to the restriction of the steel plate in the support on the deformation of the rubber layer and the restriction of the peripheral materials of the support on the core compression part (three-dimensional pressure).

When the shear strain of the bearing increases, the effective area of the core under load decreases, plus the effect of P-Δ effect, the ultimate vertical bearing capacity will decrease. However, if the axial compressive stress on bearing is constant, a large shear deformation value can be achieved when shear failure occurs. The test results of the shear failure of the larger diameter isolation bearing show that the ultimate shear strain reaches 380–450% when the axial compressive stress of the isolation bearing is 10–30 MPa. When the axial compressive stress is 10–15 MPa, the horizontal shear strain r is in the range of 350% (r is the ratio of the horizontal relative displacement D of the upper and lower plates to the total thickness nt_r of the rubber layer), and the rubber bearing will not be damaged by shear pressure. When the horizontal shear deformation displacement of the bearing is $D \leq 0.75d$ (d is the rubber diameter inside the rubber bearing), the bearing capacity of the bearing does not decrease significantly.

(3) **Vertical ultimate tensile stress**

The so-called vertical ultimate tensile stress of rubber isolation bearing is to apply only axial tensile force to the tested bearing, and load slowly or in stages until it is damaged. At the same time, draw the curve of tensile load and displacement, and determine the ultimate stress of failure according to the change trend of the curve.

When the height width ratio of the building is large, there may be large sway during the earthquake, which makes some rubber bearings in the tension state; when the vertical seismic action of the ground in the earthquake is large, coupled with the large horizontal and torsional action of the ground, may also make some rubber bearings in the tension state; in addition, when the horizontal shear deformation is large, the local area of the cross section of the bearings may produce tensile stress. Therefore, in order to ensure the stability and safety of the isolated structure under multi-dimensional strong earthquake, the rubber bearing should have a certain tensile bearing capacity.

9.1.1.4 Horizontal Performance of Rubber Isolation Bearing

The horizontal performance of rubber isolation bearing includes the initial horizontal stiffness, horizontal stiffness after yield (with lead core), equivalent viscous damping ratio and horizontal ultimate deformation.

1. **Horizontal stiffness**

The horizontal stiffness of rubber bearing includes the initial stiffness and the post yield stiffness of the bearing with lead core. The horizontal stiffness of the isolation bearing is one of the important mechanical parameters of the isolator. Its importance lies in: appropriate horizontal stiffness reasonably determines the period of the isolator, so as to achieve obvious isolation effect; sufficient initial stiffness can ensure the normal use of the isolation structure under strong wind and small earthquake and other environmental effects; appropriate initial stiffness and post yield stiffness can ensure that the isolator will not produce excessive horizontal shear and horizontal displacement.

For rubber bearings with lead core, the horizontal stiffness is mainly determined by the size of lead core and the size and shape coefficient of rubber pad. For ordinary rubber bearings, the horizontal stiffness is mainly affected by the following factors. Firstly, the mechanical properties of rubber material, the greater the hardness, elastic modulus and shear modulus of rubber material are, the greater the stiffness of rubber bearing is; secondly, the shape coefficient of rubber bearing, the horizontal stiffness of rubber pad increases with the increase of shape coefficient of rubber pad; in addition, the horizontal stiffness of rubber pad decreases with the increase of axial pressure, and decreases with the increase of shear strain in a certain range.

According to different influence conditions and theoretical calculation formula, the horizontal stiffness of rubber pad can be estimated. However, the horizontal stiffness of rubber bearing used in actual engineering must be determined according to the measured value. The test requires that under the action of design compressive stress of the product, the tested bearing shall be respectively subjected to the dynamic loading test under the shear strain $\gamma = 50\%$, $f = 0.3\,\text{Hz}$; $\gamma = 100\%$, $f = 0.2\,\text{Hz}$; $\gamma = 250\%$, $f = 0.1\,\text{Hz}$. The horizontal loading waveform is sine wave, and the loading frequency of the large-diameter specimen can be appropriately reduced. Taking the horizontal displacement corresponding to the positive shear strain γ and negative shear strain $-\gamma$ as the maximum horizontal positive displacement and negative displacement, three hysteretic curves are made continuously. Using the third hysteretic curve, calculate the horizontal stiffness of the bearing according to the following formula:

$$k_{eq} = \frac{Q^+ - Q^-}{U^+ - U^-} \quad (9.8)$$

Among them, k_{eq} is the horizontal stiffness of rubber isolation bearing; U^+ is the maximum horizontal positive displacement; U^- is the maximum horizontal negative displacement; Q^+ is the horizontal shear force corresponding to U^+; Q^- is the horizontal shear force corresponding to U^-.

For leaded core bearings, the horizontal stiffness after yield shall be determined according to the following formula according to the third hysteretic curve of the test of $\gamma = 100\%$, $f = 0.2$ Hz:

$$k_{py} = \frac{1}{2}\left(\frac{Q^+ - Q_y^+}{U^+ - U_y^+} + \left|\frac{Q^- - Q_y^-}{U^+ - U_y^-}\right|\right) \quad (9.9)$$

Among them, k_{py} is the horizontal stiffness after yielding of building isolation rubber bearing (with lead core); U_y^+ is the positive direction yield displacement; U_y^- is the negative direction yield displacement; Q_y^+ is the corresponding horizontal shear; Q_y^- is the horizontal shear force corresponding to U_y^-.

2. **Equivalent viscous damping ratio**

As an isolation rubber bearing for practical engineering application, its damping value must be obtained through the measurement of the actual rubber bearing. The test requirements and conditions are the same as the previous horizontal stiffness test. The equivalent viscous damping ratio of the tested bearing is calculated as follows:

$$\zeta_{eq} = \frac{W}{2\pi Q^+ U^+} \quad \text{or} \quad \zeta_{eq} = \frac{W}{2\pi \mathrm{k_{eq}}(\mathrm{U^+})^2} \quad (9.10)$$

Among them, ζ_{eq} is the equivalent viscous damping ratio of building isolation rubber bearing; W is the area enclosed by the hysteretic curve.

3. **Horizontal limit deformation**

Rubber isolation bearings have a great bearing capacity under axial (vertical) load. When bearing lateral displacement occurs, the effective area of the bearing decreases and the ultimate vertical bearing capacity decreases. The shear compression capacity and ultimate horizontal shear deformation capacity of rubber bearings are very large. When the design axial compressive stress is 10–15 MPa, the horizontal shear strain r is in the range of 350% (r is the ratio of the horizontal relative displacement D of the upper and lower plates to the total thickness nt_r of the rubber layer), and the rubber bearing will not be damaged by shear pressure.

9.1.2 Improved Rubber Isolation Bearing with Low Shear Modulus

9.1.2.1 Improvement of Low Shear Modulus of Rubber Isolation Bearing

The principle of isolation design is to achieve the purpose of isolation by increasing the natural vibration period of the structure, avoiding the predominant period of the earthquake and reducing the earthquake action on the upper structure, so the horizontal stiffness of the isolation bearing will directly affect the isolation effect of the structure. Therefore, in a reasonable range, properly reducing the horizontal stiffness of the isolation bearing can improve the isolation effect, reduce the cost of the isolation building, further promote the promotion of the isolation technology, and ensure the property and safety of the people from loss.

For rubber isolation bearings, the horizontal stiffness of the bearings is closely related to the shear modulus. At present, most of the isolation bearings studied or used in our country are those with shear modulus above 0.40 MPa, and the horizontal stiffness is relatively large. For this reason, the author's team proposed a low shear modulus rubber isolation bearing (LNR400) with shear modulus of 0.32 MPa from the point of view of the material characteristics of rubber itself, and its performance parameters are shown in Table 9.1 [2]. The biggest difference between the improved rubber bearing and the common bearing is its small rubber shear modulus, which will have a direct impact on the two important parameters of the isolation bearing: the vertical compression stiffness and the horizontal equivalent stiffness.

9.1.2.2 Vertical Mechanical Property Test of Isolation Bearing

1. **Test device and sample**

The test was carried out in the structural laboratory of Jiulonghu campus of Southeast University. The test is carried out on the compression shear tester. The vertical displacement was measured by dial indicator and the vertical pressure was measured by force sensor. The main test device is shown in Fig. 9.2a. The relevant parameters are: maximum vertical thrust 150,000 kN; maximum vertical tension 6000 kN; maximum horizontal thrust 2000 kN; net space: 3 m × 1.5 m × 4 m (L × W × H). The bearings used are LNR400 rubber bearings produced by Jiangyin Haida Rubber & Plastics Co., Ltd., which are numbered as #1, # 3, # 4, # 5 and # 6 (the bearing number has been compiled after Haida production, and has not been renumbered to avoid data confusion). See Table 9.2 for specific parameters. Jiangyin Haida Rubber & Plastics Co., Ltd. has carried out the physical property test table of rubber materials when the test samples leave the factory. All rubber used to make rubber isolation bearings have been tested for tensile performance, aging performance, hardness, shear performance and compression performance according to the test methods specified in the specifications. See Table 9.3 for the test results.

Table 9.1 Performance parameters of improved rubber bearing with low shear modulus LNR400

Parameter	Parameter value	Parameter	Parameter value
Product outer diameter (mm)	420	Total height of support (mm)	123.5
Effective diameter (mm)	400	Thickness of sealing plate (mm)	15
Center hole diameter (mm)	20	Total height of mould (mm)	153.5
Rubber layer thickness (mm)	2.71	Effective rubber area (mm^2)	125,349.4
Number of rubber layers	29	Rubber shear modulus G (MPa)	0.32
Total thickness of rubber layer (mm)	78.71	Rubber hardness correction coefficient K	0.913
Steel plate thickness (mm)	1.6	Vertical compression stiffness (kN/mm)	1627.8
Steel plate number	28	Horizontal equivalent stiffness (10 MPa, kN/mm)	0.5007
Total thickness of steel plate (mm)	44.8	Horizontal equivalent stiffness (12 MPa, kN/mm)	0.4970
First shape factor S_1	35	Horizontal equivalent stiffness (15 MPa, kN/mm)	0.4902
Second shape factor S_2	5.1		

2. **Loading scheme**

According to "*Rubber bearings–Part 1: Test methods for isolation rubber bearings GB/T 20688.1-2007*", the compression stress of the bearing is loaded from $0.7\sigma_0$ to $1.3\sigma_0$, as shown in Fig. 9.2b. Take the result of the third cycle, and the vertical compression stiffness is calculated by the following formula:

$$K_v = \frac{P_2 - P_1}{Y_2 - Y_1} \tag{9.11}$$

Among them, P_2 is the maximum vertical load reached at the third loading cycle, i.e. 1.3 times of P_0 (design load); P_1 is 0.7 times of design load at the third loading cycle, i.e. 0.7 times of P_0; Y_2 is the displacement corresponding to the maximum vertical load P_2 reached at the third loading cycle; Y_1 is the displacement corresponding to the vertical load P_1 at the third loading cycle.

Test the vertical compression performance of LNR400 bearing under the compression stress of $\sigma_0 = 12$ MPa and 15 MPa. See Table 9.4 for the test conditions. The specific test steps are as follows:

(1) install the test support in the compression shear tester, and ensure that the center of the support is on the center line of the compression shear tester;

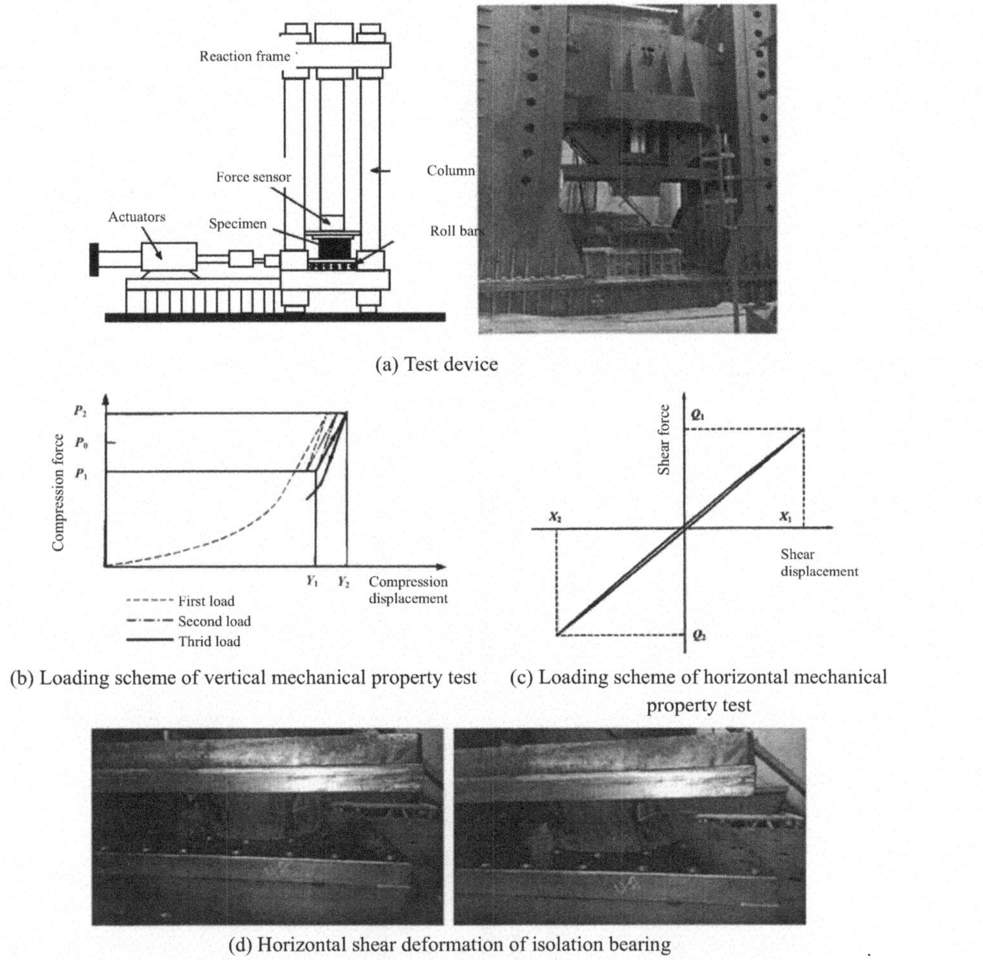

(a) Test device

(b) Loading scheme of vertical mechanical property test

(c) Loading scheme of horizontal mechanical property test

(d) Horizontal shear deformation of isolation bearing

Fig. 9.2 Vertical mechanical property test of isolation bearing

(2) install dial indicator in four directions of test bearing to measure vertical displacement;
(3) preload test: load to bearing test load, dead load for 3 min, unload to 0, stop load for 3 min, repeat the above test for three times in total;
(4) formal loading: the above compression test loading method is adopted for the loading, and the compression stress of the bearing is loaded from 0.7 to 1.3% in cycles for three times. Read the dial indicator value at 0.7 and 1.3% of the third cycle to obtain the displacement change value;
(5) unloading, end of test.

Table 9.2 Basic geometric parameters of LNR400 bearing

Bearing class	Effective diameter (mm)	Center hole diameter (mm)	Rubber layer thickness (mm)	Steel plate thickness (mm)	Thickness of sealing plate (mm)	Number of rubber layer	Number of steel plate layer	First shape factor S_1	Second shape factor S_2
LNR400	400	20	2.71	1.6	15	29	28	35	5.1

Note LNR400-#3 bearing is selected as the test bearing in this vertical compression performance test, and other bearings will be used in the horizontal shear performance test in Sect. 9.1.2.3

Table 9.3 Physical property test table of rubber for isolation bearing

Technical index		Detection value	Specification allowed value	Whether qualified
Tensile property	tensile strength (MPa)	18.3	During the production of bearing, spot check shall be conducted once a day, and the performance shall meet the requirements in Appendix B of GB20688.3-2007	Yes
	Elongation	650.5%		Yes
Aging property	Change rate of tensile strength	21.3%	±25%	Yes
	Change in elongation at break	11.7%	−50% * maximum elongation before aging	Yes
Compression performance	Compression permanent deformation	14.19%	/	Yes
Shear property	Shear modulus of elasticity	0.32	/	Yes
Hardness	Normal hardness	40	Not as main design index	Yes
	Hardness after aging	43		Yes

Table 9.4 Vertical compression test condition

Load step	Vertical surface pressure (MPa)	Shear deformation Y%	Cycle times	Vertical force (KN)
1	0–0.7 × 12	±0	\	0–1053
2	12 ± 30%	±0	3	1053–1955
3	0.7 × 12–0.7 × 15	±0	\	1053 1316
4	15 ± 30%	±0	3	1316–2444
5	0.7 × 15–0	±0	\	1316–0

3. **Analysis of test results**

When the bearing is compressed, the rubber layer bulges and the bearing displaces downward. But in the actual test process, the displacement of the bearing under the vertical load is very small, which is difficult to observe by naked eyes, and the bulge of rubber is also basically difficult to identify. The vertical compression stiffness of rubber bearing under 12 and 15 MPa can be calculated by Eq. (9.11), as shown

Table 9.5 Results of vertical compression test of isolation bearing

Compression stress (MPa)	Lower pressure of the third cycle (kN)	Higher pressure of the third cycle (kN)	Displacement change (mm)	Vertical compression stiffness test value (kN/mm)
12	1055	1960	0.5575	1623.32
15	1320	2450	0.675	1674.07

Table 9.6 Comparison between test value and theoretical value of vertical compression stiffness of rubber bearing

Bearing	Basic compressive stress (MPa)	Theoretical value of vertical compression stiffness (kN/mm)	Test value of vertical compression stiffness (kN/mm)	Deviation
LNR400-#3	12	1627.8	1623.32	−0.28%
	15		1674.07	2.84%

in Table 9.5. It should be noted that the displacement in the table is not the total displacement of the bearing under σ_0, but the relative displacement of the bearing between $0.7\sigma_0$ and $1.3\sigma_0$, the value is about 0.5–0.7 mm. Refer to the theoretical formula in Sect. 9.1.1 to calculate the vertical stiffness value of the bearing, and compare with the test value. See Table 9.6 for the results. It can be seen from the table that the vertical compression stiffness of LNR400 improved low shear modulus isolation bearing is about 1600 kN/mm, and the error between the test value and the theoretical value is very small, within 3%, which can be almost ignored; at the same time, from the test results, it can be found that the vertical compression stiffness of the bearing under 12 MPa is less than that under 15 MPa, showing the trend of increasing stiffness. This is consistent with the loading curve provided by the code and the results of other literature. The theoretical formula can not reflect this rule.

9.1.2.3 Horizontal Mechanical Property Test of Isolation Bearing

1. **Loading scheme**

The test equipment shown in Fig. 9.2a is also used for the horizontal mechanical property test of the isolation bearing. According to "*Rubber bearings–Part 1: Test methods for isolation rubber bearings GB/T 20688.1-2007*", the compression shear performance test is carried out under the vertical compressive stress, and four loading cycles are adopted, the test data of the third cycle is taken to calculate the horizontal equivalent stiffness of the bearing, as shown in Fig. 9.2c, and the calculation formula is as follows:

$$K_h = \frac{Q_2 - Q_1}{X_2 - X_1} \tag{9.12}$$

Among them, X_1 is the maximum positive horizontal displacement in the third cycle; X_2 is the maximum negative horizontal displacement in the third cycle; Q_1 is the horizontal load corresponding to X_1 in the third cycle (friction correction is required); Q_2 is the horizontal load corresponding to X_2 in the third cycle (friction correction is required). The horizontal shear performance of LNR400 bearing was tested when the compression stress is $\sigma_0 = 10$ MPa, 12 MPa and 15 MPa. See Table 9.7 for the test conditions. The specific test steps are as follows:

(1) the preparatory work is similar to that of the vertical compression test;
(2) preload test: firstly, raise the support to ±100% displacement, and test the actuator. Then, the vertical pressure was increased to σ_0, and the ±100% displacement was measured by hand;
(3) formal loading: design horizontal shear loading program, cyclic loading for 4 times, record the test data by computer, and finally take the third cyclic data;
(4) unloading, end of test.

2. **Analysis of test results**

The horizontal mechanical performance of rubber bearing is an important mechanical characteristic of the bearing, which will affect the isolation effect and safety of the isolation building to a large extent. In this test, five rubber bearings #1, #3, #4, #5 and #6 were tested for 100% horizontal deformation shear performance to study the horizontal mechanical properties and stability of this improved low shear modulus isolation bearing. During the horizontal shear process of rubber bearing, it can be clearly observed that the bearing produces great lateral displacement (Fig. 9.2d), and the load displacement curve of each vertical stress is shown in Fig. 9.3. In this test, there are 5 LNR400 rubber isolation bearings in total. From the horizontal load displacement curves under 10 MPa and 15 MPa, it can be observed that the horizontal mechanical properties of the 5 bearings are relatively stable. Using Eq. (9.12) to

Table 9.7 Horizontal shear test condition

Load step	Vertical surface pressure (MPa)	Shear deformation Y%	Cycle times	Load frequency (Hz)	Vertical force (kN)
1	0–10	±0	\	\	0–1253.5
2	10	±100	4	0.005	1253.5
3	10–12	±0	\	\	1253.5–1504
4	12	±100	4	0.005	1504
5	12–15	±0	\	\	1504–1880
6	15	±100	4	0.005	1880
7	15–0	±0	\	\	1880–0

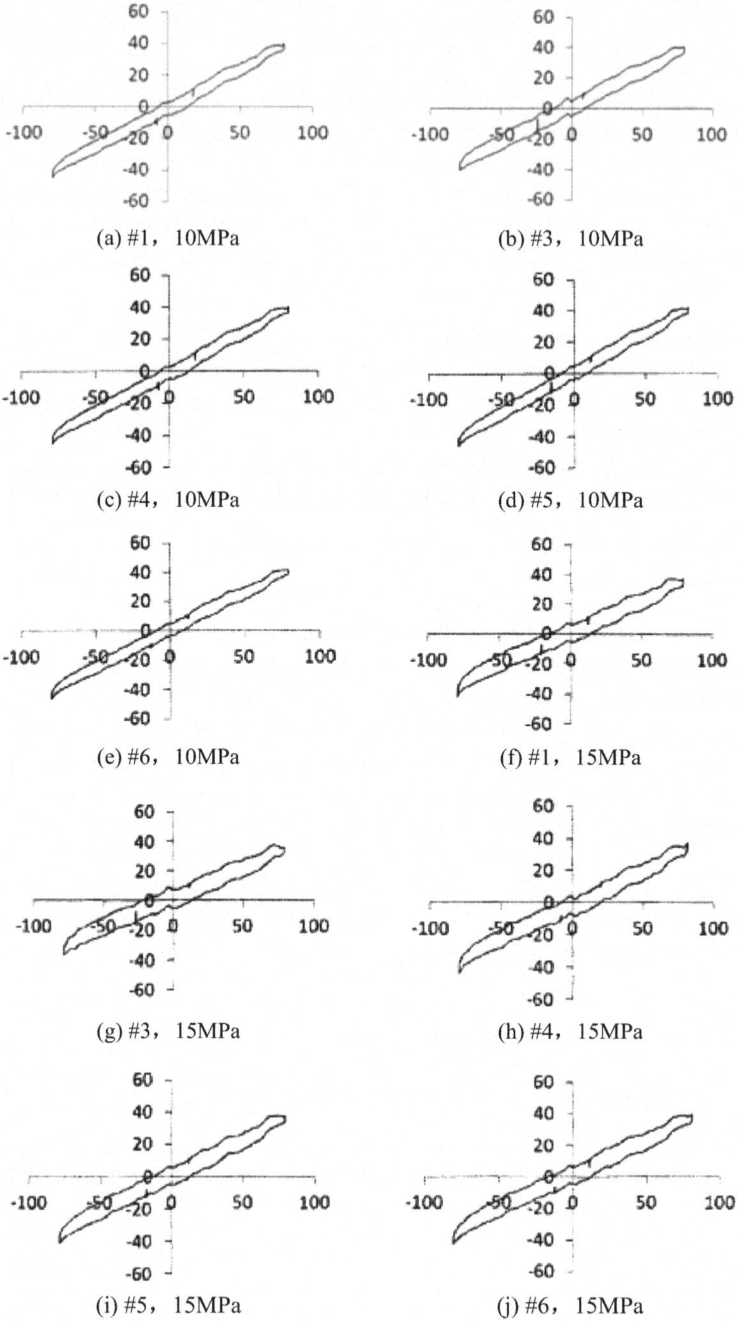

(a) #1, 10MPa
(b) #3, 10MPa
(c) #4, 10MPa
(d) #5, 10MPa
(e) #6, 10MPa
(f) #1, 15MPa
(g) #3, 15MPa
(h) #4, 15MPa
(i) #5, 15MPa
(j) #6, 15MPa

Fig. 9.3 Measured load displacement curve of isolation bearing

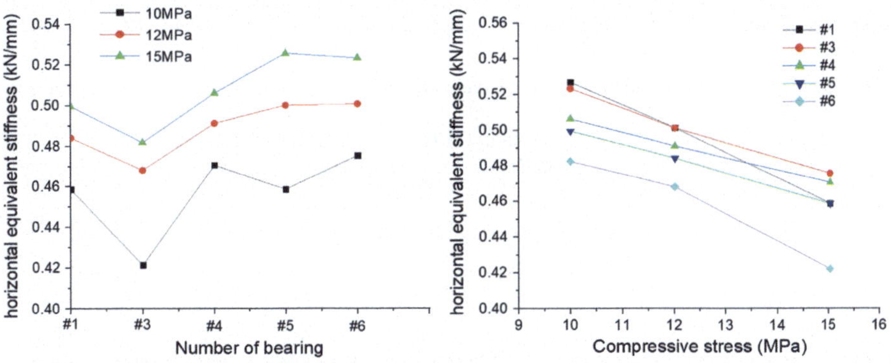

Fig. 9.4 Equivalent horizontal stiffness of isolation bearing test

calculate the corresponding horizontal equivalent stiffness, the horizontal equivalent stiffness curves of the bearings under 10, 12 and 15 MPa are shown in Fig. 9.4, in which the friction coefficient is 0.15%. It can be seen from the figure that: (1) the value of horizontal equivalent stiffness of bearing under various compressive stresses is relatively stable, and the law that the horizontal equivalent stiffness decreases with the increase of compressive stress appears under different compressive stresses. (2) under the same pressure stress, the horizontal equivalent stiffness values of the five bearings have little difference and small discreteness. The maximum deviation of the horizontal equivalent stiffness of the bearings is about 5% under 10 MPa, about 4.5% under 12 MPa, and about 8% under 15 MPa, indicating that the performance of the improved low shear modulus isolation bearings is stable. (3) the horizontal equivalent stiffness of the bearing decreases with the increase of the compressive stress, and the decreasing trend of each bearing is basically the same, with the slope of −0.01 (calculated according to the slope of the curve, only to evaluate the declining trend, regardless of the unit). This is because in the process of horizontal shear, the upper and lower planes of the support will be staggered with each other, so the vertical load will produce a bending moment on the support, play a positive role in the horizontal shear of the support, the horizontal displacement of the support increases, so the horizontal equivalent stiffness decreases, and gradually decreases with the increase of the vertical load.

The theoretical value of the horizontal equivalent stiffness of the bearing calculated according to the theoretical formula in Sect. 9.1.1 is compared with the test value. The results are shown in Table 9.8. The horizontal equivalent stiffness of the bearing is about 0.50 kN/mm, and the deviation between the test value and the theoretical value is within 2% at 10 and 12 MPa, and within 7% at 15 MPa.

Table 9.8 Comparison of test value and theoretical value of horizontal equivalent stiffness of rubber bearing

Bearing	Basic compressive stress (MPa)	Theoretical value of horizontal equivalent stiffness (kN/mm)	Test value of horizontal equivalent stiffness (kN/mm)	Deviation (%)
LNR400	10	0.5007	0.5081	1.47
	12	0.4970	0.4895	−1.51
	15	0.4902	0.4576	−6.64

9.1.2.4 Numerical Simulation of Mechanical Properties of Isolation Bearing

1. **Finite element model**

Further use the finite element software to carry out the numerical simulation of the mechanical properties of the improved low shear modulus rubber bearing, and the model is shown in Fig. 9.5a. The bearing diameter direction is divided into 24 parts, and the circumference direction is divided into 48 parts. In the processes of compression and shear of bearing, the deformations of steel plate and sealing plate are not large, and the thickness of sealing plate is large, so the sealing plate is divided into two parts along the thickness direction, while the steel plate is not divided along the thickness direction. The deformation of rubber layer is large in the process of bearing stress, which is the main influence factor of bearing mechanical properties, so it is divided into three parts along the thickness direction. In the process of making rubber isolation bearing, rubber and steel plate are laminated and vulcanized under high temperature and pressure. In the experiment, there is little peeling phenomenon between steel plate and rubber, so in the simulation, binding constraint is used between each layer of steel plate and rubber, which can greatly reduce the number of iterations needed for calculation. The fixed constraint is carried out at the bottom of the rubber bearing to limit the translation in the X, Y and Z directions and the rotation around the three axes at the bottom of the bearing. A reference point is set at the center of the top surface of the support, and the reference point and the top surface are constrained as rigid tie, so that whether the vertical compression load or the horizontal displacement load is applied on the top surface, it can be applied at the reference point. More importantly, by setting the reference point as a set and defining the output in the analysis step, the displacement and load of the corresponding working condition can be easily obtained. In order to simulate the bearing compression shear test, the rotational degrees of freedom around the shaft of reference point is fixed, keeping the top surface level, which is consistent with the actual loading situation of the bearing. It should be noted here that if the rotational degrees of freedom around the shaft are not constrained, the top surface of the support will deflect during the shear process and will no longer remain horizontal, which is inconsistent with the actual test process, so it is necessary to fix the rotational degrees of freedom around

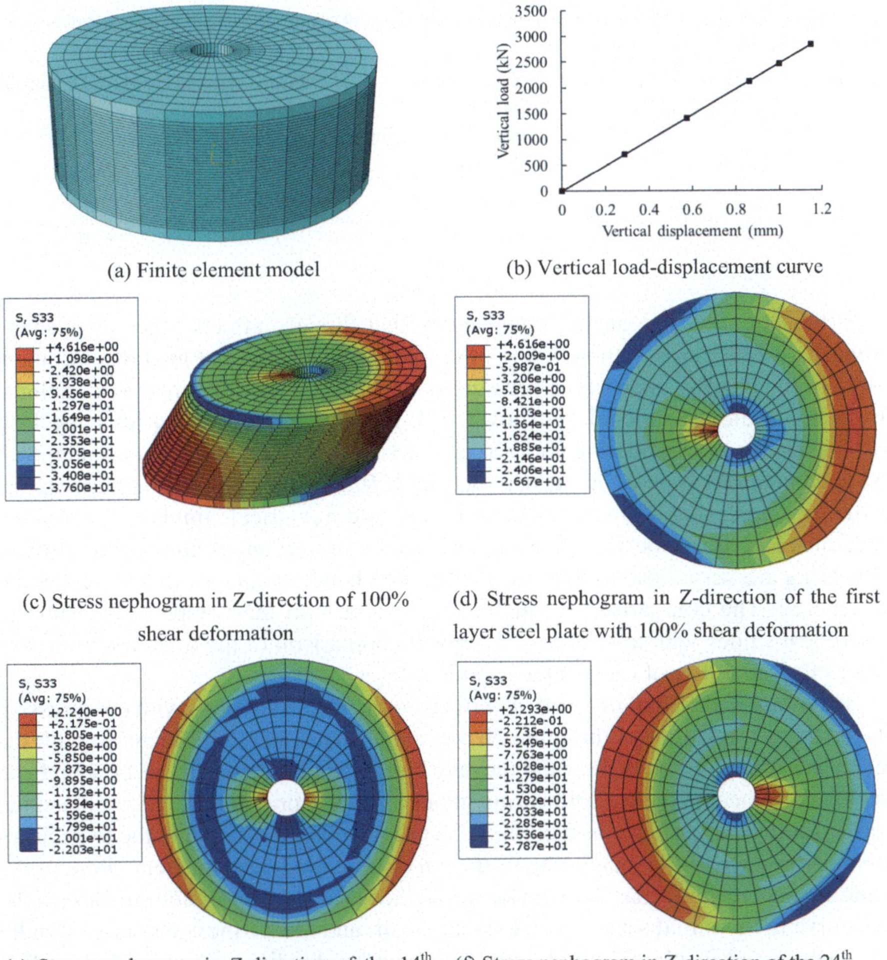

(a) Finite element model

(b) Vertical load-displacement curve

(c) Stress nephogram in Z-direction of 100% shear deformation

(d) Stress nephogram in Z-direction of the first layer steel plate with 100% shear deformation

(e) Stress nephogram in Z-direction of the 14th layer steel plate with 100% shear deformation

(f) Stress nephogram in Z-direction of the 24th layer steel plate with 100% shear deformation

Fig. 9.5 Finite element simulation of isolation bearing

the shaft. The vertical compression stress on rubber bearing is 5, 10, 12 and 15 MPa, and the horizontal shear simulation is carried out. The stress state and deformation of the bearing are extracted respectively under the horizontal shear deformation of 0%, 25%, 50%, 75%, 100% and 125%. The following is the finite element result analysis of rubber bearing.

2. **Simulation of vertical mechanical properties**

In practical engineering, rubber bearings need to bear the vertical load from the upper structure for a long time, so the vertical compression performance of rubber bearings is not only the guarantee of normal use of rubber bearings in the absence

Table 9.9 Comparison between test value and finite element result of vertical compression stiffness of rubber bearing

Bearing	Basic compressive stress (MPa)	Test value of horizontal equivalent stiffness (kN/mm)	Finite element result of horizontal equivalent stiffness (kN/mm)	Deviation
LNR400-#3	12	1623.32	1640.50	L06
	15	1674.07	1678.20	0.25

of earthquake, but also an important index that directly affects other mechanical properties of bearings in the coming of earthquake. The vertical load displacement curve of the bearing under axial pressure is obtained by the reference point defined in the finite element model, as shown in Fig. 9.5b. It can be seen that the load displacement curve of bearing under axial pressure is linear, which is consistent with the phenomenon observed in the test. See Table 9.9 for the comparison between the vertical compression stiffness obtained by test and numerical simulation. From the table analysis, it can be seen that the test results of vertical compression stiffness of bearing are very close to the results of finite element analysis, and the error can be ignored. The comparison results are consistent with the comparison results of theoretical values with test values. The result comparison of the three methods also proves the reliability of each other's results.

According to the changing law of vertical compression stiffness with compressive stress, the test results and the finite element analysis results both increase with the increase of compressive stress, and they are consistent. However, it can be seen from the load displacement curve of bearing under axial compression in Fig. 9.5b and the vertical compression test curve in Fig. 9.2b provided by the code that the two curves are not consistent at the beginning of the curve. The load displacement curve of the finite element results is basically linear, while the vertical compression stiffness of the test curve in the initial section is relatively small, and the stiffness increases rapidly after a period of displacement. At the same time, the loading section and unloading section of the test curve are not completely coincident, and the finite element analysis results can not describe this phenomenon. This problem is caused by the defect of rubber constitutive model. Mooney Rivlin constitutive model is relatively simple and can not completely describe the load displacement curve of bearing vertical compression. At the same time, the stress-strain curve of theoretical model is stable and can not describe damage, residual deformation and other phenomena. The curve will not change due to the reasons of loading and unloading. The results of bearing of the first analysis and the second analysis will not change.

3. **Simulation of horizontal mechanical properties**

The isolation effect of isolation structure depends on the horizontal shear performance of rubber bearing to a great extent. When an earthquake occurs, the isolation effect of engineering structure is realized by the horizontal deformation of rubber

Table 9.10 Comparison between test value and finite element result of vertical compression stiffness of rubber bearing

Bearing	Basic compressive stress (MPa)	Test value of horizontal equivalent stiffness (kN/mm)	Finite element result of horizontal equivalent stiffness (kN/mm)	Deviation
LNR400	10	0.5081	0.4782	−5.88
	12	0.4895	0.4658	−4.84
	15	0.4576	0.4439	−2.99

bearing. Figure 9.2b shows the change of Z-direction stress at 100% shear deformation (15 MPa). It can be seen from the figure that because of the displacement of bearing core compression area on steel plate, and the tension area appears at two sharp corners after shear deformation of bearing. Because of the existence of the central hole, the core compression area is weakened near the hole. Although the stress distribution of steel plates in different positions is different, the peak value of stress is basically the same, all of which are about 8 MPa. The peak value of the tensile stress of the steel plates at both ends of the support is not at the most edge, but about 30 mm near the edge, which is different from that the middle steel plate edge is the position of maximum tensile stress. See Table 9.10 for the comparison between the vertical compression stiffness obtained by test and numerical simulation. The test results of horizontal equivalent stiffness of bearing are very close to the results of finite element analysis, and the error is within 6%, which shows that the results are reliable. From the point of view of the decrease degree of the horizontal equivalent stiffness with the increase of the compressive stress, the decrease degree of the finite element analysis result and the test result is more consistent than the theoretical value. The comparison results show that Mooney Rivlin constitutive model is more accurate in describing the small shear deformation of bearing.

9.1.3 Honeycomb Sandwich Rubber Isolation Bearing

9.1.3.1 Design of Honeycomb Sandwich Rubber Isolation Bearing

In order to effectively reduce the horizontal stiffness of the rubber isolation bearing and maintain the overall stability of the isolation bearing, the honeycomb sandwich rubber isolation bearing is proposed from the perspective of the structure of the isolation bearing. According to the actual engineering requirements, the height and diameter of the rubber bearing with honeycomb interlayer are designed to be 280 mm and 210 mm respectively [3]. Compared with rubber bearings commonly used in engineering, the height width ratio of honeycomb sandwich rubber bearings is large, but it can still meet the requirements of stability. Its specification is: ϕ200 mm ×

Table 9.11 Dimensions of honeycomb sandwich rubber bearings

Parameter	Size (mm)	Diagram
Effective diameter R1	200	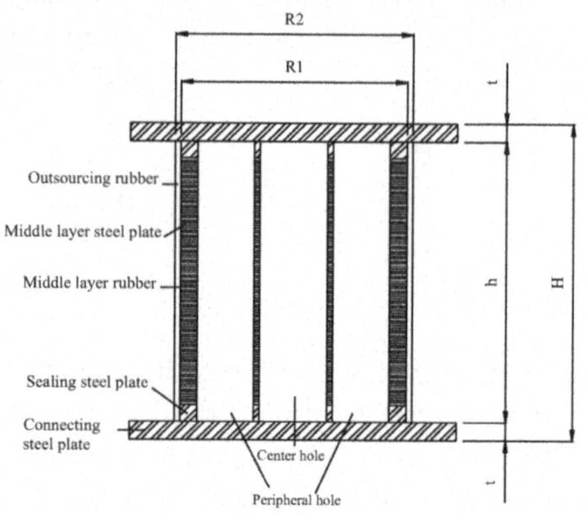
Total diameter R2	210	
Total height H	280	
Rubber layer thickness (Thickness of each layer × layer number)	2.4 × 50	
Thickness of internal steel plate (Thickness of each layer × layer number)	2 × 49	
Thickness of sealing steel plate	15	
Thickness of connecting steel plate	16	
Center hole diameter	60	
Peripheral hole diameter	50	

H300 mm, the diameter of the center hole is 60 mm, and the diameter of the four surrounding holes is 50 mm. There are 49 layers of steel plates, each layer is 2 mm thick; 50 layers of rubber, each layer is 2.4 mm thick. See Table 9.11 for specific structure and parameters.

9.1.3.2 Mechanical Property Test of Isolation Bearing

1. **Test scheme**

Four rubber bearings with honeycomb sandwich were tested with full scale model. The test equipment is MTS 458.10 hydraulic servo control system, and the test method is in accordance with "*The isolation technical code of laminated rubber bearings* (*CECS126:2001*)". In the test, two rubber bearings are superposed, and the horizontal performance is tested by the double shear method of horizontal force loading from the middle steel pull plate. The vertical design load is taken as 60 kN (Fig. 9.6).

During loading, after the vertical stiffness tests of four bearings, the horizontal performance tests of two pairs (four) bearings are carried out in two items. In the horizontal test, the static stiffness test is carried out first, and then the dynamic test is carried out according to different frequency and displacement conditions. That is to say, when the rubber bearing is under the vertical pressure of 60 kN, the dynamic loading tests with the frequency of 0.1 Hz, 0.2 Hz and 0.3 Hz and the horizontal displacement of 10 mm, 20 mm and 50 mm are carried out respectively, and the horizontal loading waveform is sine wave.

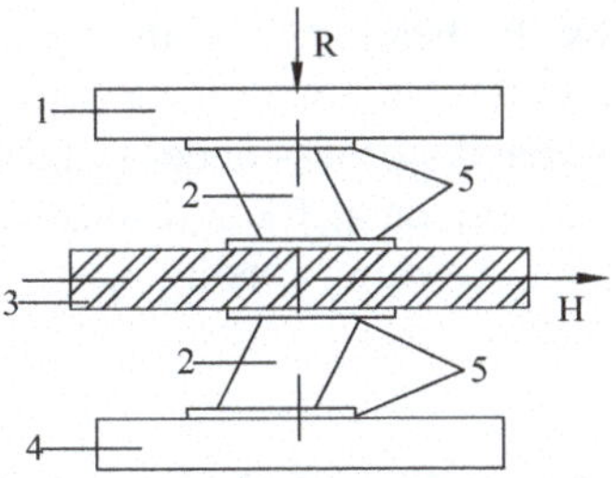

1-upper bearing plate; 2-rubber support; 3-intermediate steel pull plate; 4-lower bearing plate; 5-anti-skid friction plate

(a) Schematic diagram of double shear loading

(b) Honeycomb rubber support before and after loading

Fig. 9.6 Loading scheme for mechanical property test of isolation bearing

2. **Analysis of test results**

(1) Vertical stiffness of isolation bearing

The vertical stiffness of the honeycomb sandwich rubber bearing refers to the vertical force exerted under the unit vertical displacement of the rubber pad under the vertical pressure, which is calculated according to the following formula:

$$K_v = \frac{P}{\delta} \tag{9.13}$$

where, P is the vertical pressure of the bearing; δ is the corresponding compression deformation. Table 9.12 shows the deformation and rigidity results of each bearing under the vertical load of 60 kN.

(2) Horizontal stiffness of isolation bearing

The horizontal stiffness of the honeycomb sandwich rubber bearing refers to the horizontal force required to be exerted under the unit relative displacement of the upper and lower plate surfaces of the rubber pad, which is calculated according to the following formula:

$$K_h = \frac{Q}{X} \tag{9.14}$$

Table 9.12 Vertical stiffness of isolation bearing

Specimen	1#	2#	3#	4#
Displacement (mm)	2.50	2.52	2.50	2.51
Rigidity (kN/mm)	24.0	23.8	24.0	23.9

where, Q is the horizontal force borne by the bearing; X is the horizontal relative displacement. Figure 9.7a shows the horizontal force displacement curve under static load, and the horizontal stiffness under static load is 124.68 N/mm. Figure 9.7b–j shows the horizontal force displacement hysteretic curve of the bearing under different working conditions.

9.1.3.3 Numerical Simulation of Mechanical Properties of Isolation Bearing

1. **Mechanical properties of materials**

The honeycomb sandwich rubber bearing is mainly composed of rubber and steel plate. Rubber is a kind of hyperelastic and nearly incompressible material, which has good elasticity. It can produce large displacement and large strain under the action of external force, showing complex material nonlinearity and geometric nonlinearity. Generally, Mooney Rivlin model is used to analyze and calculate the mechanical properties of rubber materials. Because of the nonlinearity, incompressibility and large deformation of rubber material, the process of determining the mechanical property constant of rubber material by test method is complicated. Therefore, two parameters of Mooney Rivlin model of rubber material are determined as follows:

(1) according to the fitting formula from the test data of IRHD hardness H_r and E_0 of rubber material, E_0 of rubber material is calculated by known hardness H_r:

$$\lg E_0 = 0.0198H_r - 0.5432 \tag{9.15}$$

(2) E_0 has the following relationship with parameters C_1 and C_2:

$$E_0 = 6C_1(1 + \frac{C_2}{C_1}) \tag{9.16}$$

(3) finally, after determining the sum of parameters C_1 and C_2 according to the hardness, the axial compression deformation of the rubber material under different C_2/C_1 is simulated by computer to determine the C_1 and C_2 of the rubber material.

When the IRHD hardness of rubber bearing is 40°, 60° and 70°, it is the closest to the measured value when C_2/C_1 is 0.1, 0.05 and 0.02, respectively. This shows that the best C_2/C_1 is different for rubber materials with different hardness: the hardness increases and C_2/C_1 decreases. The rubber of the honeycomb sandwich rubber bearing is very soft, and its hardness is about 25°. Through the trial calculation of its axial compression deformation, it is determined that its hardness is 23°. The Mooney material constant is $C_1 = 0.1238$ MPa, $C_2 = 0.01238$ MPa, $C_2/C_1 = 0.1$.

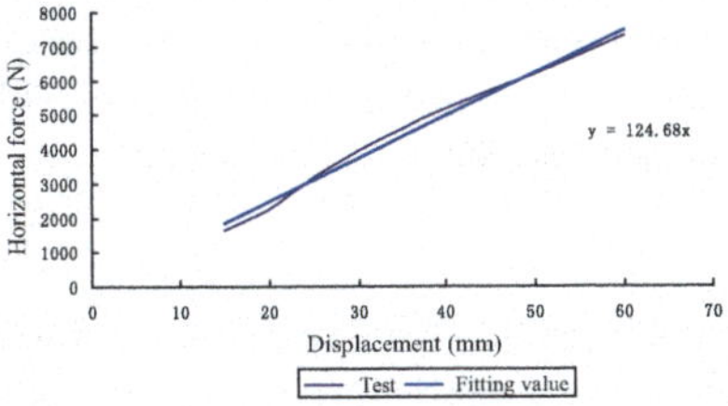

(a) Horizontal stiffness curve under static load

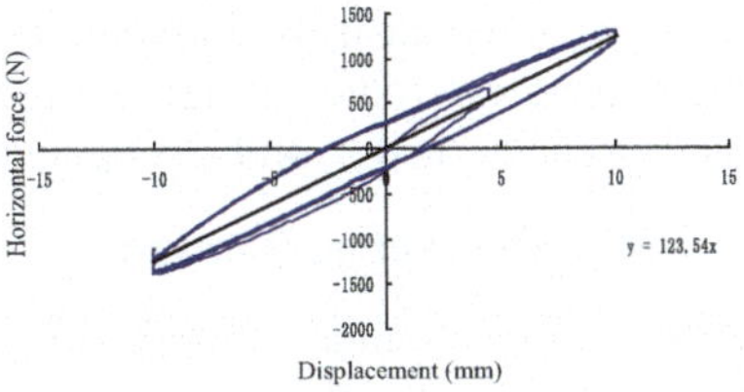

(b) 0.1 Hz-10 mm condition

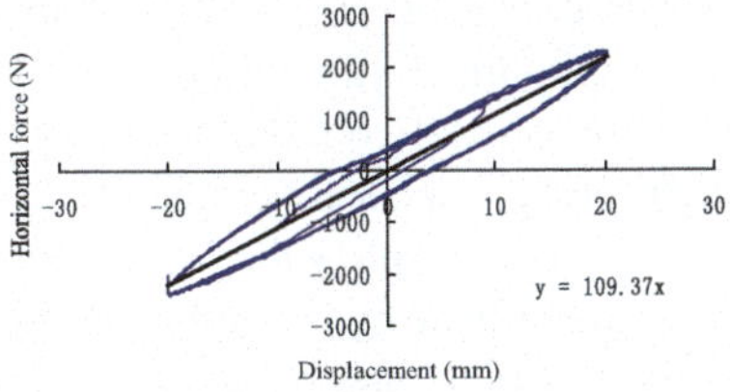

(c) 0.1 Hz-20 mm condition

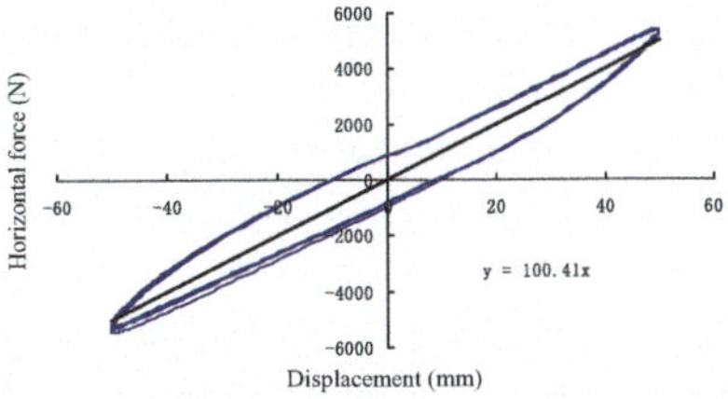

(d) 0.1 Hz-50 mm condition

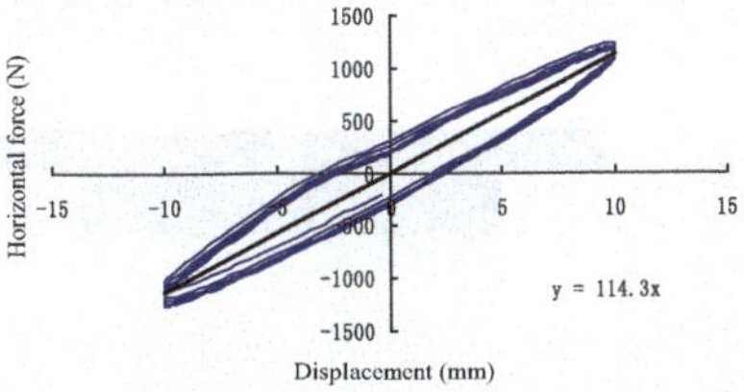

(e) 0.2 Hz-10 mm condition

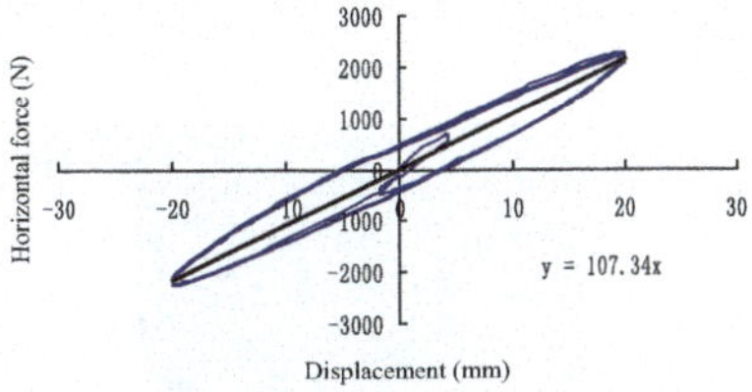

(f) 0.2 Hz-20 mm condition

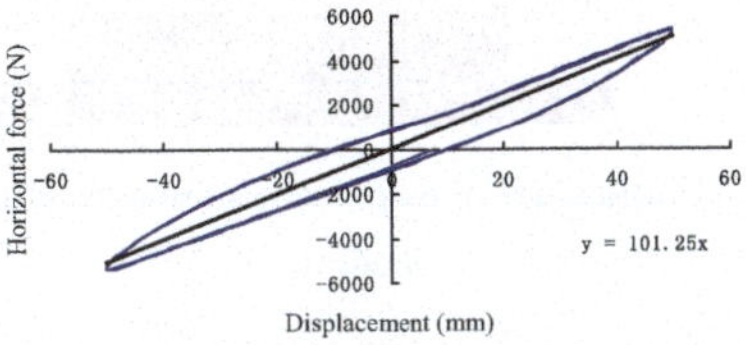

(g) 0.2 Hz-50 mm condition

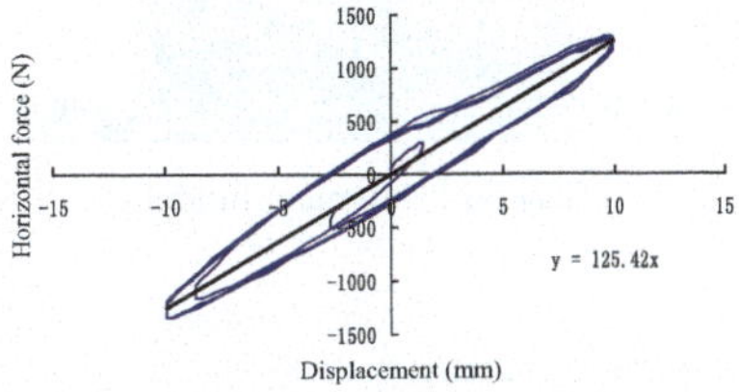

(h) 0.3 Hz-10 mm condition

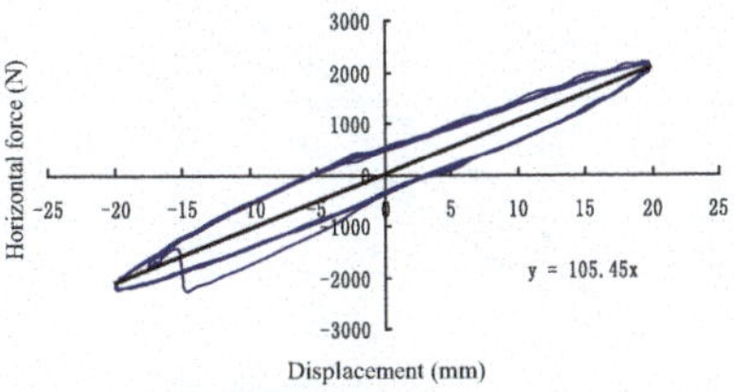

(i) 0.3 Hz-20 mm condition

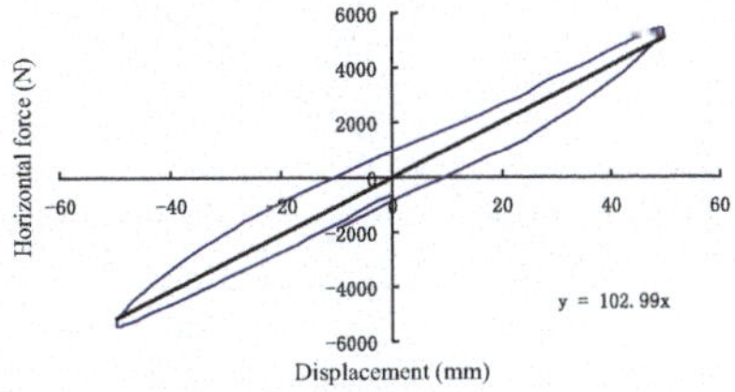

(j) 0.3 Hz-50 mm condition

Fig. 9.7 Horizontal force displacement curve of isolation bearing

Q235 steel plate is used in the support. The steel is an ideal elastic-plastic material. The bilinear isotropic strengthening model is used. The elastic modulus is 210 GPa, Poisson's ratio is 0.3, and the tangent modulus is 2100 MPa.

2. **Finite element simulation**

ANSYS finite element software is used for analysis. Solid 45 element is used for steel plate and Hyper58 element is used for rubber when establishing the model, which are three dimensional solid elements with 8 nodes. The model is shown in Fig. 9.8a. In view of that in the manufacturing process of rubber bearing, the rubber plate and thin steel plate are bonded and vulcanized under high temperature and pressure, but in actual use, the rubber and the steel plate are always closely bonded together, and the test results also verify that there is little peeling phenomenon between the steel plate and the rubber inside the rubber bearing. In order to simplify the finite element model, the joint between rubber and steel sheet in rubber bearing is combined. At the same time, fix all degrees of freedom of the lower seal plate and the Y-axis degrees of freedom of the upper seal plate, apply the Z-direction pressure, X-direction static load and X-direction cyclic reciprocating displacement load on the upper seal plate.

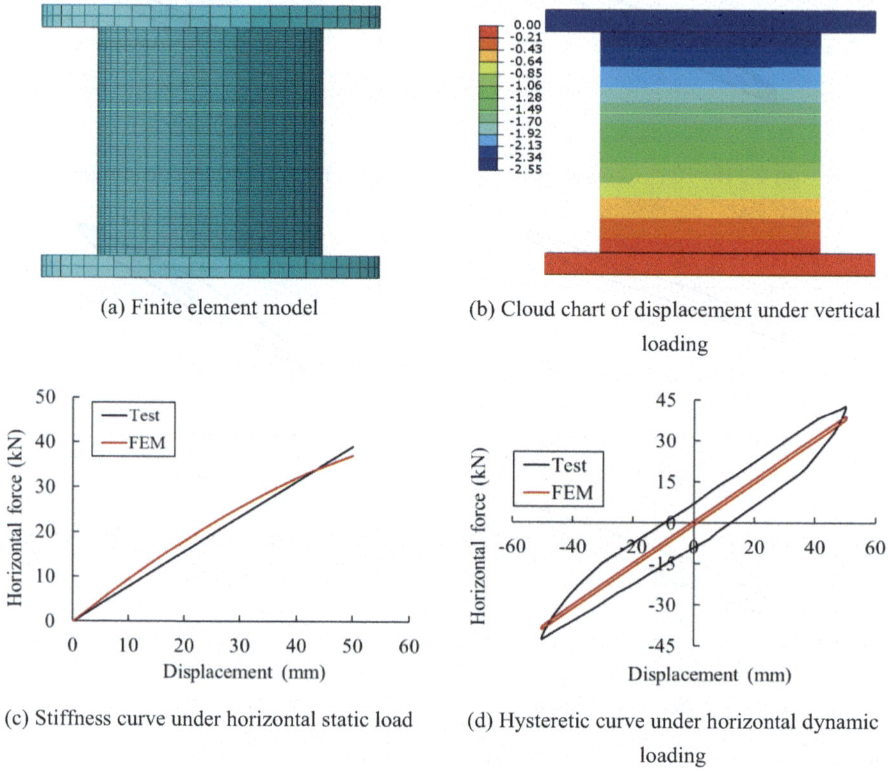

Fig. 9.8 Finite element simulation of isolation bearing

Table 9.13 Comparison of test and numerical simulation of isolation bearing stiffness

Parameter	Test value (N/mm)	Finite element simulation value (N/mm)	Error
Vertical stiffness	23,930	23,490	1.84
Horizontal static stiffness	124.68	133.33	6.94
Horizontal dynamic stiffness (0.1 Hz–50 mm condition)	100.41	93.28	7.10

In order to be able to compare with the test value, it is necessary to apply vertical compression load and lateral cyclic load to the bearing consistent with the test situation. Considering that during the test, 60 kN axial pressure was applied to the bearing firstly, then horizontal cyclic displacement was applied, so in the finite element analysis, 60 kN axial pressure was applied to the bearing firstly, then horizontal cyclic displacement was applied, and finally time history displacement load was applied for dynamic analysis. Because there are many working conditions of horizontal dynamic loading, only a group of working conditions of 0.1 Hz–50 mm is selected for finite element simulation analysis.

Through the above test analysis of the honeycomb sandwich rubber bearing, it can be concluded that the vertical stiffness of the honeycomb sandwich rubber bearing is 23.93 kN/mm, the horizontal static stiffness is 124.68 kN/mm, and the horizontal dynamic stiffness range is 100.41–123.54 N/mm. The results of finite element numerical simulation of honeycomb sandwich rubber bearing: the vertical stiffness is 23.49 kN/mm, the horizontal static stiffness is 133.33 N/mm, and the horizontal dynamic stiffness under the condition of 0.1 Hz–50 mm is 93.28 N/mm. The comparison with the test results is shown in Table 9.13, Fig. 9.8c, d. It can be seen that the finite element numerical simulation analysis method can more accurately simulate all kinds of bearing stress conditions, the error of vertical stiffness is 1.84%, the error of horizontal static stiffness is 6.94%, and the error of horizontal dynamic stiffness is 7.10%. Compared with the rubber bearing commonly used in engineering, the honeycomb sandwich rubber bearing has relatively small vertical stiffness and horizontal stiffness.

9.2 Composite Isolation Bearing

9.2.1 *Dish Spring Composite Multi-dimensional Isolation Bearing*

9.2.1.1 Design of Dish Spring Composite Multi-dimensional Isolation Bearing

Multidimensional isolation has always been a hot topic in the research of building isolation. The three-dimensional multifunctional isolation bearing designed by the author's team is shown in Fig. 9.9a, which is composed of lead rubber vibration isolation bearing, dish spring, guide rod, lower connecting plate, intermediate connecting plate, upper connecting plate, etc. [4]. A lead rubber vibration isolation support is arranged between the lower connecting plate and the middle connecting plate, a dish spring group is arranged between the middle connecting plate and the upper connecting plate, a guide device is arranged in the middle of the dish spring group, a guide rod is arranged inside the guide device, and the upper part of the guide rod is connected with the upper connecting plate. A low friction material is arranged between the contact surface of the dish spring group and the guide device to reduce the friction between the two. The middle part of the guide rod is provided with an annular flange, and the upper part of the annular flange is provided with buffer rubber. The upper part of the guide device is provided with a circular tensile baffle, the lower part of the upper connecting plate is provided with a circular groove, and the guide device can slide freely in the groove. A flange is arranged outside the groove, and the flange directly contacts with the dish spring group at the lower part to transmit the vertical load.

By analyzing the three-dimensional multi-functional vibration isolation bearing, it can be found that when the three-dimensional multi-functional vibration isolation bearing is set on the structure, the transmission order of vertical load is: (1) upper structure, (2) upper connecting plate, (3) dish spring group, (4) intermediate connecting plate, (5) lead rubber vibration isolation bearing, (6) lower connecting plate; the transmission order of horizontal load is: (1) upper structure, (2) upper connecting plate, (3) guide rod, (4) guide device, (5) intermediate connecting plate, (6) lead rubber vibration isolation support and (7) lower connecting plate.

Working mechanism of three-dimensional multi-functional vibration isolation bearing: when the three-dimensional multi-functional vibration isolation bearing is installed on the structure, the dish spring is vertically deformed, the guide device enters the reserved groove of the upper connecting plate, and the contact between the guide device and the groove is generated, which can transfer the horizontal force. Under the action of wind or small earthquake, the deformation of lead rubber bearing in the three-dimensional multi-functional vibration isolation bearing is small, so as to ensure the normal use of the structure; under the action of medium and strong earthquake, the lead rubber bearing in the three-dimensional multi-functional vibration isolation bearing has large deformation, so as to isolate the up and down transmission

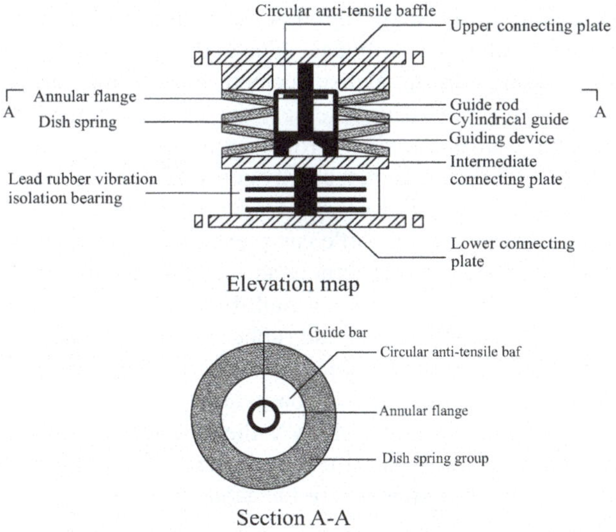

(a) Three dimensional multi-functional vibration isolation support diagram

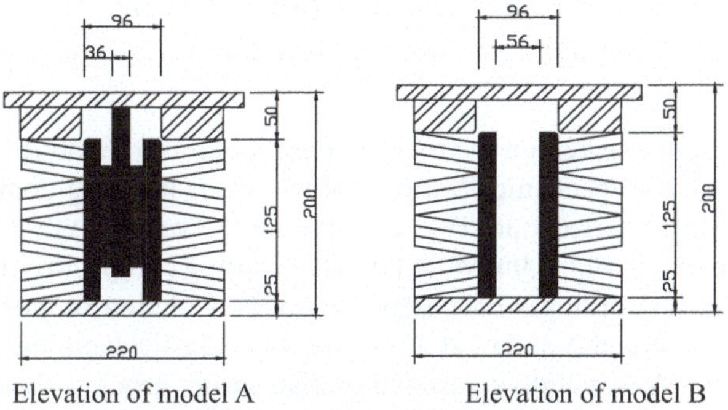

(b) Experimental model of dish spring support

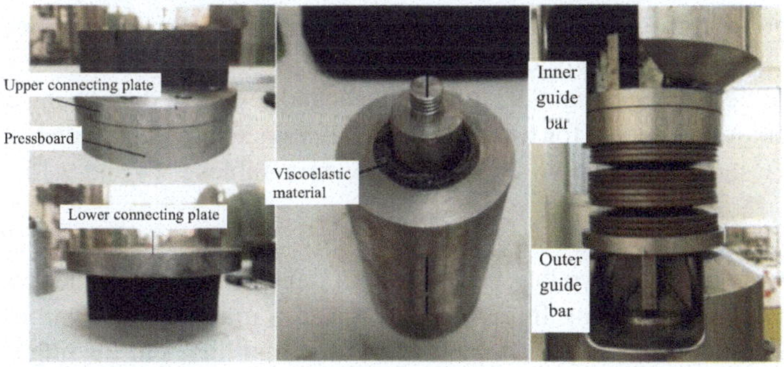

(c) Physical drawing of dish spring composite vibration isolation support

Fig. 9.9 Three dimensional multifunctional isolation bearing of vibration, and consume energy through the high damping characteristics of the lead rubber bearing; after the earthquake, due to the sufficient horizontal stiffness of the lead rubber bearing, the initial displacement state of the bearing can be restored. Under the vertical seismic action, the dish spring group has vertical deformation, which isolates the transmission of vertical ground motion to the upper structure, and the dish spring group has certain energy consumption capacity, which consumes part of the vertical seismic action energy. Under the action of horizontal and vertical earthquakes, the annular tensile baffle can prevent the excessive upward movement of the annular flange and provide certain vertical tensile capacity.

The durability of three-dimensional multifunctional vibration isolation bearing is determined by the durabilities of lead rubber vibration isolation bearing and dish spring. The existing research and practical engineering experience show that the durability of lead rubber bearing can be guaranteed. At the same time, the disc spring material is 60Si2MnA and 50CrVA, the durability can meet the use requirements. Therefore, the three-dimensional multifunctional vibration isolation bearing can meet the durability requirements of the actual project.

9.2.1.2 Mechanical Property Test of Isolation Bearing

1. **Test model**

The experimental research on the mechanical properties of lead rubber bearing has been mature, so the mechanical properties test of dish spring composite isolation bearing is mainly carried out. Two test models of vertical vibration damper are designed as shown in Fig. 9.9b. Model A is dish spring composite vibration isolation support, model B is ordinary dish spring vibration isolation support. The main difference between model A and model B is that viscoelastic damping layer is added to model A. Model A is mainly composed of dish spring group, cylindrical viscoelastic damping layer, upper and lower connecting plates, etc. The dish spring group is composed of 12 dish springs, which are combined by 3 pieces and 4 pairs. The inner and outer guide rods are respectively connected to the upper and lower connecting plates, and the cylindrical viscoelastic damping layer is arranged between the inner and outer guide rods. See Fig. 9.9c for the physical drawing of dish spring composite vibration isolation support for test. The design parameters of single dish spring and cylindrical viscoelastic damping layer are shown in Tables 9.14 and 9.15.

2. **Test scheme**

MTS electro-hydraulic servo fatigue test system is used in the test, which can provide a maximum vertical loading force of 500 kN and a maximum actuation stroke of 176 mm. Sine wave loading is adopted in the test, and different loading conditions are realized by adjusting the frequency and amplitude of sine wave. The vertical load and displacement are collected by computer.

Table 9.14 Design parameters of dish spring

Outer diameter of dish spring D (mm)	Inner diameter of dish spring d (mm)	Thickness t (mm)	Ultimate deformation h_0 (mm)	Height H_0 (mm)	Combination form
200	102	8	7	15	3 superposition, 4 involution

Table 9.15 Design parameters of viscoelastic damping layer

Diameter of outer guide rod D (mm)	Diameter of inner guide rod d (mm)	Wall thickness of outer guide rod t (mm)	Diameter of shear ring l (mm)	Length of shear ring h (mm)
96	36	20	46	100

(1) Static load test: according to the characteristics of the test equipment, the method of displacement control is used to load slowly from zero displacement to 21 mm (deformation of dish spring $f = 0.75h_0$) and 28 mm (flattening of dish spring), then unload slowly, and record the load displacement curve of the test piece.
(2) Dynamic load test: the method of displacement control is used to load the specimen repeatedly under different preload amount. The load displacement curves under different loading frequency and dynamic load amplitude are recorded. The preloading amount is 10 mm and 14 mm respectively, the loading frequency is gradually increased from 0.1 Hz, 0.2 Hz, 0.5 Hz and 1 Hz, and the displacement amplitude is gradually increased from 0.5 mm, 1 mm, 2 mm and 4 mm. The test condition number is shown in Table 9.16.

Table 9.16 Test condition number

Number	Preloading displacement (mm)	Amplitude of dynamic load (mm)	Loading frequency (Hz)
1–4	10	0.5	0.1, 0.2, 0.5, 1
5–8	10	1	0.1, 0.2, 0.5, 1
9–12	10	2	0.1, 0.2, 0.5, 1
13–16	10	4	0.1, 0.2, 0.5, 1
17–20	14	0.5	0.1, 0.2, 0.5, 1
21–24	14	1	0.1, 0.2, 0.5, 1
25–28	14	2	0.1, 0.2, 0.5, 1
29–32	14	4	0.1, 0.2, 0.5, 1

3. **Analysis of test results**

(1) Analysis of static load test results

Carry out the ultimate bearing capacity test on the dish spring composite vibration isolation bearing. The dish spring composite vibration isolation bearing starts to load slowly from the free state until the dish spring is completely flattened, and then unload slowly. Repeat three times, respectively record the initial height of the dish spring composite vibration isolation bearing and the height after unloading, and check whether the dish spring is damaged. Figure 9.10a shows the picture of dish spring when it is pressed to the limit, and Table 9.17 shows the height change of dish spring composite vibration isolation support before and after flattening. It can be seen from the table that the dish spring can return to the initial height after each loading, and there is no damage after inspection. It can be seen that the dish spring composite vibration isolation bearing has good vertical bearing capacity.

When loaded to 21 mm under static load, the load displacement curves of model A and model B are shown in Fig. 9.10b, and the vertical stiffness and vertical bearing capacity are shown in Table 9.18. Compared with model B, model A has larger vertical bearing capacity and stiffness than model B. the reason is that viscoelastic damping layer increases the vertical stiffness and bearing capacity of model A.

(a) Dish spring was pressed to limit

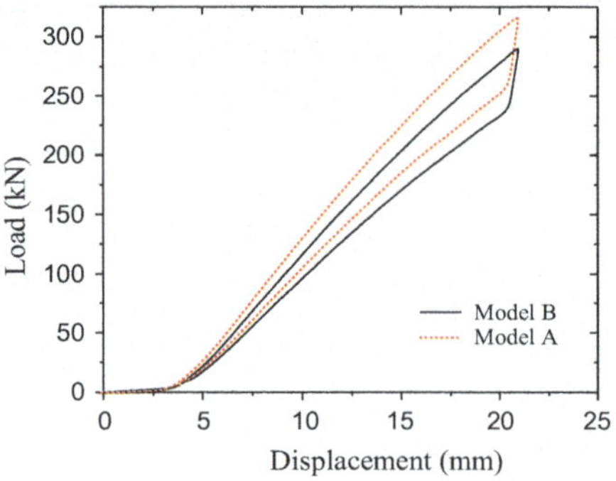

(b) Load displacement curve of bearing

Fig. 9.10 Static load test results

Table 9.17 Height of dish spring composite vibration isolation support

Test number	Initial height (mm)	Height after unloading (mm)
1	125.6	125.4
2	125.4	125.4
3	125.4	125.4

Table 9.18 Static load test results

	Vertical bearing capacity (kN)	Vertical stiffness (kN/mm)
Model A	318.02	15.16
Model B	292.36	13.92

(2) Analysis of dynamic load test results
Figure 9.10 shows the load displacement curves of the two models under various conditions. It can be seen from the figure that:

(a) Under the same loading frequency, preloading amount and dynamic load amplitude, the area of load displacement hysteresis loop of model A is significantly larger than that of model B, and more full, so the energy consumption capacity of model A is significantly better than that of model B, the reason is that viscoelastic damping layer increases the energy consumption capacity of model A.
(b) Under the same preload and dynamic load amplitude, the hysteresis curves of model A and model B change little with the increase of loading frequency. According to the analysis in Sect. 1.1, the damping force of model A includes viscous damping force, Coulomb damping force and viscoelastic damping force. Compared with model A, the damping force of model B lacks the viscoelastic damping force. Because the hysteretic curve of model B is less affected by the loading frequency, the proportion of viscous damping force of model B is smaller, and the Coulomb damping force is the main part. Since the damping force of viscoelastic damping layer is less affected by the loading frequency, the hysteretic curve of model A is also less affected by the loading frequency.
(c) Under the same preloading amount and loading frequency, with the increase of dynamic load amplitude, the fullness of hysteretic curve of model A decreases gradually. Due to different loading and unloading stiffness, the shape is obviously asymmetric. The asymmetry of the shape of hysteretic curve of model B increases obviously with the increase of dynamic load amplitude. The damping force of model B mainly comes from the friction between the dish springs, and the magnitude of the friction force is related to the pressure. Under the action of dynamic load, the pressure increases when loading, which makes the friction damping force increase, and the curve tends to be full; when unloading, with the decrease of the vertical compression deformation of the dish springs, the deformation energy decreases, and the surface friction damping between the laminated dish springs also decreases. As a result, the hysteretic curve of model B tends to be slender, which leads to the obvious asymmetry of the hysteretic curve of model B. Because the viscoelastic damping layer in model A is less

affected by the dynamic load amplitude, combined with the characteristics of the hysteresis curve of model B, the hysteresis curve of model A shows that with the increase of dynamic load amplitude, the fullness decreases gradually.

(d) Under the same loading frequency and dynamic load amplitude, the hysteretic curves of model A and model B tend to be full with the increase of preloading, indicating that the energy consumption capacity of model A and model B increases significantly with the increase of preloading. With the increase of preload, the vertical load of dish spring group increases gradually. The larger the vertical deformation of dish spring is, the larger the Coulomb friction energy consumption and deformation energy are. As a result, the hysteretic curves of model A and model B tend to be full with the increase of preload (Fig. 9.11).

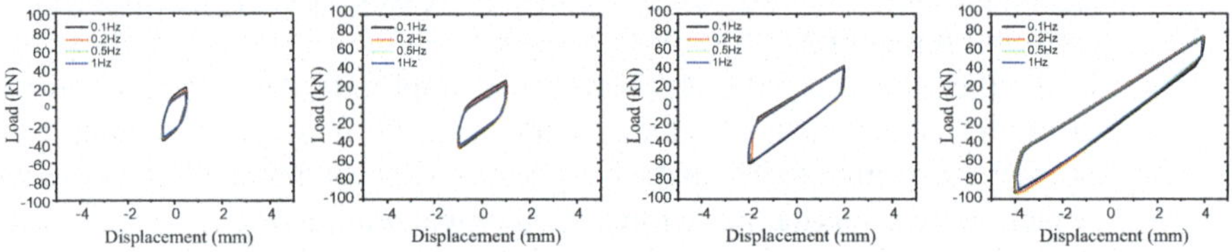

(a) Load displacement curve of dish spring of model A (preload amount of 10 mm)

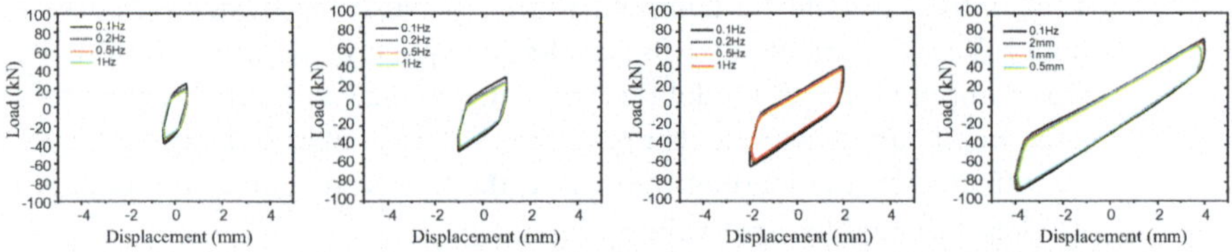

(b) Load displacement curve of dish spring of model A (preload amount of 14 mm)

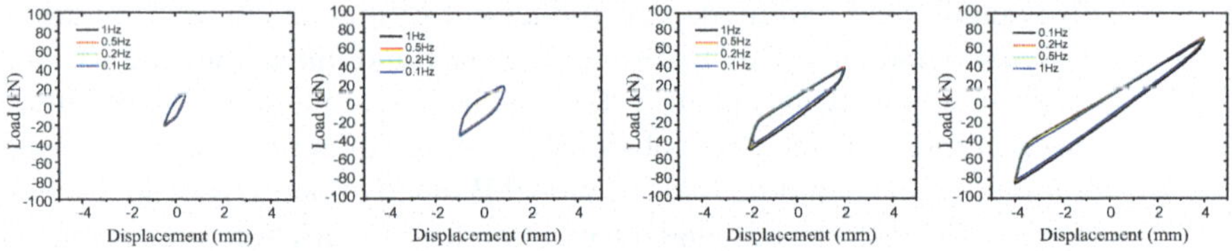

(c) Load displacement curve of dish spring of model B (preload amount of 10 mm)

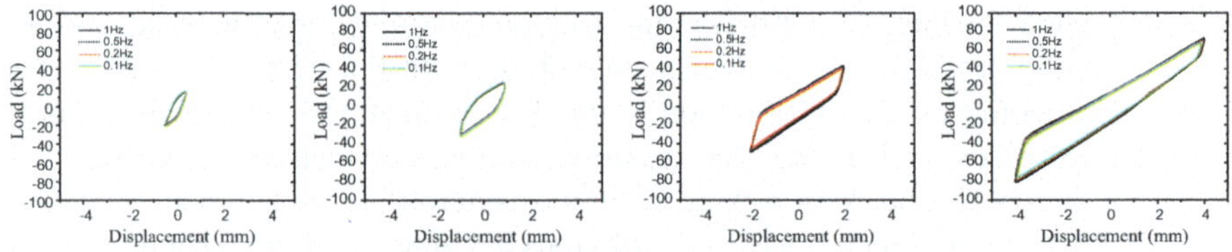

(d) Load displacement curve of dish spring of model B (preload amount of 14 mm)

Fig. 9.11 Load displacement curve of dish spring

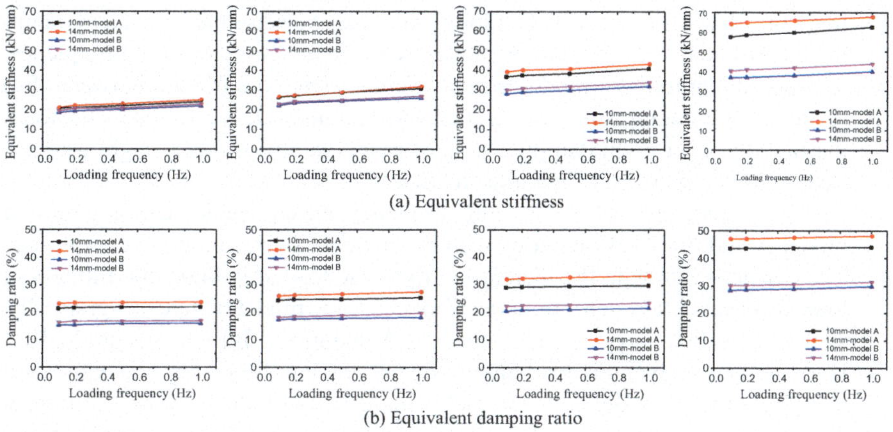

Fig. 9.12 Mechanical property parameters of dish spring

Figure 9.12a shows the equivalent stiffness of the two models under various conditions. As can be seen from the figure,

(a) Under the same preload and dynamic load amplitude, with the increase of loading frequency, the equivalent stiffness of model A and model B increases gradually, but the increase amplitude is small.
(b) Under the same preload and loading frequency, the equivalent stiffness of model A and model B increases with the increase of dynamic load amplitude. When the dynamic load amplitude is 0.5 mm, the difference between the equivalent stiffness of model B and model A is small. With the increase of the dynamic load amplitude, the difference between the equivalent stiffness of the two models increases gradually.
(c) Under the same dynamic load amplitude and loading frequency, with the increase of preloading amount, the equivalent stiffness of model A and model B increases gradually. When the preloading is 14 mm, the vertical stiffness of model A is significantly higher than that of 10 mm.
(d) Due to the addition of viscoelastic damping layer in model A, the equivalent stiffness of model A is greater than that of model B under the same loading frequency, dynamic load amplitude and preloading amount.

Figure 9.12b shows the equivalent damping of the two models under various conditions. As can be seen from the figure,

(a) Under the same preload and dynamic load amplitude, with the increase of loading frequency, the equivalent damping ratio of model A and model B increases gradually, but the increase amplitude is small. One part of the damping force of model B is viscous damping force, which is proportional to the loading frequency; the other part is Coulomb damping force, and the change of loading frequency has little effect on Coulomb damping; because the equivalent damping ratio of model B is slightly increased with the increase of loading frequency, the vertical damping of model B mainly comes from the friction between dish springs. For model A, since the viscoelastic damping is less affected by the loading rate, the equivalent damping ratio of model A also shows a slight increase trend with the increase of loading frequency.

(b) Under the same preload and loading frequency, the equivalent damping ratio of model A and model B increases with the increase of dynamic load amplitude. The reason is that with the increase of dynamic load amplitude, the fullness of load displacement hysteretic curves of model A and model B increases gradually, so the equivalent damping ratio of model A and model B increases gradually.

(c) Under the same dynamic load amplitude and loading frequency, the equivalent damping ratio of model A and model B increases with the increase of preload. The reason is that the area of load displacement hysteresis loop of model A and model B increases with the increase of preload, so the equivalent damping ratio of model A and model B increases with the increase of preload.

(d) Under the same loading frequency, dynamic load amplitude and preload, the equivalent damping ratio of model A is significantly higher than that of model B. When the dynamic load amplitude is 0.5 mm, the difference between the equivalent damping ratio of model B and model A is small. With the increase of the dynamic load amplitude, the difference between the equivalent damping ratio of the two models increases gradually.

9.2.1.3 Numerical Simulation of Mechanical Properties of Isolation Bearing

1. **Theoretical analysis**

The horizontal and vertical stiffness of the three-dimensional multi-functional vibration isolation bearing is the series connection of the stiffness of the dish spring of the upper structure of the bearing and the lead rubber vibration isolation bearing of the lower structure. The calculation formula is as follows:

$$k_V = 1/(1/k_{DV}+1/k_{RV}) \tag{9.17}$$

$$k_H = 1/(1/k_{DH}+1/k_{RH}) \tag{9.18}$$

And, k_V and k_H respectively represent the vertical and horizontal stiffness of the three-dimensional multi-functional vibration isolation bearing; k_{DV} and k_{DH} respectively represent the vertical and horizontal stiffness of the upper structure of the three-dimensional multi-functional vibration isolation bearing; k_{RV} and k_{RH} respectively represent the vertical and horizontal stiffness of the lead rubber vibration isolation bearing.

Because the vertical stiffness of the lead rubber bearing is much greater than that of the dish spring group, the vertical stiffness of the three-dimensional multifunctional vibration isolation bearing is approximately equal to that of the dish spring group, $k_V \approx k_{DV}$. Because the horizontal stiffness of the upper structure of the three-dimensional multifunctional vibration isolation bearing is much greater than that of the lead rubber vibration isolation bearing, the horizontal stiffness of the three-dimensional multifunctional vibration isolation bearing is approximately equal to the horizontal stiffness of the lead rubber bearing, $k_H \approx k_{RH}$.

The initial horizontal stiffness of the lead rubber bearing is k_i, $k_i = \alpha_0 k_d$, α_0 can be approximately 10–15. k_d is the stiffness after yield, k_d is taken by $k_d = GA/T_r$ approximately. The calculation formula of the vertical effective stiffness of the dish spring is:

$$k_{DV} = \frac{4E}{1-\mu^2} \cdot \frac{t^3}{K_1 D^3} \cdot K_4^2 \cdot \left\{ K_4^2 \left[\left(\frac{h_0}{t} \right)^2 - 3 \cdot \frac{h_0}{t} \cdot \frac{f}{t} + \frac{3}{2} \left(\frac{f}{t} \right)^2 \right] + 1 \right\} \tag{9.19}$$

where, E is the elastic modulus of the dish spring material, μ is the Poisson's ratio of the dish spring material, t is the thickness of the dish spring, K_1 and K_4 are the calculation coefficients, h_0 is the calculation value of the deformation of the dish spring at the flattening stage, f is the deformation of the single dish spring.

2. **Numerical simulation**

The numerical analysis is realized by ABAQUS software. The dish spring of dish spring composite vibration isolation support is simulated by eight node hexahedron linear reduction integral element (C3D8R), and the viscoelastic material is simulated by three-dimensional eight node hexahedron hybrid element (C3D8H).

The dish spring material is 64Si2MnA, the design value of yield strength of corresponding material is 1400 N/mm^2, the elastic modulus of material is taken as $E = 2.06 \times 10^6$ N/mm^2, and Poisson's ratio is taken as $\mu = 0.3$. The inner and outer guide rods and the upper and lower connecting plates are simulated with rigid materials. The viscoelastic materials are simulated by the five term third-order polynomial model (Polynomial). The parameters in the constitutive model are $C_{10} = 0.238$, $C_{01} = 0.0119$, $C_{20} = 0.00410$, $C_{11} = 0.00101$, $C_{02} = 2.807 \times 10^{-5}$. Poisson's ratio of viscoelastic material is 0.4997.

When the dish spring works, there will be friction between the upper and lower connecting plates, the outer guide rod and the dish spring. Considering that the contact friction between the dish spring and the outer guide rod is small, only the contact friction between the dish spring and the upper and lower connecting plates and the dish springs is considered. Based on the static load test data of the common dish spring bearing, the equivalent value of the friction coefficient between the conical surfaces of the dish spring is calculated and used for the numerical simulation of the dish spring bearing under the dynamic load. The friction coefficient between the dish spring and the upper and lower connecting plates is 0.6, and the friction coefficient between the conical surfaces of the dish spring is 0.2. The normal contact between the dish springs is simulated by the hard contact in ABAQUS, and the tangential contact between the dish springs is simulated by the Coulomb friction model in ABAQUS. The finite element model of the dish spring composite vibration isolation support is shown in Fig. 9.13a.

Figure 9.13b shows the comparison between the test results and the numerical simulation results of the load displacement hysteretic curve of model A under the vertical static load. It can be seen from the figure that under the vertical static load, the simulated value of the load displacement hysteretic curve of model A has the similar curve characteristics with the test value. The ultimate bearing capacity of the two is similar, but the stiffness changes are different at the initial load and the final load, with poor coincidence.

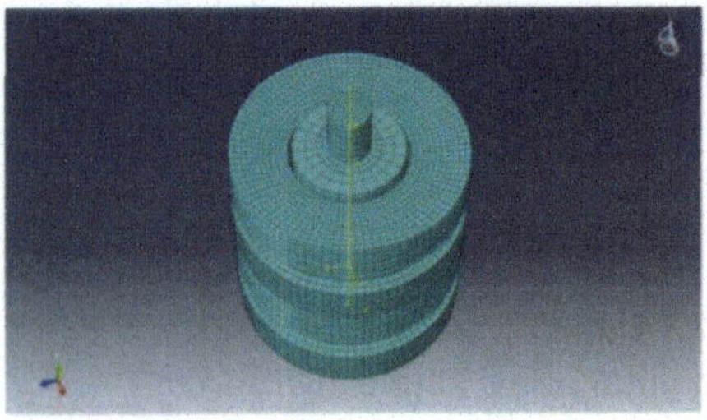

(a) Finite element model

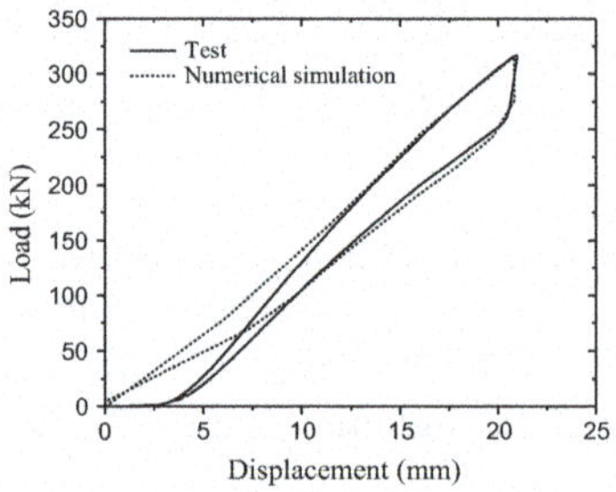

(b) Comparison of test value and simulation value under static load

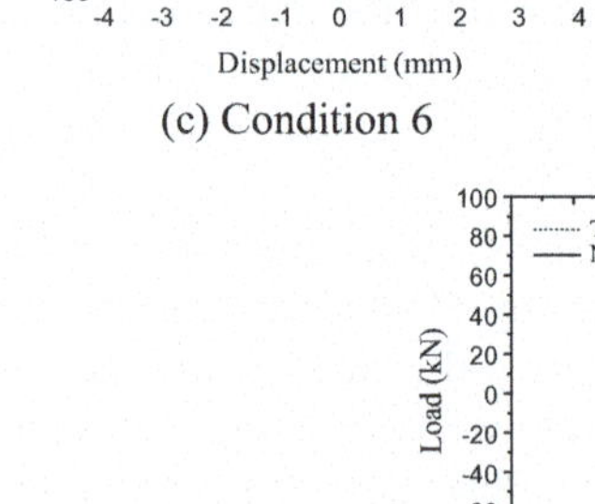

(c) Condition 6

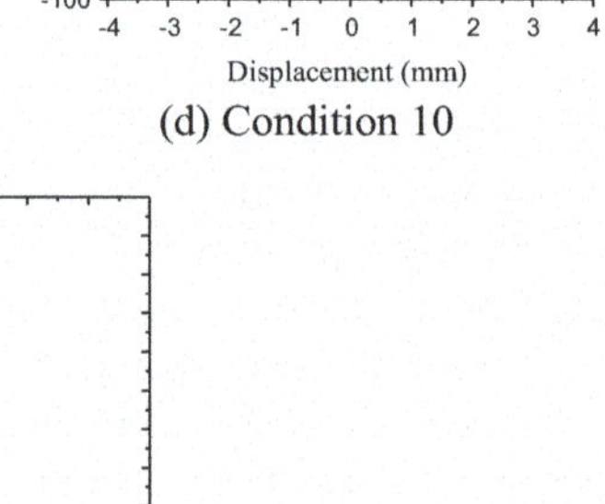

(d) Condition 10

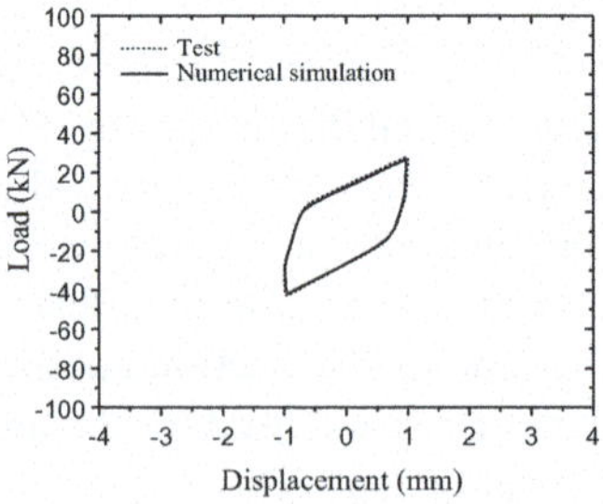

(e) Condition 24

Fig. 9.13 Numerical simulation of dish spring composite vibration isolation support

Table 9.19 Equivalent stiffness and equivalent damping ratio

Dynamic load amplitude (mm)			0.5	1	2	4
Preloading 10 mm	Equivalent stiffness (kN/mm)	Test	57.33	64.51	36.91	40.33
		Simulation	55.53	67.32	37.30	43.23
		Error ratio (%)	−3.14	4.36	1.06	7.19
	Equivalent damping ratio	Test	43.52	47.25	28.41	30.22
		Simulation	41.45	45.75	29.34	32.45
		Error ratio (%)	−4.76	−3.17	3.27	7.38
Preloading 14 mm	Equivalent stiffness (kN/mm)	Test	36.80	39.53	28.17	30.17
		Simulation	33.23	37.53	26.16	32.47
		Error ratio (%)	−9.70	−5.06	−7.14	7.62
	Equivalent damping ratio	Test	29.12	32.24	20.52	22.44
		Simulation	28.34	33.65	22.12	24.24
		Error ratio (%)	−2.68	4.37	7.80	8.02

Figure 9.13b–e shows the comparison between the test results and numerical simulation results of load displacement hysteretic curve of model A under dynamic load, and only the calculation results of condition 6, condition 10 and condition 24 are listed due to limited space. It can be seen from the figure that under the dynamic load condition, the simulated value of load displacement hysteretic curve of model A has similar curve characteristics with the test value, and the numerical simulation results are in good agreement with the test results. Table 9.19 shows the comparison between the equivalent stiffness and damping ratio of model A and model B and the numerical simulation results. It can be seen from the table that the numerical simulation results are in good agreement with the test results, and the error is within 10%.

9.2.2 *Rubber Composite Sliding Isolation Bearing*

9.2.2.1 Design of Rubber Composite Sliding Isolation Bearing

In order to realize phased passive isolation control, a new multi-functional isolation bearing is designed, which is composed of two parts in series [5, 6]. The upper part is polytetrafluoroethylene (PTFE)-stainless steel plate sliding device, and the lower part is LRB300 lead rubber isolation bearing. A transfer device is set between the upper and lower parts of the support, which is composed of four surrounded baffles, which can meet the sliding of the upper structure of 30 mm in longitudinal direction and 20 mm in transverse direction. The overall structure of the support and the processing and forming model are shown in Fig. 9.14. The specific design parameters are shown in Table 9.20.

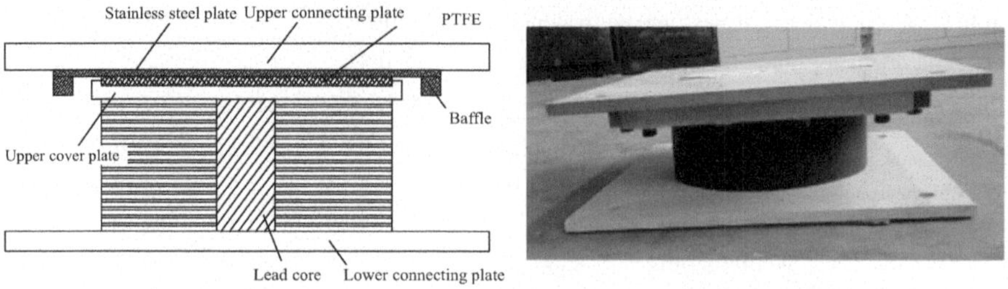

(a) Bearing structure (b) Bearing sample

Fig. 9.14 Rubber composite sliding isolation bearing

Table 9.20 Mechanical parameters of new multifunctional bearing

Parameter	Parameter value	Parameter	Parameter value
Shear modulus of rubber G (Mpa)	0.392	Single rubber thickness (mm)	3.39
Effective diameter (mm)	300	Rubber layer	18
Lead core diameter (mm)	60	First shape factor	22.12
Thickness of internal steel plate (mm)	1.5	Second shape factor	4.92
Number of layers of internal steel plate	17	Total height of support (mm)	106.5

9.2.2.2 Mechanical Property Test of Isolation Bearing

1. **Test model**

According to the design idea of the above-mentioned isolation bearings, two kinds of isolation bearings with different realization modes are designed and manufactured. A professional isolation pad manufacturer in China is entrusted to manufacture the bearing model for the test according to the relevant national standards and the parameters of the test device.

(1) B-type isolation bearing

In this kind of multi-functional seismic isolation bearing, a conversion device of "concave convex inlay" is designed to achieve the following purpose: under the action of temperature, vehicle braking and concrete creep, the required displacement is formed by the polytetrafluoroethylene sliding device along the longitudinal direction of the bridge; under the action of earthquake, when the displacement along the longitudinal direction of the bridge reaches the limit value of sliding displacement, the structure of "concave convex inlay" between the stainless steel plate with smooth surface and the upper cover plate is formed to make the lower energy dissipation isolation system play a role, thus significantly improving the seismic performance of the bridge structure. A sliding clearance of 30 mm is set along the sliding direction

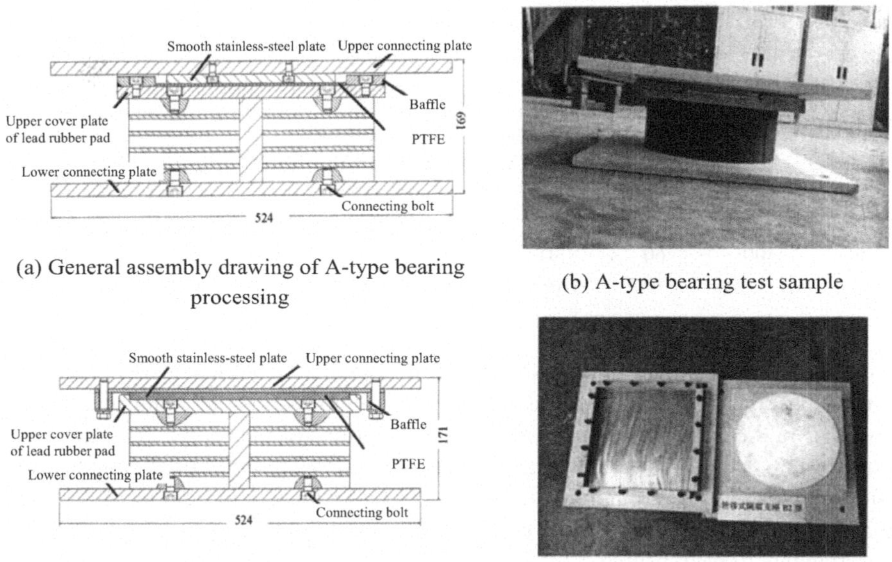

(a) General assembly drawing of A-type bearing processing

(b) A-type bearing test sample

(c) General assembly drawing of B-type bearing processing

(d) B-type bearing test sample

Fig. 9.15 Type a seismic isolation bearing

of the support, and the sliding displacement in the other direction of the support is limited. However, the contact surface between the side of the block in this direction and the side of the smooth stainless steel plate needs to be handled well, so as not to cause greater friction and affect the sliding effect of the support.

A-type isolation bearing (as shown in Fig. 9.15a) is composed of the following components: upper connecting plate, smooth stainless steel plate, polytetrafluoroethylene plate, upper top plate, limit plate, lead rubber pad and lower connecting plate. Among them, the upper connecting plate, smooth stainless steel plate and polytetrafluoroethylene plate constitute the upper structure of the bearing, which mainly solves the problem of sliding isolation of the bearing. The upper connecting plate, limit plate, lead rubber pad and the lower connecting plate constitute the substructure of the bearing, which mainly solves the problem of energy dissipation and shock absorption of the bearing under the action of large earthquake. The smooth stainless steel plate and the upper connecting plate are connected by bolts, and the limit plate and the upper top plate are connected by bolts and two fillet welds are added and welded. LRB300 type lead rubber pad is adopted, and the specific dimensions of the isolation pad are shown in Table 9.21.

(2) Type B isolation bearing

Aiming at the "embedded" sliding mechanism on the upper part of A-type seismic isolation bearing, a "outsourcing" sliding mechanism is designed and manufactured

Table 9.21 Size parameters of lead core isolation pad

Shear modulus of elasticity of rubber MPa	Effective diameter	Lead core diameter	Parameters of intermediate rubber layer (layers × thickness) mm	Parameters of middle steel plate layer (layers × thickness) mm	First shape factor	Second shape coefficient	Total height of support mm
0.392	300	60	18 × 3.39 = 61	17 × 1.5 = 25.5	22.12	4.92	106.5

to realize the performance of first sliding of the upper part and then energy consumption of the lower part. For the realization of this kind of seismic isolation bearing, considering the needs of deformation in two directions of the bridge, the functional requirements to fix one direction and release the other direction of A-type bearing can be used, or the functional requirements to release the both directions can be used to design the bearing. The sliding displacement can be set according to the specific requirements of each bridge.

The B-type seismic isolation bearing consists of the following components: upper connecting plate, smooth stainless-steel plate, polytetrafluoroethylene plate, baffle block, upper top plate, lead rubber pad and lower connecting plate. Among them, the upper connecting plate, smooth stainless steel plate, polytetrafluoroethylene plate, baffle block and upper top plate constitute the upper sliding mechanism; the upper top plate is connected with the lower lead rubber pad by bolts, and the circular polytetrafluoroethylene plate is placed in the circular groove of the square upper top plate, but its top surface shall be higher than the groove to facilitate contact and sliding with the stainless steel plate. The smooth stainless steel plate and baffle block shall be connected with the upper connecting plate by bolts in turn; the baffle block shall be arranged in a rectangular shape on the upper connecting plate, and the size of the inner rectangle formed by its enclosure shall be determined according to the design slip amount, and shall be used together with the upper roof.

The upper sliding mechanism of B-type isolation bearing can be square or rectangular, but the upper top plate of sliding mechanism is square. Such a structure complies with the design requirements, that is, when the design requires the same amount of sliding in the two principal axis directions, the upper sliding mechanism is square; when the design requires that the amount of sliding in the two principal axis directions is different, the upper sliding mechanism is rectangular. When making smooth stainless-steel plate, the quality of the product should be paid attention to ensure that the smooth surface is flat and free of burrs. Because of the weakening of baffle block caused by the connecting bolt, the cross section of the block should be used only after the strength is qualified. The upper sliding displacement of the test model is: longitudinal 30 mm, transverse 20 mm. The lead rubber pad is the same as that of type A.

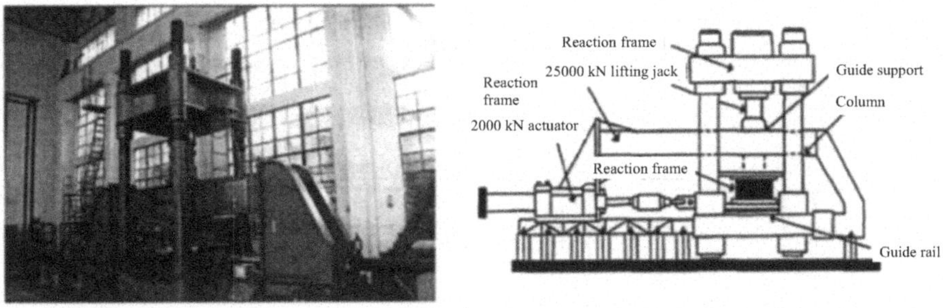

Fig. 9.16 Test loading device

2. **Test scheme**

The large compression shear tester for the test is shown in Fig. 9.16. The compression shear test system is composed of horizontal hydraulic servo loading control system, vertical loading system, computer data acquisition and analysis system and test control and detection platform. The computer data acquisition and analysis system is composed of A/D and D/A conversion hardware systems and specially designed software system, which can complete the whole process of work including loading, data acquisition, test curve drawing and giving the characteristics value of detection support. The test system can provide a maximum vertical loading force of 25 MN and a maximum horizontal loading force of 2.8 MN, with a maximum actuation stroke of ±600 mm. The testing machine adopts sine wave loading, and different loading conditions can be realized by adjusting the frequency and amplitude of sine wave. When testing the vertical mechanical properties of the bearing, a dial indicator is set at the four corners of the bearing to measure the vertical displacement of the bearing.

According to the relevant provisions of Part 1 and Part 2 of the national standard "*Rubber bearing*" of the People's Republic of China, the test conditions are designed.

(1) Vertical compression performance test

The following loading methods are adopted for A-type and B-type bearings respectively: load three times in the way of 0-P_{max}-0, and take the data of the third load for analysis. The effective minimum cross-sectional area of the test bearing is 70,686 mm^2, and the design pressure stress is 10 MPa, so Po = 70,686 × 10 = 707 kN. The loading method is shown in Fig. 9.2b.

(2) Shear performance test

A-type seismic isolation bearing

During the test, it is found that due to the design reasons, the A-type seismic isolation bearing cannot realize the preliminary assumption of "first superstructure sliding, then driving the movement of substructure". Only the test shown in Table 9.22 is carried out, and the shear performance data of the lower lead rubber pad are obtained.

Table 9.22 Shear performance test of A-type seismic isolation bearing

Condition	Shear displacement (mm)	Loading frequency (Hz)	Vertical loading force (kN)	Cycle times
1	10	0.05	707	3
2	45.5	0.2	707	3
3	67	0.1	350	3
4	67	0.1	707	3
5	67	0.2	707	3

B-type seismic isolation bearing

A series of tests on the direction of 30 mm slip of B-type seismic isolation bearing are carried out as follows: under the action of 350 kN vertical force and 0.1 Hz loading frequency, the shear performance tests under the shear displacement of 30, 50, 80, 100, 120, 130 and 140 mm are carried out respectively, with three cycles under each working condition.

A series of tests on the direction of 20 mm slip of B-type seismic isolation bearing are carried out as follows: under the action of 200 kN vertical force and 0.1 Hz loading frequency, the shear performance tests under the shear displacement of 50, 60, 80 and 100 mm are carried out respectively, with three cycles under each working condition.

(3) Shear performance test

In order to investigate the influence of different loading conditions on the shear performance of the bearing, a series of shear performance correlation tests were designed to comprehensively investigate the shear mechanical properties of the bearing. The correlation of shear performance in the direction of 30 mm slip of B-type seismic isolation bearing is mainly investigated. See Table 9.23 for the specific loading conditions of the test. See Table 9.24 for the specific loading conditions of the correlation test of shear performance in the direction of 20 mm slip of B-type seismic isolation bearing.

3. **Analysis of vertical performance test results**

See Table 9.25 for the data of vertical compression performance test of two kinds of multifunctional isolation bearings. The vertical compression stiffness of each bearing is also calculated in the table. It can be seen from the table that the vertical compression stiffness of the two kinds of bridge multi-functional isolation bearings is relatively large, which is close to that of the lead rubber bearings (900–1000 kN/mm), which shows that the upper sliding friction sliding device is nearly incompressible in the vertical direction after several cycles of compression. Compared with the vertical compression stiffness of the two kinds of isolation bearings, it can be seen that the vertical compression stiffness of A-type bearing is larger, which is caused by the difference of the two construction methods: the polytetrafluoroethylene plate used for the B-type bearing is thicker. However, the difference between the two is not significant, and there is no bulge, bias, uneven compression and other undesirable phenomena in the test process, which can meet the actual use needs.

Table 9.23 Correlation test content of shear performance of B-type seismic isolation bearing in the direction of 30 mm slip

Test item	Test loading condition	
	Condition	Change
Shear displacement correlation	350 kN, 0.1 Hz	30, 50, 80, 100, 120, 130, 140 mm
Correlation of compressive stress	50 mm, 0.1 Hz	200, 350, 707, 848 kN
Load frequency correlation	50 mm, 350 kN	0.01, 0.02, 0.1, 0.2 Hz
Low pressure deformation correlation	200 kN, 0.1 Hz	50, 100 mm
Correlation of times of repeated loading	50 mm, 0.1 Hz, 200 kN, cycle 30 times	

Table 9.24 Correlation test content of shear performance of B-type seismic isolation bearing in the direction of 20 mm slip

Test item	Test loading condition	
	Condition	Change
Shear displacement correlation	200 kN, 0.1 Hz	50, 60, 80, 100 mm
Correlation of compressive stress	50 mm, 0.1 Hz	200, 350 kN

Table 9.25 Test results of vertical compression performance of multi-functional isolation bearing

Support form	P_1 (kN)	Y_1 (mm)	P_2 (kN)	Y_2 (mm)	K_v (kN/mm)
A-type	495	0.72	919	1.19	902
B-type	495	0.75	919	1.24	865

4. **Analysis of shear performance test results**

(1) A-type isolation bearing

Figure 9.17a, b are hysteretic curves under two conditions of shear performance test of A-type seismic isolation bearing. It can be seen from the figure that the A-type bearing only reflects the shear energy dissipation capacity of the lead rubber pad, and the upper friction sliding effect is not reflected. The reason is that the vertical deformation of the type a bearing under the upper load is uneven during the test, which leads to the contact between the upper connecting plate and the small block, and the friction between the two is much greater than the friction between the Teflon plate and the stainless steel plate, which causes that the sliding device does not work. See Fig. 9.17c, d for the specific contact position. From the hysteretic curve obtained from the test, the mechanical parameters of all the lead rubber pads can also be

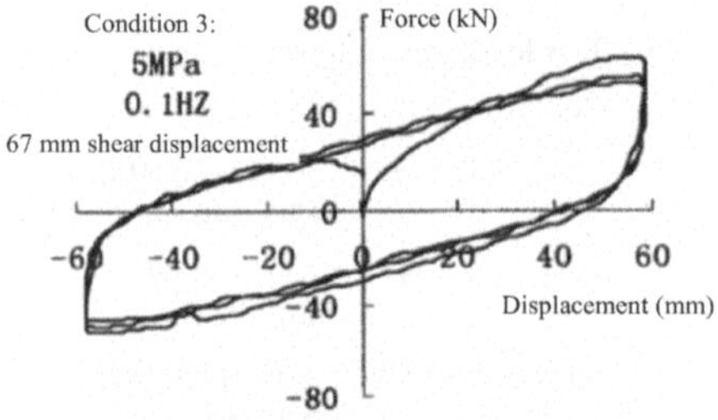

(a) Hysteresis curve of condition 3

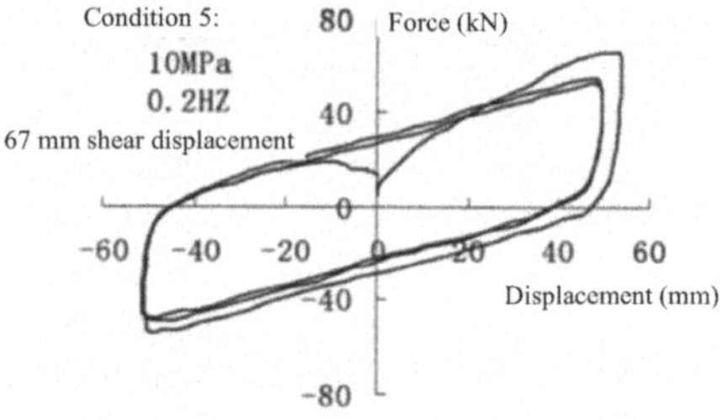

(b) Hysteresis curve of condition 5

(c) Contact position of support-block

(d) Contact position of support-upper connecting plate

Fig. 9.17 Test results of shear performance of A-type seismic isolation bearing

compared and analyzed to check the factory quality, and the mechanical parameters can also be used in the design of the following bearings.

According to the force displacement hysteretic curve of the support shown in Fig. 9.17, extract the data of the third circle, and calculate the mechanical performance parameters of the support under different working conditions. See Table 9.26 for specific values. It can be seen from the table that the bearing has good energy dissipation and isolation capacity under various working conditions (equivalent damping ratio is greater than 20%): the yield force is relatively stable, and in each working condition, the value

Table 9.26 Shear mechanical property parameters of A-type isolation bearing

Condition	Stiffness after yield (kN/m)	Equivalent horizontal stiffness (kN/m)	Yield force (kN)	Equivalent damping ratio (%)
1	987	3027	21.2	36.3
2	412	1173	28.1	26.1
3	425	899	27.4	22.4
4	475	946	25.3	22.0
5	506	1063	24.5	21.6

of the first cycle is larger than that of the last several cycles, on the one hand, it is due to the influence of Bauschinger effect, on the other hand, it is caused by the uneven processing and deformation of lead core; The equivalent horizontal stiffness decreases with the increase of shear strain. As the shear strain of condition 1 is small (about 15%), the mechanical parameters of the bearing are greatly affected by the initial stiffness of the bearing; from the comparison of the mechanical parameters of condition 4 and condition 5, it can be seen that the loading frequency has little influence on the mechanical properties of the bearing.

(2) 30 mm sliding direction of B-type isolation bearing

Figure 9.18a shows the force displacement hysteretic curve of B-type isolation bearing in the direction of 30 mm slip under 7 working conditions of the shear performance test. It can be seen from the figure that the two effects of friction sliding and energy dissipation isolation of B-type isolation bearing have been achieved. In the sliding stage, the friction of the bearing has no big jump change, the system conversion device operates well, and the energy dissipation isolation capacity of the lower lead rubber bearing can be better developed. From the curve of each working condition, the hysteretic curve of the bearing tends to be stable in the second circle.

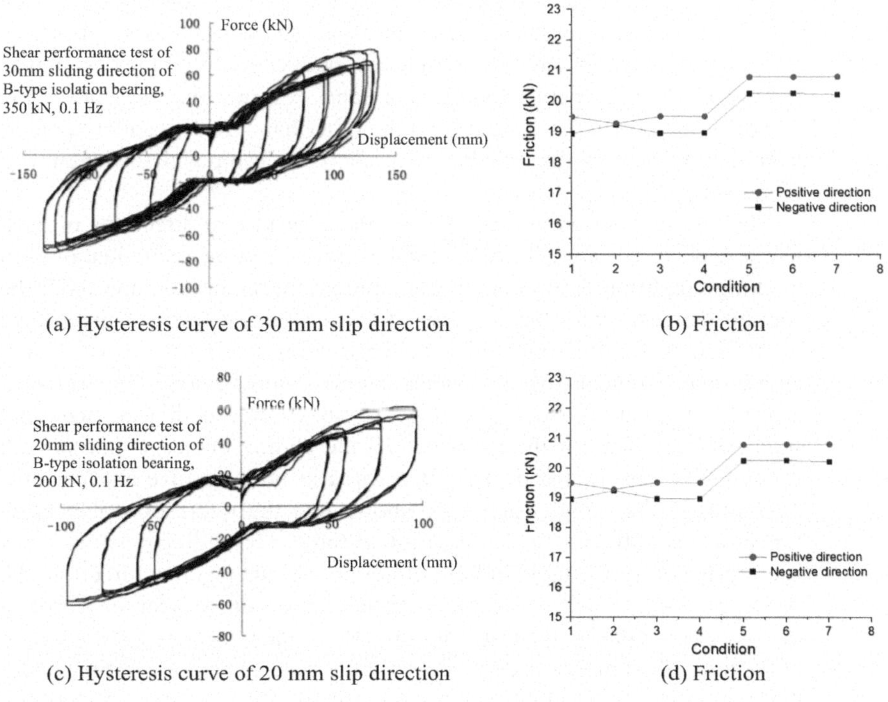

Fig. 9.18 Shear performance test of B-type seismic isolation bearing

The starting sliding point of the bearing tends to be the same, and the sliding friction of each working condition can be extracted from the test data (Fig. 9.18a), and the sliding friction coefficient of each working condition can be deduced. The calculated friction coefficient is between 0.055 and 0.06, friction increases with the increase of shear strain, and the change range of friction is small. In the case of large shear strain, the friction coefficient increases, which is because the test loading condition is from small shear strain to large shear strain gradually. When the test loading condition is from small shear strain to large strain, the temperature of friction surface increases due to friction, which leads to the increase of friction coefficient. The change trend of friction force in both positive and negative directions is the same, which shows that the bearing force is isotropic, and the realization of bearing function is not affected by the direction of force.

(3) 20 mm sliding direction of B-type isolation bearing
Figure 9.18c shows the force displacement hysteretic curve of B-type isolation bearing in the direction of 20 mm slip under 4 working conditions of the shear performance test. The two effects of friction sliding and energy dissipation isolation have also been achieved, the system conversion device operates well, the energy dissipation isolation capacity of the lower lead rubber bearing can be better developed, and the hysteretic curve of the bearing is full. When the shear displacement increases, the degradation of bearing stiffness is not obvious, which indicates that the bearing has good fatigue resistance. Compared with the 30 mm slip, both can achieve the design purpose of the bearing better. On the one hand, it shows that the conversion device of the bearing is effective, on the other hand, it shows that the realization of the bearing function has no direct relationship with the magnitude of the slip.

The sliding friction and its starting position in two directions of each working condition are extracted, and the sliding friction coefficient of each working condition is calculated. The value is between 0.06 and 0.07, the friction increases with the increase of shear strain, and the change range of friction is small. Compared with the friction coefficient in the sliding direction of 30 mm, the friction coefficient in this direction is larger. Through analysis, it is due to the processing of polytetrafluoroethylene plate and stainless steel plate of the support, and the sliding device is not smooth enough. The change trend of friction force in both positive and negative directions is the same, which also shows that the realization of bearing function is not affected by the direction of force. The sliding position of the bearing is relatively stable in both directions. In the positive direction, the sliding position slightly increases with the increase of shear strain. Through the observation during the test, it is considered that it is related to the failure of the lower lead rubber pad to recover to the vertical position in time under the large shear strain, which can also be explained by combining the change trend of the sliding position in the reverse direction.

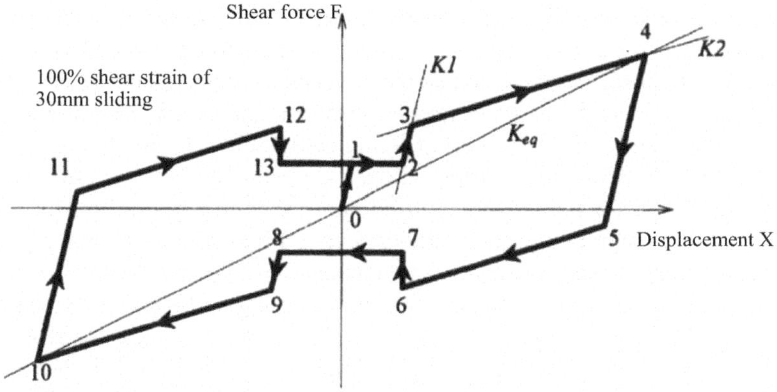

Fig. 9.19 Shear deformation performance curve of multi-dimensional isolation bearing

9.2.2.3 Numerical Simulation of Mechanical Properties of Isolation Bearing

1. **Theoretical hysteresis curve**

Figure 9.19 is the shear performance deformation curve of multi-dimensional isolation bearing. 0–13 in Fig. 9.19 are the key points of force during bearing deformation. Key point 0 is the starting point of the whole bearing stress, process 0–1 is the force displacement change line of the lower lead rubber pad controlled by its stiffness before yielding: key point 1 is the slip starting point of the upper sliding device, and the corresponding stress is the sliding friction force. When the bearing stress increases to key point 1, the upper sliding device starts to slide, and process 1–2 is the sliding range of the bearing. In this range, suppose the upper friction force is constant; key point 2 is the end point of the sliding range of the bearing. When the bearing deformation reaches this position, the "sliding-isolation" system conversion device starts to play a role, driving the movement of the lower isolation pad. Process 2–3 is the force displacement change curve of the isolation pad under the control of its stiffness before yielding. Key point 3 is the turning point of the internal lead core of the isolation pad to yield. Process 3–4 is the force displacement curve of the bearing controlled by the stiffness of the lead rubber pad after yielding; key point 4 is the mark point when the shear strain of the bearing reaches 100%; when the bearing force reaches key point 4, the bearing begins to receive the shear force in the opposite direction, and process 4–5 is the deformation curve before the bearing starts to recover its vertical state and the lower isolation pad begins to yield in the opposite direction, which is simplified as controlled by the stiffness of the lead rubber pad before yielding. Key point 5 is the turning point of the lead core inside the isolation pad to yield in the opposite direction, process 5–6 is the force displacement curve of the bearing under the opposite direction controlled the stiffness of the lead core rubber pad after yield; key point 6 is the static friction force of the sliding device, when the bearing force increases to this friction force, the upper sliding device starts to play a role, process 6–7 is that the friction coefficient of upper sliding device is changed from controlled by the static friction coefficient into the dynamic friction coefficient, that is to say, the bearing unloading phenomenon occurs in this stage; the stress of key point 7 is the dynamic friction of the sliding device, and the process 7–8 is the sliding range of the upper sliding device, which assumes that the dynamic friction coefficient of the bearing does not change.

Process 8–13 is the same as process 2–7, and the shear performance of the bearing is not affected by the bearing stress direction. When the bearing force reaches the key point 13, the next "sliding-isolation" process begins. Process 0–1, 2–3 and 4–5 are all considered to be controlled by the pre yield stiffness K_1 of the isolation pad; process 3–4 and 5–6 are all considered to be controlled by the post yield stiffness K_2 of the isolation pad; process 1–2 and 7–8 are controlled by the dynamic friction of the sliding device; process 6–7 is controlled by the change of the dynamic and static friction coefficient of the sliding device. The connection between key point 4 and key point 10 is defined as the equivalent horizontal stiffness of the bridge multi-functional isolation bearing.

2. **Finite element model**

Based on ANSYS platform, the finite element model of the new multi-functional isolation bearing is established by mapping. Due to the small thickness of sandwich steel plate and the size limitation of solid units, the total number of bearing units is more, with a total of 45132 units. See Fig. 9.20a for the finite element model of the new multi-functional isolation bearing and Fig. 9.20b for the contact pair model of the friction sliding surface of Teflon and stainless-steel plate.

In the above-mentioned finite element model, the steel plate, rubber and lead core are all simulated by Solid185 element, and the common joint simplified processing is adopted between the lead core element with the surrounding laminated rubber element and sandwich steel plate element, the top upper cover plate element with the bottom lower seal plate element. The steel plate and lead core are made of

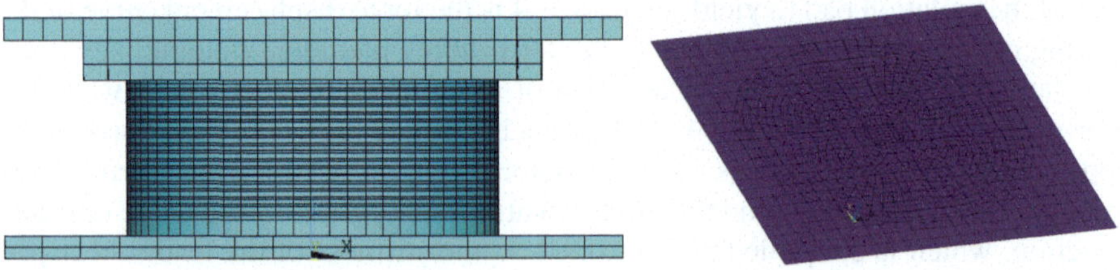

(a) Finite element model (b) Friction surface contact pair

Fig. 9.20 Finite element model of isolation bearing

equidirectional strengthened elastic-plastic materials, in which the elastic modulus of steel is 210 Gpa, the yield modulus is 2.1 Gpa, Poisson's ratio is 0.3; the elastic modulus of lead core is 16..46 Gpa, the yield strength is 10 Mpa, the modulus after yield is defined as 0, that is, the ideal elastic-plastic model, Poisson's ratio is 0.44; the stress-strain constitutive relationship of rubber is Mooney Rivlin model, and the material model can simulate accurately within 200% strain range. The modulus of elasticity E_0 of rubber material has a direct relationship with the constant C_1 and C_2 of rubber material. According to the value of modulus of rubber material in Table 9.1, the parameter of Mooney Rivlin material in this paper is set as $C_1 = 0.1782\,\text{Mpa}$, $C_2 = 0.0178\,\text{Mpa}$.

For the contact effect in the new support, such as tetrafluoroethylene plate and stainless steel plate, baffle and support body, the contact pair composed of Contact174 element and Target170 element are used. The Contact174 element is a 3D 8 node surface contact element, which is used to describe the contact and sliding state between the target surface (Target170 element) and the flexible surface defined by the element. It is suitable for the contact analysis of 3D structure and coupling field. The normal penalty stiffness and contact penetration tolerance of the contact element are taken according to the default settings of the software. The friction coefficient is taken as 0.055 according to the actual measurement, that is, the difference between the dynamic friction coefficient and the static friction coefficient is not considered. In order to prevent the sliding device from "jumping up" during loading, the contact pair of Teflon plate and stainless steel plate is set as "non separation", that is, once contacted, the program forces the contact pair not to be separated; while the contact between the baffle plate and the support body at the transverse limit is set as "standard" type, which is different from the contact pair of polytetrafluoroethylene with stainless steel plate, so it can be connected and separated normally, the rest are the same.

3. **State of stress and strain**

Figure 9.21 is the stress and strain state of each part of the support under 100 mm shear displacement. Figure 9.21a is the overall deformation of the bearing. The upper sliding device has been sliding, the upper cover plate is in contact with the upper right limit plate, and the lower lead rubber isolation bearing has been obviously sheared.

Figure 9.21b shows that the maximum shear stress of rubber reaches 0.399 Mpa, the minimum shear stress is 0.377 Mpa, and the maximum value and the minimum value appear in the inner side of rubber near the lead core. In the simplified design calculation, the shear stress in the single rubber of lead rubber bearing can be approximately considered equal, with an error of less than 5%.

Figure 9.21c is the Mises plastic strain cloud map of the lead core. On the outer surface of the lead core, the minimum value of Mises plastic strain is 0.322, and the maximum value is 0.580. After the transverse shear displacement 100 mm is applied, the lead core has great plastic deformation, which plays a good role in plastic energy

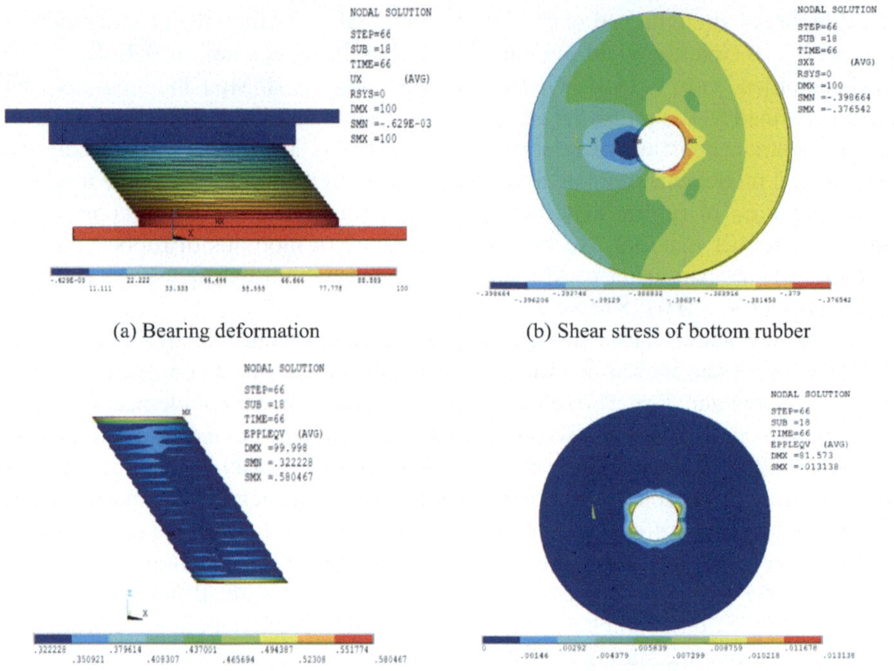

(a) Bearing deformation

(b) Shear stress of bottom rubber

(c) Von-Mises plastic strain of lead core

(d) Von-Mises plastic strain of sandwich steel plate

Fig. 9.21 Stress state of contact element between PTFE and stainless steel

consumption. However, Fig. 9.21d shows that the plastic deformation of sandwich steel plate is still only at the position closest to the lead core, and the plastic strain is not large, and the maximum value is 0.01314.

4. **Overall mechanical properties**

In order to verify the accuracy and applicability of the finite element model of the new multifunctional isolation bearing, the comparison of the hysteretic curve of the bearing is given in Fig. 9.22. Figure 9.22a is the comparison between the test results and the numerical calculation results under the vertical pressure of 350 kN and loading frequency of 0.1 Hz, and the shear displacement is 30, 50, 80, 100 mm. It can be seen from the figure that the numerical simulation results of the above-mentioned finite element model are in good agreement with the test results, and the numerical calculation can more accurately simulate the mechanical characteristics of the new isolation bearing of each stage. However, there is a certain error between the two when the bearing is in unloading state, and when the shear displacement is large, the peak value of numerical calculation results deviates greatly from the test results.

Figure 9.22b is the shear hysteresis curves under the action of the maximum shear displacement of the bearing of 50 mm, loading frequency of 0.1 Hz, vertical pressure

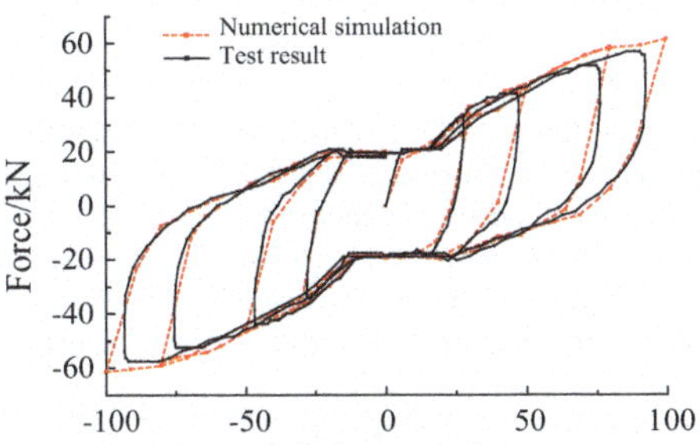

Comparison of hysteresis curves in shear strain correlation test

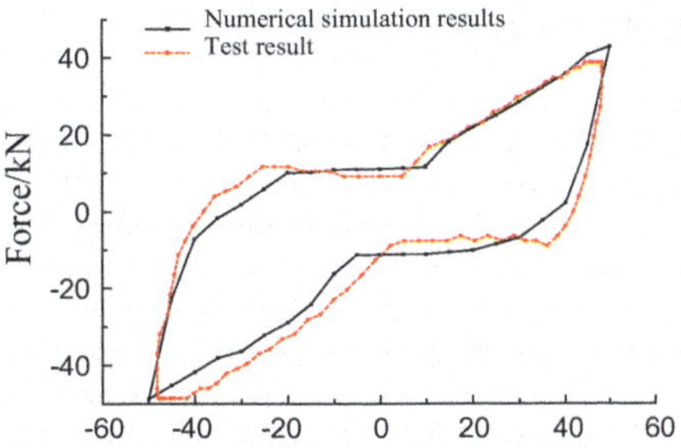

(a) Vertical pressure of 200 kN

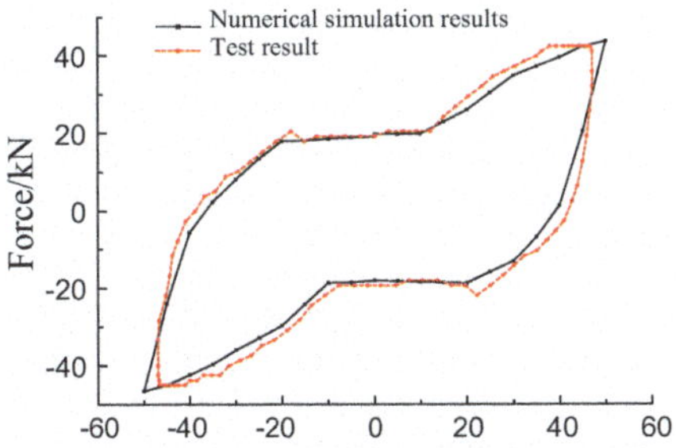

(b) Vertical pressure of 350 kN

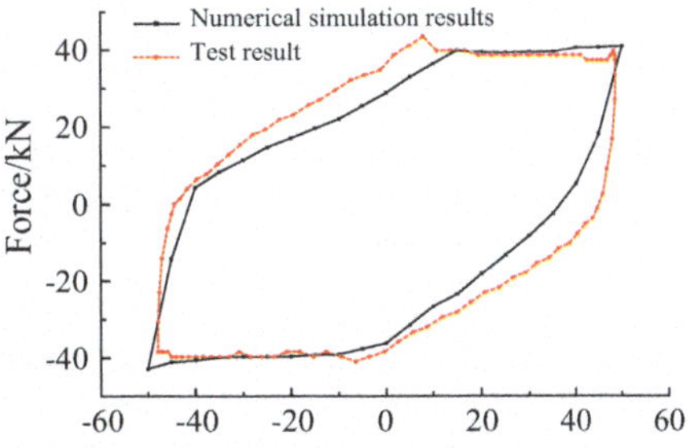

(c) Vertical pressure of 707 kN

Fig. 9.22 Comparison of pressure stress correlation hysteresis curves

of 200 kN, 350 kN and 707 kN respectively. It can be seen from Fig. 9.22a, b that when the vertical pressure is small, the new multi-functional isolation bearing can well achieve the isolation performance of "sliding first and then isolation", and the hysteretic curve obtained by numerical simulation is in good agreement with the hysteretic curve obtained by test; when the vertical pressure is 707 kN, the shear performance of the new bearing changes, and the critical force of sliding friction is greater than the yield force of the bearing, so as to reflect the characteristics of "isolation before sliding". The results of numerical simulation also verify this phenomenon.

It can be seen from Fig. 9.22 that the numerical calculation is in good agreement with the test results, but the hysteresis loop obtained from the numerical simulation is smaller than the test results. The reason for the error is that on the one hand, there is a certain error in the numerical simulation itself; on the other hand, some simplifications and assumptions are made in the simulation calculation process, such as material model, friction coefficient, contact definition, etc. However, as long as the error effect is controlled in a certain range, the numerical results can still meet the needs of the project, and can provide more local stress-strain information for the project.

References

1. Li, Aiqun. 2007. *Vibration control of engineering structure*. Beijing: China Machine Press.
2. Tang, Yonghui. 2016. *Experimental study on the mechanical properties of the improved isolation bearings of low shear modulus*. Nanjing: Southeast University. (in Chinese).
3. Zhao, Genwen. 2010. *Theoretical analysis and experimental research on the mechanical properties of the cellular laminated rubber bearing*. Nanjing: Southeast University. (in Chinese).
4. Zhao, Shuai. 2015. *Experimental study of Three-dimensional multi-function isolation bearings*. Nanjing: Southeast University. (in Chinese).
5. Jirong, Wu. 2012. *Experimental study on performance of bridge multifunctional isolation bearing*. Nanjing: Southeast University. (in Chinese).
6. Li, Xinping. 2011. *Test of new type isolation device and seismic safety analysis of Nanbo old hall*. Nanjing: Southeast University. (in Chinese).

Chapter 10
Other Damping Devices

Abstract The shape memory alloy damper and foam aluminum composite damper are discussed. About shape memory alloy damper, its damping mechanism and characteristics, super elastic mechanical property test of SMA material, constitutive model, one dimensional simplified mechanical model, mechanical model of multi-dimensional simplified SMA are introduced. Basic structure and working principle, mechanical performance test, test result analysis of tension-compression SMA damper and composite friction SMA damper are elaborated respectively. About foam aluminum composite damper and AF/PU composite damper, the preparation of foam aluminum composite damping material, damping mechanism and characteristics, deformation mechanism, phenomenological constitutive model, mechanical model of AF/PU composite material and their damper are introduced respectively.

10.1 Shape Memory Alloy Damper

Shape memory alloy (SMA) is a kind of intelligent material with many unique properties. As early as the 1990s, foreign scholars began to focus on the research of SMA damper. Author's team began to develop SMA damper for civil engineering around 2000, systematically studied the mechanical properties and constitutive model of SMA, designed a variety of SMA damper devices, and carried out corresponding theoretical and experimental research [1, 2].

10.1.1 Damping Mechanism and Characteristics of Shape Memory Alloy

10.1.1.1 Super Elastic Mechanical Property Test of SMA Material

(1) **Test scheme**

In order to study the vibration reduction mechanism of SMA, the mechanical properties of SMA were tested. The diameter of NiTi wire is 1.00 mm, the chemical composition is: Ni49.8, Ti50.2 (atomic fraction %). The length of test piece is 180 mm, its gauge distance is 100 mm, and the total number of test pieces is 96. In the designed

A. Li, *Vibration Control for Building Structures*, Springer Tracts in Civil Engineering, https://doi.org/10.1007/978-3-030-40790-2_10

test pieces, randomly select three samples for differential scanning calorimeter (DSC) test, and the phase transition temperature of the base metal is as shown in Table 10.1.

During the test, the influences of environmental temperature, loading rate, strain amplitude and cycle times on the mechanical properties of SMA are mainly considered. The mechanical properties mainly include stress-strain relationship, phase change stress, hysteretic energy consumption and material loss factor, deformation modulus, residual strain and other parameters. In order to discuss the influence of these load cases on the mechanical performance parameters, a series of test methods are developed, as shown in Table 10.2.

The test is carried out on Instron 5569 universal electronic testing machine. The temperature control is realized by the temperature control box on the testing machine, and the liquid nitrogen is used for refrigeration. During the test, according to the test method, the sine excitation method (equal displacement) is used for loading, which is automatically controlled by the computer, and the test results are automatically collected by the computer. The test device is shown in Fig. 10.1a and the loading method is shown in Fig. 10.1b.

In the test, after one cycle of loading and unloading, the typical stress-strain curve is shown in Fig. 10.1c. σ_{ms} and σ_{mf} are the stress corresponding to the beginning (point A) and the end (point B) of martensitic transformation; σ_u and ε_u are the

Table 10.1 Phase transition temperature of NiTi shape memory alloy wire base metal

Starting temperature of martensitic transformation Ms	End temperature of martensitic transformation Mf	Starting temperature of austenite transformation As	End temperature of austenite transformation Af
−18 °C	−25 °C	0 °C	8 °C

Table 10.2 Test method of each test piece

Serial number	Loading method	Loading rate (mm/Min)	Test temperature (°C)
1	1 cycle at 3%, 6%, 7.5% of the maximum strain respectively	3	15, 25, 30, 40, 50
2	3 cycles at 1%, 3%, 6% and 7.5% of the maximum strain respectively, and finally pull off	180	15, 25, 30, 40, 50
3	1 cycle at 6%, 7%, 10% of the maximum strain respectively, and finally pull off	180	15, 25, 30, 40, 50
4	50 cycles, strain amplitude 6%	30, 360	15, 20, 25, 30, 40, 50
5	3 loading rates at 6% of strain amplitude	360, 216, 72	15, 25, 30, 40, 50

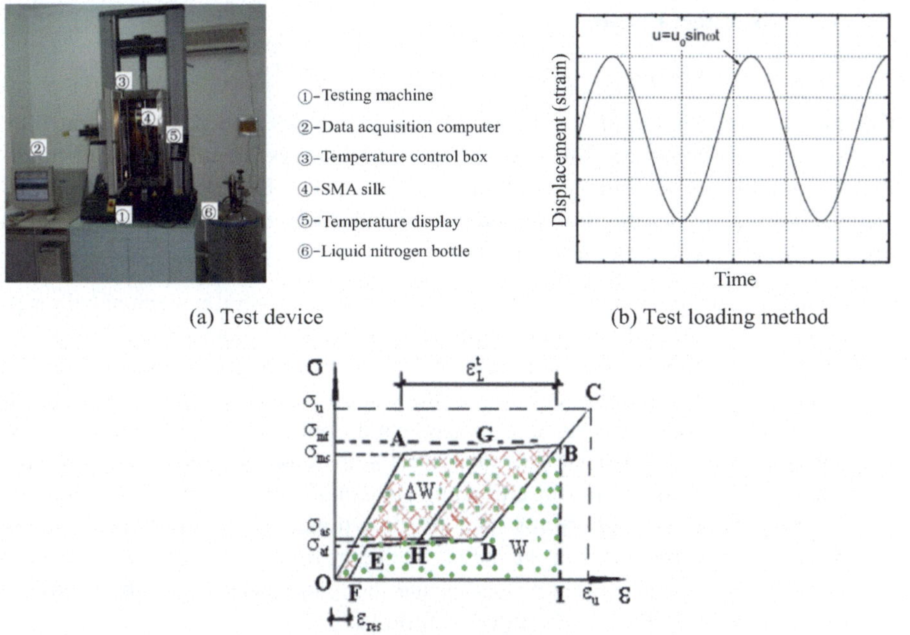

(a) Test device

(b) Test loading method

(c) Parameter specification of stress-strain curve

Fig. 10.1 SMA mechanical property test

stress and strain corresponding to SMA approaching the tensile strength (point C); σ_{as} and σ_{af} are the stress corresponding to the beginning (point D) and the end (point E) of austenite transformation; ε_L^t is the maximum transformation strain (length of yield platform) at loading; and the area ΔW enclosed by OAGBDEFO is the hysteretic energy consumption of SMA wire in one cycle of loading and unloading, and the area W enclosed by OAGBIFO is its elastic energy storage, where, $\eta = \Delta W/2\pi W$ is defined as the loss factor of SMA material; ε_{res} must be the residual strain. When loading, the OA, AGB and BC segments are in austenite, transition state from austenite to martensite and martensite respectively; when unloading, the DBC, DHE and EF segments are in martensite, transition state from martensite to austenite and austenite respectively; the slope of straight line OA and BC is the deformation modulus of austenite and martensite respectively. The slope of the line GH is the modulus of deformation at a certain loading and unloading strain amplitude value in the range of the maximum transformation strain. In the test, the value of mechanical property parameters is obtained according to the above provisions.

(2) **Analysis of test results**

1. Effect of temperature on mechanical properties of SMA

The influence of temperature on the mechanical properties of SMA is illustrated by the No. 1 in the test method (Table 10.2). Figure 10.2 shows the influences of temperature change on stress-strain curve, transformation stress, hysteretic energy dissipation and loss factor, deformation modulus, residual strain, maximum transformation stress, ultimate strength and ultimate strain.

(a) Figure 10.2a shows that the influence of temperature on the mechanical properties of SMA mainly lies in the influence on the yield plateau height of stress-strain curve, that is, the influence on four transformation stresses. With the increase of temperature, the four transformation stresses (σ_{ms}, σ_{mf}, σ_{as}, σ_{af}) all increase, and their magnitude is approximately linear with the increase of temperature (Fig. 10.2b). That is to say, with the increase of temperature, both the loading yield platform and the unloading yield platform are rising proportionally. The height of the two platforms is mainly determined by the four transformation stresses. According to the curve in Fig. 10.2b, the relationship between the magnitude of four transformation stresses and the temperature under the condition of No. 1 in Table 10.2 can be fitted as follows:

$$\sigma_{mf} = 9.20T + 229.71$$

$$\sigma_{ms} = 9.28T + 192.59$$

$$\sigma_{as} = 5.27T - 1.77$$

$$\sigma_{af} = 5.20T - 45.71$$

where, σ_{ms} and σ_{mf} are the start stress and end stress of martensitic transformation, MPa; σ_{as} and σ_{af} are the start stress and end stress of austenitic transformation, MPa; T is the ambient temperature, °C.

(b) Figure 10.2c shows the effect of temperature change on the deformation modulus of SMA during unloading. When the temperature rises from 15 to 50 °C, and the strain amplitude is 1, 3, 6, 7.5%, the corresponding SMA unloading modulus increases by 19, 16, 13, 1%. It can be seen that in the range of test temperature, the deformation modulus increases gradually with the increase of temperature, but the degree of increase is related to the magnitude of transformation strain (strain amplitude). The larger the transformation strain is, the smaller the effect of temperature on deformation modulus is. The main reason for this phenomenon is that the deformation modulus of the material under unloading is greatly affected by the content of martensite in the material, and the increase of transformation strain increases the content of martensite produced

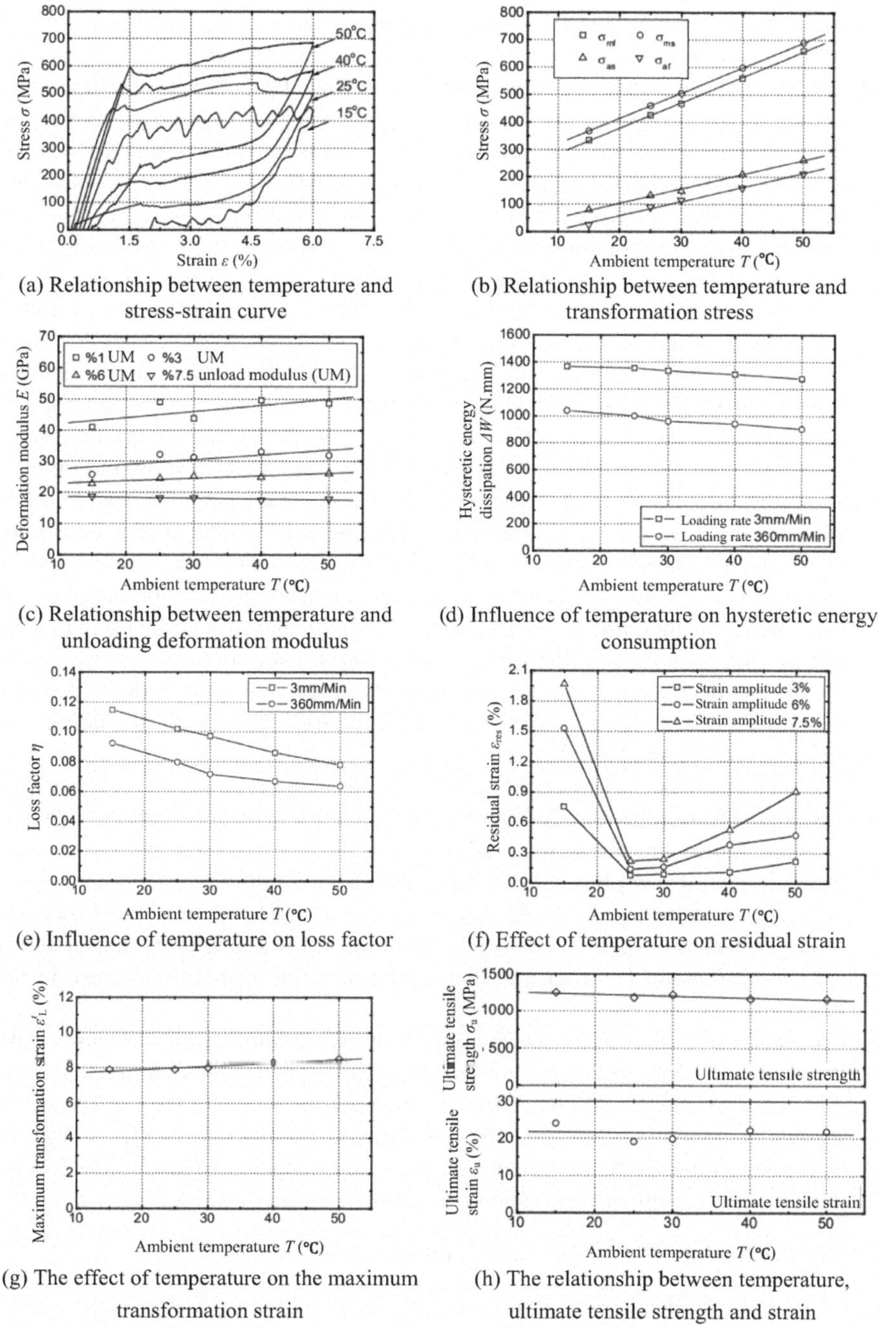

(a) Relationship between temperature and stress-strain curve

(b) Relationship between temperature and transformation stress

(c) Relationship between temperature and unloading deformation modulus

(d) Influence of temperature on hysteretic energy consumption

(e) Influence of temperature on loss factor

(f) Effect of temperature on residual strain

(g) The effect of temperature on the maximum transformation strain

(h) The relationship between temperature, ultimate tensile strength and strain

Fig. 10.2 Effect of ambient temperature on mechanical properties of SMA by stress-induced martensitic transformation, which leads to the decrease of the deformation modulus of the material. When the content of martensite reaches 100%, the deformation modulus tends to be stable. The increase of temperature results in the reverse transformation of SMA and the decrease of martensite content. When the transformation strain is large, the change of martensite content caused by temperature change is smaller than that induced by stress, so the effect of temperature on deformation modulus is not obvious. In this test, the change range of temperature is 35 °C, and the maximum change of deformation modulus is not more than 20%. So in the range of temperature change is not too large, the influence of temperature on deformation modulus can be omitted when the material is in hyperelastic state.

(c) Figure 10.2d, e respectively show the relationship between the hysteresis energy consumption, loss factor of SMA material and temperature in the test temperature range (15–50 °C). With the increase of temperature, the energy consumption capacity of the material decreases slightly (when the loading and unloading rates are 3 mm/min and 360 mm/min, the decrease is 6.67% and 13.21%, respectively). Therefore, the influence of temperature on the energy consumption capacity of the material is not the main factor. The loss factor of SMA material is the ratio of hysteretic energy consumption and elastic energy storage. With the increase of temperature, the yield platform increases, and the elastic energy storage of SMA material increases. Therefore, the corresponding loss factor decreases. It can be seen from the test results that in the hyperelastic range, the loss factor of the material decreases linearly with the increase of temperature, as shown in Fig. 10.2e.

(d) As can be seen in Fig. 10.2f, when the ambient temperature is above 25 °C, the residual strain of the material after a cycle increases linearly with the increase of temperature, but the increase amplitude is closely related to the loading and unloading strain amplitude. The larger the strain amplitude is, the greater the increase is. However, at 15 °C, the residual strain of SMA after the first cycle is larger, which is mainly because the material at this temperature also contains a certain amount of martensite, that is, the material at this time is not in the complete austenite state.

(e) At a certain loading rate, the maximum transformation strain increases with the increase of temperature, but the increase is small. This shows that after SMA reaches yield (martensitic transformation), the length of the yield platform basically remains unchanged with the change of temperature (Fig. 10.2g). In addition, within the test temperature range, the temperature has little effect on the tensile strength and corresponding ultimate strain of SMA (Fig. 10.2h).

2. Effect of loading rate on mechanical properties of SMA

Figure 10.3 shows the stress-strain curve, phase transformation stress and energy dissipation capacity of SMA under different loading rates (the first cycle is taken and the loading and unloading strain amplitude values are all 6%).

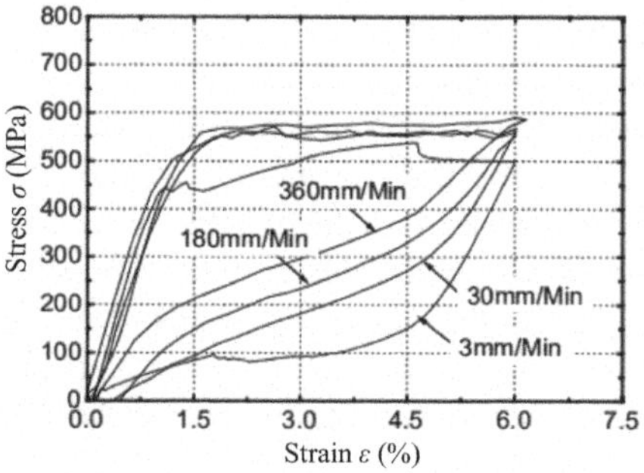

(a) Relationship between loading rate and stress-strain curve (25 °C)

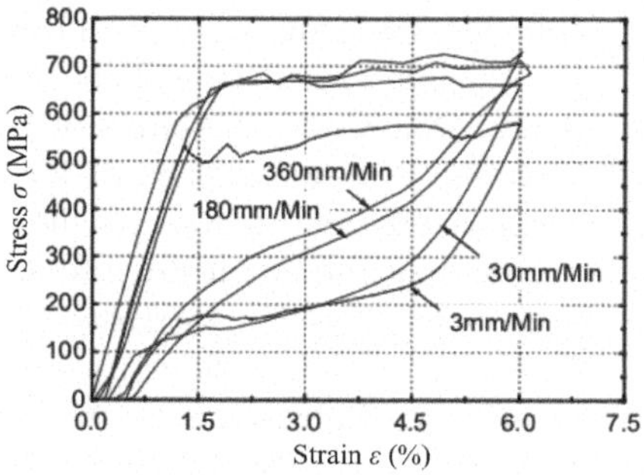

(b) Relationship between loading rate and stress-strain curve (40 °C)

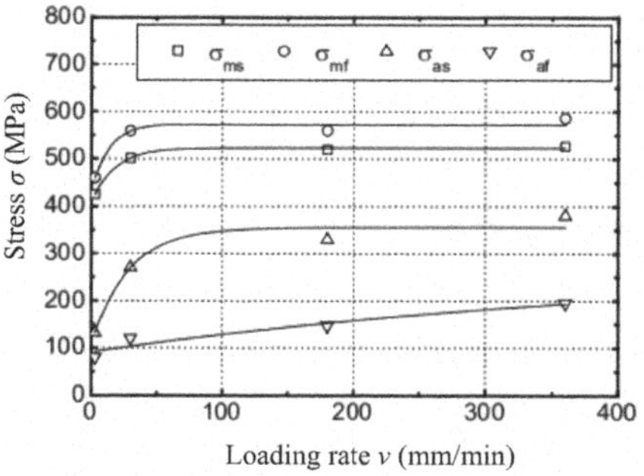

(c) Relationship between loading rate and transformation stress (25 °C)

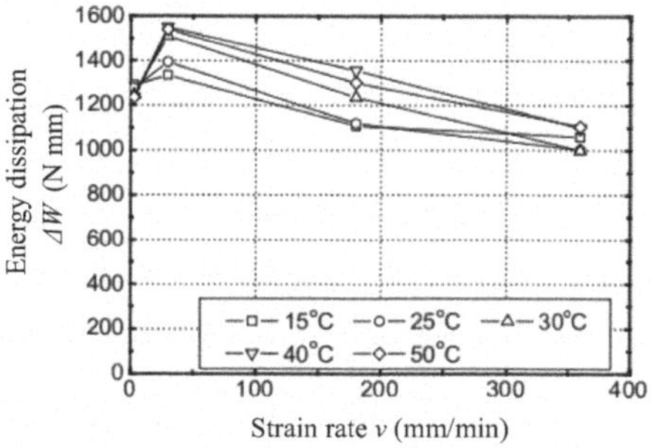

(d) Relationship between loading rate and energy consumption

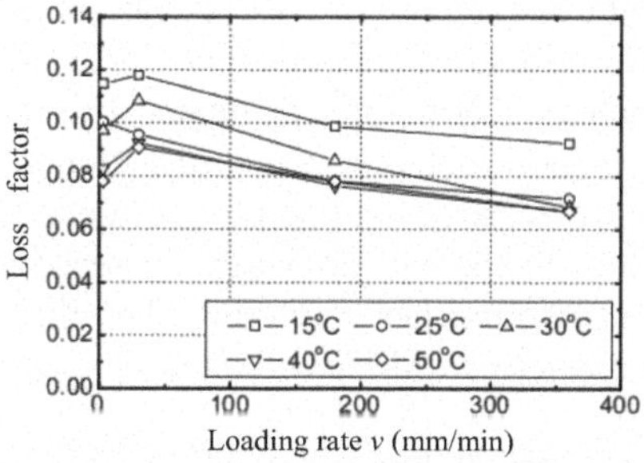

(e) Relation between loading rate and loss factor (25 °C)

Fig. 10.3 Effect of loading rate on mechanical properties of SMA

(a) As can be seen in Fig. 10.3a, b, when the loading rate is small (3 mm/min), the loading and unloading 'yield platform' of the material is very obvious, the slope of the two platforms is very small, almost close to the level, and each transformation point is easy to determine; but when the loading and unloading rate is large (360 mm/min, etc.), the loading yield platform is also obvious, while the unloading yield point (initial stress of austenite transformation σ_{ms}) and "yield platform" are not obvious, and the slope is obvious.

(b) Figure 10.3c shows the effect of loading rate on the transformation stress of materials. With the change of loading rate, the transformation stresses also change in different degrees, but the general trend is: with the increase of loading rate, the transformation stresses increase, in which the initial stress of austenite transformation σ_{as} increases greatly, while the other three transformation stresses $\sigma_{ms}, \sigma_{mf}, \sigma_{af}$ increase approximately the same. According to Fig. 10.3c, the relationship between the transformation stress and the loading speed v at 25 °C can be fitted as follows:

$$\sigma_{ms} = 20.90 \ln(v) + 413.13$$

$$\sigma_{mf} = 21.71 \ln(v) + 448.46$$

$$\sigma_{as} = 46.83 \ln(v) + 81.88$$

$$\sigma_{af} = 20.89 \ln(v) + 59.51$$

where, v is the loading rate, mm/min; σ_{ms} and σ_{mf} is the start stress and end stress of martensitic transformation, MPa; σ_{as} and σ_{af} is the start stress and end stress of austenitic transformation, MPa.

(c) Under the test conditions, when the loading rate is less than 30 mm/min, the hysteretic energy consumption and loss factor of SMA increase with the increase of loading rate; when the loading rate is greater than 30 mm/min, the hysteretic energy consumption and loss factor decrease with the increase of loading rate, and with the further increase of loading rate, the decrease amplitudes of energy consumption and loss factor will gradually decrease, as shown in Fig. 10.3d, e. When the loading rate reaches a certain value, the hyperelasticity damping energy dissipation capacity of SMA will tend to be stable.

(3) Effect of strain amplitude on mechanical properties of SMA

Figure 10.4 gives the analysis results of mechanical properties parameters of SMA at 30 °C under different strain amplitude values. The loading rates are 3 mm/min and 360 mm/min respectively.

(a) In the range of strain amplitude selected by the test, the strain amplitude has a great influence on the energy dissipation capacity of the material. When the loading rate is small (Fig. 10.4a), the height of hysteresis loop changes little with the increase of strain amplitude, but its width increases linearly; when the

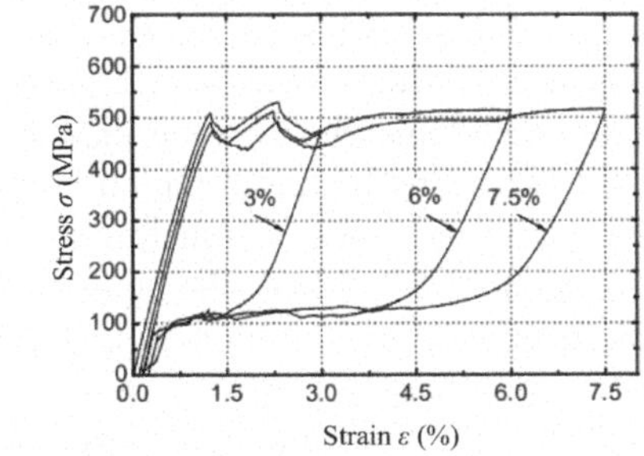

(a) Relationship between strain amplitude and stress-strain curve (3mm/min)

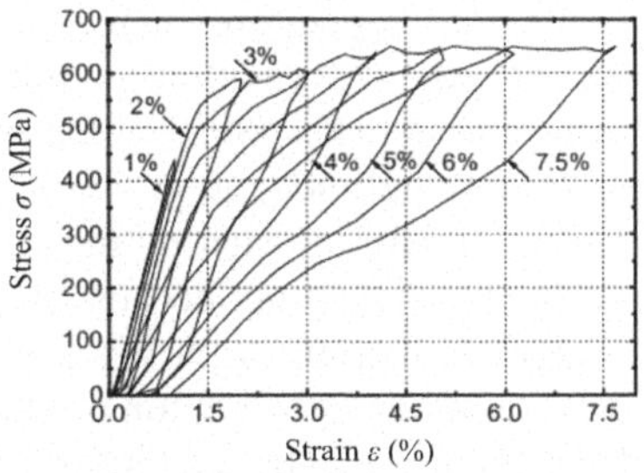

(b) Relationship between strain amplitude and stress-strain curve (360mm/min)

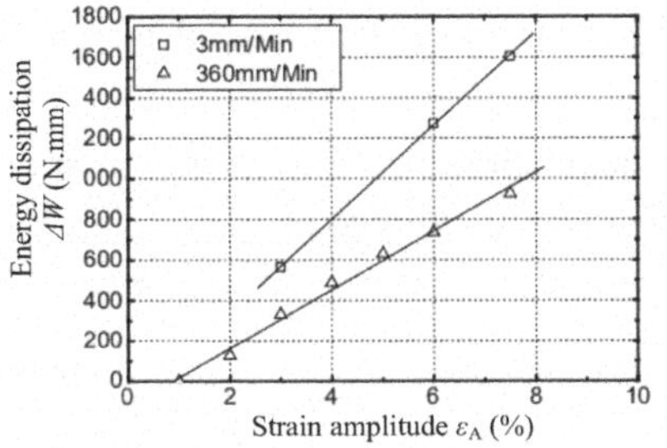

(c) Relationship between strain amplitude and energy consumption

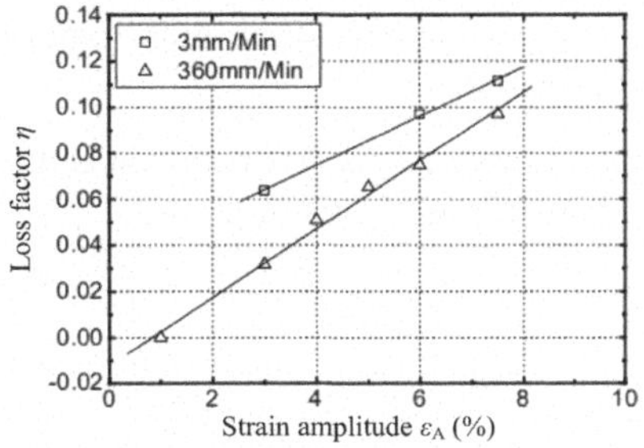

(d) Relationship between strain amplitude and loss factor

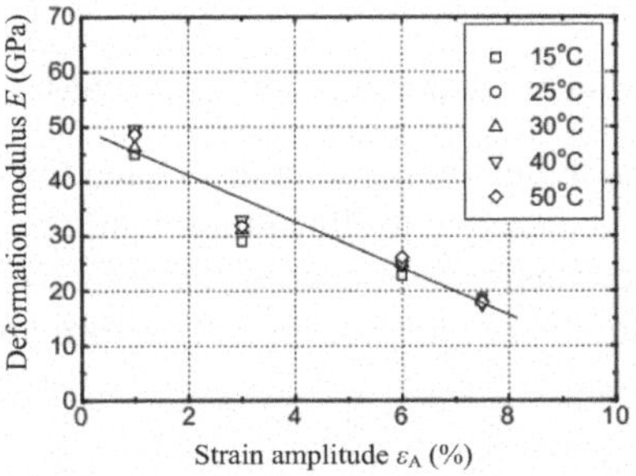

(e) Relationship between strain amplitude and deformation modulus

Fig. 10.4 Effect of strain amplitude on mechanical properties of SMA (30 °C)

loading rate is large (Fig. 10.4b), the energy consumption capacity of material also increases linearly with the increase of strain amplitude, but the increase amplitude is smaller than that of low-speed loading. Figure 10.4c, d show the relationship between the hysteretic energy consumption, loss factor and strain amplitude in the range of maximum phase change strain. It can be seen that the hysteretic energy consumption and loss factor increase linearly with the increase of strain amplitude. When the loading rate is small, the increase amplitude is large. It can be seen that the strain amplitude is the most important factor affecting the energy dissipation capacity of materials.

(b) In the test temperature range, when the strain amplitude value increases from 1 to 7.5%, the deformation modulus of the material during unloading decreases from 43.8 to 18.1 GPa, which basically decreases linearly with the increase of strain amplitude, as shown in Fig. 10.4e. Therefore, the deformation modulus is greatly affected by the strain amplitude in the range of transformation strain. This is mainly because the larger the strain amplitude is, the higher the content of martensite in stress-induced martensitic transformation is, and the corresponding deformation modulus is reduced.

(c) In addition, it can be seen from the test results that the strain amplitude has little effect on the phase transformation stress of the material, but has a certain effect on the residual strain of the material. In the hyperelastic state, with the increase of the loading strain amplitude, the residual strain of the material increases gradually.

According to Fig. 10.4e, it can be fitted that the relationship between deformation modulus E and strain amplitude ε_A at test temperature as follows:

$$E = -4.19\varepsilon_A + 48.79 \tag{10.1}$$

Among them, E is the deformation modulus under a certain strain amplitude, GPa; ε_A is the loading and unloading strain amplitude within the maximum transformation strain range, %.

4. Effect of cycle times on mechanical properties of SMA

Figure 10.5 take No. 4 (ambient temperature 25 °C, loading rate 30 mm/min) in test method (Table 10.2) as an example to illustrate the influence of cycle times on the stress-strain curve, phase transformation stress, hysteretic energy consumption, loss factor, residual strain and other main mechanical performance parameters of SMA.

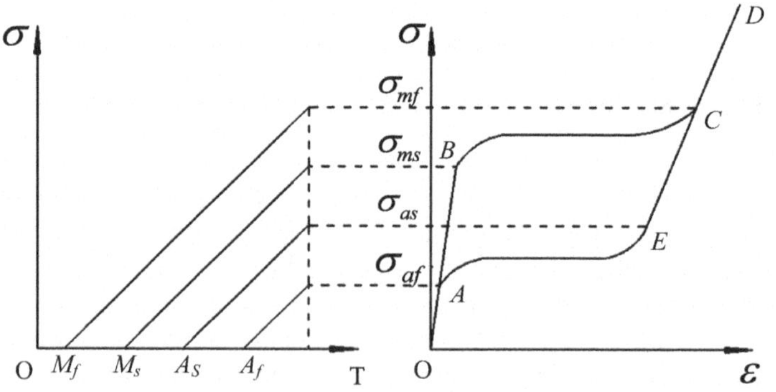

Fig. 10.5 Typical hyperelastic stress-strain-temperature relationship of SMA

(a) It can be seen from Fig. 10.5a that in the first several cycles of loading, with the increase of the number of cycles, the four transformation stresses decrease, but the decreasing amplitude is different. The decreasing amplitude of initial stress of martensitic transformation σ_{ms} is the largest while that of the other three transformation stresses σ_{mf}, σ_{as}, σ_{af} is smaller. After about the 10th cycle, the transformation stresses tend to be stable, as shown in Fig. 10.5b. According to Fig. 10.5b, the relationship between transformation stress and cycle number at 25 °C and loading rate of 30 mm/min can be fitted as follows:

$$\sigma_{ms} = -60.06\ln(N) + 489.19$$

$$\sigma_{mf} = -50.58\ln(N) + 568.04$$

$$\sigma_{as} = -13.20\ln(N) + 257.51$$

$$\sigma_{af} = -3.73\ln(N) + 73.47$$

Among them, N is the number of cycles, σ_{ms} and σ_{mf} are the start and end stresses of martensitic transformation, MPa, σ_{as} and σ_{af} are the start and end stresses of austenitic transformation, MPa;

(b) In Fig. 10.5c, the energy consumption capacity of SMA decreases with the increase of cycles. In the first few cycles, the hysteretic energy consumption of materials is greatly reduced. After about 15 cycles, the hysteretic energy consumption of materials gradually stabilizes. When the number of cycles increases from 1 to 50, the loss factor of SMA material decreases from 0.09 to 0.07, and the change of loss factor is small, as shown in Fig. 10.5d.

(c) It can also be seen from the hysteresis curve in Fig. 10.5a that, with the increase of the number of cycles, each hysteresis curve can basically remain parallel, but the slope of the yield platform changes slightly when loaded to yield (martensitic transformation). It can be seen that the number of cycles has little effect on the deformation modulus of the material.

(d) Figure 10.5e shows the change rule of the cumulative residual strain of SMA with the number of cycles. With the increase of the number of cycles, the cumulative residual strain increases. In the first few cycles, the increase of residual strain is larger, but when the number of cycles is more than 15, the increase of cumulative residual strain is smaller, and the material shows obvious complete hyperelasticity. According to Fig. 10.5e, the relationship between cumulative residual strain ε_{resN} and number of cycles N can be fitted as follows:

$$\varepsilon_{resN} = 0.1126\ln(N) + 0.3725$$

where, N is the number of cycles; ε_{resN} is the cumulative residual strain after N cycles. In addition, with the increase of the number of cycles, the maximum transformation strain of the material tends to decrease.

10.1.1.2 Common SMA Constitutive Model

(1) **Hyperelastic properties of SMA**

When SMA is under load and the ambient temperature is higher than the temperature A_f at the completion of austenite transformation, it is in hyperelastic state. The hyperelasticity of shape memory alloy is caused by the instability of martensitic transformation induced by stress. When the stress of SMA reaches a certain value, martensitic transformation begins to take place. On the macro level, it shows the yield phenomenon of materials, but the yield platform is much longer than that of ordinary metals. Because this transformation is unstable, when the stress is relieved (unloaded), the reverse transformation will make the martensite phase disappear, and the material will return to the austenite state, that is, to its shape before loading.

The typical stress-strain-temperature curve of the material in the hyperelastic state is shown in Fig. 10.5. In OAB stage, SMA material is in austenite state and shows elastic deformation; when it is loaded to point B, martensitic transformation begins, at this time, when the stress slightly increases, martensitic transformation continues until point C, and austenite is transformed into martensite. In the process of transformation, the slope of deformation curve is very small, and the material produces about 6–8% plastic deformation (transformation deformation), which is much larger than that of ordinary metal; when the stress continues to increase, the material is in the martensitic state, at this time, the stress-strain relationship is linear, but compared with the austenitic state, the young's modulus of martensite is much lower. During unloading, after the elastic recovery (martensitic state) of DCE, the starting point E of austenite transformation is reached, and after the completion of reverse transformation, the material returns to the point F of complete austenite state, and finally to the origin O.

It can be seen from the super elastic deformation characteristics of SMA that martensitic transformation is one of the main factors affecting the large deformation of materials, which is generally indicated by the volume content of martensite. In Fig. 10.5, the volume content of martensite is 0 in the fully austenitic state (OAB section) and 1 in the fully martensitic state (DCE section), while in the transformation state: loading from B to C increases the volume content of stress-induced martensite from 0 to 1, on the contrary, unloading from E to F reduces the volume content of martensite from 1 to 0. It can also be seen from Fig. 10.5 that the four critical stresses, σ_{mf}, σ_{as}, σ_{af} (i.e., the starting stress and the ending stress of martensitic transformation output, the starting stress and the ending stress of austenitic transformation output), which are the key parameters of the constitutive relation curve, and the main factor affecting the four parameters is the ambient temperature. From the test results in the previous section, they are related to the temperature with a basically linear relationship.

In a complete loading and unloading process, the super elastic SMA can return to its original state after loading and unloading for one cycle, and the residual strain is zero. Due to the large difference between the stress induced martensitic transformation and that induced austenite transformation, a significant hysteresis loop is formed on the constitutive relation curve, which shows that SMA has a significant energy dissipation effect. When applied to engineering structures, it can absorb the energy of structural vibration, reduce the vibration of structures and maintain the resilience of structures under large earthquakes.

(2) **SMA constitutive model**

Because the mechanical parameters in the model are relatively simple and easy to determine, the one-dimensional phenomenological constitutive model of SMA is more suitable for practical engineering applications, such as the model proposed by Tanaka, Brinson, Liang and Roger, etc. These models usually include the constitutive equation reflecting the stress-strain-temperature relationship and the transformation dynamic equation describing the change of crystal transformation in SMA. Among them, the temperature change and the degree of stress-induced martensitic transformation are indicated by the volume fraction of martensitic transformation. These models are briefly described below.

1. **Tanaka model**

Tanaka takes one-dimensional SMA as the research object, and its constitutive equation is as follows:

$$\sigma - \sigma_0 = E(\xi)(\varepsilon - \varepsilon_0) + \Theta(T - T_0) + \Omega(\xi)(\xi - \xi_0) \tag{10.2}$$

Among them, σ_0 and σ indicates the initial stress and the stress in a certain state; ε_0 and ε indicates the initial strain and the strain in a certain state; T_0 and T indicates the initial temperature and the temperature in a certain state; ξ_0 and ξ indicates the initial volume number of martensitic transformation and the volume number of transformation in a certain state; $E(\xi)$ is the young's modulus of the material, $E(\xi) = E_A + \xi(E_M - E_A)$; E_A and E_M are the elastic modulus in austenite and martensite respectively. Θ is thermal coefficient; ε_L^t is maximum transformation strain; $\Omega(\xi)$ is transformation coefficient: $\Omega(\xi) = -\varepsilon_L^t E(\xi)$. In the constitutive Eq. (10.2), the internal variable ξ reflecting the change of material transformation is a very important parameter. Its value is determined by the transformation dynamic equation.

When martensitic transformation occurs, i.e. transformation from austenite phase A to martensitic phase M:

$$\xi = 1 - e^{a_M(M_s - T) + b_M\sigma} \tag{10.3}$$

When austenite transformation occurs, i.e. transformation from martensitic phase M to austenite phase A:

$$\xi = e^{a_A(A_s - T) + b_A \sigma} \tag{10.4}$$

Among them, $a_A = \frac{\ln(0.01)}{A_s - A_f}$, $a_M = \frac{\ln(0.01)}{M_s - M_f}$, $b_A = -\frac{a_A}{c_A}$, $b_M = -\frac{a_M}{c_M}$; C_A and C_M are the equivalent transformation coefficients of stress and temperature when austenite transforms to martensite and martensite transforms to austenite respectively; M_s and M_f are the start and end temperatures of martensite transformation respectively; A_s and A_f are the start and end temperatures of austenite transformation respectively.

2. **Liang and Rogers model**

The Tanaka model was improved by Liang and Rogers, which mainly reflected in the kinetics equation of martensitic transformation. The cosine relationship between internal variables ξ (volume fraction of martensitic transformation) with temperature and stress was proposed. The cosine function was used in the transformation equation, and the influence the initial conditions of internal variables on martensitic transformation and austenitic transformation was also considered.

When martensitic transformation occurs, i.e. transformation from austenite phase A to martensitic phase M:

$$\xi = \frac{1 - \xi_A}{2} \cos[a_M(T - M_f) + b_M \sigma] + \frac{1 + \xi_A}{2} \tag{10.5}$$

When austenite transformation occurs, i.e. transformation from martensitic phase M to austenite phase A:

$$\xi = \frac{\xi_M}{2} \cos[a_A(T - A_s) + b_A \sigma + 1] \tag{10.6}$$

Among them, $a_A = \frac{\pi}{A_f - A_s}$, $a_M = \frac{\pi}{M_s - M_f}$, $b_A = -\frac{a_A}{c_A}$, $b_M = -\frac{a_M}{c_M}$, ξ_A and ξ_M are the initial transformation volume numbers of A → M and M → A respectively; other symbols are the same as before. The phase transformation equation is a phenomenological simulation based on the shape of the experimental curve.

3. **Brison model**

In the constitutive model proposed by Tanaka and Liang, the reorientation of martensitic phase cannot be solved at low temperature. To solve this problem, Brinson divides the volume number ξ of martensitic transformation into two parts, one is the martensitic transformation number ξ_T induced by temperature change, the other is the martensitic transformation number ξ_S induced by stress change. The constitutive equation is:

$$\sigma - \sigma_0 = E(\xi)\varepsilon - E(\xi_0)\varepsilon_0 + \Theta(T - T_0) + \Omega(\xi)\xi_s - \Omega(\xi_0)\xi_{s0} \tag{10.7}$$

Among them, $\xi = \xi_S + \xi_T$. When the material is in hyperelastic state ($T > A_f$), the phase transformation equation of the internal variable (ξ_T, ξ_T) in Eq. (10.7) is as follows:

When martensitic transformation occurs, i.e. transformation from austenite phase A to martensitic phase M:

$$\xi_S = \frac{1 - \xi_{S0}}{2} \cos\left[\frac{\pi}{\sigma_{ms} - \sigma_{mf}}(\sigma - \sigma_{mf} - c_M(T - M_f))\right] + \frac{1 + \xi_{S0}}{2} \quad (10.8)$$

$$\xi_T = \xi_{T0} - \frac{\xi_{T0}}{1 - \xi_{S0}}(\xi_S - \xi_{S0}) \quad (10.9)$$

When austenite transformation occurs, i.e. transformation from martensitic phase M to austenite phase A:

$$\xi = \frac{\xi_0}{2} \cos\left[a_A\left(T - A_S - \frac{\sigma}{c_A}\right) + 1\right] \quad (10.10)$$

$$\xi_S = \xi_{S0} - \frac{\xi_{S0}}{\xi_0}(\xi_0 - \xi) \quad (10.11)$$

Among them, the parameter with subscript 0 is the initial state; σ_{ms} and σ_{mf} are the beginning and ending stress of martensitic transformation; other parameters are the same as before.

10.1.1.3 One Dimensional Simplified SMA Mechanical Model

The Tanaka, Liang and Brison models, which are more influential at present, are introduced in the front. The establishing idea of the model is basically the same, but the kinetic equation of martensitic transformation adopted is different. Therefore, it is necessary to establish not only a constitutive equation that can correctly reflect the mechanical properties of materials, but also a transformation dynamic equation that can reflect the change rule of material transformation and has simple form, which is the basis of correctly describing the mechanical properties of SMA, and also two conditions for establishing SMA constitutive model.

Because SMA has superior super elasticity and damping performance, it can be used to design dampers, which has obvious energy dissipation and damping effect. In order to establish the super elasticity constitutive relationship suitable for the analysis of SMA damper and other composite engineering structures with SMA, the following basic assumptions are made [2]:

(1) according to the stress characteristics of SMA under the action of temperature and stress, the stress-strain relationship of each transformation state (austenite state, mutual transition state of martensite and austenite, martensite state) is simplified into four straight lines form;
(2) the working temperature T of SMA is always above the starting temperature A_f of austenite transformation, and it works under the maximum recoverable strain;
(3) the critical stress of martensitic transformation and austenitic transformation is related to temperature with a linear relationship;
(4) because the material works in hyperelastic state, its temperature range is generally about 50 °C, and the thermal expansion coefficient of SMA is in the order of 10^{-6}, so the maximum strain (0.05%) caused by thermal expansion is far less than its maximum transformation strain (6%), so the effect of thermal strain can be ignored;
(5) according to the characteristics of SMA hyperelastic transformation, the volume fraction of martensite transformation should be proportional to the transformation.

According to the above basic assumptions, the constitutive relation curve of SMA during loading and unloading is shown in Fig. 10.6. It can be seen that the magnitude of the four phase change stresses (σ_{ms}, σ_{mf}, σ_{as}, σ_{af}) is the key parameter to reflect the SMA constitutive model, which determines the level of loading and unloading yield platform on the stress-strain curve, and is also an important parameter to determine the working bearing capacity of the designed SMA damper. The transformation stress is mainly affected by the ambient temperature. As long as the ambient temperature is given, the corresponding transformation stress and the corresponding critical strain can be obtained, and the form of the stress-strain relationship is also determined.

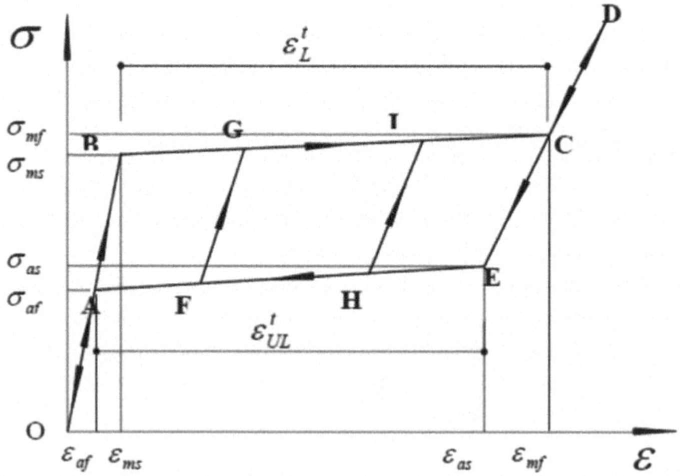

Fig. 10.6 Simplified hyperelasticity constitutive relation of SMA

Therefore, the transformation stress is an important parameter in the constitutive equation. It can be seen from the analysis of the hyperelastic properties of SMA and Fig. 10.6 that:

When loading, when the stress reaches σ_{ms}, austenite begins to transform to martensite, and ξ increases gradually from 0. In Eq. (10.5), when ξ and ξ_A are zero, the stress σ_{ms} at which austenite begins to transform to martensite can be obtained:

$$\sigma_{ms} = c_M(T - M_S) \tag{10.12}$$

When the transformation from austenite to martensite is close to the end, ξ increases gradually to 1, let $\xi = 1$, the stress σ_{mf} at the end of transformation from austenite to martensite can be obtained from Eq. (10.5):

$$\sigma_{mf} = c_M(T - M_f) \tag{10.13}$$

Similarly, in unloading, when the stress decreases to σ_{as} and the transformation from martensite to austenite starts, ξ gradually decreases from 1. In Eq. (10.5), if ξ and ξ_M are equal to 1, the stress σ_{as} when the transformation from martensite to austenite starts can be obtained:

$$\sigma_{as} = c_A(T - A_s) \tag{10.14}$$

When the transformation from martensite to austenite is close to the end, ξ decreases gradually to 0, let $\xi = 0$, the stress σ_{af} at the end of transformation from martensite to austenite can be obtained from Eq. (10.5):

$$\sigma_{af} = c_A(T - A_f) \tag{10.15}$$

From the above four formulas, we can get the relationship between the four transformation stresses on the constitutive relation curve and the ambient temperature. The limit strains ($\varepsilon_{ms}, \varepsilon_{mf}, \varepsilon_{as}, \varepsilon_{af}$) corresponding to the four transformation stresses can be determined according to the stress-strain relationship and Fig. 10.6:

$$\varepsilon_{ms} = \sigma_{ms}/E_A, \quad \varepsilon_{mf} = \varepsilon_{ms} + \varepsilon_L^t, \quad \varepsilon_{mf} = \sigma_{mf}/E_A, \quad \varepsilon_{as} = \varepsilon_{af} + \varepsilon_{UL}^t \tag{10.16}$$

In the above formula, ε_L^t and ε_{UL}^t show the maximum transformation strain of martensite and austenite respectively. Through the basic hypothesis (1), we can get the relationship between them as follows:

$$\varepsilon_{UL}^t = \varepsilon_L^t + \frac{\sigma_{ms} - \sigma_{af}}{E_A} - \frac{\sigma_{mf} - \sigma_{as}}{E_M} \tag{10.17}$$

Therefore, only one of ε_L^t and ε_{UL}^t can be determined, generally ε_L^t, the shape of the constitutive relation can be determined by calculating four transformation stresses and corresponding strains.

Based on the test results at various temperatures and loading and unloading rates, considering the reduction of SMA transformation stress after cycling, the relationships between transformation stress and ambient temperature, loading and unloading rate are as follows:

$$\sigma_{ms} = \alpha_{ms} c_M (T - M_S) + b_{ms} \ln(v) \tag{10.18}$$

$$\sigma_{mf} = \alpha_{mf} c_M (T - M_f) + b_{mf} \ln(v) \tag{10.19}$$

$$\sigma_{as} = \alpha_{as} c_A (T - A_S) + b_{as} \ln(v) \tag{10.20}$$

$$\sigma_{af} = \alpha_{af} c_A (T - A_f) + b_{af} \ln(v) \tag{10.21}$$

Among them, c_M, c_A, b_{ms}, b_{mf}, b_{as}, b_{af} are the material constants respectively. According to the test results, $c_M = 92$ MPa/°C, $c_A = 5.2$ MPa/°C, $b_{ms} = 20.89$ MPa, $b_{mf} = 21.7$ MPa, $b_{as} = 46.8$ MPa, $b_{af} = 20.9$ MPa, α_{ms}, α_{mf}, α_{as}, α_{af} are the phase change stress reduction coefficients respectively when the material performance is stable after the repeated cycle action. When the material is the base material, their values are all 1. The unit of ambient temperature is °C and the unit of loading and unloading rate is mm/min.

According to Hooke's law, the constitutive equation of SMA is:

$$\sigma = E\varepsilon^e \tag{10.22}$$

Because the total strain ε of SMA is composed of elastic strain ε^e, phase transformation strain ε^t and thermal strain $\alpha \Delta T$, then

$$\varepsilon^e = \varepsilon - \varepsilon^t - \alpha \Delta T \tag{10.23}$$

Among them, ε^e is elastic strain, ε^t is phase transformation strain, α is coefficient of thermal expansion and ΔT is temperature difference. According to the basic assumption (4), the constitutive equations of SMA can be obtained from Eqs. (10.22) and (10.23):

$$\sigma = E(\varepsilon - \varepsilon^t) \tag{10.24}$$

where E is the young's modulus of the material. According to the basic assumption (4), during the transformation, the transformation strain ε^t is:

$$\text{Loading:} \qquad \varepsilon^t = \xi \varepsilon_L^t \tag{10.25}$$

$$\text{Unloading:} \quad \varepsilon^t = \xi \varepsilon^t_{UL} \tag{10.26}$$

According to Eq. (10.23) and basic assumption (3), when phase transformation occurs (i.e. yield):

$$\text{Loading:} \quad \varepsilon^t = \varepsilon - \varepsilon_{ms} \tag{10.27}$$

$$\text{Unloading:} \quad \varepsilon^t = \varepsilon - \varepsilon_{af} \tag{10.28}$$

The internal variable ξ in the constitutive equation is the transformation volume number which reflects the change law of martensitic transformation. Its value should be determined by the corresponding transformation equation according to the law of thermodynamics, which is generally complex. In order to simplify the calculation, according to the analysis of the hyperelastic phase transition characteristics of SMA and Eqs. (3.25)–(3.28), ξ can be expressed as follows:

OAB line segment (austenite state):	$\xi = 0$	
DCE line segment (martensitic state):	$\xi = 1$	
BC line segment (transformation state from austenite to martensite):	$\xi = \frac{\varepsilon - \varepsilon_{ms}}{\varepsilon^t}$	(10.29)
EA line segment (transformation state from martensite to austenite):	$\xi = \frac{\varepsilon - \varepsilon_{af}}{\varepsilon^t_{UL}}$	

Equations (10.24)–(10.29) are the hyperelastic constitutive equations of SMA, which can be used to describe the mechanical properties of hyperelastic SMA. However, the constitutive relations of SMA in loading and unloading stages are as follows:

(1) Loading:

$$\varepsilon \leq \varepsilon_{ms} \text{ (OAB segment):} \quad \sigma = E_A \varepsilon \tag{10.30}$$

$$\varepsilon_{ms} < \varepsilon \leq \varepsilon_{mf} \text{ (BGIC segment):} \quad \sigma = \sigma_{ms} + \xi(\sigma_{mf} - \sigma_{ms}) \tag{10.31}$$

$$\varepsilon \geq \varepsilon_{mf} \text{ (CD segment):} \quad \sigma = \sigma_{mf} + E_M(\varepsilon - \varepsilon_{ms} - \varepsilon^t_L) \tag{10.32}$$

$$\begin{array}{r}\text{When loading during unloading:}\\ \text{(FG segment):}\end{array} \quad \sigma = \sigma_{af} + \xi(\sigma_{as} - \sigma_{af}) + E(\xi)(\varepsilon - \varepsilon_{af} - \varepsilon^t) \tag{10.33}$$

where, $\xi = \frac{\varepsilon - \varepsilon_{af}}{\varepsilon^t_{UL}}$, $E(\xi) = E_A + \xi(E_M - E_A)$, $\varepsilon^t = \xi \varepsilon^t_{UL}$.

(2) Unloading:

$$\varepsilon \geq \varepsilon_{as} \text{ (DCE segment):} \quad \sigma = \sigma_{as} + E_M(\varepsilon - \varepsilon_{af} - \varepsilon_{UL}^t) \tag{10.34}$$

$$\varepsilon_{af} < \varepsilon \leq \varepsilon_{as} \text{ (EA segment):} \quad \sigma = \sigma_{af} + \xi(\sigma_{as} - \sigma_{af}) \tag{10.35}$$

$$\varepsilon \leq \varepsilon_{af} \text{ (AO segment):} \quad \sigma = E_A \varepsilon \tag{10.36}$$

$$\text{When unloading during loading (IH segment):} \quad \sigma = \sigma_{ms} + \xi(\sigma_{mf} - \sigma_{ms}) - E(\xi)(\varepsilon_{ms} + \varepsilon^t - \varepsilon) \tag{10.37}$$

where, $\xi = \frac{\varepsilon - \varepsilon_{ms}}{\varepsilon_L^t}$, $E(\xi) = E_A + \xi(E_M - E_A)$, $\varepsilon^t = \xi \varepsilon_L^t$.

In order to explain the rationality of the model, the test data of some specimens are compared with the model. The relevant data measured in the test are as follows: $M_f = -25$ °C, $M_s = -18$ °C, $A_s = 0$ °C, $A_f = 8$ °C, $E_A = 51$ GPa, $E_M = 13$ GPa, $c_M = 9.2$ MPa/°C, $c_A = 5.2$ MPa/°C, $b_{ms} = 20.89$ MPa, $b_{mf} = 21.7$ MPa, $b_{as} = 46.8$ MPa, $b_{af} = 209$ MPa, $\varepsilon_L^t = 7.9\%$, $\alpha_{ms}, \alpha_{mf}, \alpha_{as}, \alpha_{af}$ all taken as 1.0. According to the above test data, the comparison between the test curve and the model curve under different ambient temperature and loading and unloading rates is shown in Fig. 10.7. It can be seen from the comparison results that the model can simulate the stress-strain relationship of SMA in hyperelastic state more accurately.

10.1.1.4 Mechanical Model of Multi Dimensional Simplified SMA

Based on the linearization of hysteretic curve and one-dimensional constitutive model in Sect. 10.1.1.3, the multi-dimensional simplified model of SMA is obtained [3]. It is assumed that the Green-Lagrange strain tensor during the transformation of SMA can be expressed by the sum of the elastic strain tensor increment and the plastic strain increment:

$$d\varepsilon_{ij} = d\varepsilon_{ij}^e + d\varepsilon_{ij}^p \tag{10.38}$$

Among them, $d\varepsilon_{ij}$ is the total strain tensor increment; $d\varepsilon_{ij}^e$ is the elastic strain tensor increment; $d\varepsilon_{ij}^p$ is the plastic strain tensor increment, and:

$$d\varepsilon_{ij}^p = \Lambda_{ij} d\xi \tag{10.39}$$

Among them, Λ_{ij} is transformation tensor, when martensitic transformation and reverse transformation occur, they have different values, which will be explained in

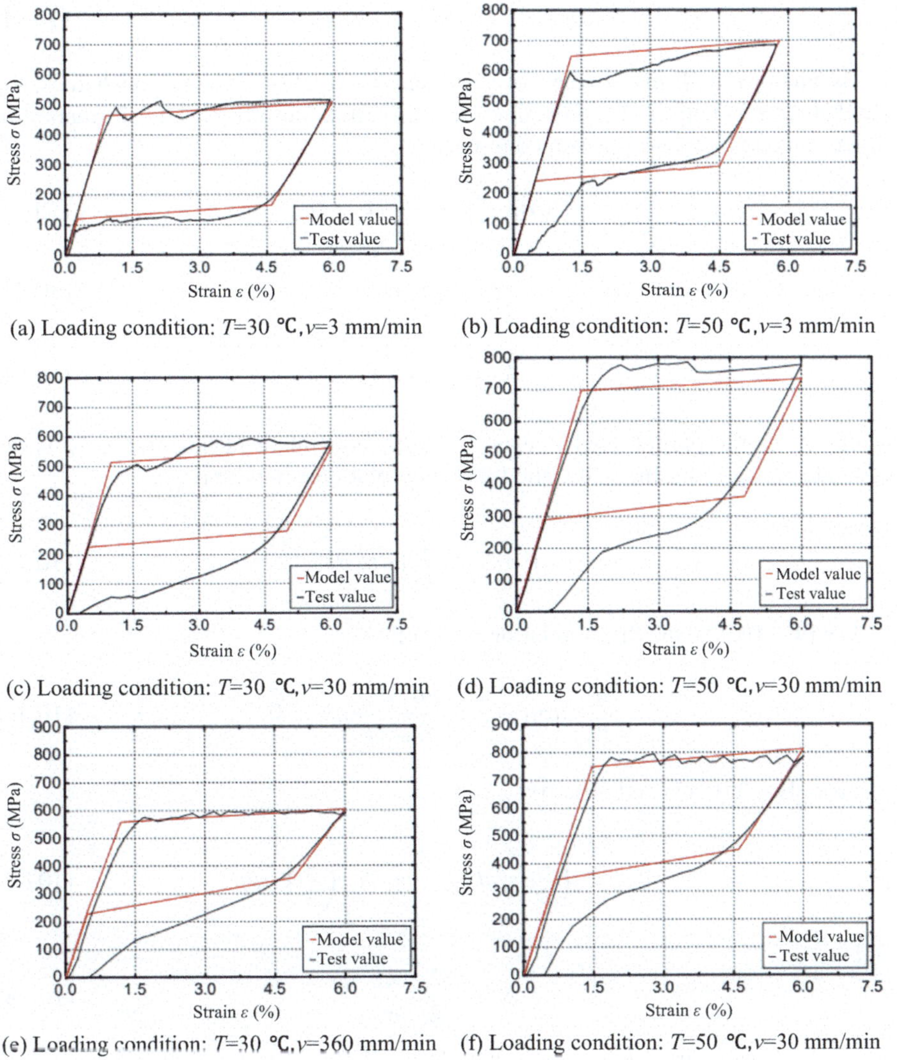

(a) Loading condition: T=30 °C, v=3 mm/min

(b) Loading condition: T=50 °C, v=3 mm/min

(c) Loading condition: T=30 °C, v=30 mm/min

(d) Loading condition: T=50 °C, v=30 mm/min

(e) Loading condition: T=30 °C, v=360 mm/min

(f) Loading condition: T=50 °C, v=30 mm/min

Fig. 10.7 Comparison between simulation results of simplified constitutive model and test values

detail later; ξ is the volume percentage of martensite. The relationship between the increment of Piola Kirchhoff stress tensor $d\sigma_{ij}$ and the increment of elastic strain tensor $d\varepsilon_{ij}^{e}$ can be expressed as follows:

$$d\sigma_{ij} = D_{ijkl}^{e} d\varepsilon_{kl}^{e} \tag{10.40}$$

where, D_{ijkl}^{e} is the elastic stiffness tensor. From Eqs. (10.38)–(10.40):

$$d\sigma_{ij} = D^e_{ijkl}(d\varepsilon_{kl} - \Lambda_{kl} d\xi) \tag{10.41}$$

By analogy with the concept of yield surface in elastic-plastic mechanics, the yield surface equation of martensitic transformation and inverse transformation of SMA induced by stress is assumed to be:

$$F(\sigma_{ij}, \varepsilon_p) = f - k = 0 \tag{10.42}$$

$$f = \frac{1}{2} s_{ij} s_{ij} \tag{10.43}$$

$$k = \frac{1}{3}\sigma_s^2(\bar{\varepsilon}_p) \tag{10.44}$$

where, s_{ij} is the partial stress tensor; σ_s is the equivalent plastic stress; $\bar{\varepsilon}_p$ is the equivalent plastic strain, $\bar{\varepsilon}_p$ has the following relationship with $d\varepsilon^p_{ij}$:

$$\bar{\varepsilon}_p = \int d\bar{\varepsilon}_p = \int \left(d\varepsilon^p_{ij} d\varepsilon^p_{ij}\right)^{\frac{1}{2}} \tag{10.45}$$

For Eq. (10.42), the differentiation is as follows:

$$\frac{\partial f}{\partial \sigma_{ij}} d\sigma_{ij} - \frac{2}{3}\sigma_s \frac{\partial \sigma_s}{\partial \bar{\varepsilon}_p} \frac{d\bar{\varepsilon}_p}{d\xi} d\xi = 0 \tag{10.46}$$

From Eqs. (10.45) and (10.46):

$$d\bar{\varepsilon}_p = \left(\frac{2}{3} d\varepsilon^p_{ij} d\varepsilon^p_{ij}\right)^{\frac{1}{2}} = \Lambda_{ij}\left(\frac{2}{3}\right)^{\frac{1}{2}} d\xi \tag{10.47}$$

and:

$$\Lambda_{ij} = \frac{3}{2}\Lambda \frac{s_{ij}}{\sigma_s} \tag{10.48}$$

Among them, Λ is the transformation constant, which has different values when martensitic transformation and reverse transformation. Substituting Eq. (2.19) into Eq. (2.18):

$$d\bar{\varepsilon}_p = \Lambda d\xi \tag{10.49}$$

Substituting Eqs. (10.39), (10.41) and (10.49) into Eq. (10.46), we can get:

$$\frac{\partial f}{\partial \sigma_{ij}} D^e_{ijkl}\left(d\varepsilon_{kl} - \frac{3}{2}\Lambda \frac{s_{ij}}{\sigma_s} d\xi\right) - \frac{2}{3}\sigma_s \frac{\partial \sigma_s}{\partial \bar{\varepsilon}_p} \Lambda d\xi = 0 \tag{10.50}$$

Then:

$$d\xi = \frac{\frac{\partial f}{\partial \sigma_{ij}} D^e_{ijkl} d\varepsilon_{kl}}{\frac{3}{2}\frac{\partial f}{\partial \sigma_{ij}} D^e_{ijkl} \Lambda \frac{s_{ij}}{\sigma_s} + \frac{2}{3}\sigma_s \frac{\partial \sigma_s}{\partial \bar{\varepsilon}_p}\Lambda} \tag{10.51}$$

Substituting Eq. (10.51) into Eq. (10.41), we can get:

$$\begin{aligned} d\sigma_{ij} &= D^e_{ijkl}\left(d\varepsilon_{kl} - \frac{3}{2}\Lambda\frac{s_{ij}}{\sigma_s} d\xi\right) \\ &= \left(D^e_{ijkl} - \frac{D^e_{ijpq} s_{pq} (\partial f/\partial \sigma_{rs}) D^e_{rskl}}{\frac{\partial f}{\partial \sigma_{ij}} D^e_{ijkl} s_{kl} + \frac{4}{9}\sigma_s^2 \frac{\partial \sigma_s}{\partial \bar{\varepsilon}_p}}\right) d\varepsilon_{kl} \end{aligned} \tag{10.52}$$

When $T \geq A_f$ and SMA is under the action of external force, the austenite will undergo elastic deformation. When the external force increases to a certain value, the austenite will start to induce martensitic transformation, that is, the austenite will gradually change into martensite under the action of external force. After unloading, because the martensite can not exist stably without external force at high temperature, the martensite will automatically gradually change into austenite, so SMA will return to the original state. These two processes can be described by Eq. (10.52).

In order to determine whether to use elastic or plastic constitutive relations, the loading and unloading criteria are defined as follows:

When $\xi = 0$, if $\frac{\partial f}{\partial \sigma_{ij}} d\sigma_{ij} > 0$, elastic loading; if $\frac{\partial f}{\partial \sigma_{ij}} d\sigma_{ij} < 0$, elastic unloading.
When $\xi = 1$, if $\frac{\partial f}{\partial \sigma_{ij}} d\sigma_{ij} > 0$, elastic loading; if $\frac{\partial f}{\partial \sigma_{ij}} d\sigma_{ij} < 0$, elastic unloading.
When $0 < \xi < 1$, if $F \neq 0$ and $\frac{\partial f}{\partial \sigma_{ij}} d\sigma_{ij} > 0$, elastic loading;

if $F \neq 0$ and $\frac{\partial f}{\partial \sigma_{ij}} d\sigma_{ij} < 0$, elastic unloading;
if $F = 0$ and $\frac{\partial f}{\partial \sigma_{ij}} d\sigma_{ij} > 0$, plastic loading, the martensitic transformation occurs;
if $F = 0$ and $\frac{\partial f}{\partial \sigma_{ij}} d\sigma_{ij} < 0$, plastic unloading, the reverse martensitic transformation occurs.

When the material is in elastic loading and unloading state, $d\sigma_{ij} = D^e_{ijkl} d\varepsilon_{kl}$; when the material is in plastic loading and unloading state, $d\sigma_{ij} = (D^e_{ijkl} - \frac{D^e_{ijpq} s_{pq} (\partial f/\partial \sigma_{rs}) D^e_{rskl}}{\frac{\partial f}{\partial \sigma_{ij}} D^e_{ijkl} s_{kl} + \frac{4}{9}\sigma_s^2 E^p(\xi_0)}) d\varepsilon_{kl}$. For SMA materials with different loading and unloading frequency, the value formula of plastic modulus $E^p(\xi_0)$ is different.

10.1.2 *Tension-Compression SMA Damper*

10.1.2.1 Basic Structure and Working Principle of Damper

The developed SMA tension-compression super elastic damper is mainly composed of cylinder, guide rod, piston and NiTi SMA super elastic wire cable, as shown in Fig. 10.8a. NiTi SMA super elastic wire cable is reliably connected with the front and rear cylinder heads by adjusting the screws and nuts. By adjusting the nuts on the screws, pre deformation can be applied to the SMA wire cable. The SMA wire cable is fixed by the slotted flat end fastening screw on the piston. The initial position of the single piston is in the middle of the cylinder block, and the initial position of the double piston is 1/3 of the cylinder block. Secure the piston to the guide rod with a nut. NiTi shape memory alloy wire cables are evenly arranged along the circumference of the cylinder block. Assemble shape memory alloy wire cable and corresponding parts into damper. Apply a little butter to the piston and the ring grooves of front and back cylinder heads. The assembly sequence is from inside to outside. Try to make the piston in the middle of the cylinder during the assembly. After each part is fixed, the shape memory alloy wire cable is pre deformed by adjusting the nuts on the screw at both ends.

The working principle of tension compression type super elastic shape memory alloy damper is to push the piston in the cylinder barrel to move back and forth under the action of earthquake (or wind) through the force on the guide rod, so that the NiTi shape memory alloy super elastic wire on both sides of the piston will produce tension or compression, produce super elastic damping, and provide full hysteresis loop and stable damping force, achieve the purpose of energy dissipation and vibration reduction. When the damper works, the guide rod is forced and the relative motion between the cylinder barrel and the piston will drive the shape memory alloy wire cable on both sides of the piston to work. In order to obtain better hysteretic cycle behavior, NiTi wire cable in shape memory golden damper is usually pre deformed to the mid point of the super elastic platform. In order to ensure that when the damper works, two groups of SMA wires can act together and stay in the super elastic platform spring all the time, so as to obtain the optimal damping performance, the maximum stroke of the damper is often limited. In the case of forward motion, the damper starts to load, and the cylinder barrel and piston move relative to each other. As shown in Fig. 10.8b, the first group of shape memory stage gold wire cable continues to be stretched from the initial position point M, and martensitic transformation continues to occur when the stress increases. When it reaches point D, the resistor begins to unload, the shape memory alloy wire cable begins to shrink, the stress decreases and reverse martensitic transformation occurs, finally back to point IV: while the second group of SMA wire cable gradually shrinks from the initial position point N, the stress decreases and reverse martensitic transformation continues to occurs, when it reaches point C, the damper begins to unload, SMA wire cable starts to be stretched, stress increases and the martensitic transformation starts, finally back to point M.

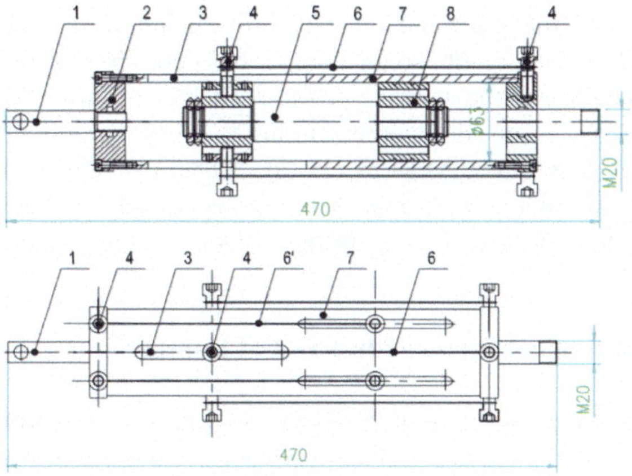

(a) Damper structure

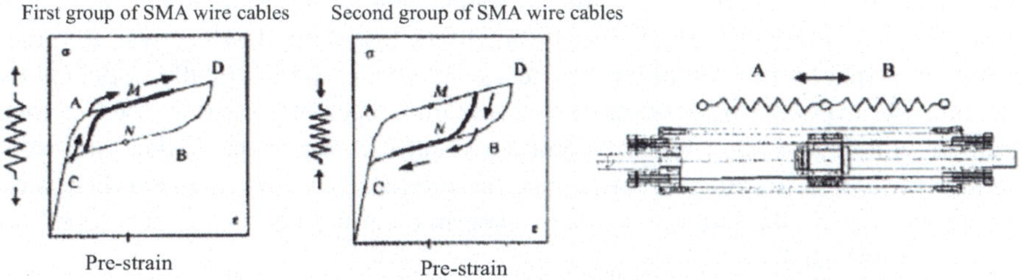

(b) Working principle

(c) Test equipment and test pieces

1-pull rod; 2 -cylinder head; 3 -slip joint; 4 -fixing bolt; 5 -shaft; 6 -SMA wire (first group); 6- SMA wire (second group) 7 -cylinder barrel; 8 -piston

Fig. 10.8 Tension compression SMA damper

In an ideal state, the upper and lower hyperelastic platforms of shape memory alloy have similar slope and length, so the damper has better hysteretic characteristics and stable restoring force in the working process. The two groups of shape memory cable are always in tension and act together, similar to the two groups of non-linear springs connected with each other. As shown in Fig. 10.8c, two sets of shape memory alloy wire cables are simulated by springs A and B respectively, and the two sets of springs are pretensioned to ensure that the spring is always in tension during the movement.

10.1.2.2 Mechanical Performance Test of Damper

The test was carried out on the INSTRON5569 universal electronic tensile machine in the testing center of Jiangyin Farson Group. The room temperature was 25 °C, and the test temperature was controlled by the temperature control box on the test machine, which was cooled by liquid nitrogen. During the experiment, the equal displacement loading was adopted according to the test method, which was automatically controlled by the computer. The test results were automatically collected by the computer every 50 ms. During the test, the influences of pre strain, loading rate, strain amplitude and other loading conditions on SMA super elastic damping were mainly considered. In order to investigate the influence of these load conditions on SMA super elastic damping, a series of experimental methods were developed, as shown in Table 10.3.

As shown in Fig. 10.8c, the effective working length of SMA wire cable in single piston damper is 150 cm, that of SMA wire cable in double piston damper is 250 cm, and four NiTi SMA wire cables are arranged along 1/4 circle.

The chemical composition of NiTi cable selected is: Ni49.8, Ti50.2 (atomic fraction%). Randomly selected samples are tested by DSC, and the phase transition temperature of base metal selected according to the American Standard F 2063-00 is shown in Table 10.4.

10.1.2.3 Test Result Analysis

Relevant literatures show that different hysteretic behavior can be obtained by applying initial pre deformation to NiTi wire in SMA damper. Figure 10.9 is the force displacement relationship curve of the tension compression type SMA super elastic damper when no pre deformation is applied, and the loading frequency is 30 mm/min. Among them, Fig. 10.9a is the force displacement relationship curve of single piston damper under six kinds of displacements: 1.5 mm, 3 mm, 4.5 mm, 6 mm, 7.5 mm and 9 mm. The strain of NiTi wire cable in damper is 1%, 2%, 3%, 4%, 5%, 6%; Fig. 10.9b is the force displacement relationship curve of single piston damper under 20 cycles with 9 mm displacement (the strain of NiTi wire cable in damper is 6%); Fig. 10.9c is the force displacement curve of the double piston damper under 5 kinds of displacements of 2.5 mm, 5 mm, 7.5 mm, 10 mm, and 12.5 mm, the strain of NiTi

Table 10.3 Test method

Serial number	Damper category	Experimental method	Loading speed
1	Single piston	Initial prestrain is applied at the end of NiTi wire cable, and each cycle is conducted under six displacements of 1.5 mm, 3 mm, 4.5 mm, 6 mm, 7.5 mm and 30 mm respectively	30
2	Single piston	Initial prestrain is applied at the end of NiTi wire cable for 10 cycles with displacement amplitude of 9 mm	30
3	Single piston	3% of initial prestrain is applied at the end of NiTi wire cable for 10 cycles with displacement amplitude of 4.5 mm	30,180,360
4	Single piston	At the end of NiTi wire cable, 3% prestrain is applied, and each cycle is conducted under six displacements of 1.5 mm, 3 mm, 4.5 mm, 6 mm, 7.5 mm and 30 mm respectively	3, 30, 180, 360
5	Double piston	Initial prestrain is applied at the end of NiTi wire cable, and each cycle is conducted under five displacements of 2.5 mm, 5 mm, 7.5 mm and 12.5 mm respectively	30
6	Double piston	Initial prestrain is applied at the end of NiTi wire cable for 10 cycles with displacement amplitude of 15 mm	30
7	Double piston	3% of initial prestrain is applied at the end of NiTi wire cable for 10 cycles with displacement amplitude of 7.5 mm	30, 180, 360
8	Double piston	At the end of NiTi wire cable, 3% prestrain is applied, and each cycle is conducted under five displacements of 2.5 mm, 5 mm, 7.5 mm and 12.5 mm respectively	3, 30, 180, 360

Table 10.4 Transformation temperature of base metal of NiTi SMA (°C)

M_f	M_s	A_a	A_f
−25	−18	0	8

wire cable in the damper is 1%, 2%, 3%, 4%, 5%; and Fig. 10.9d the force displacement curve of the double piston damper under 10 cycles of 15 mm displacement (the strain of NiTi wire cable in the damper is 6%). It can be seen that when NiTi wire cable in SMA damper is not applied with initial pre deformation, the force displacement curve of the damper is mainly distributed in the first and third quadrants, showing a small flag shape, which is similar to the restoring force model of SMA.

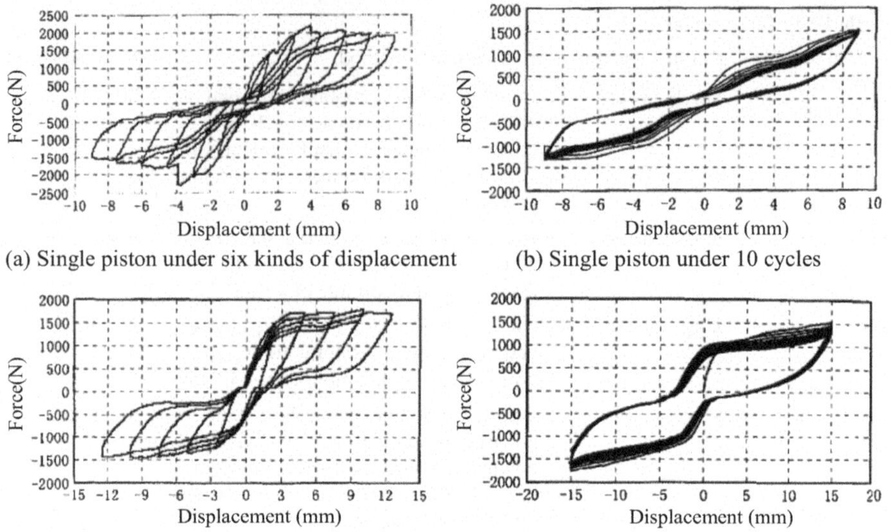

(a) Single piston under six kinds of displacement (b) Single piston under 10 cycles

(c) Double piston under six kinds of displacement (d) Double piston under 10 cycles

Fig. 10.9 Force displacement curve of damper without pre deformation

By applying initial pre deformation on the NiTi wire in the SMA damper to stay at the mid point of the super elastic platform, the damper can obtain a full hysteresis loop and a relatively stable damping force. Figure 10.10 is the force displacement

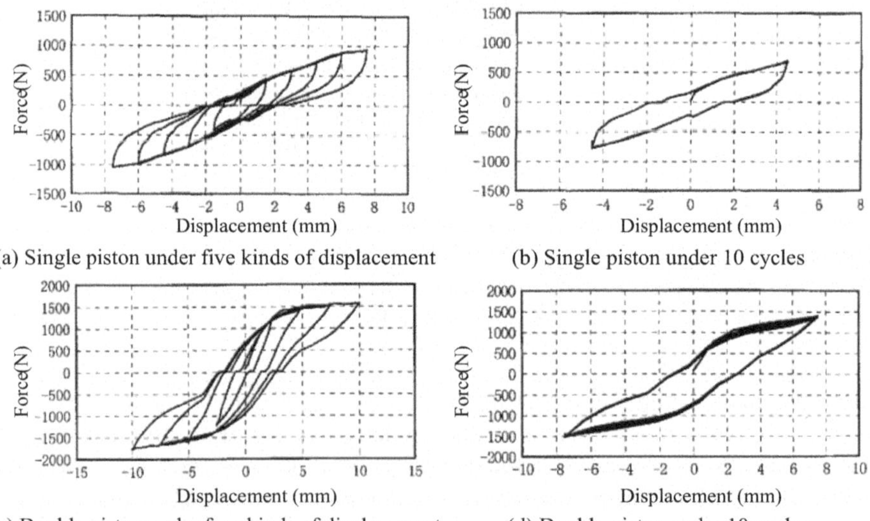

(a) Single piston under five kinds of displacement (b) Single piston under 10 cycles

(c) Double piston under four kinds of displacement (d) Double piston under 10 cycles

Fig. 10.10 Force displacement curve of damper under 3% pre deformation relationship curve of SMA tension compression type super elastic damper when 3% pre deformation is applied to NiTi wire cable in SMA damper, wherein Fig. 10.10a is the force displacement relationship curve of single piston damper under five displacements of 1.5 mm, 3 mm, 4.5 mm, 6 mm and 7.5 mm, and the strain of NiTi wire cable in damper is 1%, 2%, 3%, 4%, 5%, 6% respectively. The loading frequency is 30 mm/min; Fig. 10.10b the force displacement relation curve of single piston damper with 10 cycles of 4.5 mm displacement (the strain of NiTi wire cable in the damper is 3%), and the loading frequency is 30 mm/min; Fig. 10.10c the force displacement relation curve of double piston damper with 4 displacements of 2.5 mm, 5 mm, 7.5 mm and 10 mm, and the loading frequency is 30 mm/min, and the strain of NiTi wire cable in the damper is 1%, 2%, 3%, 4% respectively; Fig. 10.10d is the force displacement curve of the double piston damper after 10 cycles of 7.5 mm displacement (the strain of NiTi wire cable in the damper is 6%), and the loading frequency is 180 mm/min. It can be seen that when NiTi wire in SMA damper is applied with initial pre deformation to the mid point of the super elastic platform, the force displacement curve of the damper is distributed in four quadrants, which is rectangular, and the hysteresis loop is relatively full.

Figure 10.11 is the force displacement curve of single piston damper under the loading frequency of 30, 180 and 360 mm/min with initial 3% pre deformation on NiTi wire cable, and the displacement amplitude is 4.5 mm. It can be seen that with the increase of loading rate, the hysteresis loop becomes narrow and long, its surrounding area gradually decreases, the equivalent stiffness of damper slightly increases, and the equivalent damping ratio gradually decreases.

From the test results, it can be found that because the number and length of shape memory alloy wire cable of double piston damper is larger than that of single piston damper, the force displacement curve of double piston damper is fuller than that of single piston damper, and the damping performance is better than that of single piston damper. On the other hand, the connection part of SMA wire cable of double piston damping is larger than that of single piston damper, which is twice of that with single piston damper. Therefore, the probability of local stress concentration is greater than that of single piston damping. During the test, two retests were carried out because of broken wires. As shown in Fig. 10.11d, e, broken wires appeared when the damper worked at the maximum negative displacement. One case of broken wires suffered a rapid decreases in stress, stiffness and appeared hysteresis loop asymmetry.

10.1.3 Composite Friction SMA Damper

10.1.3.1 Basic Structure and Working Principle of Damper

Figure 10.12a shows the structure of the composite friction SMA damper. The damper includes left pull rod, front cover, outer cylinder, outer slide rod, inner slide rod, super elastic shape memory alloy wire, back cover, right pull rod and other components [4]. At the outer side of the inner sliding strip, four outer sliding strips are arranged

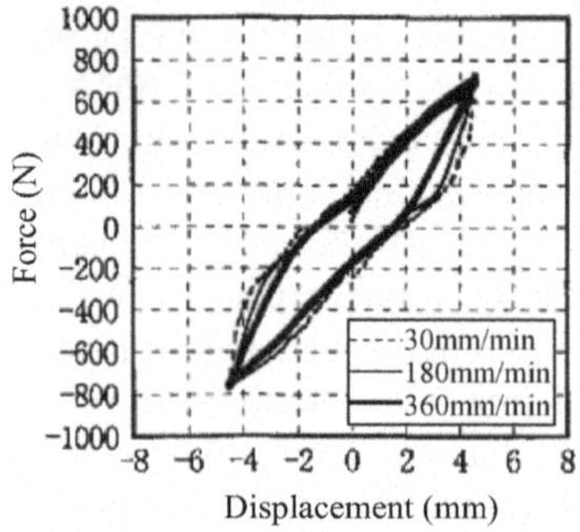

(a) Hysteretic curves under different loading frequencies

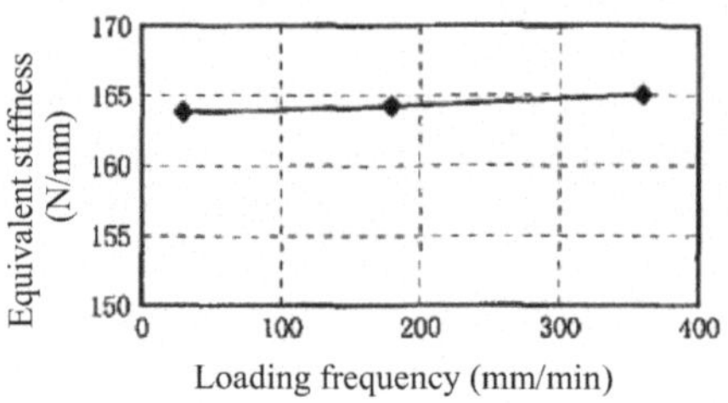

(b) Equivalent stiffness

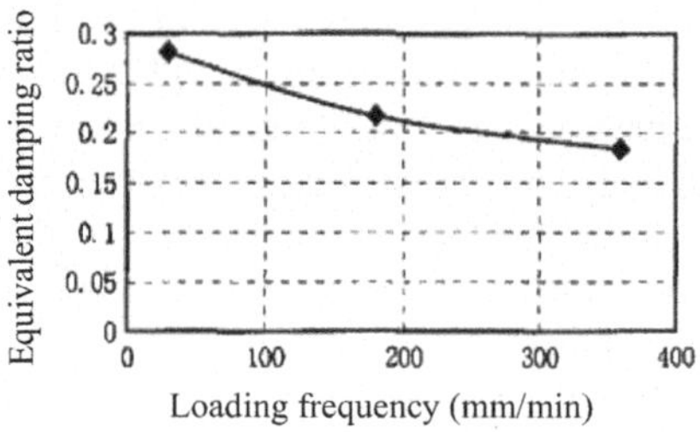

(c) Equivalent damping ratio

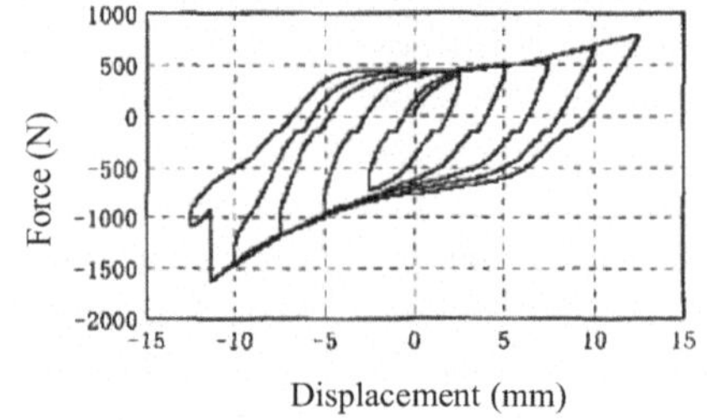

(d) Pre deformation of 3%, 3 mm/min, one wire was broken

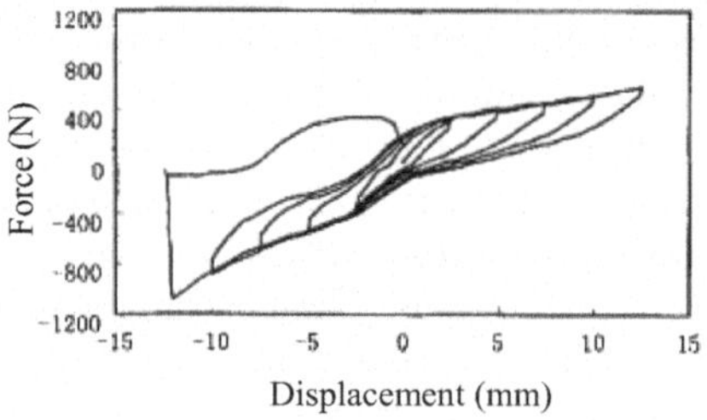

(e) Pre deformation of 3%, 180 mm/min, one wire was broken

Fig. 10.11 Influence of loading frequency on the performance of damper

symmetrically to the axis of the inner sliding strip. The outer side of the inner sliding strip and the inner side of the outer sliding strip are concave and convex waves along the axis direction, and the inner sliding strip and outer sliding strip are meshed by the concave and convex wave contact surfaces; the outer circumference of the whole composed of the inner sliding strip and the four outer sliding strips is wrapped and bound with super elastic shape memory alloy wires, and the whole body is put into the outer cylinder, and the internal structure of the damper is formed by the meshing of the super elastic shape memory alloy wire and the internal and external sliding strips, which is the main body for the damper to perform various functions; the two ends of the outer cylinder are respectively provided with front cover and back cover, which constitute the external structure of the damper, mainly for pulling and pressing,

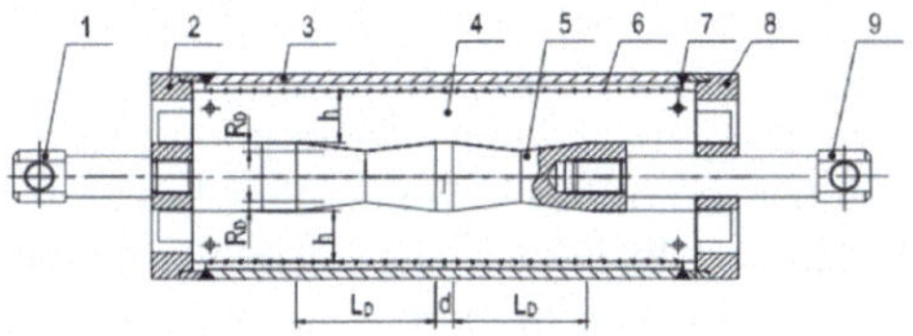

1-left pull rod; 2-front cover; 3-outer cylinder; 4-outer slide rod; 5-inner slide rod (shaft); 6-SMA wire; 7-chute locking rod; 8-rear cover; 9-right pull rod; 10-square slot

(a) Structure of damper

(b) Damper assembly

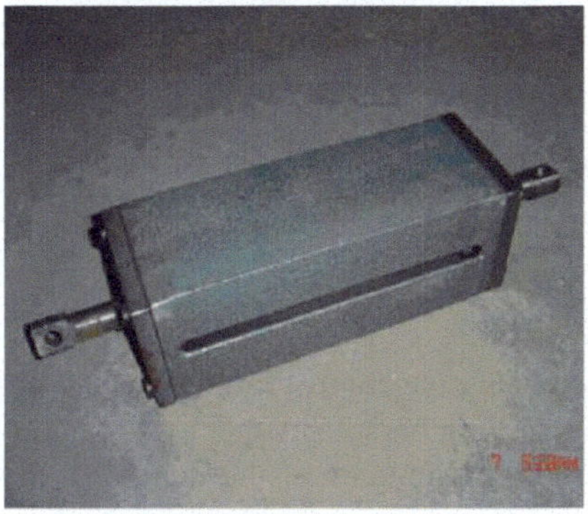

(c) Damper test piece

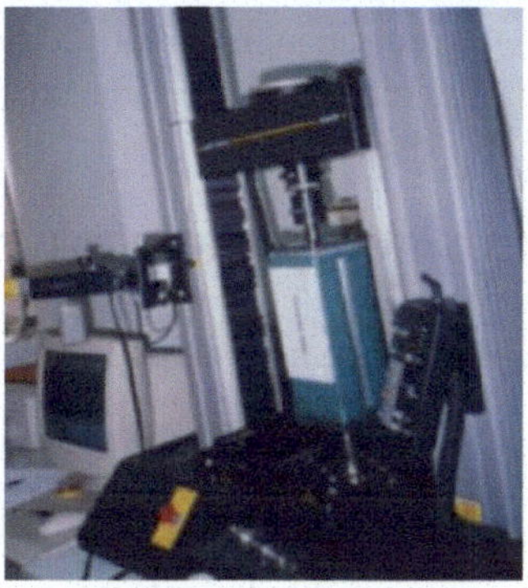

(d) Loading test of damper

Fig. 10.12 Composite friction SMA damper

protecting the internal structure and constraining the movement of the outer sliding strips; the left pull rod is fixed in the bolt hole of the front cover, the right pull rod is fixed in the bolt hole of the inner sliding strip through the guide rod hole on the rear cover; the two ends of the outer sliding strip are respectively located in the square slot on the front cover and the rear cover, and the two ends of the super elastic shape memory alloy wire are respectively fixed on the outer sliding strip. The super elastic shape memory alloy wire material restraining the inner and outer sliding strips can be wire or thin strip; the wavy concave and convex contact surfaces of the inner and outer sliding strips are rough contact surfaces, and corresponding friction materials can be set between the contact surfaces according to the vibration reduction requirements of the engineering structure.

The basic working principle of the composite friction SMA damper is: when the SMA damper is installed on the structure and the pull rods at both ends working together with the structure occur relative motion, the internal sliding strip is pushed and pulled, so that the internal sliding strip and the external sliding strip produce relative displacement along the axial and radial direction respectively. However, due to the restraint of the SMA wire to the radial displacement, the contact surface of the internal and external sliding strips produce mutual force, and the output force of damper is composed of the axial component of the force and the friction force on the contact surface. The energy dissipation mechanism of SMA damper depends on the relative displacement of the two ends of the pull rod. In the case of small displacement, the energy is mainly absorbed by the friction between the inner and outer sliding strips; in the case of large displacement, the vibration energy is dissipated by the super elastic damping of SMA wire and the friction of contact surface. Along with the energy consumption, the damper also produces a certain output control force to the engineering structure. Therefore, the damper is a SMA composite friction energy dissipation damper, its performance is not only related to the characteristics of SMA, but also to the nature of the contact surface. In addition, due to the large change of elastic modulus of SMA in different phase transformation state, the stiffness of the damper also changes greatly. Therefore, the SMA composite friction damper is a kind of damper with variable stiffness.

10.1.3.2 Mechanical Model of Damper

(1) **Basic assumptions**

The super elastic SMA damper mainly uses the passive control force provided by the interaction between the internal and external sliding bars, the super elastic damping of SMA and the friction between the internal components of the damper to dissipate energy and reduce the structural vibration. In order to establish the mechanical analysis model of the damper, the following basic assumptions are made:

1. the temperature in the outer cylinder of the damper is always above the starting temperature of austenite transformation, so as to ensure that SMA wire is in hyperelastic working state;
2. the maximum displacement stroke of damper should work under the maximum recoverable strain state of super elastic SMA;
3. in addition to considering the friction between internal and external sliding strips, the friction between other components can be ignored;
4. no other friction materials are installed between the inner and outer sliding strips, and the friction coefficient is the sliding friction coefficient between the inner and outer sliding strips;
5. the influence of loading frequency on the output force of damper is ignored.

(2) **Calculation model of super elastic SMA damper**

1. **Relationship between strain of SMA and displacement of damper**

According to the geometric relationship shown in Fig. 10.13a, it can be concluded that the length of SMA wire around the outer sliding strip is:

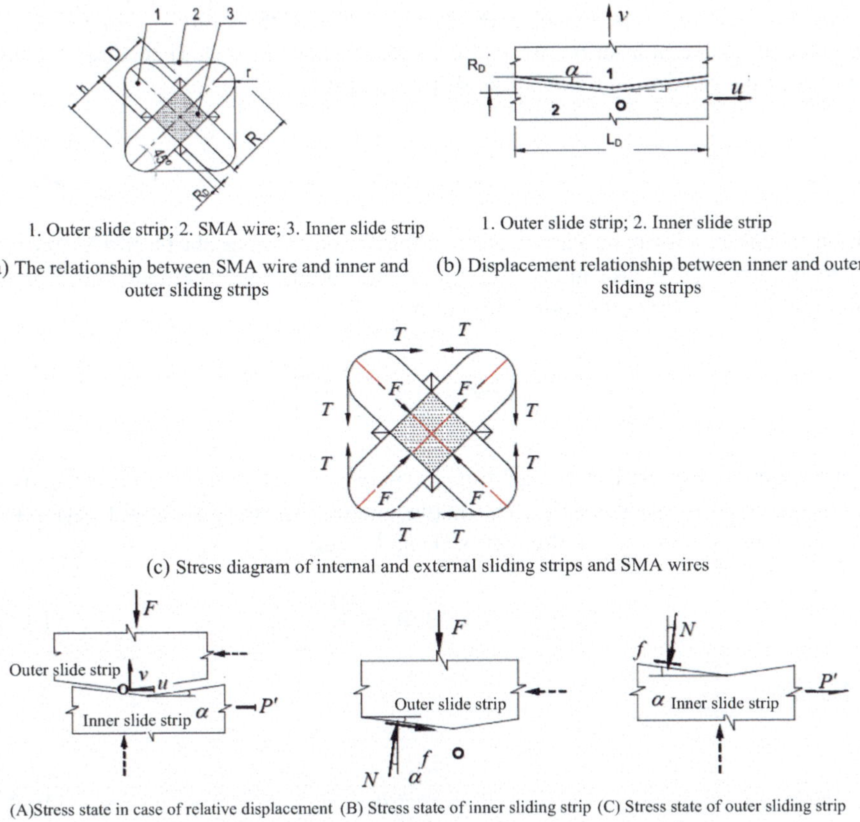

(a) The relationship between SMA wire and inner and outer sliding strips

(b) Displacement relationship between inner and outer sliding strips

(c) Stress diagram of internal and external sliding strips and SMA wires

(A)Stress state in case of relative displacement (B) Stress state of inner sliding strip (C) Stress state of outer sliding strip

(D) Interaction analysis of internal and external sliding strips ($u \geq 0$)

Fig. 10.13 Internal deformation principle of damper

$$S = 4\sqrt{2}(R - r) + 2\pi r \tag{10.53}$$

where, S is the length of a circle of SMA wire around the outer sliding strip; R is the radius of the inner circle when the inner and outer sliding strips are fully meshed; r is the circle radius of the wire surface of the outer sliding strip.

When the outer sliding strip moves v along the radial direction (as shown in Fig. 5.2), the elongation ΔS of SMA wire in one circle is:

$$\Delta S = 4\sqrt{2}v \tag{10.54}$$

Among them, ΔS is the elongation of SMA wire when it is around the outer sliding strip with one circle, and v is the radial sliding displacement of the outer sliding strip relative to the inner sliding strip.

According to the hyperelastic properties of SMA, if the strain ε_{rec} of SMA is in the range of recoverable strain, its residual strain can be ignored. Therefore, ΔS shall not exceed the maximum recoverable displacement $\varepsilon_{rec} S$:

$$\Delta S \leq \varepsilon_{\text{rec}} S \tag{10.55}$$

Among them, ε_{rec} is the maximum recoverable strain of super elastic SMA. According to the above three formulas, the maximum sliding displacement $\nu_{\max}$ of the outer sliding strip along the radial direction is:

$$\nu_{\max} = \left[(R - r) + \frac{\sqrt{2}}{4} \pi r \right] \varepsilon_{rec} \tag{10.56}$$

According to Fig. 10.13b, the relationship between the radial sliding displacement ν of the outer sliding strip and the axial displacement of the inner sliding strip (relative displacement of both ends of the damper) can be obtained:

$$\nu = utga, \quad tga = \frac{2R_D}{L_D} \tag{10.57}$$

Namely:

$$\nu = \frac{2R_D}{L_D} u \tag{10.58}$$

Among them, L_D is the length of the groove, i.e. the maximum reciprocating stroke when the damper works; R_D is the depth of the groove. To ensure that SMA works within the recoverable strain range, R_D should meet the following requirements:

$$R_D \leq \left[(R - r) + \frac{\sqrt{2}}{4} \pi r \right] \varepsilon_{rec} \tag{10.59}$$

where, u is the axial displacement of the inner sliding strip, i.e. the relative displacement of both ends of the damper; a refers to the slope angle of the wavy convex part of the outer sliding strip.

Because the strain ε of SMA wire is:

$$\varepsilon = \frac{\Delta S}{S} \tag{10.60}$$

where, ε is the strain of SMA wire. The relationship between the strain ε of SMA wire and the relative displacement u of both ends of damper is as follows:

$$\varepsilon = \frac{4\sqrt{2}R_D}{2\sqrt{2}(R-r)+\pi r} \cdot \frac{1}{L_D}u \tag{10.61}$$

2. **Relationship between the force and displacement of SMA damper**

Figure 10.13c shows the interaction between SMA wire and inner and outer sliding strips. The total tensile force T of SMA wire is:

$$T = \sigma(\varepsilon) \cdot A \tag{10.62}$$

Among them, T is the total tensile force of SMA wire; A is the total area of SMA wire; $\sigma(\varepsilon)$ is the tensile stress of SMA wire. Further, we can get:

$$F = \sqrt{2}\sigma(\varepsilon) \cdot A \tag{10.63}$$

Among them, F is the interaction between the internal and external sliding strips.

Figure 10.13d shows the forces of inner and outer sliding strips when the damper sliding strip moves to the right ($u \geq 0$) with respect to the O-point of the external sliding strip and the SMA wire produces binding force F on the external sliding strip. By using the projection equation of the outer sliding strip in the vertical direction, we can get:

$$N \cdot \cos a = f \cdot sign(\Delta|u|) \cdot \sin a + F \tag{10.64}$$

where, N is the normal interaction force on the contact surface of the inner and outer sliding strips; $\Delta|u|$ shows the increment of the absolute value of the relative displacement u of the inner and outer sliding strips along the axial direction; f is the friction force on the contact surface: $f = \mu_k \cdot N$. μ_k is the sliding friction coefficient; $sign$ is a symbolic function.

By substituting f into Eq. (10.64), we can get:

$$N = \frac{F}{\cos a - \mu_k \cdot \mathrm{sign}(\Delta|u|) \cdot \sin a} \tag{10.65}$$

By using the projection equation of the inner sliding strip in the horizontal direction, the output force P' produced by a single outer sliding strip can be obtained:

$$P' = N \sin a + f \cdot sign(\Delta|u|) \cdot \cos a \tag{10.66}$$

Among them, P' is the output force generated by the damper when a single outer sliding strip interacts with an inner sliding strip.

Namely, when $u \geq 0$,

$$P' = N[\sin a + \mu_k \cdot sign(\Delta|u|) \cdot \cos a] \tag{10.67}$$

Similarly, when the inner sliding strip of the damper moves to the left relative to the O-point of the external sliding strip ($u \leq 0$), P' is obatained as follows:

$$P' = -N \cdot [\sin a + \mu_k \cdot sign(\Delta|u|) \cdot \cos a] \tag{10.68}$$

Combining the above two formulas, in the case of $u \geq 0$ or $u < 0$, P' is:

$$P' = N \cdot sign(u) \cdot [\sin a + \mu_k \cdot sign(\Delta|u|) \cdot \cos a] \tag{10.69}$$

From Eqs. (10.65) and (10.69), the relationship between P' and F can be obtained as follows:

$$P' = F \cdot \text{sign}(u) \frac{\sin a + \mu_k \cdot sign(\Delta|u|) \cdot \cos a}{\cos a - \mu_k \cdot sign(\Delta|u|) \cdot \sin a} \tag{10.70}$$

Namely:

$$P' = F \cdot sign(u) \cdot \frac{tga + \mu_k sign(\Delta|u|)}{1 - \mu_k \cdot sign(\Delta|u|) \cdot tga} \tag{10.71}$$

Considering that the damper is equipped with four outer sliding bars, the relationship between the total output force P and F is as follows:

$$P = 4F \cdot sign(u) \frac{tga + \mu_k \cdot sign(\Delta|u|)}{1 - \mu_k \cdot sign(\Delta|u|) \cdot tga} \tag{10.72}$$

where, P is the total output force of the super elastic SMA damper.

According to Eqs. (10.57), (10.63) and (10.72), the relationship between the output force P and displacement u of SMA damper is as follows:

$$P = 4\sqrt{2} \cdot \sigma(\varepsilon) \cdot A \cdot sign(u) \frac{2R_D + \mu_k \cdot L_D \cdot sign(\Delta|u|)}{L_D - 2\mu_k \cdot R_D \cdot sign(\Delta|u|)} \tag{10.73}$$

The above formula is the mechanical calculation model of SMA composite friction damper.

10.1.3.3 Mechanical Performance Test of Damper

Due to the different components of SMA alloy materials and the different treatment processes, the mechanical properties of SMA alloy are also different. The equal atomic ratio NiTi alloy wire produced by Jiangyin Farson Company is selected. Its diameter is 1.00 mm, and its chemical composition is Ni49.8, Ti50.2 (atomic fraction,

Table 10.5 Loading method

Loading frequency (Hz)	Loading amplitude (mm)
0.02	2.0, 4.0, 8.0, 12.0
0.2	2.0, 4.0, 8.0
0.5	2.0, 4.0
1	2

%). The main paramenters of physical property of SMA wire are the same as above. Design damper test pieces are shown in Fig. 10.12b, c.

The sinusoidal excitation method is adopted in the test, and the loading system of Instron testing machine is controlled according to the input displacement $u = u_0 \sin \omega t$, which is the excitation amplitude. Test temperature: 25 °C; load frequency: 0.02–1.0 Hz; displacement amplitude range: ±1 mm, soil: (1–6%)L_0 (Note: L_0 is the effective working length around a circle of SMA wire, $L_0 = 400$ mm), 20 cycles are recorded under each working condition. The loading method is shown in Table 10.5, and the test device is shown in Fig. 10.12d. The basic test steps are as follows:

(1) installation and positioning of damper;
(2) pre cycle training of damper: after 20 cycles of stress-strain cycle with loading frequency of 0.5 Hz and displacement amplitude of 3%, adjust the damper to eliminate the loosening of damper caused by residual strain of SMA. Adjust and calibrate the loading system to test the dynamic characteristics of the damper;
(3) the sinusoidal force of a certain frequency is applied by the damper, and the input displacement amplitude is controlled step by step from small to large as shown in Table 10.5, and the corresponding force, displacement and corresponding time of 20 cycles are recorded.
(4) change the loading frequency step by step according to 0.02, 0.2, 0.5 and 1 Hz to apply sinusoidal force, input the displacement value of (3) respectively for each frequency.

10.1.3.4 Test Result Analysis

Figure 10.14 shows the comparison of model curve and test curve of load displacement curve of SMA damper under different displacement amplitude. The displacement control amplitudes of the original test are 4 mm, 8 mm and 12 mm respectively, but due to the errors of 0.5 mm (positive load) and 1.0 mm (negative load) in the manufacture and installation of the damper, the damper has certain initial displacement. When the proposed model is used for simulation, the displacement amplitudes are all reduced by 0.5 mm. It can be seen from Fig. 10.14 that there are certain differences between the model curve and the experimental curve, which is mainly caused by the gap between concave and convex mating surfaces caused by the manufacturing errors

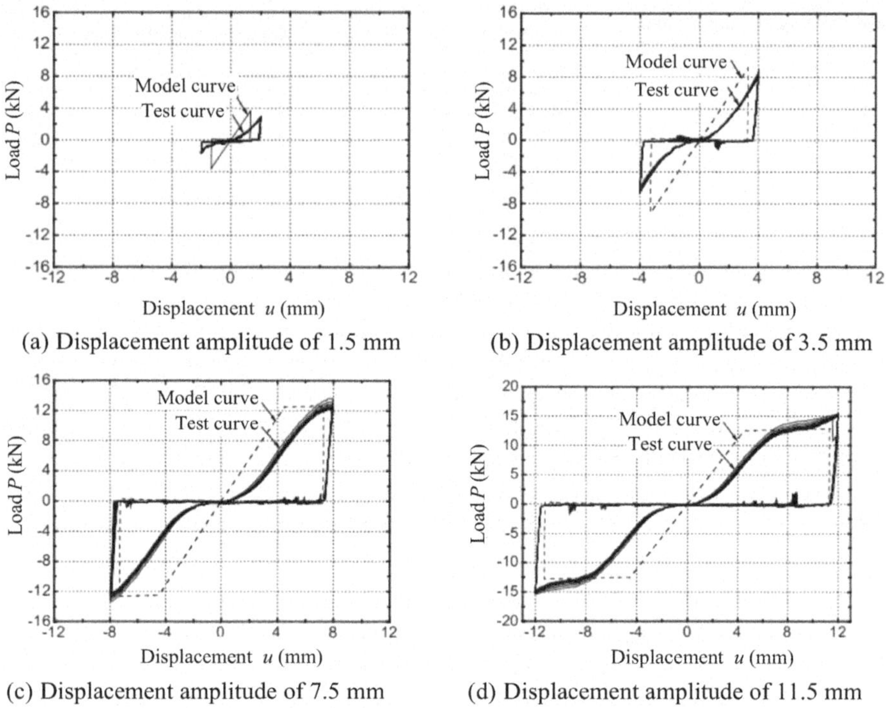

Fig. 10.14 Model curve and test curve of load-displacement of damper with different displacement amplitudes

of the inner and outer sliding strips of the damper, and the insufficient tension and uniformity of the SMA wire during winding, etc., but on the whole, the calculation model can basically reflect the mechanical properties of the SMA damper.

10.2 Foam Aluminum Composite Damper

Since the advent of foam aluminum in 1940s, due to its many excellent properties, foamed aluminum has attracted wide attention and research from academia and engineering circles at home and abroad, and has been widely applied in many fields, such as automobile, aerospace and packaging. However, the application of foam aluminum and its composite materials in civil engineering is less. It is still a relatively new material for the majority of civil engineering scholars. Although foam aluminum has been applied in the field of construction in recent years, it is not widely used. It is mainly used for outdoor decorative curtain walls, interior decoration walls, ceilings, sliding doors, mobile partitions, flooring and so on. In addition, foam aluminum plates are sometimes installed on both sides of the expressway to reduce noise and noise.

The author's team has carried out a systematic research on the preparation of foam aluminum composite damping materials and the development of shock absorbers [5, 6].

10.2.1 *Preparation of Foam Aluminum Composite Damping Material*

10.2.1.1 Obtaining of Foam Aluminum

Part of the used open cell foam aluminum with is self-made, partly purchased from Suzhou Christie's Foam Pioneer Metals Corporation. The self-made foam aluminum was prepared by solid infiltration casting method and using sodium chloride as solid pore forming grain. The whole preparation process includes the following steps (Fig. 10.15):

(a) Sieve salt (b) Salt loading and vibration compaction (c) Negative pressure seepage (d) Dissolved salt grains

Fig. 10.15 Preparation of open cell foam aluminum

(1) **Sieve salt**

(a) Salt baking. Pour the industrial salt into the crucible and put it into the heating furnace for heating. First heat to 200 °C and keep warm for 2 h, then heat to 300 °C and keep warm for 2 h. Then take out the crucible and let it cool naturally to room temperature.
(b) Salt screening. The salt is screened by a sieve with a diameter of 5, 4, 3, 2, 1.25 and 0.85 mm. The screened salt is then packed into different bags for subsequent foam aluminum production, so that the aperture of foam aluminum prepared are in different ranges.

(2) **Salt loading**

(a) Mold: use the stainless-steel mold with diameter of 14 cm and height of 20 cm.
(b) Lay a layer of fine wire mesh with the hole diameter less than 0.5 mm at the bottom of the mold, and fix the fine wire mesh with bolts and circular plates with holes. Sand the lower edge of the mold with sandpaper to ensure that there is no air leakage when pouring aluminum.
(c) Lay a layer of fine salt with the particle size of 0.5–0.9 mm on the fine wire net to ensure that the poured aluminum liquid will not percolate down too much during air extraction.
(d) Lay the salt with a certain diameter screened in front on the fine salt, add the salt while vibrating it with a machine, and stop when the salt height is installed to 10 cm.
(e) Lay a layer of fine salt with a particle size of 0.5–0.9 mm on the dense salt, and then lay a layer of wire mesh with a slightly larger hole diameter on it (to prevent the formation of eddy current during the pouring of aluminum liquid, which is adverse to the preparation). So far, the salt loading work is completed.

(3) **Preparation of foam aluminum by seepage method**

(a) Heating of salt. Put the mold with salt into the resistance furnace and heat it up to 465 °C gradually.
(b) Obtaining of aluminum liquid. Place the aluminum ingot in the furnace, heat it in the resistance furnace until it melts, and then gradually heat it to 765 °C.
(c) Use the air extraction pipe to pump air until the negative pressure is 78 kPa and stop and keep it. Place a fine wire mesh at the end of the air extraction pipe, and lay a salt with a particle size of 1.25–2 mm on it. Apply silicone oil on the connecting plate of the air extraction pipe end.
(d) Take out the salt mold from the resistance furnace, place it on the connecting plate of the air extraction port, and align it; take out the crucible containing the aluminum liquid from the resistance furnace, and pour the aluminum liquid into the mold. Then, open the air extraction valve to extract air, so that the aluminum liquid can penetrate the gap of the salt particles. Several times of air extraction ensures that the aluminum liquid can percolate down to the maximum extent.

(e) At this point, the use of percolation method for the preparation of foam aluminum has basically been completed. After waiting for a certain time until the mold is cooled to room temperature, the foam aluminum with salt grains can be removed and washed for a certain time, so that the aluminum foam can be obtained by dissolving the salt grains. So far, the work of preparing porous foam aluminum by percolation method has been completed.

The porosity of the foamed aluminum prepared by this method is between 66 and 75%, and the pore size is between 1.25 and 5.5 mm. This part of foam aluminum is mainly used in the exploratory experiment of the research group.

10.2.1.2 Acquisition of AF/PU Composite Material

(1) **Selection and preparation of polyurethane materials**

Polyurethane is a general designation of macromolecular compounds containing repeated carbamate groups (–NHCOO–) in the main chain, which is generally formed by the gradual addition and polymerization of polyisocyanate, oligomer polyol and glycol or diamine chain extender, with the characteristics of wear resistance, tear resistance, chemical corrosion resistance, good adhesion with other materials, high elasticity and strong vibration absorption. Polyurethane has good viscoelasticity, large damping coefficient, good vibration and noise reduction performance; at the same time, because polyurethane has good bonding performance with metal, good corrosion resistance, high strength, good elasticity and toughness, plus it is liquid before curing, can adapt to different structural shapes, and the curing time is moderate, so it is easy to pour and form. It is an ideal material for filling foam aluminum holes and preparing foamed aluminum composites.

Polyurethane is a typical multi block copolymer. Its molecular chain is composed of soft segments with glass transition temperature lower than room temperature and hard segments with glass transition temperature higher than room temperature. Generally, the proportion of soft segments is 50–90%, while that of hard segments is only 10–50%. The soft segment is mainly Polyacyl polyol or Polybutyric polyol, which has good flexibility and endows polyurethane with excellent elastic properties; the hard segment is composed of polyisocyanate and small molecular chain extender, which is very rigid and plays the role of strengthening filling and physical crosslinking. In general, polyurethane is prepared by adding polyisocyanate and oligomer polyol to form prepolymer (material A), then adding chain extender (material B) to carry out chain extension reaction, that is to say, polyurethane materials are generally obtained through the synthesis reaction of prepolymer and chain extension reaction of prepolymer, and in some cases, the crosslinking reaction of polyurethane at a certain temperature is required to obtain the final required polyurethane materials. Figure 10.16 gives the preparation process of polyurethane: at 80 °C, mix material A and B according to the mass ratio of 10:1, adjust the pressure of polyurethane pouring machine to make the liquid polyurethane flow out; then pour the liquid polyurethane into the polyurethane elastomer centrifugal molding machine, set the temperature of

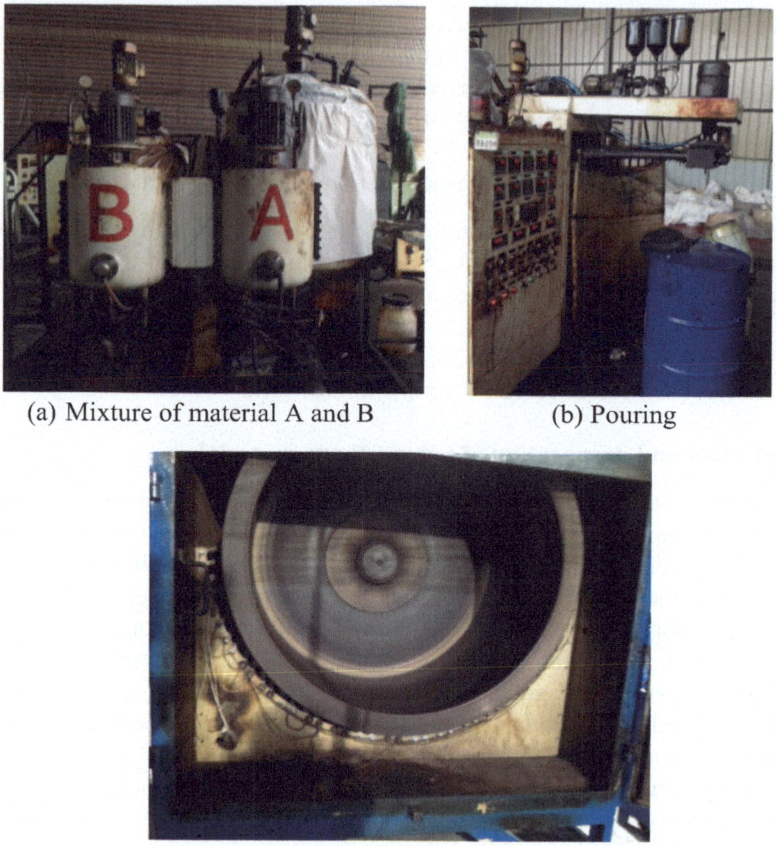

(a) Mixture of material A and B (b) Pouring

(c) Crosslinking curing

Fig. 10.16 Polyurethane preparation process

the centrifuge to 120 °C, and take it out after 4 h. At this time, the crosslinking and curing of polyurethane are completed, and the polyurethane elastomer material is obtained after demoulding. So far, the preparation process of polyurethane has been completed.

(2) **Filling technology and quality control of polyurethane material**

In order to prepare AF/PU composites, polyurethane should be filled into the pores of foam aluminum. In the experiment, a certain size mold was used to complete the operation. First, the foam aluminum sample was placed in the mold and positioned, then the polyurethane was poured into the mold, and the natural flow of the polyurethane was used to infiltrate into the pores of the foam aluminum. In order to obtain better filling effect, setting the size of the pouring mold is slightly larger than the size of the foam aluminum specimen, so that when pouring, the polyurethane can permeate into the pores from the top surface of the foam aluminum specimen and penetrate from the side of the specimen. When pouring polyurethane into the mold, it needs pouring several times and slowly pouring each time to achieve the best filling effect. Before and after filling, the quality of the foam aluminum specimens and the AF/PU composite specimens were respectively weighed, and the quality of the polyurethane filled in the pores of the aluminum foam was calculated to reflect the filling effect of polyurethane and the quality of the AF/PU composites were controlled.

(3) **Preparation of AF/PU composite material**

The foam aluminum is processed into a cube sample with a bottom dimension of 30 mm × 30 mm and a height of 25 mm by means of wire cutting, as can be seen in Fig. 10.17a. Immerse the casting mold into the mold release agent for a moment, after

(a) Cutting formed foam aluminum specimen

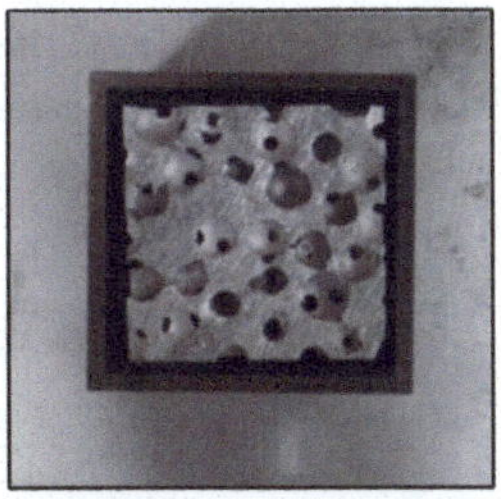

(b) Foam aluminum is placed in the mold

(c) Sample pouring in centrifuge

(d) AF/PU compositematerialsample

Fig. 10.17 Preparation of AF/PU composite material

taking out the mold, the foam aluminum sample is placed in the mold, and position and fix it as shown in Fig. 10.17b. Then the mold and the test piece are placed on the flat plate and put into the polyurethane elastomer centrifugal molding machine.

When the temperature is 80 °C, mix material A and material B according to the ratio of 10:1 to obtain liquid polyurethane, adjust the pressure of polyurethane pouring machine to make the polyurethane flow out, and use a container to catch it at the lower outlet. The polyurethane is then poured into the mold to permeate into the pores of the foam aluminum. When the polyurethane covers on the foam aluminum sample with a certain thickness, stop pouring, and the top surface of the foam aluminum is covered by the flat to ensure that the top surface of the sample is smooth, as shown in Fig. 10.17c. Set the temperature of polyurethane elastomer centrifugal molding machine to 120 °C, take it out after 4 h, at this time, the crosslinking and curing of polyurethane is completed. Take out the sample in the mold, cut off and polish off the surplus polyurethane outside the sample to obtain AF/PU composite material. The composite material sample is shown in Fig. 10.17d.

10.2.2 Damping Mechanism and Characteristics of AF/PU Composite Material

10.2.2.1 Factors Affecting Mechanical Properties

(1) **Relative density or porosity of aluminum foam**

The relative density of foam aluminum is the ratio of the apparent density of foam aluminum to the density of the parent compact aluminum. The porosity is the ratio of the total volume of the holes in the foam aluminum to the total volume of foam aluminum. The two have the same effect in reflecting the structure of the foam aluminum, and their definitions are shown as Eqs. (10.74) and (10.75) respectively. Among them, ρ_{AF} is the apparent density of foam aluminum, and, ρ_{Al} is the density of solid aluminum.

$$\text{Definition of relative density:}\quad \bar{\rho} = \frac{\rho_{AF}}{\rho_{Al}} \tag{10.74}$$

$$\text{Definition of porosity:}\quad \Phi = 1 - \bar{\rho} = 1 - \frac{\rho_{AF}}{\rho_{Al}} \tag{10.75}$$

Relative density is the most easily measured and easily obtained basic parameter for foamed aluminum. It is also the key factor determining the physical and mechanical properties of foam aluminum, such as thermal conductivity, electrical conductivity, acoustic performance and compressive strength. The relative density of aluminum foam is a very important structural characteristic parameter, whether open or closed, and its change will have a great influence on the mechanical properties of foam aluminum. High porosity foam aluminum has good sound absorption and damping properties, and high relative density foam aluminum has better mechanical properties. When the relative density of foam aluminum in AF/PU composites changes, the properties of foam aluminum change, and the properties of the composites change accordingly. Therefore, the relative density of foam aluminum is an important factor affecting the properties of composite materials.

(2) **Pore size of foam aluminum**

Pore size is one of the basic structural parameters of foam aluminum, and can be characterized by the average diameter of pores in any cross-section of foam aluminum. There is no unified conclusion about the influence of pore size on the mechanical properties of foam metal. The mechanical properties of two closed cell foam aluminum with the same density and different pore sizes under quasi-static conditions were studied by Miyoshli. It was found that the yield strength of foam aluminum with smaller pore size is higher than that of larger pore size. It is believed that the cell wall of small pore foam aluminum is smaller and thinner than that of large pore foam aluminum, and the thickness distribution is uniform, and the specific surface area is larger. The bearing capacity of cell wall is increased, the yield stress of material is increased as a whole. The research of Nieh shows that the pore size has little effect on the yield stress.

The conclusions of domestic researchers are also inconsistent. Some scholars have found that the same relative density and different pore size will lead to different dynamic mechanical properties of foam aluminum by testing: small pore size specimens have obvious sensitivity to strain rate and has high yield strength; large pore size specimens are insensitive to strain rate, yield strength and yield platform are low. Some scholars have pointed out that the yield strength, flow stress and compacting stress of open cell foam aluminum will increase along with the increase of pore size, and the strain hardening phenomenon is more significant. With the increase of pore size, the compressive energy absorption of open cell foam aluminum increases significantly, and the energy absorption rate remains unchanged. Therefore, the influence of the pore size of foam aluminum on the properties of composites remains to be studied.

(3) **Polyurethane content**

The polyurethane is filled into the pores of the foam aluminum and has a certain amount outside. As a component of the composite, the content of polyurethane will have a certain impact on the mechanical properties of the composite. The calculation method of polyurethane content in AF/PU composite is shown in Eq. (10.76). Among them, V_{total} and V_{PU} are respectively the total volume of the composite and the volume of polyurethane filled inside the composite; m_{PU} is the mass of polyurethane filled inside the composite and ρ_{PU} is the density of polyurethane.

$$PU\% = \frac{V_{PU}}{V_{total}} = \frac{m_{PU}/\rho_{PU}}{V_{total}} \tag{10.76}$$

In addition, because the preparation method is not perfect, the pores of the foam aluminum are not completely filled by polyurethane. The polyurethane filling rate of the composite specimens with the same polyurethane content is not necessarily the same, and the properties of the composites will also be different. The polyurethane filling rate is defined in Eq. (10.77).

$$(PU)\% = \frac{V_{PU}}{V_{void}} = \frac{m_{PU}/\rho_{PU}}{V_{void}} \tag{10.77}$$

Among them, V_{void} is the pore phase volume of foam aluminum.

(4) **Components of polyurethane**

It has been known from the previous definition that polyurethane is a general term for a class of materials, not a specific material. There are many kinds of synthetic raw materials, as long as changing a raw material, or changing the matching ratio of reactant, there will be differences in the structure of cross-linking, and the final polyurethane materials are also different, its physical and mechanical properties will change in a considerable range.

Therefore, polyurethane materials with different soft hardness, temperature resistance, chemical resistance and other properties can be prepared by changing the composition of raw materials and formula proportion, which will affect the mechanical properties of AF/PU composite. Therefore, the composition of polyurethane material is also an important factor affecting the mechanical properties of the composite, which needs to be further studied.

In other aspects, such as loading rate, temperature and so on will have a certain impact on the mechanical properties of the composite. In order to study the hysteretic energy dissipation performance of composite materials, the influence of loading amplitude, frequency and other factors should also be considered.

10.2.2.2 AF/PU Mechanical Property Test of AF/PU Composite

(1) **AF/PU composite test piece**

The foam aluminum used in this experiment is spherical open cell aluminum foam, purchased from Suzhou Christie's Foam Pioneer Metals Corporation. The porosity of foam aluminum is between 65 and 75% and the average pore size is 3–5 mm. The foam aluminum is processed into a cube sample of 30 mm × 30 mm × 25 mm by line cutting method, as shown in Fig. 10.18a. Use the preparation method of the composite materials mentioned above to prepare the AF/PU composite materials required for the test. The sample size is 34 mm × 34 mm × 25 mm, as shown in Fig. 10.18b. Figure 10.18c shows the filling rate of polyurethane in each composite material. It can be seen that the filling rate of most test pieces prepared in this test is more than 80%, and the filling effect is good.

(a) Foam aluminum sample

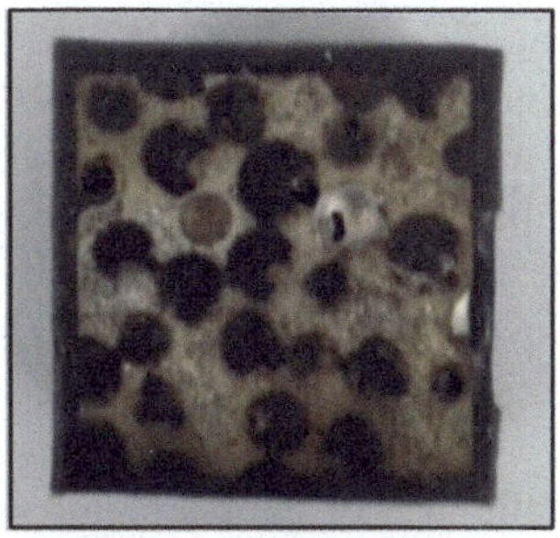

(b) AF/PU composite sample

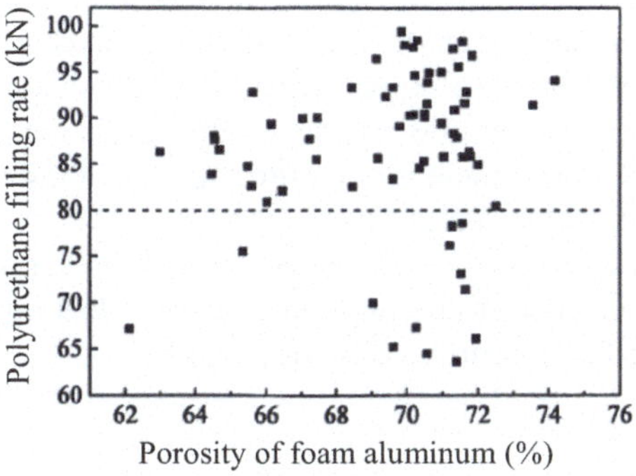

(c) Polyurethane filling rate of AF/PU composite sample

(d) Test equipment

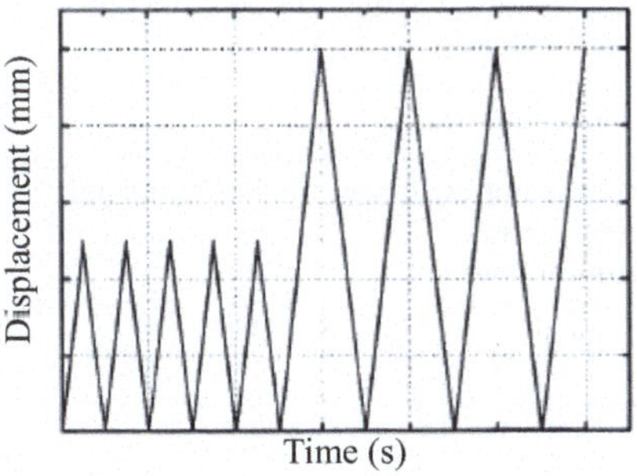

(e) Triangular wave loading mechanism

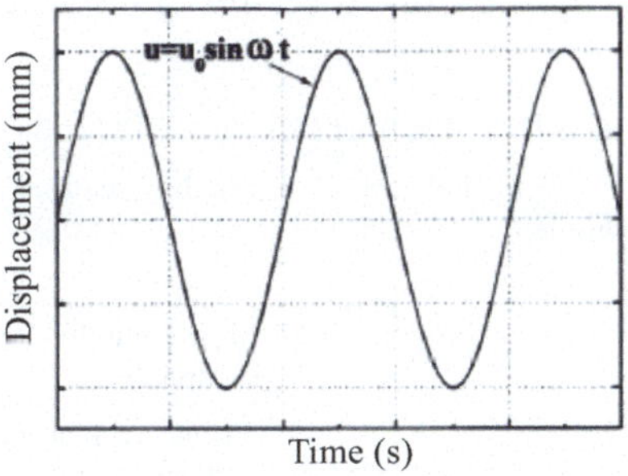

(f) Sine wave loading mechanism

Fig. 10.18 Mechanical property test of AF/PU composite

(2) **Test equipment**

The equipment used in this test is MTS810 hydraulic servo universal testing machine, School of Materials, Southeast University. The dynamic load range of the testing machine is 250 kN, the displacement range is 75 mm, the measurement accuracy of load and displacement is five thousandths of the indicated value, the maximum cycle loading frequency is 30 Hz, and the temperature range is −40 to 80 °C, which meets the requirements of this test. Displacement loading control mode is adopted in the test. During the test, triangle wave and sine wave are used for displacement loading according to the test method, which are automatically controlled by the computer. The test results are automatically collected by the sensor built in the computer. The test device is shown in Fig. 10.18d, and the loading mode is shown in Fig. 10.18e, f.

(3) **Test scheme**

The influence of the characteristics of composite materials such as relative density of foam aluminum, the content of polyurethane in composite materials, and the loading conditions such as loading frequency, strain amplitude and cycle number on the mechanical properties of AF/PU composites were mainly considered in the test. Among them, the mechanical properties mainly include stress-strain relationship, hysteretic energy consumption and material loss factor, deformation modulus and residual strain. In order to explore the influence of the above factors on the mechanical properties of composite materials, a series of test methods are developed, as shown in Table 10.6. For the specific test scheme, please refer to the appendix.

Among them, before the formal test, the test piece with test No. 4 needs to carry out 30 loading and unloading cycles under 20% strain amplitude, and the loading and unloading rate is 75 mm/min, and then carry out the subsequent test.

(4) **Analysis of test results**

1. **Effect of relative density of foam aluminum on reciprocating properties of composites**

In order to investigate the effect of relative density of foam aluminum in AF/PU composites on its reciprocating loading performance, Fig. 10.19 compared some

Table 10.6 Test methods for AF/PU composites

Experimental serial number	Loading method	Loading frequency (Hz) or rate (mm/min)	Experimental temperature (°C)
1	Monotonic loading to compaction	v = 3	25
2	5 times of loading and unloading at 5%, 10%, 15%, 20%, 25%, 30%, 40% of the maximum strain respectively, and finally compaction	v = 3.75	25
3	50 cycles at 20% strain amplitude	v = 75	25
4	3 cycles at 4%, 6%, 8% of the maximum strain respectively	f = 0.1, 0.5, 1.0, 3.0	25

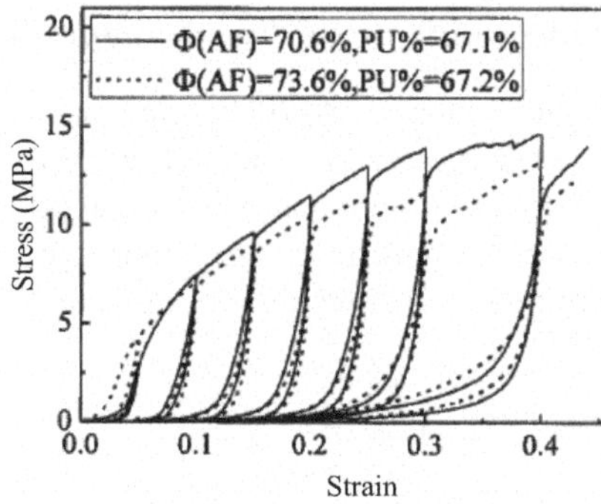

(a) Comparison of stress-strain curves

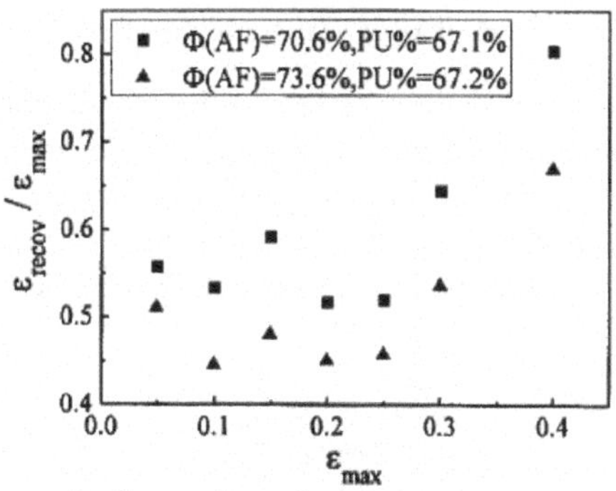

(b) Comparison of strain recovery rate

Fig. 10.19 Comparison of reciprocating loading properties of composite materials with different relative density of foam aluminum

experimental results with loading and unloading rate of 75 mm/min. Among them, Fig. 10.19a compared the stress-strain curve results of materials, and Fig. 10.19b compared the strain recovery rate of materials.

In the test, because the polyurethane covering the upper surface of some specimens is not cleared, the stress will first experience a low speed growth section, then the foam aluminum skeleton will bear the load and the curve will gradually become normal. From the comparison of the above figures, it can be seen that when the strain amplitude is small, the unloading curves and reloading curves of different composite materials are almost the same, basically coincident, and there is no significant difference in the residual strain; with the increase of strain amplitude, although the unloading curve is still not significantly different, the residual strain of the specimen with smaller relative density is larger, and the reloading curve is relatively higher, showing a smaller elastic recovery strain and larger reloading modulus. When the relative density of foam aluminum is relatively small, the properties of the composites are not very different. When the relative density of foam aluminum is relatively large, the performance difference is bigger.

Under the same condition of polyurethane content, the relative density of foam aluminum is small, and the filling rate of polyurethane in AF/PU composites is relatively small, and the unfilled pores are relatively more. Therefore, in the reciprocating loading, this part of the pores will be compressed and compacted, and there is no polyurethane filling inside, so this part of the strain can not be recovered. That is to say, the composite material with relatively small density of foam aluminum has a low recovery rate when it is reciprocating, and when reloading, the existence of this dense part will superimpose the elastic property of the dense part on the superelastic properties of the composite, showing a larger loading modulus. However, for the specimen with relatively high filling rate, due to the incompressibility of polyurethane, it can provide a large recovery force when unloading, and the strain recovery rate is correspondingly large; and because the dense part is less than the former, the modulus when reloading is also smaller than the former.

2. **Effect of polyurethane content on the reciprocating properties of composite**

In order to compare the influence of polyurethane content in AF/PU composite on its reciprocating compression performance, some test results are given in Fig. 10.20.

From the comparison of the results of pure foam aluminum and AF/PU composites of Fig. 10.20a, b, it can be seen that the slope of pure foam aluminum when unloading is larger, only a macroscopic elastic property, and the deformation can be

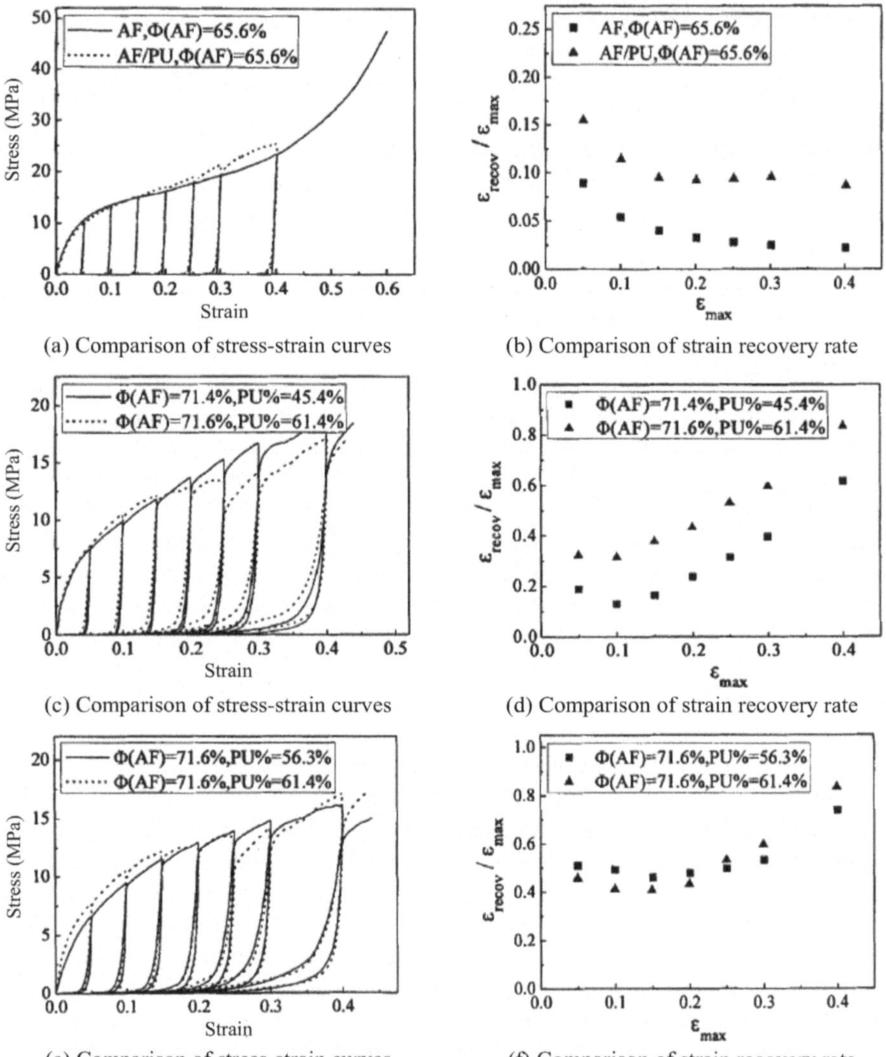

(a) Comparison of stress-strain curves
(b) Comparison of strain recovery rate
(c) Comparison of stress-strain curves
(d) Comparison of strain recovery rate
(e) Comparison of stress-strain curves
(f) Comparison of strain recovery rate

Fig. 10.20 Comparison of reciprocating loading properties of composites with different polyurethane content recovered is very little. The recoverable elastic deformation of composite materials during unloading includes two parts, namely, macro elastic recovery deformation and elastic recovery deformation related to cell edge buckling and polyurethane superelasticity. Compared with pure foam aluminum, the slope of initial unloading section of composite material is approximately the same as that of pure foam aluminum, and the restoring deformation of composite material also contains super elastic deformation recovery, so its value is more than that of pure foam aluminum. The residual deformation of foam aluminum decreases correspondingly.

The comparison of Fig. 10.20c, d shows that with the increase of polyurethane content in AF/PU composite, the total recovery deformation when unloading increases, the residual deformation decreases correspondingly, and the elastic properties of the composite are improved; at the same time, the unloading curve and reloading curve are improved. With the increase of strain amplitude, the difference of loading and unloading properties of composites with different polyurethane content tends to increase. This is because with the increase of strain amplitude, the cell edge of foam aluminum gradually breaks down and the polyurethane gradually plays the leading role: when the polyurethane content is large, the superelasticity of the composite material is obvious. It shows that the recoverable deformation increases and the reloading modulus increases; when the polyurethane content is small and the hyperelastic property is not obvious, so the gap becomes larger.

In Fig. 10.20e, f, the polyurethane content of the composite is different, but the difference is not significant, and the filling rate of polyurethane is more than 80%. It can be seen from the comparison of the results that there is little difference between the loading and unloading performance of the two. It shows that when the filling rate of polyurethane is high, the influence of its absolute value on the loading and unloading properties is not obvious.

3. **The influence of cycle times on the reciprocating properties of composites**

Figure 10.21 take the test result of Test No. 3 in Table 10.6 as an example to explain the influence of cycle times on the mechanical properties of AF/PU composite. It can be seen from Fig. 10.21a that the stress-strain curve of AF/PU composite changes greatly at the beginning of load cycle. With the increase of the number of cycles, the reloading curve gradually moves down while the unloading curve basically does not change. The above changes mainly occurred in the first 30 loading and unloading cycles, and then the stress-strain curve gradually stabilized. After about 30 cycles, with the increase of load cycles, the stress-strain curve basically does not change, the hysteresis curve tends to be stable, and the peak stress tends to be constant.

In cyclic loading, stiffness degradation Fl is a common method to evaluate material damage. Figure 10.21b shows the change process of stiffness degradation of AF/PU composite under cyclic loading. It can be seen that the degradation process of the stiffness when loaded to 50 cycles can be roughly divided into two stages, namely, the first stage, the sharp decline of F/F_0 mainly occurs in some initial cycles, because of the material softening caused by the destruction of cell edges, the subsequent peak stress loaded to the same stress is smaller than that of the previous cycle. The stiffness degradation decreases greatly in the first 30 cycles. In the second stage, the

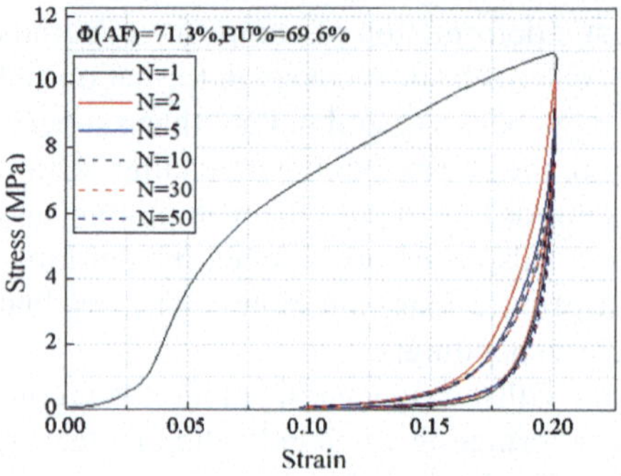

(a) Stress strain curve

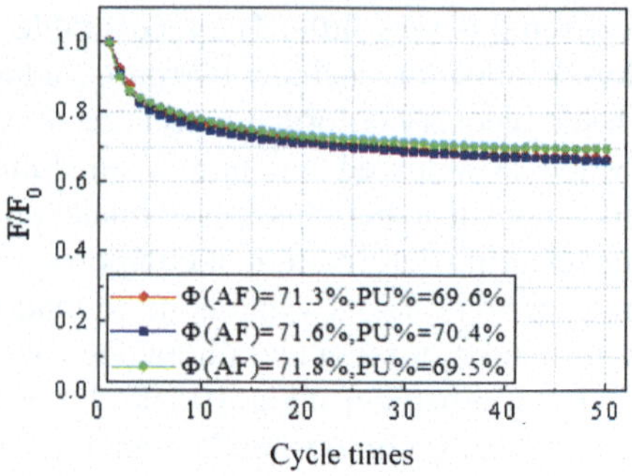

(b) Stiffness degradation

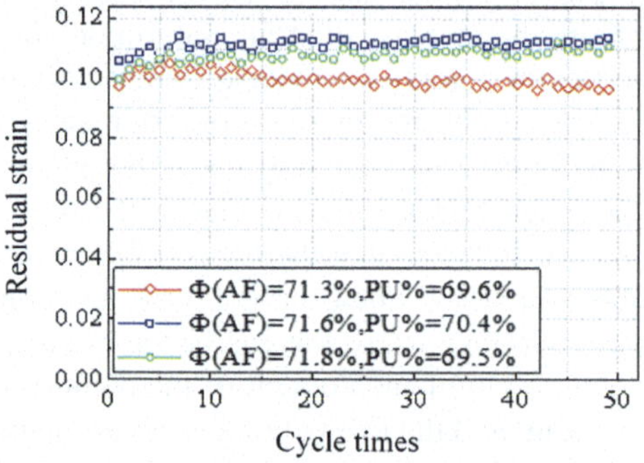

(c) Residual strain change

Fig. 10.21 The influence of cycle times on the mechanical properties of composites

degradation process of stiffness tends to be gentle, and gradually reaches a stable value. It shows that the damage in the material reaches a stable state. The above phenomenon shows that after the initial several cycles, the properties of the composite will be stable; after that, the composite will maintain its integrity and integrity, and the peak stress, deformation modulus and other parameters are no longer sensitive to the increase of the number of cycles.

Figure 10.21c shows the change law of residual strain of AF/PU composite with the number of cycles. With the increase of the number of cycles, the residual strain of the material tends to increase. In the first few cycles, the increase of residual strain is slightly larger; when the number of cycles reaches 30, the increase is small, and the residual strain basically does not change, and the stable value is about 10–11%. Although polyurethane improves the resilience recovery of foam aluminum, the residual deformation of AF/PU composites remains large. Therefore, if the composite material is directly subjected to cyclic load from its initial state without proper measures, it has the disadvantages of small deformable range and poor energy dissipation performance. Because AF/PU composite is loaded and unloaded at 20% of strain amplitude, the residual strain is about 10%, so its recoverable strain is determined as 10%.

10.2.2.3 Deformation Mechanism of Foam Aluminum and AF/PU Composites

Foam aluminum will deform in the manner shown by Fig. 10.22a under monotonic compression. The cell edge is compressed and bent under the action of external force; with the increase of external force, the weak cell edge first yields to form a plastic hinge, then the cell edge is compressed and buckled until plastic collapse occurs, making the cell hole closed and dense. With the increase of load, the yield of cell edge is gradually transferred, and the number of cell edge of yield collapse is increasing, which makes the whole specimen dense. When the test piece is unloaded in the plastic platform section. The material shows the macro elastic properties of the compressed and dense part, the elastic modulus is larger than that of the initial loading, the residual deformation is larger, and the recoverable deformation is smaller.

Due to the introduction of polyurethane as reinforcing filler, the deformation mechanism of AF/PU composites is different from that of pure foam aluminum. As shown in Fig. 10.22b, when the external force is small, the polyurethane does not contribute to the force of the composite material, and only the foam aluminum skeleton bears the load. At this stage, the performance of the composite is basically the same as that of the aluminum foam with the same structural parameters. With the increase of external force, polyurethane in the composite began to participate in the stress. Because of its incompressibility, the resistance of composite material increases when deformation, so the stress is larger than that of pure foam aluminum. At the same time, the incompressibility of polyurethane retarded the yield and buckling of the cell edge, and delayed the compactness of the composites. Therefore, the plastic

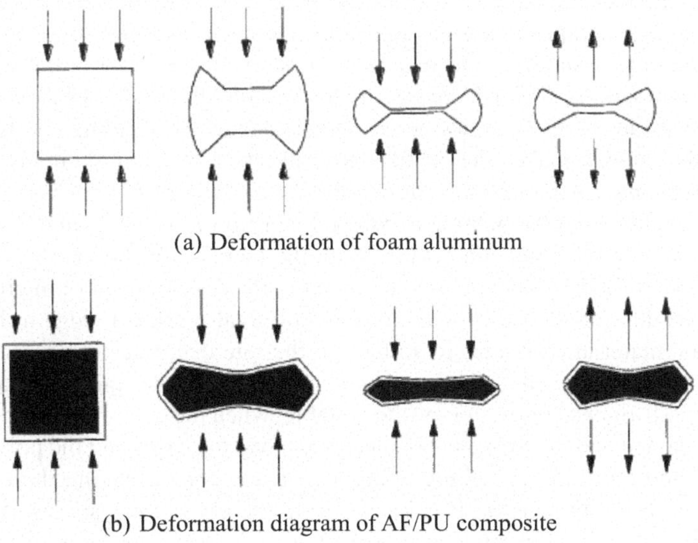

Fig. 10.22 Deformations of foamed aluminum and AF/PU composites

deformation capacity of the composite material is stronger than that of the pure foam aluminum. Because of the incompressibility of polyurethane, the transverse deformation occurs when the compression of cell pore decreases, and the cell edge is stretched and broken at the weak point. When the cell edge breaks, the pressure of polyurethane in the cell hole is released, and the stress of the material is reduced, which is shown as jitter on the stress-strain curve

After the cell edge breaks, it will continue to compress under the action of external force. When unloading, the composite will first show the macro elastic properties of the part with dense cell edges; when unloading continually, because there is no binding of cell holes, the polyurethane originally in the cell holes can give full play to its super elastic properties. Therefore, the recoverable elastic deformation of composite materials when unloading contains two parts. Besides showing the macroscopic elastic properties of foam materials, it also shows a certain superelastic deformation recovery. In contrast to the unloading process, the stress first experiences a low-speed growth period during the reloading process, showing the hyperelastic properties of the composite, and then increases at a large rate, showing the macro elastic properties of the compacted part. With the increase of the load, the phenomenon of cell edge fracture will spread from the loading end to the supporting end, while with the increase of the number of broken cell edges, the restraint of polyurethane will gradually weaken, and the hyperelastic properties of the composite will become more and more obvious. Therefore, the strain recovery rate of the composite increases with the unloading at a large strain amplitude, which is also consistent with the previous experimental analysis. The influence of the relative density of foam aluminum and the content of polyurethane in composite materials on loading and unloading properties has been discussed before.

The results of cyclic loading test can also be explained by the above deformation mechanism. The composite softens in the second cycle, and the stress value at the maximum strain is smaller than that at the first cycle. In some initial cycles, the stiffness degradation (F/F_0) decreases greatly and rapidly; in the subsequent cycles, the reduction rate of F/F_0 slows down. This is due to the yielding and fracture of the cell edge, which makes the cellular pore structure of foam aluminum undergo permanent change. Although the strain amplitude of the cell is not very large, the cell edges are not completely broken, but the foam aluminum is inevitably damaged, so the weakening phenomenon occurs. With the increase of the number of cycles, although the strain does not continue to increase, the development of micro damage and the transmission of cracks will cause the original weak cell edge to break. As a result, the material continues to soften but the rate decreases during subsequent cycles. When all the weak cell edges have been broken under this strain amplitude value, the performance of the composite is stable; when the cyclic loading continues, the polyurethane and the broken cell edges bear load together, and the polyurethane plays an important role, providing a larger recovery force, making the composite show better elastic properties. It is known from the above analysis results that the performance of AF/PU composite in this study is stable after about 30 cycles of cyclic loading under 20% strain amplitude.

In the process of edge fracture, crack propagation and micro damage development, energy will be lost, but this energy consumption mechanism is irreversible; with the decrease of the number of broken edges, this part of energy consumption will gradually reduce to zero. Therefore, during the cyclic loading process, the area surrounded by the hysteresis curve of the composite gradually decreases until it stabilizes. The energy consumption mechanism of this part of stable energy consumption will be introduced and studied later.

10.2.2.4 AF/PU Phenomenological Constitutive Model

The macroscopic phenomenological constitutive model is a kind of theory that describes the macroscopic mechanical behavior of materials on the basis of experiments. The parameters in the model are identified according to the experimental data. Because of its simplicity and practicality, it is usually the preferred model in the numerical method. Therefore, the establishment of a reasonable macroscopic phenomenological constitutive model of materials has always been the focus of scholars. Rusch firstly proposed a phenomenological constitutive model for porous foams, such as Eq. (10.78). The model is simple in form and consists of two curtain functions. LiuQ et al. proposed a phenomenological constitutive model of six parameters based on the Rusch model, which can be used to describe the tensile and compressive mechanical behavior of foams, as shown in Eq. (10.79). Among them, $A, B, C, \alpha, \beta, \gamma$ are the undetermined parameters in the model.

$$\sigma = A\varepsilon^m + B\varepsilon^n \tag{10.78}$$

$$\sigma = A\frac{e^{\alpha\varepsilon} - 1}{B + e^{\beta\varepsilon}} + e^C(e^{\gamma\varepsilon} - 1) \tag{10.79}$$

A new phenomenological constitutive model of foam materials was proposed by modifying Rusch model by Avalle M:

$$\sigma = A\left(1 - e^{(-C/A)\varepsilon(1-\varepsilon)^m}\right) + B\left(\frac{\varepsilon}{1-\varepsilon}\right)^n \tag{10.80}$$

σ and ε are the engineering stress and engineering strain of foam material respectively. In the model, A, B, C, m and n are five parameters to be determined. A, B, and C are parameters related to the relative density of foam materials. m and n are independent constants of relative density. The first part of this model is used to describe the elastic-plastic region of the stress-strain curve of the foam material, and the second one is used to describe the dense segment. In this section, this model will be used to fit the monotonic loading curve of composite materials.

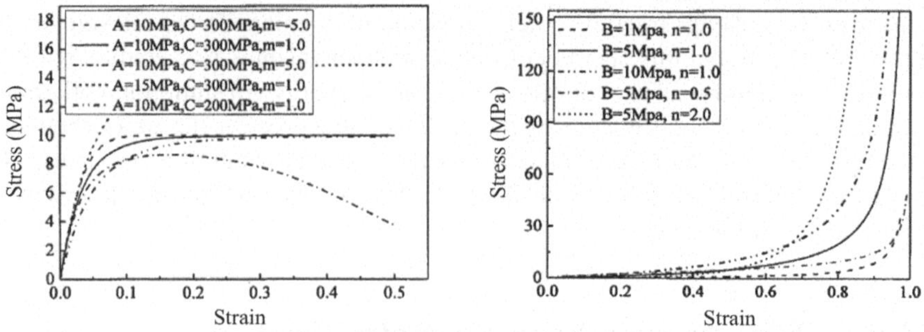

(a) Influence of A, C, m on material properties (b) Influence of B and n on material properties

Fig. 10.23 Influence of model parameters on material properties

For the part of the model describing the elastic-plastic stage, take different values of A, C and m respectively. See Fig. 10.23a for the curve. (1) when the value of parameter A is larger, the stress-strain curve is higher, and the value of platform stress is larger, indicating that parameter A affects the platform stress of the material; (2) when the value of parameter C is large, the initial slope of the curve is larger, indicating that parameter C affects the initial elastic modulus of the material; (3) when the value of parameter m is larger, the curve will soften after passing the yield point; when the value of m is small, the material will appear platform stage, and the smaller the value of m, the earlier the material will end the initial elastic segment and enter the platform stage, indicating that the parameter m affects the initial elastic deformation ability and subsequent soft and hard properties of the material. Because of the small difference of the initial elastic deformation capability of foam aluminum composites with different relative density, in order to simplify, the parameter m is regarded as constant in this study, resulting in little error. A and C are regarded as parameters varying with material parameters.

For the part of the model describing the compaction stage, take different B and n values respectively. See Fig. 10.23b for the curve. It can be seen that the larger the value of B is, the earlier the material enters the compaction section, the smaller the compaction strain is, and the faster the densification speed is. The influence of parameter n on the function is similar to that of B, that is, the larger the value of n is, the earlier the material enters the compaction, and the faster the curve changes in the compaction section. Therefore, the function of parameter n is similar to that of B. in order to simplify the simulation process, parameter n is regarded as a constant, while parameter B is regarded as a parameter varying with material parameters.

At strain $\varepsilon = 0$, Eq. (10.80) satisfies $\sigma = 0$. Therefore, the model satisfies the initial boundary conditions of loading. The derivative of stress to strain is obtained as follows:

$$\frac{\partial\sigma}{\partial\varepsilon} = \frac{\partial}{\partial\varepsilon}\left[A\left(1 - e^{(-C/A)\varepsilon(1-\varepsilon)^m}\right) + B\left(\frac{\varepsilon}{1-\varepsilon}\right)^n\right] = Ce^{(-C/A)\varepsilon(1-\varepsilon)^m}\left[(1-\varepsilon)^m\right.$$

$$-m\varepsilon(1-\varepsilon)^{m-1}\Big] + Bn\frac{\varepsilon^{n-1}}{(1-\varepsilon)^{n+1}} \tag{10.81}$$

When the strain ε approaches 0:

$$\left.\frac{\partial\sigma}{\partial\varepsilon}\right|_{\varepsilon=0} = \frac{\partial}{\partial\varepsilon}\left[Ce^{(-C/A)\varepsilon(1-\varepsilon)^m}\left[(1-\varepsilon)^m - m\varepsilon(1-\varepsilon)^{m-1}\right] + Bn\frac{\varepsilon^{n-1}}{(1-\varepsilon)^{n+1}}\right]\Bigg|_{\varepsilon=0} = C \tag{10.82}$$

According to the actual physical meaning, near the initial point of the stress-strain curve, there are:

$$\left.\frac{\partial\sigma}{\partial\varepsilon}\right|_{\varepsilon=0} = E_0 \tag{10.83}$$

where, E_0 is the slope of the stress-strain curve at the initial point, i.e. the initial elastic modulus of the material. Therefore, the physical meaning of parameter C in the model is the initial elastic modulus of the material, which is consistent with the analysis results shown in above figures.

For the first part of the model, it is noted that the range of strain e is (0, 1). In this range, as the strain increases, the value of the whole expression first increases and then decreases. When e approaches a specific value, the expression approaches a limit:

$$\lim_{\varepsilon\to\varepsilon_0}\left[A\left(1-e^{(-C/A)\varepsilon(1-\varepsilon)^m}\right)\right] = A \tag{10.84}$$

That is to say, after the strain reaches a certain value, the stress defined in this part is close to a constant, with the size of A; and from the actual physical meaning, the constant is the platform stress σ_p of the material. Therefore, the physical meaning of parameter A in the model is the platform stress of the material, which is also consistent with the previous figure.

The characteristics of this model include: (1) this model has five parameters: A, B, C, m and n. among them, A, B and C depend on the material parameters, and m and n are constants. (2) parameters A, B and C have practical physical meaning, that is, A is close to the platform stress of the material, C is the initial elastic modulus of the material, and B determines the dense strain of the material. (3) the second part of the model is the revision of Rusch model. The purpose of introducing this part is to make the curve have a vertical asymptote (corresponding to the limit of compression, i.e. strain $\varepsilon = 1$). (4) each parameter affects a specific region of the model, but it does not list an expression for each region, but has a unified expression, so it is convenient to use. (5) the exponential function is used in the model to improve the fitting effect of the model on the connection between the elastic section of the curve and the stage of the platform; as long as the value of m is appropriate, a better fitting effect can be achieved.

Table 10.7 Fitting results of different composite parameters

Specimen number	Porosity (%)	A (MPa)	B (MPa)	C (MPa)	m	n
C-1	63.0	14.0851	30.8750	367.4128	4.9217	0.9484
C-1	65.5	13.6477	23.9532	318.1636	4.5180	0.9667
C-1	66.1	10.2870	21.3084	300.8732	4.3578	0.7879
C-1	67.2	10.0271	26.5008	253.8480	5.7446	0.7669
C-1	70.5	12.2003	19.4879	151.6524	4.8424	0.9505
C-1	64.6	12.4534	25.6149	364.4697	5.1581	0.7993
C-1	66.5	9.1776	26.9626	252.4105	5.6570	0.8008

Using the above model, the monotonic loading curve of AF/PU composite is fitted, and the corresponding parameter values of each specimen are shown in Table 10.7.

According to the previous analysis, A and B should be related to the relative density of foam aluminum and the content of polyurethane, while C is related to the relative density of foam aluminum. Because the polyurethane filling rate of AF/PU composites is high here, the trace difference of the content has little effect on A and B. Therefore, the influence on the parameters of A and B is not considered here. A, B and C are all regarded as parameters controlled by relative density of aluminum foam. The corresponding fitting formula is shown in Eqs. (10.85)–(10.89), while m and n are constants, $\bar{\rho}$ is the relative density of foam aluminum in AF/PU composite. Use the following fitting formulas to fit the parameter values in Table 10.7, and get the parameter values in the fitting formula as shown in Table 10.8.

$$A = C_{1,A}\bar{\rho} + C_{2,A}\bar{\rho}^{3/2} \tag{10.85}$$

$$B = C_{1,B}\bar{\rho}^{C_{2,B}} \tag{10.86}$$

$$C = C_{1,C}\bar{\rho} + C_{2,C}\bar{\rho}^{2} \tag{10.87}$$

$$m = m_0 \tag{10.88}$$

$$n = n_0 \tag{10.89}$$

where, $m = 5.0195$, $n = 0.8574$.

Table 10.8 Fitting results of parameters in each fitting formula

	C_1	C_2
A	0	59.67
B	269.89	2.20
C	0	2605.73

Table 10.9 Prediction results of parameters of each test piece

Specimen number	Porosity (%)	A (MPa)	B (MPa)	C (MPa)	m	n
C-1	63.0	13.429	30.2852	356.7244	5.0195	0.8574
C-2	65.5	12.092	25.9650	310.1470	5.0195	0.8574
C-3	66.1	11.778	24.9819	299.4531	5.0195	0.8574
C-4	67.2	11.209	23.2332	280.3349	5.0195	0.8574
C-6	70.5	9.561	18.3990	226.7637	5.0195	0.8574
C-26	64.6	12.568	27.4786	326.5397	5.0195	0.8574
C-22	66.5	11.570	24.3380	292.4280	5.0195	0.8574

Table 10.9 gives the predicted values of the Avalle constitutive model parameters of the composite obtained by the above fitting formula parameters. Figure 10.24 is the comparison between the direct fitting results and the predicted values of the parameters of the Avalle constitutive model.

Figure 10.25 is the fitting result of monotonic loading curve of each specimen by using the predicted value of Avalle constitutive model parameters. Compared with the test results, it shows that the fitting effect is better. Therefore, the Avalle constitutive model can be used to predict the mechanical behavior of AF/PU composite under monotonic loading, which provides a reference for its application in monotonic compression.

Monotonic compression test and reciprocating loading test were carried out on AF/PU composites with different structural parameters. The deformation process of composites was observed during the test, and the stress-strain curves were obtained. The influence of relative density of foam aluminum, content of polyurethane, loading rate and cycle times on mechanical properties of composites were analyzed, and the deformation mechanism of composites was defined. The monotonic loading curve was fitted. The conclusions are as follows:

(1) The mechanical properties of AF/PU composites under monotonic loading are similar to those of pure foam aluminum, including three stages of elastic segment, plastic platform segment and dense segment. When unloading and reloading, the polyurethane plays an important role in making the composite exhibit macroscopic elastic and superelastic properties. The elastic recovery deformation has two parts, including macro elastic recovery deformation and super elastic recovery deformation, and the elastic recovery performance is improved.
(2) The relative density of foam aluminum in AF/PU composites has a great effect on the mechanical properties of the composites. The greater the relative density of foam aluminum is, the greater the initial elastic modulus of the composite is, the greater the stress of the platform is, but the shorter the length of the plastic platform segment is, the worse the plastic deformation ability is. The larger the relative density of the aluminum foam is, the smaller the residual deformation of the composite material is when unloading. The reloading modulus is also small.

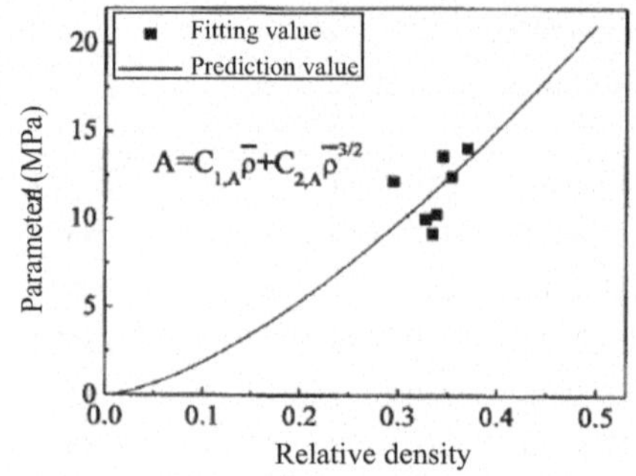

(a) Comparison of fitting value and prediction value-parameter *A*

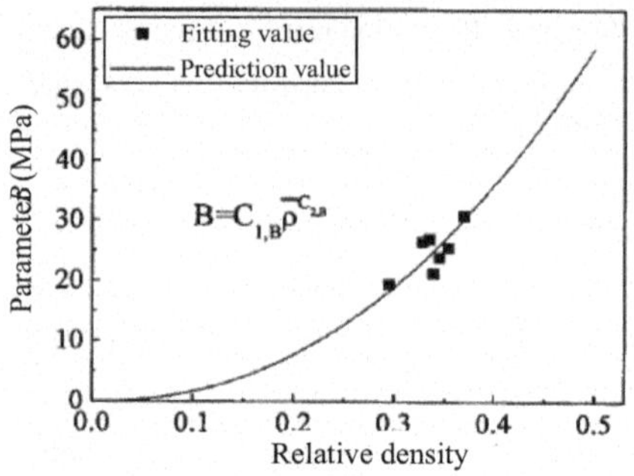

(b) Comparison of fitting value and prediction value-parameter *B*

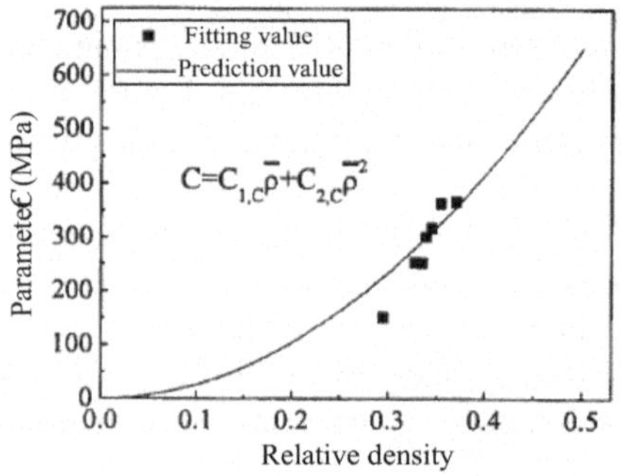

(c) Comparison of fitting value and prediction value-parameter *C*

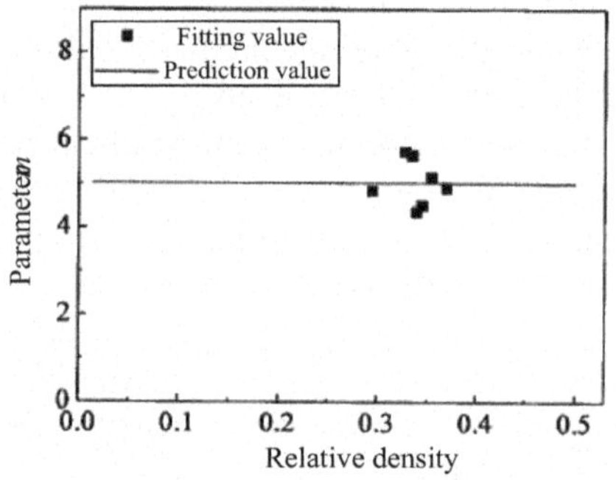

(d) Comparison of fitting value and prediction value-parameter *m*

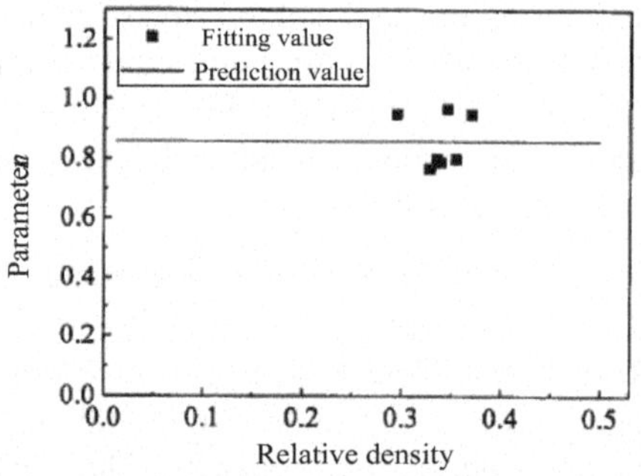

(e) Comparison of fitting value and prediction value-parameter *n*

Fig. 10.24 Comparison of fitting value and prediction value of constitutive model parameters of each specimen

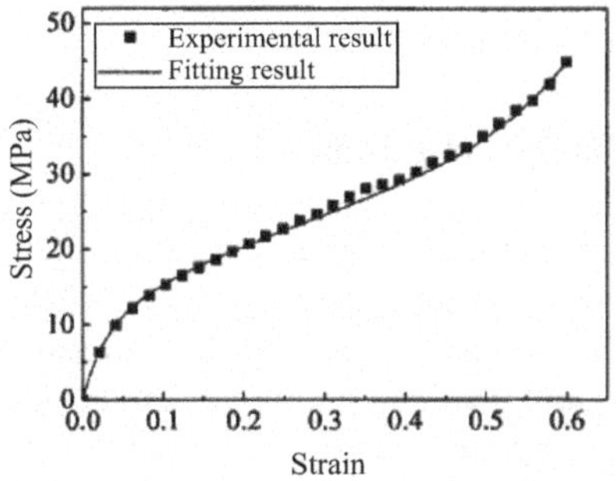

(a) Test results and fitting results-($\Phi(AF) = 63.0\%$)

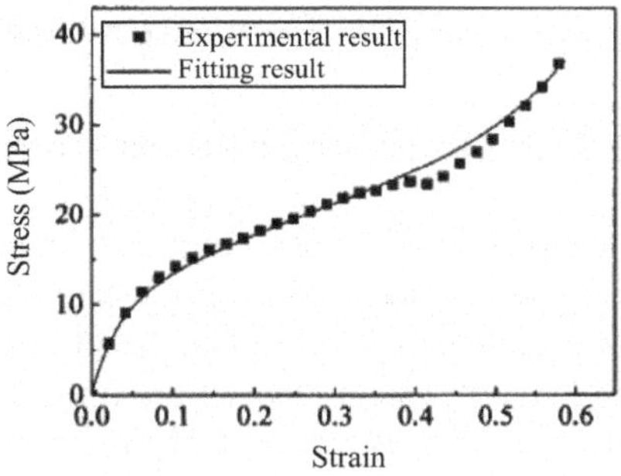

(b) Comparison between fitting value and prediction value of parameters-($\Phi(AF) = 65.5\%$)

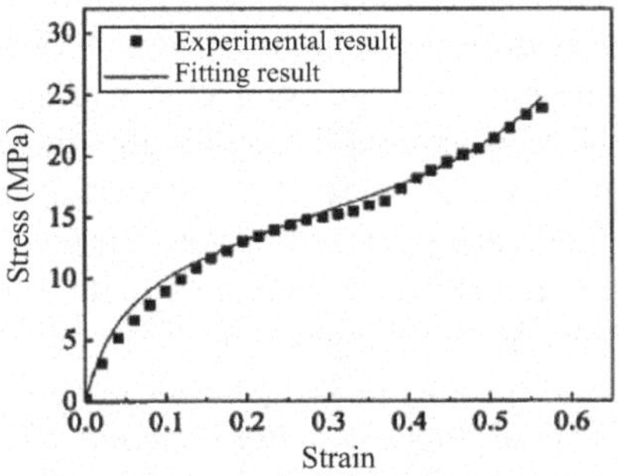

(c) Test results and fitting results-($\Phi(AF) = 70.5\%$)

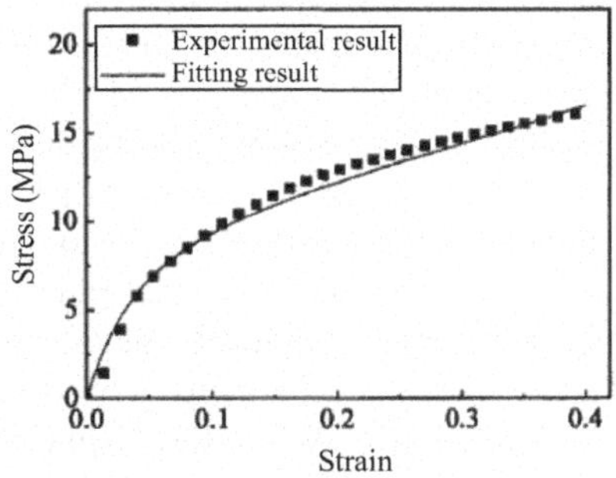

(d) Test results and fitting results-($\Phi(AF) = 71.6\%$)

Fig. 10.25 Comparison of test results and model prediction results of different composite materials

(3) The content of polyurethane in AF/PU composite has a certain influence on the stress of the composite: with the increase of polyurethane content, the platform stress of the composite becomes larger, the dense strain becomes larger, and the plastic deformation ability is improved; the larger the content of polyurethane is, the smaller the residual deformation of the composite is when unloading, and the higher the modulus when reloading is. When the filling rate of polyurethane in the composite is large, the influence of polyurethane content on the properties of the composite is not obvious.

(4) Due to the introduction of polyurethane, the deformation mechanism of AF/PU composites is different from that of pure foam aluminum. During the deformation process, the cell edges of the foam aluminum will break. With the increase of strain, the cell edge fracture will continue to pass until all the cell edges are broken, compacted, and the specimens are destroyed.

10.2.3 AF/PU Composite Damper

10.2.3.1 Basic Structure and Working Principle of Damper

The structure of AF/PU composite damper is shown in Fig. 10.26a, which is mainly composed of upper connecting steel plate 1, AF/PU composite 2, outer steel plate 3, lower connecting steel plate 4 and high-strength bolt [7]. The AF/PU composite 2 and two outer steel plates 3 are fixed by bolts, and there is no relative displacement between them; the middle of the upper connecting steel plate 1 is provided with a groove hole, so that the upper connecting steel plate 1 and the AF/PU composite 2 can move relatively; the lower connecting steel plate 4 and the two outer steel plates 3 are connected by high-strength bolts, and two AF/PU composite materials of the same thickness with the AF/PU composite 2 are used between them as a cushion block, to ensure that the load of AF/PU composite 2 is even when the preload is applied; the three steel plates and two AF/PU composite materials are connected with each other through high-strength bolts, and the preload is applied through high-strength bolts. The device uses the relative displacement between the AF/PU composite 2 and the upper connecting steel plate 1 to realize friction energy consumption. During the test, the upper connecting steel plate 1 and the lower connecting steel plate 4 are connected through the fixture device provided by MTS testing machine.

There are two mechanisms of friction and viscoelasticity in the operation of AF/PU composite damper. When the displacement is small, there is bottom friction effect between the foam aluminum and the upper connecting plate 1, and the polyurethane is shear acted to produce viscoelastic force. The resistance is assumed by the friction force and the viscoelastic force. It can be explained by the ideal visco elastoplastic model. When the displacement is larger, the resistance is mainly borne by the friction between the aluminum, polyurethane and steel plate, and the ideal rigid plastic model can be used to elaborate.

Although the structure of AF/PU friction damper is similar to that of ordinary friction damper, the energy dissipation effect of AF/PU friction damper is totally different from that of ordinary friction damper. Because in the case of small earthquake, AF/PU friction damper can give full play to the viscoelastic energy consumption of polyurethane in AF/PU; while in the case of medium earthquake and large earthquake, the friction energy consumption of aluminum material in AF/PU and the viscoelastic energy consumption of polyurethane can work together, making AF/PU friction damper like a composite damper with series viscoelastic damper and friction damper at the same time. In order to achieve this ideal energy consumption effect, it is necessary to let the surface of AF/PU friction sample keep a very thin polyurethane coating, as shown in Fig. 10.26b.

When the AF/PU friction specimen with polyurethane cover is installed in the damper, the damping force of the AF/PU friction damper can be divided into three stages with the change of external excitation. This staged performance change depends on the amplitude of external load (ΔU), the ultimate shear deformation (S) of polyurethane coating in AF/PU and the displacement (L) when polyurethane coating is completely worn out. The mechanical characteristics of AF/PU friction damper can be described in detail in combination with Fig. 10.26d.

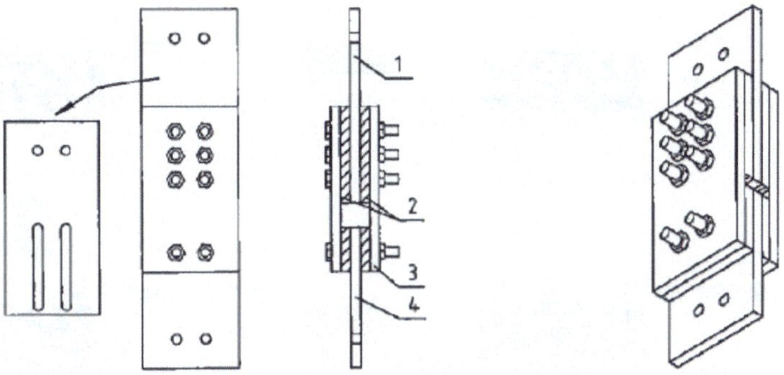

1-upper connecting steel plate; 2-AF/PU composite material; 3-outer steel plate; 4-lower connecting steel plate

(a) Construction of AF/PU composite damper

Foam aluminum

Polyurethane

Foam aluminum composite (AF/PU)

(b) AF/PU composite sample

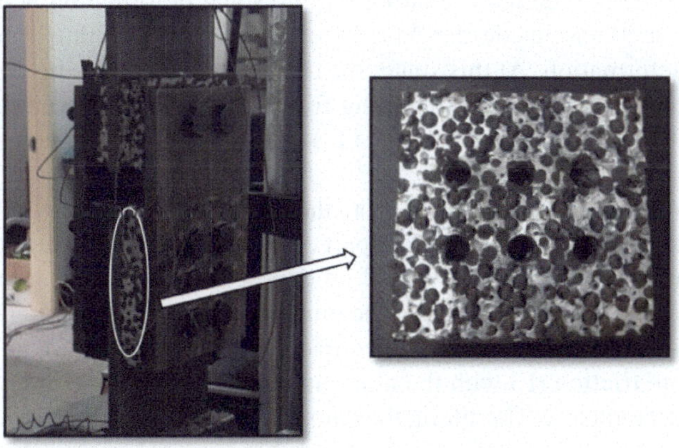

(c) AF/PU composite damper test piece

Fig. 10.26 AF/PU composite damper

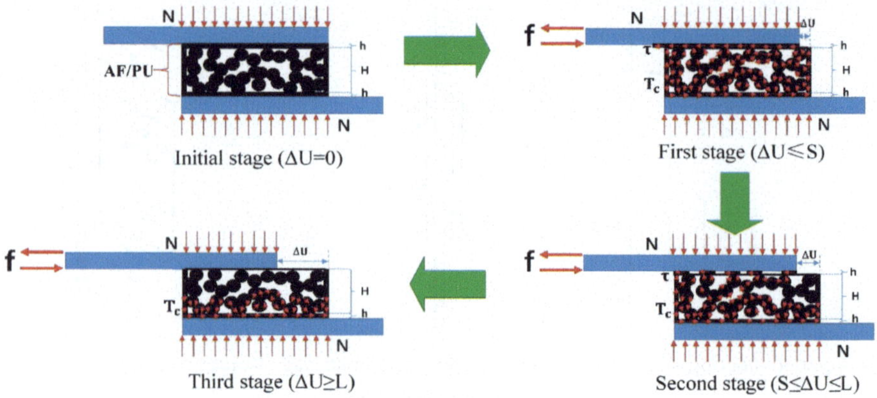

(d) Working principle

Fig. 10.26 (continued)

In the initial stage, due to the preload (N) of high-strength bolts, the aluminum skeleton in AF/PU is elastically deformed, which forces the polyurethane inside it to extrude outwards, thus increasing the thickness (h) of polyurethane coating and reducing the height (H) of AF/PU main body, which paves the way for the deformation of polyurethane coating in the next stage.

In the first stage ($\Delta U < S$): when the intermediate steel plate moves interactively, the static friction force on the contact surface between the polyurethane coating and the intermediate steel plate is greater than the shear force of the polyurethane coating, so that the polyurethane coating and the intermediate steel plate are tightly bonded together; only the polyurethane coating will do shear deformation with it, and will drive the polyurethane inside the AF/PU to extend outward, thus forming typical viscoelastic deformation. At this stage, the damping force (f) of the AF/PU friction damper, in addition to the shear damping force (τ) from the polyurethane coating, also includes the restraint damping force (T_c) of the aluminum skeleton to the internal polyurethane in the AF/PU.

In the second stage ($S < \Delta U \leq L$): as the deformation of polyurethane coating exceeds its ultimate shear deformation, part of the polyurethane coating will be torn, and the aluminum surface under it will be exposed. However, under the action of huge pre pressure (N), and because of the thin thickness (h) of polyurethane coating, the intermediate steel plate can fully contact the exposed aluminum surface, and finally generate friction (F) with the aluminum surface when the intermediate steel plate moves interactively. Therefore, the damping force (f) of AF/PU friction damper should be added with its friction force (F) at this stage. However, due to the wear of polyurethane coating, the shear damping force (τ) of polyurethane coating and the restraint damping force (T_c) of aluminum skeleton to polyurethane inside will decrease correspondingly.

In the third stage ($\Delta U \geq L$): since the polyurethane coating in this stage has been completely worn, the damping force (f) of the AF/PU friction damper in this stage is only the friction force (F) between the intermediate steel plate and the aluminum surface and the weak constraint damping force (T_c), which will gradually stabilize based on the friction force (F). At this time, the AF/PU friction damper is similar to the ordinary friction damper device.

Based on the above analysis, it can be seen that the damping force (f) of AF/PU friction damper comes from the shear damping force (r) of polyurethane covering layer, the restraint damping force (T_c) of aluminum skeleton to the internal polyurethane, and the friction force (F) between the middle steel plate and the aluminum surface of AF/PU, all of which change in stages with the change of the amplitude of the external load ((ΔU)). Due to the complex structure of AF/PU, different aluminum volume fraction and pore morphology will affect the friction performance of AF/PU. In order to make further research on AF/PU friction damper, the following two assumptions need to be put forward:

(1) when AF/PU samples have the same aluminum volume fraction, AF/PU samples have the same friction properties;
(2) when the polyurethane coating in AF/PU is very thin, the friction force (F) of AF/PU friction damper in the second stage is equal to that in the third stage.

Therefore, the variation of damping force (f) of AF/PU friction damper with external excitation can be summarized as follows:

First stage ($\Delta U < S$):

$$f(t) = \tau(t) + T_c(t) \tag{10.90}$$

Second stage ($S < \Delta U \leq L$):

$$f(t) = \tau(t) + T_c(t) + F \tag{10.91}$$

Third stage ($\Delta U \geq L$):

$$f(t) = T_c(t) + F \tag{10.92}$$

10.2.3.2 Mechanical Performance Test of AF/PU Damper

(1) Specimen design

According to the structure and working principle of the above-mentioned AF/PU friction damper, considering that the current market can not produce AF/PU of large thickness, but in order to meet the characteristics of the building structure and seismic design requirements, the AF/PU friction damper is designed in a reduced scale. Because the thickness (h) of polyurethane cover of AF/PU friction damper is very thin, the performance of damper can enter the second stage with the change of smaller load displacement amplitude (ΔU), but it needs a very large load displacement amplitude (ΔU) when its performance enters the third stage, that is, the displacement (L) of polyurethane cover completely worn out is difficult to set. In order to simplify and better design the AF/PU friction damper, and to study the transitional change of damping force in different stages, two cases, retention and wear, are used for the polyurethane cover in AF/PU. Because when the AF/PU with worn out polyurethane cover is installed in the damper, under the great preload, the foam aluminum skeleton in AF/PU will deform, forcing the polyurethane inside to be extruded, forming a small polyurethane protrusion on the surface of AF/PU, and these small protrusions are like the remaining polyurethane coating (Fig. 10.26b) of AF/PU dampers with polyurethane cover at the end of the second stage. These polyurethane protrusions can be completely worn out under small load displacement amplitude (ΔU), making the performance of AF/PU friction damper enter the third stage. Therefore, the performance of the AF/PU friction damper can be easily changed from the first stage to the second stage, and from the second stage to the third stage by setting the appropriate change of the external load displacement amplitude (ΔU).

Because the friction force (F) of the AF/PU friction damper is directly proportional to the preload (N) exerted by the high-strength bolt, it is necessary to control the preload (N) according to the compression strength of the AF/PU, so that the AF/PU specimen will not be crushed by the huge preload (N). At present, there are many ways to adjust the pre pressure (N), such as strain measurement, torque wrench and so on. Combined with the accuracy and limitation of the test, we use the torque wrench method in this test. By adjusting the torque of connecting bolt of AF/PU friction damper, different preload (N) is applied to AF/PU friction material. The torque (T) of high strength bolt and pre pressure (N) can be obtained:

$$N = \frac{T}{K \cdot d} \tag{10.93}$$

where, d is the nominal diameter of the thread and K is the torque coefficient. The torque coefficient reflects the relationship between the bolt axial preload and the tightening torque, which is a constant determined by experiments. Its value depends on the geometry of the thread pair and the friction of the thread pair. In China, the value is usually between 0.11 and 0.15. The size range of AF/PU friction material can be determined by the obtained pre pressure (N) and the compression performance test of AF/PU in the previous chapter.

Therefore, in order to study the performance of AF/PU friction dampers more intuitively, the same aluminum volume fraction was used for all AF/PU to meet the requirements of engineering assumption 1. According to the compression performance test of AF/PU, three kinds of reduced scale AF/PU friction dampers were designed: T60PU, T70PU and T70. The main design parameters are shown in Table 10.10.

Table 10.10 Main design parameters of reduced scale model of AF/PU friction damper

Part name		Damper number		
		T60PU	T70PU	T70
Thickness of intermediate steel plate (mm)		10	10	10
Section area of intermediate steel plate (mm^2)		18085	18085	18085
Thickness of outer steel plate (mm)		10	10	10
Section area of outer steel plate (mm^2)		16372	16372	16372
Bolt	Diameter (mm)	10	10	10
	Torque (N m)	60	70	70
AF/PU	Aluminum volume fraction (%)	30	30	30
	Sectional area (mm^2)	9529	9529	9529
	Height (mm)	10	20	20
	Whether there is polyurethane coating	Yes	Yes	No

(2) **Test scheme**

(a) By measuring the force displacement curve of AF/PU friction damper under different preload, loading frequency, loading displacement amplitude and cycle times, to understand the influence of these parameters on the dynamic characteristics of the damper. Among them, the dynamic characteristics mainly include force displacement relationship, hysteretic energy consumption, equivalent stiffness and equivalent damping ratio.

(b) Based on the test results, the rationality of the mechanical calculation model of AF/PU friction damper is verified.

This test is the same as the previous AF/PU material compression performance test. The equipment used in this test is MTS810 hydraulic servo universal testing machine of School of Materials, Southeast University. The test device is shown in Fig. 10.27a. The displacement loading control mode was adopted in the test, and the sine waves with variable amplitude and equal amplitude were used as the excitation for displacement loading, as shown in Fig. 10.27b, c. The test temperature is 25 °C, the loading frequency is 0.1–0.5 Hz, and the displacement amplitude range is ±2–12 mm. The specific loading method is shown in Table 10.11. The force displacement curve obtained from the test is automatically collected by the sensor built in the computer.

10.2.3.3 Test Result Analysis

Figure 10.28 shows the force displacement curve of AF/PU friction damper under various working conditions:

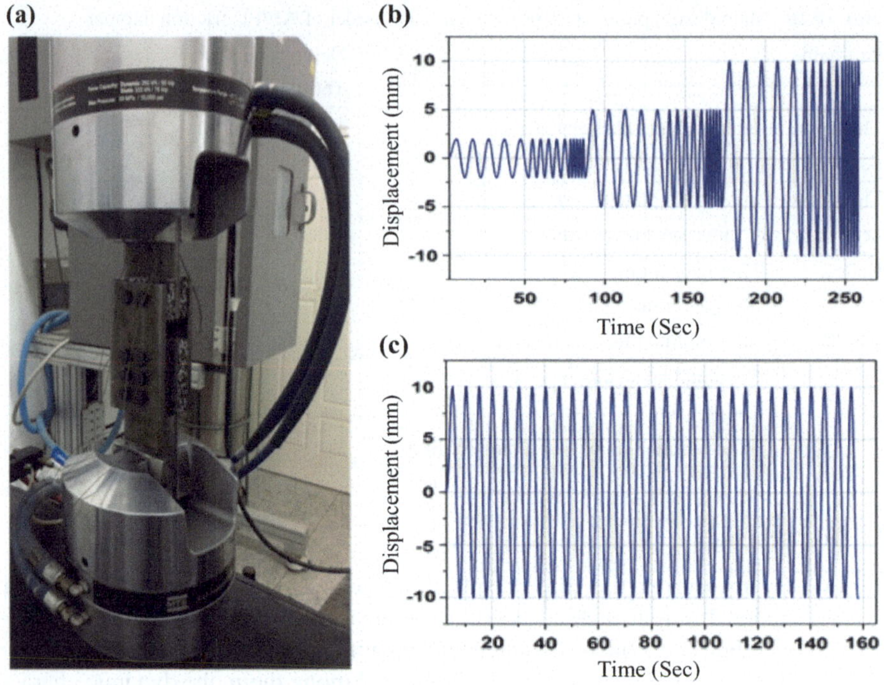

Fig. 10.27 Test device and loading scheme

Table 10.11 Loading method of AF/PU friction damper

Test serial number	Loading method	Loading frequency (Hz)	Damper number
1	5 cycles at the amplitudes of ±2 mm, ±5 mm, ±10 mm, respectively	0.1, 0.2, 0.5	T60PU T70PU
2	5 cycles at the amplitudes of ±6 mm, ±8 mm, ±12 mm, respectively	0.1, 0.2, 0.5	T70PU
3	30 cycles at the amplitude of ±10 mm	0.2	T70PU

(1) Under all loading conditions, the hysteretic effect of the force displacement curve of AF/PU friction damper is quite full, basically without pinching and sliding, and can maintain its symmetry to the center origin under different displacement amplitudes. AF/PU friction dampers have strong and stable energy dissipation capacity, and have a good prospect in the field of building seismic.

(2) For T60PU and T70PU with polyurethane coating, when the amplitude is ±2 mm, the hysteretic curve of the damper shows obvious viscoelastic hysteretic energy dissipation effect, that is, the hysteretic curve is elliptical, The mechanical performance of AF/PU friction damper is in the first stage, but

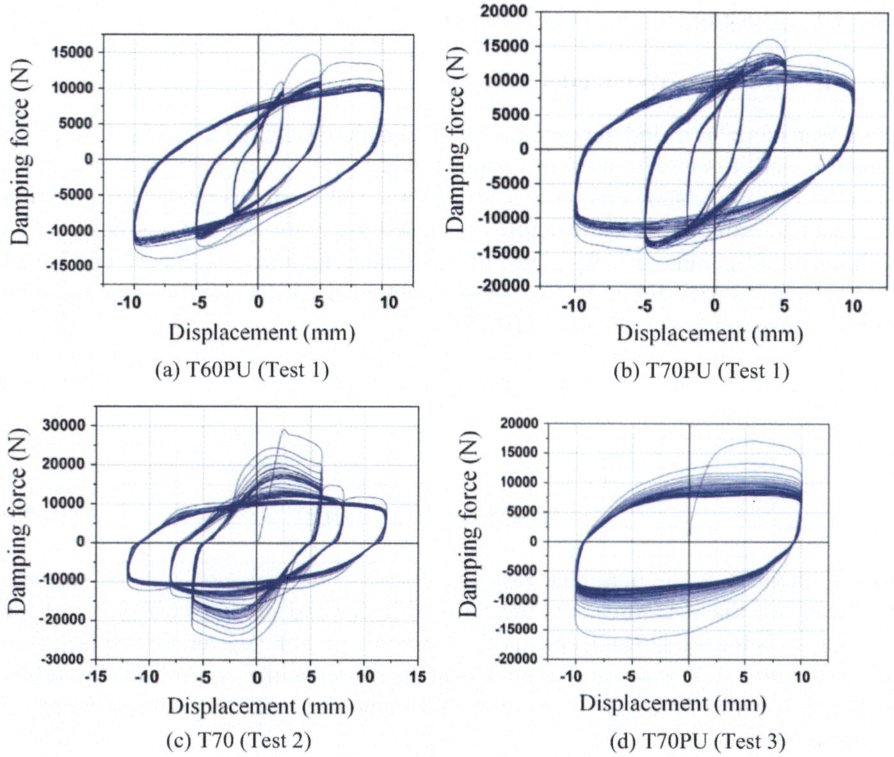

Fig. 10.28 Test force-displacement curve of AF/PU friction damper

when the amplitude is increased to ±10 mm, the rectangle shape of hysteretic curve is more significant. The energy consumption gradually changes into friction energy consumption. However, the mechanical properties of AF/PU friction dampers have been transformed into the second stage, as shown in Fig. 10.28a, b.

(3) For T70 without polyurethane coating, when the amplitude is ±6 mm, the hysteretic curve of the damper begins to be inclined rectangle, until the amplitude increases to ±12 mm, the hysteretic curve gradually falls back, and its shape becomes the hysteretic curve similar to that of ordinary friction damper. These show that the mechanical properties of AF/PU friction damper have entered the third stage, as shown in Fig. 10.28c.

10.2.3.4 Mechanical Model of Damper

(1) **Modified Bouc-Wen model**

Bouc Wen model is a kind of nonlinear hysteretic model characterized by differential equation, and it is widely used in various fields of civil engineering because of its smooth hysteretic characteristics and strong universality. In this model, the restoring force and deformation of the structure are related to a nonlinear differential equation with uncertain parameters. By choosing reasonable parameters, a large number of hysteresis loops with different shapes can be obtained. The classic Bouc-Wen model can be shown in Eqs. (10.94) and (10.95).

$$f = \alpha k u + (1 - \alpha) k z \tag{10.94}$$

$$\dot{z} = A\dot{u} - \beta|\dot{u}||z|^{n-1} - \gamma\dot{u}|z|^{n} \tag{10.95}$$

Among them, f is the nonlinear restoring force of the structure, k is the stiffness coefficient of the structure, α is the ratio of linear and nonlinear stiffness of the structure, u is the total displacement of the structure, z is the hysteretic displacement of the structure, A, β, γ are the parameters of the model, which controls the saturation degree and shape of the hysteretic curve, and n determines the smoothness degree of the hysteretic curve.

According to the performance test of AF/PU friction damper, the performance of AF/PU friction damper changes obviously in stages with the change of displacement amplitude. However, the classical Bouc-Wen model can not reflect the influences of changes of displacement amplitude and frequency on the restoring force. In order to accurately simulate the dynamic performance of AF/PU friction damper, a modified Bouc-Wen model is proposed for the law that the stiffness of AF/PU friction damper changes little with the loading frequency, but decreases with the increase of displacement amplitude. This model mainly introduces the nonlinear cumulative stiffness into the classic Bouc-Wen model. See Fig. 10.29 for the simplified model. At this time, the damping force of AF/PU friction damper is:

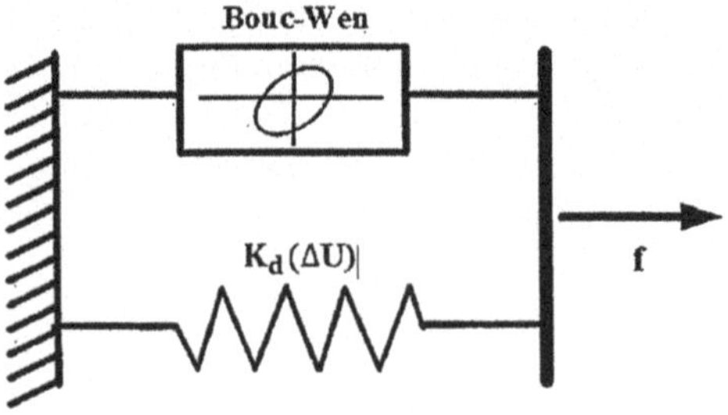

Fig. 10.29 Modified Bouc-Wen model

$$f = K_d u + \alpha_1 z \tag{10.96}$$

Among them, α_1 is the coefficient of hysteretic restoring force, which is related to the pre pressure of the damper, equivalent to the product of $(1 - \alpha)$ and k in Eq. (10.94); K_d is the cumulative stiffness of AF/PU friction damper, which is dependent on the change of displacement amplitude (ΔU). The relationship between K_d and ΔU can be assumed to be a third-order polynomial:

$$K_d = p * (\Delta U)^2 + q * (\Delta U) + K_0 = \begin{cases} > 0 \\ \leq 0, \text{ equals to } 0 \end{cases} \tag{10.97}$$

where, K_0 is the initial cumulative stiffness of the AF/PU friction damper, and p and q are the constants related to the AF/PU material, respectively. According to the previous test, the stiffness of AF/PU friction damper decreases with the increase of displacement amplitude. To reflect this feature, the constant p in Eq. (10.97) must be negative, so that the cumulative stiffness K_d may decrease to negative with the increase of displacement amplitude. However, for the AF/PU friction damper, when its first stiffness K_{d1} decreases to zero, it indicates that the performance of the AF/PU friction damper has entered the third stage, becoming a common friction damper. When Bouc-Wen model is used to simulate the damping force of friction damper, there is no need to accumulate the stiffness, that is, $K_d = 0$. Therefore, in order to meet the performance characteristics of AF/PU friction damper, the K_d calculated by Eq. (10.97) must be non negative, so Eq. (10.97) needs to be revised to:

$$K_d = p * (\Delta U)^2 + q * (\Delta U) + K_0 = \begin{cases} > 0 \\ \leq 0, \text{取} 0 \end{cases} \tag{10.98}$$

(2) **Model parameter identification**

In order to ensure the accuracy and applicability of the modified Bouc-Wen model to the AF/PU friction damper, it is necessary to identify the parameters of the modified Bouc-Wen model, so as to determine the parameters, so that under the same external excitation, the response of the numerical model with the determined parameters can be close to the real response of the AF/PU friction damper. Combined with the discussion in the previous section, there are 8 parameters to be identified in the modified Bouc-Wen model, including $A, \beta, \gamma, n, \alpha_1, p, q$ and K_0, because the response of the modified Bouc-Wen model is affected by multiple parameters, which makes it relatively difficult to identify the parameters.

In order to better identify the parameters of modified Bouc-Wen model, it is necessary to understand the influence of each parameter on the restoring force curve of AF/PU friction damper, so as to better determine the value range of each parameter. Since the AF/PU friction dampers here all use the AF/PU material of the same friction performance, the Bouc-Wen model with $n = 2$ has enough accuracy for simulating friction energy consumption. Therefore, the modified Bouc-Wen model still has seven parameters to be identified, including $A, \beta, \gamma, \alpha_1, p, q$ and K_0.

Table 10.12 Parameter identification results of modified Bouc-Wen model

Parameter	Range	T60PU	T70PU	T70
A	$0 \leq A < 10$	0.687234	0.691234	2.695099
β	$-\gamma < \beta < \infty$	9.961815	9.941815	1.15299
γ	$-\beta \leq \gamma < 0$	−8.92542	−8.98542	−1.12359
α_1	$1 < \alpha_1 < \infty$	9729.11	10929.11	1086.969
K_0	$0 < K_0 < \infty$	545.3159	245.3159	1544.068
q	$-\infty < q < \infty$	279.5579	379.5579	−562.709
p	$-\infty < p < 0$	−37.6679	−37.9609	−646.331
n	$1 \leq n < \infty$	2	2	2

According to the test data of T60PU, T70PU and T70, the parameters of modified Bouc-Wen model are identified by genetic algorithm. The final results are summarized in Table 10.12.

Taking T70PU damper as an example, all the identified parameter results of T70PU in Table 10.12 are substituted into the modified Bouc-Wen model given in the previous section, and the differential equation in the model is solved by the fourth-order Runge Kutta method. Since the performance of AF/PU friction damper is mainly affected by the displacement amplitude, and the loading frequency has little influence on the performance of AF/PU friction damper, here only uses the modified Bouc-Wen model to simulate the response of AF/PU friction damper under the fixed loading rate and different displacement amplitude. Take the case that the loading frequency of T70PU damper is 0.1 Hz as an example, draw the force displacement curve and output force time history curve of numerical model and test data respectively, as shown in Fig. 10.30.

From Fig. 10.30, it can be seen that under different displacement amplitudes, the results derived from the modified Bouc-Wen model are in good agreement with the test data, so it can be proved that it is reasonable to use the third-order polynomial related to displacement amplitudes for the cumulative stiffness in the modified Bouc-Wen model, and the obtained model curves and identification results can meet the requirements.

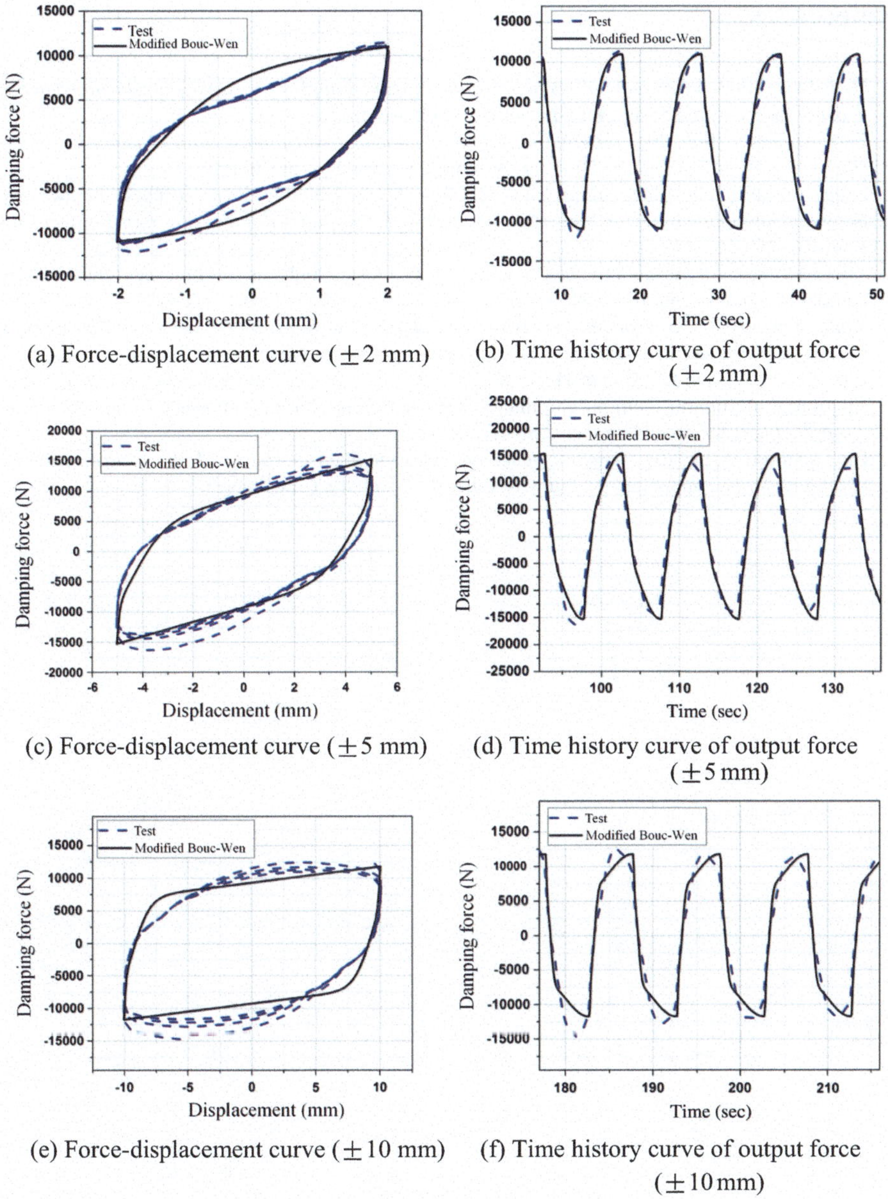

(a) Force-displacement curve (±2 mm)

(b) Time history curve of output force (±2 mm)

(c) Force-displacement curve (±5 mm)

(d) Time history curve of output force (±5 mm)

(e) Force-displacement curve (±10 mm)

(f) Time history curve of output force (±10 mm)

Fig. 10.30 Comparison between modified Bouc-Wen model and test data

References

1. Ni, Lifeng. 2004. *A new technology of SMA superelastic damping for structural vibration energy dissipation*. Nanjing: Southeast University. (in Chinese).
2. Zuo, Xiaobao. 2005. *Experimental and analytic study of vibration control of cable based on shape memory alloy damper*. Nanjing: Southeast University. (in Chinese).
3. Ding, Youliang, Xin Chen, Aiqun Li, et al. 2011. A new isolation device using shape memory alloy and its application for long-span structures. *Earthquake Engineering and Engineering Vibration* 10 (2): 239–252.
4. Zuo, XiaoBao, AiQun Li, and QingFu Chen. 2008. Design and analysis of a superelastic SMA damper. *Journal of Intelligent Material Systems and Structures* 19 (6): 631–639.
5. Xuan, Peng. 2016. *Experimental study on the damping mechanism and mechanical properties of high-performance aluminum foam*. Nanjing: Southeast University. (in Chinese).
6. Liu, Shaobo, Aiqun Li, Siyuan He, et al. 2015. Cyclic compression behavior and energy dissipation of aluminum foam-polyurethane interpenetrating phase composites. *Composites Part A-Applied Science and Manufacturing* 78: 35–41.
7. Liu, Shaobo. 2018. *Experimental study on mechanical properties and shock absorption of aluminum foam composites*. Nanjing: Southeast University. (in Chinese).

Part III
Design Method of Structural Vibration Control

With the deepening of the understanding of vibration control technology, the production and manufacture of vibration control devices are becoming more and more mature, and the engineering application of the vibration control technology of building structures is inevitable. In this respect, the perfect analysis theory and mature design method are the important theoretical basis for the application of structural vibration control technology, and the corresponding standards, regulations and atlas are the necessary prerequisites for the promotion of structural vibration reduction technology. Based on the self-developed damping device of building structure, the authors' team systematically carried out the research on the theory and design method of energy dissipation, frequency modulation and isolation, and established the perfect analysis method and design method of building structure vibration control. In this part, based on the research and development of damping device of building structure in the second part, the authors' team's research results in the field of structural damping design methods will be introduced from the perspective of structural damping design.

Chapter 11
Vibration Control Analysis Theory of Building Structure

Abstract Vibration control analysis theory and method of building structure are introduced. Dynamic model of building structure damping system, including energy dissipation structure system, frequency modulation damping structure system and isolated structure system are elaborated respectively. Analysis method of building structure vibration control, including numerical analysis method, velocity dependent damper element based on Abaqus software, variable damping viscous fluid damper element and Lead rubber bearing element based on OpenSees platform are discussed. Vibration control dynamic test of building structure, including energy dissipation, frequency modulation, isolated structure system are introduced respectively.

11.1 Dynamic Model of Building Structure Damping System

11.1.1 Dynamic Model of Energy Dissipation Structure System

In the first part of this book, the basic principle of structural vibration reduction control of single degree of freedom system has been introduced, and the dynamic model of structural vibration reduction system can be obtained by properly expanding it to multi degree of freedom system. As shown in Fig. 11.1a, according to its mechanical properties, the structure can be equivalent to a series multi degree of freedom system with n degrees of freedom. Therefore, the dynamic equation of ordinary building structure can be expressed as follows:

$$[M]\{\ddot{x}(t)\} + [C]\{\dot{x}(t)\} + [K]\{x(t)\} = \{F(t)\} \tag{11.1}$$

where, $\{F(t)\}$ is external load vector, $\{x(t)\}$, $\{\dot{x}(t)\}$ and $\{\ddot{x}(t)\}$ are the displacement, velocity and acceleration vectors of particle respectively, $[M]$, $[C]$ and $[K]$ are the mass, damping and stiffness matrices of equivalent multiple degrees of freedom respectively, which can be expressed as:

A. Li, *Vibration Control for Building Structures*, Springer Tracts in Civil Engineering, https://doi.org/10.1007/978-3-030-40790-2_11

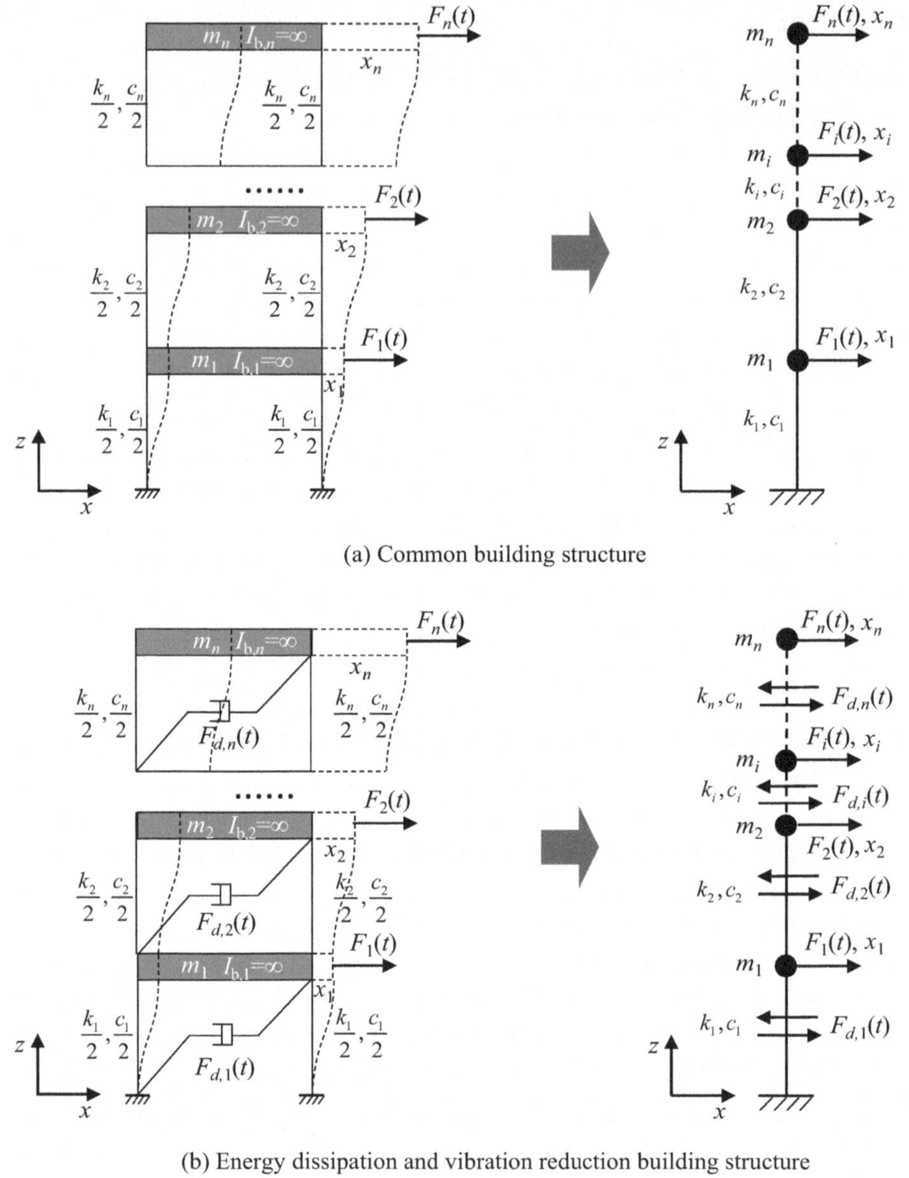

Fig. 11.1 Calculation diagram of building structure

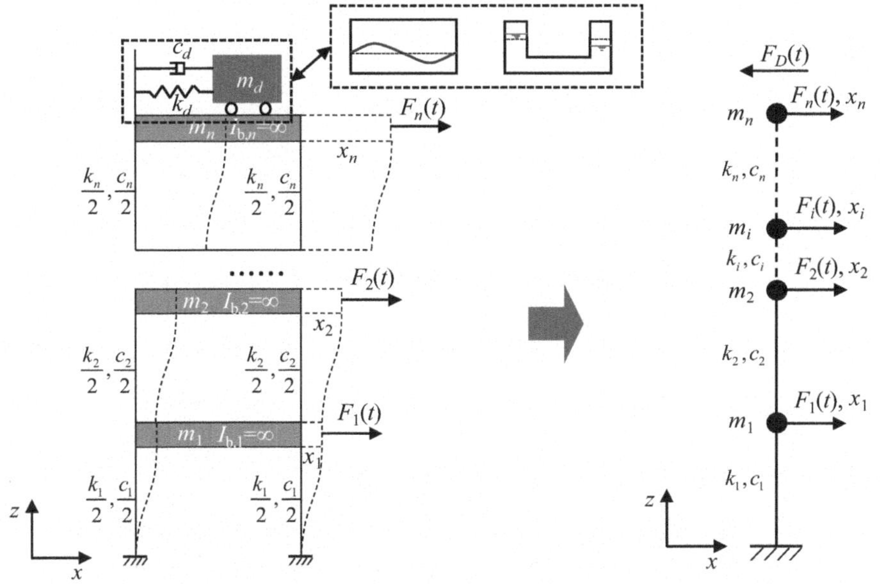

(c) Frequency modulation and vibration reduction building structure

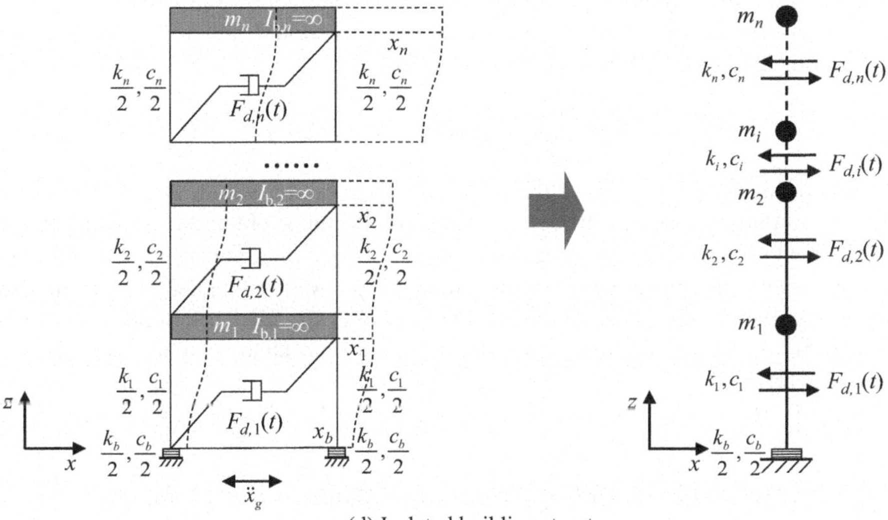

(d) Isolated building structure

Fig. 11.1 (continued)

$$[M] = \begin{bmatrix} m_1 & & & & \\ & \ddots & & 0 & \\ & & m_i & & \\ & 0 & & \ddots & \\ & & & & m_n \end{bmatrix} \tag{11.2}$$

$$[K] = \begin{bmatrix} k_1 + k_2 & -k_2 & & & & & \\ -k_2 & k_2 + k_3 & -k_3 & & & & \\ & -k_3 & \ddots & -k_i & & & \\ & & -k_i & k_i + k_{i+1} & -k_{i+1} & & \\ & & & -k_{i+1} & \ddots & k_n \\ & & & & k_n & k_n \end{bmatrix} \tag{11.3}$$

$$[C] = \alpha[M] + \beta[K] \tag{11.4}$$

Among them, α and β are Rayleigh damping coefficients. In addition to Eq. (11.4), there are other calculation methods for structural damping, but Rayleigh damping is one of the more commonly used methods to determine the damping matrix. The building structure with energy dissipation device can also be equivalent to a series multi degree of freedom system with n degrees of freedom (such as Fig. 11.1b), and its dynamic equation can be expressed as follows [1–3]:

$$[M]\{\ddot{x}(t)\} + [C]\{\dot{x}(t)\} + [K]\{x(t)\} + \{F_D(x, \dot{x})\} = \{F(t)\} \tag{11.5}$$

Among them, $\{F_D(x, \dot{x})\}$ is the damping force vector, which is determined by the type of damping device. Its calculation formula can be obtained according to the mechanical model of the damping device in the second part, generally related to the relative displacement and speed of the damper. Equations (11.1) and (11.5) can also be used to express the dynamic model when considering other structural forms such as high-rise structure and long-span structure, but only changes when the stiffness matrix is formed.

11.1.2 Dynamic Model of Frequency Modulation Damping Structure System

As shown in Fig. 11.1c, TMD, TLD or TLCD are usually set on the top of the structure, the dynamic model of the structure system with frequency modulation and vibration reduction can be expressed as follows [1]:

$$[M]\{\ddot{x}(t)\} + [C]\{\dot{x}(t)\} + [K]\{x(t)\} = \{F(t)\} + [H]\{F_D(t)\} \tag{11.6}$$

$$\text{Set TMD: } \ddot{x}_d + 2\zeta_d\omega_d\dot{x}_d + \omega_d^2 x_d = -\ddot{x}_j,\ F_{TMD}(t) = c_d\dot{x}_d + k_d x_d \tag{11.7}$$

$$\text{Set TLD: } \ddot{x}_d + 2\zeta_L\omega_L\dot{x}_d + \omega_L^2 x_d = -\ddot{x}_j \tag{11.8}$$

$$\text{Set TLCD: } \ddot{x}_d + 2\zeta_L\omega_L\dot{x}_d + \omega_L^2 x_d = -(B/L)\cdot\ddot{x}_j \tag{11.9}$$

where, j is the structure node number at the location where the frequency modulation damping device is set, $j \le n$; when the frequency modulation damping device is set at the top of the structure, $j = n.x_d$, $\dot{x}_d$ and $\ddot{x}_d$ are respectively the generalized displacement, velocity and acceleration of the damping device. For TMD, they are the relative displacement, velocity and acceleration of the mass unit; $[H]$ is the action position matrix of the frequency modulation damping device, where $H = [0,\ldots,0,1,0,\ldots,0]^{\mathrm{T}}$ (where 1 is in the jth column) indicates that the frequency modulation damping device is set at the jth node of the structure; $\{F_D(x,\dot{x})\}$ is the control force of the frequency modulation damping system, it can be obtained according to the mechanical model of the device in the first part and the second part.

11.1.3 Dynamic Model of Isolated Structure System

Figure 11.1d shows the calculation diagram of a typical base isolated structure, according to which the dynamic model of the isolated structure system can be established as follows [4–6]:

$$[M]\{\ddot{x}(t)\} + [C]\{\dot{x}(t)\} + [K]\{x(t)\} = -[M]\{\mathrm{I}\}\ddot{x}_g(t) \tag{11.10}$$

where, $\{x(t)\}$ is the displacement vector of the structural layer (including the isolation layer) relative to the ground, $\{\dot{x}(t)\}$ and $\{\ddot{x}(t)\}$ are the corresponding velocity and acceleration vectors respectively; the damping matrix and stiffness matrix can be expressed as follows respectively:

$$[C] = [C]_{FF} + \begin{bmatrix} 0 & \\ & c_b \end{bmatrix} \tag{11.11}$$

$$[K] = [K]_{FF} + \begin{bmatrix} 0 & \\ & k_b \end{bmatrix} \tag{11.12}$$

Among them, $[C]_{FF}$ and $[K]_{FF}$ are the damping and stiffness matrices of the free-free system respectively, k_b and c_b are the stiffness and damping coefficients of the isolation layer. If the isolation device is a bilinear isolation system, the above dynamic equation can be expressed as follows:

$$[M]\{\ddot{x}(t)\} + [C]\{\dot{x}(t)\} + [K]\{x(t)\} + \begin{pmatrix} 0 \\ F_b \end{pmatrix} = -[M]\{\mathrm{I}\}\ddot{x}_g(t) \tag{11.13}$$

Among them, F_b is the restraint of the isolator to the underlying structure. Assuming that the yield force of the device is Q, the system stiffness will change after yielding, F_b can be expressed as:

In the elastic phase:

$$\begin{aligned} F_b(x_1) &= (k_{b1} - k_{b2})(x_1 - x_s) + k_{b2}x_1 \\ &= k_{b1}x_1 - (k_{b1} - k_{b2})x_s \end{aligned}, \quad \text{and when} \quad \dot{x}_s = 0, \quad (k_{b1} - k_{b2})(x_1 - x_s) < Q \tag{11.14}$$

In the yield phase:

$$F_b(x_1) = k_{b2}x_1 - Q\mathrm{sgn}(\dot{x}_1), \text{ and } \dot{x}_s = \dot{x}_1 \text{ until } \dot{x}_s = 0 \tag{11.15}$$

where, $x_1(t)$ is the mass of the underlying structure;x_s is the displacement of the slider (regardless of its mass).

11.2 Analysis Method of Building Structure Vibration Control

11.2.1 Numerical Analysis Method

Due to the non-linear characteristics of the damping devices in the damping system of the building structure, the conventional frequency domain analysis method is difficult to apply. One way to solve this problem is to linearize the non-linear mechanical model of the damping device, and the other is to directly use the numerical analysis method in the time domain to solve the above-mentioned dynamic model. The latter method can get the instantaneous state of the structure, and get the peak value, mean value, variance and other key indicators of various responses of the structure, which attracts the attention of scholars and engineers. However, as for the vibration reduction system of building structure, the introduction of various new vibration reduction devices has led to the emergence of new mechanical models, which are often lack of solutions of these devices in the existing modules of existing programming languages and finite element software unit library. On the basis of the second part of the research on the damping device, on the one hand, the research group has developed the program of damping control system analysis of the simple structure; on the other hand, it has carried out the secondary development of the new damping element of the commonly used finite element software. In the process of self programming, the commonly used numerical integration methods are central difference method, Wilson-θ method, Newmark-β method, etc. when dealing with the non-linear problems caused by the damping device, Newton method and quasi Newton method can be used. This book only introduces the basic principles of the first two numerical integration methods, and other numerical analysis methods can refer to relevant professional books.

1. **Central difference method**

Central difference method is a kind of numerical integration method which is widely used in explicit method and dynamic response problem. It is based on the following formula:

First order central difference formula:

$$\{\dot{x}(t)\} = \frac{1}{2\Delta t}(\{x(t+\Delta t)\} - \{x(t-\Delta t)\}) \tag{11.16}$$

Second order central difference formula:

$$\{\ddot{x}(t)\} = \frac{1}{(\Delta t)^2}(\{x(t+\Delta t)\} - 2\{x(t)\} + \{x(t-\Delta t\}) \tag{11.17}$$

Equations (11.16) and (11.17) are substituted into the dynamic equation (11.1) of the system at time *t*, and the equation of $\{x(t+\Delta t)\}$ for solving the displacement at time $t+\Delta t$ is obtained:

$$\begin{aligned}\left(\frac{1}{(\Delta t)^2}[M] + \frac{1}{2\Delta t}[C]\right)\{x(t+\Delta t)\} &= \{F(t)\} - \left([K]\{x(t)\} - \frac{2}{(\Delta t)^2}[M]\{x(t)\}\right)\\ &\quad - \left(\frac{1}{(\Delta t)^2}[M] - \frac{1}{2\Delta t}[C]\right)\{x(t-\Delta t)\}\end{aligned} \tag{11.18}$$

It can be seen from Eq. (11.18) that in order to obtain $\{x(t+\Delta t)\}$, $\{x(t)\}$ and $\{x(t-\Delta t)\}$ must be calculated first, and it is necessary to find $\{x(-\Delta t)\}$ at the beginning. Through Eqs. (11.16) and (11.17), and $t=0$, then

$$\{x(-\Delta t)\} = \{x(0)\} - \Delta t\{\dot{x}(0)\} + \frac{1}{2}(\Delta t)^2\{\ddot{x}(0)\} \tag{11.19}$$

By using the above formula, the response of displacement, velocity and acceleration of structure under $F(t)$ can be obtained. The advantage of the central difference method is that it has high accuracy and high efficiency. The disadvantage is that the condition is stable, the time step is controlled by the stability, and the time step is:

$$\Delta t \le \Delta t_{cr}, \quad \Delta t_{cr} = \frac{2}{\omega \text{max}} = \frac{T_n}{\pi} \tag{11.20}$$

where, Δt is the time step; Δt_{cr} is the critical time step;ωmax is the maximum circular frequency; T_n is the minimum natural period.

The calculation steps of each time step $[t, t + \Delta t]$ of the central difference method are as follows:

(1) the initial displacement $\{x(t)\}$ of each step and the displacement $\{x(t - \Delta t)\}$ at the beginning of the previous step are known. They are either the end value of the previous step or the initial condition of the problem;
(2) the initial $\{x(-\Delta t)\}$ can be obtained from Eq. (11.19), the initial acceleration is

$$\{\ddot{x}(0)\} = [M]^{-1}(\{p(0)\} - [C]\{\dot{x}(0)\} - [K]\{x(0)\}) \tag{11.21}$$

(3) Equation (11.18) is used to get the displacement value $\{x(t + \Delta t)\}$ of the end point of the period;
(4) the velocity value $\{\dot{x}(t)\}$ of the starting point of the period is obtained from Eq. (11.16);
(5) the acceleration value $\{\ddot{x}(t)\}$ at the beginning of the period is calculated from Eq. (11.17).

When the fifth step is finished, the calculation of the current period is finished. To get the response of the next period, just repeat the above process and continue, all the response time history of the structure can be get.

2. **Wilson-θ method**

Wilson-θ method is essentially an extension of linear acceleration method, that is, assuming that the acceleration from t to $t + \theta\Delta t$ changes linearly, when $\theta \geq 1.37$, the method is unconditionally stable, and when θ is too large, the accuracy is poor, so commonly used $\theta = 1.4$.

Suppose $0 \leq \tau \leq \theta\Delta t$, on the assumption of linear acceleration, the acceleration vector in the interval $[t, t + \theta\Delta t]$ can be expressed as:

$$\{\ddot{x}(t + \Delta t)\} = \{\ddot{x}(t)\} + \frac{\tau}{\theta\Delta t}(\{\ddot{x}(t + \theta\Delta t\} - \{\ddot{x}(t)\}) \tag{11.22}$$

By integrating Eq. (11.22) and using the initial condition of t,

$$\{\dot{x}(t + \tau)\} = \{\dot{x}(t)\} + \tau\{\ddot{x}(t)\} + \frac{\tau^2}{2\theta\Delta t}(\{\ddot{x}(t + \Delta t)\} - \{\ddot{x}(t)\}) \tag{11.23}$$

$$\{x(t + \tau)\} = \{x(t)\} + \tau\{\dot{x}(t)\} + \frac{\tau^2}{2}\{\ddot{x}(t)\} + \frac{\tau^3}{6\theta\Delta t}(\{\ddot{x}(t + \Delta t)\} - \{\ddot{x}(t)\}) \tag{11.24}$$

Substituting $\tau = \theta\Delta t$ into Eqs. (11.23) and (11.24), then $\{\ddot{x}(t + \theta\Delta t)\}$ and $\{\dot{x}(t + \theta\Delta t)\}$ ($\{x(t + \theta\Delta t)\}$ as the basic unknown) can be solved,

$$\{\ddot{x}(t+\theta\Delta t)\} = \frac{6}{\theta^2(\Delta t)^2}(\{x(t+\theta\Delta t)\} - \{x(t)\}) - \frac{6}{\theta\Delta t}\{\dot{x}(t)\} - 2\{\dot{x}(t)\} \tag{11.25}$$

$$\{\dot{x}(t+\theta\Delta t)\} = \frac{6}{\theta\Delta t}(\{x(t+\theta\Delta t\} - \{x(t)\}) - \frac{\theta\Delta t}{2}\{\ddot{x}(t)\} - 2\{\dot{x}(t)\} \tag{11.26}$$

At the moment $t+\theta\Delta t$, the equation of motion of the system should satisfy the following equation,

$$[M]\{\ddot{x}(t+\theta\Delta t)\} + [C]\{\dot{x}(t+\theta\Delta t)\} + [K]\{x(t+\theta\Delta t)\} = \{F(t+\theta\Delta t)\} \tag{11.27}$$

The external load vector $\{p(t+\theta\Delta t)\}$ can be obtained by linear extrapolation,

$$\{F(t+\theta\Delta t)\} = \{F(t)\} + \theta(\{F(t+\Delta t)\} - \{F(t)\}) \tag{11.28}$$

From Eqs. (11.27) and (11.28),

$$\begin{aligned}&[M]\{\ddot{x}(t+\theta\Delta t)\} + [C]\{\dot{x}(t+\theta\Delta t)\} + [K]\{x(t+\theta\Delta t)\}\\ &\quad = \{F(t)\} + \theta(\{F(t+\Delta t)\} - \{F(t)\})\end{aligned} \tag{11.29}$$

Substituting Eqs. (11.25), (11.26) into (11.29),

$$\begin{aligned}&\left[\frac{6}{(\theta\Delta t)^2}[M] + \frac{3}{\theta\Delta t}[C] + [K]\right]\{x(t+\theta\Delta t)\}\\ &= \{p(t)\} + \theta(\{p(t+\theta\Delta t)\} - \{p(t)\})\\ &+ [M]\left[\frac{6}{(\theta\Delta t)^2}\{x(t)\} + \frac{6}{\theta\Delta t}\{\dot{x}(t)\} + 2\{\ddot{x}(t)\}\right]\\ &+ [C]\left(\frac{3}{\Delta t}\{x(t)\} + 2\{\dot{x}(t)\} + \frac{\theta\Delta t}{2}\{\ddot{x}(t)\}\right)\end{aligned} \tag{11.30}$$

The value of $\{\ddot{x}(t+\theta\Delta t)\}$ can be obtained by substituting $\{x(t+\theta\Delta t)\}$ (solved by Eq. (11.30)) into Eq. (11.25) and then taking $\tau = \Delta t$ (Eq. (11.22)), the value of $\{\ddot{x}(t+\Delta t)\}$ is

$$\{\ddot{x}(t+\Delta t)\} = \{\ddot{x}(t)\} + \frac{1}{\theta}(\{\ddot{x}(t+\theta\Delta t)\} - \{\ddot{x}(t)\}) \tag{11.31}$$

After obtaining $\{\ddot{x}(t+\Delta t)\}$, substitute Eqs. (11.23) and (11.24), and take $\tau = \Delta t$, $\{\dot{x}(t+\Delta t)\}$ and $\{x(t+\Delta t)\}$ can be get,

$$\{\dot{x}(t+\Delta t)\} = \{\dot{x}(t)\} + \Delta t\{\ddot{x}(t)\} + \frac{\Delta t}{2}(\{\ddot{x}(t+\Delta t\} - \{\ddot{x}(t)\}) \tag{11.32}$$

$$\{x(t+\Delta t)\} = \{\ddot{x}(t)\} + \Delta t\{\dot{x}(t)\} + \frac{(\Delta t)^2}{2}\{\ddot{x}(t)\} + \frac{(\Delta t)^2}{6}(\{\ddot{x}(t+\Delta t)\} - \{\ddot{x}(t)\}) \tag{11.33}$$

According to Eqs. (11.31), (11.32) and (11.33), the displacement, velocity and acceleration responses of structural vibration can be obtained.

The calculation steps of each time step $[t, t+\Delta t]$ of Wilson-θ method are as follows:

(1) the initial displacement $\{x(t)\}$ and velocity $\{\dot{x}(t)\}$ of each step are known, and they are either the end value of the previous step or the initial condition of the problem;
(2) calculate the initial acceleration according to Eq. (11.21);
(3) calculate the displacement value $\{x(t+\theta\Delta t)\}$ at the moment $t+\theta\Delta t$ from Eq. (11.30);
(4) calculate the acceleration value $\{\ddot{x}(t+\theta\Delta t)\}$ at the moment $t+\theta\Delta t$ from Eq. (11.27);
(5) calculate the acceleration value $\{\ddot{x}(t+\Delta t)\}$ at the moment $t+\Delta t$ from Eq. (11.31).
(6) calculate the speed $\{\dot{x}(t+\Delta t)\}$ and displacement $\{x(t+\Delta t)\}$ of the moment $t+\Delta t$ from Eqs. (11.32) and (11.33).

11.2.2 Finite Element Software and Secondary Development

11.2.2.1 Velocity Dependent Damper Element Based on Abaqus Software

1. **The principle of UEL implicit dynamic analysis**

The seismic response time history analysis of building structure belongs to dynamic analysis. The implicit dynamic analysis or explicit dynamic analysis are generally used in Abaqus. The author's team compiles subroutines using Dynamic /Implicit calculation methods and principles, using the Hilber-Hughes-Taylor (HHT) integration scheme, it is necessary to save the node velocity $\dot{u}^N$ and node acceleration $\ddot{u}^N$ of each step. These variable values are passed to UEL at the end of each calculation step. The UEL must define the parameters such as element Jacobian matrix $\tilde{K}^{NM}$, element residual vector RHS, and state variable H^α. Matrix $\tilde{K}^{NM}$ contribution will include the following three items:

$$-\frac{\partial F^N}{\partial u^M};\quad -\frac{\partial F^N}{\partial \dot{u}^M}\left(\frac{d\dot{u}}{du}\right)_{t+\Delta t};\quad -\frac{\partial F^N}{\partial \ddot{u}^M}\left(\frac{d\ddot{u}}{du}\right)_{t+\Delta t} \tag{11.34}$$

In HHT integration scheme:

$$\left(\frac{d\dot{u}}{du}\right)_{t+\Delta t} = \frac{\gamma}{\beta \Delta t}; \ \left(\frac{d\ddot{u}}{du}\right)_{t+\Delta t} = \frac{1}{\beta \Delta t^2}$$

Among them, β and γ are constants of Newmark-β integration method, parameter $\partial F^N/\partial \dot{u}^M$ is element damping matrix and $\partial F^N/\partial \ddot{u}^M$ is element mass matrix. The ABAQUS input file uses the dynamic time history analysis process defined by DYNAMIC. At this time, the dynamic implicit analysis is transmitted through the user subroutine flag array LFLAGS (1) = 11, 12. For the general normal iteration LFLAGS(3) = 1, the remaining quantity equations can be written as follows:

$$F^N = -M^{NM}\ddot{u}_{t+\Delta t} + (1+\alpha)G^N_{t+\Delta t} - \alpha G^N_t \tag{11.35}$$

The overall dynamic balance equation is:

$$-M^{NM}\ddot{u}_{t+\Delta t} + (1+\alpha)G^N_{t+\Delta t} - \alpha G^N_t = 0 \tag{11.36}$$

where, F^N is the nth node effect of the element, M^{NM} is the mass matrix, α is the HHT integral time damping constant, and the value is between −0.333 and 0. G^N_t and $G^N_{t+\Delta t}$ are respectively the total load action (excluding inertial action) at the Nth degree of freedom at the time of t, $t + \Delta t$, which is usually called static allowance. The quality matrix is defined as:

$$-\partial F^N/\partial \ddot{u}^M = M^{NM} \tag{11.37}$$

The total stiffness matrix defined by implicit dynamic analysis is:

$$AMATRX = N^{MN}\left(\frac{d\ddot{u}}{du}\right) + (1+\alpha)C^{NM}\left(\frac{d\dot{u}}{du}\right) + (1+\alpha)K^{NM} \tag{11.38}$$

Among them, $C^{NM} = -\partial G^N_{t+\Delta t}/\partial \dot{u}^M$, $K^{NM} = -\partial G^N_{t+\Delta t}/\partial u^M$. In addition, it is necessary to define RHS = F^N and update the status variable H^α.

In addition, the HHT integration method needs to define the half step residual vector of the relevant degrees of freedom to determine whether the iteration converges. Through LFLAGS(3) = 5 transfer half step margin calculation, when ABAQUS calculates the half step margin, the time step increment will be adjusted to meet the requirements of $F^N_{1/2}$ < HAFTOL. For HHT criterion, half step margin is defined as:

$$F^N_{1/2} = -M^{NM}\ddot{u}_{t+\Delta t/2} + (1+\alpha)G^N_{t+\Delta t} - \frac{\alpha}{2}(G^N_t + G^N_{t-}) \tag{11.39}$$

where, G^N_{t-} is the static margin at the beginning of the previous incremental step. Define RHS = $F^N_{1/2}$ and update the state variable $H^\alpha_{t+\Delta t/2}$.

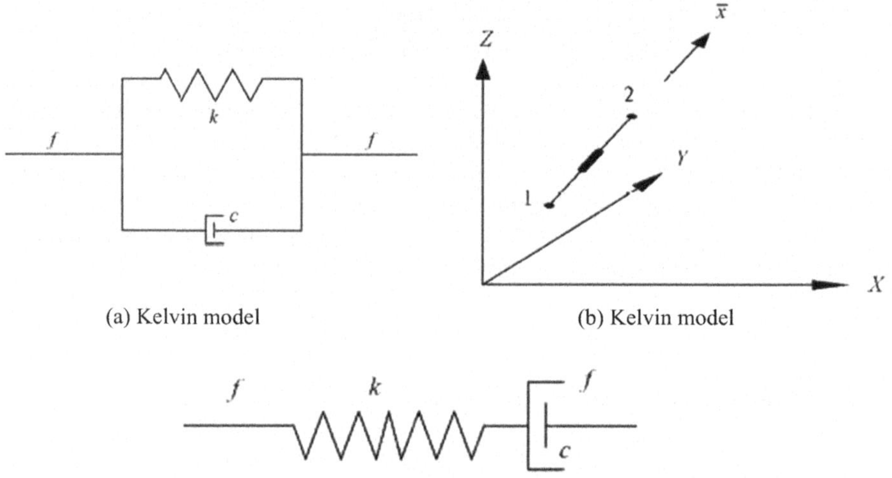

(a) Kelvin model (b) Kelvin model

(c) Maxwell model

Fig. 11.2 Mechanical model of velocity dependent damper

2. **Element characteristics of viscoelastic damper**

The calculation model of viscoelastic damper is usually Kelvin model, which is composed of a spring and a viscoelastic pot in parallel. As shown in Fig. 11.2a, the mechanical equation of Kelvin model is as follows:

$$f = c\dot{u} + ku \tag{11.40}$$

where, k is the spring stiffness; c is the damping coefficient of the pot; u is the displacement load acting on the model; f is the output force of the damper. Figure 11.2b is a viscoelastic element with inclined support in the space. The two end points 1 and 2 of the damper are respectively installed between the upper and lower adjacent layers of the structure, and the joints are generally treated as hinged joints. Let the local coordinate be $\bar{x}$, in the local coordinate system, the displacements of nodes 1 and 2 are $\bar{u}_1$ and $\bar{u}_2$, respectively, the velocity are $\dot{\bar{u}}_1$ and $\dot{\bar{u}}_2$, and the corresponding forces at both ends are $\bar{P}_1$ and $\bar{P}_2$. In the whole coordinate system XYZ, let the displacement of any node be u, v and w respectively, and the node force corresponding to u, v and w be P_x, P_y and P_z. Let the direction cosine of viscoelastic element be:

$$c_x = \cos(\bar{x}, X), c_y = \cos(\bar{y}, Y), c_z = \cos(\bar{z}, Z)$$

Let $[B] = \left[-c_x \ -c_y \ -c_z \ c_x \ c_y \ c_z\right]$, $\{U\} = [u_1 \ w_1 \ u_2 \ v_2 \ w_2]^T$, the local coordinates are as follows:

$$\Delta\bar{u} = [B]\{U\} \tag{11.41}$$

$$\Delta\dot{\bar{u}} = [B]\{\dot{U}\} \tag{11.42}$$

The force along the local coordinate $\bar{x}$ direction is:

$$\Delta\bar{P} = \bar{P}_2 - \bar{P}_1 = K\Delta\bar{u} + C\Delta\dot{\bar{u}} = K[B]\{U\} + C[B]\{\dot{U}\} \tag{11.43}$$

Let $\{P\}^e = [P_{x1}, P_{y1}, P_{z1}, P_{x2}, P_{y2}, P_{z2}]^T$, then the force at both ends of the node in the overall coordinate system is:

$$\begin{aligned}\{P\}^e &= [B]^T\Delta\bar{P} = [B]^T(K[B]\{U\} + C[B]\{\dot{U}\}) \\ &= K[B]^T[B]\{U\} + C[B]^T[B]\{\dot{U}\}\end{aligned} \tag{11.44}$$

The energy consumption increment of damper is calculated as follows:

$$\Delta E_d = \int_{t_i}^{t_{i+1}} \Delta\bar{P}\cdot\Delta\dot{\bar{u}}dt \tag{11.45}$$

The energy consumption increment of damper of the $i+1$ time step can be obtained by the numerical integration of trapezoid formula:

$$\begin{aligned}\Delta E_d^{(i+1)} &= \int_{t_i}^{t_{i+1}} \Delta\bar{P}(\tau)\cdot\Delta\dot{\bar{u}}(\tau)dt \\ &= \frac{1}{2}(\Delta\bar{P}(t_i)\cdot\Delta\dot{\bar{u}}(t_i) + \Delta\bar{P}(t_{i+1})\cdot\Delta\dot{\bar{u}}(t_{i+1}))\cdot\Delta t\end{aligned} \tag{11.46}$$

Overall stiffness matrix of viscoelastic damper element is:

$$\begin{aligned}AMATRX &= (1+\alpha)K[B]^T[B] \\ &\quad + (1+\alpha)C[B]^T[B](d\dot{u}/du)\end{aligned} \tag{11.47}$$

The residual vector of each degree of freedom of the node is:

$$RHS_{i+1}^j = RHS_i^j - P^j \tag{11.48}$$

3. **Element characteristics of viscous damper**

The calculation model of viscous damper is usually Maxwell model, which consists of a spring and a viscous pot in series, as shown in Fig. 11.2c. When the actual frequency is less than the cut-off frequency 4 Hz, the influence of spring stiffness on displacement can be ignored. Therefore, the relationship between the damping force F of viscous damper and its relative velocity U at both ends is as follows:

$$F = -c < \dot{U} >^m = \begin{cases} -c\dot{U}^m, (\dot{U} \geq 0) \\ c(-\dot{U})^m, (\dot{U} < 0) \end{cases} \quad (11.49)$$

Among them, c is the damping coefficient; when the flow index is $m = 1$, it is the linear viscous damping; when m $\neq$ 1, it is the nonlinear viscous damping; the flow index of the nonlinear viscous damper is between 0.1 and 1.0.

In general nonlinear analysis, ABAQUS uses Newton-Raphson iteration method. For the nonlinear viscous damper, the damping matrix C is constantly changing, and the tangent slope at the end of iteration is generally used to measure the incremental characteristics of nonlinear damping. In programming, only the element characteristics of viscous damper are described, and the element characteristics of other structures do not need to be concerned, and the main program transfers the displacement, velocity and acceleration to the subprogram after calculation, while the damping matrix of viscous damper is related to the velocity, so the tangent method is not used to deal with it, and Eq. (11.49) is derived as follows:

$$F = \begin{cases} -c\dot{U}^{m-1} \cdot \dot{U}, (\dot{U} \geq 0) \\ c(-\dot{U})^{m-1} \cdot (-\dot{U}), (\dot{U} < 0) \end{cases} \quad (11.50)$$

and

$$c_d = \begin{cases} -c\dot{U}^{m-1}, (\dot{U} > 0) \\ c(-\dot{U})^{m-1}, (\dot{U} < 0) \end{cases} \quad (11.51)$$

Because $0 < m < 1$, so $c_d = \infty$, when $\dot{U} = 0$, it is obviously meaningless, the above formula will lead to errors in the subroutine, so it is necessary to modify the incremental damping coefficient or damping force near the origin of velocity. The velocity $\dot{U}_0$ at a point near the origin is used, and the secant stiffness of the point is used to replace the damping coefficient near the velocity origin.

In the calculation and analysis, $\dot{U}_0 = 1.0\,\text{mm/s}$ (when the model is difficult to converge, it can be taken as 5.0 mm/s. The larger the value is, the larger the error will be, but the better the convergence will be) is used to deal with the non-linear segment near the origin. In $\left[-\dot{U}_0, \dot{U}_0\right]$, the overall stiffness matrix AMATRX is expressed in Eq. (11.47), only $K = 0$ is needed. For the non-linear segment, the derivation is as follows:

In the local coordinate system $\overline{xyz}$, the damping force produced by the viscous damper along the $\bar{x}$ direction is as follows:

$$\Delta\bar{P} = -c(\dot{\bar{u}}_2 - \dot{\bar{u}}_1)^m = -c([B] - \{\dot{U}\})^m \quad (11.52)$$

Then the force at both ends of the node in the overall coordinate system is:

$$\{P\}^e = [B]^T \Delta\bar{P} = -c[B]^T([B]\{\dot{U}\})^m \quad (11.53)$$

The above formula is transformed into:

$$\{P\}^e = -c([B]\{\dot{U}\})^{m-1}[B]^T[B]\{\dot{U}\} \ = [C_{DP}]\{\dot{U}\} \tag{11.54}$$

where

$$[C_{DP}] = -c([B]\{\dot{U}\})^{m-1}[B]^T[B] \tag{11.55}$$

which is the nonlinear damping matrix. The overall stiffness matrix of viscous damper element is as follows:

$$AMATRX = (1+\alpha)[C_{DP}](d\dot{u}/du) \tag{11.56}$$

The energy consumption of viscous dampers and the residual vector of joints are calculated according to Eqs. (11.46) and (11.48), respectively.

4. **Example verification**

The shaking table test model of 4-story steel frame structure carried out by the author's team is used for verification. According to the prototype structure, the plane size is 9 m × 4.5 m, the storey height is 3 m, the beam section is H320 × 130 × 15 × 9.5, the column section is H300 × 150 × 15 × 10, the line load of the bottom beam is 2.5 kN/m, and the line load of the remaining layers is 6.0 kN/m. The plan is shown in Fig. 11.3a. Viscoelastic damper and viscous fluid damper are respectively arranged between axis 1 and axis 2 of column A and B, and 8 dampers are respectively arranged in the form of diagonal connection. For this steel frame, the El-Centro-WE wave (peak acceleration 210 gal) is input along the X direction (length direction). In this case, the 8.5° frequent earthquake action is simulated, and the peak acceleration of the seismic wave is adjusted to 110 gal. The time history analysis of the structure without damper and with additional damper is carried out by ABAQUS and its subroutines Udamper.for and SAP2000 respectively. The SAP2000 model has passed the inspection and verification through the shaking table test of the scale model (see Sect. 11.3 for details).

Firstly, the simulation accuracy of viscoelastic damper element is compared and analyzed. The equivalent stiffness and damping of damper are taken as K = 8000 N/mm and C = 700 Ns/mm respectively. The structural displacement response is shown in Fig. 11.3b, c. It can be seen from the figure that the error between the calculated vertex displacement and interlayer displacement of SAP2000 and the result calculated by the subroutine udmper compiled in this paper is very small, which shows the correctness of this program. The interlayer displacement is very different before and after the shock absorption. The interlayer displacement before the shock absorption reaches 25 mm, and only 6 mm after the shock absorption. It

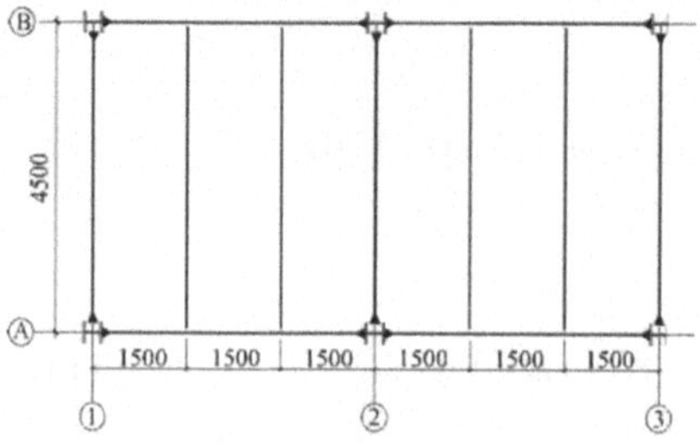

(a) Plan of steel frame

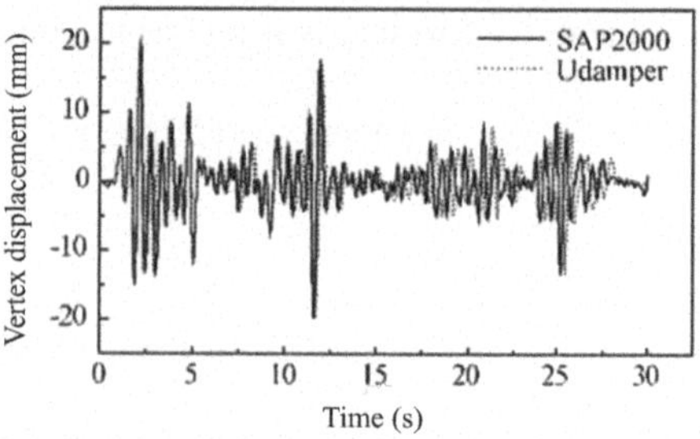

(b) Vertex displacement of viscoelastic damping structure

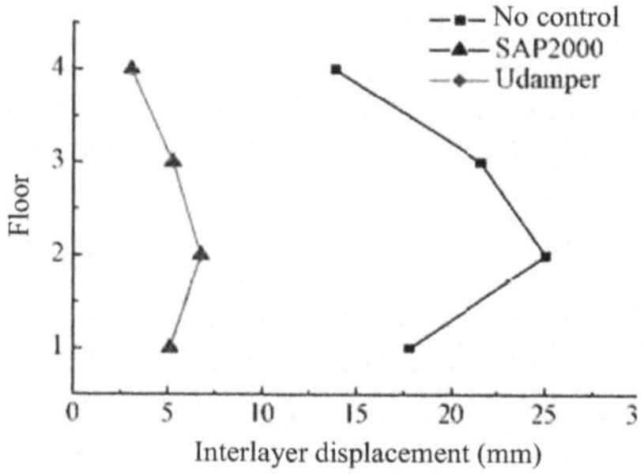

(c) Floor displacement of viscoelastic damping structure

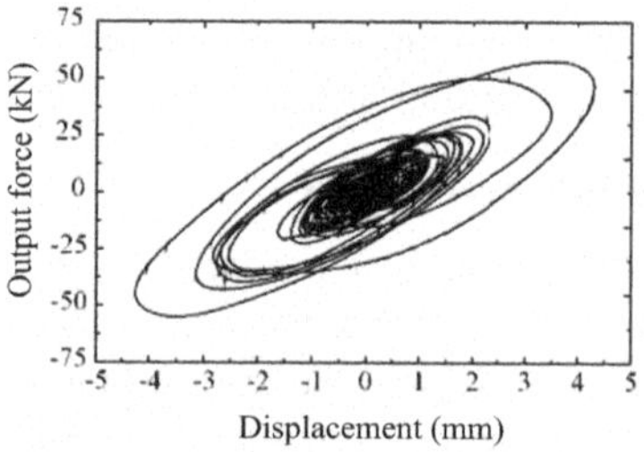

(d) Hysteretic curve of viscoelastic damper

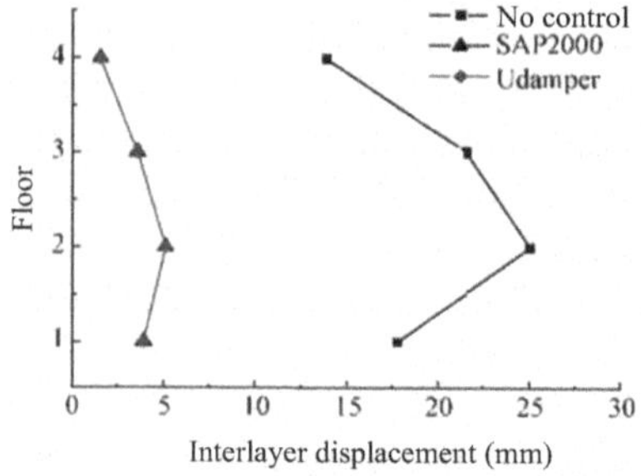

(e) Floor displacement of viscous fluid damping structure

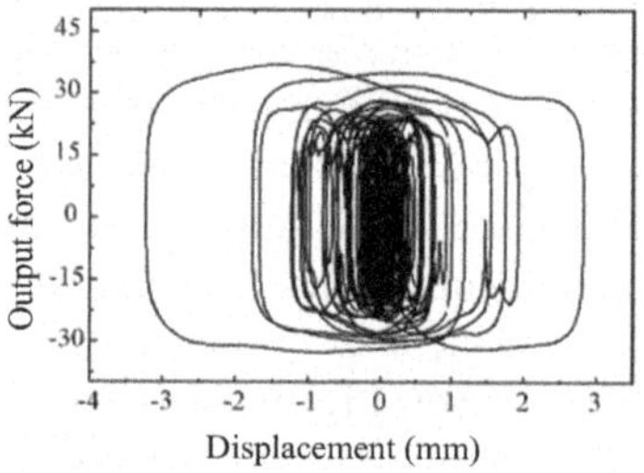

(f) Hysteretic curve of viscous fluid damper

Fig. 11.3 Comparison of numerical examples

shows that the viscoelastic damper element selected has the effect of shock absorption. In order to better reflect the energy dissipation performance of the viscoelastic damper in Udamper. for, select a viscoelastic damper hysteretic curve on the first floor as shown in Fig. 11.3d. It can be seen that the hysteretic curve is full, and the shape is approximately elliptic with stiffness, which is consistent with the theory and experimental research results of the authors' team.

Then, the simulation accuracy of viscous damper element is compared and analyzed. The damper damping coefficient $c = 100$ kN/(m/s) $^{0.3}$ and damping index $m = 0.3$ are taken. The structural displacement response is shown in Fig. 11.3e. It can be seen from the figure that the interlayer displacement calculated by SAP2000 is close to the result calculated by the subroutine Udamper.for. The interlayer displacement of the structure before the vibration reduction is 25 mm, and only 5 mm after the vibration reduction, which shows that the selected viscous damper element has the effect of vibration reduction. As shown in Fig. 11.3f, it can be seen that the hysteretic curve of a viscous damper on the first floor is full, and its shape is approximately a rectangle without stiffness, which is consistent with the theory and experimental results of the authors' team.

11.2.2.2 Variable Damping Viscous Fluid Damper Element Based on OpenSees Platform

1. **Parameter setting and development of variable damping viscous fluid damper**

According to the test results, the main working indexes of the variable damping viscous damper are the distribution of damping index α and damping coefficient C, among which the distribution of damping coefficient is determined by the length of small displacement segment h_1, damping coefficient C_1 of small displacement segment and damping coefficient growth slope k of large displacement segment. Although the physical meaning of k is clear, on the one hand, considering that the parameter interface given in the design can intuitively control and express the size of damping coefficient; on the other hand, the theoretical analysis results show that if the working condition displacement is large enough, the damping coefficient will be a fixed value after the damper piston completely enters the large displacement straight round hole section. Therefore, the damping coefficient C_1 in the small displacement section, the damping coefficient C_d in the pure large displacement section and the damping coefficient changing section length Δl are used instead of the slope, $k = \frac{C_d - C_1}{\Delta l}$.

where, $C_2 = C_1 + kx = C_1 + \frac{C_d - C_1}{\Delta l}x$, is damper displacement.

The simulation of the damper in OpenSees is realized by defining the viscous damping material, and then giving the material characteristics to the element. Because the mechanical calculation of the material is on the micro level of stress-strain, it is necessary to have a certain element length and area A to convert the stress-strain into the element force and displacement, and then correspond to the restoring force model of the damper. To sum up, C_1, C_d, h_1, Δl, α, l and A are selected as the basic input parameters of the material element of the variable damping viscous damper in the OpenSees calculation program.

Based on the original OpenSees damper simulation material source code, the parameter interface used to describe the working performance of VSD is added, and the new simulation material is embedded in the source code through Microsoft Visual

Studio (VS) and recompiled to generate the new OpenSees computing program, which includes both the damper simulation original material and the VSD simulation new material. The technical route of is shown in Fig. 11.4a.

2. **Comparison and selection of OpenSees viscous damping materials**

There are three kinds of material elements that can be used to simulate viscous fluid damper in OpenSees material element library: Maxwell material, viscous material

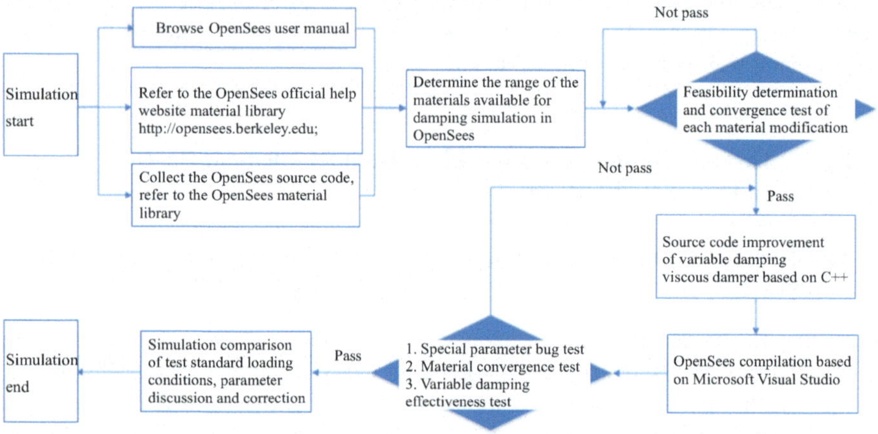

(a) Technical route of simulation of variable damping viscous fluid damper

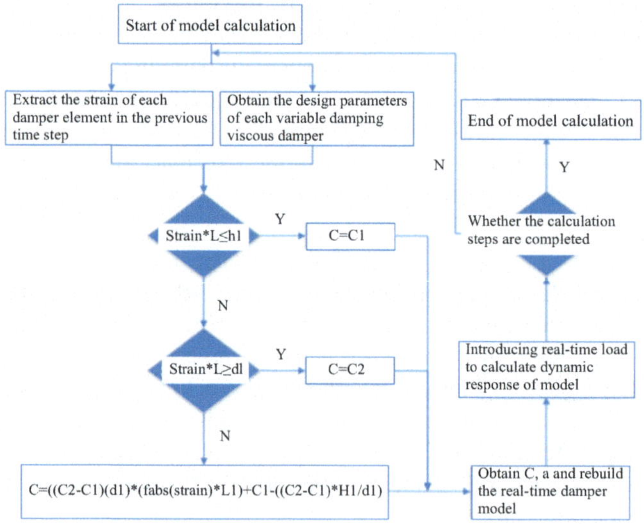

(b) Flow chart of structural time history analysis

Fig. 11.4 General idea of development of viscous fluid damper element with variable damping and ViscousDamper material. Among them, Maxwell material is the source code compiled by Professor Sarven Akcelyan of McGill University. The restoring force model adopted is Maxwell model. Therefore, the source code of variable damping viscous damper can be directly improved on the basis of this material element.

The viscous material prepared by Professor Mehrdad Sasani of Northeastern University of Boston can be used to simulate the damper material with the following stress-strain relationship. The core formula for calculating the restoring force of the damper is the deformation of the physical experiment result formula $F = CV^{\alpha}$. However, in the actual calculation process, the fitting convergence of viscous material to the nonlinear damper is not good, so it can be mainly used to simulate the linear damper.

$$\sigma = C\left(\frac{d\varepsilon}{dt}\right)^{a} \tag{11.57}$$

Among them, σ is the axial normal stress of damper element, ε is strain, C is damper damping coefficient and a is damping index.

ViscousDamper material is an improvement of Viscous material. This material element introduces an adaptive iterative algorithm to solve the problem of poor convergence of Viscous material. However, due to the introduction of this material of from OpenSees2.5.0 version, PEER only disclosed the source code before OpenSees2.3.0, so the introduction of new materials in this paper cannot be modified with this material unit.

Considering that the quasi convergence of Viscous material is not good due to the existence of non integer exponential operation, in the actual source code modification process, in addition to the above two material elements, through the formula $\mathrm{F} = \mathrm{C}V^{\alpha}$ to carry out the following Taylor expansion and other deformation to eliminate the non integer exponential calculation in the restoring force calculation, it is expected to improve the convergence of viscous material. To sum up, the new material element of variable damping viscous damper is based on Maxwell material, viscous material and Taylor expansion viscous material for improvement and embedding of new parameter interface. The convergence test results and feasibility modification feasibility of each material are summarized in Table 11.1.

Table 11.1 Selection of simulation materials for dampers of OpenSees

Material category	Open source or not	Astringency	Resilience computing core	Whether to choose
Maxwell material	Yes	Excellent	Maxwell model	Yes
Viscous	Yes	Poor	$\mathrm{F} = \mathrm{CV}^{\alpha}$	Yes
Viscous damper	No	Excellent	$\mathrm{F} = \mathrm{CV}^{\alpha}$	No
Taylor expansion viscous	Self compiling	To be tested	Taylor expansion of $\mathrm{F} = \mathrm{CV}^{\alpha}$	Yes

$$F = CV^{\alpha} = C\left(\frac{du}{dt}\right)^{a} = C\left(\frac{d\varepsilon}{dt}\right)^{a} l^{\alpha} \Rightarrow \sigma = C\left(\frac{d\varepsilon}{dt}\right)^{a} \frac{l^{\alpha}}{A} \tag{11.58}$$

McLaughlin expansion on the above formula:

$$\begin{aligned}\sigma &= C\left(\frac{d\varepsilon}{dt}\right)^{a} \frac{l^{\alpha}}{A} \\ &= C\frac{l^{\alpha}}{A}\left[1 + \left(\frac{d\varepsilon}{dt} - 1\right) + \frac{1}{2}\left(\frac{d\varepsilon}{dt} - 1\right)^{2}\right. \\ &\quad \left. + \frac{1}{6}\left(\frac{d\varepsilon}{dt} - 1\right)^{3} + \cdots \frac{1}{n!}\left(\frac{d\varepsilon}{dt} - 1\right)^{n} + Rn(x)\right]\end{aligned} \tag{11.59}$$

3. **Development of variable damping material element**

From the source code of OpenSees2.3.0 version, the files of Maxwell material.h, Maxwell material.cpp, Viscous.h and Viscous.cpp are found. Based on the compiling platform of Microsoft Visual Studio, the code of each material is modified by C++ language. The modification idea is to add parameter interface $C1$, Cd, $h1$ and Δl, without changing the original formal parameter and actual parameter, before calling the stress calculation function, First, the damping coefficient of this time step is calculated by using the parameter interface $C1$, Cd, $h1$ and Δl through adding conditional statement and the structural dynamic response of the previous time step. The calculation result is given to the damping coefficient parameter C, which is used for the subsequent stress calculation. The calculation process after the reassignment of C is the same as that of the conventional energy dissipation structure with damper. The calculation process of each time step is shown in Fig. 11.4b.

Among them, according to the calculation result that the damping coefficient C increases linearly after the loading test piston enters into the hole diameter changing section, the calculation method of the finite element damping coefficient C is as follows: when the displacement of the damper piston is less than the length of the small displacement section of the damper rod, the damping coefficient is taken as C_1; when the displacement of the piston is between $h1$ and Δl, the damping coefficient C is obtained by linear interpolation of $(h_1 C_1)$ and $(\Delta l c_d)$; When the piston displacement is greater than Δ, the damper damping coefficient is taken as C_d.

After completing the source code modification of each simulation material element, add the source codes of three groups of new materials (including class body file. cpp and class declaration file. h) into the OpenSees uniaxial material library, and add the corresponding class declaration in the relevant material declaration, call and assembly file to ensure that the three groups of new materials can be declared and called normally. After embedding the new material, recompile OpenSees with Microsoft Visual Studio and generate the OpenSees calculation program containing the new material (Table 11.2).

Table 11.2 Input of material parameters

Conventional damper	Definition of Maxwell material model	UniaxialMaterial Maxwell $matTag $K $C $a $Length
	Definition of viscous model	UniaxialMaterial Viscous $matTag $C $alpha
Variable damping viscous damper	Variable damping Maxwell material model	UniaxialMaterial VarMaxwell $matTag $K $C1 $Cd $a $h1 $dl $Length
	Variable damping viscous model	UniaxialMaterial VarViscous $matTag $C1 $Cd $alpha $h1 $dl
	Variable damping Taylor expansion viscous model	UniaxialMaterial VarTaylorViscous $matTag $C1 $Cd $alpha $h1 $dl

Note $matTag indicates the self named material name, $C1 is the input damping coefficient of small displacement segment, $Cd is the constant damping coefficient of large displacement segment, $h1 is the length of small displacement segment, $dl is the length of damping coefficient changing segment

After compiling, different loading conditions are applied to debug the failure and test the convergence of the simulation materials. The test results show that there are still some problems in the convergence of the variable damping viscous model and the variable damping Taylor expansion viscous model, which can only be used to simulate the linear damper with a damping index of 1; the variable damping Maxwell material model has good convergence, and can successfully simulate a fuller "dog bone type hysteretic curve", which can be used to carry out the analysis and simulation of energy dissipation and vibration reduction of engineering structures with variable damping viscosity dampers.

4. **Test verification**

After the compilation of the new OpenSees program, the successfully compiled Maxwell material model with variable damping is transformed into a closed matlab function file, and the standard loading condition of sinusoidal simple harmonic displacement wave of the variable damping viscous damper is simulated by matlab software.

The performance test data of variable damping viscous fluid damper conducted by the team is used to verify the developed element, and the comparison results are shown in Fig. 11.5. It can be seen that the simulation results of Maxwell material model with variable damping obtained from SJ2 calculation are similar to the test results, which shows obvious "dog bone" variable damping characteristics and frequency correlation under the condition of large displacement. Compared with the conventional damper, the variable damping viscous damper can still provide large damping force after entering into large displacement. It can be seen from Fig. 11.5 that in general, the MATLAB fitting value based on the variable damping Maxwell material model is close to the test result, and the difference is no more than 10% of the

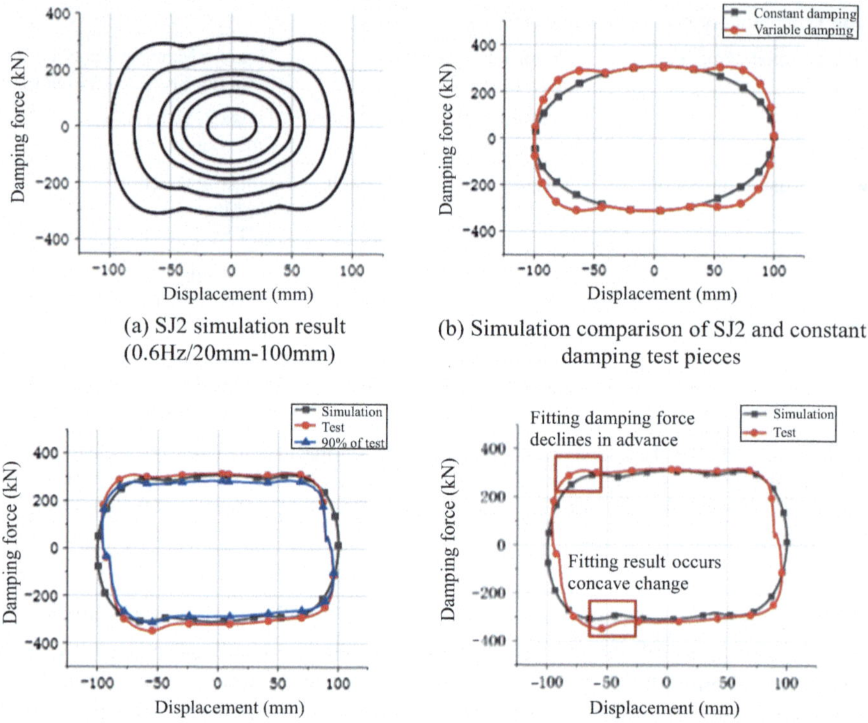

(a) SJ2 simulation result (0.6Hz/20mm-100mm)

(b) Simulation comparison of SJ2 and constant damping test pieces

(c) Comparison of SJ2 test and simulation values

(d) SJ2 test and simulation value comparison analysis (0.6Hz/100mm)

Fig. 11.5 Verification of variable damping viscous fluid damper element development

test value, so it can be considered that the variable damping Maxwell material model can simulate the working performance of the variable damping viscous damper more accurately in the finite element analysis.

Compared with the software simulation and test results, it can be found that the test results at the end of the positive displacement are close to the fitting results, and the test values at the negative displacement end are obviously larger than the fitting values. Therefore, it can be inferred that the displacement loss caused by the MTS support looseness or the piston itself does not accurately recover to the balance position before the loading under the loading condition of 0.6 Hz-100 mm, and the hysteresis curve of the damper test of SJ2 appears slight "big and small head" phenomenon. The fitting damping force of the damper is significantly reduced before the "dog bone" lifting section of the hysteretic curve compared with the test results, and the fitting results show an obvious "concave" shape at this position. It can be seen from the SJ2 damping coefficient distribution diagram obtained from the data processing of the test results that the distribution rule of the damping coefficient of the variable damping viscous damper test piece is not strictly linear change in the damping coefficient change section, but the damping coefficient value of the varying damping coefficient section defined in the simulation material element is obtained by linear interpolation. This algorithm is not completely consistent with the test results, so it causes the different damping coefficient change section of the two hysteresis loops. The fitting damping force starts to decrease obviously when the piston displacement reaches 80 mm, while the test damping force starts to decrease linearly when the displacement is close to the peak value. The reasons are as follows: the phase difference between the output and displacement of the damper is not exactly 90° in the test loading, but slightly greater than 90°. However, the influence of this tiny phase difference is not considered in the source code of the varying damping Maxwell material model in the simulation of the standard working condition, and the change of the damping force and the displacement still show a strict 90° phase difference. Therefore, the position of the peak value of the damping force in the test results has a little lag compared with the fitting result.

11.2.2.3 Lead Rubber Bearing Element Based on OpenSees Platform

1. **Development of lead rubber bearing element**

As shown in Fig. 11.6, the developed element model is a three-dimensional macro model, including 2 nodes and 12° of freedom. Based on the principle of coordinate transformation, the basic unknown quantity (displacement and force) in the element will be transformed from the basic coordinate system to the local coordinate system of the element, and finally to the overall coordinate system of the element. The mechanical properties of the element in each direction in the basic coordinate system will be described in detail below.

(a) **Horizontal shear performance**

The lead rubber bearing shows nonlinear hysteretic behavior in the horizontal direction. Park et al. extended the one-way Bouc-Wen model to obtain the two-way park model. The model can consider the two-way coupling effect of the bearing in the horizontal plane, and has been widely used in programs such as 3D-BASIS and SAP2000. However, the Park model has inconsistent loading and unloading rules in

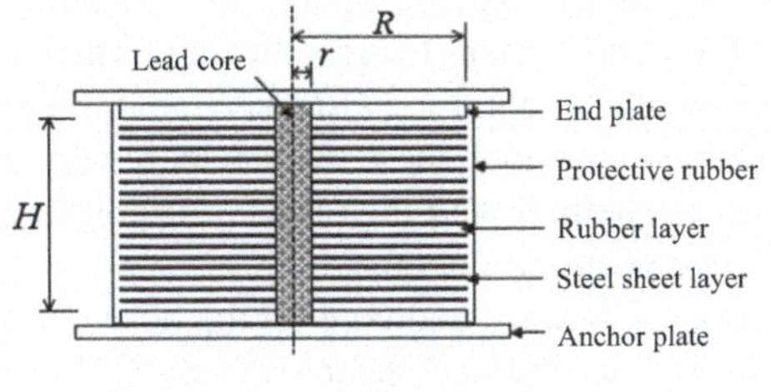

(a) Support diagram

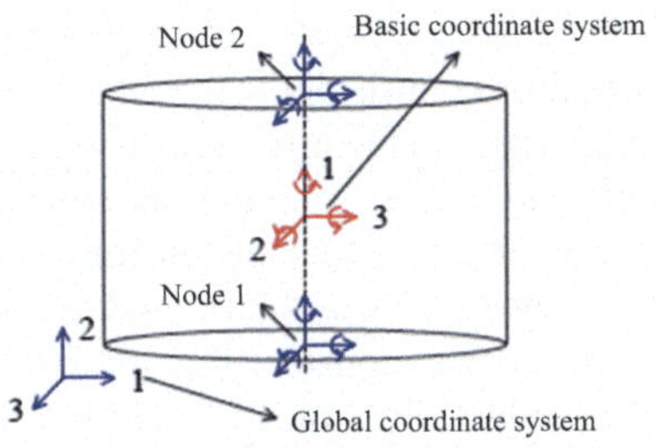

(b) Element coordinate system

Fig. 11.6 Element model of lead rubber bearing

two orthogonal directions, which violates the assumption of circular section bearing isotropy. Casciati improved the loading and unloading conditions in two orthogonal directions to meet the requirements of this assumption, so this book uses Casciati model to consider the horizontal two-way coupling effect of the support.

In the Casciati model, the restoring forces F_x and F_y in x and y directions satisfy the following relations:

$$\begin{bmatrix} F_x \\ F_y \end{bmatrix} = K_{d,0}\begin{bmatrix} U_x \\ U_y \end{bmatrix} + Q_d\begin{bmatrix} Z_x \\ Z_y \end{bmatrix} + c_d\begin{bmatrix} \dot{U}_x \\ \dot{U}_x \end{bmatrix} \tag{11.60}$$

where, $K_{d,0}$ is the stiffness of the bearing after initial yield; U_x and U_y are the deformation of the bearing in X and Y directions, respectively; Q_d is the characteristic strength of the bearing; c_b is the viscous damping of the rubber material, which is generally ignored; and $Z = [Z_x \;\; Z_y]^T$ is the hysteretic displacement as a function of U_x and U_y, which satisfies the following first-order nonlinear differential equation:

$$Y\begin{bmatrix} \dot{Z}_x \\ \dot{Z}_y \end{bmatrix} = \left(AI - \chi\begin{bmatrix} Z_x^2 & Z_x Z_y \\ Z_x Z_y & Z_y^2 \end{bmatrix}\right)\begin{bmatrix} \dot{U}_x \\ \dot{U}_y \end{bmatrix} \tag{11.61}$$

And:

$$\chi = \|Z\|^{\eta-2}[\delta + \beta sign(\dot{U}_x Z_x + \dot{U}_y Z_y)] \tag{11.62}$$

Among them, the yield displacement of the bearing is $Y = Q_d/((\tau - 1)K_d)$, parameter τ is the stiffness ratio before and after yield, usually taken as 10; parameter η controls the smoothness of the transition from elastic to plastic state of the restoring force curve; parameter I is the second-order unit matrix; parameter δ, β and A jointly control the shape and size of the restoring force curve; the range of the hysteretic displacement Z is as follows: $||z|| \le \sqrt[\eta]{A/(\beta+\delta)}$. When $A = 1$, $\delta + \beta = 1$, Eq. (11.61) represents circular yield surface, and $||z||$ is between $\pm$ 1. Based on the existing research, the values of the above parameters are as follows: $A = 1, \beta = 0.5, \delta = 0.5$ and $\eta = 2$. *Sign* () is a sign function, which is different from the $sign(\dot{U}_i Z_i)$, $i = x, y$, which is used in Park model. Cassiati model adopts the improved $sign(\dot{U}_x Z_x + \dot{U}_y Z_y)$ discriminant, which makes the loading and unloading conditions consistent in the two orthogonal directions, so as to meet the requirements of isotropy. In the process of solving the internal element, Eq. (11.61) is transformed into a nonlinear system of incremental equations, and then the Z of each incremental step is obtained by Newton Raphson iterative method, and then the corresponding horizontal stiffness matrix and restoring force vector are obtained.

In order to consider the influence of the existence of lead core on the horizontal mechanical properties of bearing, the concept of effective shear stiffness G_{eff} of bearing is introduced:

$$G_{eff} = \frac{K_{d,0}T_r}{A_b} \tag{11.63}$$

where, A_b is the cross-sectional area of the support; T_r is the total rubber thickness ($T_r = n \times t_r$, where n is the number of rubber layers; t_r is the thickness of single rubber layer). Based on the theory of double spring model, the influence of vertical pressure load on the horizontal stiffness of bearing is considered:

$$K_d = K_{d,0}\left[1 - \left(\frac{P}{P_{cr,0}}\right)^2\right] \tag{11.64}$$

where, P is the vertical pressure load and $P_{cr,0}$ is the initial critical bearing capacity of the bearing.

(b) **Axial performance**

Due to the small horizontal stiffness of the bearing, the isolation layer often produces a large horizontal displacement under the action of ground motion, and the vertical stiffness and critical bearing capacity of the bearing decrease accordingly; in addition, under the combined action of the vertical seismic component and the overturning moment of the upper structure, the bearing may produce tensile deformation. Therefore, the influence of the above factors is considered in this element model, so that the mechanical properties of the bearing can be described more accurately. When the internal and external radii of the bearing section are γ and R respectively, the initial compression stiffness $K_{v,0}$ is calculated as follows:

$$K_{v,0} = \frac{E_c A_b}{T_r} \tag{11.65}$$

Among them, the compression modulus E_c of rubber is:

$$E_c = \frac{1}{\left(\frac{1}{6G_{eff}S_1^2\lambda} + \frac{4}{3K}\right)} \tag{11.66}$$

and

$$\lambda = \frac{R^2 + r^2 - \frac{R^2 - r^2}{\ln(R/r)}}{(R - r)^2} \tag{11.67}$$

where, the first shape coefficient is $S_1 = R/(2t_r)$; K is the bulk modulus of elasticity of rubber, usually 2 GPa.

The verified piecewise linear method is used to consider the influence of the horizontal displacement u_h on the vertical stiffness K_v of the bearing:

$$\frac{K_v}{K_{v,0}} = \begin{Bmatrix} 1 - 0.4(u_h/R), & u_h/R \le 2 \\ 0.2, & u_h/R > 2 \end{Bmatrix} \tag{11.68}$$

In order to effectively distinguish the failure mode of bearing, it is very important to accurately calculate its vertical critical bearing capacity. The initial critical load $P_{cr,0}$ under zero lateral displacement of the bearing is calculated as follows:

$$P_{cr,0} = \sqrt{P_s P_E} = \frac{\pi\sqrt{GAE_b I}}{T_r} \tag{11.69}$$

and

$$E_b = \left(\frac{1}{2G_{eff}S_1^2\lambda_b} + \frac{10}{7K}\right)^{-1} \tag{11.70}$$

The overlapping area method is used to calculate the critical bearing capacity P_{cr} under the condition of lateral displacement of bearing:

$$P_{cr} = P_{cr,0}\left(\frac{A_r}{A_b}\right) \tag{11.71}$$

In the existing structural analysis programs, the tensile properties of bearings are often ignored or the linear stiffness model is used to describe the tensile mechanical behavior of bearings. In this book, the tensile phenomenological model proposed by Kumar et al. is used to consider the tensile properties of bearings. It should be pointed out that the existence of lead core has little effect on the tensile properties of bearing. As shown in Fig. 11.6, after the bearing enters the tensile state, before the tensile load reaches the initial damage strength ($F_c = 3GA_b$), it is in the linear elastic stage and the tensile rigidity is consistent with the compression rigidity. Thereafter, as the tensile load increases, the tensile capacity is calculated as follows:

$$F = F_c\left[1 + \frac{1}{kT_r}(1 - e^{-k(u_v - u_c)})\right] \tag{11.72}$$

Among them, u_v is the current tensile displacement value; u_c is the tensile deformation value corresponding to the initial damage strength F_c, and $u_c = F_c/K_v$; parameter k is used to reflect the tensile damage degree of the bearing, the greater the value of k, the greater the reduction degree of bearing capacity after the bearing damage.

When the tensile displacement exceeds u_c and then unload, the unloading path is different from the original loading path; when the load is loaded again, it will first follow the previous unloading path, this loading process ends until u_v exceeds the experienced maximum displacement u_{max}. When the load is further increased, the subsequent process still follows Eq. (11.72); when the load is unloaded again, a new unloading path is formed again. The unloading path of each cycle is approximately along the straight line between (u_{max}, F_{max}) and (u_{cn}, F_{cn}), and both of them are constantly changing during the loading process, where F_{cn} is the current damage strength, and the degree of reduction depends on the maximum deformation value u_{max} that has been experienced at present:

$$F_{cn} = F_c\left(1 - \phi\text{max}\left[1 - e^{-\alpha\left(\frac{u_{\text{max}} - u_c}{u_c}\right)}\right]\right) \quad (11.73)$$

Among them, α is the strength degradation parameter and ϕ_{max} is the maximum damage parameter that can be predicted. Based on the existing research, the above parameters are as follows: $k = 10$, $\phi_{max} = 1$ and $\alpha = 0.75$.

(c) **Bending and torsion performance**

The bending and torsion characteristics of the bearing have little influence on its overall performance, so the linear model is adopted in this element to simplify the consideration of the bending and torsion performance of the bearing:

$$K_b = 2GS_1^2 I_s\left(1 - \frac{3GS_1^2}{K}\right)\frac{H}{T_r} \quad (11.74)$$

$$K_r = \frac{2GI_s}{T_r} \quad (11.75)$$

where, K_b is the bending stiffness; K_r is the torsional stiffness; I_s is the moment of inertia of the support section; H is the total height of the steel plate layer and rubber layer inside the support.

2. **Test verification**

In order to verify the validity of the element model, this section carries out simulation analysis for the static and dynamic tests in the relevant literature; considering that the size of the bearing used in the tests with specific research purposes is small, the influence of the bearing specification parameters on the mechanical properties in all directions considered by the element is studied; in addition, the variation effect of the compression stiffness and the critical bearing capacity with the lateral displacement on the performance of bearing is considered.

(a) **Static horizontal two-way displacement control test simulation**

In order to verify the accuracy of the element model for the simulation of the horizontal two-way coupling effect of the bearing, the simulation analysis of the two-way displacement control test of the lead rubber bearing in the literature is carried out. The parameters and specifications of the bearing are as follows: the outer diameter of the bearing is 167.9 mm, the inner diameter is 29.97 mm, the total thickness of the rubber layer is $19 \times 3.175 = 60.325$ mm, the total thickness of the steel plate layer is $18 \times 1.905 = 35.29$ mm, the total height of the bearing is 151.77 mm, the shear modulus of the rubber is $G = 0.64$ MPa, the characteristic strength value of the bearing is $Q_d = 6.786$ kN, and the rigidity after initial yield is $K_{d,0} = 0.1775$ kN/mm.

Figure 11.7 compares the results of test and simulation, the support shows strong two-way coupling effect in the horizontal plane, when the x-direction horizontal shear strain reaches 100% and remains unchanged, with the increase of Y-direction displacement, the X-direction restoring force decreases to about half of the amplitude of one-way restoring force (Fig. 11.7a), the element model can relatively reasonably

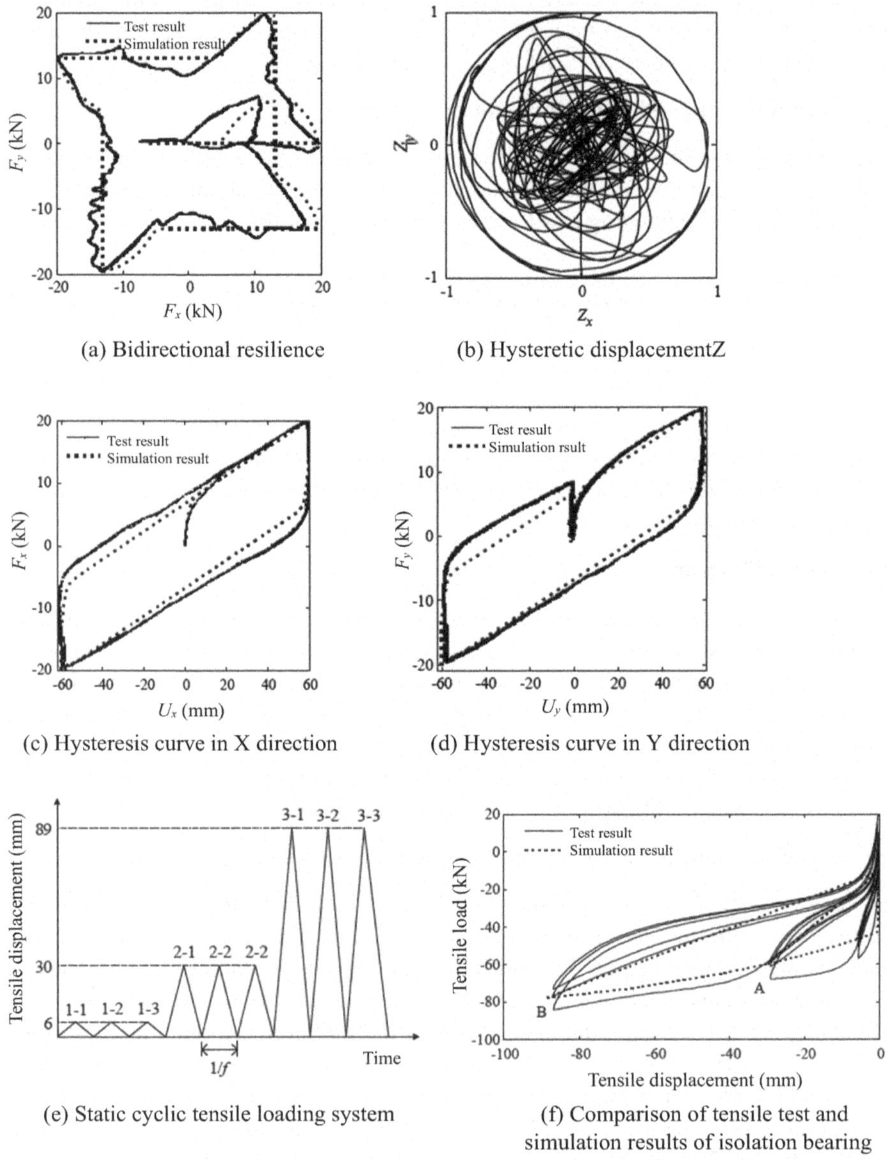

(a) Bidirectional resilience

(b) Hysteretic displacementZ

(c) Hysteresis curve in X direction

(d) Hysteresis curve in Y direction

(e) Static cyclic tensile loading system

(f) Comparison of tensile test and simulation results of isolation bearing

Fig. 11.7 Comparison of test and simulation results of isolation bearing simulate the two-way coupling behavior of restoring force. The time history results of the hysteretic displacement Z are shown in Fig. 11.7b. It can be seen that it is a circular yield surface, and it is between $\pm$ 1. As shown in Fig. 11.7c–d, the experimental and simulation results of the hysteresis curves in X and Y directions are basically the same in the given range of horizontal shear strain.

(b) **Static cyclic tensile performance test simulation**

The cyclic tensile test of lead rubber bearing in the literature is simulated and analyzed. The parameters of the test bearing are as follows: the outer diameter is 152 mm, the inner diameter is 30, 3 mm single rubber, 20 layers in total, 3 mm single steel plate, total bearing height is 167 mm, the shear modulus of rubber is 0.82 MPa, characteristic strength Q_d is 7 kN, the stiffness after initial yield $K_{d,0}$ is 0.398 kN/mm. The static cyclic tensile loading system is shown in Fig. 11.7e. The displacement amplitudes are 6, 30 and 89 mm respectively. three times of loading for each displacement amplitude level, and the loading frequency is 0.01 Hz.

Figure 11.7f compares the test and simulation results of the bearing tensile load-displacement. During the loading cycle 1-1, 1-2 and 1-3, the tensile displacement amplitude is 6 mm, the bearing is in the linear elastic state as a whole, the tensile stiffness is basically unchanged, and the loading and unloading path is consistent. In the process of tensile cycles 2-1, 2-2 and 2-3, when the tensile displacement reaches 30 mm for the first time (in cycle 2-1), the displacement amplitude exceeds the initial hole displacement u_c, and the inner rubber layer appears hole damage, which causes the bearing's tensile stiffness to decrease (point A). The permanent damage of rubber layer shows nonlinear energy dissipation characteristics, so that the unloading follows a new path; when it is loaded again (cycles 2-2 and 2-3), the loading and unloading are conducted along the unloading path of cycle 2-1. In the process of tensile cycle 3-1, the bearing first follows the original path to point A, and then, based on the loading path indicated by Eq. (11.72), it reaches point B, and the unloading also follows the new unloading path. The loading and unloading processes of cycle 3-2 and cycle 3-3 follow the unloading path of tensile cycle 3-1. It can be seen from Fig. 11.7 that this element model well simulates the mechanical characteristics of permanent damage and strength degradation of bearing under cyclic tensile load.

11.3 Vibration Control Dynamic Test of Building Structure

In order to further verify the correctness of the theory and method of the vibration reduction analysis of the building structure facing the developed vibration reduction device, so as to lay a theoretical foundation for the further design method and engineering application, the dynamic tests of the structure energy dissipation and damping structure system, frequency modulation and damping structure system and the isolation structure system are carried out systematically [1, 7–9].

11.3.1 Dynamic Test of Energy Dissipation and Damping Structure System

11.3.1.1 Viscous Fluid and Mixed Damping Vibration Control Test

1. **Test model**

The design goal of shaking table test is to keep the elastic state of steel structure frame in the test process, to keep the elasticity of soft steel damper in the small earthquake, to provide the rigidity for the structure, to start to yield in the middle earthquake, and to yield deeply with energy consumption in the large earthquake. The prototype structure of shaking table test is proposed to be built in the area with seismic fortification intensity of 8.5°, and the design basic seismic acceleration is 0.3 g. The model is designed as an under artificial mass model. According to the performance of the shaking table and laboratory conditions, it is determined that the length similarity ratio of the shaking table model frame and the prototype structure is S_l= 1/3, the materials used for the frame are the same as the prototype structure, and Q235 steel is used, and the similarity ratio of the elastic modulus is 1. According to the similarity relationship discussed above, the similarity ratio of the main physical quantities of the model is determined as shown in Table 11.3.

The test model is a four story spatial steel frame (Fig. 11.8), with three-dimensional dimensions of 3 m × 1.5 m × 4 m. H-shaped steel is used for frame column, with section size of H100 × 100 × 6×8, I-shaped steel is used for frame beam, with section size of I100 × 68 × 4.5 × 7.6. The herringbone support for the installation of mild steel damper is in the form of double angle steel, and the section size of angle steel is ∠50 × 4.5 mm thick steel plate shall be laid on each floor to install artificial mass block. The artificial mass is calculated according to the similarity relationship. 3.32 t of mass block is set in the 1st–3rd floor and 1.4 t of mass block is set in the 4th floor.

2. **Test equipment and test conditions**

In the shaking table test, the one-way shaking table of Civil Engineering Laboratory of Jiulonghu campus of Southeast University is used as the loading platform (as shown in Fig. 11.9a). Its main performance and technical parameters are as follows: table size: 4 m × 6 m; frequency range: 0.1 ~ 50 Hz; maximum model mass: 25 t; maximum displacement: X direction: ± 250 mm; maximum speed: X direction: 600 mm/s; maximum acceleration: X direction: 3.0 g (no load), 1.5 g (load 25 t). On the shaking table and in the middle of the beam on the first, second, third and fourth floors, a pull type displacement meter and an accelerometer are arranged respectively along the moving direction of the shaking table.

During the test, there are three models: one is pure frame model I; one is model II with two parabolic mild steel dampers on the bottom of the frame and two viscous fluid dampers on the second and third floors of the frame; one is model III with two viscous fluid dampers on the second and third floors of the structure. The loading system of these three framework models is identical. For the convenience of description, the pure frame model is recorded as model I, the frame with mild

Table 11.3 Model similarity (model/prototype)

Physical quantity	Length	Density	Stress	Time	Displacement	Velocity	Acceleration	Frequency	Mass
Similarity relation	1/3	2.29	1	0.50	1/3	0.66	1.31	2	0.085

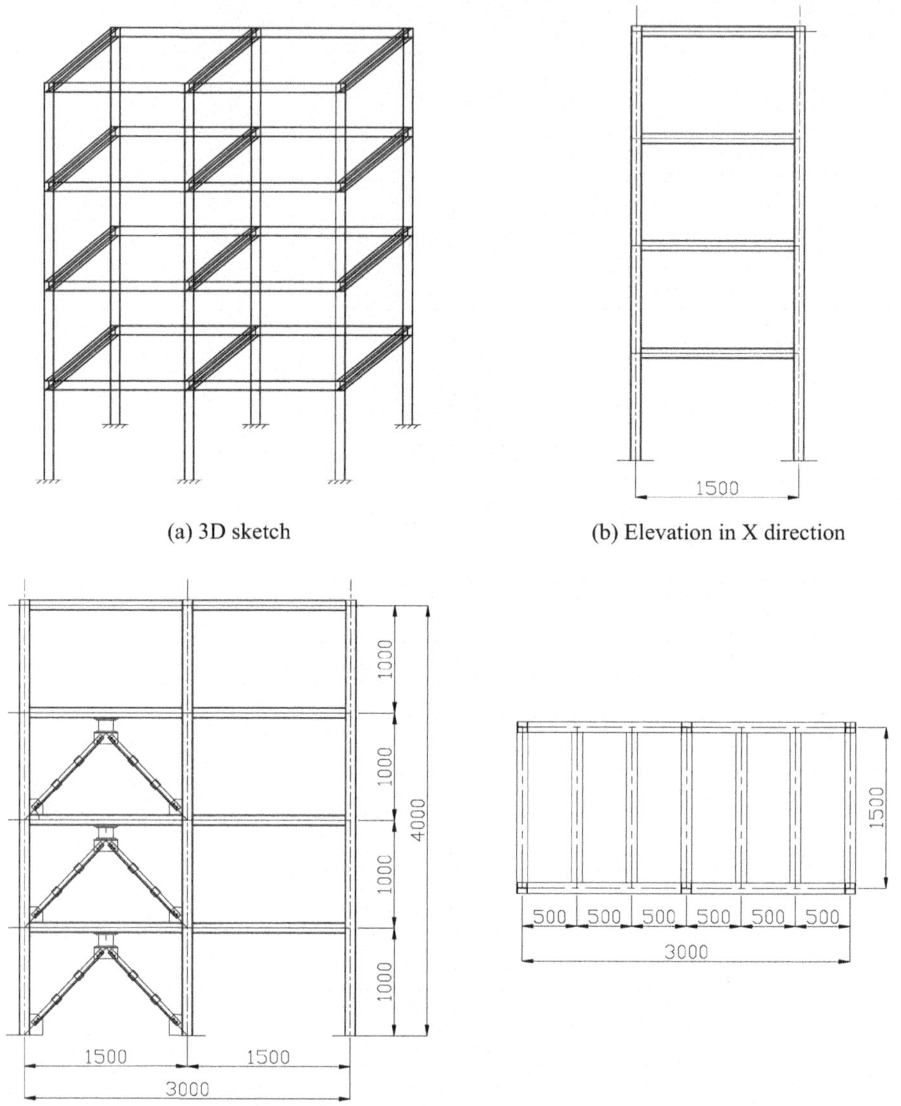

(a) 3D sketch (b) Elevation in X direction

(c) Elevation in Y direction (d) Plan layout

Fig. 11.8 Experimental model design

steel damper and viscous fluid damper as model II, and the frame with only viscous fluid damper as model III. Model I, model II and model III are loaded according to the loading sequence shown in Table 11.4. First, the shaking table test of model I, then the shaking table test of model II, and finally the shaking table test of model III (Fig. 11.10).

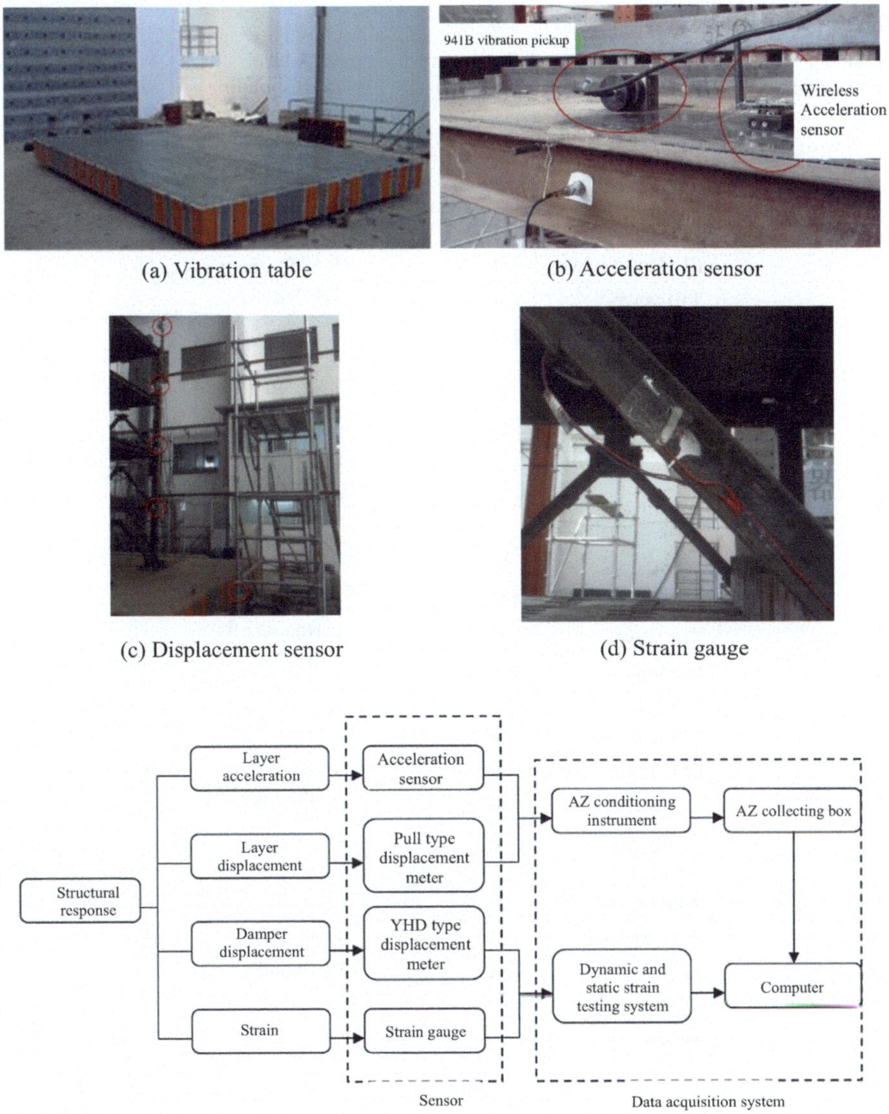

(a) Vibration table

(b) Acceleration sensor

(c) Displacement sensor

(d) Strain gauge

(d) Data acquisition system

Fig. 11.9 Test equipment

3. Acceleration response of structure

Under the same working condition, the ratio of the maximum value of acceleration response of each floor of the model structure to the maximum value of acceleration applied in the corresponding direction on the base is the acceleration amplification coefficient of acceleration response of each floor of the model under this working

Table 11.4 Loading system of shaking table test

Loading sequence	Seismic wave	Peak of wave (gal)	Risk level corresponding to the original full-scale structure	Peak value of wave corresponding to the original full-scale structure (gal)
1	White noise	50		
2	El-Centro wave	180	8.5° frequent Earthquake	110
3	Taft wave	180		
4	Washington wave	180		
5	White noise	50		
6	El-Centro wave	330	8° fortification earthquake	200
7	Taft wave	330		
8	Washington wave	330		
9	White noise	50		
10	El-Centro wave	490	8.5° fortification earthquake	300
11	Taft wave	490		
12	Washington wave	490		
13	White noise	50		
14	El-Centro wave	660	8° rare earthquake	400
15	Taft wave	660		
16	Washington wave	660		
17	White noise	50		
18	El-Centro wave	840	8.5° rare earthquake	510
19	Taft wave	840		
20	Washington wave	840		

condition. After data processing, we can get the acceleration amplification coefficient of the steel frame model in the vibration direction of the shaking table under the same level earthquake, and the acceleration amplification coefficient of the model under the same seismic record and different level earthquake. Under the earthquake action of different seismic waves and different risk levels, the acceleration amplification coefficient is shown in Tables 11.5, 11.6 and 11.7:

When the peak value of applied seismic wave is relatively small, for example, when the peak value of planned applied seismic wave is 180 gal, due to the installation error of damper, installation error of measuring device, acquisition accuracy of experimental instrument and other problems, the accuracy of experimental data

(a) Steel frame model II

(b) Steel frame model III

(c) Mild steel damper

(d) Viscous fluid damper

Fig. 11.10 Test model

may be low. For example, under the effect of Washington wave with a peak value of 180 gal, the acceleration amplification coefficient of the first floor of model I is 1.36 The acceleration amplification coefficient of the first floor of model III with viscous fluid damper is 1.48, which is larger than the acceleration amplification coefficient of the first floor of model I, which is not consistent with common sense and may be abnormal data; however, with the increase of seismic wave strength, the experimental data obtained is more accurate. Through the test data analysis of Tables 11.5, 11.6 and 11.7, the following conclusions can be obtained:

(1) from the acceleration data in Table 11.5, it can be seen that the acceleration amplification coefficient of different floors is different. The higher the floor is, the larger the acceleration amplification coefficient will be. The acceleration amplification coefficient of the top floor is the largest. Due to the difference of spectrum characteristics, the amplification coefficient of the top floor acceleration is also very different under the action of different seismic waves. After the

Table 11.5 Acceleration amplification factor of pure frame

Name of wave	Peak value of planned applied wave (gal)	Peak value of actual table acceleration (gal)	Acceleration amplification coefficient of the first floor	Acceleration amplification coefficient of the second floor	Acceleration amplification coefficient of the third floor	Acceleration amplification coefficient of the fourth floor
El-Centro wave	180	202	1.26	1.54	1.75	1.99
El-Centro wave	330	317	1.49	1.94	2.12	2.49
El-Centro wave	490	448	1.49	1.93	2.01	2.66
El-Centro wave	660	601	1.51	1.89	1.95	2.63
El-Centro wave	840	840	1.58	1.98	1.78	2.82
Washington wave	180	176	1.36	1.76	2.09	2.47
Washington wave	330	313	1.48	2.15	2.52	3.49
Washington wave	490	408	1.54	2.35	2.87	4.06
Washington wave	660	529	1.56	2.24	2.97	4.31
Washington wave	840	710	1.62	2.01	2.98	4.16

Table 11.6 Acceleration amplification factor of frame with mild steel damper and viscous fluid damper

Name of wave	Peak value of planned applied wave (gal)	Peak value of actual table acceleration (gal)	Acceleration amplification coefficient of the first floor	Acceleration amplification coefficient of the second floor	Acceleration amplification coefficient of the third floor	Acceleration amplification coefficient of the fourth floor
Washington wave	180	218	1.30	1.43	1.46	2.19
Washington wave	330	372	1.32	1.36	1.30	1.98
Washington wave	490	459	1.29	1.32	1.25	1.81
Washington wave	660	525	1.29	1.28	1.20	1.64

Table 11.7 Acceleration amplification factor of frame with viscous fluid damper

Name of wave	Peak value of planned applied wave (gal)	Peak value of actual table acceleration (gal)	Acceleration amplification coefficient of the first floor	Acceleration amplification coefficient of the second floor	Acceleration amplification coefficient of the third floor	Acceleration amplification coefficient of the fourth floor
Washington wave	180	168	1.48	1.57	1.58	1.88
Washington wave	330	282	1.34	1.39	1.40	1.68
Washington wave	490	398	1.25	1.39	1.36	1.61
Washington wave	660	467	1.22	1.36	1.39	1.69

El centro wave is applied to the frame, the acceleration amplification coefficient of the top floor is large, about 2.5, the acceleration amplification coefficient of the top layer is about 4 after the Washington wave is applied to the frame, which shows that the acceleration response of the frame is greatly affected by the spectrum characteristics of the seismic wave.

(2) from the data in Table 11.6 it can be seen that when soft steel dampers are installed at the bottom of the empty frame and viscous fluid dampers are installed at the second and third floors of the empty frame, the acceleration response of each floor is significantly smaller; because there is no damper on the top floor, the acceleration response is much larger than that of the bottom three floors, but compared with the undamped pure frame, the acceleration response is still reduced. Under the action of small earthquake, the control effect of acceleration is not obvious. Under the action of medium earthquake and large earthquake, the acceleration response is reduced to about 40% of that under the condition of pure frame, which effectively controls the acceleration response of the structure.

(3) It can be seen from the data in Table 11.7 that after adding viscous fluid damper to the second and third floors of the pure frame, the acceleration response is significantly smaller; since there is no damper on the top floor, the acceleration response is much larger than that of the bottom three floors, but compared with the empty frame without damper, the acceleration response is much smaller; Through the damping effect of viscous fluid damper, the acceleration response is reduced to about 50% of the case of pure frame.

(4) From the data in Tables 11.6 and 11.7, it can be seen that under the small earthquake action, the damping effect of only installing viscous fluid damper is better, while under the large earthquake action, the damping effect of installing soft steel damper and viscous fluid damper is better.

(5) for the frame structure under small and medium earthquakes, according to the magnitude of interlayer displacement angle, dampers can be installed in the layer with large interlayer displacement angle to achieve the goal of vibration

reduction; however, this is not necessarily the best way, because under the action of large earthquakes, the floor acceleration without dampers is easy to produce amplification, so different seismic risk levels should be considered. For example, from the perspective of floor acceleration response, although the acceleration response of the top floor has been well controlled, it is still relatively large compared with the bottom three floors. Under the large earthquake action, it may be better to add a certain amount of dampers to the top floor. When the structure enters the elastic-plastic state, the acceleration response of the story without damper may be significantly enlarged. Therefore, it is very necessary to consider different seismic risk levels for energy dissipation and damping structures, and to carry out elastic-plastic analysis of damping structures under large earthquakes.

In order to observe the change of acceleration of the structure after adding the damper more vividly and intuitively, the acceleration time history curve of some working conditions under the action of Washington wave is given in Fig. 11.11 stands for the pure steel frame, 2 stands for the steel frame with soft steel damper and viscous fluid damper, 3 stands for the steel frame with viscous fluid damper only; the abscissa of the chart represents the time, and the ordinate represents the acceleration amplification coefficient of the top layer.

From the time history response curve of acceleration, it can be seen that the acceleration amplification coefficient of the top floor is significantly reduced after the damper is added, which effectively controls the response of the structure, at the same time reduces the floor shear force and overturning moment of the structure, so as to protect the structure well; soft steel damper is added to model II, which increases the rigidity of the structure, but at the same time adds the viscous fluid damper, which increases the damping of the structure, and the floor acceleration control effect of the structure is significant finally, which is not inferior to the floor acceleration control effect of model III only with viscous fluid damper, and the acceleration control effects of model II and model III are similar.

4. **Damper response**

The hysteretic curve of the damper in the test is given in Fig. 11.12. It can be seen from the hysteretic curve of the force displacement relationship of the soft steel damper that the hysteretic curve of the force displacement relationship of the soft steel damper is relatively full. On the one hand, the soft steel damper adds a part of damping to the structure, and reduces the response of the structure through energy dissipation, so as to protect the structure; on the other hand, the soft steel damper provides a part of stiffness for the structure, which can reduce the displacement response of the structure through the stiffness effect. From the hysteretic curve of force displacement relationship of viscous fluid damper, it can be seen that the hysteretic loop of viscous fluid damper is relatively full, absorbing a considerable part of the energy input to the structure from the earthquake, so as to effectively protect the structure and reduce the damage of the structure.

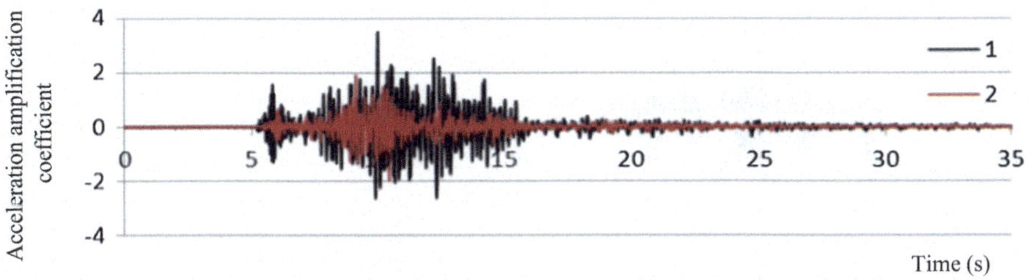

(a) Comparison between model I and model II (Washington wave, 330gal)

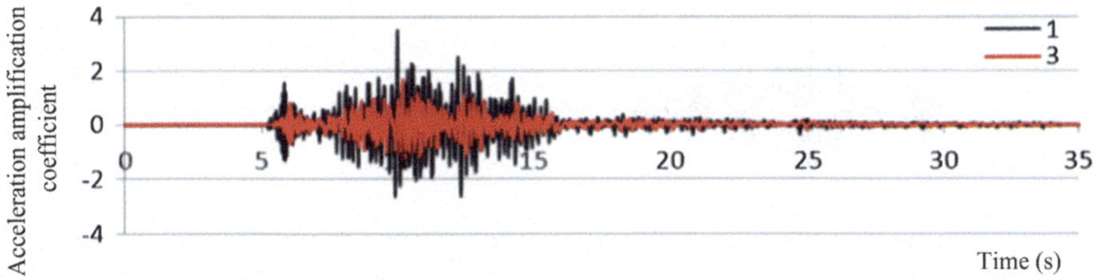

(b) Comparison between model I and model III (Washington wave, 330gal)

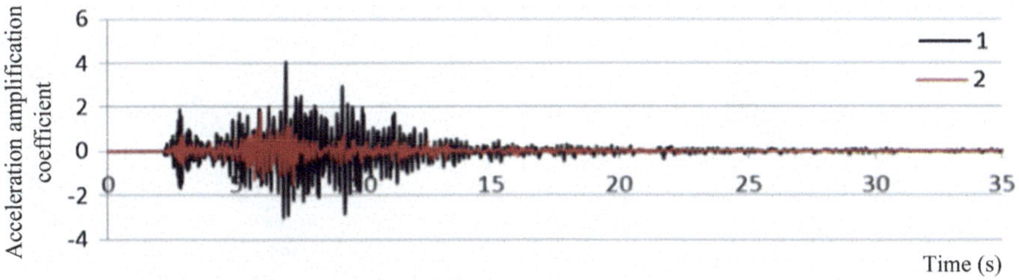

(c) Comparison between model I and model II (Washington wave, 4900gal)

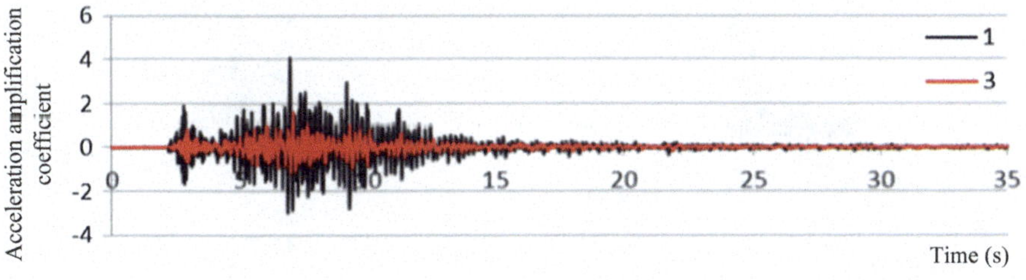

(d) Comparison between model I and model III (Washington wave, 490gal)

Fig. 11.11 Time history curve of acceleration amplification coefficient of top floor under Washington wave action

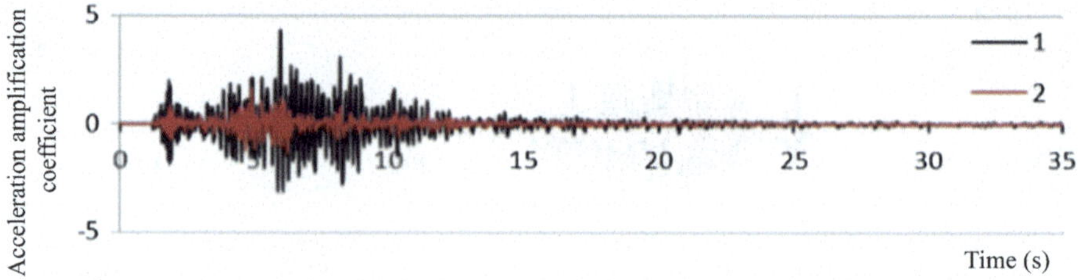

(e) Comparison between model I and model II (Washington wave, 660gal)

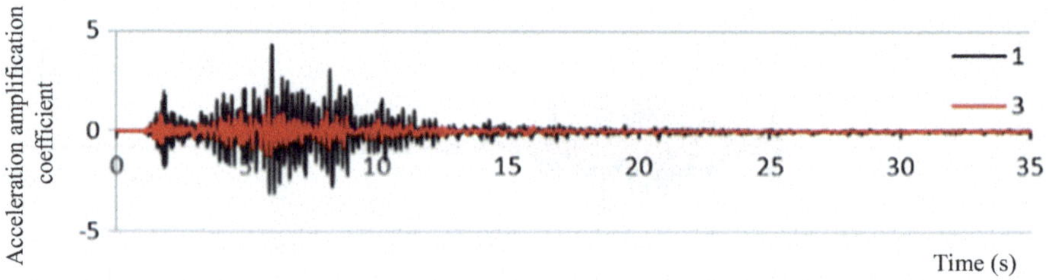

(f) Comparison between model I and model III (Washington wave, 660gal)

Fig. 11.11 (continued)

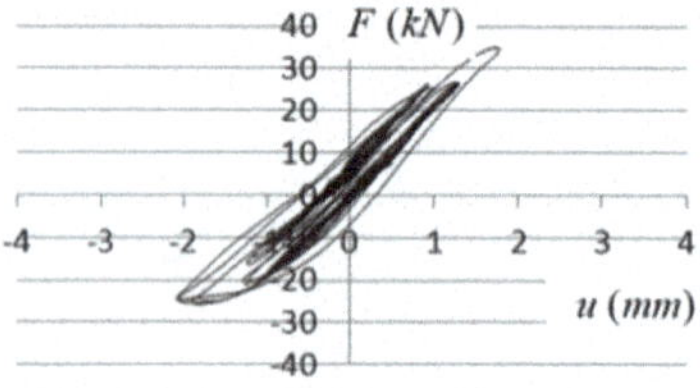

(a) Model II, mild steel damper (Taft wave)

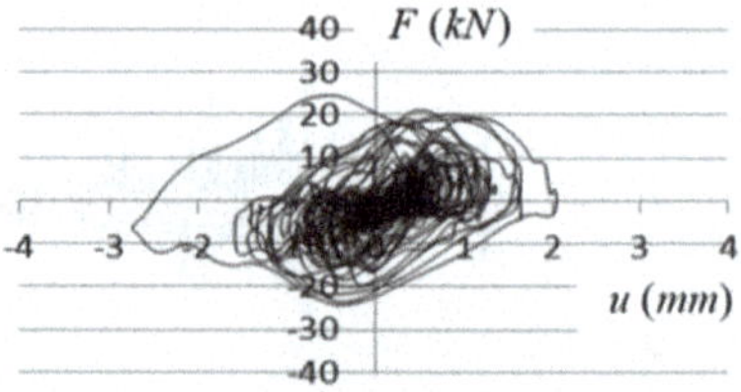

(b) Model II, viscous fluid damper (Taft wave, 2nd layer)

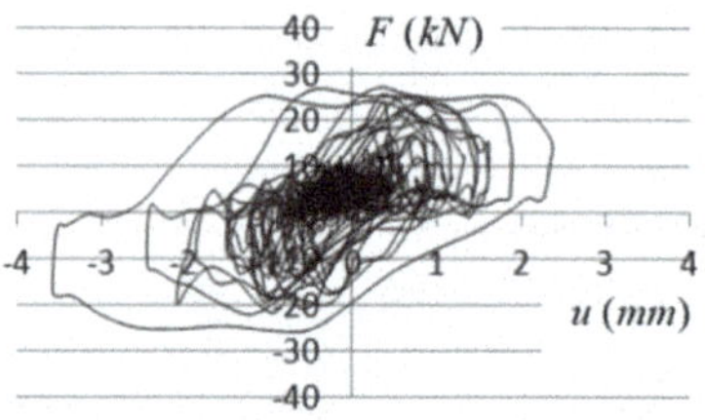

(c) Model III, viscous fluid damper (Washington wave, 2nd layer)

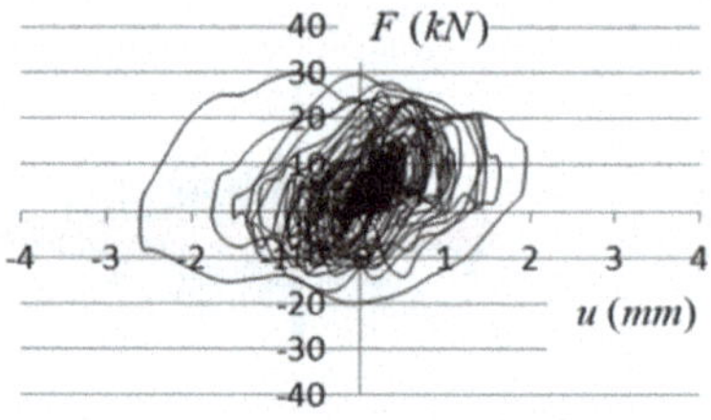

(d) Model III, viscous fluid damper (Washington wave, 3rd layer)

Fig. 11.12 Damper response

11.3.1.2 Viscoelastic Damping Test

1. **Test model**

The design of viscoelastic damping wall is determined by the modal strain energy method. The test temperature is kept at about 23 °C (±1 °C), and the damping ratio of the structure is determined to be 15% by the seismic design response spectrum. The design objective of this test is to keep the steel frame (the same as Sect. 11.3.1.1) in elastic state all the time. In order to meet the structural performance requirements, the inter story displacement angle is less than or equal to 0.4%, so that the thickness of single layer of viscoelastic damping wall is 4 mm, and the effective dimension of viscoelastic material layer is 100 mm × 100 mm. The specific design is shown in Fig. 11.13a. The viscoelastic damping wall (as shown in Fig. 11.13b) is connected to the support by welding, and then connected to the upper and lower flanges of the steel frame beams by bolts, and arranged in a single span along the long axis of the steel frame and close to the actuator. The test model of steel frame after the installation of viscoelastic damping wall is shown in Fig. 11.13c.

2. **Test conditions**

In this test, multiple loading method is adopted, and the loading wave direction is unidirectional along the long span direction of the frame, mainly divided into the following two conditions:

(1) **Structural vibration modal test**: the white noise excitation method is used in structural vibration modal test. When the acceleration input value is changed, the white noise excitation is applied to the whole structure through the exciter to obtain the dynamic characteristics of the frame system. The vibration modal parameters of the frame are determined according to the frequency response function.
(2) **Seismic dynamic response loading**: for the dynamic response loading of structures, three seismic waves (El Centro wave, Taft wave and Washington wave) are selected in this test, and the vibration amplitude is 240, 500 and 700 gal. Firstly, the controlled structure with viscoelastic damping wall was loaded, and then the empty frame was loaded after the viscoelastic damping wall was removed.

3. **Structural dynamic characteristics**

The white noise frequency sweep of El Centro wave, Taft wave and Washington wave before and after 700 gal acceleration is carried out for the controlled structure (taking 1234 type combined mode as an example) and uncontrolled structure respectively. The spectrum characteristic curve of the acceleration measuring point at the first floor is obtained, and the structure vibration modes under various working conditions are analyzed. Refer to Table 11.8 for the first three natural frequencies of the structure. It can be seen from the table that during the whole loading process, the natural

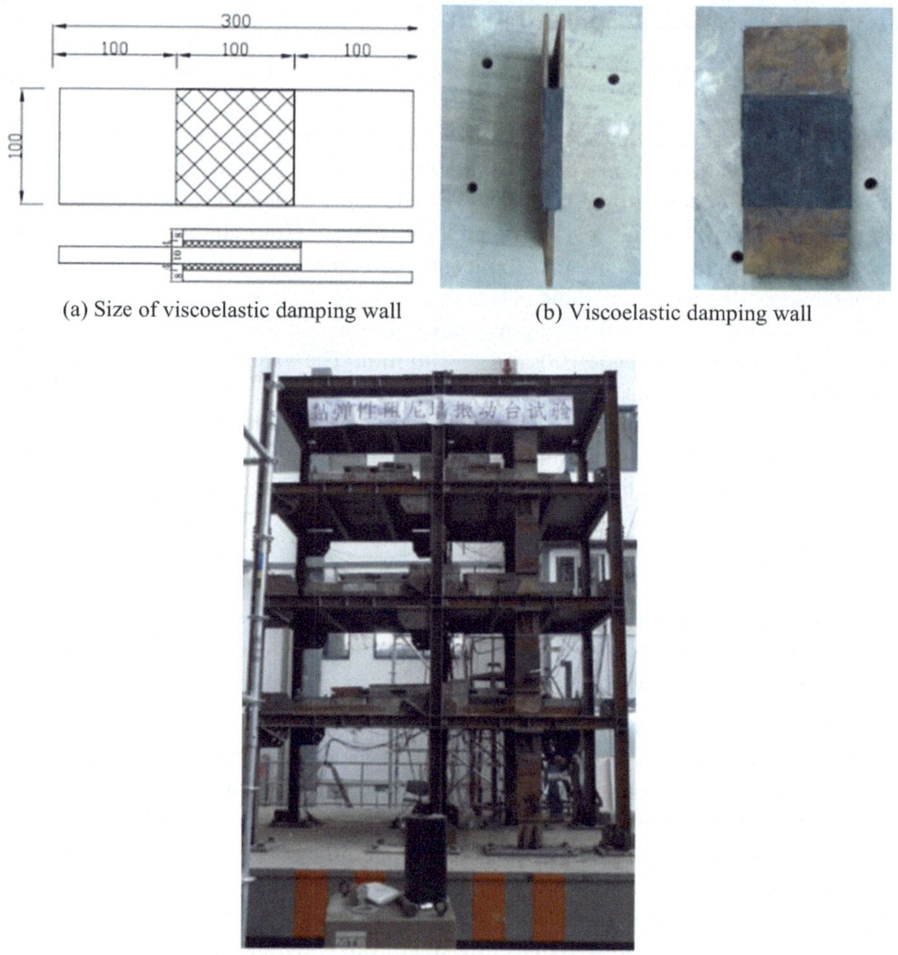

(a) Size of viscoelastic damping wall

(b) Viscoelastic damping wall

(c) Viscoelastic damping structure system model

Fig. 11.13 Viscoelastic damping test model

vibration frequency of the structure remains stable, the up and down floating is very small, and the steel frame always maintains the elastic state, which conforms to the design objective of this test. Compared with the uncontrolled structure, the controlled structure improves the natural frequency of the structure under the additional stiffness provided by the viscoelastic damping wall.

4. **Acceleration response of structure**

Table 11.9 shows the maximum acceleration response and its attenuation rate of the first ~ fourth floors of the frame under the condition of 700 gal acceleration. Figure 11.14a–c shows the comparison curve of the acceleration duration response of the middle of the first floor under the earthquake duration of the uncontrolled

Table 11.8 Natural frequency of the vibration mode of the structure swept by X-direction white noise

Loading condition	First order (Hz)	Second order (Hz)	Third order (Hz)
First white noise sweep	5.00	14.13	24.81
Frequency sweep of white noise after 700 gal El wave (controlled/uncontrolled)	5.13/4.94	14.88/14.06	26.88/24.75
Frequency sweep of white noise after 700 gal of Taft wave (controlled/uncontrolled)	5.31/4.88	15.32/14.69	27.37/24.69
Frequency sweep of white noise after 700 gal of Wa wave (controlled/uncontrolled)	5.25/5.00	15.44/14.01	27.69/24.75

Table 11.9 Comparison of maximum acceleration of each layer of controlled structure and uncontrolled structure (unit: m/s^2)

Loading condition	First floor	Second floor	Third floor	Fourth floor
El wave (uncontrolled/controlled)	9.26/6.37 (31.2%)	10.95/6.96 (36.4%)	13.13/7.77 (40.8%)	16.60/9.32 (43.9%)
Taft wave (uncontrolled/controlled)	8.96/7.16 (20.1%)	12.04/10.21 (15.2%)	14.52/13.56 (6.61%)	18.92/16.51 (12.7%)
Wa wave (uncontrolled/controlled)	10.82/7.34 (32.2%)	13.08/10.21 (21.9%)	20.52/12.23 (40.4%)	23.91/15.09 (36.9%)

Note the acceleration attenuation rate is in brackets, and its calculation formula is $\frac{\text{Acceleration of the uncontrolled structure} - \text{Acceleration of the controlled structure}}{\text{Acceleration of the uncontrolled structure}} \times 100\%$

structure and the controlled structure (taking 1234 type combination mode as an example). It can be seen from the table that the maximum acceleration of each layer of controlled structure and uncontrolled structure appears at the top floor, and the damping form of viscoelastic damping wall provides a better damping performance for the controlled structure, which makes the maximum acceleration of each layer of the steel frame attenuated, with the attenuation range of 6.61–43.9%, among which the El wave has the most obvious damping effect, with the damping range of 31.2–43.9%, Washington wave secondly, the damping range is 21.9–40.4%, and Taft wave is the smallest.

5. **Displacement response of structure**

Table 11.10 shows the maximum interstorey displacement response and its attenuation rate of the first ~ fourth floors of the frame under the three kinds of waves of 700 gal acceleration. Figure 11.14d–f shows the comparison curve of the displacement duration response of the middle of the first floor under the earthquake duration of the uncontrolled structure and the controlled structure (taking 1234 type combination mode as an example). It can be seen from the table that under the action of 700 gal

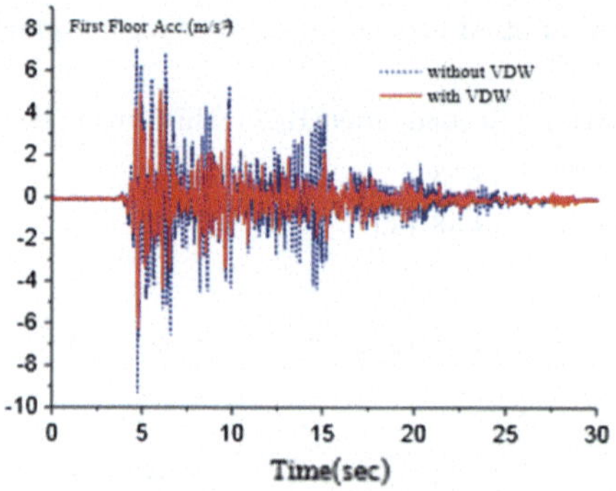

(a) Acceleration of the first floor under El Centro wave

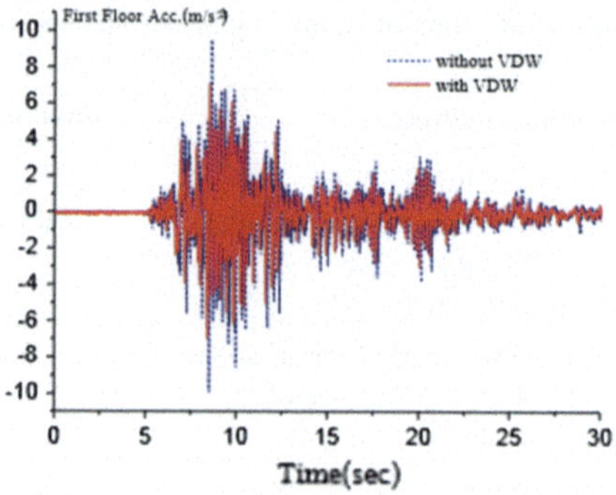

(b) Acceleration of the first floor under Taft wave

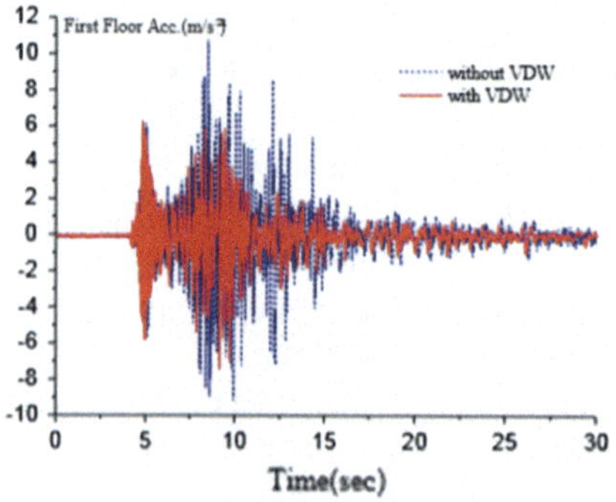

(c) Acceleration of the first floor under Washington wave

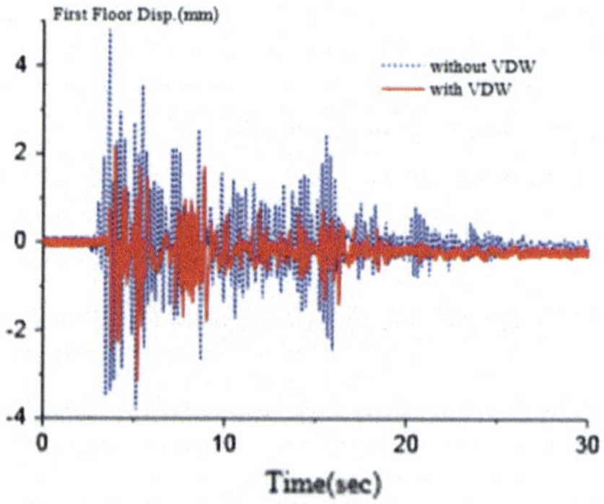

(d) Interlayer displacement of the first floor under El Centro wave

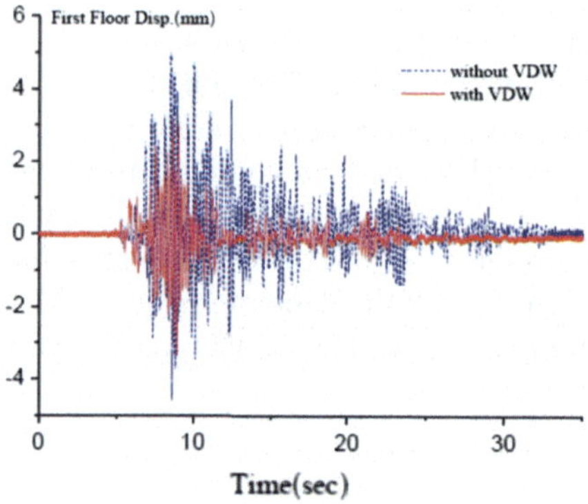

(e) Interlayer displacement of the first floor under Taft wave

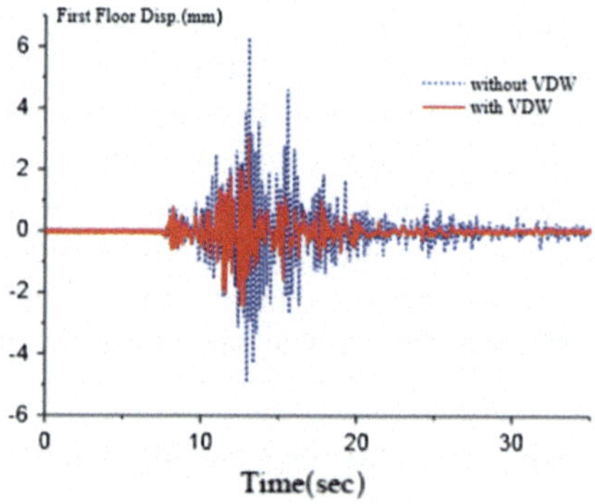

(f) Interlayer displacement of the first floor under Washington wave

Fig. 11.14 Comparison of time history curves of structural seismic response

Table 11.10 Comparison of maximum inter story displacement of each floor of controlled structure and uncontrolled structure (unit: mm)

Loading condition	First floor	Second floor	Third floor	Fourth floor
El wave (uncontrolled/controlled)	4.82/3.13 (35.1%)	4.39/2.96 (32.6%)	4.20/2.86 (31.9%)	4.03 /2.66 (34.0%)
Taft wave (uncontrolled/controlled)	5.01/3.21 (35.9%)	4.86/3.06 (51.4%)	4.56/1.99 (56.4%)	4.03/1.82 (54.8%)
Wa wave (uncontrolled/controlled)	6.31/3.12 (50.6%)	5.32/2.98 (44.0%)	4.68/2.63 (43.8%)	3.98/2.19 (45.0%)

Note the interlayer displacement attenuation rate is in brackets, and its calculation formula is $\frac{\text{Interlayer displacement of the uncontrolled structure} - \text{Interlayer displacement of the controlled structure}}{\text{Interlayer displacement of the uncontrolled structure}} \times 100\%$

seismic wave, the displacement of each floor of the uncontrolled structure under the three kinds of waves exceeds the elastic displacement limit, and the maximum value of the interlayer displacement angle has reached 0.006 within the elastic range, far exceeding the specification limit of 0.004. The controlled structure with viscoelastic damping wall can effectively control the interlayer displacement angle, and the interlayer displacement is controlled within 4 mm, greatly improving the side shifting safety and comfort of the steel frame. The interstory displacement of uncontrolled structure and controlled structure decreases in turn from the first floor to the top floor, showing a shear deformation mode. In this mode, the first floor interstory displacement of the frame is the largest, the shear deformation of viscoelastic damping wall is also the largest, and the effective stiffness can play an excellent state. Therefore, at the first floor, El wave and Wa wave have the best interlayer displacement control effect, with the attenuation rate of 35.1 and 50.6% respectively. In contrast, the effect of viscoelastic damping wall on the first floor displacement response is better than that of acceleration response.

6. **Comparative analysis of numerical simulation**

The time history analysis model of the controlled structure with 1234 viscoelastic damping wall and empty frame is established by SAP2000 analysis software. The acceleration response and displacement response are compared with the test results. In this calculation, the viscoelastic damping wall element adopts the Maxwell model, its parameters are: viscoelastic material layer number $n = 2$, material thickness $t = 4$ mm, effective shear area $A = 10{,}000$ mm^2, spring rigidity 10.5×10^6 N/m, damping coefficient 0.8×10^6 N/m. Figure 11.15a–d shows the comparison diagram of SAP2000 simulation analysis and test results of empty frame and 1234 type controlled structure under 700 gal El Centro wave. Through comparative analysis, it can be seen that the simulation results and test results are in good agreement as a whole, and the calculation results can reflect the dynamic mechanical behavior of steel frame under the action of this seismic wave, which provide a more reliable reference for the follow-up further design and practical application of viscoelastic damping wall. Among them, the controlled structure is higher than the uncontrolled structure in the consistency between the calculation results and the test results.

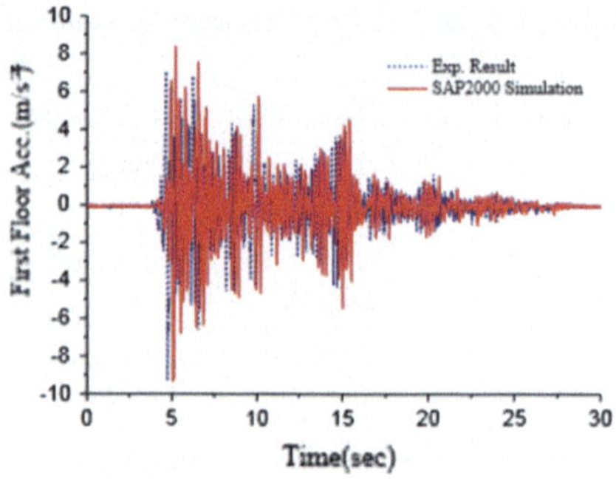

(a) Acceleration of the first floor of uncontrolled structure

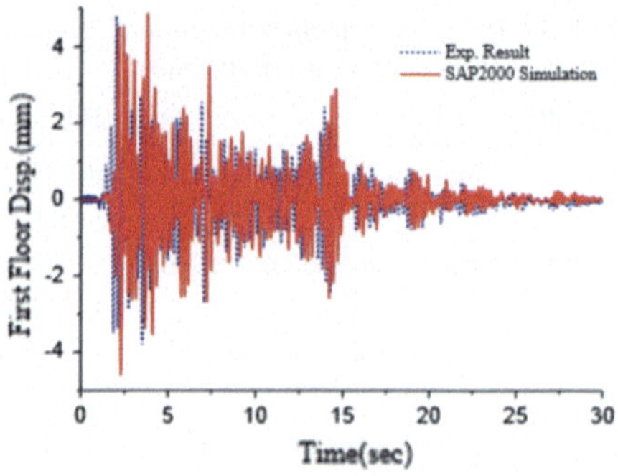

(b) Displacement of the first floor of uncontrolled structure

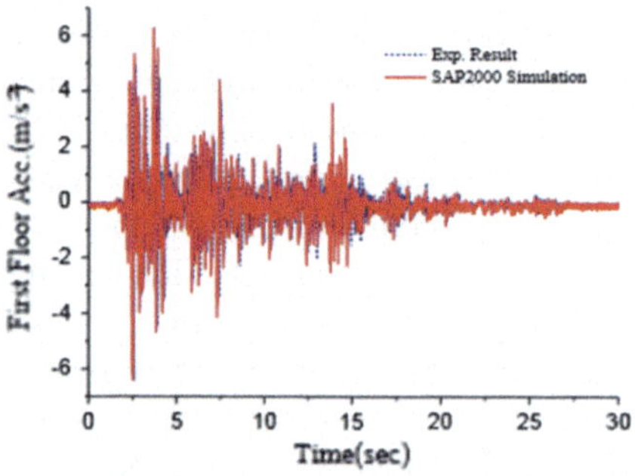

(c) Acceleration of the first floor of controlled structure

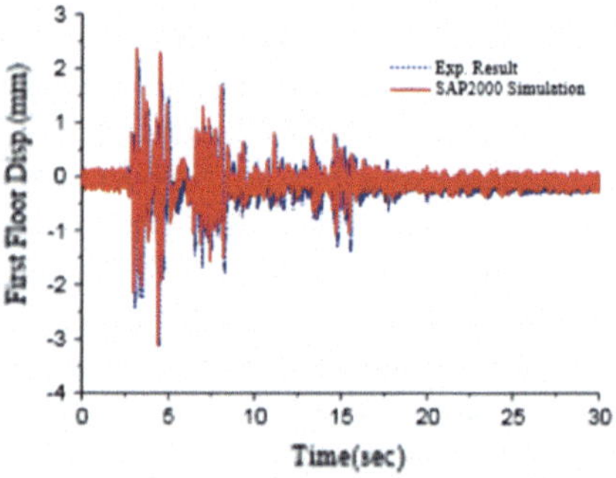

(d) Displacement of the first floor of controlled structure

Fig. 11.15 Comparison of numerical simulation and test results

11.3.1.3 Metal Damping Vibration Control Test (Parabola Shaped Mild Steel Damper)

1. **Test model**

The same structural model as Sect. 11.3.1.1 is adopted. The parabolic mild steel damper (such as Fig. 11.16b) is made of 6 mm thick Q235 steel. The damper is symmetrically arranged in 1–3 layers and connected to the structure and support by bolts. See Fig. 11.16c for specific dimensions. The test model with parabolic shape mild steel damper is shown in Fig. 11.16a. The finite element model of the damper prototype and model is established by ANSYS software. The elastic stiffness, plastic stiffness, yield displacement, yield load and the second stiffness coefficient of the model and prototype damper are calculated, and the similarity ratio of each physical quantity is obtained. Compared with the theoretical similarity ratio, see Table 11.11. It can be seen that the stiffness similarity ratio is $S_K = S_l = 1/3$, and the yield displacement and load are consistent with the theoretical similarity ratio.

(a) Mild steel damping structure model

(b) Parabolic shape mild steel damper

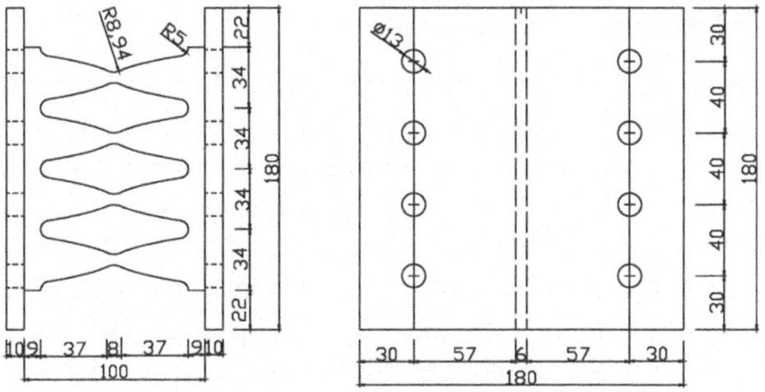

(c) Dimensions of parabolic shape mild steel damper

Fig. 11.16 Experimental model of damping and vibration reduction of parabola shaped mild steel

Table 11.11 Damper model similarity (model/prototype)

Damper parameters	Model	Prototype	Calculated similarity ratio	Theoretical similarity ratio
Elastic stiffness K_e (kN/mm)	51.56	154.67	0.33	0.33
Plastic stiffness K_p (kN/mm)	1.76	5.33	0.33	0.33
Yield displacement D_y (mm)	0.39	1.18	0.33	0.33
Yield load F_y (kN)	20.24	182.03	0.11	0.11
Second stiffness coefficient $a = K_p/K_e$	0.03	0.03	0.99	1.00

According to the above similarity ratio, the similarity ratio of the second stiffness coefficient is 1, that is to say, the stiffness after yield also meets the similarity ratio requirements.

2. **Test conditions**

The shaking table test adopts one-way loading. Four working conditions were designed, including three working conditions with damper and one working condition without damper. First of all, the shaking table test is carried out after installing dampers on the first, second and third floors of the structure (case 1). In order to compare the effect of damper's installation position on the damping effect, the first and third story dampers (case 2) and the first and second story dampers (case 3) were tested respectively. Finally, the frame structure without dampers was tested. The white noise is used to scan between the seismic waves of each magnitude, and the frequency change of the structure is monitored in real time. Three seismic waves, El Centro, Taft and Washington, are selected. The amplitude values of the loaded seismic waves are 150 gal, 410 gal, 700 gal and 850 gal respectively.

3. **Structural dynamic characteristics**

After each magnitude of seismic wave is loaded, white noise scanning is carried out to determine the frequency, damping ratio and formation of the structure. Table 11.12 lists the first and second frequencies of different working conditions. It can be seen from the table that for the same working condition, the natural frequency of the structure does not change much during the test, indicating that the main frame structure is not damaged in the test. The mild steel damper provides additional stiffness for the structure and increases the frequency of the structure. When mild steel dampers are installed on the first, second and third floors (case 1), the natural frequency of the structural system is the largest. For the case of two-layer dampers, the impact of dampers installed on different floors is different. The natural frequency of the

Table 11.12 Analysis results of structural dynamic characteristics

Case	Frequency (Hz)	Table input acceleration (gal)				
		Before test	150	410	700	850
Case 1	First order	7.81	7.48	7.56	7.47	7.43
	Second order	16.94	16.64	16.67	16.47	16.59
Case 2	First order	5.58	5.92	5.97	5.94	/
	Second order	16.61	16.67	16.65	16.79	/
Case 3	First order	7.14	6.46	6.67	6.41	/
	Second order	15.89	15.47	15.71	15.36	/
Case 4	First order	5.17	5.23	5.30	5.51	/
	Second order	14.48	14.60	14.72	14.68	/

structural system with dampers installed on the first and second floors (case 3) is higher than that installed on the first and third floors (case 2).

4. **Acceleration response of structure**

In order to study the influences of damper energy dissipation on the acceleration of the structure system, the maximum acceleration response of 1–4 floors of the frame under different working conditions of the three seismic waves under a large earthquake (700 gal) are analyzed. The results are shown in Fig. 11.17. It can be seen from the figure that: (1) compared with the working condition without damper (working condition 4), the three arrangement types of dampers reduces the acceleration response, especially the top acceleration; (2) different damper arrangements and different seismic wave actions have different control effects. In case 1, three-layer dampers are arranged, with the maximum reduction of acceleration; in case 3 and case 2, two-layer dampers are arranged, with the overall control of acceleration slightly worse than that in case 1; in case 3, the control effect is better than that in case 2; in case 3, the seismic wave is different from that in case 1, and the control effect is also different. Under the action of El Centro wave and Taft wave, the control effect of condition 3 is similar to that of condition 1, even better than that of some floors. Under the action of Washington wave, the control effect of condition 3 is slightly worse than that of condition 1, but the acceleration of the table is also greater under condition 3.

5. **Displacement response of structure**

The storey displacement of the structure can be calculated directly by the stay wire type displacement meter. Because the range of the stay wire type displacement meter is large, but the storey displacement of this test is small, there is a certain error, but it does not affect the analysis of the damping performance. Table 11.13 gives the maximum inter story displacement of three seismic waves under four conditions

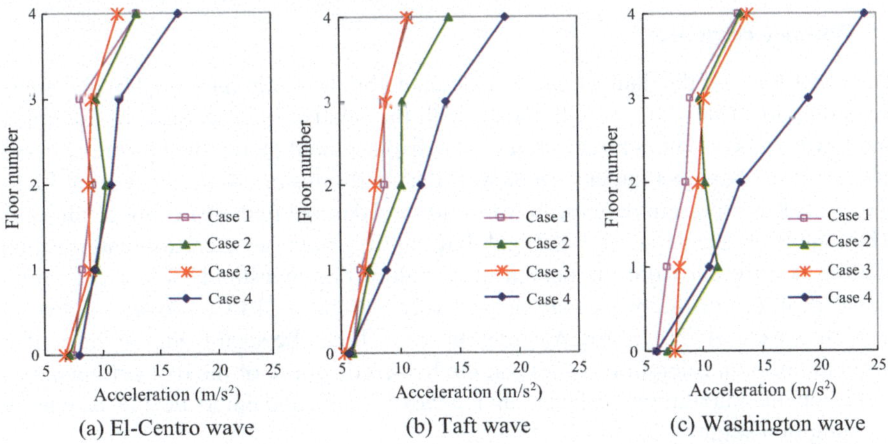

Fig. 11.17 Acceleration response of structural system (peak acceleration 700 gal)

Table 11.13 Maximum interstorey displacement and damping ratio of structure (peak acceleration of 700 gal)

Seismic wave	Floor number	Interstorey displacement (mm)				Damping ratio (%)		
		Case 4	Case 1	Case 2	Case 3	Case 1	Case 2	Case 3
El-Centro wave	1	4.71	1.80	2.44	1.60	61.84	48.21	65.90
	2	4.47	1.61	2.86	1.34	63.98	36.09	70.13
	3	3.27	1.04	1.34	1.94	68.27	58.92	40.63
	4	5.99	1.41	1.78	1.90	76.49	70.20	68.27
Taft wave	1	5.11	1.35	2.41	1.78	73.54	52.85	65.18
	2	5.46	1.58	3.45	1.39	71.06	36.82	74.48
	3	4.11	1.04	1.55	2.12	74.73	62.19	48.35
	4	3.12	1.79	2.26	2.10	42.75	27.61	32.68
Washington wave	1	7.49	1.46	2.63	1.64	80.47	64.89	78.14
	2	7.51	1.96	3.04	1.70	73.86	59.47	77.39
	3	5.44	1.12	1.51	2.81	79.43	72.26	48.28
	4	5.89	1.90	1.99	2.51	67.80	66.21	57.40

when the peak acceleration is 700 gal. It can be seen from the figure that: (1) the three damper arrangements effectively reduce the interlayer displacement, with a reduction rate of 27 ~ 80%. (2) Different arrangement of dampers has different control effect on the displacement between floors, and the control effect of condition 1 is the best; for condition 3 and 2, the arrangement position of dampers is different, and the control effect of each floor is different. For the first floor, the control effect of condition 3 is better than that of condition 2, while for the fourth floor, the control effect is different. (3) The control effect of the three arrangements is different for different seismic waves. The displacement control effect of the three arrangements is the best for Washington wave, followed by Taft wave, and slightly worse for El Centro wave.

6. **Damper response**

The shear force of the damper can be calculated by the strain measured by the angle steel support of the damper. Combined with the relative displacement of the upper and lower ends of damper measured by the displacement meter, the hysteretic curve of the soft steel damper under various working conditions can be drawn. Figure 11.18 shows the hysteretic curve of the bottom soft steel damper under working condition 1 when the peak acceleration is 850 gal, but the maximum acceleration under actual loading is different from the design input value. The measured table acceleration values of three seismic waves are respectively 880 gal for El Centro wave, 670 gal for Taft wave and 750 gal for Washington wave. It can be seen from the figure that when the peak acceleration is 850 gal, the hysteretic curve of the damper appears a yield step, the maximum displacement reaches 1.7 mm, and enters the plastic energy consumption stage.

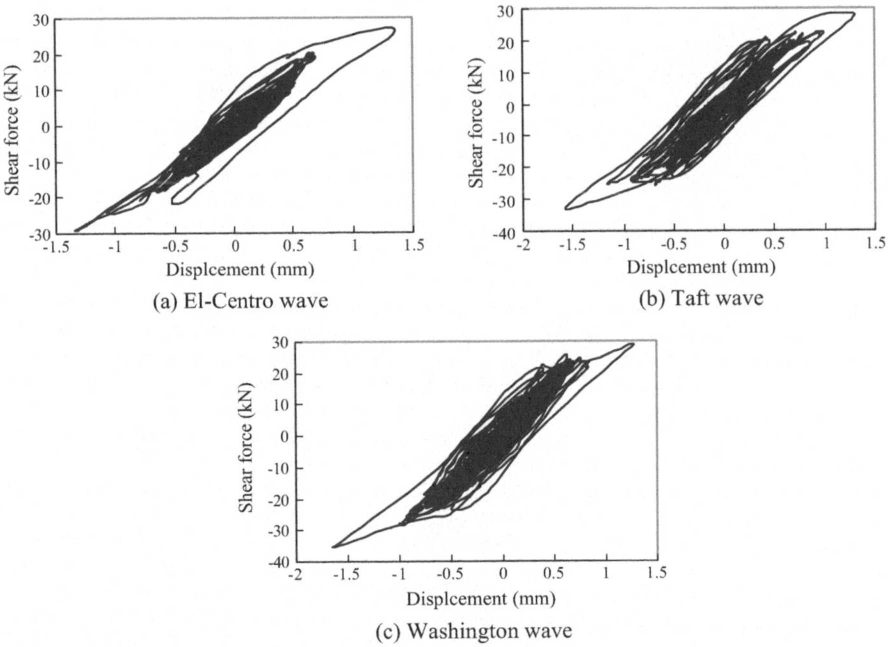

Fig. 11.18 Hysteretic curve of damper at the bottom of frame under condition 1 (850 gal)

7. **Comparative analysis of numerical simulation**

The finite element model of the test model is established, and the measured interlayer displacement is compared with the calculated value. Figure 11.19 shows the absolute displacement time history curve of the first floor of the structure with the peak seismic acceleration of 850 gal under condition 1. It can be seen that: (1) the results of finite element calculation are basically consistent with the measured displacement time history curve; (2) in the last stage of seismic wave action, there are deviations between the two groups of data, but they do not affect the engineering design and analysis.

11.3.1.4 Metal Damping Test (Curved Steel Plate Damper)

1. **Test model**

The same structural model as Sect. 11.3.1.1 is still used. The curved steel plate damper (Fig. 11.20b) is made of 5 mm thick Q235 steel. The damper is symmetrically arranged in 1–3 layers and connected with the structure and support by bolts. See Fig. 11.20c, d for specific dimensions. The test model with curved steel plate damper is shown in Fig. 11.20a. The finite element model of damper prototype and

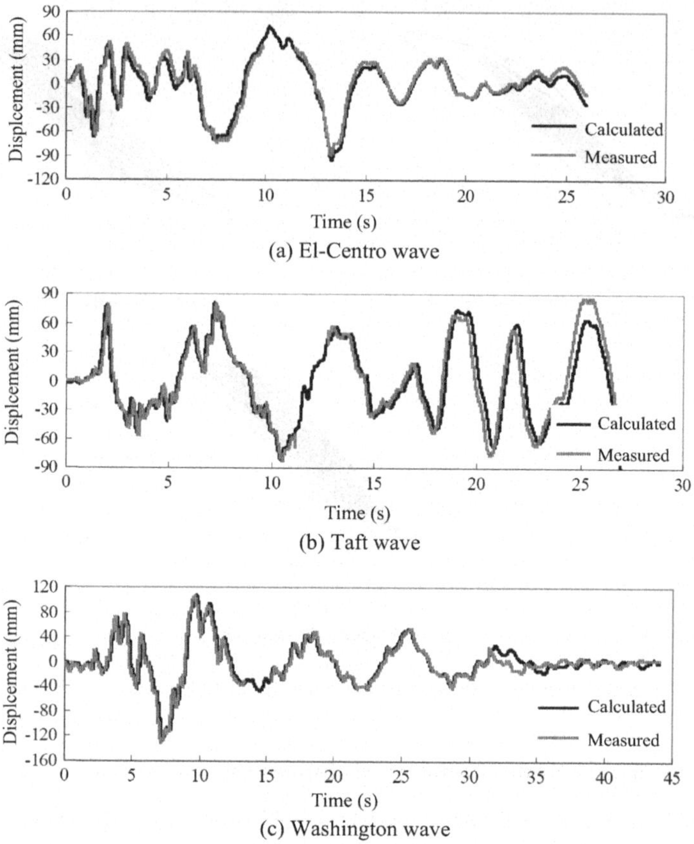

Fig. 11.19 The displacement of the first layer of different seismic waves under condition 1

model is established by ANSYS software. The elastic stiffness, plastic stiffness, yield displacement, yield load and second stiffness coefficient of the model and prototype damper are calculated, and the similarity ratio of each physical quantity is obtained.

2. **Test conditions**

Two kinds of working conditions are designed in the experiment. Working condition 1 is each floor from the first to the third floor of the structure model is equipped with two curved steel plate dampers, a total of 6 dampers. In case 2, there is no damper for the structure model. Three seismic waves, El Centro, Taft and Washington are selected. The amplitude values of the loaded seismic waves are 150 gal, 400 gal, 680 gal and 830 gal respectively. The corresponding peak acceleration values in the working conditions are respectively the maximum values when the seismic fortification intensity is 8° (0.3 g) and that of frequent earthquake, fortification earthquake,

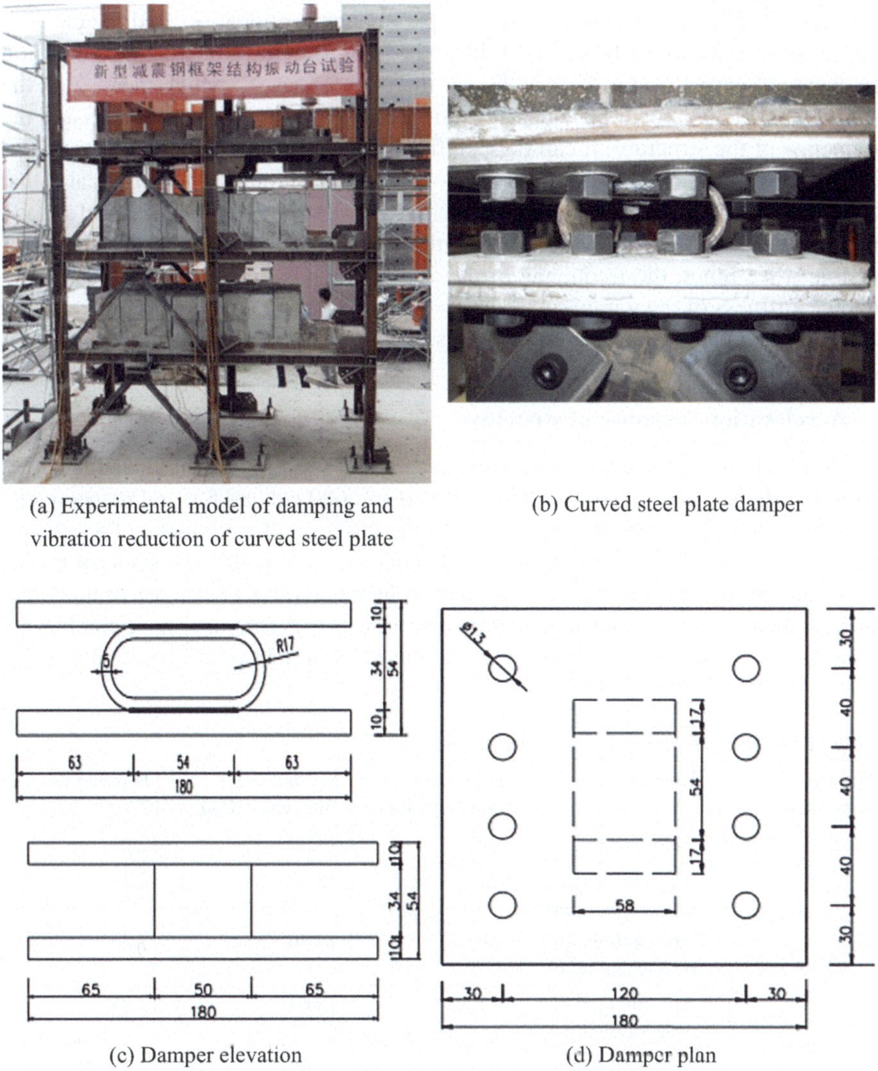

(a) Experimental model of damping and vibration reduction of curved steel plate

(b) Curved steel plate damper

(c) Damper elevation

(d) Damper plan

Fig. 11.20 Test on damping structure system of curved steel plate

rare earthquake and rare earthquake when the seismic fortification intensity is 9°. In order to monitor the frequency variation of the structural model, the white noise of 50 gal is used before and after the seismic wave of each magnitude is loaded.

3. **Structural dynamic characteristics**

Before and after the test, the structure was scanned for white noise. Using the time domain identification method of modal parameters to identify the acceleration response of the structure, the frequency and damping ratio of the structure can be obtained. The STD method is used for parameter identification, and the calculation results are shown in Table 11.14. It can be seen from the table that the structural frequency of case 1 is significantly higher than that of case 2, which means that the curved steel plate damper provides a certain additional stiffness, thus increasing the frequency of the structure. It can be seen from the frequency obtained under condition 2 that the frequency of the structure does not change much, which indicates that the damage of the steel frame during the test is very small. From the first and second frequency of case 1, it can be seen that with the increase of the peak acceleration of seismic wave input, the frequency of the structure tends to decrease, which shows that the stiffness of the damper will decrease in addition to the slight damage of the structure. A damping ratio of about 8% is added to the structure with curved steel plate damper (case 1).

4. **Acceleration response of structure**

Figure 11.21 shows the acceleration response of the test model with the peak acceleration of three seismic waves of 150 gal, 400 gal, 680 gal and 830 gal respectively. It can be seen from the figure that: (1) when there is a small earthquake of 8° (0.3 g), the maximum acceleration of the structure under the two conditions is not much different, and the damper has not yet played the role of damping. (2) In case of moderate earthquake of 8° (0.3 g), the acceleration response of the structure under condition 1 is smaller than that under condition 2, and the reduction amount is basically larger

Table 11.14 Dynamic characteristics of structure

Working condition	Explain	First order frequency (Hz)	Second order frequency (Hz)	Damping ratio (%)
Case 1	Before test	6.89	15.93	12.02
	After 150 gal of peak value input	6.72	15.82	12.39
	After 400 gal of peak value input	6.69	15.70	12.86
	After 680 gal of peak value input	6.52	15.64	12.88
	After 830 gal of peak value input	6.41	15.59	12.93
Case 2	Before test	4.43	13.25	4.55
	After 150 gal of peak value input	4.57	13.66	4.42
	After 400 gal of peak value input	4.53	13.72	4.66
	After 680 gal of peak value input	4.57	13.77	4.75
	After 830 gal of peak value input	4.60	13.69	4.95

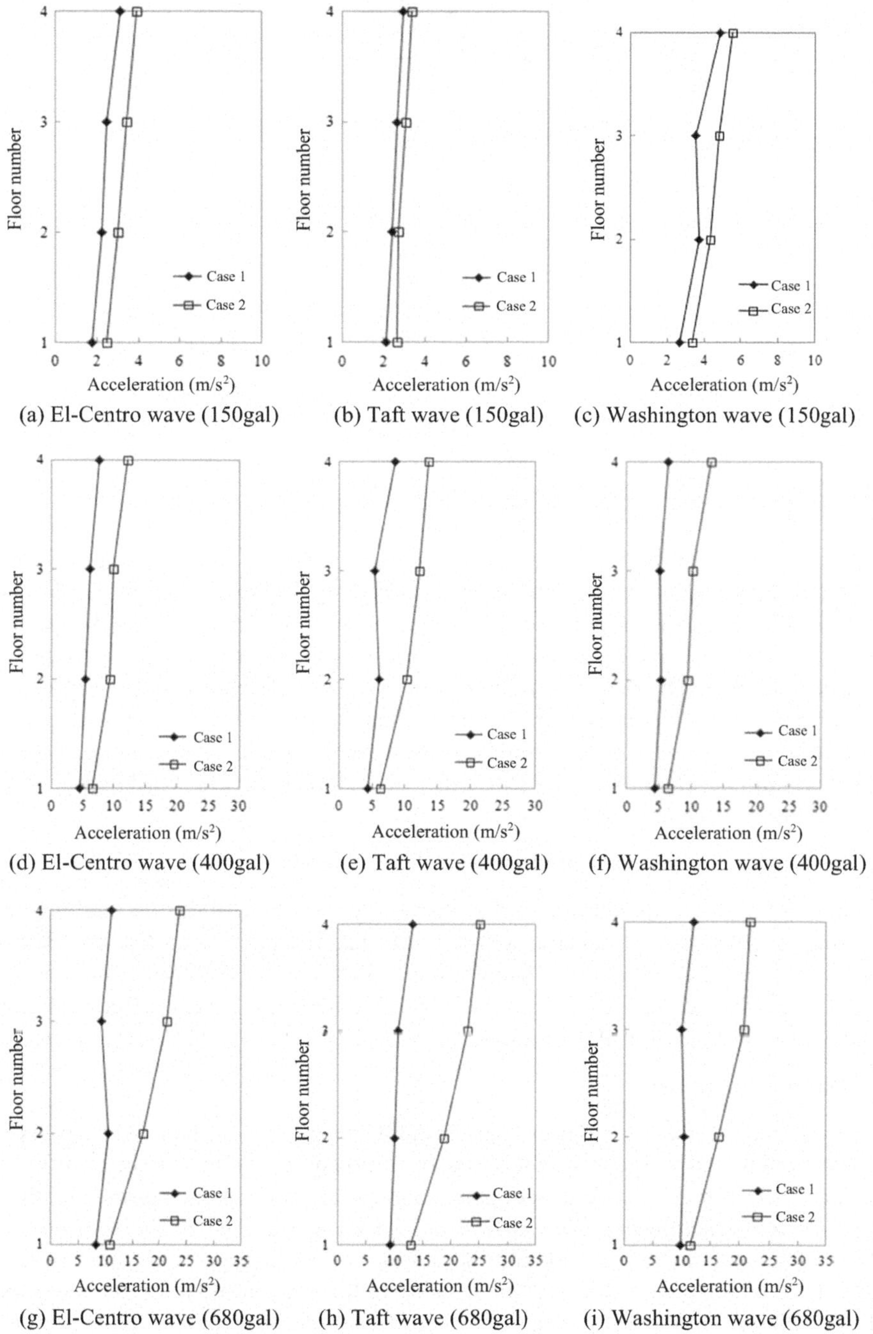

Fig. 11.21 Acceleration response of structure

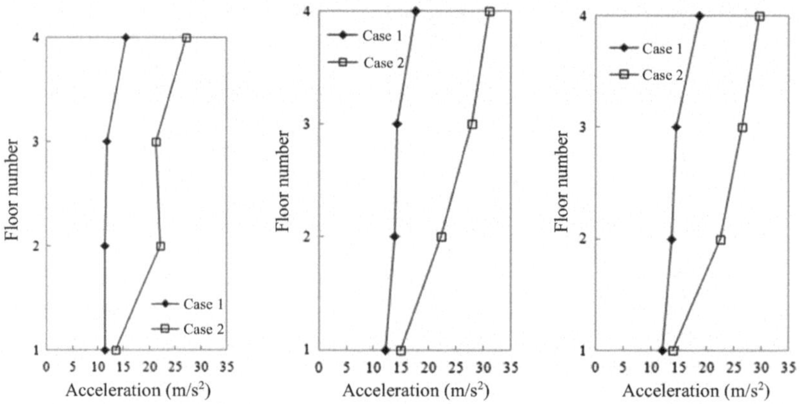

(j) El-Centro wave (830gal) (k) Taft wave (830gal) (l) Washington wave (830gal)

Fig. 11.21 (continued)

with the increase of the floor number. The reduction rate of the maximum acceleration under El Centro wave, Taft wave and Washington wave is 36.31, 55.46 and 50.90%. (3) When the earthquake is 8° (0.3 g), the control effect of the damper on the acceleration is obvious, and with the increase of the floor number, the reduction is basically larger. The reduction rate of the maximum acceleration under El Centro, Taft and Washington waves is 52.96, 55.87 and 51.46%. (4) The control trend of acceleration response of 9° earthquake is the same as above. The reduction rate of the maximum acceleration under El Centro wave, Taft wave and Washington wave is 48.00, 48.89 and 45.29%.

5. **Displacement response of structure**

Table 11.15 gives the maximum displacement between layers of the test model when the peak acceleration of seismic wave is 680 gal. It can be seen that the damping ratio under the action of El Centro wave, Taft wave and Washington wave is 39.63 ~ 53.78%, 45.52 ~ 52.24% and 47.63 ~ 57.48% respectively under 8° (0.3 g) earthquake, with obvious damping effect.

6. **Damper response**

In case 1, curved steel plate dampers are installed on the first, second and third floors of structural model. In order to analyze the operation of the dampers more intuitively, the hysteretic curves of curved steel plate dampers installed on the bottom and third floors of the structure under the action of El Centro wave when the acceleration peaks of seismic wave are 400 gal, 680 gal and 830 gal respectively are given in Fig. 11.22. It can be seen that with the increase of the number of layers, the lower the plasticity degree of the damper is, the smaller the hysteretic area is; with the increase of the peak acceleration of the ground motion, the plasticity degree of the damper is gradually increased, and the hysteretic curve tends to be full.

Table 11.15 Maximum displacement between layers of structural model and damping ratio (680 gal)

Seismic wave	Floor number	Interlayer displacement (mm)		Damping ratio (%)
		Case 1	Case 2	
El-Centro wave	1	4.22	9.13	53.78
	2	4.43	9.09	51.27
	3	3.9	6.46	39.63
	4	4.07	7.3	44.25
Taft wave	1	4.40	9.11	51.70
	2	4.80	10.05	52.24
	3	4.22	8.15	48.22
	4	4.50	8.26	45.52
Washington wave	1	3.9	8.85	55.93
	2	5.2	12.23	57.48
	3	4.32	8.23	47.63
	4	4.21	8.21	48.72

7. **Comparative analysis of numerical simulation**

Using SAP2000 to establish the numerical model corresponding to the test model, carry out structural modal analysis and dynamic time history analysis, and compare with the test results. Table 11.16 gives the comparison of the first two vibration frequencies of the structure. It can be seen that compared with the test model, the error of the numerical model is between 4.00 and 7.45%, which can reflect the dynamic characteristics of the test model. Figure 11.23 shows the acceleration and displacement response of the structure system with curved steel plate damper when the peak ground acceleration is 680 gal. the comparison shows that the numerical model is in good agreement with the test results, which can reflect the real response of the structure after the installation of curved steel plate damper.

11.3.2 Dynamic Test of Frequency Modulation Damping Structure System

11.3.2.1 Horizontal TMD Damping Test

1. **Test model**

In order to study the effectiveness of TMD, TLD, TLCD and other frequency modulation and vibration reduction devices designed by the team, as well as the reliability of frequency modulation and vibration reduction analysis method, taking a 90 m

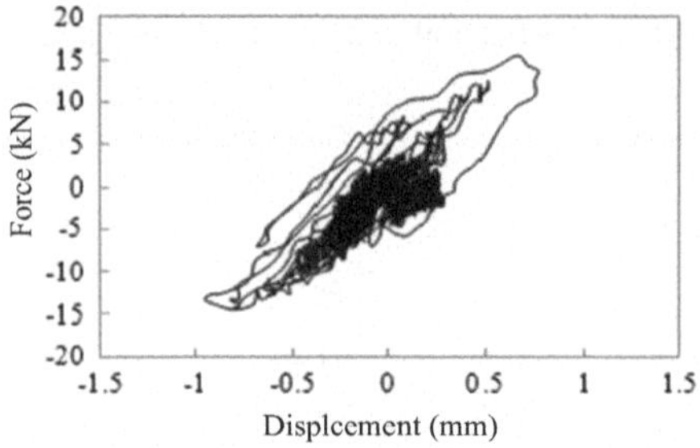

(a) First floor (El Centro wave, peak 400gal)

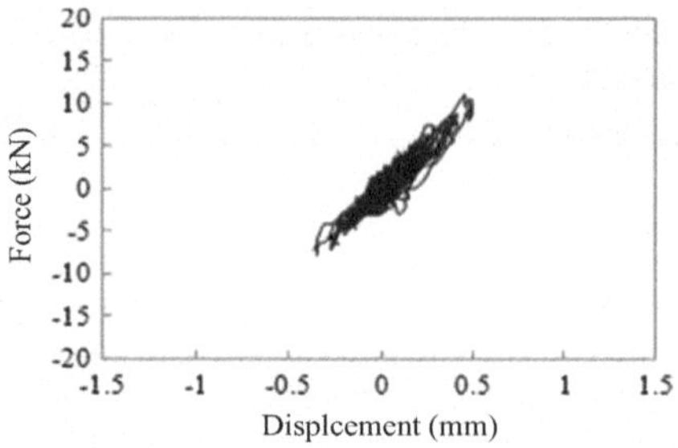

(b) Third layer (El Centro wave, peak 400gal)

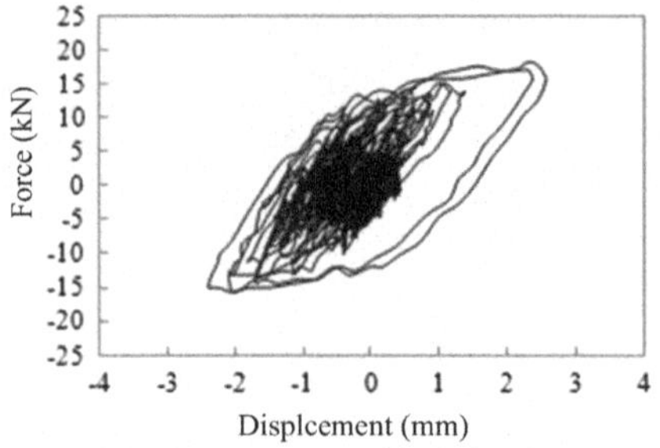

(c) First floor (El Centro wave, 680gal peak)

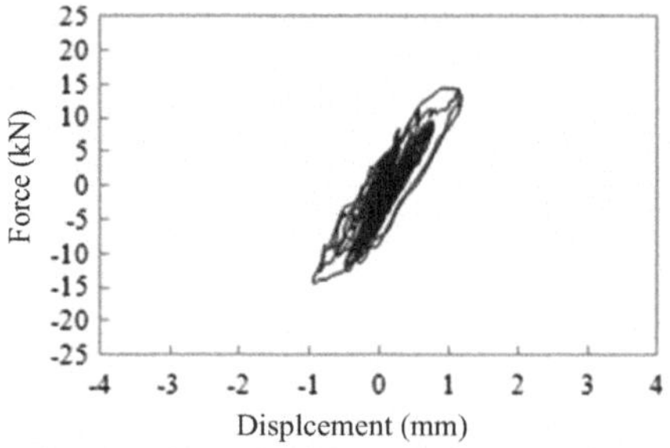

(d) Third layer (El Centro wave, peak 680gal)

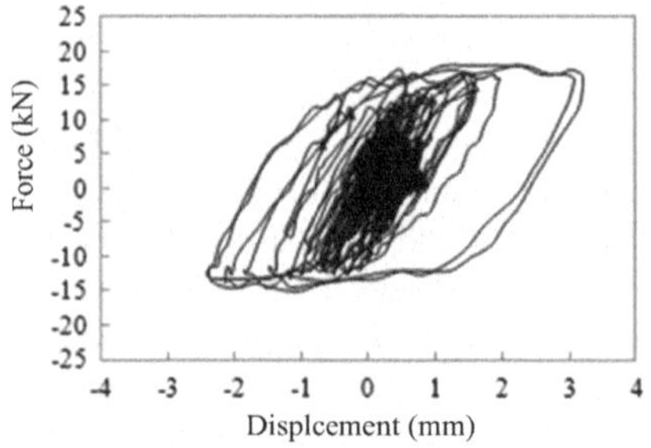

(e) First floor (El Centro wave, 830gal peak)

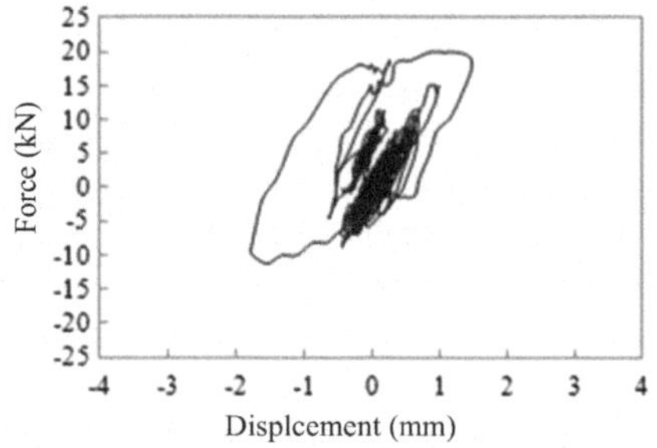

(f) Third layer (El Centro wave, peak 830gal)

Fig. 11.22 Hysteretic curve of damper

Table 11.16 Comparative analysis of frequency

Working condition	Explain	Test value	Calculated value	Error (%)
Case 1	First order frequency (Hz)	6.89	7.33	6.39
	Second order frequency (Hz)	15.93	16.97	6.53
Case 2	First order frequency (Hz)	4.43	4.76	7.45
	Second order frequency (Hz)	13.25	13.78	4.00

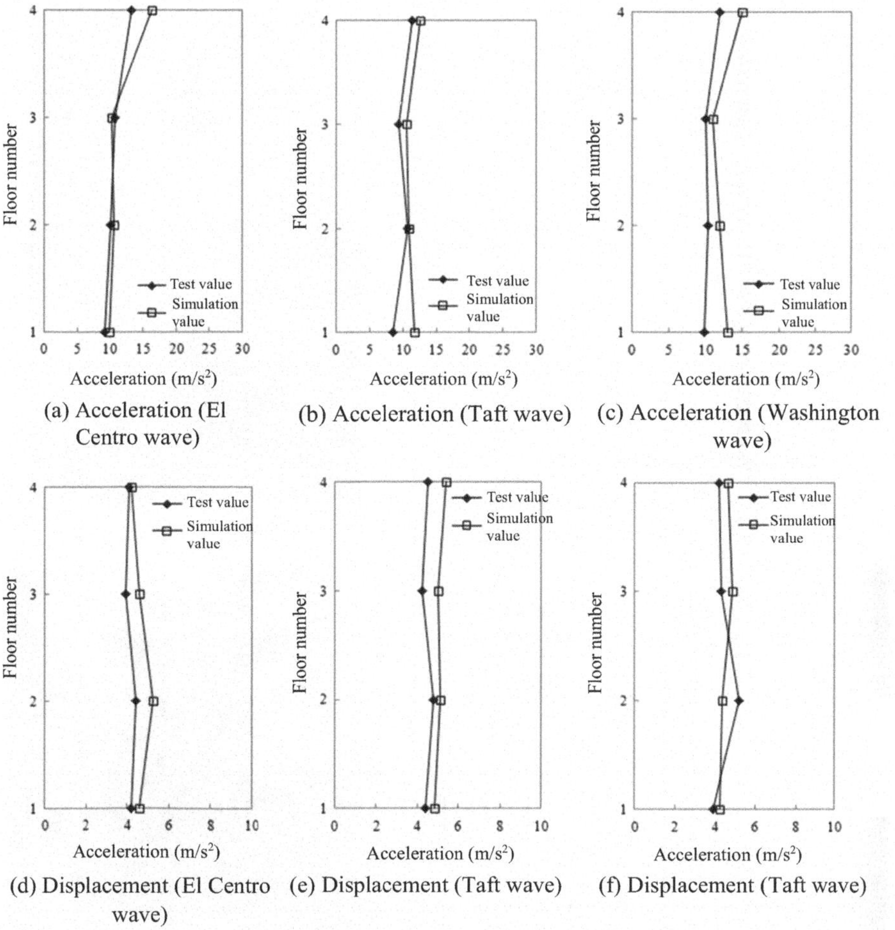

(a) Acceleration (El Centro wave) (b) Acceleration (Taft wave) (c) Acceleration (Washington wave)

(d) Displacement (El Centro wave) (e) Displacement (Taft wave) (f) Displacement (Taft wave)

Fig. 11.23 Comparative analysis of numerical simulation of damping system of curved steel plate

high steel chimney as the prototype, the frequency modulation and vibration reduction test model of high-rise structure is designed as shown in Fig. 11.24, the inner diameter of the chimney is 0.244 m, the outer diameter of the chimney is 0.25 m, the height of the chimney is 9 m, the thickness of the extension plate is 0.014 m, and the structure is provided with stiffeners; the model is divided into nine parts. Each segment is 1 m high, and 8 high-strength bolts with a diameter of 20 mm are used to connect the segments. The bottom segment is connected with the laboratory through four high-strength bolts with a diameter of 30 mm. The counterweight of the top segment of the model is 50 kg, and those of the rest segments are 100 kg, with a total of 9 mass blocks, which are made into a circular iron block and connected to the extension plate by bolts. The material is Q235 steel. The final prototype height is

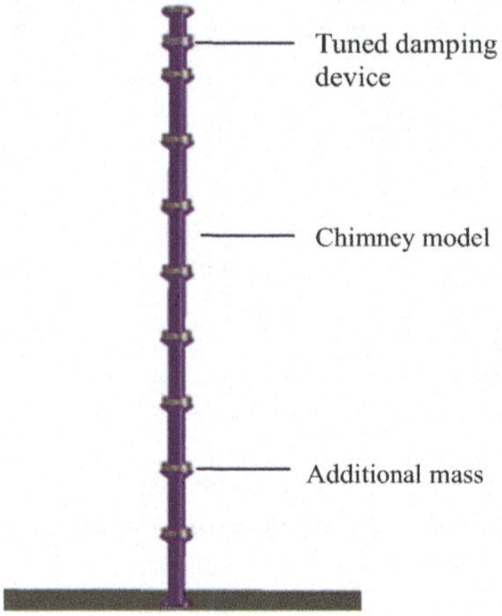

(a) Three dimensional diagram of test model

(b) Physical photographs

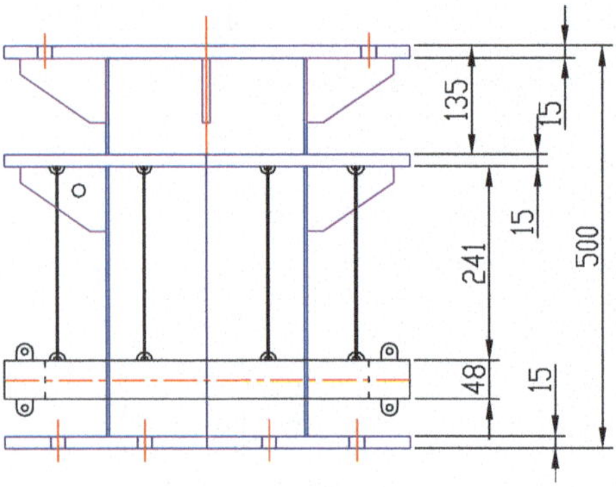

(c) Design drawing

(d) Physical photographs

Fig. 11.24 Test model of frequency modulation and vibration reduction for high-rise structure

90 m, the outer diameter is 2.5 m, the wall thickness is 0.03 m, the material is steel, the elastic modulus is $E = 2.06\text{e} + 11\ \text{N/m}^2$, the density is $\rho = 7850\ \text{kg/m}^3$.

Select the suspension ring TMD (RSR-TMD, as shown in Fig. 11.24c, d) for vibration reduction test. The TMD mass is 26.86 kg (the first mode mass ratio is 0.1), and the spring stiffness is 1254 N/m (the frequency ratio is 0.9, mainly considering the influence of increasing TMD on the structural flexibility). The inner diameter of the mass block is 0.4 m, the outer diameter is 0.5 m, the height is 0.048 m, the suspension height is 0.265 m, and the diameter of the suspension cable is 2 mm, totally 8 pieces.

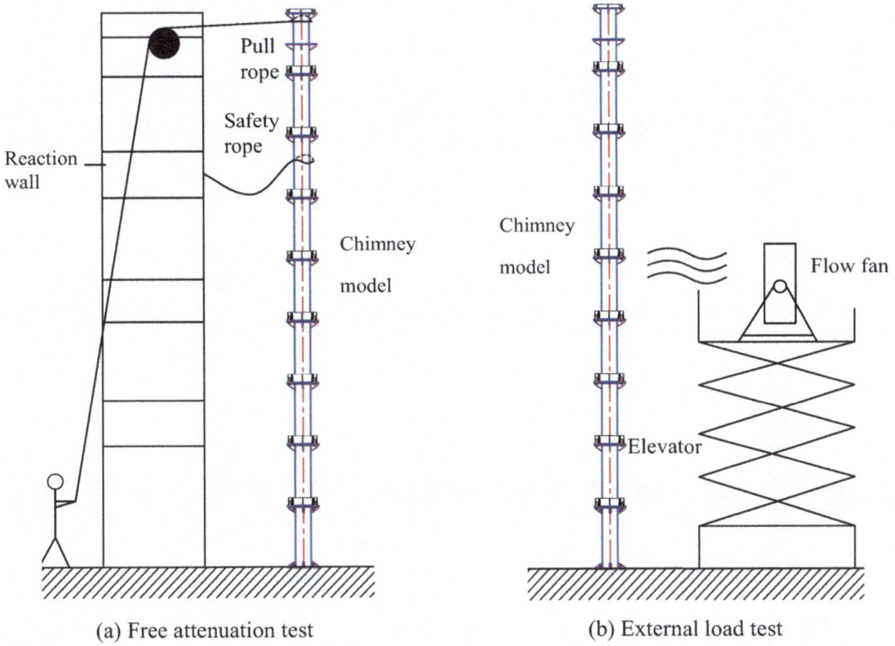

Fig. 11.25 Test scheme

2. **Test loading scheme and data acquisition**

In order to accurately measure the increase of the equivalent damping ratio of the structure after the installation of the damping device, so as to reflect the control effect of the damping device on the dynamic response of the chimney structure, the dynamic test research of the free attenuation of the structure is carried out. The test method is shown in Fig. 11.25a: by applying a certain force on the top of the model, make the displacement of the reference point reached the expected value, and then the pull rope is loosened to make the model occur a motion of free attenuation. At the same time, in order to study the effect of damping device under the action of wind load to a certain extent, the large flow fan is used for dynamic test under the action of external load. The test method is shown in Fig. 11.25b: place the flow fan on the large lift, control the height of the flow fan through the lift, and the flow fan outputs the external load that makes the model vibrate.

The arrangement of the test points is shown in Fig. 11.26: the acceleration sensor points mainly correspond to the first four largest vibration modes; the strain measurement points are arranged at the bottom and the middle of the model. The arrangement positions are the maximum strain positions of the model when the first-order vibration is the main vibration and the second-order vibration is the main vibration respectively. Strain gauges are pasted on both sides along the moving direction axis, with

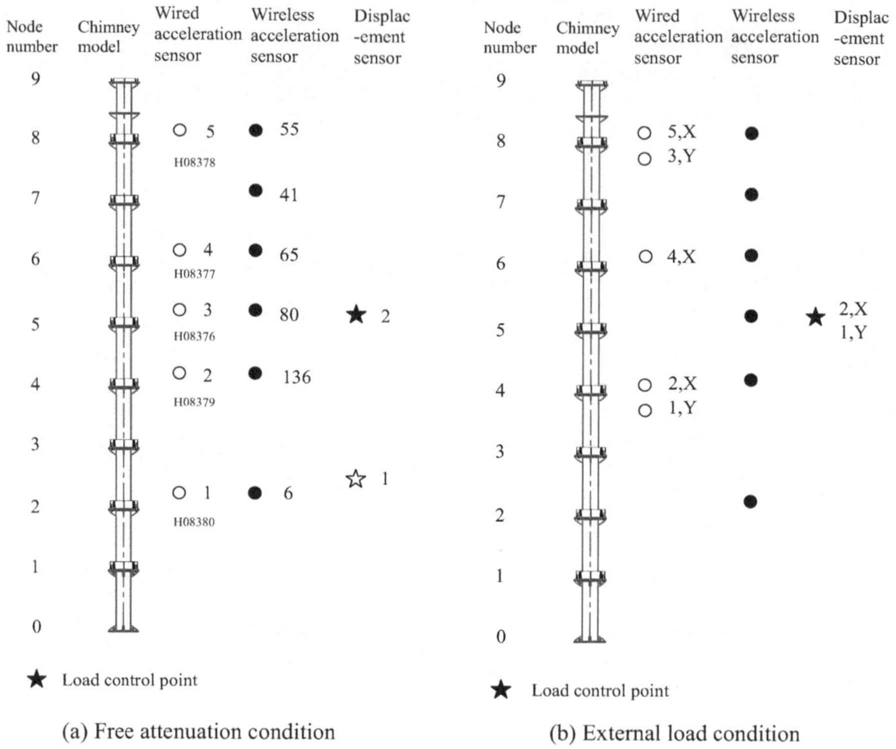

Fig. 11.26 Test point layout of test model

a total of 8 strain gauges. The displacement sensor is arranged in the middle and the bottom of the structure, it is mainly used to control the application of external load and check the acceleration sensor. It requires 2 displacement sensors, 5 acceleration sensors, 6 wireless sensors and 8 strain gauges. See Table 11.17 and Fig. 11.27 for the test instruments used in the test: the wireless sensor and its acquisition system adopt a series of products developed by Professor B.F. Spensor of UIUC; the main performance parameters of the flow fan are: air volume: 25,000 m^3/h, rotating speed: 1000r/min, power rate: 3 kW, voltage: 380 V, total pressure: 340 Pa.

According to the designed test scheme, the vibration control of high-rise steel chimney was carried out on the civil traffic test platform of Jiulonghu campus of Southeast University. See Table 11.18 for main test conditions.

3. **Structural dynamic characteristics and model modification**

In this section, the dynamic test is carried out for the model without any damping device, and the numerical model is modified. First of all, analyze the data obtained from the test, verify the correctness of the data obtained from each sensor, so as to lay

Table 11.17 List of test instruments

Device function	Equipment	Name	Number
Acquisition device	Wired acceleration sensor	941B vibration pickup	5
	Wireless acceleration sensor	Imoto2-SHM-A	6
	Strain gauge		8
	Displacement meter	YHD50 displacement sensor	2
Acquisition system	Strain collector	TST3827 dynamic and static strain test system	2
	Acceleration acquisition instrument	AZ308 8-channel data acquisition box	1
	Amplifier	AZ808 8 channel signal conditioner	1
	Personal computer		3
Software	Strain acquisition	TST3827 dynamic and static signal test and analysis system	
	Displacement acquisition	TST3827 dynamic and static signal test and analysis system	
	Wired acceleration acquisition	Anzheng Cras vibration and dynamic signal acquisition and analysis software V7.0	
	Wireless acceleration acquisition		
Incentive equipment	Flow fan	New energy saving low noise flow fan (SF 7-4)	1
Auxiliary equipment	Elevator		1

a foundation for further data analysis. A representative part of the test data is selected for comparison and analysis. Limited to space, in this section, the test data at the control loading point of free attenuation test when the displacement of the loading control point is approximately 15 mm (corresponding to the wired acceleration sensor measuring point 3, wireless acceleration sensor No. 80 and displacement measuring point 2) is discussed as an example. Figure 11.28a shows the acceleration time history collected by wired acceleration sensor and wireless acceleration sensor. It can be seen from the comparison that the values and waveforms of the two are basically the same. Therefore, the acceleration response of the test model can be obtained accurately whether wired sensor or wireless sensor is used. Because wireless sensor does not need wiring, installation and acquisition are more convenient than wired sensor, and it has greater advantages when the measurement conditions are limited and the installation of measuring points is difficult. But compared with wired sensor, it needs additional energy, and it is difficult to ensure long-term stable work when the current battery technology has not made great progress. For the wireless sensor used in this section, it uses three No. 5 batteries, and usually needs to be

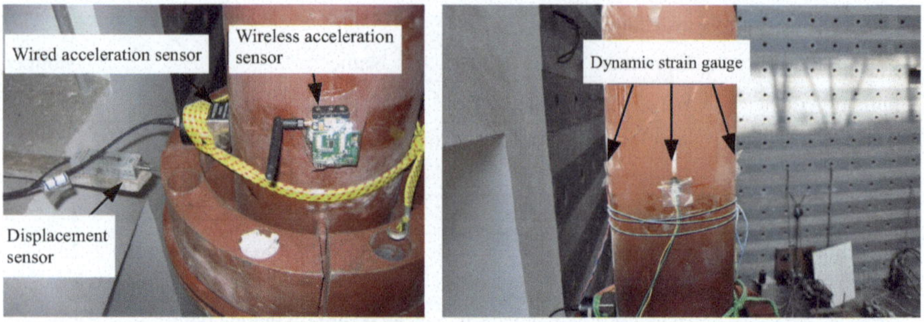

(a) Acquisition device

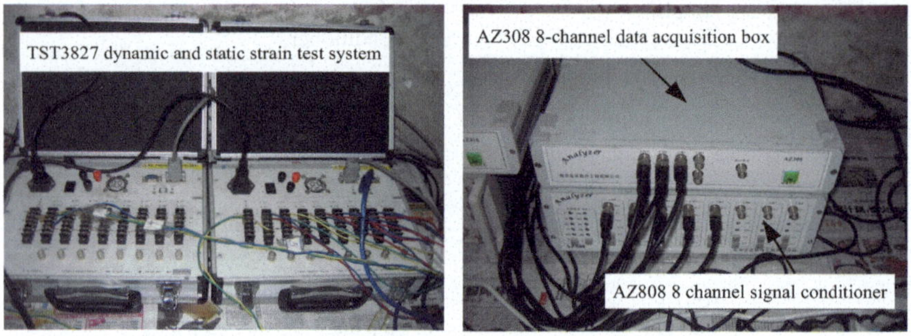

(b) Acquisition system

Fig. 11.27 Some experimental equipment

Table 11.18 Main test conditions

Working condition name	Working condition number	Working condition description
Free attenuation condition	FD-1	After the control point is stretched for 5 mm, it is suddenly relaxed
	FD-2	After the control point is stretched for 10 mm, it is suddenly relaxed
	FD-3	After the control point is stretched for 15 mm, it is suddenly relaxed
	FD-4	After the control point is stretched for 20 mm, it is suddenly relaxed
External load condition	W-1	The flow fan blows in the 8th segment of the model
	W-2	The flow fan blows between 8th segment and 9th segment of the model

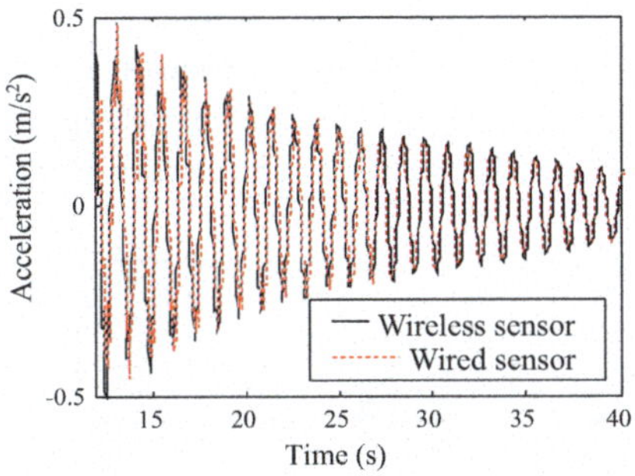

(a) Data obtained by wired and wireless sensors

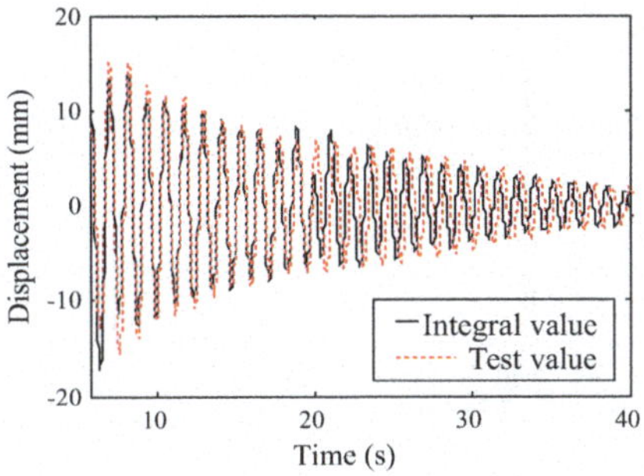

(b) Data obtained by acceleration sensor and displacement sensor

Fig. 11.28 Test data comparison and verification

replaced in about 2 h. Figure 11.28b shows the free attenuation displacement signal obtained by displacement sensor and the integration of acceleration sensor signal. The comparison shows that the difference between them is very small, which indicates that both acceleration sensor and displacement sensor work normally in the test process, the collected data is accurate and can be used for further analysis.

For the vibration control test, in addition to the concern of the model response itself, the modal parameters of the model (such as frequency, damping ratio, etc.) are also important. These parameters can be easily extracted by using the test data collected under different working conditions, with the help of spectrum transformation method and the definition of damping ratio. Considering that there are many factors affecting the modal parameters in the test process, the original structure was tested three times in this test, and the response signals under each test condition were collected. Using the data of free attenuation condition in three tests for analysis, Fig. 11.29a gives the linear spectrum (RMS) of acceleration response when the reference point tensile displacement is 20 mm. By picking up the extreme point, the first three modal frequencies of the original structure in different test processes can be obtained (Model 1 represents the first test, Model 2 represents the second test, Model 3 represents the third test). During of the second test, further tightening of the connecting bolts reduces the influence of the connection between the segments, and the structural frequency becomes larger; while the third test is to measure again after one day after tightening, and it is found that the value of the first two frequencies decreases, but the third frequency changes slightly; it can be seen that the tightening degree of the bolts has a greater impact on the dynamic characteristics of the model due to the more segments. Figure 11.29b shows the damping ratio (M1P1, representing Model 1 and measuring point 1) calculated by three tests at different

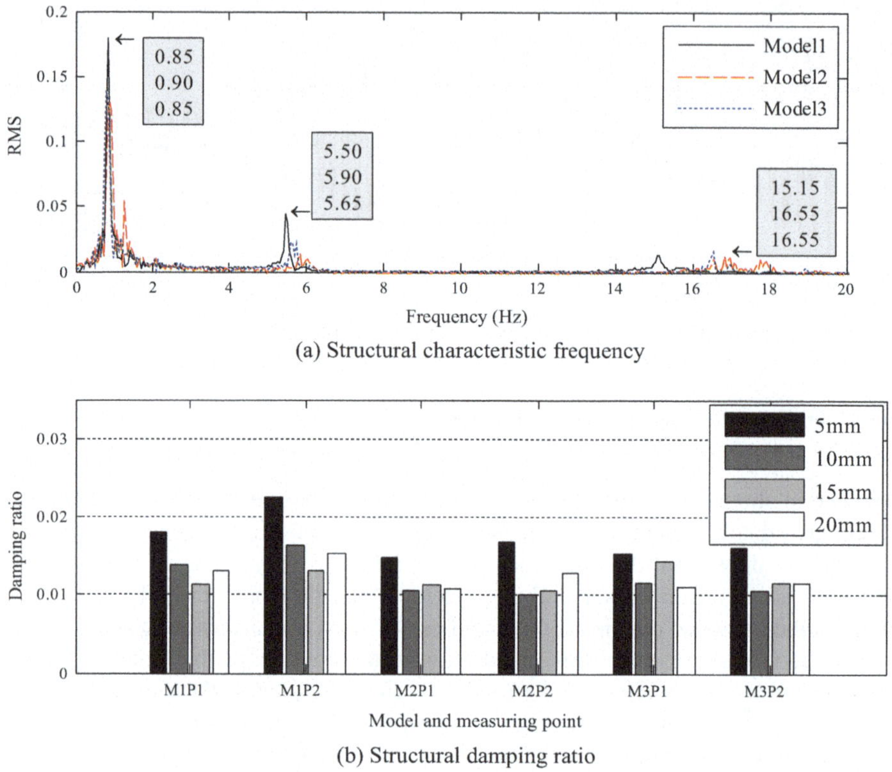

Fig. 11.29 Structural modal parameters

displacement measuring points and under different working conditions, with the mean value of 0.0134 and root variance of 0.003. It can be seen that the damping ratio is relatively large when the tensile distance is small, but generally speaking, the damping ratio obtained under each working condition has a small difference, so the mean value can be taken as a parameter for further analysis.

Through the comparison between the results of the experimental model and the numerical model, it can be seen that there is a certain error between the real value of the dynamic characteristics of the model and the initial calculated value. The main reasons for this situation are analyzed as follows:

Reason expression	Connecting bolt	Bottom constraint	Additional mass
Actual phenomena	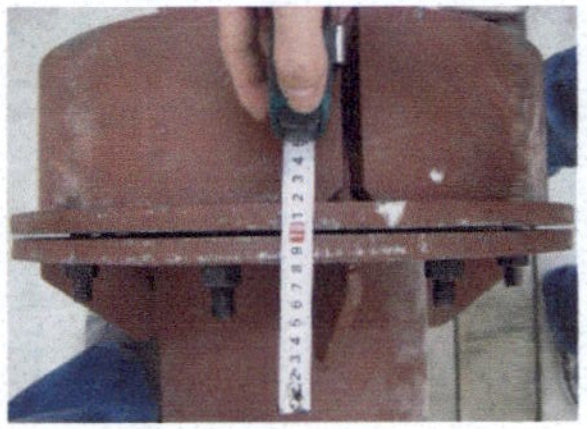		
Influence mechanism	Due to the tightening degree of the connecting bolts and the small warpage of the connecting plate, there is a certain gap between the connecting plates after the installation, and the structure produces multiple "flexible mezzanine", which makes the rigidity between the concentrated masses smaller and the structure as a whole more flexible	Due to the limitation of the ground anchor hole distance in the anchor hole area of the laboratory, the integrity of the bottom plate connection is affected, the bottom is changed from fixed connection to elastic connection, and the structure is flexible	Due to the acceleration sensor and the influence of the thickness of the connecting plate (between 15 and 18 mm), the additional mass used in the test is not exactly the same as that used in the analysis, which affects the mass matrix of the structure
Correction method	Correction of stiffness between concentrated masses	Increase the elastic rotation constraint at the bottom and correct its rigidity	Through analysis, it is found that the influence is small and no correction is needed

With the help of the above analysis, the modified variables can be determined to modify the numerical model, and the optimization based model modification method (Fig. 11.30a, in which the genetic algorithm is used in the process of variable updating) is adopted. Select the first three frequencies from the first test as the target, and construct the target function as follows:

$$J = \min\left(\frac{\alpha_1|f_1 - f_{1,m}|}{f_{1,m}} + \frac{\alpha_2|f_2 - f_{2,m}|}{f_{2,m}} + \frac{\alpha_3|f_3 - f_{3,m}|}{f_{3,m}}\right) \tag{11.76}$$

Among them, α_1, α_2 and α_3 are weight coefficients. For the flexible cantilever structure such as test model, the first-order mode controls its vibration, so the three weight coefficients are 0.6, 0.3 and 0.1 respectively; f_1, f_2 and f_3 are the first three frequencies of the model obtained by numerical simulation; $f_{1,\mathrm{m}}$, $f_{2,\mathrm{m}}$ and $f_{3,\mathrm{m}}$ are the first three frequencies of the model actually measured. Figure 11.30b shows the change of the objective function with the number of iterations in the process of model modification. It can be seen that after 50 generations of inheritance, the objective function basically tends to the optimal value. After modification, the first three frequencies of the numerical model are 0.85, 5.52 and 15.73 respectively. Compared with the measured results, it can be seen that the maximum error of the first three frequencies of the numerical model is 3.8% (the third frequency), of which the first order frequency is almost the same. In order to verify the model and further verify the static characteristics of the model, take the displacement results when the model was stretched to 20 mm in the first test, the test start time is 7.40 s, the displacement corresponding to control measurement point 2 is 21.05 mm, and the displacement corresponding to displacement measurement point 2 is 7.32 mm; when the displacement of control measurement point (i.e. measurement point 2) is 21.05 mm, the displacement of displacement measurement point 2 is calculated as 7.55 mm, which is 3.1% different from the measured result. Figure 11.30c shows the comparison of the displacement time history curve at the displacement measuring point 2 under this working condition. It can be seen that the difference between the calculated value and the measured value is small both in value and phase. To sum up, the model modification method based on genetic algorithm can better modify the numerical model, and the modified model basically meets the analysis requirements.

4. **Structural dynamic response and analysis**

This section mainly introduces the test of RSR-TMD structure system, and establishes the model for analysis to verify the correctness of the above analysis theory and the preparation of the program. In order to determine the parameters of TMD, the dynamic test of TMD was carried out before and after the structural system test of RSR-TMD to determine the dynamic parameters of TMD. The test results of TMD are given in Fig. 11.31. PSD represents the power spectral density and numbers 1 and 2 represent two orthogonal vertical directions respectively. After the dynamic test of the original structure, considering that the model is relatively flexible in design, the suspension height of the damper is increased in the possible range. However, due to

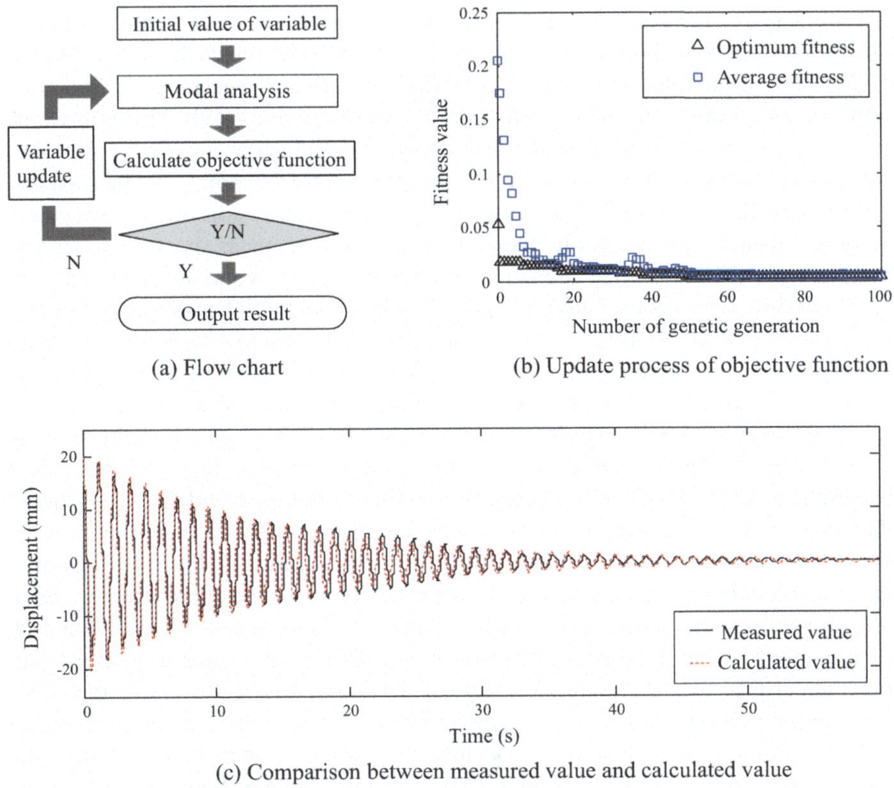

Fig. 11.30 Revision of numerical model

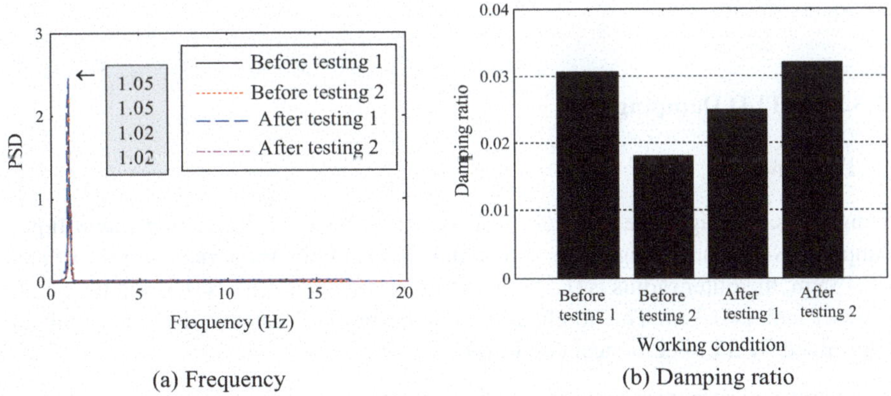

Fig. 11.31 TMD test results

the processing of the damper is before the test, the reserved space is not enough to adjust to the theoretical optimal frequency; Fig. 11.31b calculates the damping ratio of TMD, the mean value of four tests is 0.0264, and the root variance is 0.0063.

Figure 11.32 shows the test results after the installation of RSR-TMD, in which Fig. 11.32a, b respectively show the displacement time history and linear spectrum of displacement measurement point 2 when stretching for 20 mm. It can be seen that after the installation of TMD, the structure becomes slightly soft, the displacement response attenuation is accelerated, and the peak value of linear spectrum is significantly reduced. Figure 11.32c compares the results of analysis and measurement, and the data of displacement measurement point 2 is still selected. The comparison shows that the numerical simulation can represent the real situation to a certain extent. With the increase of time, there is a slight phase error, which is mainly caused by the nonlinearity caused by the error of segment connection and bottom connection in the real model. Figure 11.32d, e show the two-way acceleration distribution and one-way acceleration time history curve of the fourth node (corresponding to displacement measurement point 2 and acceleration measurement points 1 and 2) of the model when the flow fan is blowing in the eighth section of the model (W-1 working condition). At this time, the acceleration response is small and the displacement response can be ignored, but it can be seen that the acceleration response of the structure has a large attenuation after TMD is installed. Figure 11.32f shows the structural damping ratio calculated from the displacement response of two displacement measurement points under four free attenuation conditions. The comparison shows that the structural damping ratio is basically the same under each condition, with the mean value of 0.034, root variance close to 0, and the mean value greater than 0.0134 of the original structure. It can be seen that TMD can improve the damping ratio of the whole structure system, so as to effectively control the dynamic response of steel chimney. The program of dynamic response analysis compiled in this section has certain credibility.

11.3.2.2 TLD Damping Test

1. **Test model**

Using the same high-rise structure test model as Sect. 11.3.2.1, ring tuned liquid damper (RS-TLD) is designed as shown in Fig. 11.33: the inner radius of water tank is 0.135 m, the outer radius is 0.225 m, the inner diameter ratio is 0.6, and the height of water tank is 0.3 m. The height and damping ratio of water in the water tank are determined by the specific test conditions.

2. **Structural dynamic response and analysis**

This section introduces the dynamic test of RS-TLD structure system with different water depth and different damping methods: under the condition that the TLD section cannot be changed, by adjusting the water depth to change the TLD tuning frequency, so as to achieve the purpose of affecting the control effect. This section carried out

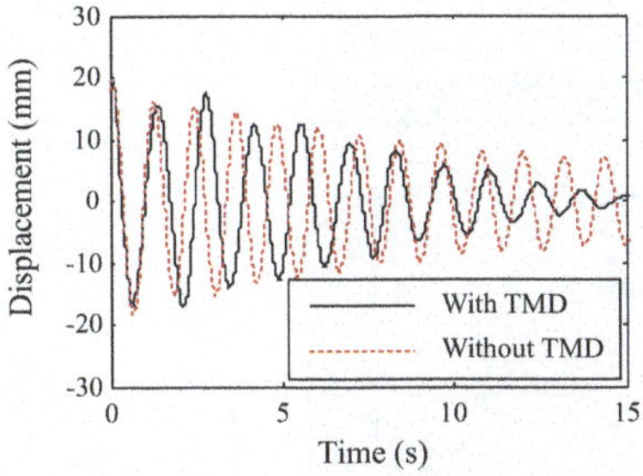

(a) Time history curve of displacement measuring point 2 under FD-4 working condition

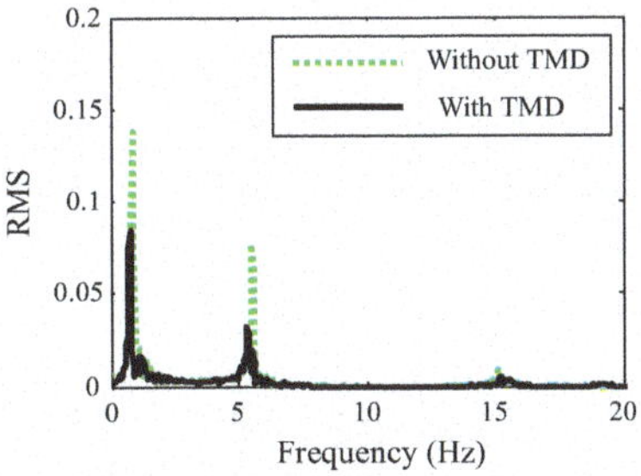

(b) Linear spectrum of acceleration measuring point 3 under FD-4 working condition

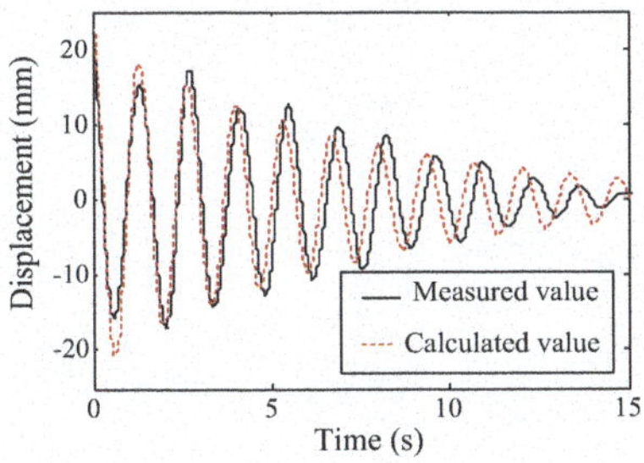

(c) Comparison between analysis and measurement

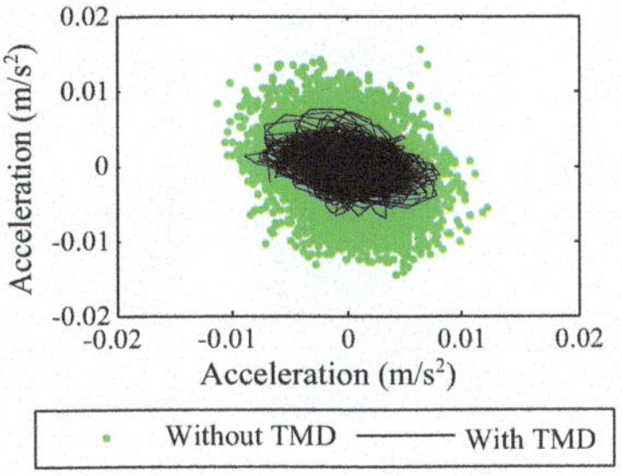

(d) Acceleration in two orthogonal directions of node 4 under W-1 condition

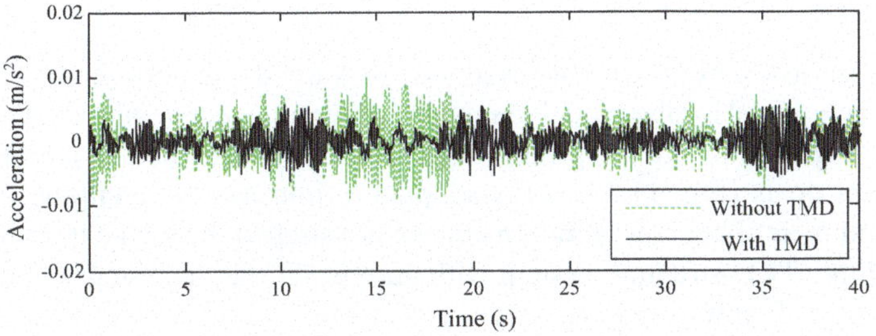

(e) Acceleration time history in one direction of node 4 under W-1 condition

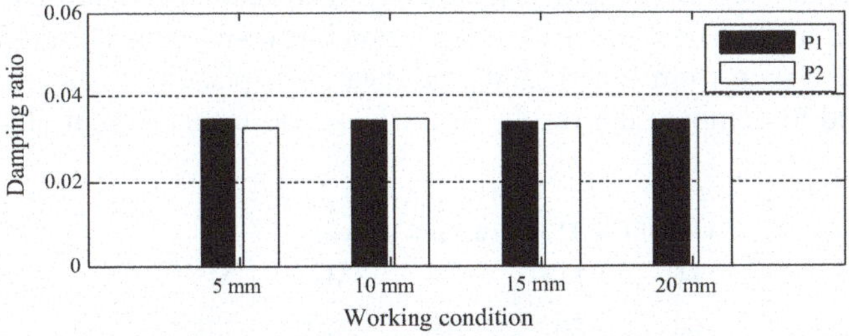

(f) Damping ratio of RSR-TMD structure system

Fig. 11.32 Analysis of test results of RSR-TMD structural system

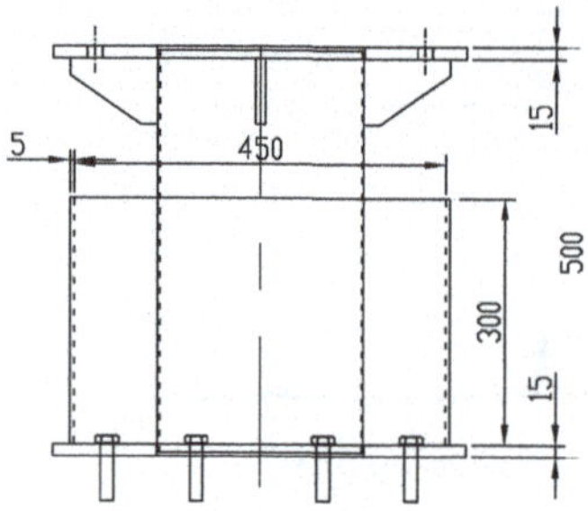

(a) Design chart

(b) Physical photograph

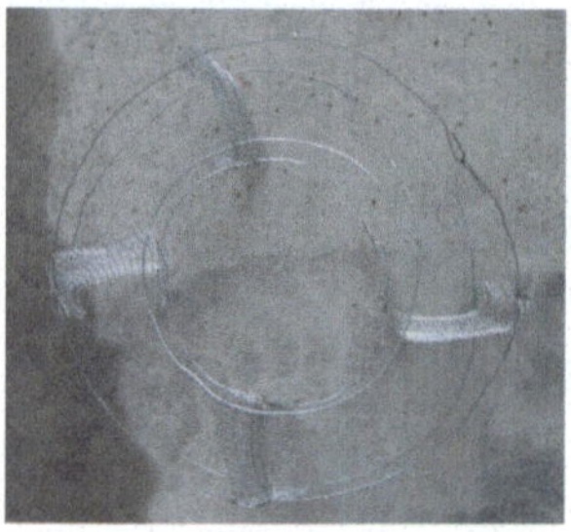

(c) Wire mesh

(d) Big foam

(e) Small foam

Fig. 11.33 RS-TLD for test

the dynamic test of RS-TLD structure system with water depth between 0.06 m and 0.18 m (see Table 11.19 for the model number of different water depths); at the same time, considering the low damping ratio of pure water, the method of adding wire mesh and foam floats (Fig. 11.33c–e) is adopted to improve the damping ratio of pure water, and their effects on vibration control are studied through experiments.

First of all, the TLD structure system with pure water only is tested without any damping additives. Figure 11.34 gives some test results for model M2 and the damping ratio measured by each model. It can be seen from Fig. 11.34a, b: after TLD is installed, the structure becomes slightly flexible, the displacement response attenuation is accelerated, and the peak value of acceleration linear spectrum is significantly reduced. As can be seen from Fig. 11.34c, the data of displacement measurement point 2 is selected to compare the results of analysis and measurement. The error

Table 11.19 Experimental model of RS-TLD structural system

Model number	M1	M2	M3	M4	M5	M6	M7
Depth of water/mm	60	80	100	120	140	160	180
TLD tuning frequency/Hz	0.672	0.765	0.841	0.904	0.956	0.998	1.033

Note for pure water and barbed wire only, all models were tested; for additional foam, only M3, M4 and M5 models were tested

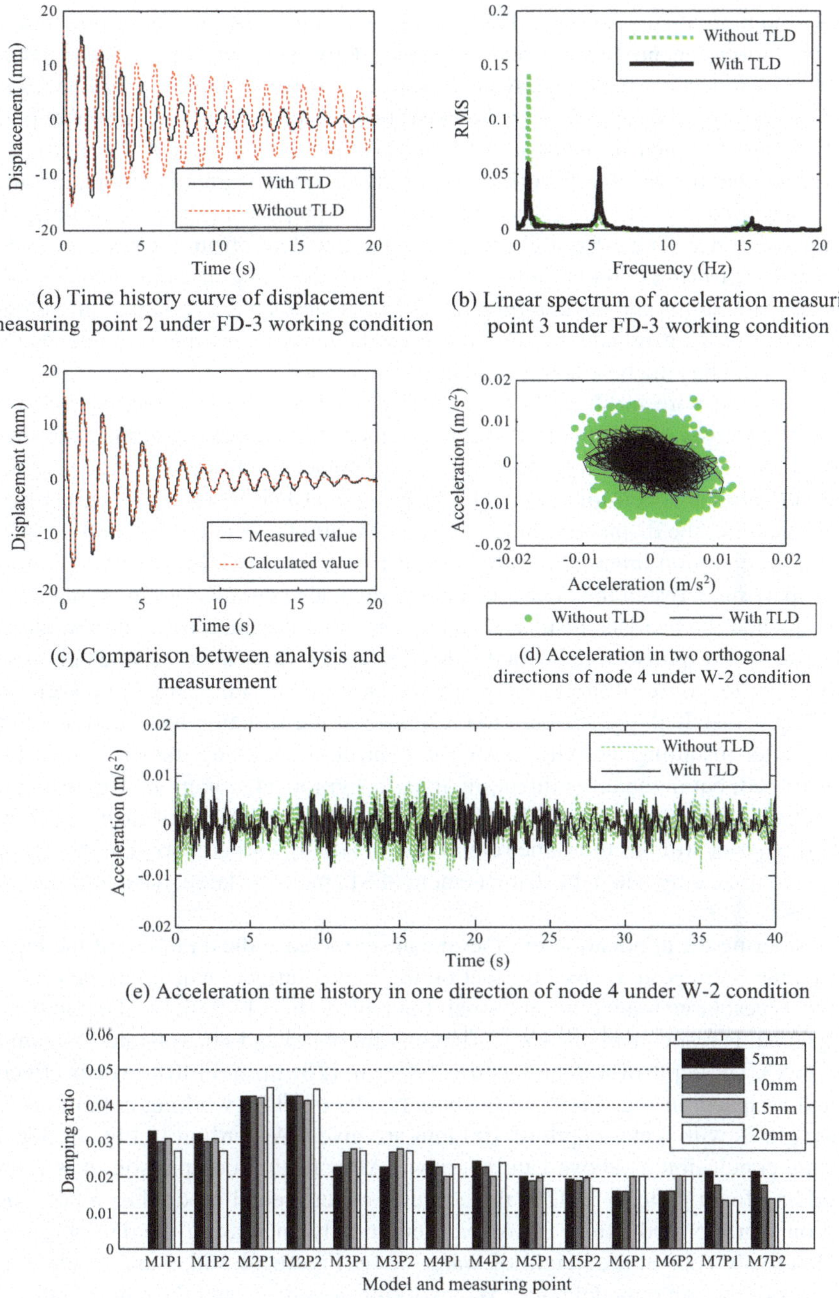

(a) Time history curve of displacement measuring point 2 under FD-3 working condition

(b) Linear spectrum of acceleration measuring point 3 under FD-3 working condition

(c) Comparison between analysis and measurement

(d) Acceleration in two orthogonal directions of node 4 under W-2 condition

(e) Acceleration time history in one direction of node 4 under W-2 condition

(f) Damping ratio of RS-TLD structure system

Fig. 11.34 Analysis of test results of RS-TLD structural system (pure water) between numerical simulation and measurement results is small. Therefore, numerical simulation can predict the real response of the structure to a certain extent. Figure 11.34d, e show the two-way acceleration distribution and one-way acceleration time history curve of node 4 (corresponding to displacement measurement point 2 and acceleration measurement points 1 and 2) of the model under W-2 condition. It can be seen that the acceleration response of the structure is greatly attenuated after TLD is installed. Under such external loads, TLD can play a role in controlling the structure response. The structural damping ratio calculated from the results of each model and measuring point is shown in Fig. 11.34f, the damping ratio of each model is basically the same, and the damping ratio of each model has a certain improvement compared with the original structure, reaching the maximum value in model M2, at this time, TLD has the best control effect on the model.

Then, install wire mesh in TLD and carry out relevant experimental research. Figure 11.35 gives part of the test results of model M2 and the damping ratio measured by each model. It can be seen from the displacement time history curve and acceleration linear spectrum in Fig. 11.35a, b: after TLD is installed, the structure is slightly flexible, the displacement response attenuation is accelerated, and the peak value of acceleration linear spectrum is significantly reduced. Figure 11.35c shows the comparison between the measured data of displacement measurement point 2 of the model and the analysis results. It can be seen that the numerical simulation can predict the real response of the structure to a certain extent. Under the W-2 condition, the dynamic response of the model is shown in Fig. 11.35d, e. It can be seen that after TLD is installed, the acceleration response of the structure has a certain attenuation. After installing the wire mesh, the control effect along the stress direction has increased, but in the other direction, it has a significant reduction. The structural damping ratio (as shown in Fig. 11.35f) calculated from the results of each model and measuring point still has the same rule as that of adding barbed wire, but the overall value decreases, only when the displacement of M2 model is large, the damping ratio increases.

Considering the application of other means to increase the damping of the liquid in TLD, the foam plate is used to float on the liquid surface, which can be divided into two types: large foam plate and small foam plate (Fig. 11.33d, f). The number of arrangement is respectively 8, 16, 32 (large foam plate) and 80, 160 and 320 (small foam plate). The depth of water is taken as 100 mm, 120 mm and 140 mm respectively. Limited to space, in Fig. 11.36, only some results of FD-3 working conditions for 8 foam plates with water depth of 100 mm are given. According to Fig. 11.36a–c, the same conclusion as above can be drawn. After further comparison, it is found that adding foam plate to TLD at this scale of experimental model has no obvious effect on vibration reduction. However, theoretically, increasing the damping ratio can effectively control the height of water wave. Figure 11.36d, e are acceleration responses of the same model under W-2 working condition, and the control effect is obvious. Figure 11.36f calculated the average value of damping ratio under different water depth and different number of foam plates under tensile test conditions. It can be seen that with the increase of water depth, the damping ratio decreases, and the number of foam plates slightly affects the damping ratio, but the change is not

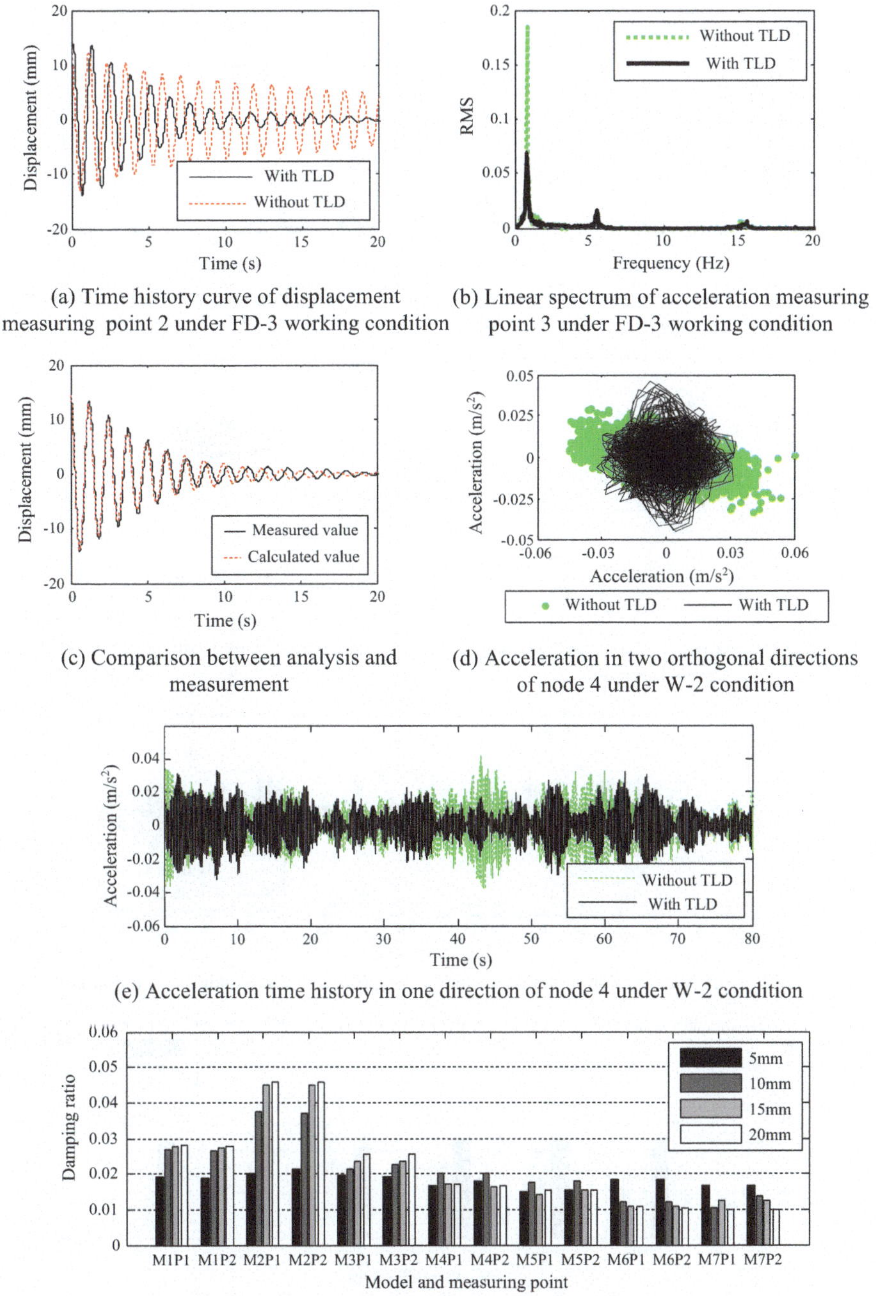

(a) Time history curve of displacement measuring point 2 under FD-3 working condition

(b) Linear spectrum of acceleration measuring point 3 under FD-3 working condition

(c) Comparison between analysis and measurement

(d) Acceleration in two orthogonal directions of node 4 under W-2 condition

(e) Acceleration time history in one direction of node 4 under W-2 condition

(f) Damping ratio of RS-TLD structure system

Fig. 11.35 Analysis of test results of RS-TLD structural system (Wire mesh)

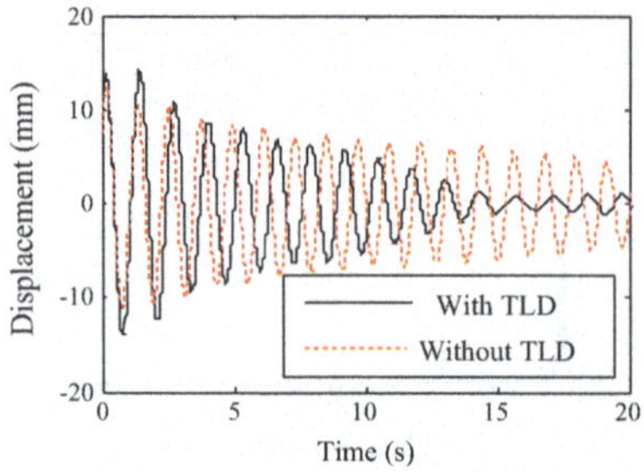

(a) Time history curve of displacement measuring point 2 under FD-3 working condition

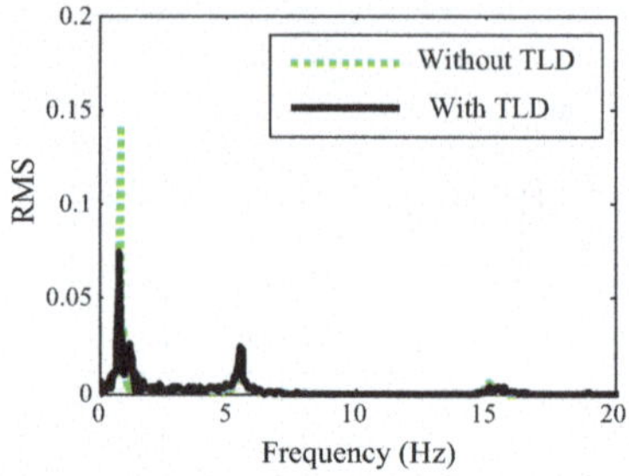

(b) Linear spectrum of acceleration measuring point 3 under FD-3 working condition

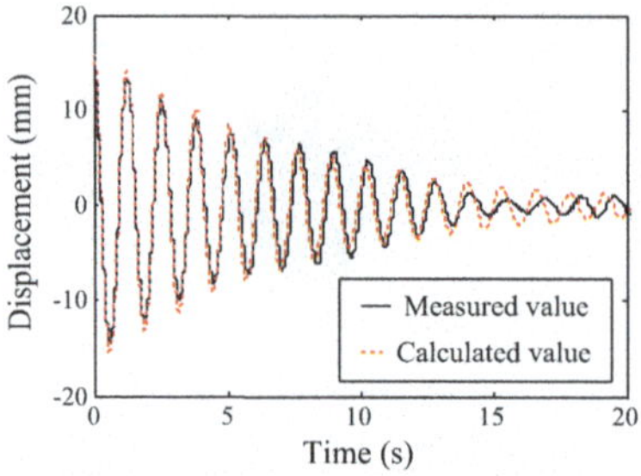

(c) Comparison between analysis and measurement

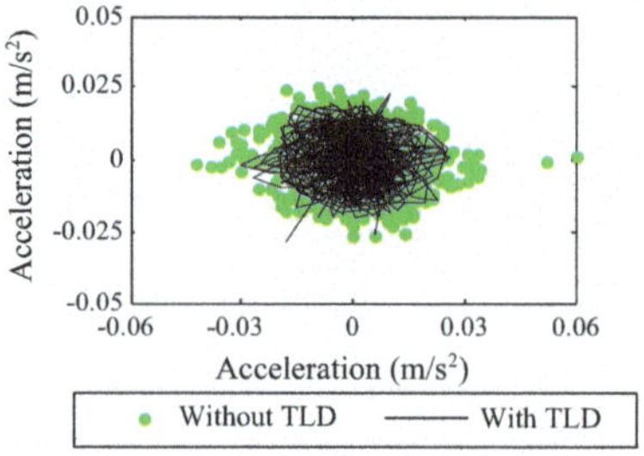

(d) Acceleration in two orthogonal directions of node 4 under W-2 condition

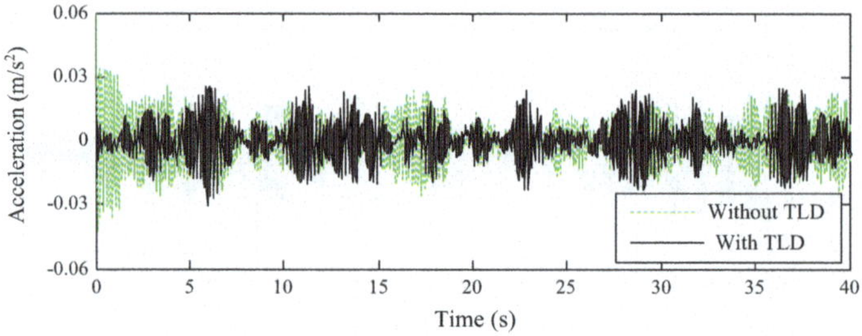

(e) Acceleration time history in one direction of node 4 under W-2 condition

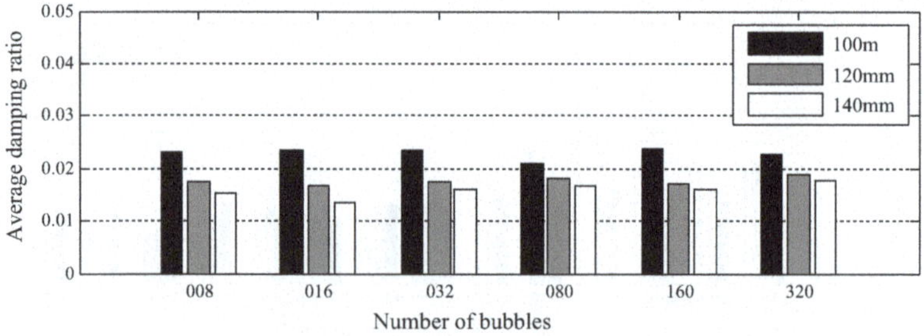

(f) Average damping ratio of RS-TLD structure system

Fig. 11.36 Analysis of test results of RS-TLD structural system (foam plate) obvious. Compared with pure water, because the foam plates restrict the movement of water waves, the control effect is reduced to some extent, but the specific effect on water wave height control needs further study.

To sum up, TLD can effectively control the dynamic response of steel chimney structure; adjusting its tuning frequency reasonably can greatly increase the overall damping ratio of the structure and achieve better damping effect; adopting measures to increase the damping ratio of water does not necessarily improve the control level, but through the observation of field test, we can see that the measures can better inhibit the water wave height. The specific quantitative effect needs further study.

11.3.2.3 TLCD Damping Test

1. **Test model**

The same high-rise structure test model as Sect. 11.3.2.1 is still used. The design of ring tuned liquid column damper (RS-TLCD) is shown in Fig. 11.37: the pipe is a square section of 0.1 m, the center radius of the water tank is 0.18 m, the height of the water tank is 0.25 m, $A_0/A = 0.5$, and the height of the water in the water pipe is determined by the test conditions.

2. **Structural dynamic response analysis**

In this section, the test results of RS-TLCD structure system are introduced and compared. The dynamic tests of the structure system under different water depth are carried out. See Table 11.20 for the water depth and model number. Figure 11.38a, b show the time history curve and power spectral density of displacement measurement point 2 when the control displacement of model M3 is 15 mm. After TLCD

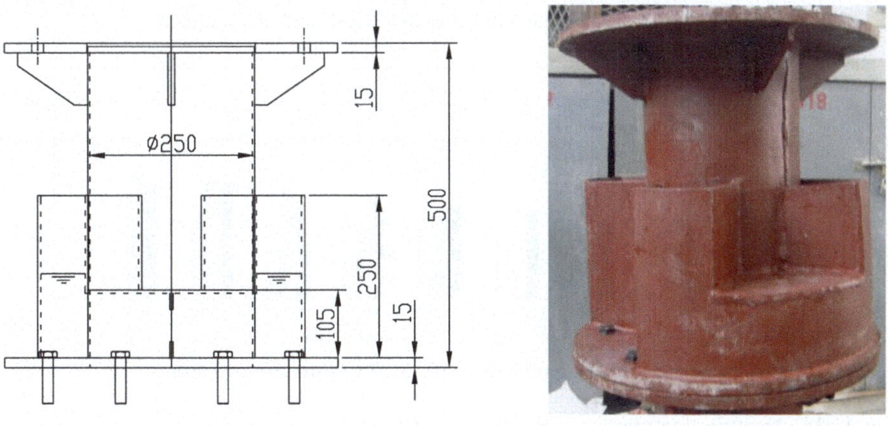

(a) Design chart (b) Physical photograph

Fig. 11.37 RS-TLCD for test

Table 11.20 RS-TLCD Experimental model of RS-TLCD structural system

Model number	M1	M2	M3	M4
Water depth/mm	160	180	200	220
TLCD tuning frequency/Hz	0.994	0.956	0.923	0.893

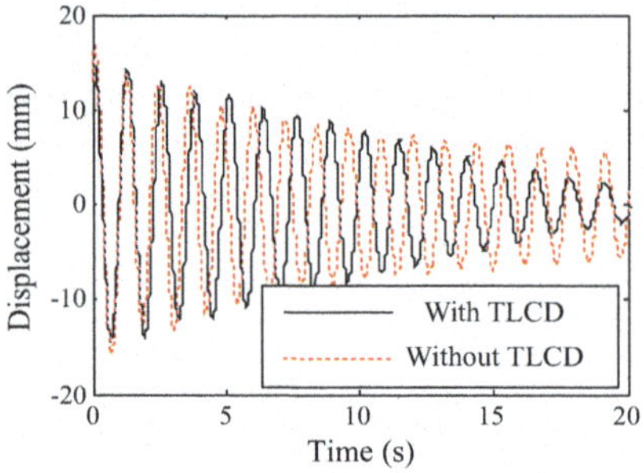

(a) Time history curve of displacement measuring point 2 when the control displacement is 15 mm

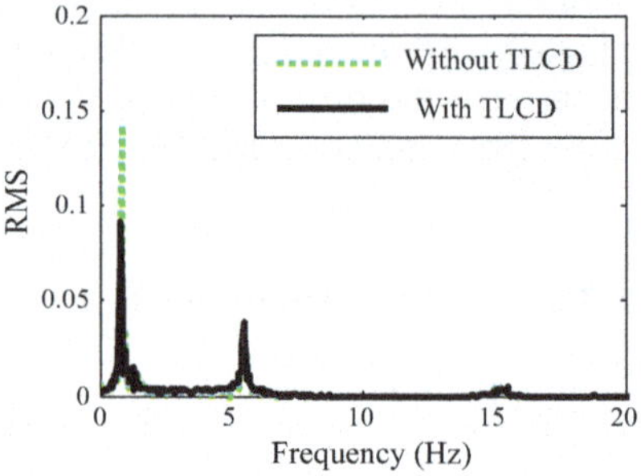

(b) Linear spectrum of acceleration measuring point 3 when the control displacement is 15 mm

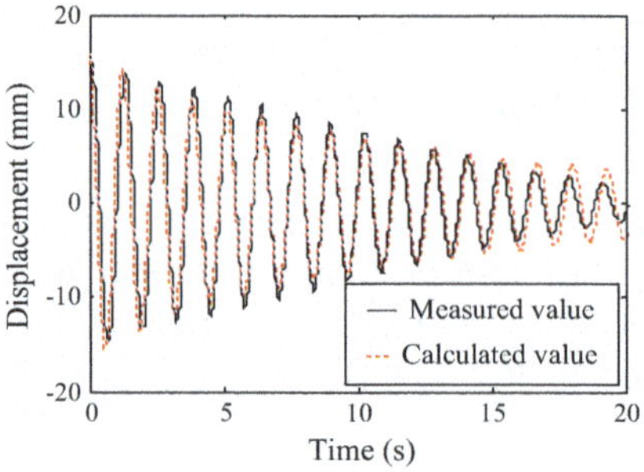

(c) Comparison between analysis and measurement

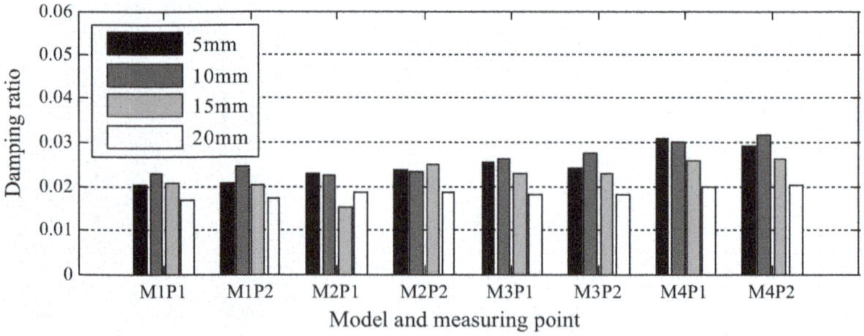

(d) Damping ratio of RS-TLCD structure system

Fig. 11.38 Analysis of test results of RS-TLCD structural system is installed, the structure becomes flexible as a whole, the displacement response decays faster, the peak value of power spectral density becomes smaller, and the peak position slightly shifts to the low frequency. Figure 11.38c compares the measured and numerical simulation results, and the numerical difference between them is small. Because the experimental model is not completely ideal, the installation problem leads to some non-linear situations, which are not considered in the numerical model, so the phase and numerical error increase with the increase of time. In order to further study the influence of TLCD on the vibration response of the model, the damping ratios of four models are calculated respectively, as shown in Fig. 11.38d. The analysis shows that: the damping ratios obtained from different measuring points of the same model are basically the same; the mean value of the damping ratio of model M1 is 0.0205, and the root variance is 0.0025; the mean value of the damping ratio of model M2 is 0.212, and the root variance is 0.0034; the mean value of the damping ratio of model M3 is 0.0231, The root variance is 0.0034; the mean value of damping ratio of model M4 is 0.0267, and the root variance is 0.0046; with the increase of water depth, the tuned frequency of TLCD is more and more close to the first-order natural frequency of the structure, and the damping ratio of RS-TLCD structure system is increased, and the damping effect is gradually improved. It can be seen from the above comparative study that the installation of RS-TLCD can effectively improve the damping ratio of steel chimney and suppress the dynamic response of steel chimney, and its control effect is closely related to the frequency of RS-TLCD itself.

11.3.2.4 Vertical TMD Damping Test

1. **Test model**

The long-span roof structure is also one of the structural forms for which the frequency modulation and vibration reduction devices are most widely used. For this reason, the team carried out frequency modulation and vibration reduction test for the long-span roof structure based on vertical TMD. Steel concrete composite structure is adopted as the test component. The structural arrangement is shown in Fig. 11.39, with spans of 4700 mm and 7700 mm respectively. The plane size of the slab is 5000 × 8000 mm, the thickness of the slab is 60 mm, and the concrete grade is C30. The main beam is hot-rolled common I20b I-beam and I22b I-beam, and the secondary beam is hot-rolled common I12.6 I-beam, as shown in Table 11.21.

The test members simulate the typical beam slab structure in normal use. The design of components shall meet the requirements of bearing capacity and normal use stage. The TMD model is shown in Fig. 11.39d. It can be seen from the figure that the damper consists of eight springs, a viscous damper and a mass block. It is suspended in the mid span of the test floor. When the response test under the single walking load before the vibration reduction is conducted, it is locked by the iron bars around the TMD to limit its displacement as the additional mass of the structure, and

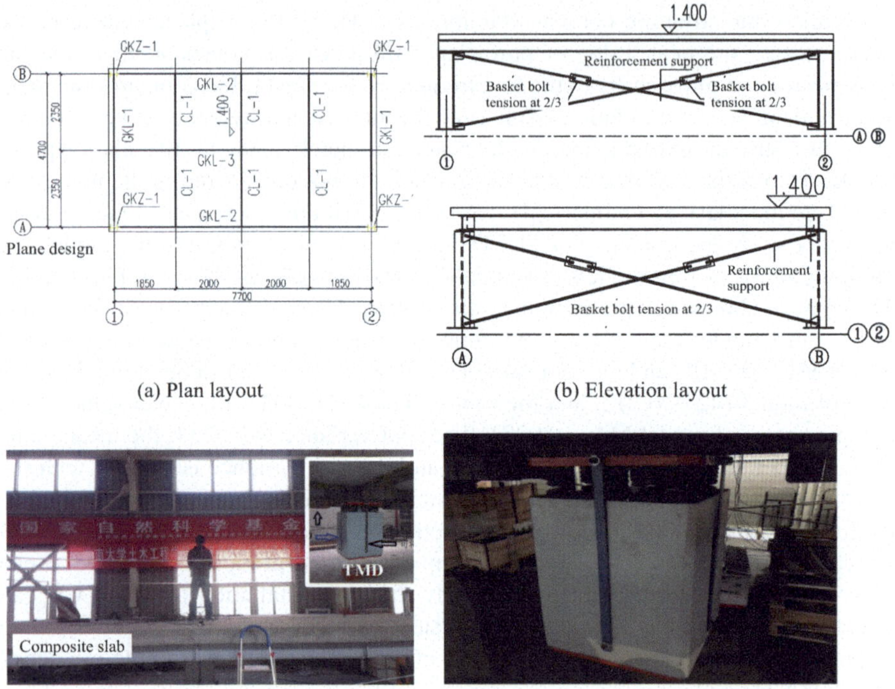

(a) Plan layout (b) Elevation layout

(c) Floor model (d) TMD model

Fig. 11.39 Test model of vertical TMD damping system

Table 11.21 Material list of test components

Code name	Specification	Arrangement plan of stud with cylindrical head
GKZ-1	H300 × 200 × 6×10	
GKL-1	I20b	Ø10 × 120@200 (protruding out of structural layer)
GKL-2	I20b	Ø10 × 120@200 (protruding out of structural layer)
GKL-3	I22b	Ø10 × 40@200
GL-1	I12.6	Ø10 × 40@200

the response test after the vibration reduction is required, only four limit devices are removed, TMD can work normally and participate in the structural vibration control.

2. **Structural dynamic response**

In order to verify the vibration reduction effect of TMD under the excitation of single person walking on the test floor, the experimenter carried out the single person walking test under five working conditions for the TMD locked state (uncontrolled structure) and the TMD opened state (controlled structure). Figure 11.40 gives the acceleration time history curve under the excitation of single person walking under

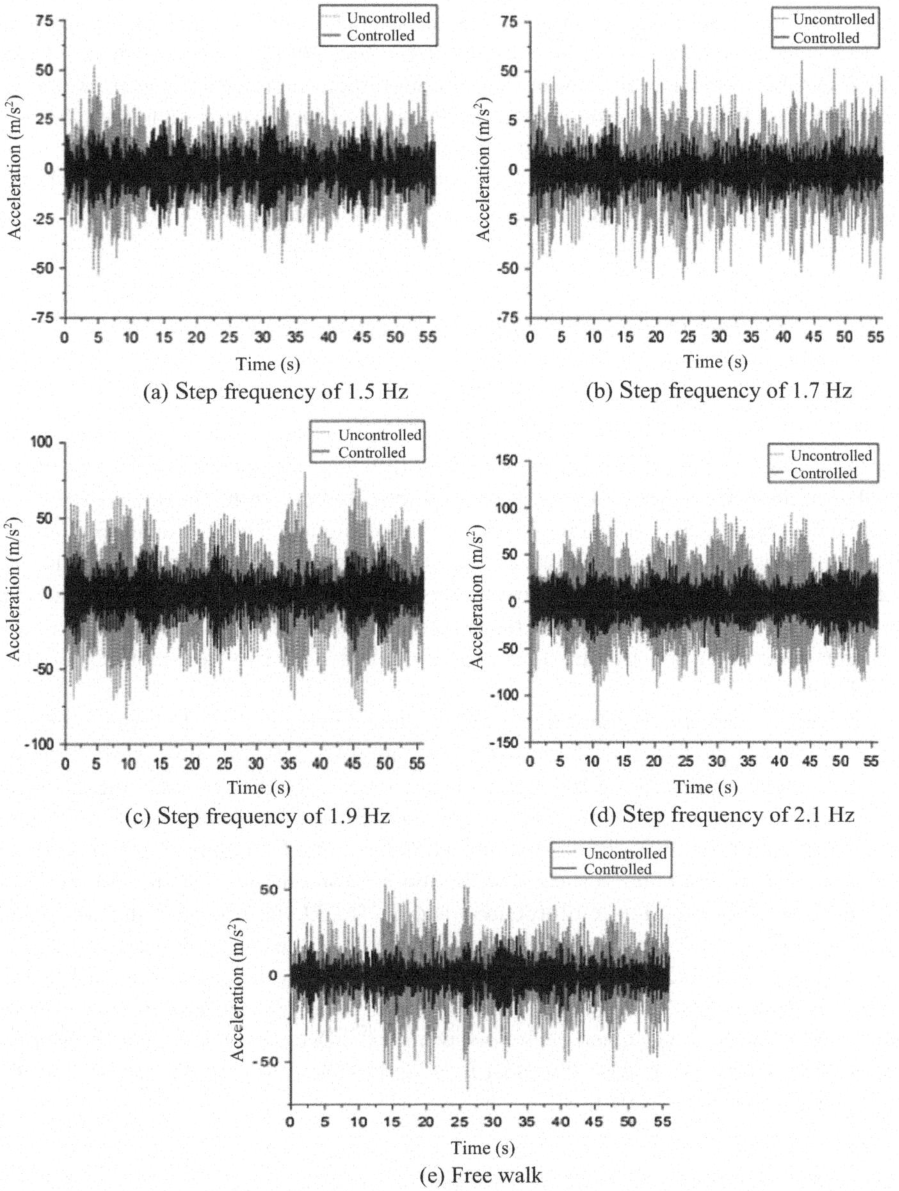

Fig. 11.40 Comparison of acceleration in middle of structure before and after vibration reduction under single walking excitation

each walking condition, here only gives the acceleration time history comparison before and after damping at the middle of the test floor (measurement point No. 3). It can be seen from the figure that although there is a certain gap between the frequency of TMD and the fundamental frequency of the structure, and the vibration control effect cannot be fully exerted, the vibration acceleration of the structure is still significantly reduced under the action of TMD, and with the increase of step frequency, the vibration reduction effect is more obvious.

11.3.3 Dynamic Test of Isolated Structure System

11.3.3.1 Test Model

In order to study the effectiveness of the isolation bearing and the correctness of the isolation analysis method, a four story prefabricated concrete shear wall structure with a new type of assembled horizontal joint is designed. The fortification intensity is 8°, the design basic seismic acceleration is 0.2 g, the seismic grade is grade II, and the site category is class II. In the design of the model, considering the size of the vibration table, load, model material (superstructure material, isolation layer, support) and other constraints, the value of the test similarity constant is shown in Table 11.22.

The test model and standard floor plan are shown in Fig. 11.41. The wall thickness of each layer of the model is 35 mm, the storey height is 750 mm, the plane dimension of the model is 1400 mm × 1400 mm, and the cross section dimension of the coupling beam is 100 mm × 100 mm. The design strength grade of concrete is C20, and the main reinforcement and stirrup are HPB235. The longitudinal reinforcement of concealed column and vertical distribution reinforcement of the model areφ2.2, the horizontal distribution reinforcement is ϕ 2.2@120, the stirrup of the concealed column is ϕ 2.2@60, the longitudinal reinforcement of the coupling beam is ϕ 3, and the stirrup is ϕ 2.2@50. The floor reinforcement is double-layer two-way ϕ 4@100. After the completion of construction, the base weighs 2.43 t, the wall of each floor of the model weighs 0.315 t, and each floor weighs 0.245 t. According to the similarity relationship, 0.6 t counterweight must be added to each floor of the model, and the total structure mass is 7.07 t.

Table 11.22 Similarity coefficient between model and prototype

Physical quantity	Similarity relation	Physical quantity	Similarity relation
Mass	$S_m = 1/16$	Displacement	$S_U = 1/4$
Length	$S_L = 1/4$	Volume	$S_V = 1/64$
Modulus of elasticity	$S_E = 1$	Time	$S_t = 1/2$
Acceleration	$S_a = 1$	Frequency	$S_f = 2$
Velocity	$S_V = 1/2$	Force	$S_F = 1/16$

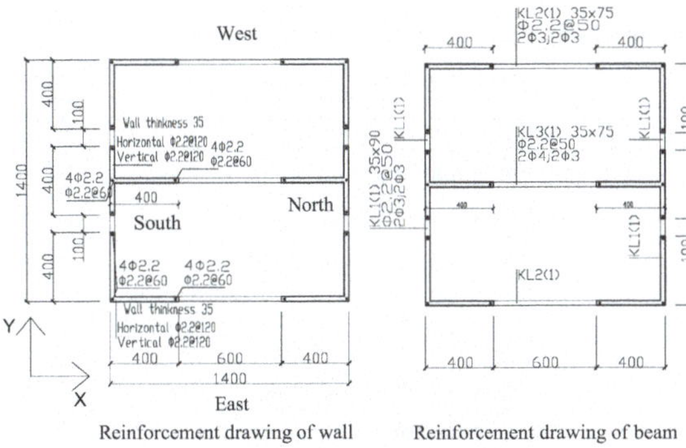

(a) Standard floor plan

(b) Bottom fixed structure

(c) Isolated structure

(d) High damping isolation bearing

Fig. 11.41 Test model of isolated structure system

The low hardness and high damping rubber isolation bearing is used as the isolation bearing for the test. Considering the stability of the mechanical properties of the rubber bearing, the total weight of the model structure and the requirements of the design parameters, the bearing with the diameter of 100 mm is selected. The basic parameters are shown in Table 11.23, and the appearance shape is shown in Fig. 11.41d. High damping isolation bearing has small horizontal stiffness and certain energy dissipation capacity. The bilinear model is used to simulate the hysteretic curve of the high damping isolation bearing. From Table 11.23, it can be seen that the initial stiffness of the high damping isolation bearing is large, which is 0.384 kN/mm, ensuring that the high damping isolation bearing can keep stable under wind load and small earthquake. When the high damping isolation bearing is bent, the equivalent horizontal stiffness of the high damping isolation bearing decreases gradually, which makes the high damping isolation bearing produce large horizontal deformation, dissipate part of seismic energy, and protect the safety of superstructure. When the shear strain of the high damping isolation bearing reaches more than 200%, the

Table 11.23 Design value of basic parameters of high damping isolation bearing

Basic dimension parameter	Value	Mechanical property parameter	Value
Effective diameter d_0 (mm)	100	Rubber shear modulus G (N/mm^2)	0.4
Rubber layer thickness t_r (mm)	1.5	First shape factor S_1	17
Total thickness of rubber T_r (mm)	21	Second shape factor S_2	4.95
Thickness of steel plate layer t_s (mm)	4	Equivalent horizontal stiffness (kN/mm)	0.164
Total thickness of steel plate (mm)	52	Equivalent damping ratio	0.11
Total height of support (mm)	89	Initial stiffness (kN/mm)	0.384
		Stiffness after yield (kN/mm)	0.156

Note the equivalent horizontal stiffness and equivalent damping ratio are parameters when the shear deformation is 100%

high damping isolation bearing enters the hardening stage, which can avoid the large horizontal deformation of the isolation structure under the rare earthquake action and ensure the stability of the structure.

According to "*Standard for test methods of concrete structures*" (GB/T50152-2012), the reserved concrete test blocks are tested. The test blocks are 150 mm × 150 mm × 150 mm cubes. The average compressive strength of concrete cubes measured is 25.63 MPa (first layer), 24.76 MPa (second layers), 25.63 MPa (third layers), 26.1 MPa (fourth layers), 33.76 MPa (base). The measured mechanical properties of iron wire are shown in Table 11.24.

Bolt is used to fix the structure directly on the vibration table, that is to form a fixed structure. The high damping isolation bearing is set between the structure and the vibration table top. The high damping isolation bearing is anchored with the vibration table top and the model base respectively by the anchor rod, which forms the isolation structure. In the isolation structure, the model base is also used as the isolation layer. Four high damping isolation bearings are used for the isolation layer, which is located at the lower part of L-shaped wall limb. Before hoisting the isolation model, first install the high damping isolation bearing on the vibration table top, then drop the model to make the vertical pressure of each bearing as uniform as possible, and tighten the bolts after all the screw holes are adjusted in place. Before the installation of the model, the displacement sensor and acceleration sensor needed in the test should be calibrated, so that the components can work normally during the test.

Table 11.24 Mechanical properties of iron wire

Wire type	Yield strength /MPa	Ultimate strength/MPa	Elongation rate/%	Modulus of elasticity/MPa
φ1.6	249.12	335.38	19.83	1.89×10^5
φ2.6	245.32	330.97	20.66	1.90×10^5
φ3	240.53	315.78	20.28	1.88×10^5

11.3.3.2 Test Loading Scheme and Test Point Layout

Six acceleration sensors and six pull type displacement sensors are arranged in the test model. The acceleration sensor is distributed in the isolation floor and each floor of upper structure. The test load is unidirectional and the loading direction is Y direction as shown in Fig. 11.41a.

According to the seismic fortification requirements and site category (class II), Castaic (CA) wave, Taft (TA) wave and artificial (RG) wave are selected for the test. Since the period similarity constant is 1/2, three seismic waves are compressed in the proportion of 1/2 and input. In the input stage of the same peak acceleration, according to the order of acceleration spectrum value of each seismic wave at the basic period of the structure from small to large, the input order of seismic wave is CA wave, TA wave and RG wave. Before and after the change of peak acceleration of each earthquake, white noise is input.

In the test process, 96 working conditions of isolated structure and fixed structure were carried out. Table 11.25 is the working condition sequence, which is as follows:

In the first step, a high damping isolation bearing is set up under the base of the model to form the isolation structure, and carry out seismic wave loading with peak acceleration of 25, 45 and 65 gal;
In the second step, the high damping isolation bearing under the base of the isolation structure is removed, and the model is fixed on the vibration platform to form a fixed structure, and carry out seismic wave loading with peak acceleration of 25, 45 and 65 gal;
In the third step, install the high damping isolation bearing again, and load the seismic wave with the peak acceleration of 65, 85, 110, 140, 180, 220, 270, 320, 400 gal;
In the fourth step, remove the high damping isolation bearing at the lower part of the isolation structure to form a fixed structure, and load the seismic wave with the peak acceleration of 65, 85, 110, 140, 180, 220, 270, 320, 400 gal.

11.3.3.3 Test Phenomena and Damage Status

During the test, the vibration phenomena of isolated structure and fixed structure are observed. The main test phenomena are summarized as follows:

(1) In the fixed structure, the vibration of the structure is very severe and the damping of the structure is very small. After the input of seismic wave, the vibration of the structure still has a great response, and it takes tens of seconds to completely stop the vibration. In the isolated structure, the vibration of the structure is obviously weakened, and the displacement response of the isolation layer is large. After the input of seismic wave, the vibration of the upper structure of the isolation layer stops within a few seconds, and the isolation layer stops the vibration completely in tens of seconds.

Table 11.25 Test condition

Working condition number	Seismic wave	Peak acceleration/gal	Remarks
2, 3, 4	CA, TA, RG	25	Isolated structure
6, 7, 8	CA, TA, RG	45	Isolated structure
10, 11, 12	CA, TA, RG	65	Isolated structure
15, 16, 17	CA, TA, RG	25	Fixed structure
19, 20, 21	CA, TA, RG	45	Fixed structure
23, 24, 25	CA, TA, RG	65	Fixed structure
28, 29, 30	CA, TA, RG	65	Isolated structure
32, 33, 34	CA, TA, RG	85	Isolated structure
36, 37, 38	CA, TA, RG	110	Isolated structure
40, 41, 42	CA, TA, RG	140	Isolated structure
44, 45, 46	CA, TA, RG	180	Isolated structure
48, 49, 50	CA, TA, RG	220	Isolated structure
52, 53, 54	CA, TA, RG	270	Isolated structure
56, 57, 58	CA, TA, RG	320	Isolated structure
60, 61, 62	CA, TA, RG	400	Isolated structure
65, 66, 67	CA, TA, RG	85	Fixed structure
69, 70, 71	CA, TA, RG	110	Fixed structure
73, 74, 75	CA, TA, RG	140	Fixed structure
77, 78, 79	CA, TA, RG	180	Fixed structure
81, 82, 83	CA, TA, RG	220	Fixed structure
85, 86, 87	CA, TA, RG	270	Fixed structure
89, 90, 91	CA, TA, RG	320	Fixed structure
93, 94, 95	CA, TA, RG	400	Fixed structure

Note the discontinuous part of working condition number is white noise working condition

(2) It can be seen from the observation of the test phenomenon that the horizontal displacement of the isolation layer is relatively obvious, and the maximum displacement of the high damping isolation bearing is about 20 mm.

For the shear wall limb in the loading direction, with the increasing of seismic intensity, the first crack appears at the end of coupling beam in the isolation structure and the fixed structure, and then the horizontal crack appears at the bottom of shear wall limb. See Fig. 11.42 for the final crack diagram of isolated structure and fixed structure. It can be seen from the figure that the cracks of the isolation structure are obviously less than that of the fixed structure, and the high damping isolation bearing has a good isolation effect. There are only cracks at the end of coupling beam of each floor and the bottom of shear wall limb of the first floor in the isolation structure, while there are cracks at the end of coupling beam and the bottom of shear wall limb of each floor in PCSW fixed structure.

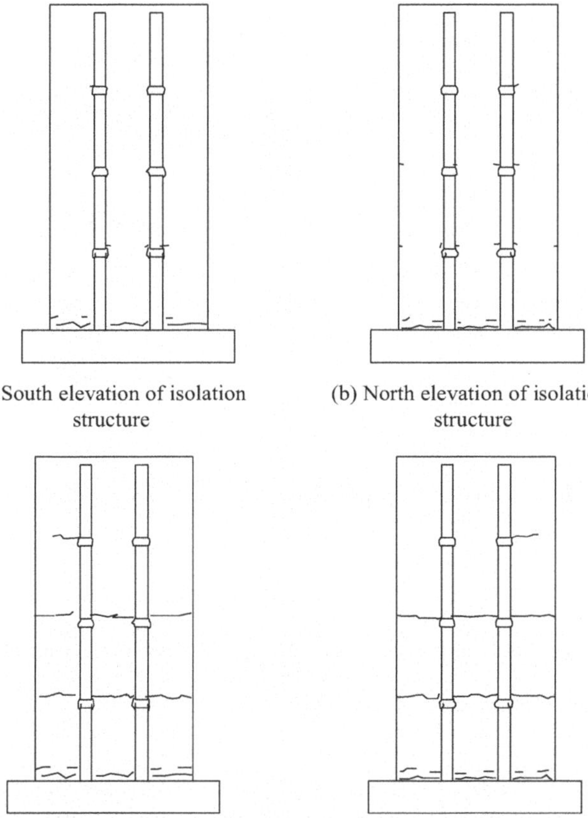

(a) South elevation of isolation structure

(b) North elevation of isolation structure

(c) South elevation of fixed structure

(d) North elevation of fixed structure

Fig. 11.42 Structural crack distribution diagram

11.3.3.4 Dynamic Characteristics

1. **Natural frequency and damping ratio**

The first and second order natural frequencies and damping ratios of isolated structure and fixed structure at different stages are measured in the test. Figure 11.43a, b show that: (1) with the increase of peak acceleration of seismic wave, the first-order frequency of the isolated structure changes little, maintaining at about 2.2 Hz; when the peak acceleration of local seismic wave does not exceed 65 gal, the second-order frequency of the isolated structure changes little, maintaining at about 24 Hz; when the peak acceleration of seismic wave is greater than 65 gal, the second-order frequency of the isolated structure decreases gradually, and when the peak acceleration of seismic wave is 400 gal, the second frequency of the isolated structure is reduced to 10.75 Hz. (2) With the increase of the peak value of seismic wave acceleration,

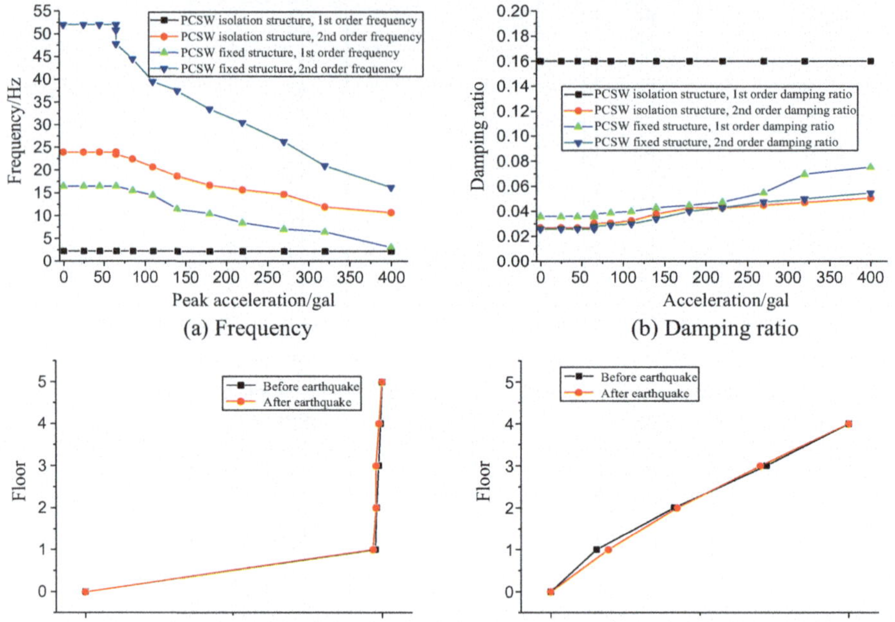

(a) Frequency

(b) Damping ratio

(c) First order mode shape of isolated structure

(d) First mode shape of fixed structure

Fig. 11.43 Dynamic characteristics of test model

the first-order damping ratio of the isolated structure does not change much, maintaining at about 16%; when the peak value of local seismic wave acceleration does not exceed 65 gal, the second-order damping ratio of the isolated structure does not change much, maintaining at about 2.7%; when the peak value of local seismic wave acceleration is greater than 65 gal, the second-order damping ratio of the isolated structure increases gradually, and when the peak value of local seismic wave acceleration is 400 gal, the second damping ratio of PCSW isolation structure increases to 5%. (3) With the increase of the peak value of seismic wave acceleration, the first and second order frequencies of the fixed structure change obviously. When the peak value of local seismic wave acceleration does not exceed 65 gal, the first order frequency is maintained at about 16.5 Hz, and the second order frequency is maintained at about 52 Hz; when the peak value of local seismic wave acceleration is greater than 65 gal, the first and second order frequencies of the fixed structure gradually decrease, and when the peak value of local seismic wave acceleration is 400 gal, the first and second frequencies of the fixed structure are reduced to 3.12 and 16.25 Hz. (4) With the increase of the peak value of seismic wave acceleration, the first and second order damping ratio of the fixed structure changes obviously. When the peak value of local seismic wave acceleration does not exceed 65 gal, the first order damping ratio is basically maintained at about 3.6%, and the second order damping ratio is maintained at about 2.6%; when the peak value of seismic wave acceleration is greater than 65 gal, the first and second order damping ratio of the fixed structure increases gradually, and when the peak value of seismic wave acceleration is 400 gal, the first and second damping ratios of the fixed structure are 7.54 and 5.47%.

With the increasing of the peak acceleration of seismic wave, the change of frequency and damping ratio of isolated structure is small; the change of frequency and damping ratio of fixed structure is large, the frequency is decreasing, and the damping ratio is increasing. The reason is that the frequency and damping ratio of the isolation structure are mainly determined by the characteristics of the isolation layer. Because the high damping isolation bearing has better anti fatigue performance, the frequency and damping ratio of the isolation structure with high damping isolation bearing have little change with the increase of the earthquake intensity. Under the rare earthquake, the damage of the fixed structure is accumulating, which leads to the decrease of the frequency and the increase of the damping ratio of the structure.

2. **Mode shape**

The vibration modes of the isolated structure and the fixed structure are measured by the test. The change comparison diagram of the vibration modes before and after cracking is given in Fig. 11.43c, d. It can be seen from the figure that the first mode of vibration of the isolated structure is the translation of the isolation layer, and the upper structure moves approximately as a rigid body; the first mode of vibration of the fixed structure before and after cracking is bending.

11.3.3.5 Structural Dynamic Response

1. **Acceleration response**

When the peak acceleration of local seismic wave is 400 gal, the absolute acceleration response envelope of each layer of the model is shown in Fig. 11.44a. It can be seen from the figure that the absolute peak acceleration of each layer of the isolated structure is far less than that of the fixed structure. The absolute acceleration response of each layer above the isolation layer of the isolation structure is close to that of the rigid body. The damping effect of the absolute acceleration of the top layer of the isolation structure is the most obvious, which is 63–71%. The absolute acceleration damping effect of the first layer of the isolation structure is relatively poor, which is 38–59%.

Figure 11.44b shows the absolute acceleration response envelopes of isolated structure vertex and fixed structure vertex. It can be seen from the figure that the envelope value of the absolute acceleration of the isolation structure vertex is far less than that of the absolute acceleration of the fixed structure vertex, and the trend is more obvious with the increase of the peak value of the seismic wave acceleration. With the increase of the peak acceleration of seismic wave, the envelope value of the absolute acceleration of the top of the fixed structure increases nonlinearly, and the increase trend of the absolute acceleration of the top tends to slow down. The reason is that the internal damage accumulation of the structure causes the decrease of the structure stiffness and the increase of the damping ratio. With the increase of peak acceleration of seismic wave, the envelope value of absolute acceleration at the top of isolation structure approximately increases linearly. The reason is that the upper structure of isolation layer approximately moves as a rigid body, and the damage of upper structure is small.

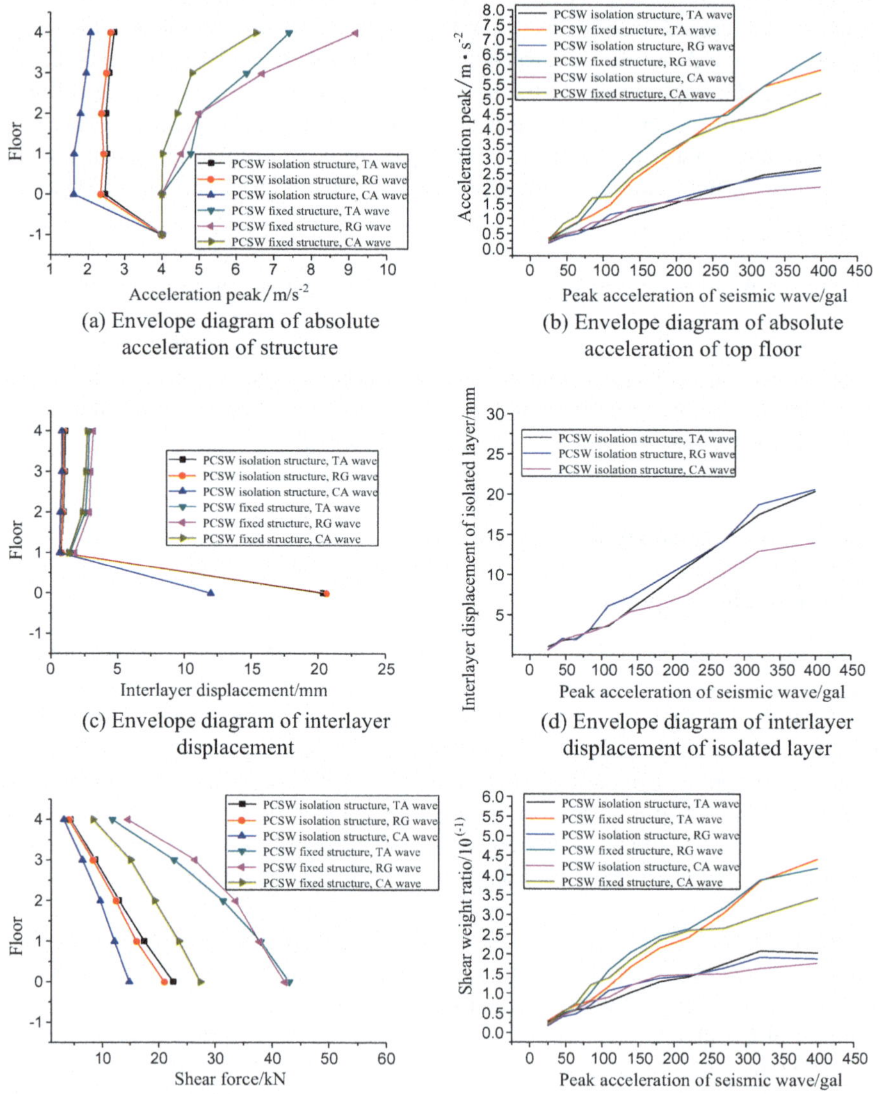

(a) Envelope diagram of absolute acceleration of structure

(b) Envelope diagram of absolute acceleration of top floor

(c) Envelope diagram of interlayer displacement

(d) Envelope diagram of interlayer displacement of isolated layer

(e) Envelope diagram of interlayer shear

(f) Shear weight ratio

Fig. 11.44 Dynamic response of isolated structure model

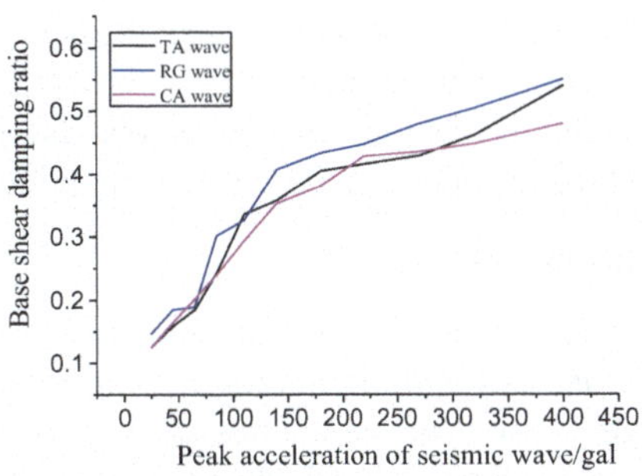

(g) Isolation rate of base shear

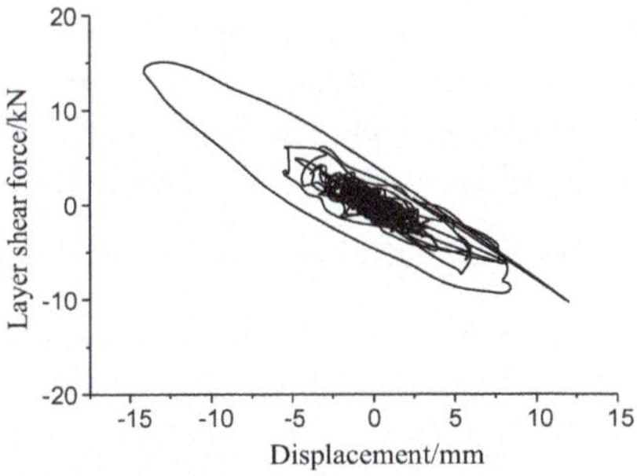

(h) Hysteresis curve of isolation layer (PGA 400gal, CA wave)

Fig. 11.44 (continued)

The damping ratio of absolute acceleration at the top of PCSW isolation structure increases with the increase of peak value of seismic wave acceleration. When the peak value of local seismic wave acceleration is small, the damping ratio of absolute acceleration at the top is about 25%. With the increase of peak value of seismic wave acceleration, the damping ratio of absolute acceleration at the top is about 65%.

2. **Interlayer displacement response**

Figure 11.44c is the envelope diagram of displacement between layers under the action of seismic wave with peak acceleration of 400 gal. It can be seen from the figure that the interlayer displacement of the isolation structure is larger, and the interlayer displacement of the upper structure of the isolation layer is smaller. In contrast, the interlayer displacement of the isolation layer is smaller under the action of Ca wave, and larger under the action of TA wave and RG wave. The interlayer displacement of the upper story of the fixed structure is much larger than that of the corresponding story of PCSW isolation structure. Figure 11.44d is the envelope diagram of the interlayer displacement of the isolation structure in each stage of the test. It can be seen from the diagram that the interlayer displacement of the isolation structure increases approximately linearly with the increase of the peak acceleration of the seismic wave.

3. **Interlayer shear response**

Under the action of base seismic wave, the maximum nominal interlayer shear force $F_i(t)_{\max}$ of the ith floor at time t can be calculated approximately according to the following formula, $F_i(t)\max = \left| -\sum_{j=i}^{n} m_j a_j(t) \right|_{\max}$, where n is the total number of floors, a_j is the acceleration of the jth floor, and m_j is the concentrated mass of the jth floor.

Figure 11.44e is the envelope diagram of the interlayer shear response of the isolated structure and the fixed structure under the action of the seismic wave with the peak acceleration of 400 gal. It can be seen from the diagram that the interlayer shear of the isolated structure is far less than the interlayer shear of the fixed structure under the action of the rare earthquake with 8°. Among them, under the action of CA wave, the interlayer shear of isolated structure and fixed structure is relatively small, and under the action of TA wave and RG wave, the interlayer shear is relatively large.

4. **Shear weight ratio and base shear isolation ratio**

Figure 11.44f is a comparison diagram of shear weight ratio of isolated structure and fixed structure. It can be seen from the figure that with the increase of the peak acceleration of seismic wave, the shear weight ratio of isolated structure and fixed structure increases continuously, but the shear weight ratio of isolated structure is smaller than that of fixed structure.

Figure 11.44g is the base shear isolation rate of the isolated structure. It can be seen from the figure that with the increase of the peak acceleration of the seismic wave, the base shear isolation rate of the isolated structure increases continuously, but its increase amplitude is smaller and smaller. The reason is that the initial stiffness of the high damping isolation bearing is large, when the peak value of seismic wave acceleration is small, the deformation of the high damping isolation bearing is small, and the isolation effect is not obvious. When the peak value of seismic wave acceleration is large, the equivalent horizontal stiffness of the isolation bearing is small and the deformation is large, and the isolation bearing has a good isolation effect. When the peak acceleration of seismic wave is 400 gal, the base shear isolation rate of isolated structure reaches 48–54%.

5. **Hysteretic curve of isolation layer**

Compared with the traditional non isolation structure, the isolation structure mainly dissipates energy through the large deformation of the isolation layer, and the hysteretic energy dissipation performance of the isolation layer reflects the damping performance of the structure. Figure 11.44h is the hysteretic curve of the isolation layer. It can be seen from the figure that under the action of earthquake, the hysteretic curve of the isolation layer is full and shuttle shaped, with good energy consumption capacity.

References

1. Chen, Xin. 2012. *Theoretical and Experimental Study on Vibration Control of High-rise Steel Chimneys Under Wind Load*. Nanjing: Southeast University. (in Chinese).
2. Li, Aiqun. 2007. *Vibration Control of Engineering Structure*. Beijing: China Machine Press.
3. Sun, Guangjun. 2010. *Research on Analysis Methods of Random Earthquake Response and Reliability of Base Isolation Structures and Seismic Reduction Structures*. Nanjing: Southeast University. (in Chinese).
4. Li, Aiqun. 1991. Simplified Solution of Motion Equations for Base Vibration Isolator. *Journal of Southeast University (Natural Science Edition)* 21 (04): 130–135. (in Chinese).
5. Mao, Lijun. 2004. *Research on Building Structures of Sliding Base-isolation*. Nanjing: Southeast University. (in Chinese).

6. Wu, Yifeng, Hao Wang, and Aiqun Li. 2019. The inelastic displacement spectra and its utilization of DDB design for seismic isolated bridges subjected to near-fault pulse-like ground motions. *Earthquake Spectra* 35 (3): 11091140.
7. Xu, Yanhong. 2013. *Theoretical and Experimental Research on the New Mild Steel Dampers Used in the Structure*. Nanjing: Southeast University. (in Chinese).
8. Zheng, Jie. 2015. *Theoretical and Experimental Research on a Curved Steel Damper Used in the Structures*. Nanjing: Southeast University. (in Chinese).
9. Xu, Junhong. 2015. *Experimental Study on the New Kind of Viscoelastic Damping Wall*. Nanjing: Southeast University. (in Chinese).

Chapter 12
Vibration Control Design Method of Building Structure

Abstract Design method of vibration control in building structure, including energy dissipation control, frequency modulation vibration control and isolation control are introduced. General design frame, design process and examples of the above-mentioned vibration control of building structure are given respectively.

12.1 Performance Level of Building Structure and Quantification

The performance level of a building is used to describe the damage degree of a building under certain ground motion. It is a combination of structural performance, non structural component and system performance as well as indoor facility performance. Generally divided into: structural performance level and non structural performance level. Before the vibration control design of building structure, it is necessary to define the performance level of the structure. The research reports of FEMA273, SEAOC Vision2000, ATC-40 of the United States and the Architectural Research Institute of the Ministry of Construction of Japan divide the building performance level, and the specific division method is shown in Table 12.1. In the "*Code for seismic design of buildings*" (GB 50011-2010) in China, the performance level and deformation reference value of all kinds of buildings are divided by reference to the

Table 12.1 Foreign division of performance standards

FEMA273	SEAOC Vision2000	ATC-400	Japan research report
–	Completely normal use	–	–
Normal use	Normal use	Normal use	Normal use
Immediate occupancy	–	Immediate occupancy	Easy to repair
Life safety	Life safety	Life safety	Life safety
Prevent collapse	Approaching collapse	Structural stability	–

A. Li, *Vibration Control for Building Structures*, Springer Tracts in Civil Engineering, https://doi.org/10.1007/978-3-030-40790-2_12

"*Standard for classification of earthquake damage level of buildings*". Generally, they can be divided into five grades. See Table 12.2 for details.

Table 12.2 Performance level and deformation reference value of building structure

Name	Failure description	Possibility of continued use	Deformation reference value
Basically intact (including intact)	Bearing members are in good condition; some non bearing members are slightly damaged; accessory members are damaged to different degrees	Generally, it can be used without repair	$<[\Delta u_e]$
Slight damage	Some load-bearing members are slightly cracked (residual deformation for steel structure members), and some non load-bearing members are obviously damaged; auxiliary members are damaged to different degrees	It can still be used without repair or slight repair	$(1.5 \sim 2)[\Delta u_e]$
Moderate damage	Most load-bearing members have slight cracks (or residual deformation), some obvious cracks (or residual deformation); some non load bearing members are seriously damaged	It needs general repair and can be used properly after taking safety measures	$(3 \sim 4)[\Delta u_e]$
Serious damage	Most load-bearing members are seriously damaged or partially collapsed	Overhaul and partial removal	$< 0.9[\Delta u_p]$
Collapse	Collapse of most load-bearing members	Need to be demolished	$>[\Delta u_p]$

Note 1. Individual refers to less than 5%, part refers to less than 30%, and most refers to more than 50%

2. For the deformation reference value of moderate failure, the average value of the standard elastic and elastic-plastic displacement angle limit value is roughly taken, and for the slight damage, 1/2 of average value is taken

Table 12.3 Performance level of non structural components

Name	Damage situation
Functional integrity	Non structural components can maintain function
Can continue to live	Under the premise of meeting the requirements of building structure performance level, the non structural part will have limited damage, but the basic use and personal safety function will not be lost
Life safety	The non structural damage which is serious or causes great economic loss occurs, non structural components will not collapse or fall down, so as to threaten the life safety of people inside and outside the building
Risk reduction	The non structural parts are seriously damaged, but the larger and heavier components such as low wall, panel, heavy gypsum ceiling, etc. will not fall and cause life danger to many people

For the level of non structural components (building decorative components, partition, ceiling, HVAC system, piping system, fire protection and lighting system, etc.), FEMA356 has made corresponding division, see Table 12.3 for details.

The test and theoretical study show that there is a good correlation between the performance of building structure in each stage and its deformation index, that is to say, the quantitative relationship can be established between the structural performance and its deformation index. The main reason for the collapse of building structure under the action of large earthquake is due to its insufficient deformation capacity and energy consumption capacity. Therefore, the inter story displacement angle index is the key parameter in the seismic design of the structure.

FEMA273 suggests that the inter story displacement angle limit of reinforced concrete frame structure can be divided into vibration process deformation and permanent deformation, that is to say, the inter story displacement angle limit of the structure is 1/100, 1/50 and 1/25 when the structure is in immediate occupancy, life safety and collapse prevention performance level in the process of vibration deformation; in Vision2000 Committee of SEAOC in the United States, it is suggested that limit values of inter story displacement angle of the reinforced concrete frame structure are 1/500, 1/200, 1/67 and 1/40 for the performance levels including should not be damaged, repairable, unrepairable and severely damaged respectively. Among them, when the structure has three corresponding performance levels, the limit values of the inter story displacement angle of the structure of 1/100 and 1/25 are respectively negligible.

According to the seismic code for buildings in China, the limit value of elastic interlayer displacement angle of reinforced concrete frame structure is 1/550 under the action of frequent earthquake and 1/50 under the action of rare earthquake.

Therefore, it is more reasonable and convenient for engineering design to adopt the interlayer displacement angle as the quantitative index of structural performance level in the process of structural vibration reduction design. In order to correspond to the three levels of seismic fortification targets in the code of seismic design of buildings (GB 50011-2010), and referring to the provisions of "limit value of elastic-plastic interlayer displacement angle of energy dissipation and damping structures should meet the expected deformation control requirements, and should be appropriately reduced compared with non energy dissipation and damping structures" in "*Code for seismic design of buildings*" (GB 50011-2010), the book puts forward the three levels of energy dissipation and damping reinforced concrete frame structure, namely, the limit values of interlayer displacement angles are 1/550, 1/200 and 1/80 respectively under well used for good, life safety and collapse prevention.

12.2 Design Method for Energy Dissipation and Vibration Control of Buildings

12.2.1 General Frame for Energy Dissipation and Vibration Control Design of Buildings

It can be seen from the contents of Chap. 11 of this book that, due to the strong nonlinear characteristics of vibration reduction devices, it needs more in-depth professional knowledge and simulation skills to carry out refined numerical analysis, which brings great inconvenience to the popularization and application of vibration reduction technology. Therefore, many scholars put forward some practical simplified design methods in combination with specifications, which greatly improves the design efficiency. Among them, the design method based on the equivalent damping ratio, with clear concept, simple operation and close combination with specifications, is a design method accepted by the majority of designers. Equivalent damping ratio refers to the damping effect of vibration reduction devices on the structure, which can be equivalent to a part of the damping applied to the original structure, so as to reduce the dynamic response of the structure. By using the equivalent damping ratio to analyze the vibration control system of the structure, the nonlinearity and the increase of the degree of freedom caused by the installation of the damping device can be eliminated, and the motion equation form of the original structure can be kept unchanged, which effectively simplifies the calculation process.

The general frame of energy dissipation and vibration reduction design of buildings is to use the equivalent damping ratio to complete the design of energy dissipation and vibration reduction structure system by modifying the existing design method [1–3]. The key of the design method lies in the determination of the equivalent damping ratio. The methods to determine the equivalent damping ratio usually include energy method, forced decoupling method and modal strain energy method. Among them, when the non orthogonal term of the damping matrix cannot be ignored, the energy method can only be used to solve it. The energy method is based on the principle that the energy consumed by a linear damping system is equivalent to the energy consumed by a nonlinear damping system. It is suitable for the calculation of the damping ratio of linear and nonlinear vibration reduction devices. In China's aseismic specifications, there are provisions for calculating the additional equivalent damping ratio of energy dissipation components by using this method:

$$\zeta_e = W_d/(4\pi W_s) \tag{12.1}$$

Among them, W_d is the energy consumed by the vibration damping device in the structure for one cycle under the expected displacement of the structure; W_s is the total strain energy of the structure with the vibration damping device under the expected displacement. For the regular multi-storey building structure, when the torsional effect can be ignored, the total strain energy of the structure under the external load can be estimated as follows:

$$W_s = (1/2)\{x_j\}^{\mathrm{T}}\{F_j\} \tag{12.2}$$

where, $\{F_j\}$ is the standard value vector of the j-mode horizontal load; $\{x_j\}$ is the displacement vector of the j-mode under the action of the standard value of the horizontal load. The energy consumed by the speed dependent energy dissipation and damping device in one cycle of reciprocating cycle under the action of wind load is calculated by the following formula:

$$W_d = \left(2\pi^2/T_j\right)\{x_j\}^{\mathrm{T}}[H]\left[C_{deq}\right]\{x_j\} \tag{12.3}$$

where, $\left[C_{deq}\right]$ is the equivalent local damping matrix of the damping device; $[H]$ is the position matrix of the damping device; T_j is the periodic value of the j-mode. The energy consumption of displacement related energy dissipation device in one cycle of reciprocating cycle under wind load can be estimated as follows:

$$W_d = A \tag{12.4}$$

where A is the area of the resilience hysteresis loop. The effective stiffness of energy dissipation and damping device can be taken as secant stiffness of hysteresis loop. Under the action of general external load, the displacement can be expressed as the product of the maximum displacement of the vertex and the vibration mode, $\{x_j\} = x_{j,\max}\Phi_j$, which can be substituted into the above formula as follows:

$$W_s = (1/2)x_{j,\max}\Phi_j^{\mathrm{T}}\{F_j\} \tag{12.5}$$

$$W_d = \left(2\pi^2/T_j\right)x_{j,\max}^2\Phi_j^{\mathrm{T}}[H]\left[C_{deq}\right]\Phi_j \tag{12.6}$$

It should be noted that the additional equivalent damping ratio calculated by the above method shall not exceed 25%. When it exceeds 25%, it shall still be calculated as 25%. The total equivalent damping ratio of vibration reduction structure can be calculated by Eq. (12.1). As shown in Fig. 12.1, the design method of wind-induced response of energy dissipation and vibration reduction structure system of high-rise steel chimney can be seen from the figure that the simplified method actually

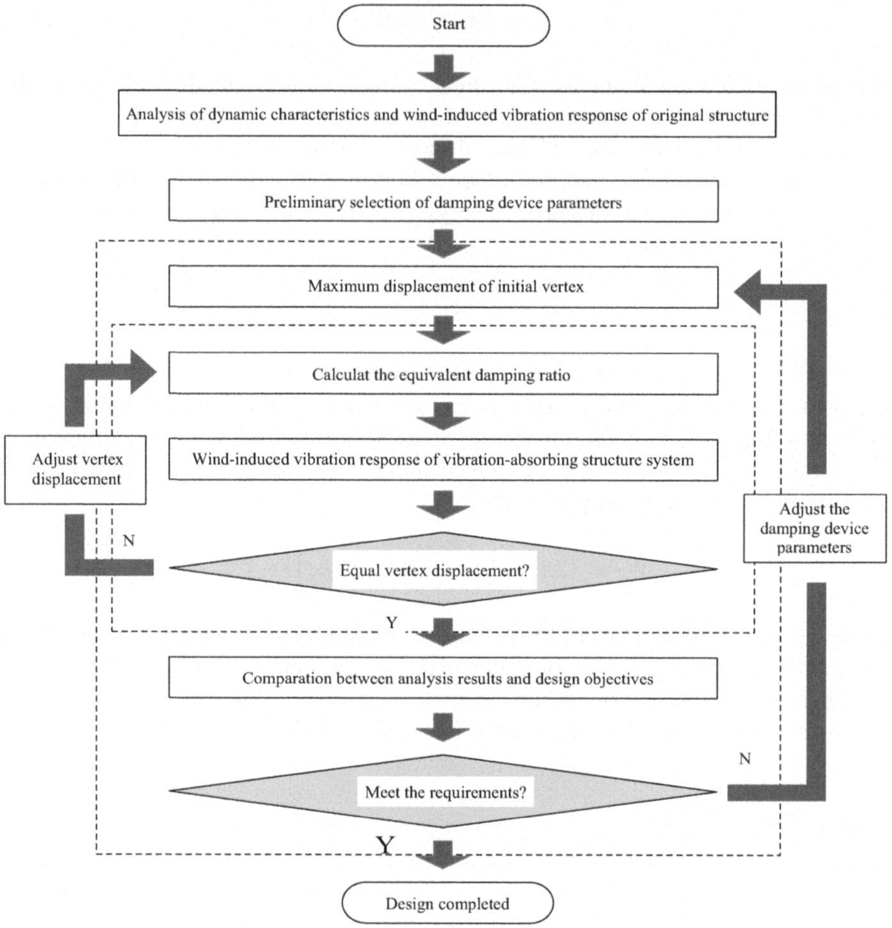

Fig. 12.1 General frame of energy dissipation and vibration control design of building structure

includes two circulation processes of calculation cycle and design cycle: (1) calculate the dynamic characteristics and wind-induced response of the original structure, and preliminarily select the parameters of the vibration reduction device; (2) select the initial peak maximum displacement, calculate the equivalent damping ratio, and then analyze the wind-induced vibration response of the vibration reduction structure system, compare the vertex displacement, and when it is not satisfied, continue to iterate according to the method shown in the figure until the expected displacement and the calculated displacement error meet the requirements; (3) compare the analysis results with the design objectives, if it meets the requirements, if it does not meet the requirements, adjust the parameters according to the rules of the vibration damping device parameter analysis, and carry out step (2) again in this way, it continues to cycle until it meets the design requirements. The above is the general framework of energy dissipation and vibration reduction design of building structures. The following two typical energy dissipation and vibration reduction technologies, viscous fluid damping and metal damping, are taken as examples to introduce the specific design process of velocity dependent and displacement dependent dampers respectively.

12.2.2 Viscous Fluid Damping Design of Building Structure

The displacement based vibration reduction design method is to first express the actual structure as an equivalent single degree of freedom system with "alternative structure", and then use the secant stiffness corresponding to the target displacement and the assumed equivalent damping of the structure after the additional energy dissipation damper to represent the actual structure. According to the displacement response spectrum, the period corresponding to the target displacement can be obtained from the pre-determined target displacement and additional damping ratio. Finally, the base shear force can be calculated and distributed along the height of the structure to obtain the horizontal seismic action of each floor. According to the combination of seismic effect and other effects, the structural section and reinforcement are modified, and appropriate energy dissipation dampers are selected according to the additional damping ratio.

The specific flow of displacement based vibration reduction structure design is shown in Fig. 12.2.

The specific steps and parameters of the design process are determined:

1. The preliminary design of the structure includes determining the arrangement of shear walls and columns, calculating the mass m_i of each layer of the structure, assuming that the viscous damping ratio attached to the structure by the viscous energy dissipation damper, the viscous damping ratio can be selected according to the economic situation.
2. The seismic fortification target of shock absorption structure should be higher than that of conventional structure. The target displacement shall be set according to the purpose, importance of the specific building structure and the owner's requirements, and the equivalent mass M_{eff} and equivalent displacement δ_{eff} of equivalent SDOF shall be calculated according to Eqs. (12.7) and (12.8);

$$M_{eff} = \sum m_i c_i = \left(\sum m_i \delta_i\right) / \delta_{eff} \tag{12.7}$$

$$\sum F_i \delta_i = V_b \delta_{eff}, \quad \delta_{eff} = \frac{\sum F_i \delta_i}{V_b} = \frac{\sum m_i \delta_i^2}{\sum m_i \delta_i} \tag{12.8}$$

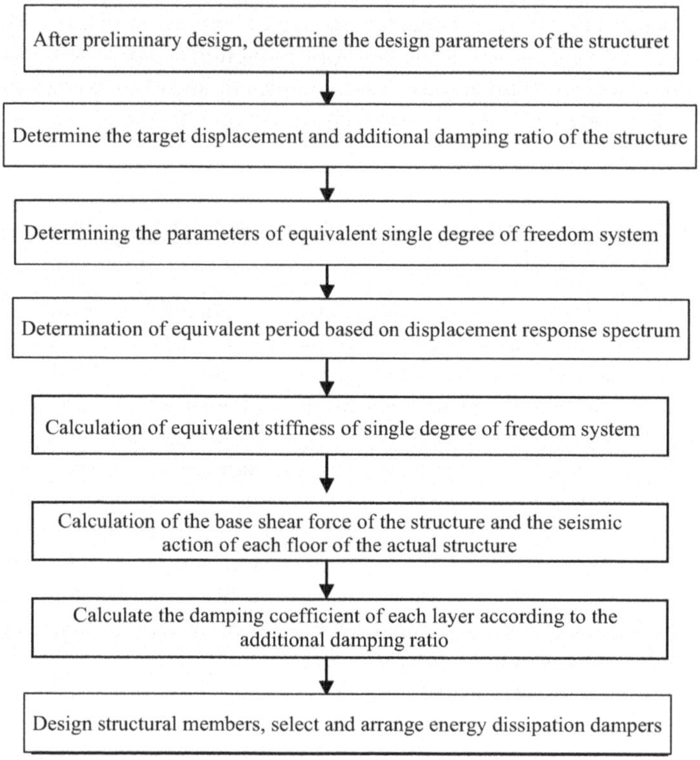

Fig. 12.2 Design process of viscous fluid damping vibration control for building structure

3. The elastic displacement spectrum is established according to the following formula:

Straight up stage:

$$T^2[0.45 + 10(\eta_2 - 0.45)T] = \frac{4\pi^2 S_d}{\alpha_{\max} g} \quad (T \leq 0.1\,\text{s}) \tag{12.9}$$

Level stage:

$$T = 2\pi \sqrt{\frac{S_d}{\eta_2 \alpha_{\max} g}} \quad (0.1 \leq T \leq T_g) \tag{12.10}$$

Curve descent stage:

$$T = \left(\frac{4\pi^2}{T_g^{\gamma}} \cdot \frac{S_d}{\eta_2 \alpha_{\max} g} \right)^{\frac{1}{2-\gamma}} \quad (T_g \leq T \leq 5T_g) \tag{12.11}$$

Straight descent stage:

$$T^2\left[0.2^{\gamma}\eta_2 - \eta_1\left(T - 5T_g\right)\right] = \frac{4\pi^2 S_d}{\alpha_{\max} g} \quad \left(5T_g \le T \le 6.0\,\mathrm{s}\right) \tag{12.12}$$

Upper middle: $\gamma = 0.9 + \dfrac{0.05 - \xi}{0.3 + 6\xi}$ $\eta_1 = 0.02 + \dfrac{0.05 - \xi}{4 + 32\xi}$ $\eta_2 = 1 + \dfrac{0.05 - \xi}{0.08 + 1.6\xi}$

where, γ is the attenuation coefficient at the falling end of the curve; η_1 is the slope adjustment coefficient at the falling end of the straight line, which is 0 when it is less than 0; η_2 is the damping adjustment coefficient, which is 0.55 when it is less than 0.55; $\alpha_{\max}$ is the maximum value of the seismic influence coefficient; ξ is the damping ratio.

4. Determine the equivalent damping ratio of the structure according to the following formula;

$$\xi_d = \frac{\sum_j W_j}{4\pi W_k} = \frac{\sum_j \lambda C_j \phi_{rj}^{1+\alpha} \cos^{1+\alpha}\theta_j}{2\pi w^{2-\alpha} A^{1-\partial} \sum_i m_i \phi_i^2} \tag{12.13}$$

$$\xi_{eff} = \xi_0 + \xi_d \tag{12.14}$$

Among them, ξ_{eff} is the overall effective damping ratio of the system (including viscous energy dissipation damper device); ξ_0 is the inherent damping ratio of the system itself; ξ_d is the damping ratio contributed by viscous damper device.

5. According to the determined equivalent damping ratio and equivalent displacement, the equivalent period T_{eff} of the structure is calculated by displacement spectrum;
6. The equivalent stiffness K_{eff} of equivalent SDOF is determined by Eq. (12.15);

$$K_{eff} = \frac{4\pi^2 M_{eff}}{T_{eff}^2} \tag{12.15}$$

7. The distribution of horizontal seismic force is determined by Eq. (12.16);

$$F_i = \frac{m_i \delta_i}{\sum m_i \delta_i} V_b \tag{12.16}$$

8. The damping coefficient assigned to each layer is calculated by the assumed additional damping ratio provided to the structure:

It is defined that the displacement difference is $\Delta u_i = (u_{i0} - u_i)$, u_{i0} is the displacement of the first layer i of the original structure. Considering that ξ_d is distributed according to the displacement difference between damping layers, therefore:

$$\xi_{di} = \frac{\Delta u_i}{\sum_i \Delta u_i} \xi_d = \frac{u_{i0} - u_i}{\sum_i u_{i0} - u_i} \xi_d \tag{12.17}$$

Special attention should be paid in the design. This method is only used to determine the number and parameter optimization of energy dissipation dampers needed in the early stage of design. Finally, the finite element time history analysis method should be used for modification before it can be used in the actual project.

12.2.3 Metal Damping Design of Building Structure

In the design of multi particle damping structure, the energy dissipators obtained by single degree of freedom can be simply distributed according to the characteristics of the stiffness distribution of the main structure layer, and the number of energy dissipators in each layer can be determined. In this case, the equivalent period and damping ratio of single mass point system and multi mass point system can be equal, and the prediction results of seismic response based on performance curve can be applied to multi mass point system. The design process is shown in Fig. 12.3, and the specific steps are as follows:

1. According to the characteristics of the structure, the component model or shear layer model of the main structure is established;
2. Based on the static elastic calculation, the natural vibration period and stiffness distribution of the main structure are analyzed;
3. The maximum inter story displacement angle of the main structure is calculated by using the response spectrum of the input seismic wave;
4. Based on the desired inter story displacement angle θ_g and the maximum inter story displacement angle θ_0, the target displacement reduction rate R_d ($R_d = \theta_g/\theta_0$) is calculated;
5. In the simple prediction method, the number of the required metal energy dissipaters (expressed as the stiffness ratio $k_\mathrm{d}/k_\mathrm{s}$ relative to the main structure) is obtained by the single mass point model according to the target value of the maximum seismic response and the performance curve. The energy dissipaters are arranged in each layer of the multi mass point system according to the above method;
6. The time history response analysis of the vibration reduction structure is carried out to test the vibration reduction effect and determine the seismic performance of the structure.

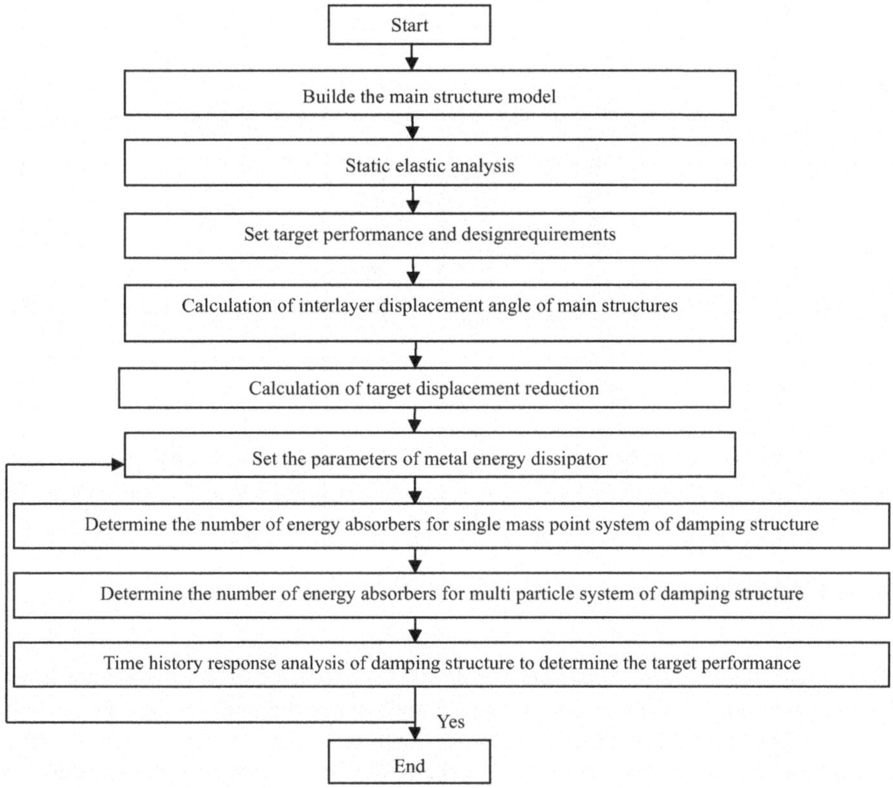

Fig. 12.3 Design process of metal damping and vibration control of building structure

12.2.4 Example of Energy Dissipation and Vibration Control Design of Buildings

1. **Project overview**

Taking the viscous fluid damping design of building structure as an example, the application of the above design method is illustrated. The structure shown in Fig. 12.4 is a frame core tube structure with 20 floors. The fortification intensity of the area where the structure is located is 8° (0.3 g), site category III, design group I, characteristic period $Tg = 0.45$ s, structural type is steel reinforced concrete frame core tube structure, the height of the first floor is 5.7 m, the height of the second and third floors is 4.2 m, and the height of other floors are 4 m. The steel used is Q345, and the reinforcement is Q345.

2. **Calculation design under the action of frequent earthquake**

The Etabs software is used to analyze the response spectrum of the structure, and the parameters such as floor mass and inter floor displacement angle are obtained, as shown in Table 12.4. The first vibration mode period is $T = 1.663$ s. According to

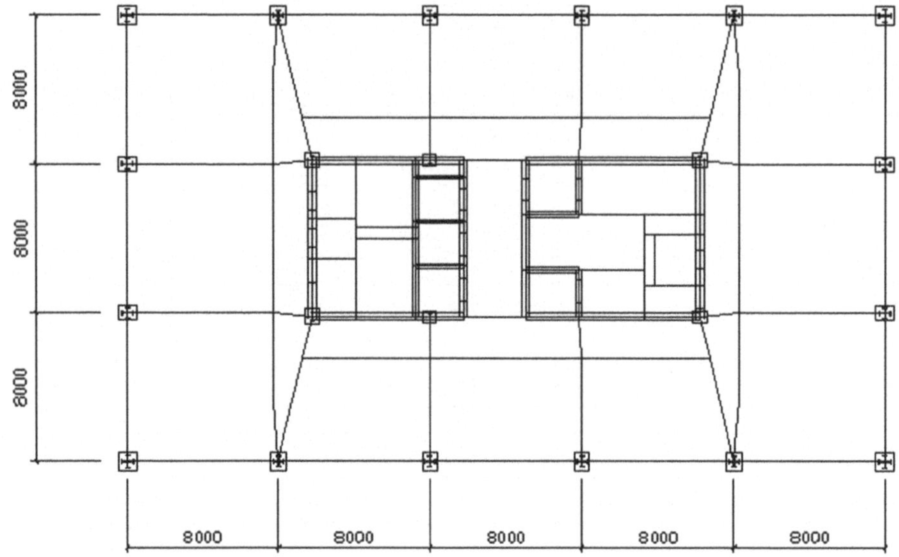

Fig. 12.4 Structural plan layout

"*Code for seismic design of buildings*" (GB 50011-2010), the limit value of interlayer displacement angle of frame core tube structure under frequent earthquakes action is $\theta_e = 1/800$, and the maximum value of interlayer displacement angle is 1/662 in response spectrum analysis, which exceeds the limit value of specifications. In this paper, the target story displacement angle of 10–13 stories is taken as 1/800 according to the requirements of the specifications, and the distribution of other stories is distributed according to the proportion of story displacement angle under the response spectrum analysis. See Table 12.5 for structural calculation parameters and Table 12.6 for equivalent single degree of freedom system parameters.

From the displacement angle and displacement response spectrum of the target story, the additional damping ratio required by the structure can be calculated to be about 12.24%. In this paper, 12% is taken for calculation. The equivalent displacement $\delta_{eff} = 0.0573$ m and the equivalent mass $M_{eff} = 28{,}048.62$ kN s^2/m are calculated.

This structure is a steel reinforced concrete structure. In the elastic stage, the inherent damping ratio of the structure itself is taken as $\xi_0 = 0.04$, the additional damping ratio of the viscous energy dissipation damper added to the structure is taken as $\xi_d = 0.12$, the maximum inter story displacement angle of the structure is 1/800, and the structure is in the elastic state. So the equivalent damping ratio of the single degree of freedom system is $\xi_{eff} = 0.16$. Then we can get: $\gamma = 0.813$, $\eta_1 = 0.0079$, $\eta_2 = 0.673$.

Then the equivalent period $T_{eff} = 1.337$ s is obtained. By substituting M_{eff} and T_{eff} into Eq. (12.15), $K_{eff} = 616{,}682.52$ kN/m is obtained, so the base shear force $V_b = 34{,}335.91$ kN. The base shear is distributed according to Eq. (12.16), as shown in Table 12.7.

Table 12.4 Structural parameters

Floor	Mass (kN s^2/m)	Analysis of displacement angle between lower layers by response spectrum
20	1122.8	1/825
19	2034	1/796
18	2020	1/761
17	2002	1/734
16	2002	1/711
15	2029	1/692
14	2014	1/679
13	2002	1/669
12	2002	1/663
11	2002	1/662
10	2034	1/666
9	2016	1/677
8	2005	1/699
7	2005	1/726
6	2005	1/767
5	2010	1/824
4	2059	1/933
3	2188	1/1093
2	2032	1/1387
1	2008	1/2554

According to the combination of horizontal seismic effect and gravity load effect obtained in Table 12.7, the internal force design value of structural members is obtained, and the section is adjusted according to the internal force design value. At the same time, the additional damping ratio is allocated, the damping coefficient of each layer is calculated, and the energy dissipator is selected. See Table 12.8 for damping ratio of each layer.

According to the distribution of additional damping ratio of each layer in Table 12.8, 4 energy dissipation dampers are arranged for each layer of 1–5 and 17–20 layers, and 8 energy dissipation dampers are arranged for each layer of the rest layers, so as to determine the type of energy dissipation damper. The arrangement of energy dissipation dampers for each floor is shown in Tables 12.9 and 12.10.

3. **Calculation design under rare earthquake**

According to "*Code for seismic design of buildings*" (GB 50011-2010), the storey displacement angle limit of frame core tube structure under frequent earthquakes is $\theta_e = 1/100$. In this paper, the maximum storey displacement angle of the project is taken as 1/150 according to the owner's requirements, the same distribution is

Table 12.5 Design parameters of shock absorption based on displacement

Floor	Mass ($kN\ s^2/m$)	Analysis of displacement angle between lower layers by response spectrum	Target inter story displacement angle	Interlayer displacement (m)	Floor displacement (m)
20	1122.8	1/825	1/997	0.00401	0.08497
19	2034	1/796	1/962	0.00415	0.08096
18	2020	1/761	1/919	0.00435	0.07681
17	2002	1/734	1/887	0.00451	0.07246
16	2002	1/711	1/859	0.00466	0.06795
15	2029	1/692	1/836	0.00478	0.06329
14	2014	1/679	1/820	0.00487	0.05851
13	2002	1/669	1/800	0.005	0.05364
12	2002	1/663	1/800	0.005	0.04864
11	2002	1/662	1/800	0.005	0.04364
10	2034	1/666	1/800	0.005	0.03864
9	2016	1/677	1/813	0.00492	0.03364
8	2005	1/699	1/845	0.00473	0.02872
7	2005	1/726	1/877	0.00456	0.02399
6	2005	1/767	1/926	0.00432	0.01943
5	2010	1/824	1/995	0.00402	0.01511
4	2059	1/933	1/1127	0.00355	0.01109
3	2188	1/1093	1/1320	0.00318	0.00754
2	2032	1/1387	1/1676	0.00251	0.00436
1	2008	1/2554	1/3086	0.00185	0.00185

divided according to the proportion of storey displacement angle under the response spectrum analysis, and the maximum storey displacement angle is taken as 1/150 for floors 10–13. The damping ratio of viscous energy dissipation damper added to the structure is equal to $\xi_d = 12\%$ in the case of frequent earthquakes. See Table 12.11 for structural calculation parameters and Table 12.12 for equivalent single degree of freedom system parameters.

According to Eq. (12.16), the equivalent displacement $\delta_{eff} = 0.3168$ m can be calculated. At the same time, the equivalent mass $M_{eff} = 28{,}551.12$ kN s^2/m. In this state, the displacement ductility coefficient of the structure is $\mu = 5$, the structure is steel reinforced concrete structure, the inherent damping ratio of the structure itself in the elastic stage is $\xi_0 = 0.04$, and the equivalent damping ratio of the original structure without additional energy dissipation damper is:

$$\xi_{eq} = 0.04 + \frac{1 - (1-\gamma)/\sqrt{u} - \gamma\sqrt{u}}{\pi} = 0.193 \tag{12.18}$$

Table 12.6 Parameter calculation of single degree of freedom system

Floor	Mass m_i (kN s^2/m)	Floor displacement δ_i (m)	$m_i\delta_i$	$m_i\delta_i^2$
20	1122.8	0.0847	95.10116	8.055068252
19	2034	0.08096	164.67264	13.33189693
18	2020	0.07681	155.1562	11.91754772
17	2002	0.07246	145.06492	10.5114041
16	2002	0.06795	136.0359	9.243639405
15	2029	0.06329	128.41541	8.127411299
14	2014	0.05851	117.83914	6.894768081
13	2002	0.05364	107.38728	5.760253699
12	2002	0.04864	97.37728	4.736430899
11	2002	0.04364	87.36728	3.812708099
10	2034	0.03864	78.59376	3.036862886
9	2016	0.03364	67.81824	2.281405594
8	2005	0.02872	57.5836	1.653800992
7	2005	0.02399	48.09995	1.153917801
6	2005	0.01943	38.95715	0.756937425
5	2010	0.01511	30.3711	0.458907321
4	2059	0.01109	22.83431	0.253232498
3	2188	0.00754	16.49752	0.124391301
2	2032	0.00436	8.85952	0.038627507
1	2008	0.00185	3.7148	0.00687238
Σ			1607.74716	92.1560842

The additional damping ratio of the viscous energy dissipation damper to the structure is taken as $\xi_d = 0.12$. So the equivalent damping ratio of the single degree of freedom system is $\xi_{eff} = 0.313$. Then we can get: $\gamma = 0.779$, $\eta_1 = 0.0012$, $\eta_2 = 0.55$; through the above three parameters, we can get the equivalent period $T_{eff} = 2.57$ s. From M_{eff} and T_{eff}, $K_{eff} = 170{,}654.07$ kN/m, so the base shear $V_b = 54{,}063.21$ kN. See Table 12.13 for the distribution of base shear.

12.3 Design Method of Building Frequency Modulation and Vibration Control

12.3.1 General Frame for Frequency Modulation and Vibration Control Design of Buildings

For the general building structure, a frequency modulation vibration reduction device is usually set on the top. The dynamic equations of each frequency modulation vibration reduction system are compared and analyzed. Under the dynamic load,

Table 12.7 Floor lateral force and shear force

Floor	Equivalent displacement δ_{eff} (m)	Equivalent mass M_{eff} (kN s^2/m)	Floor lateral force F_i (kN)	Floor shear force V_i (kN)
20	0.0573	34,335.91	2031.031	2031.031
19			3516.837	5547.868
18			3313.598	8861.466
17			3098.084	11,959.55
16			2905.256	14,864.806
15			2742.508	17,607.314
14			2516.635	20,123.949
13			2293.421	22,417.37
12			2079.641	24,497.011
11			1865.862	26,362.873
10			1678.49	28,041.363
9			1448.363	29,489.726
8			1229.786	30,719.512
7			1027.248	31,746.76
6			831.989	32,578.749
5			648.621	33,227.37
4			487.661	33,715.031
3			352.329	34,067.36
2			189.208	34,256.568
1			79.335	34,335.903
Σ			34,335.91	

the building structure with tuned vibration reduction device can be described by the following dynamic equations [1]:

$$[M]\{\ddot{x}(t)\} + [C]\{\dot{x}(t)\} + [K]\{x(t)\} = \{F(t)\} + [H]F_D(t) \tag{12.19}$$

$$B_1\ddot{x}_D(t) + C_D\dot{x}_D(t) + K_Dx_D(t) = -B_2[H]^{\mathrm{T}}\{\ddot{x}(t)\} \tag{12.20}$$

$$F_D = -\left(B_2\ddot{x}_D(t) + B_3[H]^{\mathrm{T}}\{\ddot{x}(t)\}\right) \tag{12.21}$$

Among them, $\ddot{y}_D$, $\dot{y}_D$ and y_D are the acceleration, velocity and displacement of the moving mass of the vibration control device respectively; $[H]$ is the action position matrix of the vibration control device, $H = [0, \ldots, 0, 1, 0, \ldots, 0]^{\mathrm{T}}$ (1 in column i), i is the number of the installation node of the vibration control device; $\{P(t)\}$ is the external load. B_1, B_2 and B_3 are the calculation parameters of the vibration control device, and different vibration control devices have different values (Table 12.14 gives the

Table 12.8 Damping ratio distribution table

Floor	Additional damping ratio	u_{i0} (m)	u_i (m)	Δu_i (m)	ξ_{di}
20	0.16	0.00485	0.00401	0.00084	0.00784
19		0.00503	0.00415	0.00088	0.00784
18		0.00526	0.00435	0.00091	0.008
17		0.00545	0.00451	0.00094	0.008
16		0.00563	0.00466	0.00097	0.0088
15		0.00578	0.00478	0.001	0.0096
14		0.00589	0.00487	0.00102	0.0096
13		0.00598	0.005	0.00098	0.0096
12		0.00603	0.005	0.00103	0.0096
11		0.00603	0.005	0.00103	0.0096
10		0.00604	0.005	0.00104	0.0096
9		0.00591	0.00492	0.00099	0.0088
8		0.00572	0.00473	0.00099	0.0088
7		0.00551	0.00456	0.00095	0.008
6		0.00551	0.00432	0.00119	0.0104
5		0.00485	0.00402	0.00083	0.0064
4		0.00428	0.00355	0.00073	0.0064
3		0.00384	0.00318	0.00066	0.00512
2		0.00303	0.00251	0.00052	0.0048
1		0.00223	0.00185	0.00038	0.0032
Σ				0.01788	

Table 12.9 Type of energy dissipation damper

Type of energy dissipation damper	Damping index a	Damping coefficient C (kN m/s)	Quantity (unit)
A	0.25	1100	16
B	0.25	1000	88
C	0.25	900	20

Table 12.10 Floor arrangement of energy dissipation damper

Floor	X direction		Y direction	
	Model	Number/layer	Model	Number/layer
1–5	C	2	C	2
6–16	B	4	B	4
17–20	A	2	A	2
				Total: 124

Table 12.11 Displacement based damping design

Floor	Mass (kN s^2/m)	Analysis of displacement angle between lower layers by response spectrum	Target inter story displacement angle	Interlayer displacement (m)	Floor displacement (m)
20	1122.8	1/825	1/186	0.021	0.4677
19	2034	1/796	1/180	0.022	0.4467
18	2020	1/761	1/172	0.0232	0.4247
17	2002	1/734	1/166	0.0241	0.4015
16	2002	1/711	1/157	0.0255	0.3774
15	2029	1/692	1/154	0.0259	0.3519
14	2014	1/679	1/153	0.0261	0.326
13	2002	1/669	1/150	0.0266	0.2999
12	2002	1/663	1/150	0.0266	0.2733
11	2002	1/662	1/150	0.0266	0.2467
10	2034	1/666	1/150	0.0266	0.2201
9	2016	1/677	1/152	0.0263	0.1935
8	2005	1/699	1/156	0.0256	0.1672
7	2005	1/726	1/158	0.0253	0.1416
6	2005	1/767	1/166	0.0241	0.1163
5	2010	1/824	1/174	0.0229	0.0922
4	2059	1/933	1/191	0.0209	0.0693
3	2188	1/1093	1/210	0.02	0.0484
2	2032	1/1387	1/314	0.0134	0.02438
1	2008	1/2554	1/578	0.00986	0.00986

parameter expressions of TMD, ring TLD and ring TLCD, and other frequency modulation vibration control devices can refer to); C_D and K_D are the equivalent damping and equivalent stiffness of the vibration control device respectively.

The coordinate transformation of Eq. (12.19) by mode superposition method can be obtained as follows:

$$\ddot{q}_j(t) + 2\zeta_j\omega_j\dot{q}_j(t) + \omega_j^2 q_j(t) = f_j(t) + f_{D,j}(t) \tag{12.22}$$

$$\ddot{x}_D(t) + 2\zeta_D\omega_D\dot{x}_D(t) + \omega_D^2 x_D(t) = -B_2\Phi_{ji}\ddot{q}_j(t) \tag{12.23}$$

where, $f_j = \Phi_j^{\mathrm{T}}\{P(t)\}/M_j$, which is the generalized load of the j-mode of the structure; $f_{D,j}(t) = \Phi_j^{\mathrm{T}}[H]F_D/M_j$ is the generalized control force of the vibration

Table 12.12 Calculation parameters of single degree of freedom system

Floor	Mass m_i (kN s²/m)	Floor displacement δ_i (m)	$m_i\delta_i$	$m_i\delta_i^2$
20	1122.8	0.4677	525.13356	245.604966
19	2034	0.4467	908.5878	405.8661703
18	2020	0.4247	863.348	368.9949352
17	2002	0.4015	803.803	322.7269045
16	2002	0.3774	755.5548	285.14638
15	2029	0.3519	714.0051	251.25839
14	2014	0.326	656.564	214.03986
13	2002	0.2999	600.3998	180.0599
12	2002	0.2733	547.1466	149.53517
11	2002	0.2467	493.8934	121.8435
10	2034	0.2201	447.6834	98.535116
9	2016	0.1935	390.096	75.483576
8	2005	0.1672	335.236	56.051459
7	2005	0.1416	283.908	40.201373
6	2005	0.1163	233.1815	27.119008
5	2010	0.0922	185.322	17.086688
4	2059	0.0693	142.6887	9.8883269
3	2188	0.0484	105.8992	5.1255213
2	2032	0.02438	57.7088	1.6389299
1	2008	0.00986	30.12	0.4518
Σ			9080.2797	2876.658

damping device of the j-mode; $M_j = \Phi_j^T[M]\Phi_j$ is the generalized mass of the j-mode of the structure; ω_j is the circular frequency of the j mode of the structure; Φ_{ji} is the mode value of the i point of the j-mode of the structure; ζ_D and ω_D are the damping ratio and tuning frequency of the vibration reduction device. Then,

$$\ddot{q}_j(t) + 2\zeta_j\omega_j\dot{q}_j(t) + \omega_j^2 q_j(t) = f_j(t) - \Phi_{ji}\big(B_2\ddot{y}_D(t) + \Phi_{ji}B_3\ddot{q}_j(t)\big)/M_j \tag{12.24}$$

Let $y_D = \Phi_{ji}g$ be further replaced by:

$$\begin{bmatrix} 1+\mu_3 & \mu_2 \\ \mu_2 & \mu_1 \end{bmatrix}\begin{Bmatrix} \ddot{q}_j \\ \ddot{g} \end{Bmatrix} + \begin{bmatrix} 2\zeta_j\omega_j & 0 \\ 0 & 2\mu_1\zeta_D\omega_D \end{bmatrix}\begin{Bmatrix} \dot{q}_j \\ \dot{g} \end{Bmatrix} + \begin{bmatrix} \omega_j^2 & 0 \\ 0 & \mu_1\omega_D^2 \end{bmatrix}\begin{Bmatrix} q_j \\ g \end{Bmatrix} = \begin{Bmatrix} f_j \\ 0 \end{Bmatrix} \tag{12.25}$$

Among, $\mu_1 = \Phi_{ji}^2 B_1/M_j$, $\mu_2 = \Phi_{ji}^2 B_2/M_j$, $\mu_3 = \Phi_{ji}^2 B_3/M_j$.

Table 12.13 Floor lateral force and shear force

Floor	Equivalent displacement δ_{eff} (m)	Equivalent mass M_{eff} (kN s²/m)	Floor lateral force F_i (kN)	Floor shear force V_i (kN)
20	0.3168	28,551.12	3118.066	3118.066
19			5381.127	8499.193
18			5070.246	13,569.439
17			4745.438	18,314.877
16			4449.362	22,764.239
15			4200.964	26,965.203
14			3855.51	30,820.713
13			3511.789	34,332.502
12			3182.815	37,515.317
11			2853.842	40,369.159
10			2565.226	42,934.385
9			2211.252	45,145.637
8			1877.957	47,023.594
7			1564.964	48,588.558
6			1268.445	49,857.003
5			990.863	50,847.866
4			744.347	52,592.213
3			539.305	53,131.518
2			292.164	53,923.682
1			123.73	54,063.311
Σ			54,063.21	

Table 12.14 Parameter expression of tuned damping device

Parameter	TMD	RS-TLD	RS-TLCD
B_1	m_d	$M_L F_1 d_1' \left[1 + k' R_1^1 (k' \sigma_1)\right]$	$2\rho AL$
B_2	m_d	$M_L F_1 d_1' \left[1 + k' R_1^1 (k' \sigma_1)\right]$	$2\rho AB$
B_3	m_d	$M_L \left(1 + k'^2\right)$	$2\rho AL$
ζ_D	$c_{d,eq}/(2\omega_d m_d)$	ζ_L	$\sqrt{\frac{1}{2gL}} \frac{\xi \lvert \dot{y}_d(t) \rvert}{4}$
ω_D	$\sqrt{k_d/m_d}$	$\sqrt{\frac{g}{a}\sigma_1 \tanh\left(\sigma_1 \frac{h}{a}\right)}$	$\sqrt{2g/L}$

Let f_j be a simple harmonic input, that is, substitute Eq. (12.25) to obtain the frequency response function of the vibration damping structure system as follows:

$$H_{qj}(i\omega) = \frac{1}{D}\left(-\mu_1\omega^2 + 2\mu_1\zeta_D\omega_D(i\omega) + \mu_1\omega_D^2\right) \tag{12.26}$$

$$H_D(i\omega) = \frac{1}{D}\left(\mu_2\omega^2\right) \tag{12.27}$$

Among,
$$\begin{aligned} D = & \left[\mu_1(1+\mu_3) - \mu_2^2\right]\omega^4 - \left[(1+\mu_3)\mu_1 \cdot 2\zeta_D\omega_D + \mu_1 \cdot 2\zeta_j\omega_j\right]i\omega^3 \\ & - \left[\mu_1(1+\mu_3)\omega_D^2 + \mu_1\omega_j^2 + 4\zeta_j\omega_j\zeta_D\omega_D\mu_1\right]\omega^2 \\ & + \left[\mu_1 \cdot 2\zeta_D\omega_D\omega_j^2 + 2\zeta_j\omega_j\omega_D^2\mu_1\right]i\omega + \mu_1\omega_j^2\omega_D^2 \end{aligned}.$$

In this way, the method of random vibration can be used to compare the expression of undamped structure, and the equivalent damping ratio of the vibration reduction device to the structure can be obtained:

$$\zeta_e = \frac{1}{D_e}\left[\gamma\zeta_D\lambda(\zeta_D + \zeta_j\lambda)\right] \tag{12.28}$$

Among, $\lambda = \omega_D/\omega_j$, $\gamma = \mu_2^2/\mu_1$,

$$\begin{aligned} D_e = & \left[(1+\mu_3)^2\zeta_D\right]\lambda^4 + \left\{\zeta_j\left[4(1+\mu_3)\zeta_D^2 + \gamma\right]\right\}\lambda^3 \\ & + \left\{\zeta_D\left[4\zeta_j^2 + 4(1+\mu_3)\zeta_D^2 + \gamma - 2(1+\mu_3)\right]\right\}\lambda^2 \\ & + 4\zeta_j\zeta_D\lambda + \zeta_D \end{aligned}.$$

The equivalent damping ratio of the vibration reduction structure can be expressed as:

$$\zeta_{je} = \zeta_j + \zeta_e \tag{12.29}$$

In order to achieve the best control effect of the vibration control device, it is necessary to find the optimal parameters, among which there are more TMD structural systems, and relatively less TLD structural systems. For high-rise steel chimney structure, its damping ratio is far less than that of general structure, so the influence of structural damping ratio can be ignored to simplify the problem. However, it is still a complex constrained optimization to find the optimal parameters of the vibration control device. In order to facilitate the design and application, a practical optimization method for tuning and vibration control of high-rise steel chimneys is proposed. The specific process is shown in Fig. 12.5. Firstly, the type of tuned vibration reduction device is selected according to the engineering requirements, and the modal mass ratio is determined, usually between 0.03 and 0.08. Then, if there is no limit to the displacement of the vibration reduction device, only $\zeta_j = 0$ in Eq. (12.28), and the vibration control of the frequency ratio and damping ratio of the damping device is calculated, and then the simultaneous equation is solved, then the optimal vibration control device can be obtained without constraints parameter:

$$\lambda_{opt} = \left(\omega_D/\omega_j\right) = (1+\mu_3 - 0.5\gamma)^{0.5}/(1+\mu_3) \tag{12.30}$$

$$\zeta_{opt} = 0.5\left\{\gamma(1+\mu_3 - 0.25\gamma)/\left[(1+\mu_3)(1+\mu_3 - 0.5\gamma)\right]\right\}^{0.5} \tag{12.31}$$

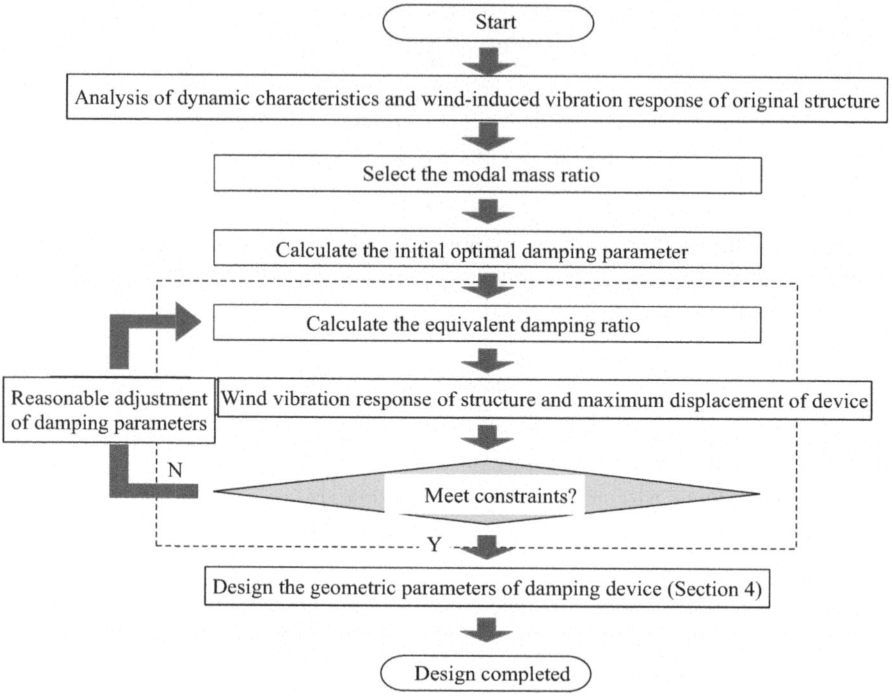

Fig. 12.5 Design process of tuned vibration reduction system for high-rise steel chimney

Then, the results of the above formula are substituted into Eq. (12.28) to calculate the equivalent damping ratio, and the equivalent damping ratio and the structural damping ratio are added as the damping ratio when calculating the structural response according to the specification to obtain the response of the structure and the vibration control device. The response of the vibration control device is compared with its limit value. If the limit value requirements are met, the design is completed. If the limit value requirements are not met, the change of parameters and results will be considered. The trend is to adjust the vibration control parameters appropriately, and then substitute the parameters into the calculation response, so that the cycle can be repeated until the constraint requirements are met. Among them, the calculation and change trend of the response of the vibration control device can be seen in Sect. 5.

12.3.2 *Example of Structure Frequency Modulation and Vibration Control Design*

1. **Project overview**

A high-rise steel chimney is 50 m in height and 2.3 m in diameter. Q235B is used as structural steel and Q345C as foundation anchoring material. The construction site is of class B roughness, the basic wind pressure is 0.6 kN/m^2, the section parameters are shown in Table 12.15, and the mass of wire is 183.8 kg/m. The preliminary analysis shows that the characteristic frequency of the steel chimney is 0.85 Hz, and the height diameter ratio is 21.7. It can be seen that the structure has the characteristics of light weight, large slenderness ratio and small damping. It has strong wind sensitivity and is easy to produce large response under the action of wind load.

2. **Design of vibration control device**

Based on the relevant contents of China's code, the wind resistance analysis and vibration control design are carried out for the project. The modal mass ratio is about 0.05. The specific value needs to be adjusted according to the characteristics of the vibration control device, and the other parameters are designed according to the design method in Sect. 12.3.1. The analysis shows that for a 50 m high steel chimney with a damping ratio of 0.01, there is little difference between the crosswind and crosswind dynamic displacement responses, so the crosswind wind vibration is selected for design first, and then checked in crosswind direction. Table 12.16 gives the results of vibration control design of high-rise chimneys, in which n in RS-TLD is the number of dampers; n in RS-TLCD is 8 TLCD for each layer, and N is 5 sets (i.e. 5 layers), which are all set at the top, and the damping ratio in brackets indicates the opening rate. It can be seen from the analysis that TMD and RS-TLD can achieve better vibration control effect, while RS-TLCD can not; Due to the need to control the stroke of the vibration control device, the damping ratio of the device is

Table 12.15 Structural section parameters

Height (m)	Section thickness (mm)
0–4.8	18
4.8–12	16
12–21.6	14
21.6–50	12

Table 12.16 Design parameters of damping device

Vibration damping device	RS-TMD	RS-TLD	RS-TLCD
Basic parameters	$m_{TMD} = 458.6$ kg $k_{TMD} = 11.3$ k N/m $C_{TMD} = 721$ N/m	$R_d = 0.3$ m $k' = 0.7$ $h = 1$ m $\zeta_L = 0.085$ $n = 2$	$R_d = 1.26$ m $r = 0.1$ m $H_w = 0.25$ m $n = 8$ $N = 5$
Modal mass ratio	0.05	0.064	0.054
Frequency ratio	0.966	0.975	0.935
Damping ratio	0.106	0.085	0.121 (0.5)
Additional equivalent damping ratio	0.0488	0.0555	0.0113
Equivalent damping ratio of structure	0.0588	0.0655	0.0213

relatively large; Through careful analysis of the final design parameters of RS-TLD, it is found that the inner diameter is only 0.3 m, and the geometry can not meet the size of the steel chimney. Therefore, the design of RS-TLD is not applicable to all self-supporting high-rise structures. The analysis shows that RS-TLD is applicable to self-supporting high-rise structures with larger slenderness ratio.

In order to further discuss the effectiveness of the equivalent damping ratio method, with the help of a self-made analysis program, the Equivalent Damping Method (EDM) and the Time History Methods (THM) are used to analyze the downwind wind vibration response, and the results are compared. It can be seen from Fig. 12.6 that the calculation results obtained by the two methods are similar, so EDM can be used for vibration reduction analysis of self-supporting high-rise structure; in the power spectral density curve, there is only one extreme value for EDM curve, while there are two obvious extreme values for THM curve in the case of small damping (RS-TLD). Figure 12.6d shows the displacement root variance of each vibration reduction system obtained by two methods. The comparison shows that: the additional vibration reduction device can better attenuate the dynamic response of the steel chimney, and the average attenuation rate of the displacement of the vertex root variance of each vibration reduction system is 25%; the vibration reduction effect of RS-TMD and RS-TLD is relatively good, and the damping effect of the other two vibration reduction systems is relatively poor, which is similar to that of the other two damping systems The results of equivalent damping ratio are consistent with each other; the average error of EDM and THM is 5.37%, which meets the needs of engineering analysis, so the method of equivalent damping ratio can be used for vibration reduction design.

The equivalent damping ratio of the structure obtained in Table 12.16 can be directly substituted into the specification method to obtain the displacement response of the chimney top designed according to the Chinese specification (see Fig. 12.7, where NoD represents the structure without any vibration reduction device). The analysis shows that: For the downwind response, the total response has a certain attenuation after the installation of damping device, with an average of more than 15%, up to 22%; the response caused by the fluctuating wind has an average attenuation of more than 35%, up to 46%; the response caused by the static wind will not change, so the vibration reduction device mainly controls the displacement caused by the fluctuating wind; For the response caused by crosswind, the installation of vibration reduction device can reduce the response by 60% on average, and the maximum attenuation value is about 85%; Compared with the results of crosswind and crosswind, the vibration control technology has a better control effect on the response caused by crosswind of self-supporting high-rise structure.

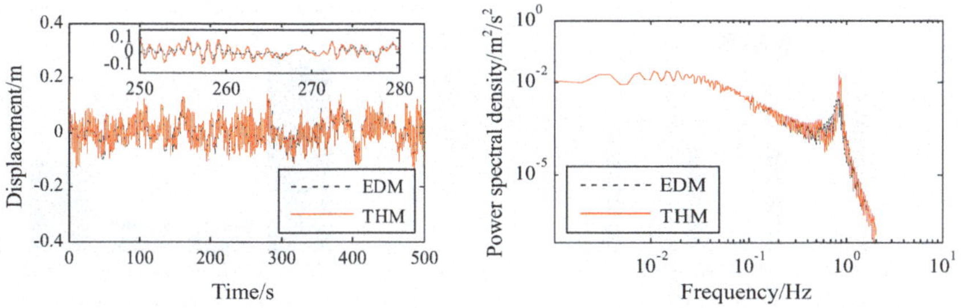

(a) Peak displacement time history and power spectral density of RS-TMD

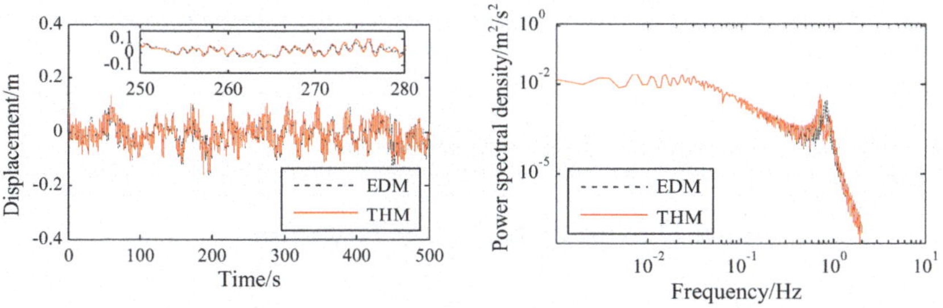

(b) Peak displacement time history and power spectral density of RS-TLD

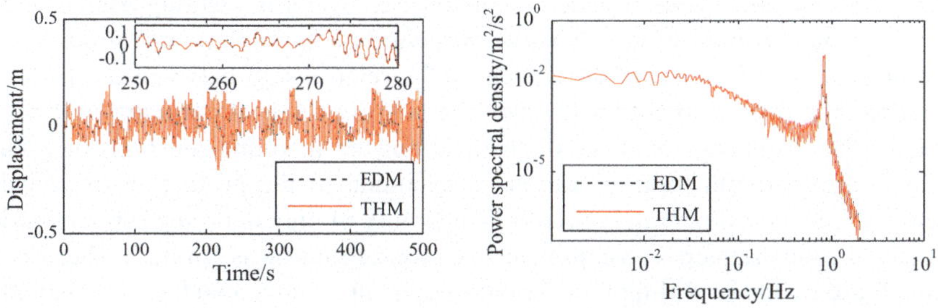

(c) Peak displacement time history and power spectral density of RS-TLCD

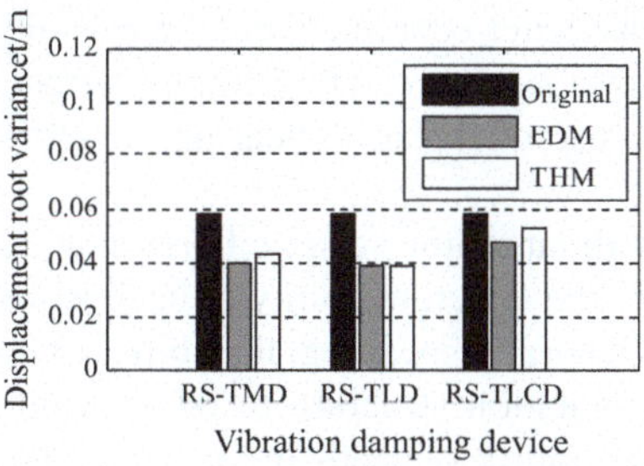

(d) Variance displacement of top root of chimney

Fig. 12.6 Comparison of the results of time history analysis of vertex displacement by different methods

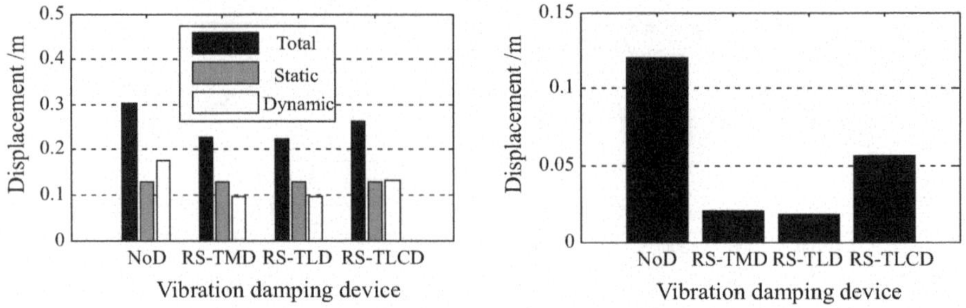

(a) Along wind displacement response (b) Crosswind displacement response

Fig. 12.7 Comparison of wind-induced vibration response of structure before and after installation of damping device

12.4 Design Method of Building Isolation

12.4.1 Conceptual Design of Building Isolation

1. **Selection of isolation system**

For common isolation systems, such as rubber base isolation, sliding base isolation and hybrid base isolation, the dynamic behavior of the isolation system is very different because of the great difference of the relationship between the isolation layer force and displacement hysteresis. The bearing force of common rubber pad is linear with displacement, which can filter out the medium and high frequency components of ground motion, which is beneficial to the protection of valuable instruments and equipment in buildings; The ability of high damping rubber bearing to filter the high frequency components of ground motion is limited; There is no definite basic period of sliding base isolation structure, and it also has good isolation effect in weak site. However, due to the strong nonlinearity of the isolation layer, the high frequency response of the isolation system is significant; In general, parallel base isolation structure has better economy, and the seismic response of the system is similar to sliding isolation structure; The isolation effect of series base isolation structure is good, but the use of isolation bearings is more and the cost is higher [4, 5].

In addition, in areas with large non seismic horizontal action such as wind load, attention should be paid to the different ability of each isolation system to resist the initial horizontal action. Generally speaking, the ability of sliding isolation structure and lead rubber cushion isolation structure to resist wind load and microseismic action is larger, while the ability of lead-free ordinary laminated rubber cushion isolation structure to resist such microseismic action is poor. The current specification stipulates that the total horizontal force produced by the standard value of horizontal load for wind load and other non earthquake action of rubber pad isolation structure shall not exceed 10% of the total gravity of the structure.

Therefore, on the basis of fully understanding the seismic response of each isolation system and considering the actual conditions and requirements of the project, the design of isolation structure should reasonably select the type of isolation system.

2. **Setting of isolation layer position**

The isolation layer should be set below the first floor of the structure, such as the top surface of the foundation or the top surface of the basement. When there is no basement or other special circumstances in the house, the isolation layer can also be set in other parts. For example, for the brick concrete structure system at the upper part of the bottom frame, the isolation layer can be set at the junction of the bottom frame and the upper brick concrete. There are already such projects at home and abroad. Another example is Japan Dacheng headquarters building (16 story reinforced concrete structure, 49 m high). Its isolation layer is set on the 8th floor with abrupt stiffness, so that the storey shear force of the whole building (upper and lower structures) is reduced by 40–80%.

When the isolation layer is set in other parts, detailed design analysis shall be carried out and reliable measures shall be taken.

3. **Selection of construction site and foundation**

Hard soil site is more suitable for isolated buildings; weak site filters out the medium and high frequency components of seismic wave, and prolonging the period of isolated structure will increase rather than decrease its seismic action. The earthquake in Mexico is a typical example. According to the isolation standard of Japan, rubber base isolation structure is only applicable to class I and class II sites. In most areas of China (the first group), the design characteristic periods of class I, II and III sites are small, so they are suitable for rubber pad base isolation structure except for class IV site. The sliding isolation structure is not sensitive to the frequency spectrum characteristics of ground motion and has strong adaptability to the site, but there is no corresponding specification for this.

The uneven settlement of the foundation will lead to the redistribution of the weight of the upper structure borne by each isolation bearing, resulting in the actual bearing capacity of some bearings exceeding the design value, which poses a threat to the stability of the structure. Therefore, the isolation structure should have good foundation.

4. **Requirements for building shape**

The requirements of building shape include height, number of floors, ratio of height to width, regularity of plane and elevation, and ratio of basic period in two directions of structure.

The base isolation technology of rubber pad is suitable for the bottom and multi-storey structures, and the best isolation effect is that the basic period is less than 10 s without isolation. With the increase of the height and the number of stories, the basic period of the structure increases. In order to ensure the safety and achieve the isolation effect, it is bound to require the use of larger specifications, high-quality isolation bearings, with a larger bearing capacity, a longer period of natural vibration and a larger deformation capacity, but also requires a more reasonable and reliable analysis and design of the isolation structure. However, with the development of technology, in recent years, the height and number of floors of newly built or renovated isolation buildings in foreign countries are higher, and there are more than 100 m isolation buildings.

Generally, the tensile property of base isolation bearing is poor. Therefore, when the height width ratio of the house is large, the anti overturning checking calculation of the house under the rare earthquake should be carried out to prevent the bearing from buckling or tensile stress, so as to ensure the overall stability of the house under the rare earthquake.

For the case of irregular plane or vertical of building structure, conceptually speaking, the stiffness center of the isolation layer and the mass center of the superstructure can be basically the same by adjusting the layout of the components of the isolation layer, so as to avoid structural torsional damage. However, the floor above the isolation layer must bear the shear transfer caused by horizontal shear redistribution, and the vibration deformation of each part caused by high-order vibration mode is still inconsistent. Therefore, for the irregular building structure, it is necessary to select the structural calculation model which is in line with the actual situation for more accurate seismic analysis, and take necessary strengthening measures for the local components of the structure.

It is generally believed that the characteristic period of the design response spectrum of the same site in any horizontal direction is the same. If the difference between the basic periods of two directions is too large, the isolation effect of two directions will be greatly different, so it is necessary to limit the difference between the basic periods of two directions not to exceed 30% of the smaller value.

5. **Integrity requirements of isolation layer**

In order to ensure the overall coordination of the isolation layer, the top of the isolation layer should be equipped with a beam slab system with sufficient rigidity in the plane, such as the use of cast-in-place reinforced concrete floor. When the prefabricated reinforced concrete floor is adopted, in order to make the vertical and horizontal beam system transmit the vertical load and coordinate the distribution of the transverse shear force in each isolation support, the vertical and horizontal beam system above the support shall be cast-in-place. In order to increase the in-plane stiffness of the top beam plate of the isolation layer, the section size and reinforcement of the girder should be added. The stress state of the beam and column near the isolation support is complex, and they will be impacted by the punching force during the earthquake, so the stirrup should be increased, and the net reinforcement should be provided if necessary. Considering that the isolation layer has no isolation effect on the vertical earthquake, the seismic structural measures of the superstructure should keep the requirements related to the vertical resistance.

6. **Arrangement of isolation bearing**

The arrangement of the isolation bearing of the isolation layer shall make the center of the rigidity of the isolation layer coincide with the center of mass of the superstructure, so as to reduce the torsional effect of the system; when selecting the isolation bearing of various specifications, attention shall be paid to give full play to the bearing capacity and horizontal deformation capacity of each isolation bearing; the wind resistant device set in the isolation layer shall be arranged symmetrically and dispersedly around the building.

In addition to the above points, we should also pay attention to control the maximum displacement limit of the isolation layer and deal with the structural requirements of the isolation layer. For example, the isolation support shall be reliably connected with the upper and lower structures, and the vertical pipeline passing through the isolation layer shall be provided with flexible joints, etc.

12.4.2 Requirements and Methods of Building Isolation Structure Design

1. **Concept and value of horizontal damping coefficient**

The horizontal damping coefficient describes the degree to which the seismic action of the isolated building is lower than that of the non isolated building. Its value can be calculated by Eqs. (12.32) and (12.33).

$$\psi = \frac{(\psi_i)\max}{0.7} \tag{12.32}$$

$$\psi_i = \frac{Q_{gi}}{Q_i} \tag{12.33}$$

Among them, ψ is the horizontal damping coefficient; $(\psi_i)_{\max}$ is the maximum value of the shear ratio between each floor under the fortification intensity when the structure is isolated or not; ψ_i is the shear ratio between the i story and the i story when the structure is isolated; Q_{gi} is the shear force between the i story when the structure is isolated under the fortification intensity; Q_i is the shear force between the i story when the structure is not isolated under the fortification intensity. When the ratio of the maximum story shear force of each story after the structure is isolated to the maximum story shear force of the corresponding story when the structure is not isolated is not greater than the values of the first row and columns in Table 12.17, the horizontal damping coefficient can be determined according to the table.

The value of horizontal damping coefficient should not be less than 0.25, and the total horizontal seismic action after isolation should not be less than the total horizontal seismic action of non isolation structure in the 6° fortification. In Table 12.17, the horizontal seismic damping coefficient is the maximum inter story shear ratio divided by 0.7, so the horizontal seismic action of the structure above the isolation

Table 12.17 The corresponding relationship between the maximum ratio of story shear and horizontal damping coefficient

Maximum inter story shear ratio	0.53	0.35	0.26	0.18
Horizontal damping coefficient	0.75	0.50	0.38	0.25

layer is only 70% of the horizontal seismic action of the structure corresponding to the damping coefficient. This means that there are about 0.5 safety reserves for horizontal seismic action, seismic checking calculation and member bearing capacity of the structure above the isolation layer. Therefore, for class C buildings, the corresponding structural requirements can also be reduced.

2. **Calculation method of horizontal damping coefficient**

The time history analysis method should be used to calculate the story shear force of isolated and non isolated cases. For masonry structure and structure with the same basic period as masonry structure, the horizontal damping coefficient can be determined by the following methods.

(1) Horizontal damping coefficient of multistory masonry structure is

$$\psi = \sqrt{2}\eta_2(\frac{T_{gm}}{T})^{\gamma} \tag{12.34}$$

Among them, ψ is the horizontal damping coefficient; η_2 is the damping adjustment coefficient of the seismic influence coefficient, which is determined according to the requirements of the seismic influence coefficient curve according to the equivalent damping ratio of the isolation layer; γ is the attenuation index of the curve falling section of the seismic influence coefficient, which is determined according to the requirements of the seismic influence coefficient curve according to the equivalent damping ratio of the isolation layer; T_{gm} is the design characteristic period of the masonry structure when the isolation scheme is adopted. The division of the design characteristic period in this area is determined according to the seismic code, but it is adopted as 0.4 s when it is less than 0.4 s; T_1 is the basic period of the isolated structure, which should not be greater than the larger value of 2.0 s and 5 times of the characteristic period.

(2) Horizontal damping coefficient equivalent to the period of masonry structure is

$$\psi = \sqrt{2}\eta_2(\frac{T_g}{T})^{\gamma}(\frac{T_0}{T_g})^{0.9} \tag{12.35}$$

Among them, T_0 is the calculation period of the non isolated structure, when it is less than the characteristic period, the value of the characteristic period value should be used; T_1 is the basic period of the isolated structure, which should not be more than 5 times the characteristic period value; T_g is the site characteristic period.

(3) Masonry structure and the structure corresponding to its basic period. The basic period of the structure after isolation is calculated according to the following formula:

$$T_1 = 2\pi\sqrt{\frac{G}{K_h g}} \tag{12.36}$$

Among them, G is the representative value of the gravity load of the structure above the isolation layer; K_h is the horizontal dynamic stiffness of the isolation layer under the frequent earthquake action; g is the gravity acceleration.

Under the action of frequent and rare earthquakes, the horizontal dynamic stiffness K_h and the equivalent viscous damping ratio ζ_{eq} of the isolation layer can be calculated according to the following formula:

$$K_h = \sum K_j \tag{12.37}$$

$$\zeta_{eq} = \sum \frac{K_j \zeta_j}{K_h} \tag{12.38}$$

Among them, ζ_{eq} is the equivalent viscous damping ratio of the isolation layer; ζ_j is the equivalent viscous damping ratio determined by the test of j isolation bearing, and the damper set separately shall include the corresponding damping ratio of the damper; K_j is the horizontal dynamic stiffness determined by the test of j isolation bearing (including the damper).

The horizontal dynamic stiffness and equivalent viscous damping ratio of the isolation layer are the design parameters determined by the test. The specific test conditions are as follows: under the action of design value of compression bearing capacity, the horizontal stiffness and equivalent viscous damping ratio with horizontal loading frequency of 0.3 Hz and shear deformation of isolation bearing of 50% should be adopted for checking calculation of frequent earthquakes; For the calculation of rare earthquake, the horizontal dynamic stiffness and equivalent viscous damping ratio of the isolation bearing with the diameter less than 600 mm should be adopted when the horizontal loading frequency is 0.1 Hz and the shear deformation of the isolation bearing is not less than 250%; The horizontal dynamic stiffness and equivalent viscous damping ratio of the isolation bearing with a horizontal loading frequency of 0.2 Hz and a shear deformation of 100% can be used for the isolation bearing with a diameter of no less than 600 mm.

12.4.3 Design of Isolation Layer

The design of the base isolation structure includes the arrangement of the isolation layer, the checking calculation of the compression bearing capacity of the isolation layer, the checking calculation of the displacement of the isolation bearing, the checking calculation of the anti overturning of the isolation structure, the checking calculation of the anti wind device and the corresponding structural measures.

1. **Arrangement of isolation layer**

The isolation layer can be composed of isolation support, damping device and anti wind device. The damping device and the anti wind device can be integrated with the isolation support, and can also be set separately. When the displacement of the isolation layer is large, the limit device can be set, and the adverse impact of collision should be avoided. When the isolation bearing has larger horizontal deformation capacity and larger damping, and has reliable connection with the upper and lower structures, the maximum displacement of the isolation layer is generally not too large, and the limit device may not be set separately.

In order to reduce the overall torsional effect of the isolated structure, the wind resistant device should be symmetrically and dispersedly arranged around the building, and the center of the stiffness of the isolation layer should coincide with the mass center of the superstructure as much as possible.

The plane layout of the isolation bearing should correspond to the plane position of the vertical load-bearing members in the upper structure and the lower structure. The bottom surface of the isolation bearing should be arranged at the same elevation, or at different elevation if necessary.

The bearing capacity and horizontal deformation capacity of each isolation bearing should be fully exerted when multiple specifications of isolation bearing are selected for the same building. When multiple isolation bearings are selected at the same support, the clear distance between them shall be greater than the space size required for installation and replacement. The distance between the isolation bearings under the shear wall should not be greater than 2.0 m.

2. **Checking calculation of compression bearing capacity of isolation layer**

The checking calculation of the bearing capacity of the isolation layer is the guarantee of the reliability and stability of the isolation layer. During the checking calculation, it should be ensured that the design value of the total bearing capacity of the isolation layer is greater than 1.1 times of the representative value of the total gravity load of the superstructure; The design value of the bearing capacity of each isolation bearing is greater than the representative value of the gravity load transferred from the superstructure to the isolation bearing; When checking the overturning of the superstructure, the representative value of the gravity load transferred from the superstructure to the isolation bearing should consider the added value caused by the overturning moment. Under the rare earthquake, the isolation bearing should not

Table 12.18 Average compressive stress limit of rubber isolation bearing (MPa)

Building category	Class A building	Class B building	Class C building
Average compressive stress limit	10	12	15

Note 1. For the structure to be checked for overturning, the design value of average compressive stress shall include the combination of horizontal seismic action effect; for the structure to be calculated for vertical seismic action, the design value of average compressive stress shall include the combination of vertical seismic action effect
2. When the second shape coefficient of rubber bearing is less than 5.0, the average compressive stress limit shall be reduced; when it is less than 5 but not less than 4, it shall be reduced by 20%; when it is less than 4 but not less than 3, it shall be reduced by 40%
3. For rubber bearings with effective diameter less than 300 mm, the average compressive stress limit is 12 MPa for class C building

have tensile stress. If the isolation bearing is inevitably in the state of tension, its tensile stress should not be greater than 1.2 MPa; When considering the vertical seismic action of the superstructure, the representative value of the gravity load transferred from the superstructure to the isolation bearing increases. 20 and 40% of the representative value of the gravity load of the superstructure can be taken respectively at 8 and 9°; the design value of the compression bearing capacity of the isolation bearing shall conform to the provisions of Table 12.18.

According to the average compressive stress limit listed in Table 12.18, the safety factor of the design value of the compressive stress of the isolation bearing is 6–9, which can ensure the bearing capacity and stability when the isolation layer reaches the maximum horizontal displacement, and can preliminarily determine the diameter of the isolation bearing.

It is stipulated that the tensile stress should not occur in the isolation bearing, mainly considering the internal damage after the rubber tension, which reduces the elastic performance of the bearing; at the same time, the tensile stress of the bearing in the isolation layer means that the superstructure is in danger of overturning.

3. **Displacement checking calculation of isolation bearing**

A large number of test results at home and abroad show that the laminated rubber pad bearing with good quality and reliable connection between upper and lower parts has a shear strain of more than 400% and a maximum displacement of more than 0.65d in case of horizontal shear (or instability) failure while maintaining a constant design compressive stress. Considering that the engineering design safety factor is generally not less than 1.2, the Seismic Specification stipulates that the maximum horizontal displacement of each isolation bearing in the isolation layer under rare earthquake shall meet the following requirements:

$$u_{\max} < 0.55d \tag{12.39}$$

$$u_{\max} < 3t_{\mathrm{r}} \tag{12.40}$$

Among them, u_{max} is the maximum horizontal displacement of the isolation bearing when considering the torsional effect under the rare earthquake; d is the diameter of the isolation bearing; t_r is the total thickness of the rubber pad of the isolation bearing.

4. **Checking calculation of elastic resilience and anti wind device of isolation bearing**

The horizontal bearing capacity of the wind resistant device of the isolation layer needs to be checked and calculated. At the same time, in order to ensure that the isolation bearing still has good reset performance after multiple earthquakes, the elastic recovery capacity of the isolation bearing must be greater than 1.4 times of the design value of the shear bearing capacity of the wind resistant device (or the design value of the horizontal yield load of the isolation bearing) under the seismic fortification intensity (shear strain of the isolation bearing is 100%). The following requirements shall be met:

(a) **Checking calculation of anti wind device**

$$\gamma_w V_{wk} \leq V_{Rw} \tag{12.41}$$

where, V_{Rw} is the design value of horizontal bearing capacity of wind resistant device. When the wind resistant device is an integral part of the isolation support, the design value of the horizontal yield load of the isolation support is taken; when the wind resistant device is set separately, the horizontal bearing capacity of the wind resistant device can be determined according to the design value of the material yield strength; γ_w is the partial coefficient of the wind load; V_{wk} is the standard value of the horizontal shear force of the isolation layer under the wind load.

(b) **Checking calculation of elastic resilience of isolation bearing**

$$K_{100} t_r \leq 1.40 V_{Rw} \tag{12.42}$$

Among them, K_{100} is the horizontal stiffness of the isolation bearing when the horizontal shear strain is 100%.

5. **Checking calculation of anti overturning of isolated structure**

For the structure with large height width ratio, the anti overturning checking calculation under rare earthquake should be carried out. The anti overturning checking calculation includes the checking calculation of the overall anti overturning of the structure and the checking calculation of the bearing capacity of the isolation bearing. When checking the overall anti overturning of the structure, the overturning moment shall be calculated according to the rare earthquake action, and the anti overturning checking calculation shall be calculated according to the representative value of the gravity load of the upper structure. The safety factor of anti overturning checking calculation shall be greater than 1.2. The precautions for anti overturning checking calculation of the isolation bearing are shown in the above checking calculation contents of the isolation layer.

6. **Structural measures**

The structural measures of the isolation layer shall be strengthened according to the following principles: the isolation support has reliable connection with the upper structure and the lower structure; the beams, columns and piers connected with the isolation support shall consider horizontal shear and vertical local pressure, and take reliable structural measures, such as densifying stirrups or configuring mesh reinforcement; When the component reinforcement is used as the lightning rod, the flexible conductor shall be used to connect the upper and lower structural reinforcement.

The vertical pipeline passing through the isolation layer shall meet the following requirements: for the flexible pipeline with smaller diameter, the extension length shall be reserved at the isolation layer, and its value shall not be less than 1.2 times of the maximum horizontal displacement of the isolation layer under the rare earthquake action; Flexible materials or flexible joints should be used for pipes with larger diameter in the isolation layer; flexible joints should be used for important pipes and pipes that may leak harmful or combustible.

When the isolation layer is set in the use space with fire resistance requirements, the isolation support and other components shall take corresponding fire protection measures according to the fire resistance rating of the use space.

The gap formed by the isolation layer can be sealed and filled with flexible materials according to the functional requirements. With waterproof or lime sand intrusion, it can also prevent rats and insects from entering and ensure its function is not affected. The sealing material used must be flexible so as not to affect the isolation effect.

The isolation layer should have space for easy observation and replacement of the isolation bearing. In fact, the isolation bearing generally does not need to be replaced within the service life of the building. The purpose of this measure is to facilitate the inspection in case of abnormal conditions and the transformation of the structure in case of changes in the use function of the house.

The superstructure and isolation layer components shall be separated from the surrounding fixings. The separation distance from the horizontal fixed object shall not be less than 1.2 times of the maximum displacement of the isolation layer under the rare earthquake, and shall not be less than 200 mm; the separation distance from the vertical fixed object shall be 1/25 of the maximum thickness of the rubber pad in the adopted isolation support plus 10 mm, and shall not be less than 15 mm.

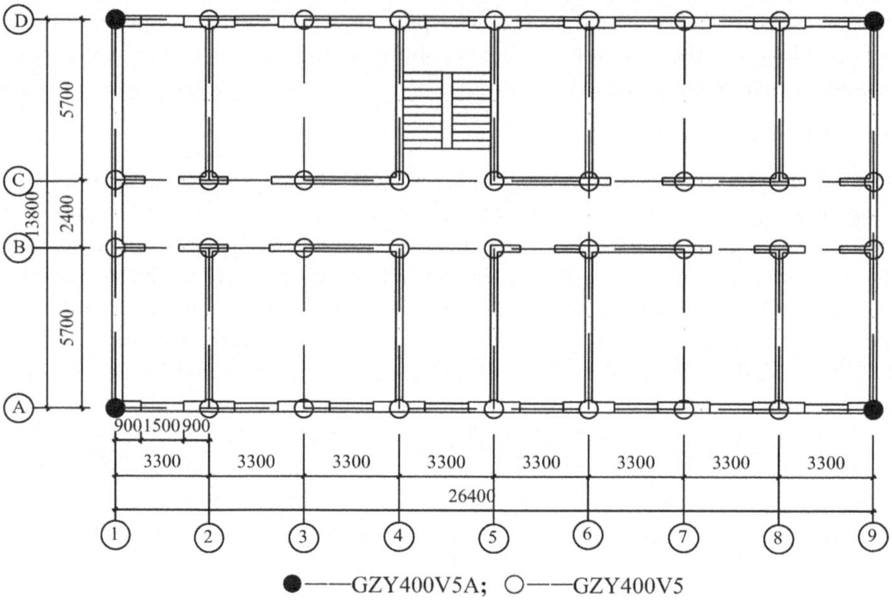

Fig. 12.8 Plane layout of isolation device

12.4.4 Example of Building Structure Isolation Design

1. **Project overview**

A building is a masonry structure (bearing transverse wall, common clay brick, 370 mm thick external wall, 240 mm thick internal wall). See Fig. 12.8 for the layout plan of its bottom layer and standard layer, six layers in total, 2.8 m high, 17.7 m high in total, and the maximum height width ratio is 1.28; class C building, fortification intensity: 8°; class II site; the design earthquake group is the first group.

2. **Preliminary design**

(a) The isolation scheme can be adopted for the building:

- When the building is not isolated, the basic period of the building is 0.3 s, less than 1.0 s;
- The total height of the building is 17.7 m and the number of floors is 6, which conforms to the relevant provisions of "Code for seismic design of buildings";
- The construction site is class II without liquefaction;
- The horizontal load of wind load and other non seismic actions shall not exceed 10% of the total gravity of the structure.

All of the above items meet the requirements of the code for the isolation scheme of buildings.

(b) Determine the location of the isolation layer:
The isolation layer is located at the top of the basement, and the rubber isolation bearing is set at the place with large stress. Its specification, quantity and distribution are determined by calculation according to the requirements of vertical bearing capacity, lateral stiffness and damping. The isolation layer should be stable under rare earthquake and should not be deformed irrecoverably. Under the rare earthquake action, the rubber bearing of the isolation layer should not have tensile stress.

(c) Total gravity above the isolation layer $G = 28{,}420\,\text{kN}$, $G_1 = 5000\,\text{kN}$, $G_2 = G_3 = G_4 = G_5 = 4840\,\text{kN}$, $G_6 = 4060\,\text{kN}$.

3. **Selection and arrangement of isolation bearing**

The axial force on each support is calculated from the superstructure. According to the corresponding requirements of the seismic code, the average compressive stress limit of the isolation bearing of class C building shall not be greater than 15 MPa, so the diameter of each bearing is determined (see Fig. 12.8 for the plane layout of the isolation device). Through trial calculation, 32 GZY350V5 lead core isolation bearings and 4 GZY350V5A lead core isolation bearings are selected. The basic parameters of lead core isolation bearing are as Table 12.19.

4. **Calculation of horizontal damping coefficient ψ**

In case of frequent earthquake, the shear deformation of the isolation bearing is 50% of the horizontal stiffness and the equivalent viscous damping ratio.

By Eq. (12.37): $K_\text{h} = \sum K_j = 1.55 \times 32 + 2.38 \times 4 = 59.1\,\text{kN/mm}$;

By Eq. (12.38): $\zeta_\text{eq} = \sum \frac{K_j \zeta_j}{K_\text{h}} = \frac{4\times2.38\times0.27+1.55\times32\times0.25}{59.1} = 25.32\%$;

Base period of structural isolation from Eq. (12.36) $T = 2\pi\sqrt{G/K_\text{h}\cdot g} = 2\pi\sqrt{28{,}420/59{,}100\cdot 9.8} = 1.3918\,\text{s}$.

According to the calculation formula in the "*Code for seismic design of buildings*", it is concluded that:

$$\eta_2 = 1 + \frac{0.05 - \zeta_\text{eq}}{0.06 + 1.7\zeta_\text{eq}} = 0.5857 > 0.55,\ \eta_2 = 0.5857,$$

$$\gamma = 0.9 + \frac{0.05 - \zeta_\text{eq}}{0.5 + 5\zeta_\text{eq}} = 0.7849$$

According to regulations, structural design characteristic period: $T_{gm} = 0.4\,\text{s}$.

Therefore, from Eq. (12.34): $\psi = \sqrt{2}\eta_2(\frac{T_{gm}}{T_1})^\gamma = \sqrt{2} \times 0.5857 \times (\frac{0.4}{1.3918})^{0.7849} = 0.3113$.

5. **Superstructure calculation**

(a) Standard value of horizontal seismic action

$$\alpha_1 = \psi\alpha_\text{max} = 0.3113 \times 0.16 = 0.0498;$$

Table 12.19 Basic parameters of lead core isolation bearing

Model	Bearing diameter (mm)	Total thickness of rubber layer (mm)	Design bearing capacity (kN)	Horizontal deformation (50%)		Horizontal deformation (250%)		Total	Second shape factor
				Horizontal stiffness (kN/mm)	Damping ratio (%)	Horizontal stiffness (kN/mm)	Damping ratio (%)		
GZY350V5	350	100.42	1400	1.55	25	0.84	10	32	5.15
GZY400V5A	400	102.58	1800	2.38	27	1.18	13	4	5.83

$$F_{EK} = \alpha_1 G = 0.0498 \times 28{,}420 = 1415.5\,\text{kN}$$

(b) The standard value of interlaminar shear force after isolation is calculated according to the following formula, $F_{ik} = \frac{G_i}{\sum_{k=1}^{n} G_k} F_{Ek}\quad (i = 1, \ldots, n)$.
See Fig. 12.9 for calculation diagram of horizontal seismic action and horizontal shear diagram of structure (Table 12.20).

(c) According to the requirements of the seismic code, the fortification intensity is 8° and the horizontal damping coefficient is not more than 0.5, so the vertical seismic action should be calculated (omitted).

6. **Checking calculation of horizontal displacement of isolation layer**

In case of rare earthquake, the shear stiffness and equivalent viscous damping ratio when the shear deformation of the isolation bearing is not less than 250%.

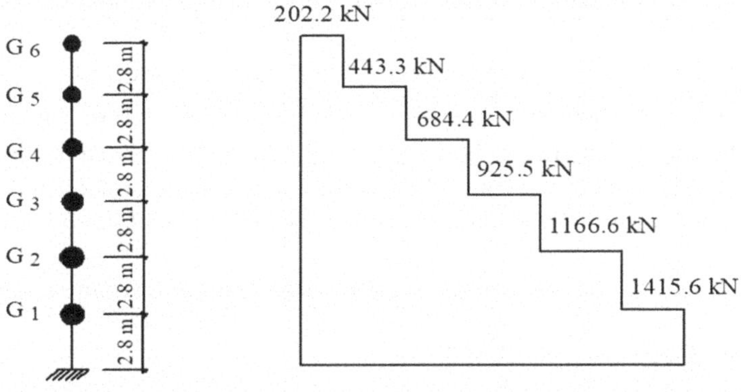

(a) Calculation diagram of horizontal seismic action of structure

(b) Distribution of horizontal shear force between floors of structure

Fig. 12.9 Calculation diagram of horizontal seismic action and distribution of horizontal shear force of structure

Table 12.20 Superstructure calculation results

Layer number	G_i (kN)	$\sum G_i$ (kN)	F_{EK} (kN)	F_{ik} (kN)	V_{ik} (kN)
6	4060	28,420	1415.5	202.2	202.2
5	4840			241.1	443.3
4	4840			241.1	684.4
3	4840			241.1	925.5
2	4840			241.1	1166.6
1	5000			249.0	1415.6

(a) Calculation of eccentricity of isolation layer e:
The structure and the isolation device are arranged symmetrically with eccentricity $e = 0$.

(b) Calculation of horizontal displacement at the center of mass of isolation layer:
According to site conditions, characteristic period $T_g = 0.35 + 0.05 = 0.4\,\text{s}$.
The horizontal dynamic stiffness of the isolation layer under rare earthquake action is as follows: $K'_{\text{h}} = \sum K'_j = 0.84 \times 32 + 1.18 \times 4 = 31.6\,\text{kN/mm}$
Equivalent viscous damping ratio of isolation layer under rare earthquake: $\zeta'_{eq} = \sum \frac{K'_j \zeta'_j}{K'_{\text{h}}} = 10.45\%$.
The basic period of structural isolation is obtained: $T'_1 = 2\pi\sqrt{G/K'_{\text{h}} \cdot g} = 2\pi\sqrt{28{,}420/31{,}600 \cdot 9.8} \doteq 1.9\,\text{s}$.
With the above parameters, it can be calculated: $\gamma = 0.9 + \frac{0.05-\zeta'_{\text{eq}}}{0.5+5\zeta'_{\text{eq}}} \doteq 0.85$, $\eta_1 = 0.02 + \frac{0.05-\zeta'_{\text{eq}}}{8} = 0.0132$, $\eta_2 = 1 + \frac{0.05-\zeta'_{\text{eq}}}{0.06+1.7\zeta'_{\text{eq}}} \doteq 0.77$.
The fortification intensity is 8°, rare earthquake, $\alpha_{\max} = 0.9$, then, $\alpha_1(\zeta'_{\text{eq}}) = \left[\eta_2 0.2^{\gamma} - \eta_1(T - 5T_{\text{g}})\right] \alpha_{\max} = \left[0.77 \times 0.2^{0.85} - 0.0132 \times (1.9 - 5 \times 0.4)\right] \times 0.9 = 0.1787$.
If $\lambda_{\text{s}} = 1.0$, then, $u_{\text{c}} = \frac{\lambda_{\text{s}}\alpha_1(\zeta'_{\text{eq}})G}{K'^{\text{h}}_{\text{h}}} = \frac{1.0 \times 0.1787 \times 28{,}420}{31.6} = 160.71\,\text{mm}$.

(c) Checking calculation of horizontal displacement (the most unfavorable bearing)
There is no eccentricity in the isolation layer of the project, and the opposite support is $\beta_i = 1.15$, so the horizontal displacement of the side support is:

$$u_i = \beta_i u_c = 1.15 \times 160.71 = 184.82\,\text{mm}.$$

(1) Checking bearing GZY350V5:

$$\begin{aligned}[\mu] &= \min\{0.55 \text{ times effective diameter, } 3 \text{ times of the total} \\ &\quad \text{thickness of each rubber layer of the support}\} \\ &= \min\{0.55 \times 350 = 192.5\,\text{mm}, \quad 100.42 \times 3 = 301.26\,\text{mm}\} \\ &= 192.5\,\text{mm}\end{aligned}$$

$\mu_i = 184.8\,\text{mm} < [\mu] = 192.5\,\text{mm}$. Therefore, the bearing deformation meets the requirements.

(2) Checking calculation of the right most upper corner support GZY400V5A:

$$\begin{aligned}[\mu] &= \min\{0.55 \text{ times effective diameter, } 3 \text{ times of the total} \\ &\quad \text{thickness of each rubber layer of the support}\} \\ &= \min\{0.55 \times 400 = 220\,\text{mm}, \quad 102.58 \times 3 = 307.74\,\text{mm}\} \\ &= 220\,\text{mm}\end{aligned}$$

$\mu_i = 184.8\,\text{mm} < [\mu] = 220\,\text{mm}$. Therefore, the bearing deformation meets the requirements.

7. **Calculation of the lower part of the isolation layer**

The shear force of each isolation bearing is distributed according to the horizontal stiffness.

(a) Calculation of horizontal shear force of isolation layer under rare earthquake: The horizontal shear force of masonry structure under rare earthquake is $V_c = \lambda_s \alpha_1(\zeta'_{eq})G = 1.0 \times 0.1787 \times 28{,}420 = 5078.7\,\text{kN}$.
(b) The total stiffness of the isolation layer is 31.6 kN/mm. The horizontal shear force of each GZY350V5 isolation pad is 135.0 kN; the horizontal shear force of each GZY400V5A isolation pad is 189.6 kN.

References

1. Chen, Xin. 2012. *Theoretical and experimental study on vibration control of high-rise steel chimneys under wind load*. Nanjing: Southeast University. (in Chinese).
2. Sun, Guangjun. 2010. *Research on analysis methods of random earthquake response and reliability of base isolation structures and seismic reduction structures*. Nanjing: Southeast University. (in Chinese).
3. Yang, Cantian, Linlin Xie, and Aiqun Li. 2019. Ground motion intensity measures for seismically isolated RC tall buildings. *Soil Dynamics and Earthquake Engineering* 125: 105727.
4. Zhang, Fuyou. 2003. *Research on reliability of base isolation and series and parallel isolation*. Nanjing: Southeast University. (in Chinese).
5. Mao, Lijun. 2004. *Research on building structures of sliding base-isolation*. Nanjing: Southeast University. (in Chinese).

Chapter 13
Intelligent Optimization Method of Building Structure Vibration Control

Abstract Intelligent optimization method of vibration control in building structure are introduced. General framework for intelligent optimization design of building structure are given. Basic calculation flow and method of intelligent optimization design of building structure based on comprehensive objective method and Pareto optimization, including genetic algorithm, pattern search and hybrid algorithm are discussed respectively.

13.1 General Framework for Intelligent Optimization Design of Building Structure

With the continuous development of structural vibration control technology, it is gradually applied to practical projects. How to make the structural design with damping devices achieve the optimal efficiency has become one of the focuses of designers and researchers. However, due to the high complexity of the building structure itself and the strong nonlinearity of the damping device, this problem is quite different from the general small-scale optimization problems in design variables, optimization objectives, calculation scale, solution spatial distribution, etc. Traditional optimization algorithms, such as linear programming, nonlinear programming, integer programming and dynamic programming, are more complex, have more constraints and have poor global optimization ability. They are generally only suitable for solving small-scale optimization problems, and are difficult to be used in the scheme optimization of the complex system of building structure damping system. In the mid-1950s, people got rid of the shackles of some classical mathematical programming methods and were inspired by the process of biological evolution. They proposed to adopt the structural characteristics, evolutionary laws, thinking structure and behavior mode of foraging process of simulating human, nature and other biological populations. According to the natural mechanism, they intuitively constructed the calculation model to solve the optimization problem. To a certain extent, it solves the complex problems of large space, nonlinear, global search, combinatorial optimization and so on. Subsequently, many intelligent algorithms continue to emerge, such as genetic algorithm, particle swarm optimization, simulated annealing algorithm and so on.

A. Li, *Vibration Control for Building Structures*, Springer Tracts in Civil Engineering, https://doi.org/10.1007/978-3-030-40790-2_13

In recent years, intelligent algorithm has been applied more and more in the field of civil engineering. It has made great progress in structural optimization design, sensor optimization, vibration control and other fields. The key to solve the optimization problem lies in the establishment of a reasonable optimization mathematical model and the selection of an appropriate and efficient optimization algorithm. A general optimization problem can be described as a design variable whose objective function is minimal (or maximal) under certain constraints. Its standard mathematical model can be written as follows:

$$\begin{aligned} \text{Seeking } X:\ & X = (x_1, x_2, \cdots, x_i, \cdots, x_n)^{\mathrm{T}} \\ min\quad F(X) &= min[f_1(x), f_2(x), \cdots, f_m(x)]^{\mathrm{T}} \\ \text{s.t.}\quad g_j(X) &\le 0 \quad (j = 1, 2, \cdots, m) \\ g_j(X) &= 0 \quad (j = 1, 2, \cdots, m) \end{aligned} \tag{13.1}$$

where, X is the design variable; x_i is the ith design parameter; $F(X)$ is the objective function; $g_j(X) \le 0$ is the inequality constraint; $g_j(X) = 0$ is the equality constraint. Aiming at the generalized optimization design of building structure vibration reduction, the author team has established the general framework of intelligent optimization design of building structure, as shown in Fig. 13.1:

(1) According to the characteristics of the actual engineering project of the building structure, combined with the performance objectives of the structure and non structure, the design objectives and constraints are determined to reflect the particularity of the design of the building structure and damping device.

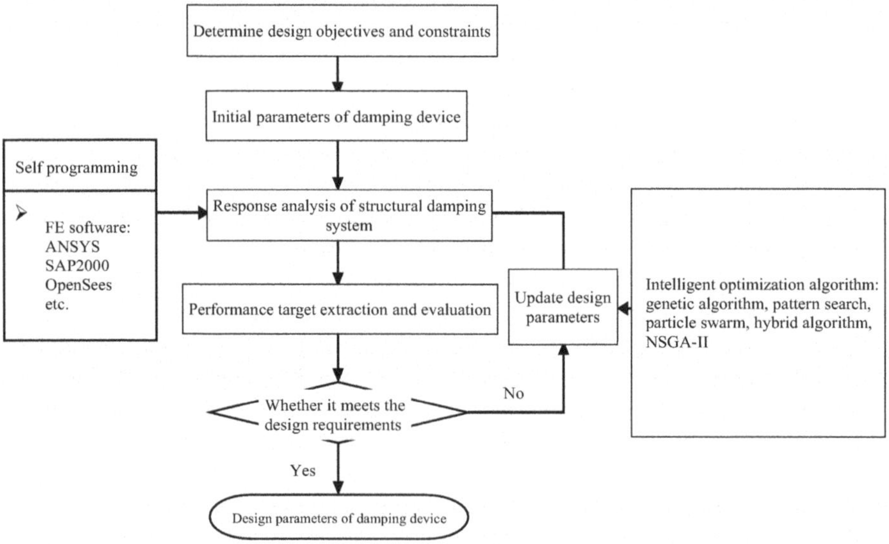

Fig. 13.1 General framework of intelligent optimal design of building structure

(2) According to the vibration reduction design method of the building structure introduced in Chap. 12, the design parameters of the vibration reduction device, such as installation position, quantity, performance parameters, etc., are preliminarily determined.
(3) In order to analyze the response of the vibration reduction system of the building structure under the dynamic load, the time-domain or frequency-domain analysis method can be selected according to the design objective, and the analysis can be carried out with the help of the self-made analysis program or professional finite element software.
(4) According to the pre-set performance objectives, evaluate the design results. If the requirements are met, the design parameters of the damping device will be obtained. If not, proceed to step (5).
(5) The optimization algorithm is used to update the design parameters, and the intelligent optimization algorithm (genetic algorithm, etc.) is introduced to guide the direction of rapid update of design parameters.
(6) Repeat steps (3) and (4) until satisfactory design parameters are obtained.

Based on the above framework, the author team developed the optimization design program of building structure vibration reduction, which includes the self-made structural vibration reduction analysis module, the finite element software API interface module, the structural performance evaluation module and the intelligent optimization algorithm module. Then, a series of research on intelligent optimization method of building structure vibration control is carried out.

The vibration control of building structure usually includes many requirements (such as displacement, velocity, acceleration, device output, realization cost, etc.), which are expressed in the optimization model as the existence of multiple optimization objectives, and ultimately summed up in the category of multi-objective decision-making in mathematics. The most significant characteristics of multi-objective decision-making are the incommensurability and contradiction between objectives. The commonly used method is to give weight to each objective and convert it into a single objective optimization problem to solve, that is, the so-called weight coefficient transformation method. However, in practice, with the increase of the number of objectives, the selection of specific weight values will become more and more difficult. At present, there are three ways to deal with multi-objective optimization problems: one is to search first and then make decisions (that is, to find the Pareto solution set of the problem first and then select the appropriate results according to the specific requirements), the other is to make decisions first and then search (such as the weight coefficient transformation method mentioned above), and the other is to search while making decisions (various interactive methods). In the following, some research results of the team in this field are introduced from the comprehensive objective method of first decision and then search and then decision-making and intelligent optimization design based on Pareto optimization.

13.2 Intelligent Optimization Design of Building Structure Based on Comprehensive Objective Method

13.2.1 Intelligent Optimization Design of Building Structure Based on Genetic Algorithm

13.2.1.1 Basic Principle of Genetic Algorithm

Genetic algorithm (GA) is a kind of random search algorithm proposed by Professor John Holland of the University of Michigan in 1975, which is based on the natural selection and natural genetic mechanism of biology. As early as the 1990s, it has been introduced into the structural optimization in the field of civil engineering. In the field of structural vibration control, GA can achieve good results in terms of the optimal design of the vibration reduction control system for seismic action and wind load.

The idea of GA is derived from Darwin's evolutionism, and it is a computational model to simulate the evolutionary process of genetic selection and natural elimination in the biological world. It takes all individuals in the population as objects, and uses randomization technology to search the encoded parameter space efficiently. Its basic genetic operations include selection, crossover and mutation. The basic unit of biological heredity is chromosome. Crossover operation is to exchange two chromosome segments, and mutation operation is to randomly change a chromosome segment. New chromosomes are produced by crossing and mutation between chromosomes, and those with strong adaptability survive while those with weak adaptability are eliminated. After many generations of genetic process, it can produce chromosomes with higher adaptability, and its process is very similar to the biological genetic process. The steps of basic genetic algorithm are shown in Fig. 13.2:

(1) Several individuals with the characteristics of the problem are randomly generated, each of which is formed by some encoding (binary encoding, decimal encoding, etc.);
(2) The fitness of the objective function of each individual in the group is evaluated and arranged in the order from the best individual to the worst individual;
(3) There are three ways to generate new offspring individuals: (A) two parents form offspring individuals by crossing; (B) parents form offspring individuals by mutation; (C) excellent parents copy directly to offspring individuals to ensure survival in iterative operation;
(4) Using the new population to carry out the next genetic iteration, and repeat step 2 and step 3 until the pre-set iteration stop conditions (iteration number, fitness function limit, running time, etc.) are reached;
(5) Return the best result in the current population.

From the above description of the principle and process, selection, crossover and mutation are the three most important operations of genetic algorithm which are different from the general algorithm. The so-called selection is to select the operation with strong individual adaptability in the genetic process. Because the selection is a

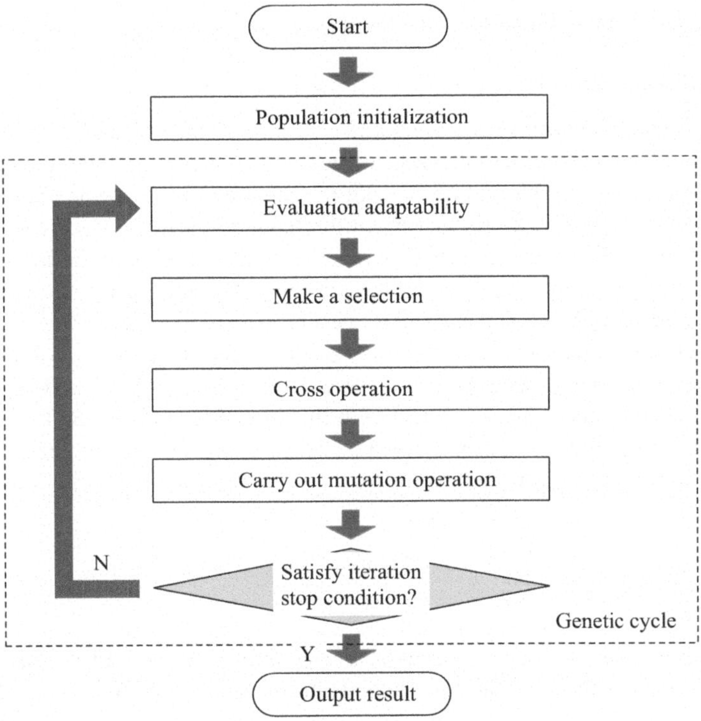

Fig. 13.2 Basic calculation flow of genetic algorithm

random process, the individuals with poor genes will not necessarily be eliminated, but their probability of being eliminated is relatively large, which is the same as the law in nature. There are many ways to choose, such as Roulette Wheel Selection, Stochastic Tournament, random League selection, and optimal reservation selection. Crossover is to randomly select two individuals from the current population according to a certain probability, randomly determine the location of gene exchange, and exchange a part of genes, thus forming two new individuals. As the next generation of chromosomes, there are also many methods, such as single point crossover, two point crossover, uniform crossover, cyclic crossover, sequential crossover, etc.

Variation is a process in which one of the individual genes changes with a certain probability. By introducing such a disturbance process, local extremum can be avoided. It is an important factor for organisms to present individual differences and evolution. Common methods include basic mutation, probability self-adjusting mutation, uniform mutation, effective gene mutation, etc.

13.2.1.2 Intelligent Optimization Design

(1) **Optimization model**

Taking a large-span maintenance hangar (Fig. 13.3a) as the object, based on the seismic response characteristics of this kind of structure, this paper studies the reasonable passive control technology and carries out the location optimization analysis of the damping device [1]. The roof structure of the hangar is a space grid structure, with a span of 176 + 176 m, a depth of 110 m and a height of 30 m at the bottom chord. The roof structure adopts the scheme of combining the inclined pyramid three-layer steel pipe grid with the steel truss at the opening side of the gate.

The energy dissipation support is arranged between the support columns of the hangar. One end of the support is connected to the lower chord node of the grid structure, and the other end is connected to the embedded part of the lower support column. If the number of dampers is set as m, x_j represents the jth gene value, and the constraints can be expressed as follows:

$$\sum_{j=1}^{p} x_j = x_1 + x_2 + \cdots + x_p = m \tag{13.2}$$

In order to consider the constraints, the penalty term should be added to the fitness function to transform the constrained optimization problem into unconstrained optimization problem.

$$F'(x) = \begin{cases} F(x) & x \text{ meet constraints} \\ F(x) - P(x) & x \text{ constraints not met} \end{cases} \tag{13.3}$$

In order to achieve the established damping effect and improve the damping efficiency of energy dissipation support, this paper attempts to consider the adaptability function of the number of dampers. The improved adaptability function is as follows:

$$J_i(p_i, q_i) = \frac{p_i(x, y) + \alpha q_i(x)}{1 + \alpha} \tag{13.4}$$

Among them, $p_i(x, y) = \frac{E_i(x,y)}{E_0} \times 100\%$, it refers to the numerical ratio of a response index after and before the structural vibration reduction. The higher the value is, the worse the energy consumption effect is. On the contrary, the better the energy consumption effect is. $q_i(x) = \frac{N_i(x)}{N_{Total}} \times 100\%$, it refers to the percentage of the number of dampers arranged by an individual in the total position of the dampers that can be arranged. The higher the value is, the more the number of dampers is, the higher the cost is, and the worse the economy is. This function considers the energy dissipation effect and the number of dampers into the fitness function. α is the weight coefficient to adjust the relationship between them.

In the construction of the fitness function, considering that there is a big difference between the energy consumption of dampers and the number of dampers, we first

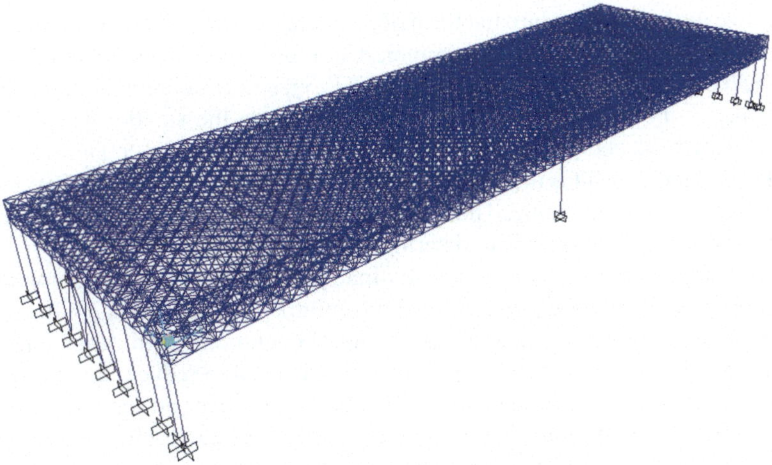

(a) Finite element analysis model

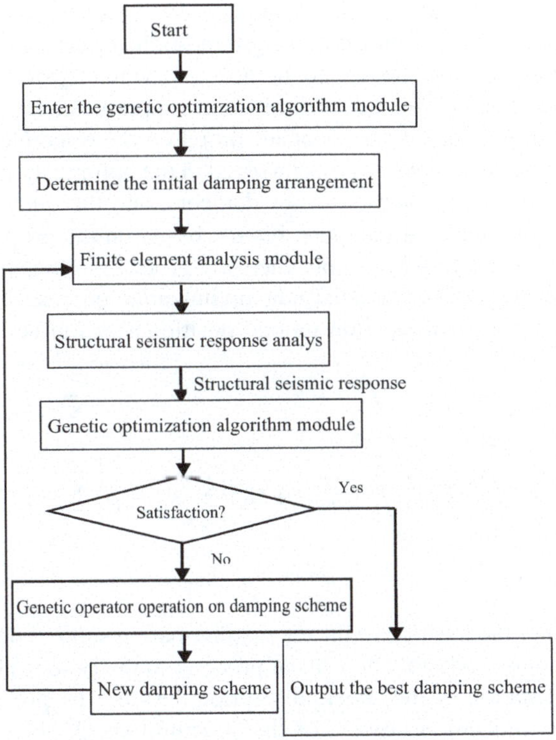

(b) Optimization process of vibration reduction

Fig. 13.3 Optimal design of viscous fluid damping and damping for large span hangar standardize the two factors into the form of percentage, so that they can interact in the same function. In the determined structure, p_i is a function of the number of dampers x and the location of dampersy, expressed as $p_i(x,y)$; $q_i(x)$ is a function of the number of dampers x, increasing with the increase of x. Since the smaller the $p_i(x,y)$ and $q_i(x)$, the better the energy consumption effect and the better the economy, it can be considered that the result is the best when the objective function of the above formula is taken as the minimum value. Therefore, the key to the optimal synchronization of damper position and number is to determine the appropriate value of α.

If there are N positions of energy dissipation support between the columns of hangar structure, the search space of location optimization analysis is 2^N. In addition, the hangar structure is huge, and the workload of operation optimization analysis is huge. In this paper, the method of group coding is used to reduce the optimization space reasonably. The specific method is as follows: group the arrangement positions of energy dissipation supports between the columns of hangar, each group has several dampers. When the corresponding position of this group is 1 in the genetic code, it means that there are dampers in these positions, and the value is 0, it means that there are no dampers in these positions. When grouping the damper positions, since the hangar structure is about the axial symmetry, the symmetrical positions on both sides of the central axis are grouped into a group, so the optimization results ensure that the final damper arrangement is about the axial symmetry of the structure.

Firstly, the locations of the energy dissipation supports among the columns that can be arranged in the hangar are grouped. Based on the principle of symmetrical and uniform arrangement, the positions involved in the optimization are divided into 15 groups, including two groups with six dampers, and the other 13 groups with four dampers in total have 64 dampers. Firstly, based on the proportion of energy dissipation of damper to the whole input energy of ground motion, the corresponding fitness function is constructed to satisfy the optimization process. The optimization fitness function of energy dissipation support position is as follows:

$$g(X) = \begin{cases} 1 - \sum\limits_{j=1}^{m} E_j x_j \Big/ E_{input} & \sum\limits_{j=1}^{m} x_j = p \\ 1 - \sum\limits_{j=1}^{m} E_j x_j \Big/ E_{input} + 0.5\left|\sum\limits_{j=1}^{m} x_j - p\right| & \sum\limits_{j=1}^{m} x_j \neq p \end{cases} \quad (13.5)$$

(2) **Optimal design**

The optimization of the location of the energy dissipation support between the long-span hangar columns is a highly non-linear process, so the genetic algorithm is used to optimize the location of the shock absorption device. The process is shown in Fig. 13.3b: the optimization process of the structural shock absorption using the improved genetic algorithm includes the structural seismic response analysis and the genetic algorithm optimization. In this paper, based on the use of large-scale general finite element analysis software to analyze the seismic response of the hangar structure, through the optimization program to call the finite element analysis results,

Table 13.1 Optimization results of energy dissipation support position in hangar

Damper arrangement position limit	Number of dampers	Optimal placement position vector	Energy consumption rate (%)
1	4	[0,0,0,0,0,0,0,0,1,0,0,0,0,0,0]	10.91
2	10	[1,0,0,0,0,0,0,0,1,0,0,0,0,0,0]	18.54
3	14	[1,0,0,0,0,0,0,1,1,0,0,0,0,0,0]	24.66
4	18	[1,0,0,0,0,0,0,1,1,1,0,0,0,0,0]	30.37
5	22	[1,0,0,0,0,0,1,1,1,1,0,0,0,0,0]	35.04
6	26	[1,0,0,0,0,0,1,1,1,1,0,0,1,0,0]	38.95
7	30	[1,0,0,0,0,0,1,1,1,1,0,0,1,1,0]	43.66
8	34	[1,0,0,0,0,0,1,1,1,1,1,0,1,1,0]	45.58
9	38	[1,0,0,0,0,0,1,1,1,1,1,1,1,1,0]	47.85
10	44	[1,0,0,0,0,0,1,1,1,1,1,1,1,1,1]	49.97
11	48	[1,0,0,1,0,0,1,1,1,1,1,1,1,1,1]	51.36
12	52	[1,0,0,1,0,1,1,1,1,1,1,1,1,1,1]	52.46
13	56	[1,0,0,1,1,1,1,1,1,1,1,1,1,1,1]	53.20
14	60	[1,0,1,1,1,1,1,1,1,1,1,1,1,1,1]	54.06
15	64	[1,0,0,1,1,1,1,1,1,1,1,1,1,1,1]	54.66

so as to achieve the interaction between the optimization program and the finite element program. The binary coding based on {0, 1} symbol set is the basic coding method of genetic algorithm. In this optimization model, the location parameters of damping device can be binary coded. If there are p positions where the damping devices can be arranged in the damping structure, the corresponding code length is p. The value of the J gene is 1, which means that the damping device is set at the corresponding structural position of the gene; on the contrary, 0 means that the damping device is not set at the corresponding position. All results of optimization of energy dissipation support position of hangar are listed in Table 13.1.

On the basis of the above analysis, the penalty function which limits the number of dampers is cancelled, and the fitness function is modified according to Eq. (13.4), and the arrangement scheme of energy dissipation support is optimized with different α values. See Table 13.2 for some results.

The energy consumption rate of a single damper can be obtained by dividing the total energy consumption rate and the number of dampers. The larger the value is, the higher the efficiency of the damper will be. The relation curve between the energy consumption rate of a single damper and the weight coefficient α of the number of dampers is shown in Fig. 13.4.

The value of α determines whether the optimization results pay more attention to the overall effect or the total number of dampers. In order to better analyze the reasonable value range of α value, divide the total energy consumption rate by the total number of dampers to get the energy consumption rate of a single damper. It can

Table 13.2 Optimization results of energy dissipation support location considering economic factors in hangar

Weight coefficient of damper number α	Number of dampers (unit)	Optimal location vector of damper	Energy consumption rate (%)	Energy consumption rate of single damper (%)
0.05	64	[1,1,1,1,1,1,1,1,1,1,1,1,1,1,1,1]	54.66	0.854
0.1	60	[1,0,1,1,1,1,1,1,1,1,1,1,1,1,1,1]	54.06	0.901
0.2	48	[1,0,0,1,0,0,1,1,1,1,1,1,1,1,1,1]	51.36	1.070
0.3	44	[1,0,0,0,0,0,1,1,1,1,1,1,1,1,1,1]	49.97	1.136
0.4	34	[1,0,0,0,0,0,1,1,1,1,1,0,1,1,1,0]	45.58	1.341
0.5	30	[1,0,0,0,0,0,1,1,1,1,0,0,1,1,1,0]	42.66	1.422
0.6	22	[1,0,0,0,0,0,1,1,1,1,0,0,0,0,0,0]	35.04	1.593
0.7	22	[1,0,0,0,0,0,1,1,1,1,0,0,0,0,0,0]	35.04	1.593
0.8	18	[1,0,0,0,0,0,0,1,1,1,0,0,0,0,0,0]	30.37	1.687
0.9	14	[1,0,0,0,0,0,0,1,1,0,0,0,0,0,0,0]	25.66	1.833
1.0	10	[1,0,0,0,0,0,0,0,1,0,0,0,0,0,0,0]	19.54	1.954
1.1	10	[1,0,0,0,0,0,0,0,1,0,0,0,0,0,0,0]	19.54	1.954
1.2	4	[0,0,0,0,0,0,0,0,1,0,0,0,0,0,0,0]	10.91	2.728
1.3	4	[0,0,0,0,0,0,0,0,1,0,0,0,0,0,0,0]	10.91	2.728
1.4	4	[0,0,0,0,0,0,0,0,1,0,0,0,0,0,0,0]	10.91	2.728
1.5	4	[0,0,0,0,0,0,0,0,1,0,0,0,0,0,0,0]	10.91	2.728
1.6	4	[0,0,0,0,0,0,0,0,1,0,0,0,0,0,0,0]	10.91	2.728

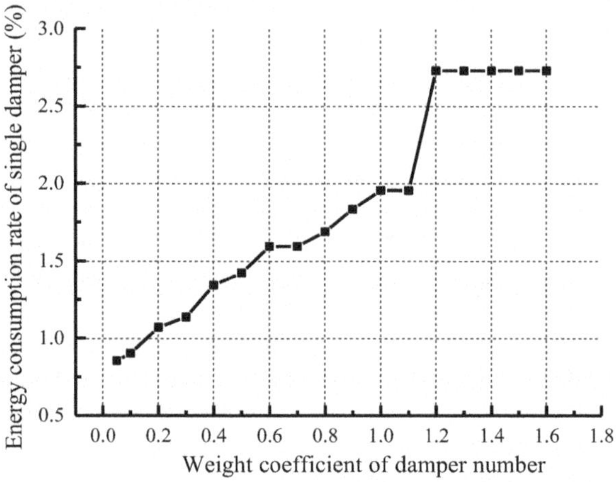

Fig. 13.4 The curve of energy dissipation rate of a single damper with the weight coefficient α of damper quantity

be seen that when α exceeds 1.2, the energy consumption ratio of a single damper hardly changes, but when α increases from 1.1 to 1.2, the energy consumption ratio of a single damper increases sharply, which shows that the number of dampers in this value range is very sensitive to the change of α value. When α is less than 1.0, the energy consumption ratio of a single damper changes linearly with the increase or decrease of α. It can be seen that, on the premise of determining the overall damping effect target, the energy consumption efficiency of a single damper can be improved by introducing the number of dampers into the fitness function.

13.2.2 Intelligent Optimization Design of Building Structure Based on Pattern Search

13.2.2.1 Basic Principle of Pattern Search Algorithm

Pattern search algorithm is a special subset family of direct search algorithm. The basic idea is to extract the value of objective function from the special direction set. By comparing the size of these objective functions, we can find the direction of descent and solve the optimization problem. The basic process is as follows [1, 2]:

(1) Determine the base point and pattern vector set $\{V_i\}$;
(2) According to the pattern vector set, the grid is generated, that is, the position of each time point searched by PS algorithm, thus forming a grid array in space;
(3) The grid points are voted to find the minimum points according to the basic points of the current iteration and the newly generated point set as the base point of the next iteration;
(4) Define the expansion factor $\alpha_e(>1)$ and reduction factor $\alpha_c(<1)$. When the voting is successful, expand the grid scale, and enter step (5). When the voting fails, reduce the grid scale, and return to step (3);
(5) Meet the convergence requirements, output the results; if not, return to step (2), where the stop conditions of the pattern search algorithm mainly include: grid size tolerance, maximum number of iterations, total evaluation times of the objective function, distance between the two elected points (variable tolerance), change of the objective function (function tolerance), etc.

13.2.2.2 Intelligent Optimization Design

(1) **Optimization model**

In the optimal design of TLCD control for a self-supporting high-rise structure, the geometric parameters (such as pipe diameter, device radius, liquid column height, etc.) and performance parameters (such as vibration frequency, damping, etc.) of the device are mainly involved. According to the calculation model of the ring TLCD, there is a unique corresponding relationship between the geometric parameters and performance parameters, which can be directly optimized by using geometric parameters. Therefore, in view of The multi-objective satisfactory optimization model for this problem is as follows:

$$\begin{aligned} &\text{Seeking } X \quad X = [R_d, r, \xi_L, H_w]^{\mathrm{T}} \\ &min \quad S(X) \\ &\text{s.t.} \quad g_j(X) \le 0 \qquad (j = 1, 2, \cdots, m) \\ &\qquad g_j(X) = 0 \qquad (j = 1, 2, \cdots, m) \\ &\qquad X_1 \le X \le X_2 \end{aligned} \tag{13.6}$$

where, X is the design variable; $S(X)$ is the composite satisfaction function; R_d is the TLCD ring radius; $g_j(X) \le 0$ is the inequality constraint condition, $g_j(X) = 0$ is the equality constraint condition.

In this example, under the action of wind load, the influence of structural stress on the safety of high-rise steel chimney is mainly considered, which is directly related to its vertex displacement. The velocity term and acceleration term are more related to the comfort degree of general building structure. Therefore, the displacement of high-rise steel chimney is one of the optimization objectives in this paper. In addition, considering the rationality and economy of the application of ring TLCD in practical engineering, the smaller the quality of the device, the smaller its cost and the smaller the impact on the main structure, the mass ratio is another optimization objective; the shape of the device has a certain impact on the size and characteristics of the wind load on the main structure, so the smaller the windward area is, the better, so it is considered as the third optimization Goals. Based on the above reasons, the objective function is defined as follows:

$$\begin{cases} f_1(X) = 1 - |x_{n,w}|_{\mathrm{std}} / |x_{n,wo}|_{\mathrm{std}} \\ f_2(X) = 10 m_{tlcd} / M_{s,gen} \\ f_3(X) = 2(R_d + r)(H_w + r + w_{max})/10 - 1 \end{cases} \tag{13.7}$$

Among them, $|x_{n,wo}|_{\mathrm{std}}$ is the variance of displacement at the top of the original structure; $|x_{n,w}|_{\mathrm{std}}$ is the variance of displacement at the top of the damping structure; w_{max} is the maximum displacement of the liquid column away from the equilibrium position; m_{tld} is the liquid mass of the ring TLD, $M_{s,gen}$ is the first-order modal mass of the structure. It can be seen that $f_1(X)$ is the attenuation rate of the displacement at the top of the structure, which reflects the control objective of the response of the structure. The larger the value is, the higher the satisfaction is. The parameter $a = -20$, $b = 1$ in the Sigmoid function is selected. $f_2(X)$ is the ratio of modal mass ratio to 0.1, reflecting the influence of device mass on the structure. 0.1 refers to the actual application of tuned mass damper (TMD) in the project, and the modal mass ratio is 0.1; the smaller the function value is, the higher the satisfaction is, and the parameter a = 20, b = 1 in the sigmoid function is selected. $f_3(X)$ is the normalized value of

Table 13.3 Optimization model constraints

Number	Expression	Explain
1	$R_d + r < 1.5R_{s,out}$	Control the maximum outer diameter of TLCD
2	$R_d - r > R_{s,out} + 0.005$	Control TLCD minimum outside diameter
3	$0 < r < 0.1$	Restricted diameter
4	$H_w - 2r > 0$	Define liquid column height greater than 0
5	$0 < \xi_L < 18.1$	When the damping hole area ratio is 0.55, the liquid damping coefficient is 18.1

the windward area of the device, reflecting the influence of the size of the device on the wind load, of which 10 is the expected smaller windward area; the smaller the value of the function, the higher the satisfaction, select the parameter a = 20, b = 1 in the Sigmoid function.

$$\begin{cases} S_1 = \frac{1}{1+\exp(-20(f_1/f_{10}-1))} \\ S_2 = \frac{1}{1+\exp(20(f_2/f_{20}-1))} \\ S_3 = \frac{1}{1+\exp(20(f_3/f_{30}-1))} \end{cases} \quad (13.8)$$

Among them, S_1, S_2 and S_3 are independent satisfaction functions of f_1,f_2 and f_3 respectively; f_{10},f_{20} and f_{30} are thresholds of three objective functions respectively. The linear weighting method is used to obtain the composite satisfaction of the problem

$$S = w_1 S_1 + w_2 S_2 + w_3 S_3 \quad (13.9)$$

Among them, w_1, w_2 and w_3 are the weight coefficients of three satisfaction functions, and the sum of them is equal to 1.

Considering that the ring TLCD is designed in accordance with the shape of the structure, items 1–3 in Table 13.3 constrain the geometry of TLCD; considering that TLCD can achieve the expected performance, item 4 in Table 13.3 limits the height of liquid column; considering that too small damping holes will reduce the ability of liquid sloshing in TLCD, item 5 in Table 13.3 controls that the area ratio of damping holes in TLCD is less than 0.55. The constraints of the optimization model are shown in Table 13.3, where $R_{s,out}$t is the outer diameter of the structure.

(2) **Optimal design**

The threshold value of objective function is 0.4, 0.1 and 0.05 respectively, and the combination of weight coefficient is [1/3 1/3 1/3]. In the optimization design, the maximum number of iterations is 30, the grid acceleration technology is used in pattern search, and the convergence stop tolerance is set as 1e−3 (too small tolerance is of little significance in practical engineering applications, and the number of reanalysis iterations is wasted).

The process of pattern search is shown in Fig. 13.5a. It can be seen that: (1) the satisfactory optimization of TLCD of self-supporting high-rise steel structure with pattern search algorithm converges fast, and the optimal solution can be obtained only by about 5 iterations; (2) combining with the analysis in Fig. 13.5, even if the weights are different, the calculation can converge within 10 times for this example; (3) This method combines the characteristics of fast convergence in the early stage of pattern search algorithm and relatively lower requirements of satisfaction for global optimization. Compared with other algorithms with general function as the goal, it improves the iteration speed relatively. Figure 13.5b–d respectively give the displacement time history, acceleration time history and displacement power spectral density when taking the optimal value. At this time, (1) the optimal parameters R_d, r, ξ_L and H_w are respectively 1.216 m, 0.061 m, 3.311 m and 2.167 m, and the frequency ratio of TLCD to structure is 1.01; (2) the objective functions f_1, f_2 and f_3 are respectively 0.392, 0.091 and 0.037, which basically meet the threshold requirements. At the same time, the acceleration is basically 0.392, 0.091 and 0.037. The damping ratio of the structure increases from 0.003 to 0.01, and the first natural frequency changes from 0.293 to 0.292. It can be seen that the additional equivalent damping ratio of the device to the structure is 0.007. At the same time, only considering the

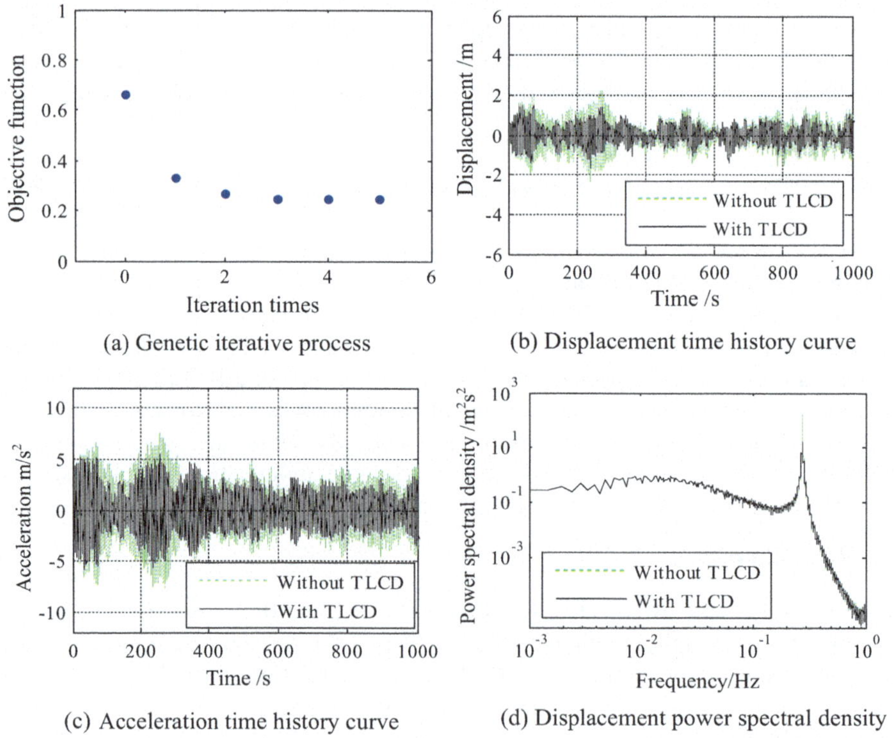

(a) Genetic iterative process

(b) Displacement time history curve

(c) Acceleration time history curve

(d) Displacement power spectral density

Fig. 13.5 Optimal design of ring TLCD

weight of liquid has little effect on the dynamic characteristics of the main structure; (3) The method proposed in this paper can be used to calculate the optimal solution of TLCD parameters and control the corresponding objective function to meet the limit value set in engineering design.

13.2.3 Intelligent Optimization Design of Building Structure Based on Hybrid Algorithm

13.2.3.1 Basic Principle of Hybrid Algorithm

This kind of intelligent algorithm, genetic algorithm, has high global convergence in the early stage and fast convergence speed, but it is easy to fall into local optimum in the later stage, and the convergence speed is slow, which leads to a great increase in the number of reanalysis. In order to solve this problem, it is an important direction in the field of engineering optimization to combine two or more algorithms, make full use of their advantages and overcome their shortcomings. Genetic Algorithm (GA) has better global convergence, but the convergence speed is slow in the later stage; Pattern Search (PS) has faster convergence speed, but it is easy to fall into local optimum. Therefore, we can consider the combination of the two, learn from each other, so as to form a hybrid optimization algorithm that takes into account both the optimal solution and the search efficiency, so as to achieve better performance in both the calculation efficiency and the optimization result [3].

The basic principle and process of the two algorithms are analyzed, and the algorithm is combined with two ideas: one is to use genetic algorithm for global optimization in the early stage of optimization, and use pattern search algorithm for local optimization in the later stage. In fact, for the optimization problem of general vibration control, the early genetic search can determine the scope of global optimization, and a large number of repeated global optimization in the later stage largely Therefore, in the later stage, using pattern search algorithm in the global optimal range for local optimization can improve the calculation efficiency on the premise of obtaining the global optimal solution; The other is to use genetic algorithm in the grid optimization of pattern search algorithm. In this way, genetic algorithm is used in every iteration of pattern search algorithm, which ensures the global optimization of the direction of the pattern search process and ensures that the optimal solution of pattern search is the global optimization. The basic flow of the two hybrid optimization algorithms is shown in Fig. 13.6a. Hybrid Algorithm A (HA) corresponds to the first hybrid method, which is a serial combination method; Hybrid Algorithm B (HB) corresponds to the second hybrid method, which is a method of inserting genetic algorithm as a parameter search in the grid point voting process of the pattern search algorithm.

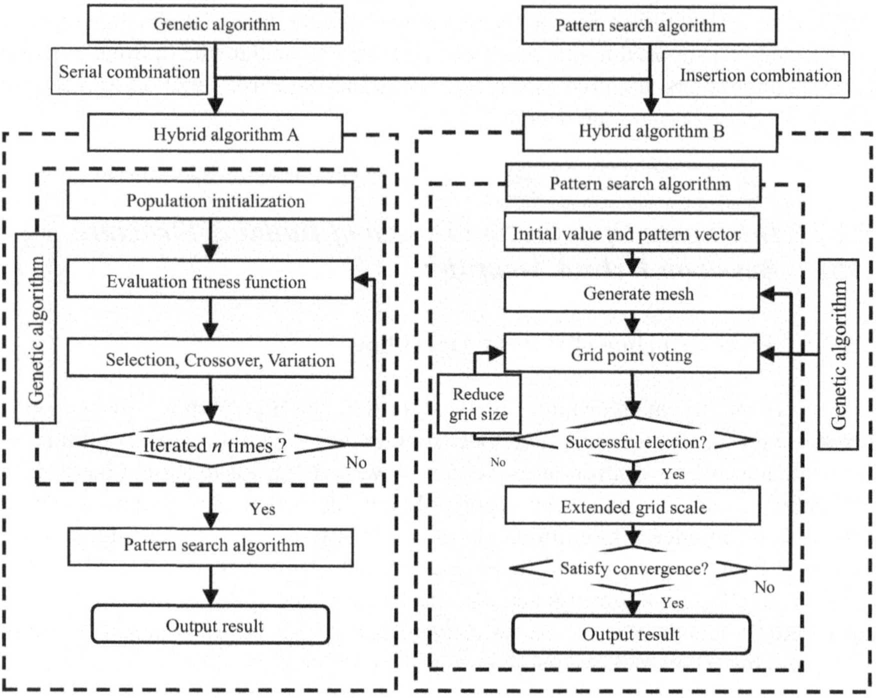

(a) Basic flow of hybrid algorithm

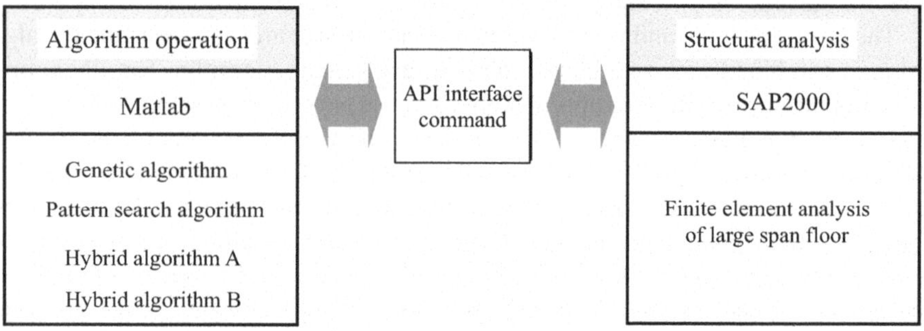

(b) The implementation of the optimal search process

Fig. 13.6 Basic principle and implementation of hybrid algorithm

13.2.3.2 Intelligent Optimization Design

(1) **Optimization model**

A commercial mezzanine (Fig. 13.7a) is designed at 17 m above the carriageway and landing platform (elevation 9 m) of a high-speed railway station building. The mezzanine floor adopts steel truss structure system, the main beam and secondary

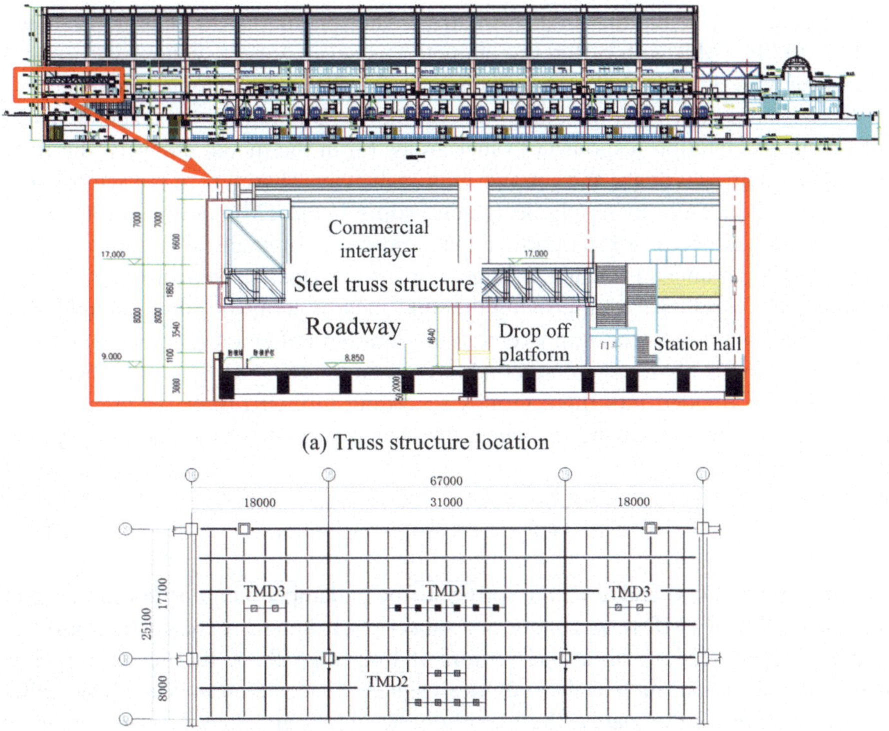

(a) Truss structure location

(b) Structure layout and MTMD layout

Fig. 13.7 A high-speed railway station building business platform

beam are steel pipe truss, and the floor is profiled steel composite floor. As shown in Fig. 13.7b, the platform is 67 m in longitudinal direction and 25.1 m in transverse direction. The longitudinal ends are supported on the columns on both sides. Four steel reinforced concrete columns are arranged at the edge and middle of the floor. The floor has the characteristics of large span, irregular structure layout, small stiffness and damping, and it is easy to gather large density of people to form large crowd load according to the requirements of building function. Therefore, it is necessary to analyze the floor vibration under the action of pedestrian load and investigate its normal performance.

TMD is usually arranged at the place with larger vibration mode of control mode, and the number is selected according to the influence of total mass of TMD on the static displacement of structure. According to the above analysis of modal and acceleration response, TMD is determined to be arranged in the three areas shown in Fig. 13.7b; the total mass of floor structure is 3.46×106 kg, and the total mass of TMD is controlled to be no more than 0.5% to reduce its impact on the performance of the structure, 0.46%. Therefore, considering the symmetry of the structure and the convenience of the actual production, 16 TMDs are set up, each TMD has a mass of

1000 kg, and the actual mass ratio is 0.46%. The specific layout scheme is shown in Fig. 13.7b. TMD with different parameters is set in zone 1, zone 2 and zone 3 respectively to form the MTMD damping system of the whole floor structure.

Because the floor is an integral structure, the TMD parameters in each area will affect the acceleration response of other areas. From the modal analysis, the floor modal distribution is dense and coupled with each other. Therefore, it is obviously not the optimal parameter to simply design according to the TMD optimization design formula, so it is necessary to optimize the parameters. In practical engineering, too many TMD parameters are not conducive to its production and installation, and although it is conducive to improving the robustness in terms of vibration reduction effect, there is no essential improvement in vibration reduction efficiency, and less TMD grouping is obviously more conducive to engineering application. Therefore, 16 TMDs are divided into three groups according to their respective areas. If the parameters of each group are the same, the design variable of the damping system is:

$$X = [k_1, C_1, \alpha_1, k_2, C_2, \alpha_2, k_3, C_3, \alpha_3]^{\mathrm{T}} \tag{13.10}$$

Among them, k, C and α are the spring stiffness, damper damping coefficient and damping index of TMD respectively, and subscripts 1, 2 and 3 represent their groups.

For the problem of vibration comfort of long-span floor, the design target is usually acceleration, such as the root mean square of acceleration or the peak value of acceleration. In this paper, the maximum acceleration values of nodes in areas 1, 2 and 3 are selected as the design objectives. Considering that they are both comfort indicators, each objective has the same importance, and the weight is taken as 1/3:1/3:1/3. Therefore, the objective function of optimization is defined as:

$$f_1(X) = \sum_{i=1}^{3} \beta_i \left(\frac{a_{i,w}}{a_{i,wo}} \right) = \frac{1}{3} \sum_{i=1}^{3} \frac{a_{i,w}}{a_{i,wo}} \tag{13.11}$$

where, $a_{i,w}$ and $a_{i,wo}$ are the peak acceleration of the node with the largest acceleration in area i when MTMD is installed and when MTMD is not installed, respectively; β_i are the weight coefficients.

To sum up, the optimization model of long-span floor in this paper can be expressed as follows:

$$\begin{aligned} &\text{Seeking } X \quad X = [k_1, C_1, \alpha_1, k_2, C_2, \alpha_2, k_3, C_3, \alpha_3]^{\mathrm{T}} \\ &min \quad f(X) \\ &\text{s.t.} \quad g_j(X) \leq 0 \quad (j = 1, 2, \cdots, m) \\ &\qquad g_j(X) = 0 \quad (j = 1, 2, \cdots, m) \\ &\qquad X_1 \leq X \leq X_2 \end{aligned} \tag{13.12}$$

where, $g_j(X) \leq 0$ is the inequality constraint; $g_j(X) = 0$ is the equality constraint. The acceleration response is obtained by the finite element model.

According to the above optimization algorithm and optimization model, the optimization design method of MTMD vibration reduction of complex long-span floor under pedestrian load is established. The specific steps are as follows:

(1) The structural analysis model is established for modal analysis, pedestrian load simulation and dynamic response analysis. The finite element software SAP2000 is used for this step;
(2) According to the results of modal analysis and the law of acceleration response distribution, combined with the actual factors such as floor self weight, fabrication and installation, the number and location of TMD are determined. This step is mainly realized by expert experience and theoretical calculation;
(3) The TMD is grouped and the mathematical model of the optimization problem is established;
(4) The optimization algorithm described in Sect. 13.2.1 is used to search the optimal variables of TMD and obtain the optimal design parameters, as shown in Fig. 13.6b. SAP2000 and MATLAB are used to realize this step.

(2) **Optimal design**

Based on four kinds of optimization algorithms (GA, PS, HA, HB), the optimization design is realized by programming, and the advantages and disadvantages of the two hybrid optimization algorithms are examined from two aspects of optimization ability and calculation efficiency. Figure 13.8 shows the optimal value, time-consuming and times of reanalysis of the optimization problem with Eq. (13.11) as the goal after running four algorithms, among which the basic parameters are set in Table 13.4. For comparison, in the hybrid algorithm, except the number of iterations and population, other parameters are the same as GA and PS. It can be seen from the comparison that: (1) the optimal value obtained by HA and HB algorithm is the smallest, and

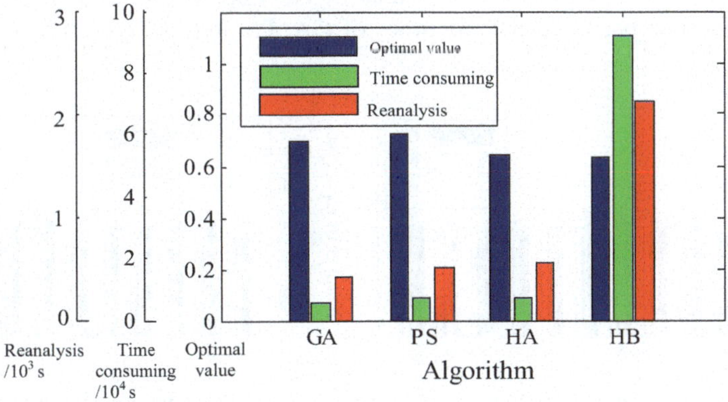

Fig. 13.8 Performance comparison of four algorithms

Table 13.4 TMD parameters of each optimization scheme

Programme	Position	Damping coefficient N/(m/s)$^{\alpha}$	Damping index	Rigidity (N/m)
STMD	whole	801.04	0.80	1,606,562.84
MTMD1	Area 1	5351.72	0.96	795,775.50
	Area 2	800.00	0.93	1,723,013.88
	Area 3	800.00	0.43	1,062,692.00
MTMD2	Area 1	13,205.03	1.00	1,008,436.29
	Area 2	7129.05	1.00	1,602,940.89
	Area 3	16,304.92	1.00	1,629,583.86

the value obtained by PS algorithm is the largest; (2) the number of reanalysis is mainly determined by the set parameters, but HB algorithm needs to call GA to search for each sub step, so the number of reanalysis required is far greater than other algorithms, and the calculation time is relatively large; (3) the optimal results of HA and HB algorithm are similar, but the consumption of HA is relatively large The time is far less than HB, so the HA algorithm is more suitable for the structure with long time of single analysis, such as complex long-span floor.

The analysis results of the intelligent algorithm are usually closely related to the initial value selection, parameter setting and other aspects, and the results of each analysis are discrete. When GA, PS and HA are used to optimize the stability of MTMD design of long-span floor, the number of reanalysis of the three algorithms is about 600 by adjusting the parameters, and the stability of the optimal results and analysis time is compared. The analysis results are shown in Fig. 13.9: (1) Fig. 13.9a gives the optimal value results calculated by each optimization method 10 times. The average values of GA, PS and ha algorithm are 0.699, 0.720 and 0.653 respectively, and the standard deviation is 0.013, 0.027 and 0.019 respectively. It can be seen that the results of HA algorithm are the best, the stability is equivalent to GA, and the results and stability of PS algorithm are poor; (2) Fig. 13.9b shows the number of reanalysis times calculated by each optimization method 10 times. The average number of reanalysis times of GA, PS and ha algorithm is 620, 586

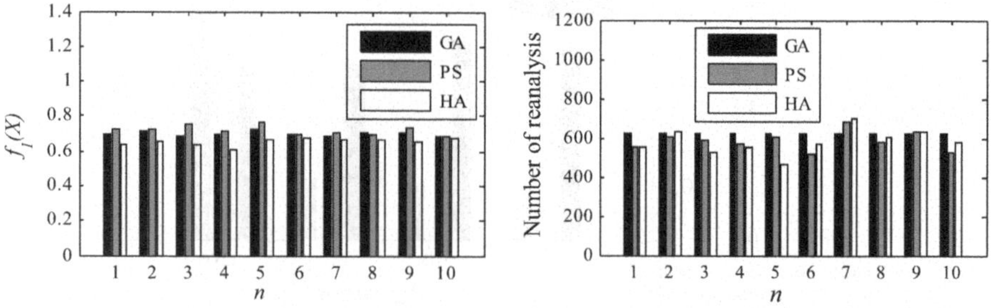

(a) Optimal value (b) Number of reanalysis

Fig. 13.9 Algorithm optimization stability

and 580 respectively, and the standard deviation is 0, 49 and 64 respectively. Due to the principle of the algorithm, GA algorithm takes the most time, while HA algorithm takes the least time; (3) in general, ha algorithm can get relatively better optimization results in less computation time, which improves the efficiency of GA and PS algorithm to a certain extent.

The working TMD is often limited by its space of use and the stroke of built-in damper. When it is subjected to crowd load for a long time, the strength and fatigue life of TMD spring is also one of the important factors to be considered in the design, and these factors can be attributed to the stroke of TMD. Therefore, it is necessary to take the TMD stroke as one of the design objectives when designing the MTMD of long-span composite floor. Based on this, the normalized TMD stroke term is added to Eq. (13.11) to get the following formula:

$$f_2(X) = \sum_{i=1}^{3} \beta_i \left(\frac{a_{i,w}}{a_{i,wo}} \right) + \beta_4 \left(\frac{d_{TMD,max}}{0.2} \right) \tag{13.13}$$

where, $d_{TMD,max}$ is the maximum displacement of TMD and β_4 is the weight coefficient.

The HA algorithm is used to optimize the design with the targets of f_1 and f_2 respectively. Table 13.4 shows that when the weight coefficients in f_2 are all taken as 0.25 (at this time, it is considered that the travel of TMD is of the same importance as the comfort degree of each area), the TMD parameters of each design scheme, and the acceleration time history and power spectral density of nodes in each area after the MTMD parameters are optimized are given in Fig. 13.10. Among them, WoMTMD represents the original structure without TMD, STMD represents Table 16 group TMD when it is designed with the same parameters and f_1 as the goal, MTMD1 and MTMD2 represent the results when it is designed with f_1 and f_2 as the goal respectively. It can be seen from the figure that (1) after TMD is installed, the acceleration response of the floor is significantly reduced; (2) when STMD is adopted, the acceleration response of area 1 and area 2 is greater than that of MTMD; (3) when f_2 is taken as the objective optimization, the acceleration response of the floor is greater than that of f_1, and the control effect is reduced after controlling the travel of MTMD. Figure 13.11 shows the displacement response of one TMD in the two MTMD optimization schemes. It can be seen that the vertical displacement of TMD obviously attenuates when f_2 is taken as the objective of optimization, so as to meet the space requirements of TMD and improve the long-term performance of TMD.

Figure 13.12 shows the acceleration response and the attenuation rate distribution (attenuation rate = (structural response—damping structural response)/original structural response) of the floor after adopting the mtmd1 and mtmd2 schemes. The comparison shows that, (1) Before and after the installation of MTMD, the value of acceleration response decreases, and the response distribution of mtmd1 scheme is similar to that of the original structure, while the peak response node of area 2 in mtmd2 scheme moves inward from the edge of the floor; (2) The maximum attenuation rate node of mtmd2 scheme does not exactly correspond to the TMD layout

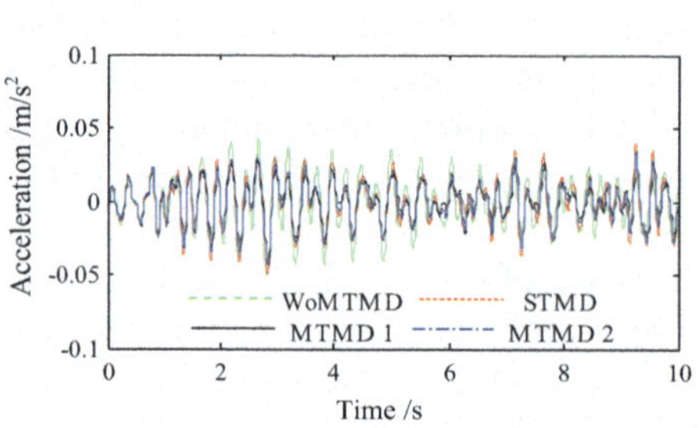

(a) Zone 1 time history

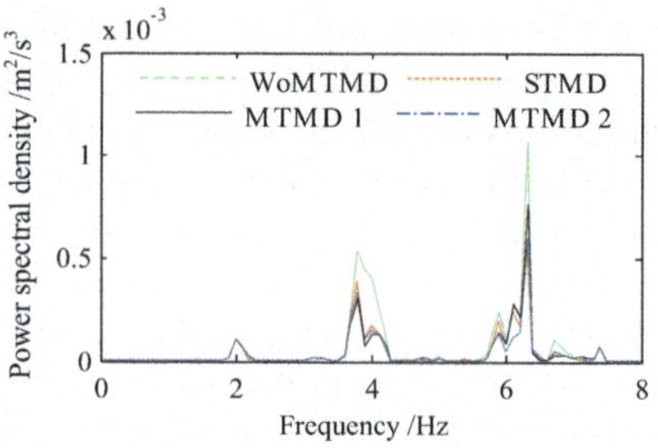

(b) Area 1 power spectral density

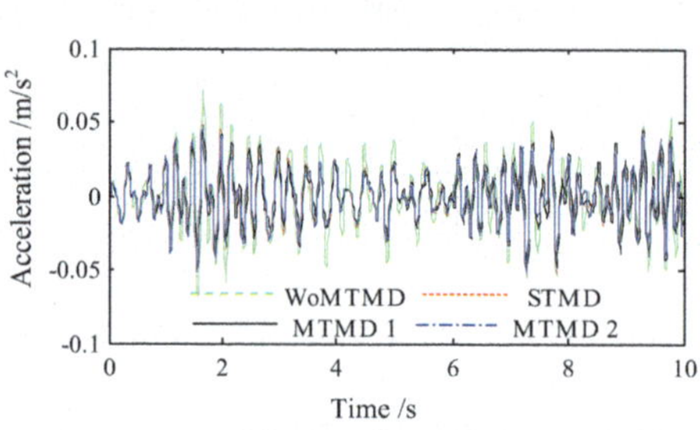

(c) Zone 2 time history

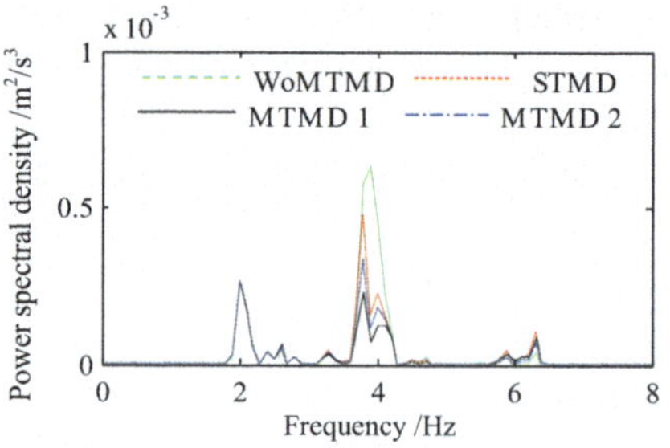

(d) Area 2 power spectral density

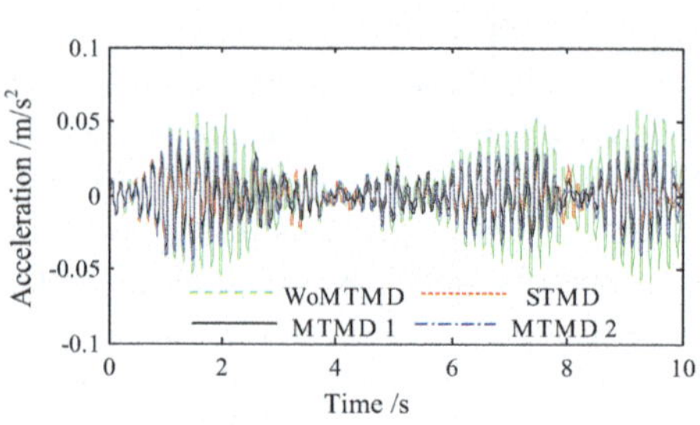

(e) Zone 3 time history

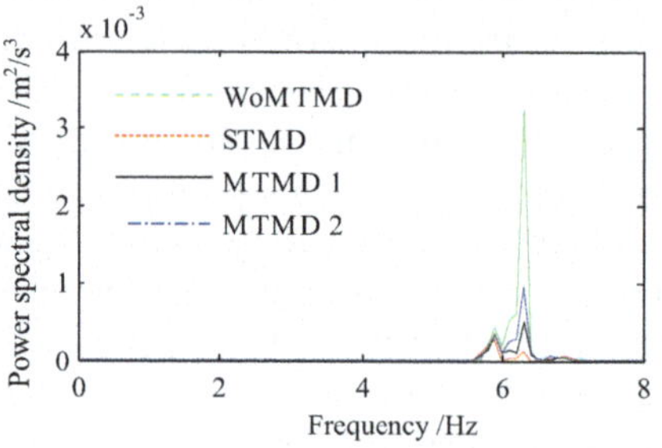

(f) Area 3 power spectral density

Fig. 13.10 Peak acceleration time history and power spectral density

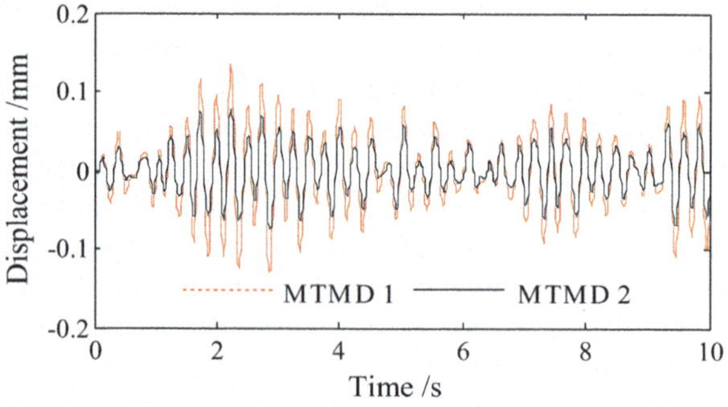

Fig. 13.11 TMD displacement response

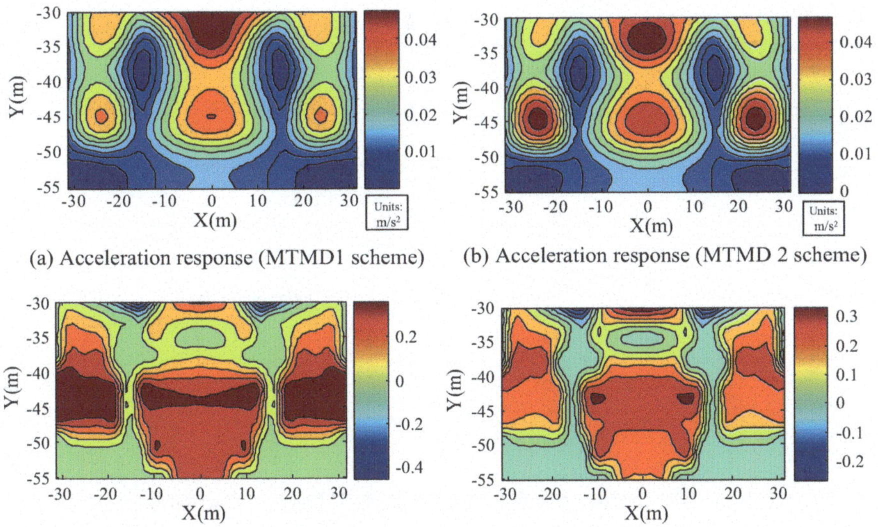

(a) Acceleration response (MTMD1 scheme)

(b) Acceleration response (MTMD 2 scheme)

(c) Attenuation rate (MTMD 1 scheme)

(d) Attenuation rate (MTMD 2 scheme)

Fig. 13.12 Distribution of acceleration and attenuation rate of floor under mtmd2 scheme

position, but is near it. However, generally speaking, the attenuation rate distribution of the two schemes is close to the response distribution of the original structure, which is caused by the design mainly aiming at reducing the response of the peak node in the original structure; (3) In Fig. 13.12c, d, acceleration amplification occurs near the main truss of the cantilever edge of the floor. Compared with the acceleration response of the original structure at the corresponding location, the acceleration here is relatively small in the original structure. After the installation of MTMD, the acceleration distribution law of the floor changes, resulting in a slight increase in the acceleration here.

13.3 Intelligent Optimization Design of Building Structure Based on Pareto Optimization

13.3.1 NSGA-II Basic Principles

Generally, it is difficult to get one solution to make all the objectives optimal, and multi-objective evolutionary algorithm can provide multiple alternative solutions, so it has a high practical value, and has been developed rapidly in recent years. NSGA-II is one of the most representative and widely used algorithms. It is an improved non dominated sorting genetic algorithm based on NSGA, which adopts fast non dominated sorting process, elite retention strategy and non parametric niche

Fig. 13.13 Multiobjective optimization problem

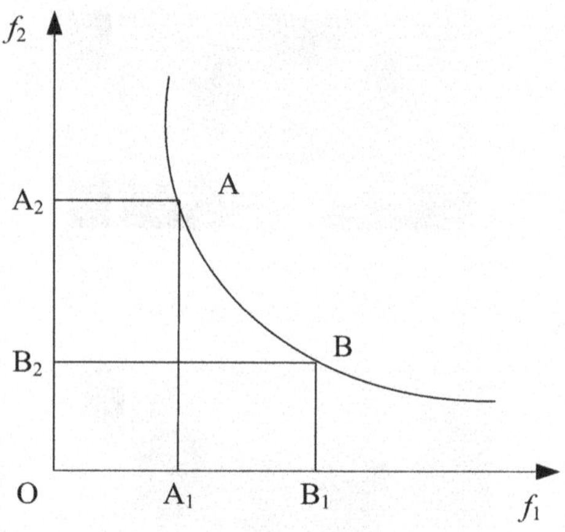

operator. Numerical experiments show that NSGA-II has some advantages over other multiobjective evolutionary algorithms. Before the introduction of NSGA-II, this paper first introduces multi-objective optimization and some basic concepts used in NSGA-II.

(1) **Some basic concepts in NSGA-II**

Pareto optimal solution: In the optimization problem shown in Fig. 13.13, the objective functions f_1 and f_2 are contradictory, because $A_1 < B_1$ and $A_2 > B_2$, that is to say, the improvement of one objective function needs to be at the cost of the reduction of another objective function, that is, such solution A and B are non-inferior solutions, or Pareto optima. The purpose of multi-objective optimization algorithm is to find these Pareto optimal solutions. The condition that the solution of multi-objective optimization problem is Pareto optimal solution is that the value of any objective function of the solution can not be further improved without worsening the value of other objective functions. Obviously, there is more than one Pareto optimal solution. In fact, in general multi-objective optimization problems, Pareto optimal solutions are often continuous and infinite, which constitutes the concept of Pareto frontier. The final solution of optimization problem is to choose an optimal compromise solution from all the optimal solutions.

Dominate and Non-inferior: in a multi-objective problem, if at least one objective of individual p is better than that of individual q, and all the objectives of individual p are not worse than that of individual q, then individual p dominates individual q, or individual q is dominated by individual p, that is to say, individual p is not inferior to individual q.

Rank and front: if p dominates q, then the order of p is lower than q. If p and q do not dominate each other, or p and q are not inferior to each other, then p and q have the same order value. The individuals with order value 1 belong to the first front end, the individuals with order value 2 belong to the second front end, and so on. Obviously, in the current population, the first front end is completely independent, and the second front end is subject to the individual in the first front end. In this way, individuals in the population can be divided into different front ends by sorting.

Crowding distance: crowding distance is used to calculate the distance between an individual in a front-end and other individuals in the front-end, to represent the crowding degree between individuals. Obviously, the larger the crowding distance, the less crowding between individuals, and the better the diversity of population. It should be noted that only individuals in the same front end need to calculate congestion distance, and it is meaningless for individuals in different front ends to calculate congestion distance.

Pareto fraction: the Pareto fraction is defined as the proportion of individuals in the population in the optimal front end, that is, the number of individuals in the optimal front end = min {Pareto fraction × population size, and the number of existing individuals in the front end}.

(2) **The basic principle and improvement of NSGA and NSGA-II**

The difference between NSGA and normal GA is only in selection method, and there is no difference between crossover and mutation. Before NSGA is selected, it is sorted by the Non-dominated individual, and then the non dominated solution is given a large Dummy fitness. The virtual fitness values of all non dominated solutions are the same, which can give all non dominated solutions the same reproduction ability. At the same time, in order to maintain the diversity of population, these individuals and their virtual fitness values are shared. Sharing is done by reducing the fitness value (the original quantity divided by a quantity that is proportional to the number of individuals around), so that multiple optimization points can exist together without being eliminated. After sharing, these non dominated solutions are temporarily put aside, and other parts of the population are treated in the same way to find the second batch of non dominated solutions, namely Pareto front. The virtual fitness assigned to this solution set is always smaller than that of the previous Pareto front. Cycle through the process until the entire population is divided into several fronts. The population is propagated according to the virtual fitness value. Because the first frontier individual has the largest fitness value, their offspring are more than other individuals. This strategy has a fast convergence speed, and the sharing method can ensure extensive search in the region. NSGA emphasizes the non dominated solution, and its efficiency depends on the virtual fitness function of sorting process using the non dominated solution. NSGA-II is improved on the basis of NSGA. In NSGA-II, except for non dominated sorting, the operation rules completely different from NSGA are used. In NSGA-II, the concept of archive is introduced in an all-round way: it makes the progress and expansion of Pareto frontier more reliable than NSGA; because the parent exploration population is generated from archive according to the elimination selection, exerting a large selection pressure on the individuals with high Pareto superiority; finally, it improves the advance ability of Pareto frontier greatly. As an alternative method of fitness sharing in NSGA, the methods of "crowding distance" and "crowding distance sorting" are introduced to avoid that the search is focused on a part of Pareto frontier, and that the individual groups at the end of Pareto frontier are eliminated together. In the non dominated sorting process of NSGA-II, we need

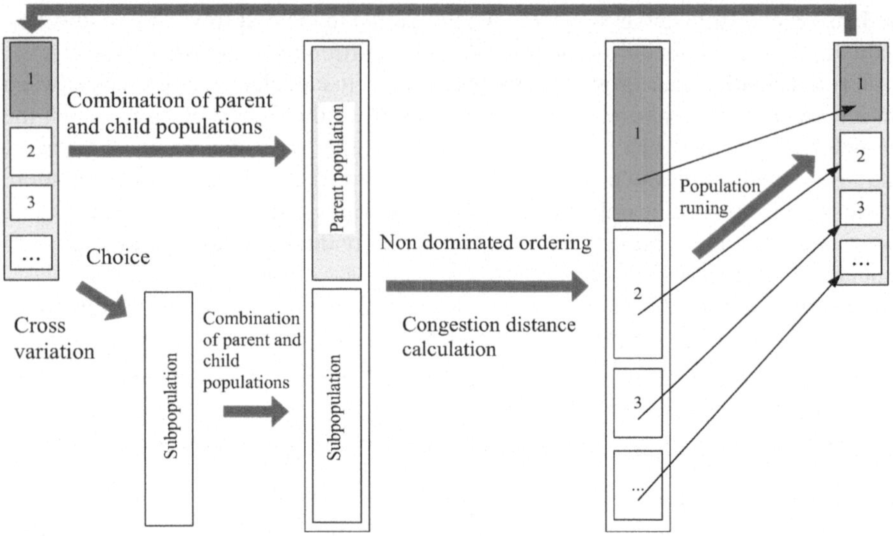

Fig. 13.14 A detailed explanation of the single step evolution process of improved NSGA-II

to cross and mutate the current parent population to get the sub population, and then merge the two populations to get the new population. In order to improve the search efficiency of NSGA-II, the concept of optimal front-end individual coefficient is introduced to prune the population. The specific process is shown in Fig. 13.14. The basic analysis process of NSGA-II after improvement is shown in Fig. 13.15. It can be seen that the difference between NSGA-II and common GA is also the process of selection.

13.3.2 *Intelligent Optimization Design*

(1) **Optimization model**

The multi-objective genetic algorithm NSGA-II is used to optimize the parameters of the damping system of high-rise structure. The specific analysis flow is shown in Fig. 13.16. It can be seen from the figure that the optimization design method of wind-induced vibration control based on NSGA-II mainly includes three steps: optimization modeling, structure modeling and multi-objective genetic search: optimization modeling mainly includes the determination of design variables, optimization objectives, constraints and variations The structure modeling mainly includes two parts: establishing analysis model and wind load simulation. The multi-objective genetic search has more processes of population merging, sorting and pruning than the general genetic search, and also has different details in the specific steps.

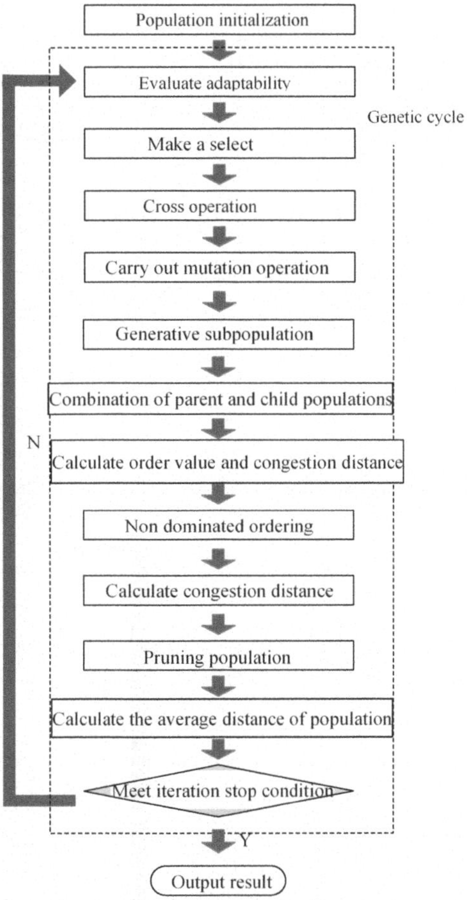

Fig. 13.15 Improving the calculation flow of NSGA-II

The optimization model is established according to the specific structure and vibration control methods. Different control technologies must have different design variables, constraints and initial values of variables, sometimes with different design objectives. For the optimal design of ring TLD control of high-rise structures, the geometric parameters (such as internal and external diameters) are closely related to the performance parameters (such as vibration frequency), so there are two kinds of optimization processes: one is that the geometric parameters are necessarily the only performance parameters, so the geometric parameters can be directly used for optimization; the other is to optimize the performance parameters first, and then the geometric parameters The optimal design approaches the optimal performance parameters. The second method is more commonly used, but it has two cycles. At the same time, since the latter cycle may not be able to find the geometric parameters that meet the conditions, it may need multiple reciprocating operations, that is, there is a large cycle outside, with low efficiency. Therefore, the first method is adopted in this paper. The optimization model can be expressed as follows [2]:

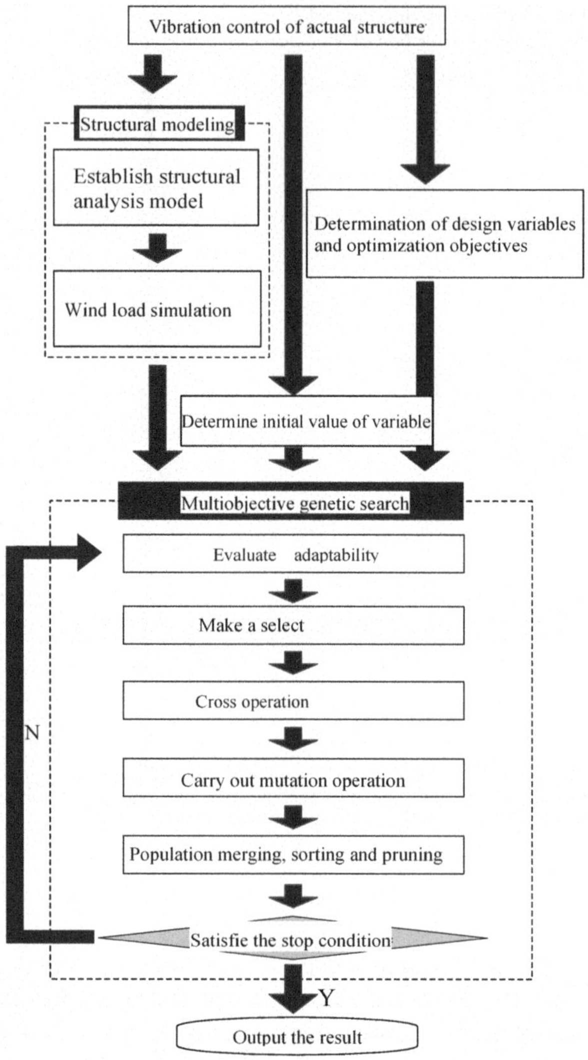

Fig. 13.16 Optimal design method of wind vibration control based on NSGA-II

$$
\begin{aligned}
&\text{Seeking } X \quad X = [R_{in}, R_{out}, \xi_L, h]^{\mathrm{T}} \\
&min \quad F(X) = min[f_1(X), f_2(X)]^{\mathrm{T}} \\
&\text{s.t.} \quad g_j(X) \le 0 \qquad (j = 1, 2, \cdots, m) \\
&\qquad g_j(X) = 0 \qquad (j = 1, 2, \cdots, m) \\
&\qquad X_1 \le X \le X_2
\end{aligned}
\tag{13.14}
$$

Among them, X is the design variable; R_{in} is the inside radius of the water tank; R_{out} is the outside radius of the water tank; ξ_L is the damping ratio of liquid vibration;

$F(X)$ is the objective function; $f_1(X)$ is the structural response control objective, this paper selects the structural displacement of vibration reduction/the original structural displacement; $f_2(X)$ is the limiting objective of the vibration reduction device, this paper selects the RS-TLD liquid vibration wave height; Inequality constraint $g_j(X) \leq 0$, equality constraint $g_j(X) = 0$ and $X_1 \leq X \leq X_2$ are all determined according to the characteristics of the structure and RS-TLD.

(2) **Optimal design**

There is no comfort in high-rise steel chimneys. The influence of structural stress caused by wind load is mainly considered in the design, which is directly related to the vertex displacement. However, the ring TLD is greatly influenced by the geometric dimension in practical application, which has certain requirements for the wave height of liquid vibration. The excessive wave height will increase the size of the device, make the design difficult and increase the cost. Therefore, the objective function is defined as follows:

$$f_1(X) = \left|y_{n,w}\right|_{\text{std}} / \left|y_{n,wo}\right|_{\text{std}} \tag{13.15}$$

$$f_2(X) = |\eta|_{\text{std}}/h \tag{13.16}$$

Among them, $\left|y_{n,wo}\right|_{\text{std}}$ is the variance of displacement at the top of the original structure; $\left|y_{n,w}\right|_{\text{std}}$ is the variance of displacement at the top of the damping structure; $|\eta|_{\text{std}}$ is the variance of wave height.

Considering the actual project, the given constraints are as follows: the inner radius of the water tank is larger than the outer radius of the chimney; the outer radius of the water tank is assumed to be smaller than 1.5 times of the outer radius of the chimney; the modal mass ratio of TMD is 0.1 when considering the practical application of the calculation example, and the mass m_{d} of the water tank is limited to be smaller than 0.1 times of the first-order modal mass of the structure; the damping ratio of the water in the water tank varies from 0.005 to 0.09. These constraints are realized by adding linear and nonlinear constraints to the optimization problem.

The optimization results of NSGA-II are closely related to population number, genetic algebra and other factors, so it is necessary to analyze the influence of these factors. Figure 13.17 shows the evolution of Pareto front in different evolution algebras when the optimal front-end ratio is 0.3 and the population number is 10, 30 and 50 respectively. It can be seen that in the process of NSGA-II evolution, the Pareto front of the population evolves continuously to the optimal Pareto front; when the population number is 10, the concentration elements of the solution obtained are too few to form the Pareto front, and the evolution efficiency is low; When the population number is 50, a better Pareto solution set can be obtained when the evolution algebra is 10; in general, for the optimization of TLD parameters of high-rise structures, a

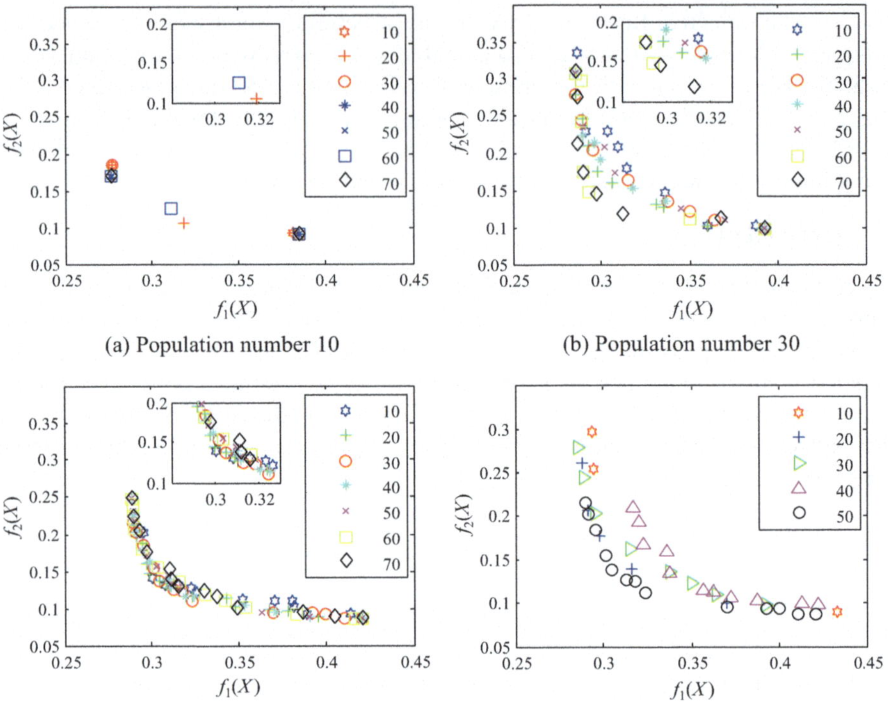

(a) Population number 10

(b) Population number 30

(c) Population number 50

(d) The influence of population number on Pareto frontier

Fig. 13.17 Pareto frontier evolution chart

Pareto solution satisfying the engineering requirements can be obtained when more than 30 generations of genetic evolution is selected.

Figure 13.17d shows the Pareto frontier of different population numbers when the optimal front-end ratio is 0.3 and the genetic age is 30 generations. It can be seen that: when the genetic evolution algebra is the same, the more population, the more satisfactory the Pareto optimal solution; too few population will make the concentration elements of Pareto solution too few, which is not conducive to the engineering decision-making after optimization; for the optimization problem in this paper, it is more appropriate to select more than 30 population; When the population number is 40, the result is slightly worse than that of 30, which is mainly due to the randomness of NSGA-II operation. However, from the perspective of engineering application, both of them are satisfactory optimal solution sets. In general, the larger the population number and the more the genetic algebra, the better the stability of the algorithm, but the calculation cost increases at the same time. Therefore, in the optimization design, it is necessary to select the specific parameters reasonably according to the specific problems.

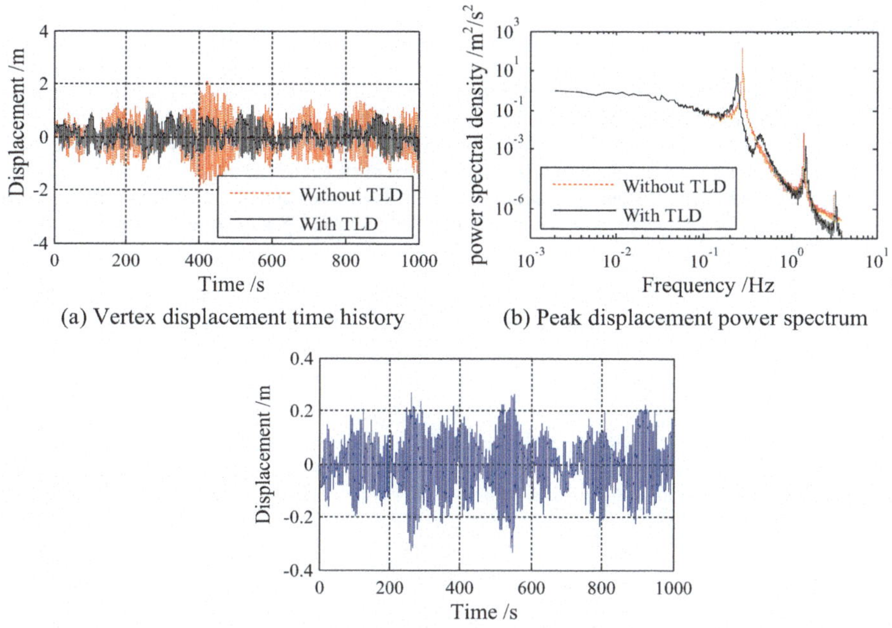

(a) Vertex displacement time history (b) Peak displacement power spectrum

(c) Liquid wave height

Fig. 13.18 Wind vibration response of structure

From the analysis of Pareto solution set, it can be seen that the vibration reduction effect of the displacement of the structure vertex is better. Therefore, from Pareto solution set, select a group of better solutions for wave height control. The design parameters of ring TLD are: inner diameter 1.160 m, outer diameter 1.353 m, liquid height 0.988 m, liquid damping ratio 0.085. After analysis, the peak displacement and liquid wave height of the structure are shown in Fig. 13.18: at this time, f_1 is 0.421, f_2 is 0.087, which effectively suppresses the wind-induced vibration response of the structure; at this time, in order to achieve better attenuation effect, the first-order modal mass ratio reaches 0.0948, which is close to the upper limit of 0.1 set; In order to achieve the damping effect and limit the relative wave height of the liquid, Fig. 13.18b the longitudinal coordinates of the peak tips on both sides of the first-order frequency of the original structure are not equal, and the liquid damping ratio is close to the set upper limit.

References

1. Qingyang, Xu. 2008. *Research on vibration control and optimization analysis of long-span hangar structure*. Nanjing: Southeast University. (in Chinese).
2. Chen, Xin. 2012. *Theoretical and experimental study on vibration control of high-rise steel chimneys under wind load*. Nanjing: Southeast University. (in Chinese).
3. Chen Xin, AiQun Li, ZhiQiang Zhang, et al. 2017. Hybrid optimization of the multiple tuned mass dampers in long-span floor for human comfort. *Journal of Vibration Engineering* 30 (5): 827–836. (in Chinese).

Part IV
Engineering Practice of Vibration Control for Building Structures

Since the 1990s, the vibration reduction theory and technology of building structures have been developed rapidly and matured gradually. The corresponding technical standardization construction has been continuously improved. The pilot applications of new technology have been tested by engineering practices, and the society's acceptance of building structure damping technology is increasing. Especially since 2008, many major earthquakes such as the Wenchuan earthquake have brought tragic losses and heavy lessons to the country and the people. Research and practice have shown that improving the resilience of building structures is the most effective means of ensuring the safety of buildings under disasters. In the new century, the society has become more and more aware of the concept, investment, policy and technical system of building structure disaster prevention and mitigation. In February 2014, the Ministry of Housing and Urban-Rural Development of the People's Republic of China (MOHURD) issued *A number of opinions on the promotion and application of seismic isolation technology for building construction projects (provisional)*. The wide application of China's building structure control technology has entered a new stage. Based on the systematic research of basic theory and applied technology, the author team further solved a series of problems of technology industrialization and engineering, and presided over the engineering practices of structural vibration control for more than 100 buildings, including the first domestic active mass damper (AMD), viscoelastic dampers and large-span floor stack load vibration control (M-TMD).

Chapter 14
Vibration Control Engineering Practice for the Multistory and Tall Building Structure

Abstract Project cases of vibration control engineering practice for the multistory and tall building structure are introduced. Case 1 is a high-rise office building in high intensity zone with the seismic fortification intensity degree 8. Case 2 is an office building in high intensity zone with the fortification intensity of degree 9. Case 3 is a middle school library with the seismic fortification intensity degree 7. Case 4 is tall residential building located in near-fault zone with the seismic fortification intensity degree 8.5. Viscous fluid dampers, viscoelastic dampers, metal dampers and rubber isolaters are used in above mentioned four projects for earthquake resistance respectively. In every case, project overview, structural energy dissipation design, structural analysis model, seismic wave selection and analysis of structural vibration absorption performance are introduced respectively.

14.1 High-Rise Office Building 1 in High Intensity Zone (Viscous Fluid Damper, Earthquake)

14.1.1 Project Overview

The project is a high-rise office building of octave area with an overall structure of cast-in-place reinforced concrete frame-shear wall structure system. The building area is 33,349.82 m^2 and the total height of the building is 89.5 m. There are 22 floors above ground and 1 floor of basement. The first floor is 5.0 m in height and the second floor is 4.5 m in height. The other layers are 4.0 m in height.

The engineering design base period is 50 years, the seismic fortification intensity is degree 8, the design basic seismic acceleration value is 0.3 g, the seismic fortification category is general seismic fortification category C, the site category is Class III, and the design earthquake group is the first group. The site characteristic period T_g = 0.45 s. The basic wind pressure during the 50-year return period is 0.45 kN/m^2, the ground roughness is B, and the building body factor is 1.3.

A. Li, *Vibration Control for Building Structures*, Springer Tracts in Civil Engineering, https://doi.org/10.1007/978-3-030-40790-2_14

14.1.2 Structural Energy Dissipation Design

The SATWE calculation results of the main structure under frequent earthquakes, in the case of meeting the functional requirements and maximally arranging the shear wall and the frame column, indicate that the displacement angles corresponding to the maximum floor displacements in the X and Y directions reached to 1/627 and 1/618, respectively, which exceed the limit of 1/800 allowed by the specification. At the same time, the main structure has many over-reinforcement phenomenon at the main beam and coupling beam under the frequent earthquakes, which is difficult to be adjusted by design. Moreover, the performance of the structure is more difficult to meet the requirements under the large earthquake. In addition, the rigidity of the structural body will be further increased if adopting the method of increasing the thickness of the shear wall section, and the design requirements cannot be met. Therefore, it is difficult to meet the seismic design requirements of the project relying solely on the traditional design of "hard resistance".

In order to improve the seismic capacity of the project, the non-linear viscous fluid dampers were installed in the structure. After optimum design, a certain number of viscous fluid dampers were set up along the two main axes in the 1–22 layers. Among them, the arrangement of first floor dampers is shown in Fig. 14.1a, and the assembly of the 7-axis damper is shown in Fig. 14.1b. The parameters and specific setting number of dampers are shown in Tables 14.1 and 14.2, respectively.

14.1.3 Structural Analysis Model

14.1.3.1 Finite Element Model

The thickness of the main shear wall of the structure is 400 mm, the thickness values of the shear wall near the elevator room include 300 and 200 mm. The main column cross-section size is 800 mm * 1000 mm and 1000 mm * 1000 mm; the main beam cross-section size is 300 mm * 600 mm; the floor plate thickness is 120 mm; the concrete strength grade of beams, columns, walls and slabs is between C30 and C60. The distribution of concrete strength grades in each floor is shown in Table 14.3.

The model of elastic analysis under frequent earthquakes was established independently by the professional finite element software CSI ETABS, as shown in Fig. 14.2a; the model of elastic-plastic time history analysis under fortification intensity and rare earthquakes was established independently by general finite element software MSC. MARC, as shown in Fig. 14.2b. In the elastic analysis model of ETABS, the beam and column elements were simulated by spatial bar elements, the shear walls and connecting beams were simulated by shell elements, the floors were simulated by membrane elements, and the viscous fluid dampers were simulated by damper elements. In MSC. MARC elastic-plastic analysis model, the THUFIBER subroutine based on the principle of fiber model was developed using the secondary development

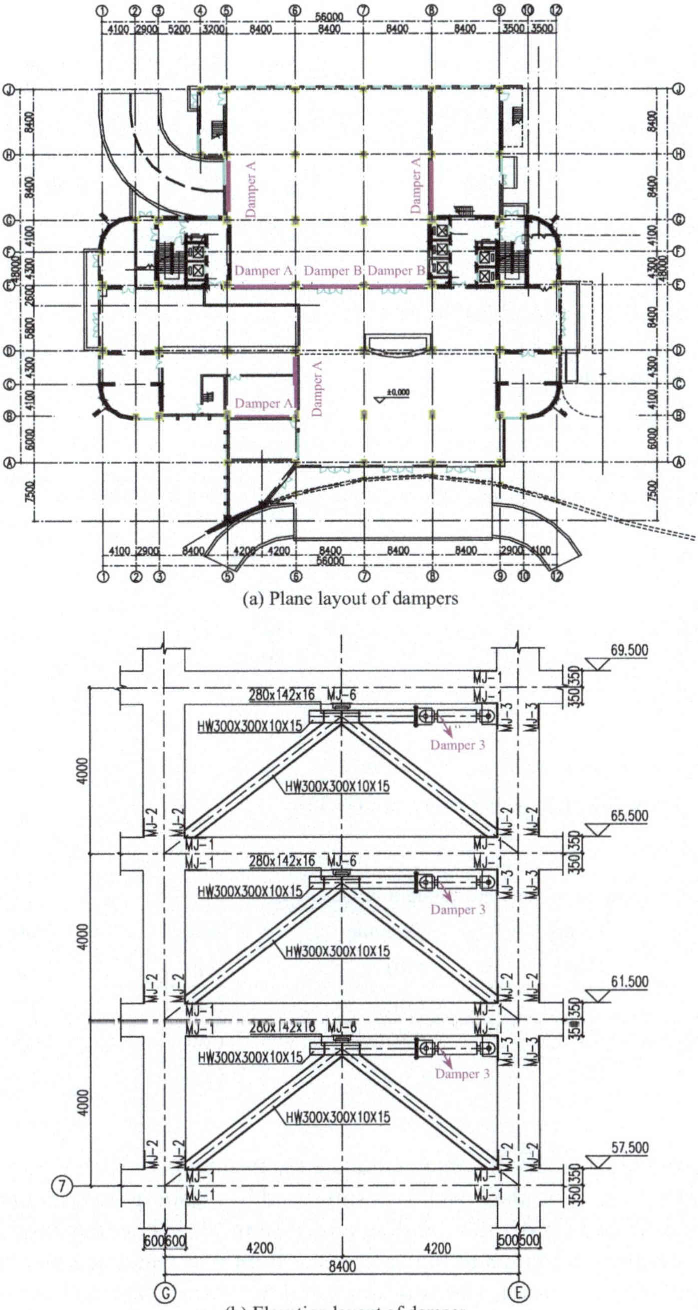

(a) Plane layout of dampers

(b) Elevation layout of damper

Fig. 14.1 Layout scheme of viscous fluid dampers

Table 14.1 Viscous fluid damper parameters

Type	Damping index (α)	Damping coefficient C (kN m/s)	Maximum stroke (mm)	Maximum damping force (kN)
A	0.20	1200	±50	850
B	0.20	1500	±40	1050
C	0.20	1600	±40	1100

Table 14.2 Statistics of damper distribution by floor

Floor	Type and quantity of damper in X-direction	Type and quantity of damper in Y-direction
18–22	C/2	C/2
11–17	C/2	C/5
8–10	C/3	C/4
7	C/3	B/3 + C/2
6	C/4	B/4 + C/2
5	C/3	B/2
4	C/4	B/2
3	C/3	B/2
2	A/3	A/2
1	A/4	A/3
Subtotal	A/12 + B/13 + C/111	
Total	57 (X-direction) + 79 (Y-direction) = 136	

Table 14.3 Distribution of concrete strength grades by floor

Floor	Beam	Column	Wall	Plate
21, 22	C40	C40	C40	C30
13–20	C40	C40	C40	C40
8–12	C40	C50	C50	C40
1–7	C40	C60	C60	C40

platform provided by the software to simulate the beam and column components, as shown in Fig. 14.2c. The shear wall was simulated by elastic-plastic laminated shell element, which can consider the coupling effects of in-plane bending, in-plane shear, out-plane bending. As shown in Fig. 14.2d, the floor was simulated by elastic shell element, and the viscous damper was simulated by setting the attributes of springs between nodes.

In order to ensure the accuracies of the established models, the characteristic period, the base shear force and the overall quality of the model calculated by the above two models were compared with that of the PKPM model. As

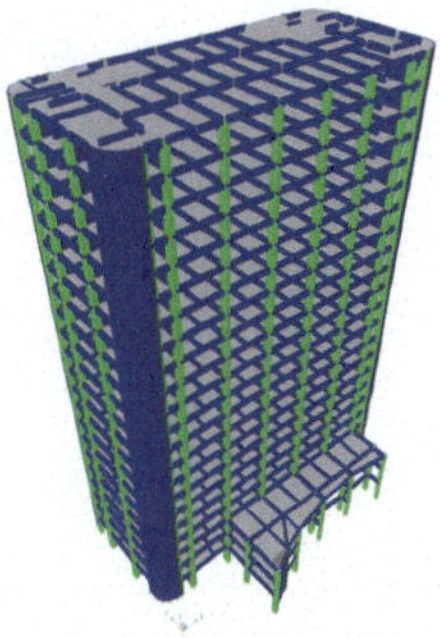

(a) ETABS model

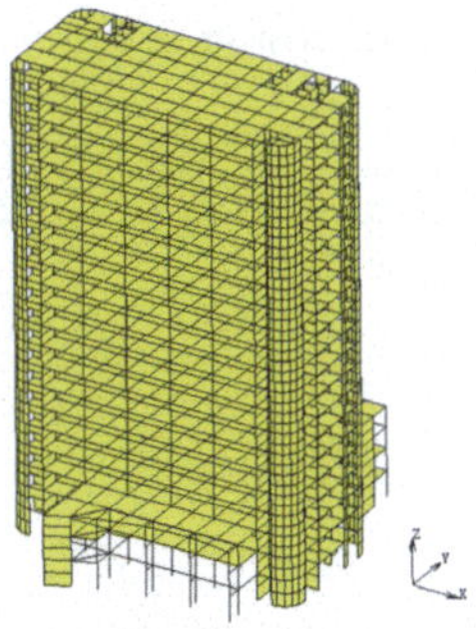

(b) MARC model

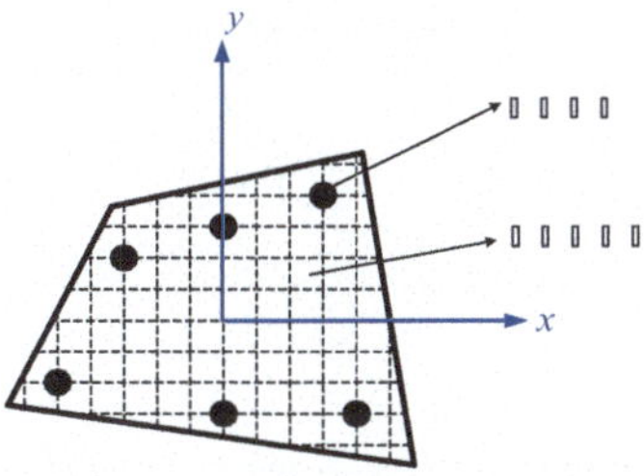

(c) Fiber element of beam and column

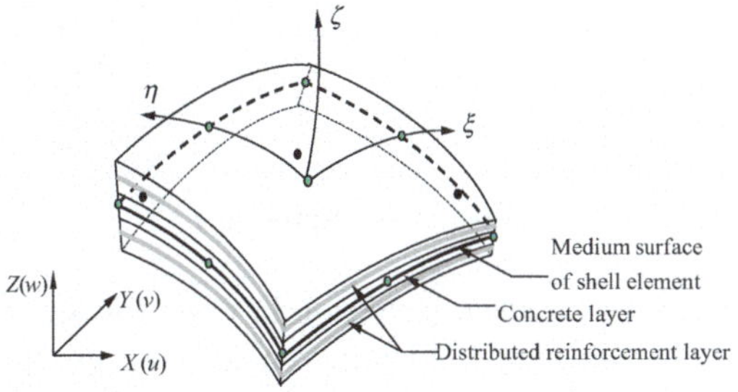

(b) Layered shell element of shear wall

Fig. 14.2 Structural analysis model

shown in Tables 14.4, 14.5 and 14.6, the error percentage 1 = (SATWE result—ETABS result)/SATWE result; the error percentage 2 = (SATWE result—MARC result)/SATWE result. The elastic model and elastic-plastic finite element model independently established have little difference with the PKPM model in many calculation indexes, which can ensure the accuracy of the model and can be used for further structural response analysis and shock absorption design.

Table 14.4 Comparison of structural characteristic period

Mode shape	Order					
	1	2	3	4	5	6
SATWE/s	1.863	1.799	1.369	0.529	0.481	0.378
ETABS/s	1.838	1.684	1.3400	0.500	0.461	0.373
MARC/s	1.786	1.693	1.288	0.465	0.447	0.349
Error percentage 1 (%)	1.38	6.86	2.15	5.81	4.27	1.09
Error percentage 2 (%)	4.33	6.31	6.29	6.97	7.51	7.95

Table 14.5 Comparison of shear force of structural basement based on response spectrum method

Principal axis	X-direction (kN)	Y-direction (kN)
SATWE	37,138	38,506
ETABS	38,570	39,140
MARC	36,758	38,289
Error percentage 1 (%)	−3.86	−1.65
Error percentage 2 (%)	1.02	0.56

Table 14.6 Comparison of overall quality of structure

FE Model	SATWE	ETABS	MARC
Mass/t	54,158	54,960	54,045
Error percentage 1	−1.48%	Error percentage 2	0.21%

14.1.3.2 Seismic Wave Selection

Based on the clear provision in *Code for Seismic Design of Buildings* (*GB50011-2010*) of China: "When seven or more groups of time history curves are taken, the average value of time history method and the larger value of mode decomposition response spectrum method can be taken as the calculation results."; "In elastic time history analysis, the bottom shear force calculated by each time history curve should not be less than 65% of that calculated by mode decomposition response spectrum method, and the average value of bottom shear force calculated by multiple time history curves should not be less than 80% of that calculated by mode decomposition response spectrum method". Seven seismic waves including lwd-00, el, lwd-90, nrg, sfy, user 587 (artificial wave), and user 362 (artificial wave) were determined by calculation. The comparison of the time history analysis results of the undamped structures and those of the response spectra of PKPM and ETABS is shown in Table 14.7. The comparison of the response spectra of seven selected seismic waves under 8-degree frequent earthquakes (110gal) and the standard response spectra at 5% damping ratio is shown in Fig. 14.3, which reflects that the selected seismic wave can satisfy the condition of wave selection given in the code and can be used for dynamic time history analysis of this project.

14.1.4 Analysis of Structural Shock Absorption Performance

14.1.4.1 Elastic Analysis of Structural Response Under Frequent Earthquakes

The comparisons between the interlayer shear force and the interlayer displacement angle before and after shock absorption under frequent earthquakes are shown in

Table 14.7 Base shear force of structure under frequent earthquakes

Seismic wave	X-direction		Y-direction	
	Shear force (kN)	Ratio (%)	Shear force (kN)	Ratio (%)
lwd-00	28,990	78	35,170	91
el	33,180	89	32,130	83
lwd-90	30,500	82	35,980	93
nrg	34,280	92	35,800	93
sfy	32,390	87	38,270	99
587	30,070	81	31,990	83
362	36,130	97	35,790	93
Average value	32,220	87	35,019	91
PKPM	37,138	100	38,506	100
ETABS	38,570	–	39,140	–

Note Ratio = Base shear force under earthquake wave/Base shear force of response spectrum

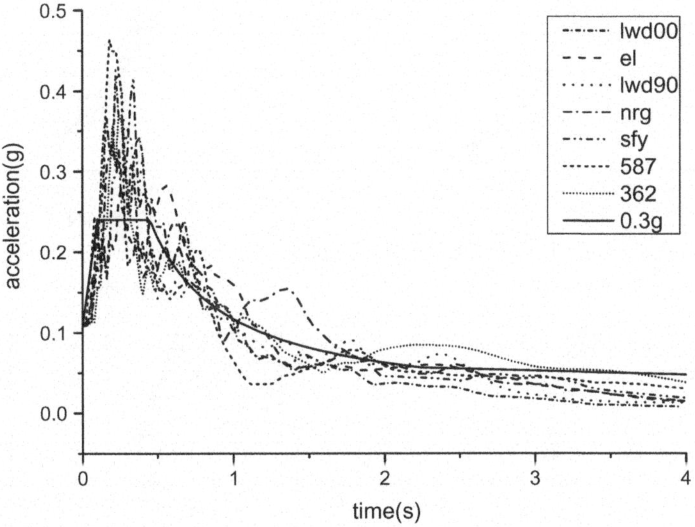

Fig. 14.3 Comparison of response spectra of design earthquake motion (110gal) and that of code at 5% damping ratio

Fig. 14.4. It can be seen from Fig. 14.4c, g that the interlayer displacement angle of the structure without shock absorption under the frequent earthquakes cannot fully meet the specifications, and the maximum interlayer displacement angles in the X and Y directions reached to 1/698 and 1/703, respectively, and breaking the limit of 1/800. However, as can be seen from Fig. 14.4d, h that the interlayer displacement angle after damping can fully meet the requirements, and the maximum damping

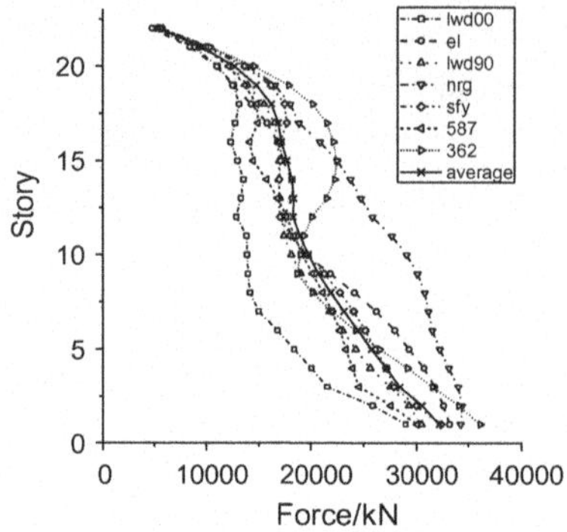

(a) Interlayer shear force in X-direction before shock absorption

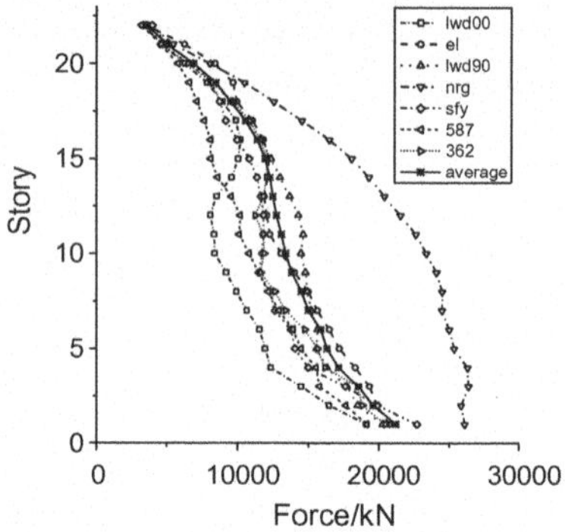

(b) Interlayer shear force in X-direction after shock absorption

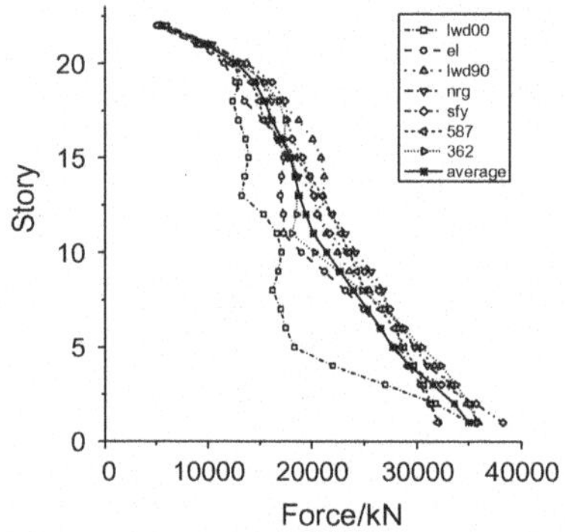

(c) Interlayer shear force in Y-direction before shock absorption

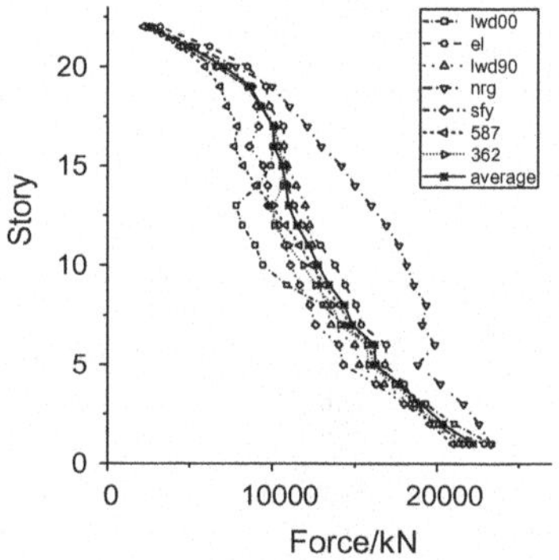

(d) Interlayer shear force in Y-direction after shock absorption

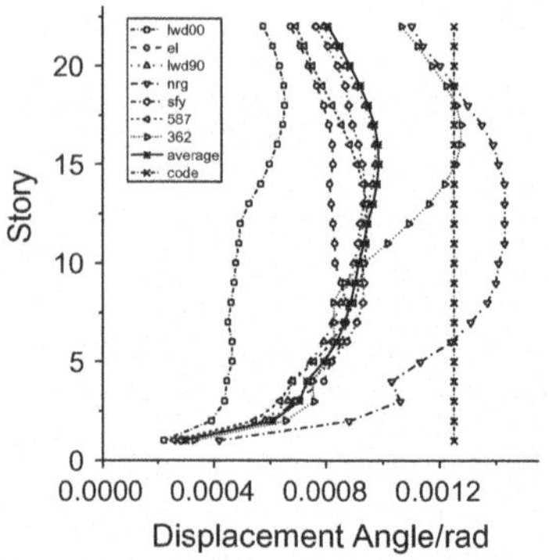

(e) Interlayer displacement angle in X-direction before shock absorption

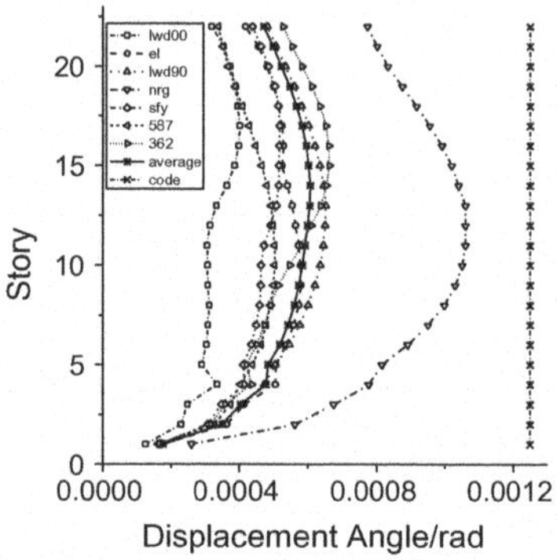

(f) Interlayer displacement angle in X-direction after shock absorption

Fig. 14.4 Comparison of structural response before and after shock absorption under each seismic wave

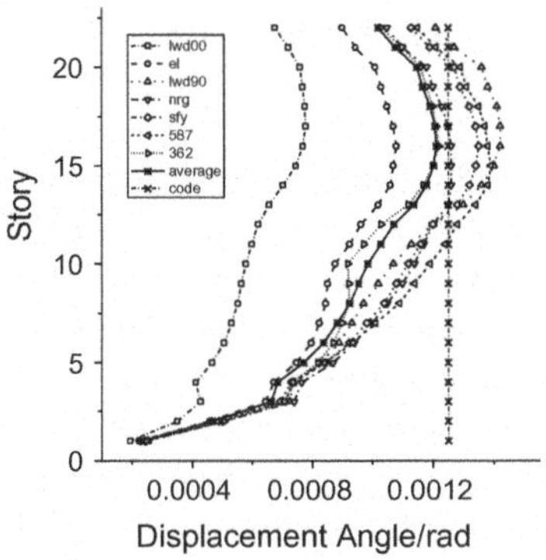

(g) Interlayer displacement angle in Y-direction before shock absorption

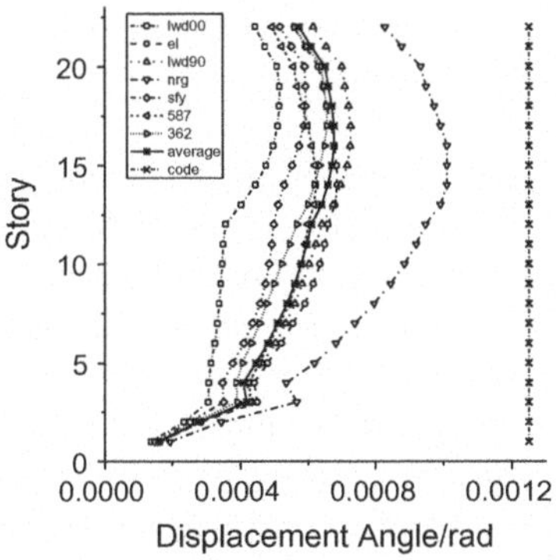

(h) Interlayer displacement angle in Y-direction after shock absorption

Fig. 14.4 (continued)

rates in the X and Y directions reached to 38% and 37%, respectively. The interstory shear force also has a large seismic reduction rate. The average seismic reduction rates of base shear force in X and Y directions were 34.05% and 36.42%, respectively. Therefore, the proposed seismic reduction scheme has good seismic reduction effect considering the action of frequent earthquakes, which can effectively suppress the seismic response of the structure and improve the overall seismic performance of the structure.

14.1.4.2 Elastoplastic Analysis of Structural Response Under Fortification Earthquakes

The interstory displacement angles of the structure without shock absorption exceed the limit of 3/800 stipulated in the code under the action of earthquake with 8-degree fortification intensity (300gal), and all the inter-story displacement angles satisfy the stipulation after shock absorption. The average values of the maximum seismic reduction rates of the seven seismic waves reach to 28% and 29% in the X and Y directions, respectively. Only a small number of coupling beams yield, but the rebars in shear walls and frame columns hardly yield. The seismic design achieves the goal of keeping the shear elasticity and bending resistance of the main structure basically under the action of fortification intensity earthquake, and the overall structure has good seismic performance.

14.1.4.3 Elastoplastic Analysis of Structural Response Under Rare Earthquakes

The maximum interlayer displacement angle of individual seismic wave exceed the limit of 1/100 stipulated in the code under the action of 8-degree rare earthquake (500gal). All of the interlayer displacement angle after shock absorption meet the stipulations. The average value of the maximum seismic reduction rates of the seven seismic waves reach to 25% and 29% in the direction of X and Y, respectively. The inter-story shear force of the seismic absorption structure is much smaller than that of the structure without seismic absorption. The average seismic reduction rates of the base shear force are 14.96% and 16.29% in the X and Y directions, respectively. The distribution of plastic hinges in X and Y directions of the whole structure under rare earthquakes before and after shock absorption is shown in Fig. 14.5, taking the lwd-00 wave as an example. The plastic hinges are shown in red. The number and development degree of plastic hinges of the structure after shock absorption under 8-degree rare earthquakes (0.3 g) are reduced compared with that of the structure without shock absorption, especially the yield number and degree of steel bars in shear walls and frame columns are obviously reduced, which fully guarantees that the damage degree of the main structure can be effectively controlled under rare earthquakes. The development degree of plasticity is small, which makes the whole structure have good seismic performance, and is more conducive to the realization of the fortification goal of "no collapsing with strong earthquake".

14.1.4.4 Performance of Damper

The hysteretic curves of the damper installed on the 3rd, 9th, 15th and 18th floors of the structure in the X direction under the rare earthquake of 8-degree (0.3 g) and the lwd_90_nor wave, are shown in Fig. 14.6. It can be seen that the hysteretic curves show full non-linear viscous characteristics, which can well simulate the non-linear viscous fluid dampers. The maximum output forces of the four dampers are less than 700 kN under rare earthquakes, and the maximum displacements are less than 30 mm, which are all within the maximum design performance range of damper.

14.1.4.5 Economic Analysis of the Shock Absorption Scheme

The direct economic comparison between the proposed energy dissipation scheme and the original design scheme is shown in Fig. 14.8. The comparison shows that the scheme could save RMB 1.3632 million yuan by saving 3408 m^3 of concrete, which was 400 yuan per cubic meter; save RMB 3.396 million yuan by saving 566 tons of steel bars, which was 6000 yuan per ton, and save RMB 560,000 yuan by saving 35 piles, which was 16,000 yuan per bar. And 136 sets of dampers were installed, each set was calculated at an average of RMB 36,000 yuan, a total of RMB 4.896 million yuan was needed. Finally, the direct economic savings of RMB 423,200

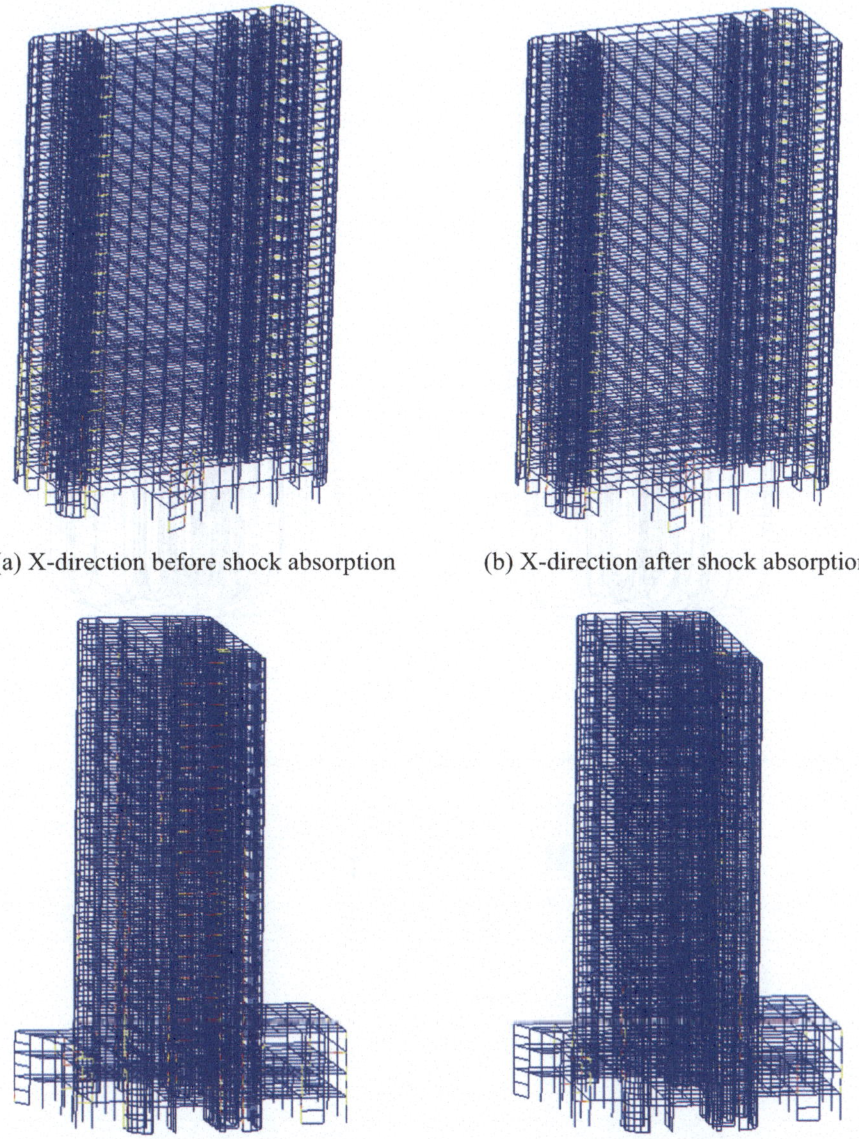

(a) X-direction before shock absorption (b) X-direction after shock absorption

(c) Y-direction before shock absorption (d) Y-direction after shock absorption

Fig. 14.5 Structural plastic development under rare earthquakes

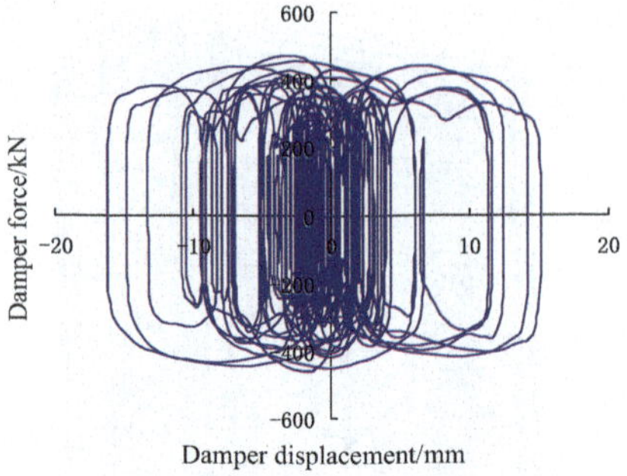

(a) 3rd floor

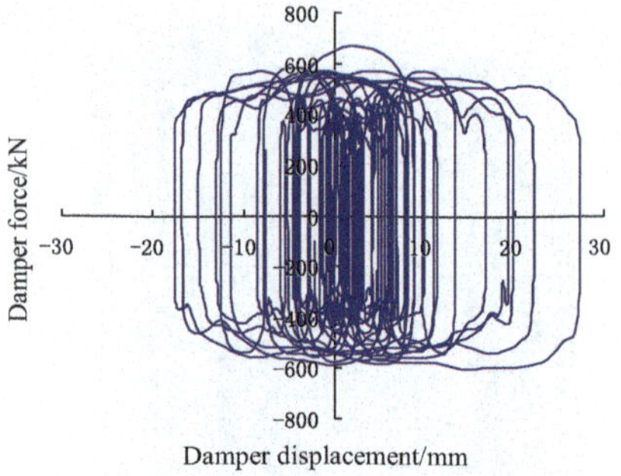

(b) 9th floor

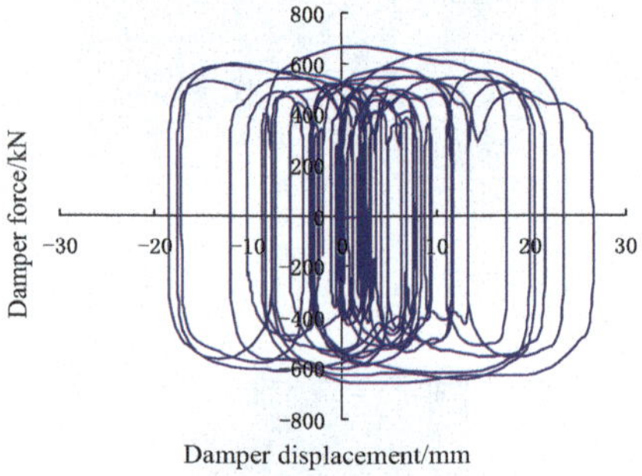

(c) 15th floor

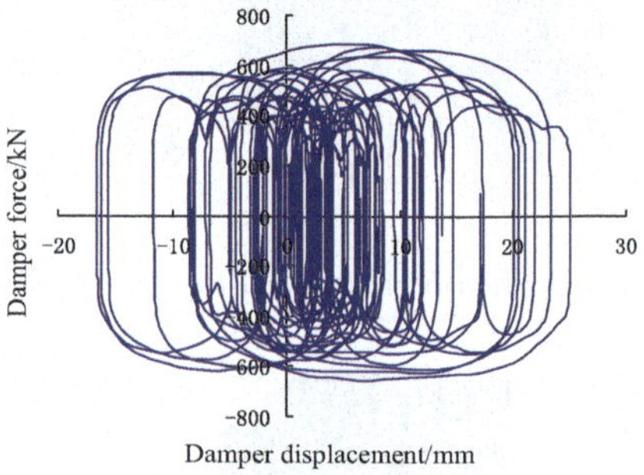

(d) 18th floor

Fig. 14.6 Force-displacement relationship of partial dampers under rare earthquakes

Photos

Fig. 14.7 An office building in 9-degree district

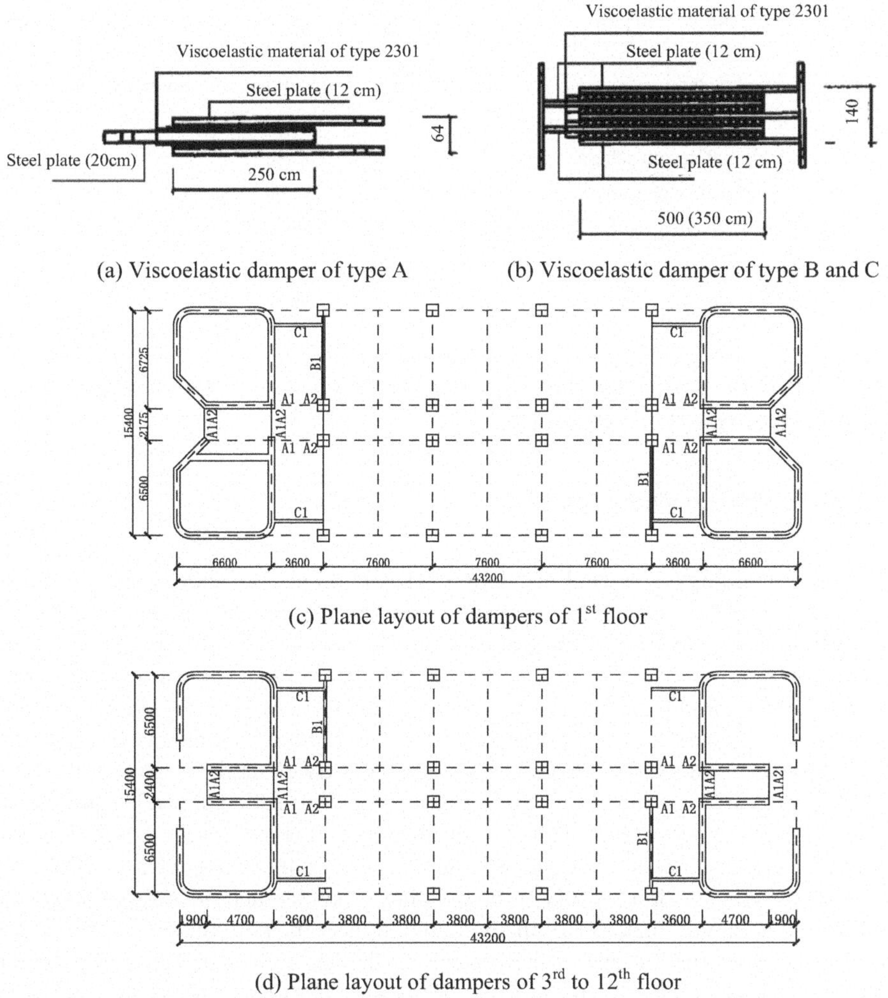

Fig. 14.8 Viscoelastic damping scheme

yuan was achieved by adopting the proposed technical scheme of shock absorption (Table 14.8).

Table 14.8 Economic comparison between the energy dissipation scheme and original design scheme

Order	Scheme type	Material type	Quantity	Reduction rate (%)
1	Energy dissipation scheme	Concrete	12,440 m^3	21.50
	Original design scheme		15,848 m^3	
2	Energy dissipation scheme	Steel bar	2395 t	19.10
	Original design scheme		2961 t	
3	Energy dissipation scheme	Pile foundation	205	14.60
	Original design scheme		240	

14.2 Office Building 2 in High Intensity Zone (Viscoelastic Damper, Earthquake)

14.2.1 Project Overview

The office building (shown in Fig. 14.7a) is a key building and belongs to the second-class high-rise building [1]. The whole building is equipped with the central air conditioning system. The building area is 14,188.5 m^2 and the total height is 57.8 m. The main building has 13 floors above ground, 16 floors of towers on both sides, and 1 floor of basement. The height values of 1st–12th floors and 14th–16th floors are 3.6 m. The height of 13th floor is 5.0 m. The reinforced concrete frame-shear wall structure is adopted, with the fortification intensity of degree 9, structural safety grade of degree 2, classification standard of seismic fortification of level 2, and site of type II. The thickness of tube wall of the original structure is 450 mm, and the bottom frame column section is 800 mm * 800 mm, and the frame beam height is 800 mm. The plane and elevation layouts of the building is regular, and the ratio of length to width and height to thickness of the building all meet the requirements of the code. Due to the construction age of the project, all the analysis in this section is based on the relevant provisions in the *Code for Seismic Design of Building Structures (GBJ11-89)*.

14.2.2 Structural Energy Dissipation Design

The original seismic design scheme could not meet the requirements under frequent and rare horizontal earthquakes based on the seismic calculations, and the reinforcement ratios of beams and columns were very high, so the construction was difficult. Therefore, the vibration reduction technology was considered to reduce the seismic response of the structure and improve the structural design scheme. After optimum design, the types and numbers of dampers arranged on each floor are shown in Table 14.9, in which the configurations of dampers are shown in Fig. 14.8a, b and

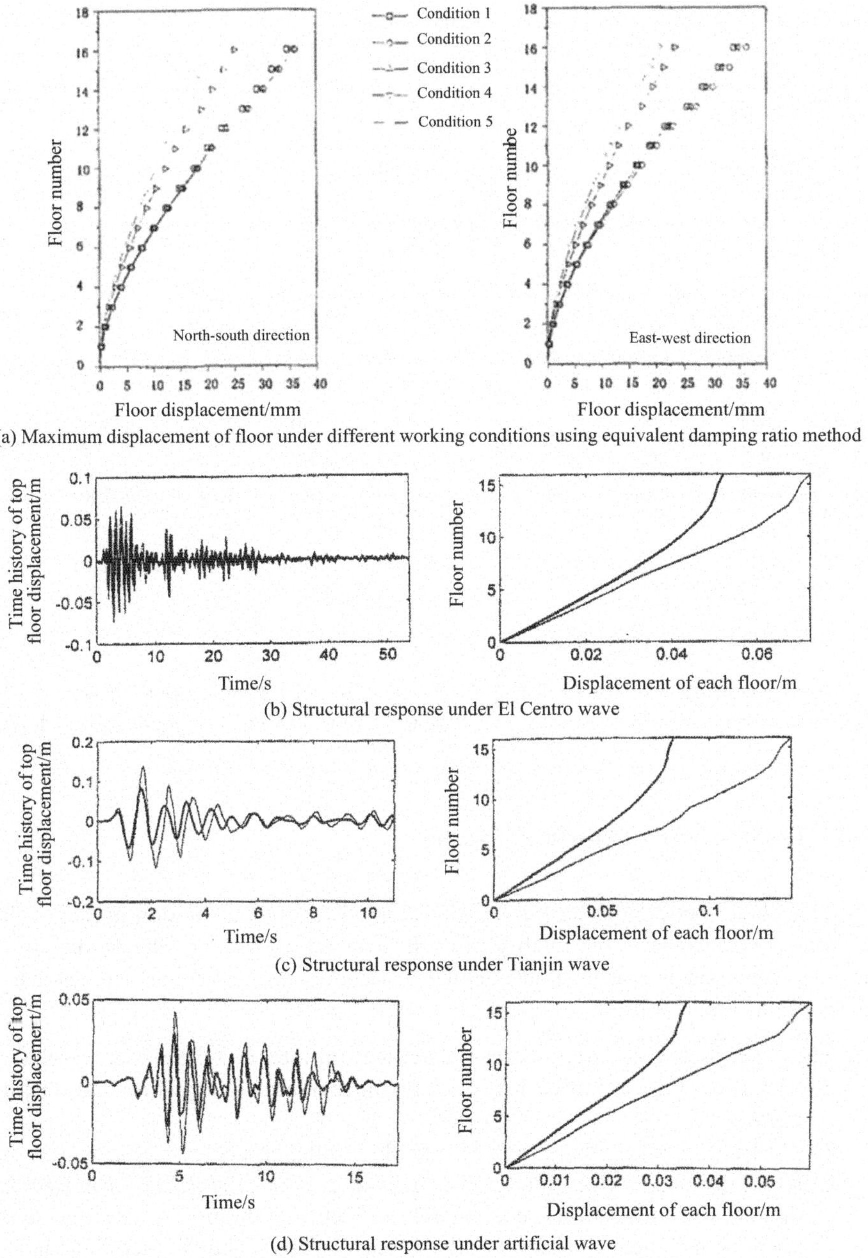

(a) Maximum displacement of floor under different working conditions using equivalent damping ratio method

(b) Structural response under El Centro wave

(c) Structural response under Tianjin wave

(d) Structural response under artificial wave

Fig. 14.9 Elastic time history analysis under frequent earthquakes

Table 14.9 Type and quantity of the viscoelastic dampers

Floor	Type A1	Type A2	Type B1	Type B2	Type C1	Type C2
1st floor	8	8	0	0	0	0
2nd floor	8	8	2	0	0	0
3rd–12th floors	8 × 10	8 × 10	2 × 10	0	4 × 10	0
13th floor	8	8	0	2	0	4

Table 14.10 Summary of the parameters of energy dissipators of each layer

Floor	Height (h/mm)	Width (w/mm)	Thickness (t/mm)	Elastic stiffness (kN/m)	Yield load (kN)	SR	
						Transverse	Longitudinal
6	300	300	3	154,000	60	2.42	1.86
5	300	300	3	154,000	60	2.29	1.73
4	450	450	5	261,000	141	3.84	2.89
3	450	450	5	261,000	141	3.82	2.88
2	600	600	8	412,000	320	6.13	4.59
1	600	600	8	412,000	320	8.00	6.00

the plane locations are shown in Fig. 14.8c, d. A total of 208 dampers of type A, 24 dampers of type B, and 44 dampers of type C were used in the project (Table 14.10).

14.2.3 Structural Analysis Model

The inter-layer bending shear model was adopted for the calculation model of the structure to concentrate the quality of each layer on each floor. The whole structure was regarded as a series multi-degree-of-freedom cantilever system. The basic assumptions are as follows:

1. The stiffness of the slab in its own plane was infinite, and there was no relative deformation of the top of each vertical member in each layer within the seismic joint section;
2. The stiffness center of the house coincided with the center of mass, and the structure did not produce a torsion deformation under the action of horizontal earthquakes. In order to more accurately reflect the vibration of the reinforced concrete frame-shear wall or cylinder-frame structure, the adopted calculation model considered both inter-layer bending deformation and shear deformation.

The viscoelastic damper installed in the reinforced concrete structure can provide both stiffness and damping. In practice, the two were combined to form a control force of the viscoelastic damper on the structure. In the elastoplastic analysis, the degraded three-fold line restoring force model with a descending section was used to simulate the plastic properties of the floor. After the model was established, the reinforced concrete cylinder-frame elastoplastic seismic response analysis program EPRD developed by the Center for Seismic and Shock Absorption of Building Engineering of Southeast University was used to analyze the dynamic time history of the structure after the viscoelastic damper was installed. The El Centro wave, Tianjin wave, and artificial wave were used in the analysis, with the predominant periods of 0.455 s, 1.149 s, and 0.337 s, respectively. The predominant period of artificial wave is consistent with the predominant period of the engineering site soil, while the predominant period of Tianjin wave is close to the natural vibration period of the structure. Considering the effects of frequent earthquakes and rare earthquakes, the peak acceleration of ground motion was selected as 0.14 g and 0.55 g, respectively. The durations of El Centro wave, Tianjin wave, and artificial wave were 53.78 s, 10.94 s, and 17.425 s, respectively.

14.2.4 Analysis of Structural Seismic Absorption Performance

14.2.4.1 Elastic Analysis of Structural Response Under Frequent Earthquakes

The TAT program developed by the CAD Department of the Institute of Structural Engineering of the Chinese Academy of Architectural Sciences was used to preliminarily calculate and analyze the effect of viscoelastic damping under frequent earthquakes. Five working conditions were considered in the calculation. The first one was the condition when the damping ratio of the structure was 5% before adjusting the member cross-section of uncontrolled structure; the second one was the condition when the damping ratio of the structure was 5% after adjusting the member cross-section of uncontrolled structure; the third one was the condition when the total damping ratio of the structure was 11% after adjusting the cross-section of structural members considering the shock absorption effect of viscoelastic damper; the fourth one was the condition when only steel braces with equal stiffness were installed at the same position after adjusting the section of structural members, and the structural damping ratio was 5%. The first four working conditions were calculated according to the seismic fortification intensity of degree 9. The fifth condition was that when the damping ratio of the structure was 5% after adjusting the member section of uncontrolled structure, and the seismic fortification intensity was calculated at degree 8. The response spectrum mode combination method (SRSS method) was used to calculate all kinds of working conditions, and the combination of the first six mode shapes were considered. Under the action of frequent earthquakes, the maximum floor displacements of the structure under various working conditions in the north-south and east-west directions are shown in Fig. 14.10a.

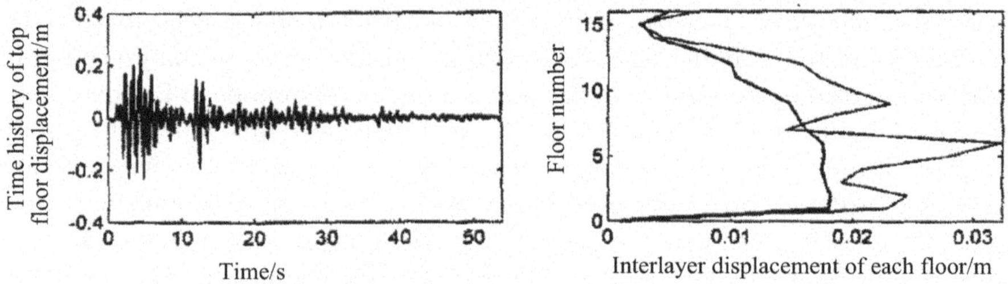

(a) Structural response under El Centro wave

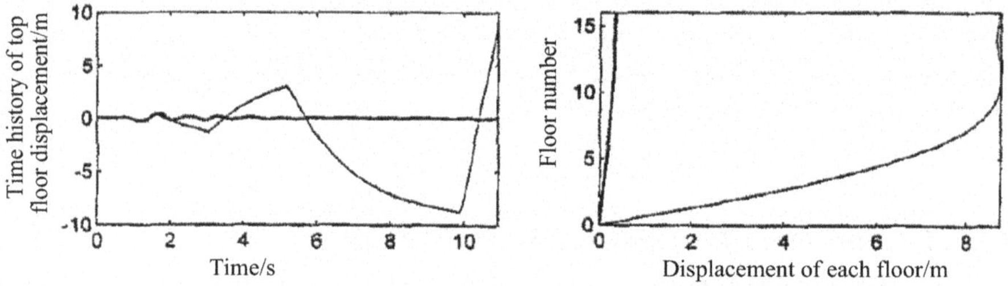

(b) Structural response under Tianjin wave

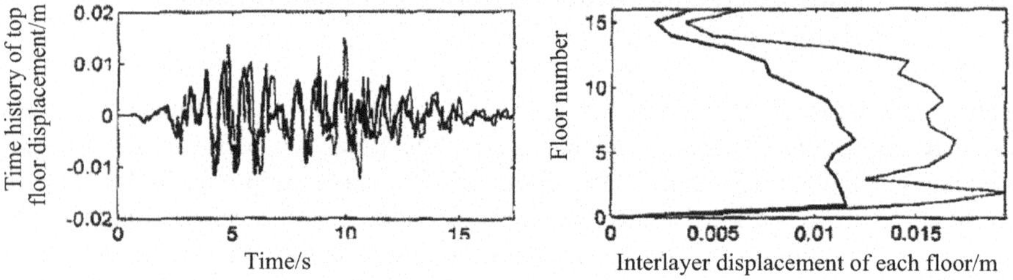

(c) Structural response under artificial wave

Fig. 14.10 Elastic time history analysis under rare earthquakes

It can be seen from the Fig. 14.9 that the maximum displacement of each floor under the first, second and fourth working conditions is close to each other. The maximum displacement of the fourth one is slightly smaller than that under the second working condition. That is to say, the maximum displacement of each floor decreases when steel braces with equal stiffness were installed in the uncontrolled structure, but the displacement of each floor decreases slightly because the damping of the structure does not increase. The maximum displacement of each floor under the third condition is much smaller than that under the second and fourth conditions, which indicates that the displacement of each floor decreases obviously when the damping ratio of the structure increases to 11%. At the same time, the maximum displacement of each floor under the third condition is smaller than that under the fifth condition, namely, the displacement of each floor is close to that of uncontrolled structures calculated by seismic fortification intensity of degree 8 after adding viscoelastic dampers.

Further, assuming that the ambient temperature is 20 °C, the elastic time history analysis of the structure under frequent earthquakes was carried out using EPRD. The elastic time history analysis results of the structure in the north-south direction are shown in Fig. 14.9b–d. It can be seen that the maximum interlayer relative displacements under El Centro wave of the uncontrolled structure and the controlled structure are 1/590 and 1/783, respectively; the maximum interlayer relative displacements under Tianjin wave of the uncontrolled structure and the controlled structure are 1/347 and 1/637, respectively; the maximum relative displacements under artificial wave of the uncontrolled and controlled structures are 1/766 and 1/1200, respectively,.

14.2.4.2 Elastoplastic Analysis of Structural Response Under Rare Earthquakes

By adjusting the peak value of ground motion acceleration to 0.55 g, the dynamic elastoplastic analysis of the structure at ambient temperature of 20 °C was carried out. The elastoplastic time history analysis results of the structure in north-south direction are shown in Fig. 14.10a–c. It can be seen that most of the uncontrolled structures have entered the plastic state under El Centro wave. The seismic weak layer is the sixth floor, and the maximum inter-story displacement is 32 mm. The largest inter-story displacement of the first floor of the controlled structure is 18 mm, that is, the inter-story displacement angle is 1/199. Under the action of Tianjin wave, because the predominant period of ground motion is close to the natural vibration period of the uncontrolled structure, the uncontrolled structure resonates and the displacement diverges. All the structures are in plastic state and are seriously damaged. When the viscoelastic damper is installed, the stiffness and damping of the structure change, and the displacement converges rapidly. The maximum inter-story displacement angle is 1/120. Under the action of artificial wave, the weak layer of uncontrolled structure is the first layer, the maximum interlayer displacement is 19 mm, and the maximum interlayer displacement of the controlled structure is 12 mm, that is, the interlayer displacement angle is 1/303. Under rare earthquakes, the maximum inter-story displacement angle of the controlled structure is less than the limit of 1/50 stipulated in the *Code for Seismic Design of Building Structures* (*GBJ11-89*), and the safety performance of the structure is significantly improved.

14.2.4.3 Economic Analysis of Shock Absorption Scheme

The cross-sections of structural members were greatly adjusted after the installation of viscoelastic damping braces, the tube wall was reduced to 400 mm, the cross-section of bottom frame columns was reduced to 700 mm * 700 mm, and the height of frame beams was reduced to 700 mm. The basic frequency of natural vibration of the structure was increased from 1.1 to 1.16 Hz, that is, the lateral stiffness of the structure was slightly smaller than that of the original structure. The engineering budget estimates show that, the building area was increased by nearly 90 m^2 after the section adjustment, and the concrete consumptions of shear walls, columns, beams, and slabs were saved 192 m^3, 118 m^3, 153 m^3, and 125 m^3, respectively, with a total of 586 m^3. In addition, the design of cross-section was conducted considering the seismic fortification intensity of degree 8, the reinforcement ratios of shear walls, columns, beams, and other structural components were greatly reduced, and nearly 200 tons of steel bars were saved after estimation roughly. It can still save more than RMB 1 million yuan after deducting the cost of steel support, viscoelastic damper, and connecting steel components, that is, nearly 10% of the cost of the main structure was saved.

14.3 A Middle School Library (Metal Damper, Earthquake)

14.3.1 Project Overview

This project is a seismic strengthening project of a middle school library. The main building has 6 floors above ground, 18.9 m in height. The bottom floor is 3.9 m in height, and the standard floor is 3 m in height. This project is a reinforced concrete frame structure. The cross-sectional dimensions of standard floor frame beams and columns are 200 mm * 500 mm and 400 mm * 400 mm, respectively. The column spacing is 6 m. The concrete strength grade of the beam, column, and slab of each floor is C30, the HRB400 is used for the longitudinal reinforcement and stirrup.

The design life of the structure is 50 years. The safety grade of the building structure is class 2, the fire prevention grade of the building structure is class 1, and the design grade of the foundation is class 1. The seismic fortification intensity is degree 7, the design basic seismic acceleration is 0.10 g, the site soil is type III, the seismic design belongs to the second group, the characteristic period of site soil is $T_g = 0.55$ s. The basic wind pressure of the structure is $w_0 = 0.40$ kN/m^2 in the 50-year recurrence period. The ground roughness of the wind load is class B, and the shape coefficient of the wind load is 1.3.

According to the relevant provisions of the *Code for Seismic Design of Buildings* (*GB50011-2010*), there are mainly the following overrun situations in this project: (1) the plane of the standard floor is polygonal, which belongs to irregular plane shape; (2) the maximum torsional displacement ratio is 1.30 under the accidental eccentric seismic load, which belongs to irregular torsion; (3) the minimum inter-story displacement angle under frequent earthquakes is 1/472, which does not meet the code limit; (4) the building is upgraded from the standard fortification category to the fortification category, and the structural measures do not meet the requirements.

As the building has been put into use, if the traditional seismic reinforcement method is adopted, it will greatly affect the normal use of the library and the building function. Therefore, considering various factors, using energy dissipation technology for seismic reinforcement can not only effectively improve the seismic capacity of the building, but also greatly reduce the construction noise and shorten the construction cycle.

14.3.2 Structural Energy Dissipation Design

According to the design process under frequent earthquakes, the size parameters of anti-buckling resistance shear steel plate dampers were selected and listed in Table 14.11. The structural quality and vertical stiffness of the project are evenly distributed along the vertical direction, so the steel plate dampers were evenly arranged along the vertical direction. The plane and elevation layouts of the dampers are shown in Fig. 14.11. This layout scheme can not only effectively reduce the displacement response of each layer, but also reduce the response difference of each layer to avoid the occurrence of weak layer.

Table 14.11 Structural base shear force under frequent earthquakes

Ground motion	X-direction		Y-direction	
	Shear force (kN)	Ratio	Shear force (kN)	Ratio
GOPESHWAR	816.9	0.725	807.6	0.718
KATAKHAL	1200	1.066	1200	1.067
SH2	969.5	0.861	973.8	0.866
Average value	995.5	0.884	993.8	0.883
Response spectrum	1126	/	1125	/

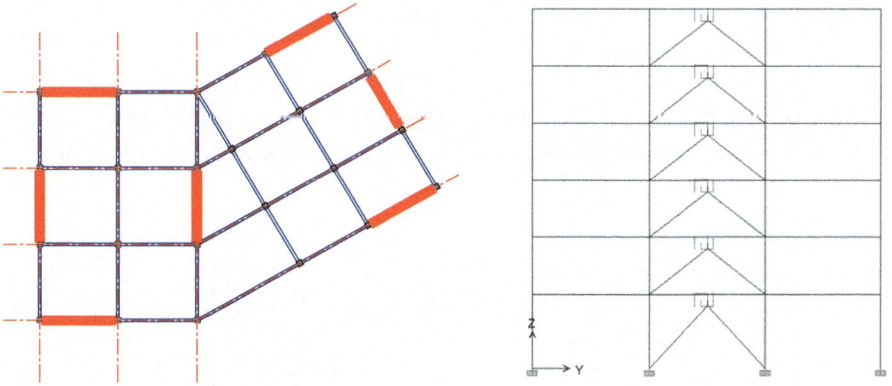

(a) Plane layout (b) Elevation layout

Fig. 14.11 Layout of damper

14.3.3 Structural Analysis Model

14.3.3.1 Finite Element Model

The CSI ETABS software was used to conduct the structural elasticity analysis. The calculation model of the steel frame was established according to structural design drawings, as shown in Fig. 14.12a. When calculating seismic action, the inter-layer displacement and lateral stiffness K_f of each floor were obtained by static or elastic time history analysis, considering the effects of accidental eccentric seismic action,

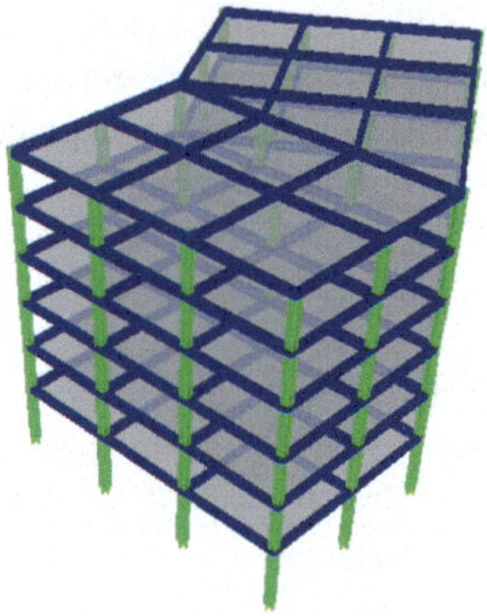

(a) ETABS model

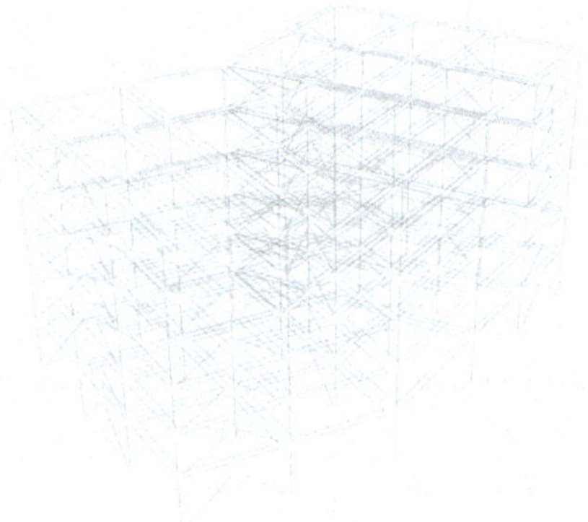

(b) Perform3D model

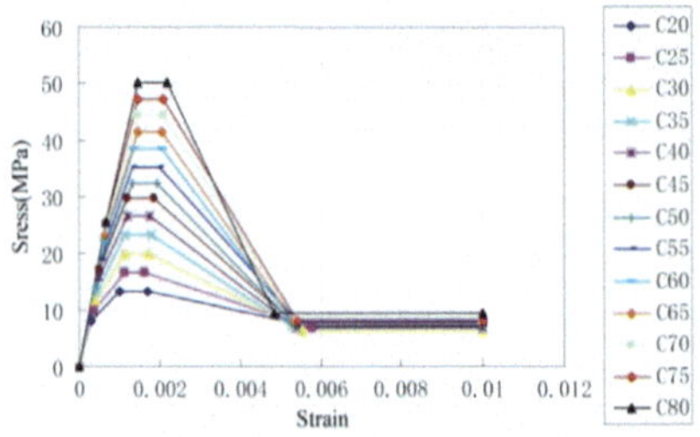

(c) Constitutive relationship of concrete

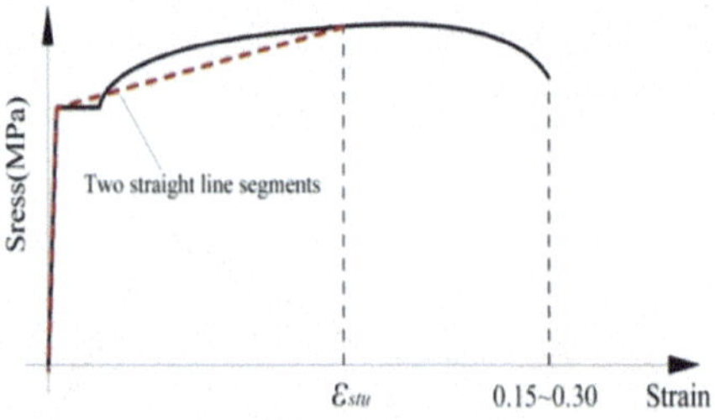

(d) Constitutive relationship of steel

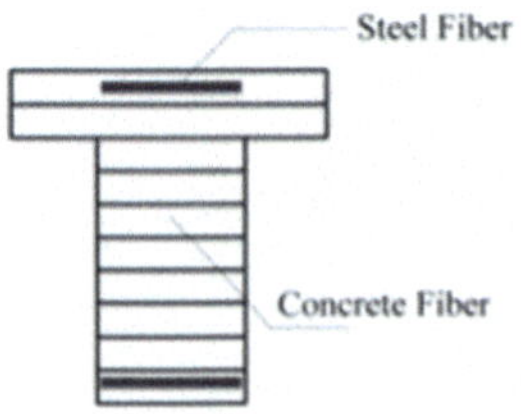

(e) Fiber section division of beam

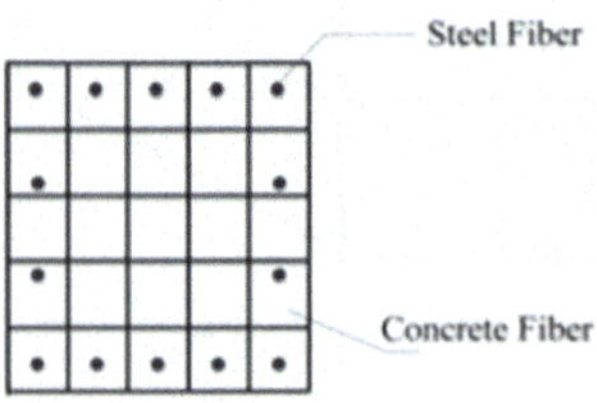

(f) Fiber section division of column

Fig. 14.12 Structural analysis model

bidirectional seismic action, torsional coupling, and simulated construction loading. The CSI Perform-3D software was used for structural elastoplastic analysis. The software is based on the basic concepts of structural engineering and takes the settings of mechanical properties of structural components as the premise. The seismic performance evaluation of the whole structure was obtained by means of structural analysis, which was in line with the understanding of structural performance by engineers. The analysis results were easy to be verified using the structural concepts and tests, which had been approved by the engineering community. The elastoplastic analysis model is shown in Fig. 14.12b. The skeleton curve was simplified and obtained according to the "Y-U-L-R-X" five-fold constitutive model, using the uniaxial stress-strain relationship of concrete provided in Appendix C of *Code for Design of Concrete Structures* (*GB50010-2010*), as shown in Fig. 14.12c. The bilinear constitutive model was adopted to simulate the steel bars. The stiffness ratio after yielding was $\gamma = 0.02$, as shown in Fig. 14.12d. At the same time, the one-dimensional fiber element was used to simulate the beams, and the two-dimensional fiber element was used to simulate the columns. The fiber division of cross-section is shown in Fig. 14.12e, f. The non-linear shear plate model of infill panel element was used to simulate the shear plate damper, which only had the shear stiffness and shear strength, while the bending stiffness and out-of-plane stiffness were zero.

14.3.3.2 Seismic Wave Selection

In order to make the results of time history analysis truly reflect the seismic performance of the structure, the seismic waves should be reasonably selected. The GOPESHWAR wave and KATAKHAL recorded by actual strong earthquakes were used in the design of this project, and SH2 wave was selected as the artificial wave. The peak acceleration of each earthquake was 35 cm/s^2 for small earthquakes, 100 cm/s^2 for medium earthquakes, and 220 cm/s^2 for large earthquakes. In time history analysis, the overall damping ratio of the structure was 5%. China's *Code for Seismic Design of Buildings* (*GB50011-2010*) stipulates that in elastic time-history analysis, the bottom shear force calculated by each time-history curve should exceed 65% of that calculated by mode decomposition response spectrum method, and the average value of bottom shear force calculated by multiple time-history curves should exceed 80% of that calculated by mode decomposition response spectrum method. The base shear force of each working condition is listed in Table 6.1. The selected ground motion meets the requirements of the code.

Table 14.12 Comparison of natural vibration periods of the structure

Natural vibration period	T_1	T_2	T_3	T_4	T_5	T_6
Frame structure	1.3056	1.2674	1.1553	0.4099	0.3993	0.3649
Shock dissipation structure	0.7236	0.6513	0.2558	0.1063	0.0802	0.0691

14.3.4 Analysis of Structural Shock Absorption Performance

14.3.4.1 Elastic Analysis of Structural Response Under Frequent Earthquakes

The natural vibration periods of the pure frame structure and shock dissipation structure are shown in Table 14.12. The natural vibration period of the shock absorption structure is smaller than that of the pure frame structure, because the considerable lateral stiffness is provided by the anti-buckling shear plate dampers attached to the structure. The first three modes of the two systems are X-directional translation, Y-directional translation, and Z-directional torsion, but the torsion effect of the shock dissipation structure is smaller than that of the pure frame structure, which shows that the arrangement of metal dampers improves the torsion effect of the structure.

The comparison of interlayer displacement angles (IDA) between pure frame structure and shock dissipation structure under frequent earthquakes is shown in Fig. 14.13. The inter-story displacement was effectively reduced by the setting of metal dampers, and the inter-story displacement angle of each floor after shock absorption meets the requirements of the code limit.

Taking the GOPESHWAR wave as an example, the base shear force of structure under frequent earthquakes before and after shock dissipation is shown in Fig. 14.14. It can be seen that the shear force at the bottom of column of the structure after shock dissipation is generally smaller than that of the pure frame structure, which shows that the setting of metal dampers can effectively reduce the base shear force under rare earthquakes, thus playing the role of protecting the bottom column. However, the decrease of the peak shear force of the base is not obvious, mainly because the seismic action on the damper will eventually pass to the bottom of the bottom column. Further comparisons show that the interlayer shear forces of every floors decrease after the damper is installed, and the internal force of components generally does not increase, so the original structure does not need to be strengthened.

14.3.4.2 Elastic Analysis of Structural Response Under Rare Earthquakes

The maximum elastoplastic inter-story displacement angle of each story of the structure under rare earthquakes is shown in Figs. 6.43 and 6.44. The maximum elastoplastic displacement angle in X direction is 1/182 and that in Y direction is 1/154,

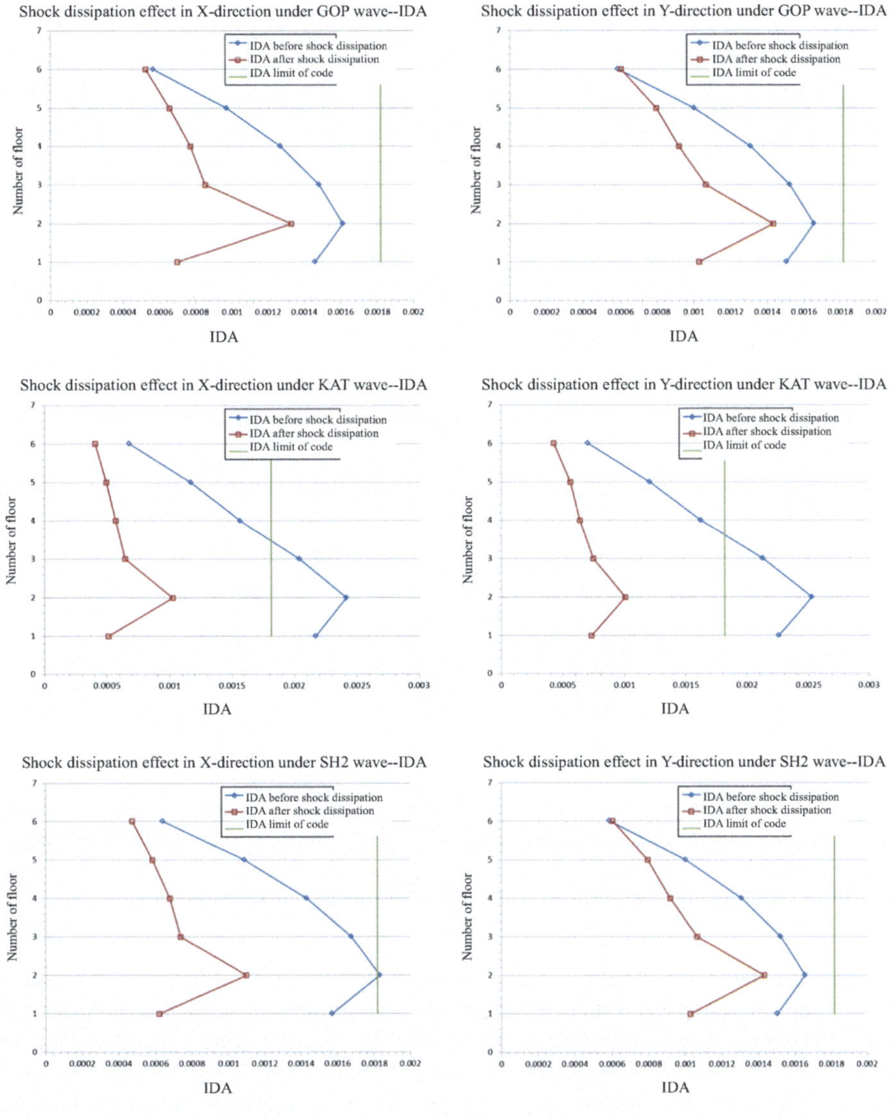

Fig. 14.13 Interlayer displacement comparison under frequent earthquakes

which are far less than the limit value of elastoplastic displacement angle of reinforced concrete frame structure stipulated in the *Code for Seismic Design of Buildings* (*GB50011-2010*). This shows that the setting of anti-buckling shear steel plate damper can effectively reduce the inter-story displacement of the structure. The structure has a higher safety reserve (Fig. 14.15).

Taking the X-direction input of GOPESHWAR wave as an example, the structural deformation checking under rare earthquakes was carried out. The definitions of

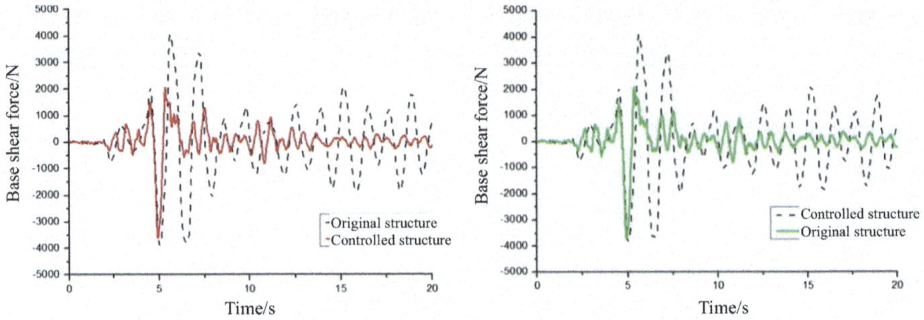

(a) Comparison of X-direction base shear force before and after seismic reduction

(b) Comparison of Y-direction base shear force before and after seismic reduction

Fig. 14.14 Comparison of base shear force before and after seismic reduction under GOPESHWAR wave

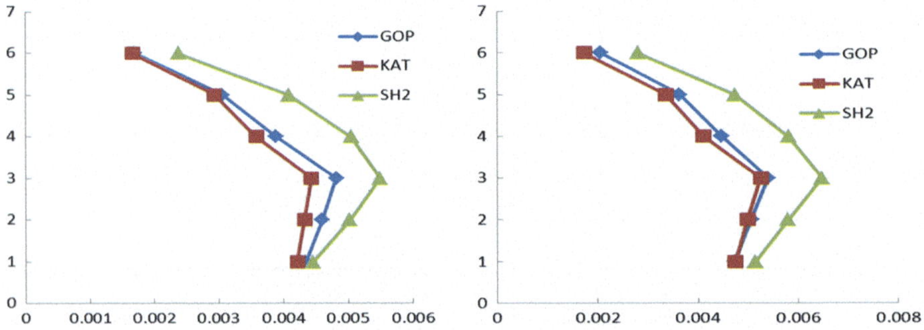

(a) X-direction interlayer displacement angle

(b) Y-direction interlayer displacement angle

Fig. 14.15 Interstory displacement angle of the structure under rare earthquakes

component states are shown in Table 14.13. Before seismic absorption, 53 beams and 14 columns entered IO state, 53 beams and 14 columns entered LS state, 53 beams and 14 columns entered CP state, 40 beams and 18 columns approached IO state, 40 beams and 17 columns approached LS state, 22 beams and 8 columns approached CP state. After seismic absorption, 14 beams and 5 columns approached IO state, 14 beams and 5 columns approached LS state, 7 beams and 2 columns approached CP state. It can be seen that the use of anti-buckling shear steel plate damper for seismic

Table 14.13 Selection of performance objectives of reinforced concrete beams and columns

Component type	Plastic corner at different stages (rad)		
	IO	LS	CP
Frame beam	0.005	0.005	0.01
Frame column	0.003	0.012	0.015

reinforcement can effectively delay the time when frame beams and columns enter the plastic state under rare earthquakes, thereby improving the seismic capacity of the structure.

14.4 Tall Residential Building (Rubber Isolator, Earthquake)

14.4.1 Project Overview

The project consists 31 seismically isolated RC tall buildings used as office and residential building [2, 3]. The aerial view of the project is shown in Fig. 14.16. These buildings are located in a near-fault region with the seismic fortification intensity degree 8.5 according to Chinese code for seismic design of buildings (GB50011-2010, 2010), which means the corresponding peak ground acceleration (PGA) values of the service level earthquake (SLE), design basis earthquake (DBE), and maximum considered earthquake (MCE) are 110, 300, and 510 cm/s^2 with a 63%, 10%, and 2% probability of exceedance in 50 years, respectively. The site condition belongs to Site Class III and second group in GB50011-2010; the characteristic period of the site is 0.55 s. The closest distance from these buildings to the rupture plane (i.e., R) is 7.5 km. When R is less than 10 km and greater than 5 km, a near-field influence coefficient with a minimum value of 1.25 is introduced to consider the near-field effect. Because the seismic design load of these buildings is extremely large and the occupancy importance is very high, a special advisory committee (referred to as "the committee" hereafter) was established to guide and monitor the seismically isolated design. The near-field influence coefficient was recommended to be 1.25 through a

Fig. 14.16 Aerial view of the seismically isolated RC tall buildings project

detailed discussion by the committee. This means that the PGA values of SLE, DBE, and MCE will be increased to 137.5, 375.0, and 637.5 cm/s^2, respectively.

The characteristic properties of these tall buildings are listed in Table 14.14. The number of floors of the 31 tall buildings ranges from 17 to 21 for superstructures and

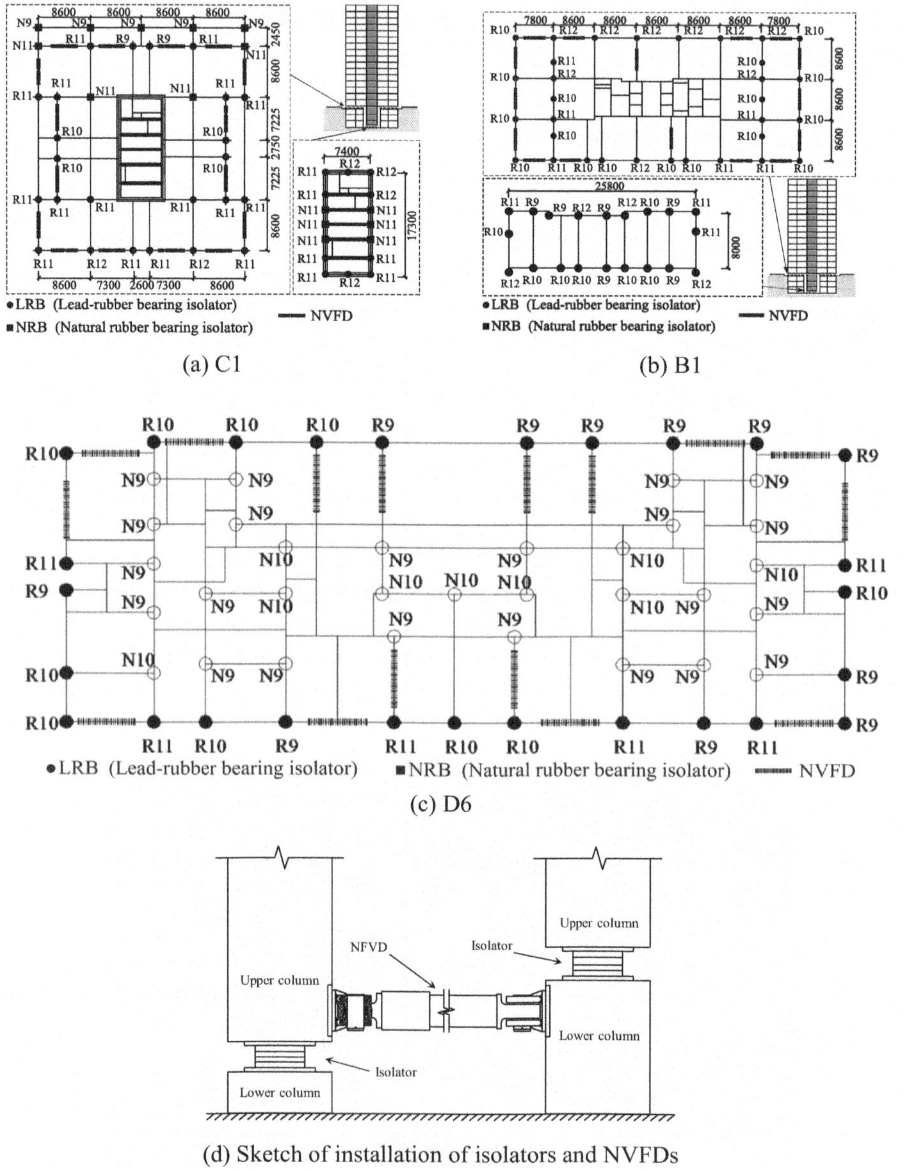

Fig. 14.17 Isolation system of typical buildings

Table 14.14 Characteristic properties of 31 seismically isolated RC tall buildings

Building notation	Total height (m)	Height-to-width ratio	Number of stories		Structural system
			Above ground	Underground	
B1, B2, B3	61.5	2.38	17	4	Frame-shear wall
B4, B5, B6	61.5	2.38	17	4	Frame-shear wall
C1	79.2	2.30	22	4	Frame-core tube
C2	79.2	2.30	22	4	Frame-core tube
C3	72.6	2.11	21	4	Frame-core tube
C4	65.8	1.91	17	4	Frame-core tube
C5	65.6	1.91	17	4	Frame-core tube
C6, C7, C8	68.3	1.98	20	4	Frame-core tube
D1, D2, D3	68.5	3.59	21	4	Shear wall
D4, D5, D6, D7	70.4	3.69	23	4	Shear wall
E1, E2	75.1	2.90	22	5	Frame-shear wall
E3	76.8	2.98	22	5	Frame-shear wall
G1, G2	73.4	2.84	21	5	Frame-shear wall
G3, G4	68.3	2.82	20	5	Frame-core tube
H1	70.0	1.98	20	5	Frame-core tube
H2	71.7	2.03	21	5	Frame-core tube

4–5 for basement, with total hight ranges from 61.5 to 79.2 m. The structure system utilized in this project including shear wall structure, frame-shear wall structure and frame-core tube structure. The three-dimensional views of typical buildings and typical floors are shown in Figs. 14.18 and 14.19. The thickness of the outer shear walls and inner shear walls are 350–600 mm and 200–400 mm, respectively. The sectional dimensions of the columns are 700 × 700 mm–900 × 900 mm. The thickness of the floors are 120–300 mm (Fig. 14.17).

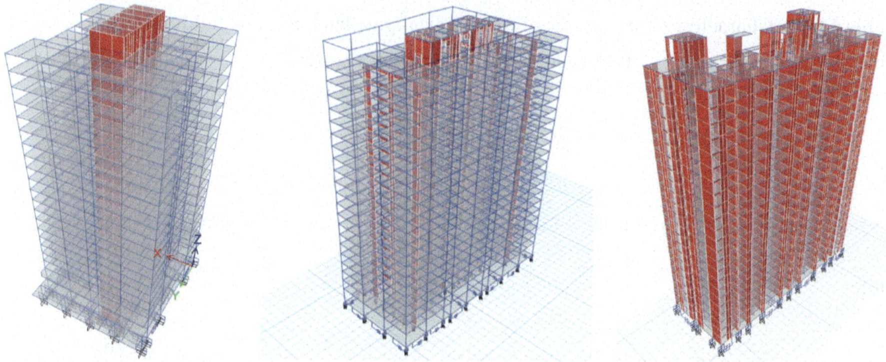

(a) Frame-core tube structure (b) Frame shear wall structure (c) Shear wall structure

Fig. 14.18 Three-dimensional views of typical buildings

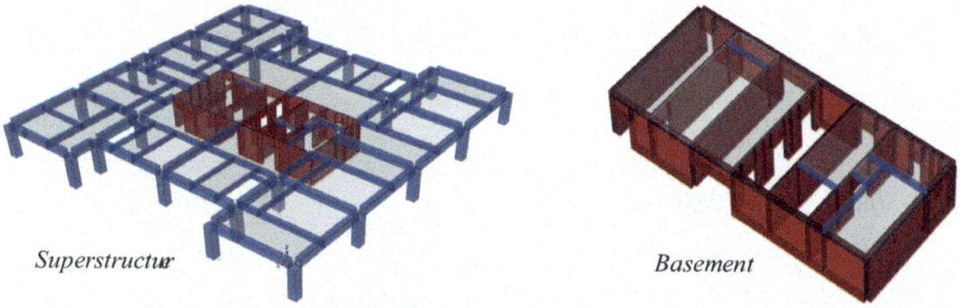

(a) Frame-core tube structure

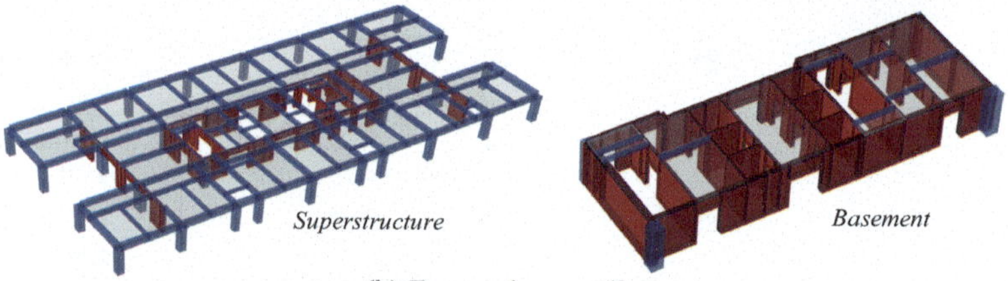

(b) Frame shear wall structure

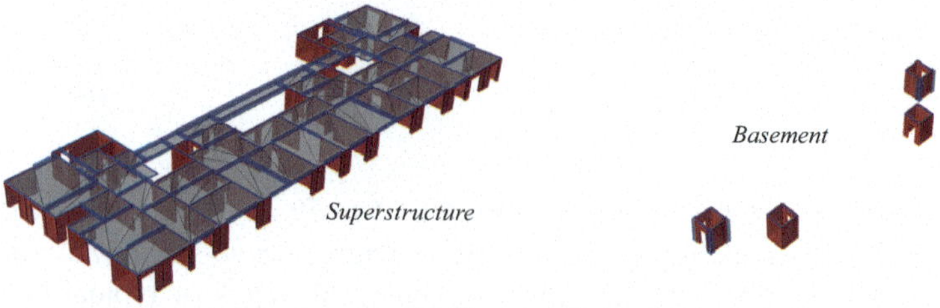

(c) Shear wall structure

Fig. 14.19 Three-dimensional view of typical floors of typical buildings

14.4.2 Structural Isolation Design

The CBI scheme, which isolates the superstructure in an elevation of ±0 m, cannot satisfy the architectural design requirements. Meanwhile, the base-isolated scheme, which isolates the entire structure at the bottom of the basement, leads to a total height of up to approximately 95 m, thus leading to unacceptable tensile stress in the isolators. Hence, a PBI scheme as shown in Fig. 14.17 is recommended by the committee and adopted for these four buildings. Specifically, the columns are isolated in an elevation of ±0 m, while the core tubes or shear walls are isolated at the bottom of the basement.

Because of the width limit of the isolation gap, which is recommended to be 600 mm by the committee, nonlinear fluid viscous dampers (NFVDs) are adopted to control the maximum bearing displacement (MBD) under MCE. The layout of the lead rubber bearing (LRB) isolators, natural rubber bearing (NRB) isolators, and NFVDs of typical buildings are schematically shown in Fig. 14.17a–d. The schematic diagram for the installation of NFVD are presented in Fig. 14.17e. The characteristic parameters of all isolators are listed in Table 14.15.

Table 14.15 Characteristic parameters of the isolators

Type	NRB900	NRB1100	LRB900	LRB1000	LRB1100	LRB1200
Notation	N9	N11	R9	R10	R11	R12
Effective diameter (mm)	900	1100	900	1000	1100	1200
Total rubber thickness (mm)	176	216	176	197	216	235
Vertical stiffness (kN/m)	3,630,000	4,519,000	4,168,000	4,639,000	5,550,000	5,940,000
Equivalent stiffness at 100% shear strain (kN/m)	1110	1358	2070	2300	2450	2600
Post-yield stiffness (kN/m)	/	/	1070	1190	1310	1470
Horizontal yield force (kN)	/	/	238	294	355	410
Rubber shear modulus (N/mm^2)	0.32	0.32	0.32	0.32	0.32	0.32

Table 14.16 Critical design parameters and indices of typical buildings

Indices	C1	B1	D6
T_f (s)	1.59	1.43	1.56
T_{is} (s)	4.44	4.13	3.88
β	0.36	0.35	0.37
MBD (mm)	488	495	486
σ_{max}^{g} (MPa)	12.38	13.27	13.82
σ_{max}^{p} (MPa)	29.07	26.16	28.54
σ_{max}^{t} (MPa)	0.00	0.00	0.10

14.4.3 Analysis of the Isolation Structure

The critical design parameters and indexes of the prototype buildings are listed in Table 14.16. The fundamental period of the corresponding fixed-base structure is denoted as T_f. The isolation period, T_{is}, is calculated using the equivalent stiffness of the isolators when the horizontal shear strain of isolator is 100% according to the GB50011-2010.

The horizontal seismic absorbing coefficient, β, which indicates the efficiency of the isolation system, is one of the most important indexes in the design of seismically isolated structures. It is defined as the maximum shear force ratio of the isolated structure to that of a fixed-base structure in each story subjected to the DBE. For seismically isolated building with NFVDs, β is required to be no more than 0.38 according to the GB50011-2010.

The MBD under the MCE is another important design index that determines the width of the isolation gap (i.e., D_g). Specifically, MBD = $\max\{\eta_i \cdot u_c\}$, where η_i is the torsion influence coefficient (i.e., a scale factor to consider the effects of torsion) of ith isolator with a value of 1.15 for the prototype buildings, and u_c is the maximum displacement of the mass center of the isolation system under the MCE. For each isolator, MBD is required to be no more than the smaller of the two values of 0.55 times the effective diameter of the isolator and 3.0 times the total rubber thickness. In addition, D_g is required to be no less than 1.2 times the MBD. As for the prototype buildings, D_g is 600 mm, which is larger than 1.2 times the MBD.

The maximum compressive and tensile stresses of the isolators are also important design indexes for the seismically isolated structures. According to the GB50011-2010, the compressive stress of isolators under gravity load (i.e., σ_{max}^{g}) should be no more than 15 MPa. Furthermore, the compressive and tensile stresses of isolators under MCE (i.e., σ_{max}^{p} and σ_{max}^{t}) should be no more than 30 MPa and 1 MPa, respectively. The typical results presented in Table 14.16 indicate that the design of the typical buildings can meet the abovementioned requirements.

References

1. Chang, Yejun. 2003. *Research and application of viscoelastic damper and energy dissipation structure*. Nanjing: Southeast University. (in Chinese).
2. Aiqun, Li, Cantian Yang, Linlin Xie, et al. 2017. Research on the rational yield ratio of isolation system and its application to the design of seismically isolated reinforced concrete frame-core tube tall buildings. *Applied Sciences-Basel* 7 (11): 1191.
3. Yang, Cantian, Linlin Xie, and Aiqun Li. 2019. Ground motion intensity measures for seismically isolated RC tall buildings. *Soil Dynamics and Earthquake Engineering* 125: 105727.

Chapter 15
Engineering Practice of Vibration Control for Tall Structures

Abstract Project cases of vibration control engineering practice for tall structure are introduced. Case 1 is Beijing Olympic Tower with the height 248.6 m and TMD are used for wind resistance. Case 2 is the Nanjing TV Tower with the height 310.1 m and AMD (Active TMD) are used for wind resistance. Case 3 is Beijing Olympic Multi-functional Broadcasting Tower with the height 135 m, TMD and variable damping viscous damper are used for wind resistance. Case 4 is the proposed Hefei TV Tower with the height of 339 m, and TMD are designed for earthquake and wind resistance. In every case, project overview, structural energy dissipation devices, structural analysis model, analysis of structural vibration absorption performance and field test and analysis are introduced.

15.1 Beijing Olympic Tower (Wind Vibration, TMD)

15.1.1 Project Overview

The Beijing Olympic Tower (Fig. 15.1) is located in the central area of the Olympic Forest Park [1]. The main building is adjacent to the central axis landscape avenue. The building area is about 6600 m^2, the above ground area is 4946.50 m^2, and the underground area is 13,030,00 m^2. The tower is one of the symbolic buildings of the Olympic Park. It is composed of five independent towers with different heights and layers. The middle tower is the highest, the height of the building top is 248.6 m, and the four small towers around the middle main tower are connected with it through corridors. The plane of the tower is hexagonal and circular, and several lower floors are the standard floors. After reaching a certain height, the plane of each tower gradually enlarges and the whole tower is mushroom-shaped. The main tower body of this structure is composed of outer shell and core frame tube. There are 16 concrete filled steel tubular columns in the outer shell, which are supported by I-beam and H-beam. The body of auxiliary tower is composed of shell, which consists of six concrete filled steel tubular columns supported by I-shaped steel beams and column supports. In order to enhance the overall lateral stiffness of the structure, a connecting truss is set up between five single towers to form a connected structure, which greatly

A. Li, *Vibration Control for Building Structures*, Springer Tracts in Civil Engineering, https://doi.org/10.1007/978-3-030-40790-2_15

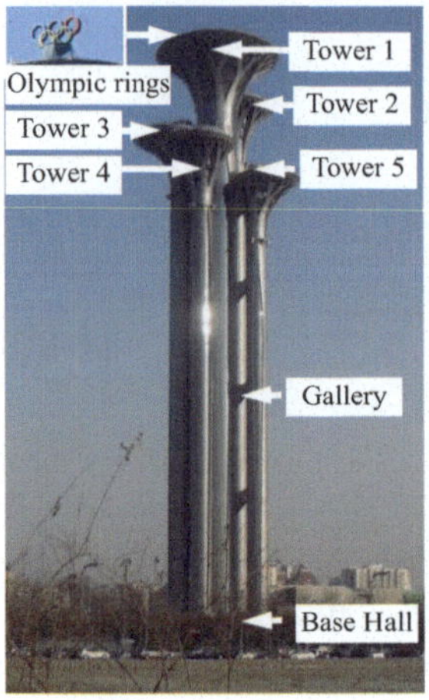

(a) Architectural Scene

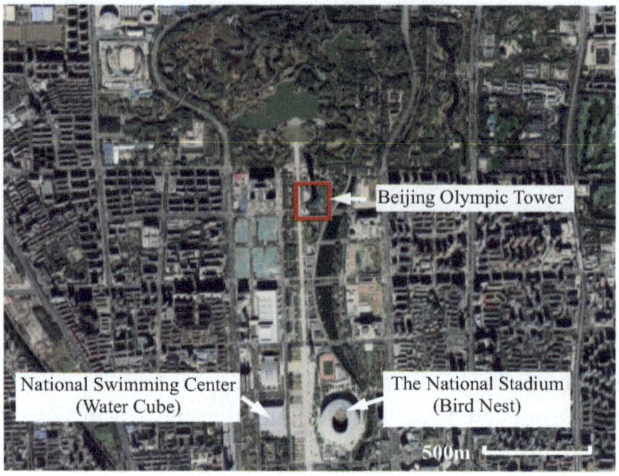

(b) Architectural orientation

Fig. 15.1 Beijing Olympic Tower

Table 15.1 Structural dimensions of Beijing Olympic Tower

Tower number	Tower 1	Tower 2	Tower 3	Tower 4	Tower 5
Building contour diameter/m	16.20	9.60	9.60	8.30	8.30
Axis diameter of structure/m	14.00	7.30	7.30	6.00	6.00
Total roof height/m	244.35	228.00	210.00	198.00	186.00
Aspect ratio	17.45	31.23	28.77	33.00	31.00
Height of top/m	31.35	21.00	21.00	21.00	21.00
Diameter of roof part/m	51.20	33.60	32.40	30.00	26.40
Distance to tower 1/m	–	19.17	15.75	15.10	16.90

improves the anti-overturning ability of the structure. The height and width of the connecting truss are about 3.0 and 2.7 m, respectively, and there are axillaries on one side of the main tower. The detailed structural dimensions are shown in Table 15.1.

15.1.2 Structural Vibration Reduction Design Using TMD

Considering the influences of architecture and technology, TMD could only be installed in tower 1. Therefore, the water tank of equipment layer at 232.5 m elevation of the structure was directly used as the quality unit considering no additional mass of the structure. As shown in Fig. 15.2, the TMD was in the fire water tank in the middle of tower 1 on the plane. Both theoretical calculation and experimental analysis show that when the frequency of TMD is close to the natural frequency of

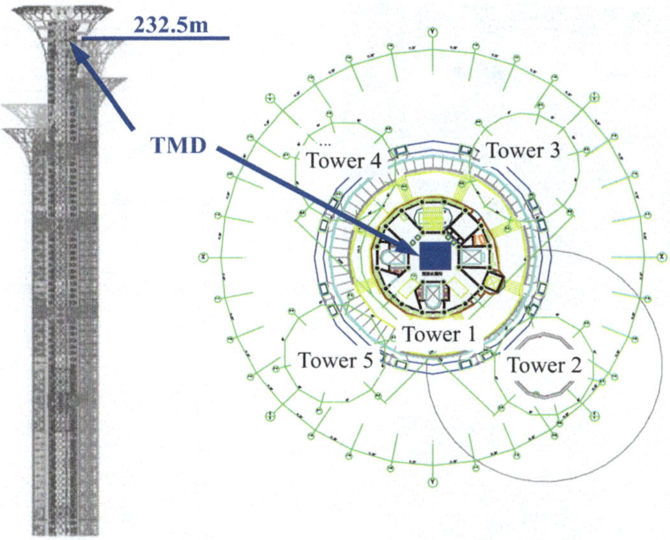

(a) Elevation position of TMD (b) Plane position of TMD

Fig. 15.2 Installation location of TMD

the structure and the damping ratio is within a certain range, the vibration reduction effect is better. In order to design TMD accurately, the wind-induced responses of the structure before and after TMD installation were analyzed using finite element model and three-dimensional wind field based on wind tunnel test, and the frequency and damping ratio of TMD were optimized. The design parameters for TMD were obtained as shown in Table 15.2.

The TMD used for high-rise structures needs to meet three structural requirements in engineering practice: long period, limited space, and multi-directional. In order to achieve these characteristics, the water tank was lifted by cables to form a TMD structure of suspension pendulum. Thus, the similar dynamic performance of TMD in many directions was achieved. At the same time, considering the contradiction between long period and limited space, the concept of two-stage suspension pendulum was adopted to design the TMD. As shown in Fig. 15.3, the TMD consisted of cables, suspension frame, fire water tank, water tank bracket, viscous damper, and other components.

Table 15.2 Parameter of TMD system

Direction	X	Y
Stiffness (N/m)	64,000	64,000
Mass (t)	50	50
Damping index	1	1
Damping coefficient (N s/m)	13,576	13,994

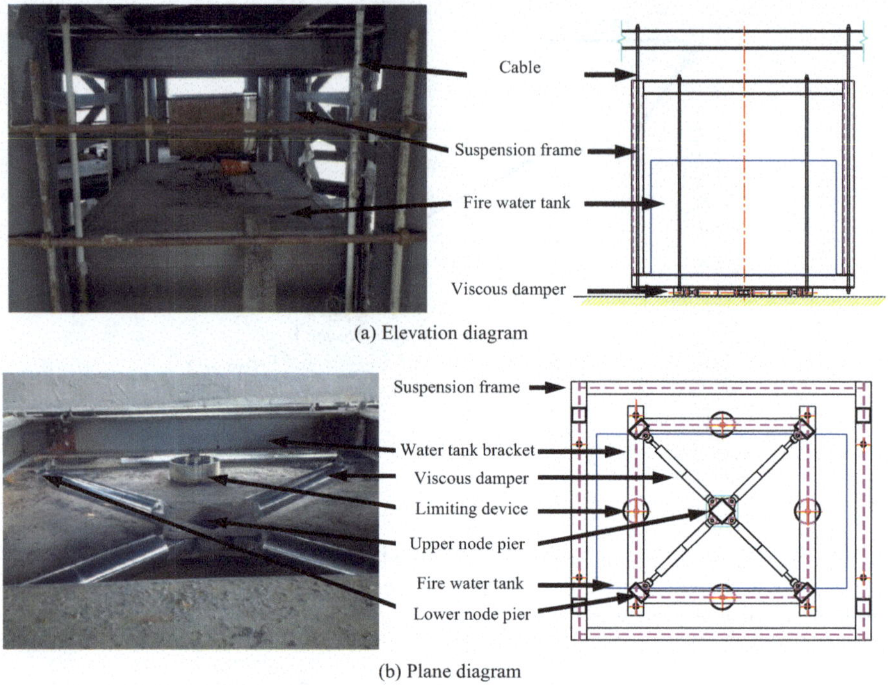

Fig. 15.3 Construction of two-stage suspended pendulum TMD

As a quality unit, the weight of empty fire water tank was 32.8 t, and the total weight was 50 t. The effective length of the cable was 8600 mm, and the theoretical length was 8350 mm. The water tank was mounted on the water tank bracket, and was suspended on the external frame through the water tank bracket. The external frame was suspended on the main structure through the cables. A two-stage suspension system was formed, and a long-term suspension TMD was realized in the limited space (4.80 m * 4.20 m * 4.55 m). In order to provide reasonable damping for TMD, four viscous dampers were installed at the bottom of the tank. One of the ends of the damper was connected to the floor through the lower node pier, and the other end was connected to the water tank through the upper node pier. The dampers and cables were adjusted within a certain range according to field test results to ensure better control effect.

15.1.3 *Structural Analysis Model*

According to the structural design information, a three-dimensional finite element model was established as shown in Fig. 15.4, in which the spatial beam element was adopted to simulate the steel beams, columns, and trusses; the quadrilateral spatial

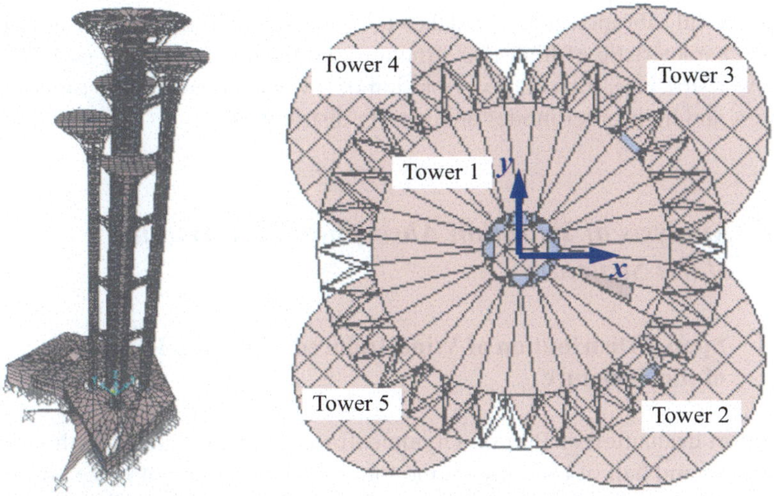

(a) 3D diagram (b) Plane diagram

Fig. 15.4 Structural FE model

shell element was adopted to simulate the reinforced concrete slabs and walls; in order to consider the restraint effect of the base on the tower, the model of the base was established according to the actual situation, and the nodes at the bottom of base were fixed.

The frequency and description of the first six modes of the structure were obtained by analyzing the dynamic characteristics of the structure as shown in Table 15.3. Because the corridor stiffness is large and the towers and tower 1 are well connected as a whole, the first several modes show the mode of global vibration, of which the first and second modes are global bending and the third modes is global torsion;

Table 15.3 Natural frequency of the structure

Order	Frequency/Hz	Description
1	0.176	First-order global bending (45°)
2	0.181	First-order global bending (135°)
3	0.227	Antisymmetric first-order bending of towers 2 and 4 (45°) + antisymmetric first-order bending of towers 3 and 5 (135°) (first-order global torsion)
4	0.512	Second-order global bending (45°)
5	0.559	Second-order global bending (135°)
6	0.598	Antisymmetric second-order bending of towers 2 and 4 (45°) + antisymmetric second-order bending of towers 3 and 5 (135°) (second-order global torsion)

Note The 45° and 135° in the table indicate the direction of vibration in the X-Y plane with 45° and 135° angles to the X-axis the first natural vibration period reaches to 5.687 s, which is a typical long-period dynamic characteristic of high-rise structure; the torsion cycle ratio is 0.775. As a multi-tower structure, the coupled vibration of towers is unavoidable for connected structures, and high-order modes are densely distributed.

15.1.4 Analysis of Vibration Absorption Performance of the Structure

15.1.4.1 Spatial Distribution of Wind-Induced Vibration Response of the Structure

The comfort evaluation of Beijing Olympic Tower is mainly aimed at the observation platform and hall of each tower (the roof of the tower crown is the observation platform, and the next floor is the observation hall). The distribution of acceleration response of the structure at wind direction angles of 45° and 135° is shown in Fig. 15.5, (1) the accelerations of towers 2 and 4 are relatively large, while the acceleration responses of towers 1, 3, and 5 are relatively small; (2) the acceleration response of the tower crown is larger than that of the tower body, and there is a vertical acceleration response at the top of the tower crown, which is greater than 0.14 m/s^2, due to the cantilever shape; (3) the maximum total acceleration of the three-dimensional synthesis appears at the edge of the tower crown of tower 5, reaching to 0.23 m/s^2 under the wind of 45° and 0.25 m/s^2 under the wind of 135°, the local comfort cannot meet the requirements; (4) after the installation of TMD, the spatial distribution of the response is similar to that of the original structure, and the horizontal and three-dimensional acceleration responses decrease in a certain range. Because the horizontal TMD is installed, the control effect of vertical acceleration is relatively weak. Overall, the acceleration response distribution of the structure under different wind directions basically follows the law of gradually increasing from bottom to top and from inside to outside.

15.1.4.2 Wind Direction Distribution of Wind-Induced Vibration Response

Based on the calculation results under 8 main wind directions, the effects and laws of TMD control under different wind directions were discussed. The horizontal acceleration responses of the viewing platform under the wind loads of different wind directions are shown in Fig. 15.6. Among them, the mean and maximum values are the average and maximum values of the peak acceleration of each node in the layer, respectively. The analysis shows that: (1) the maximum acceleration response usually occurs at the edge of the tower crown, and the restraints of the structure on these points are relatively weak, so the TMD has a good control effect on the maximum acceleration value relative to the mean value; (2) the vibration reduction effect is

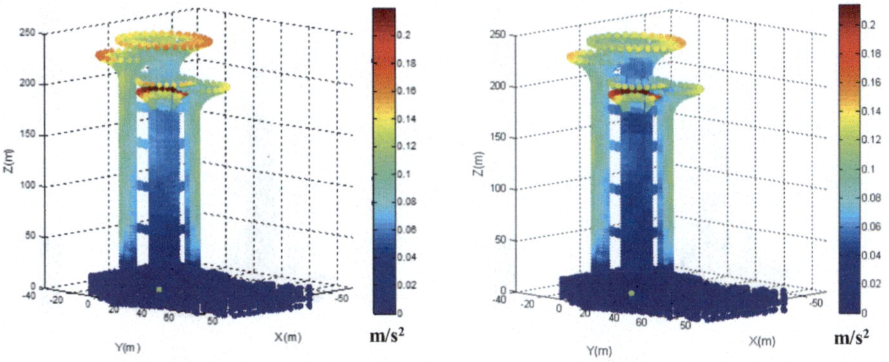

(a) Horizontal direction (45° wind direction angle, the left figure represents original structure, the right figure represents shock absorption structure)

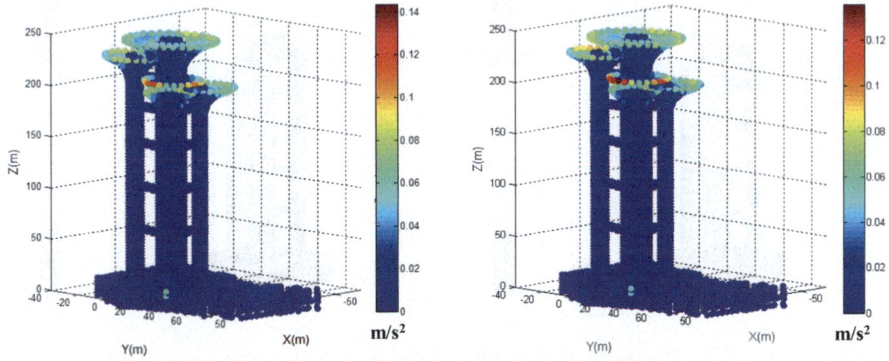

(b) Vertical direction (45° wind direction angle, the left figure represents original structure, the right figure represents shock absorption structure)

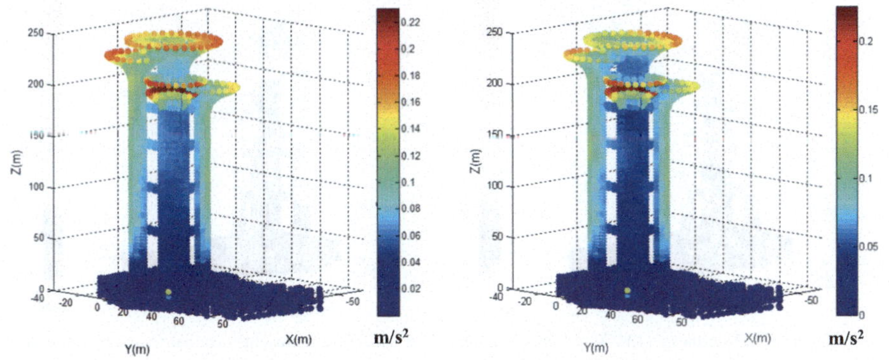

(c) Tridirectional synthesis (45° wind direction angle, the left figure represents original structure, the right figure represents shock absorption structure)

Fig. 15.5 Spatial distribution of structural acceleration response

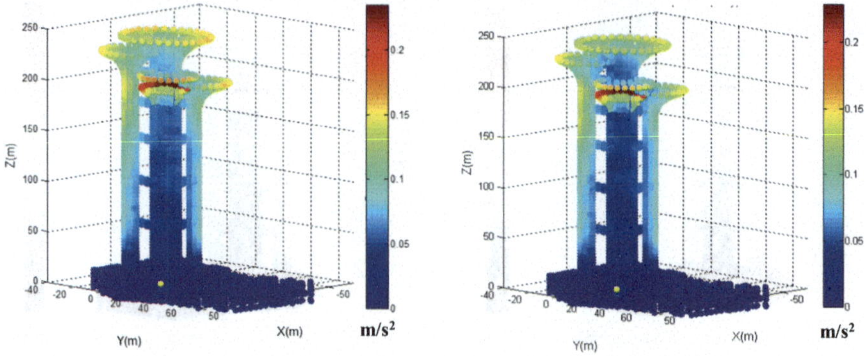

(d) Horizontal direction (135° wind direction angle, the left figure represents original structure, the right figure represents shock absorption structure)

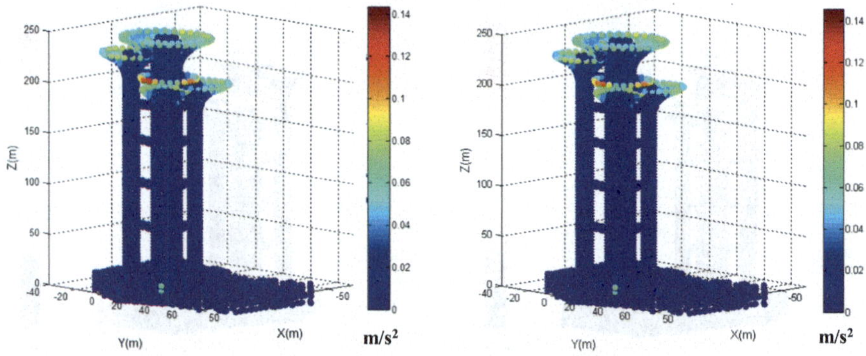

(e) Vertical direction (135° wind direction angle, the left figure represents original structure, the right figure represents shock absorption structure)

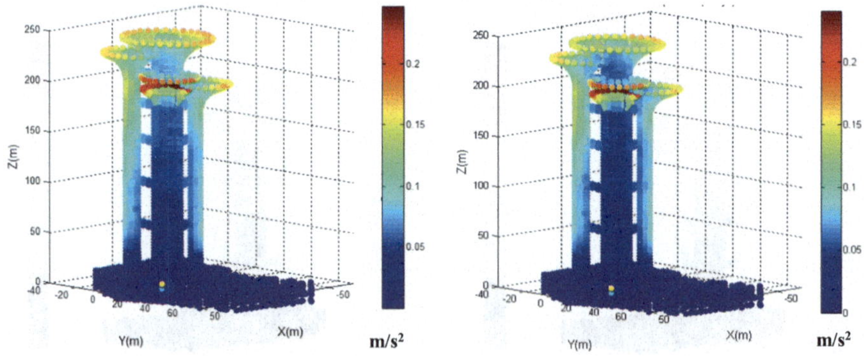

(f) Tridirectional synthesis (135° wind direction angle, the left figure represents original structure, the right figure represents shock absorption structure)

Fig. 15.5 (continued)

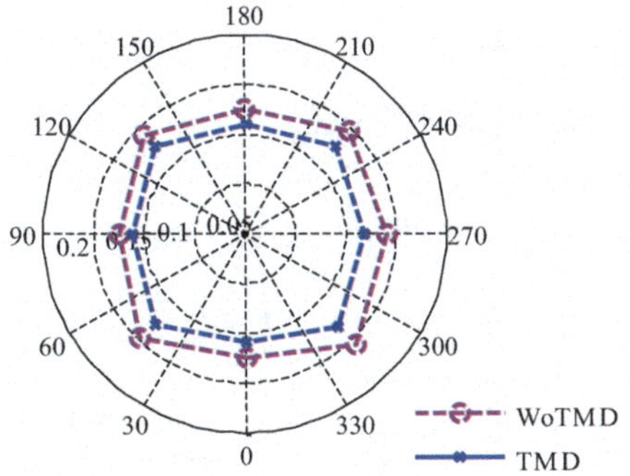

(a) Mean value of tower 1

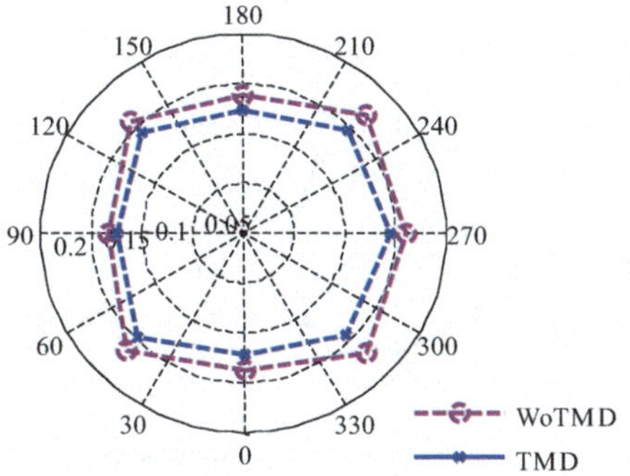

(b) Maximum value of tower 1

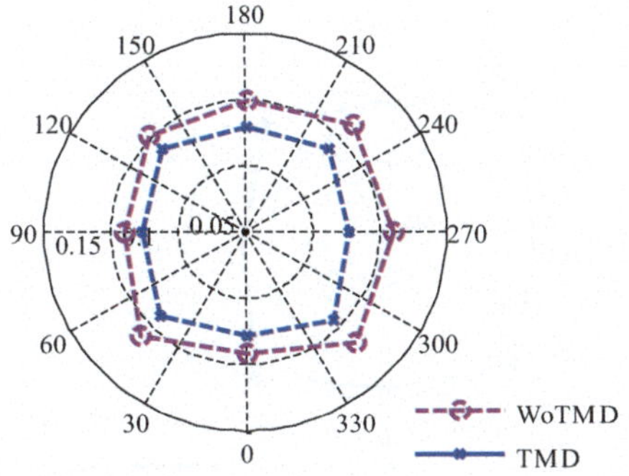

(c) Mean value of tower 2

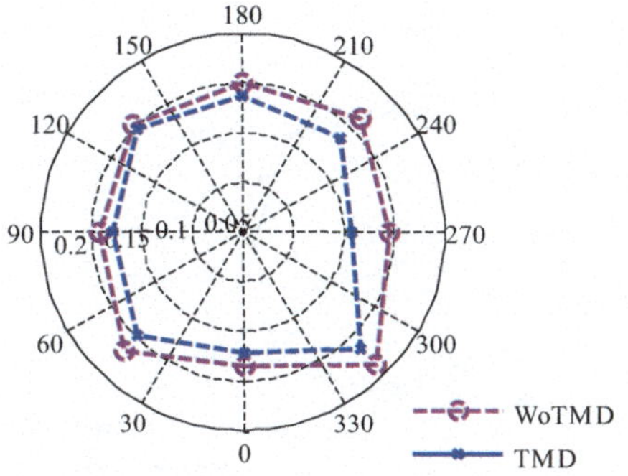

(d) Maximum value of tower 2

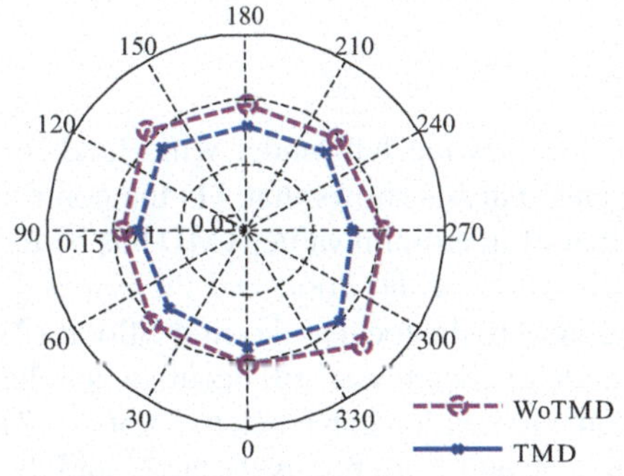

(e) Mean value of tower 3

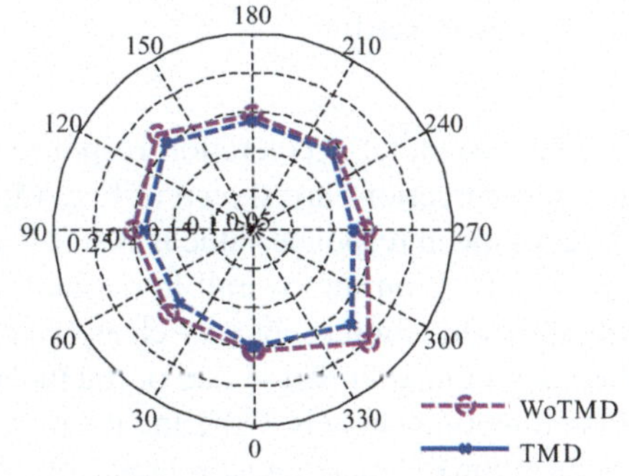

(f) Maximum value of tower 3

Fig. 15.6 Horizontal acceleration response of viewing platform of each tower

the best when the wind direction angle of each tower is 270°, the attenuation rates of tower 1, 2, and 3 can reach to 16.44%, 30.05%, and 22.63%, respectively; (3) the acceleration response of tower 5 decreases slightly, and the TMD installed in the main tower has poor effect on the vibration reductions of some towers, and the distribution of vibration reduction effect of each tower is different.

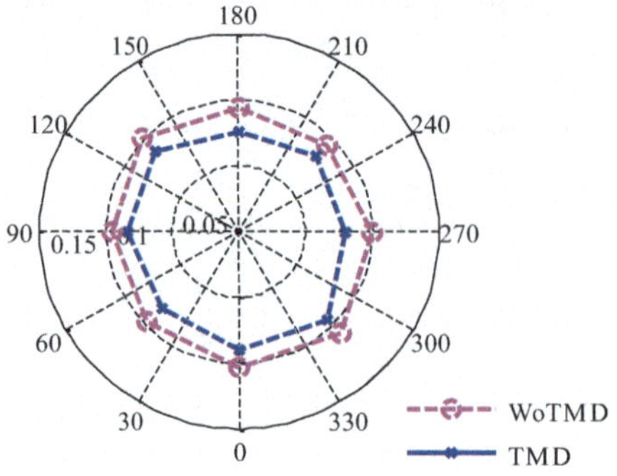

(g) Mean value of tower 4

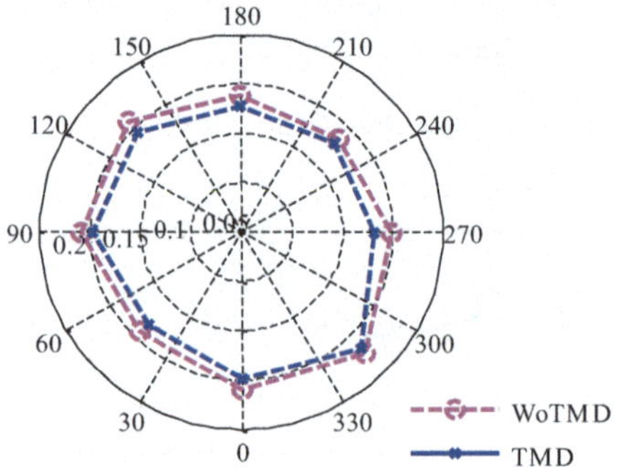

(h) Maximum value of tower 4

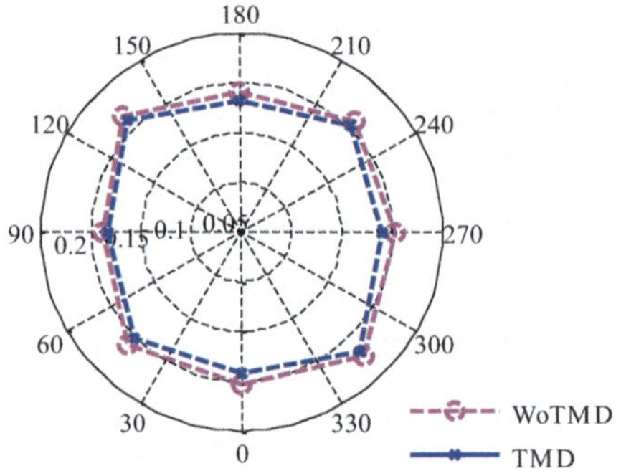

(i) Mean value of tower 5

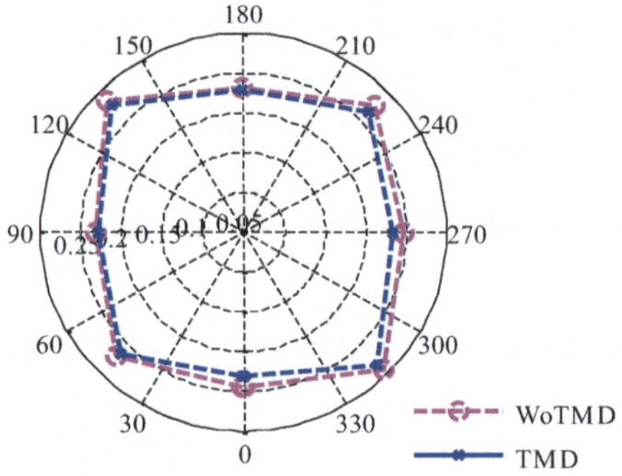

(j) Maximum value of tower 5

Fig. 15.6 (continued)

The horizontal acceleration responses of the viewing hall under wind loads of different directions are shown in Fig. 15.7. The analysis shows that: (1) the control effect of mean response value is better than that of maximum value, TMD has better control effect on the overall horizontal acceleration of the floor, and the effect of vibration reduction is not very clear when it comes to the local nodes in the floor; (2) the maximum attenuation rate of the mean value of each tower still occurs when the wind direction angle is 270°, and the maximum attenuation can reach to 24.56%; (3) for each tower, the attenuation rates of tower 1, 2, and 3 are relatively large, and the response attenuation of tower 5 is the smallest.

In conclusion, the TMD control effect for the horizontal acceleration response is the best when the wind direction angle is 270°. The vibration reduction effect of tower 5 is relatively poor, mainly because the connection between tower 1 and tower 5 is relatively weak, and the control effect of TMD installed in tower 1 on tower 5 is very limited. For this kind of complex high-rise structure, the vibration characteristics are more complex, which leads to different distribution law of wind-induced responses along different wind directions of different tower. In order to obtain reliable results, the detailed multi-condition analysis is needed in the design.

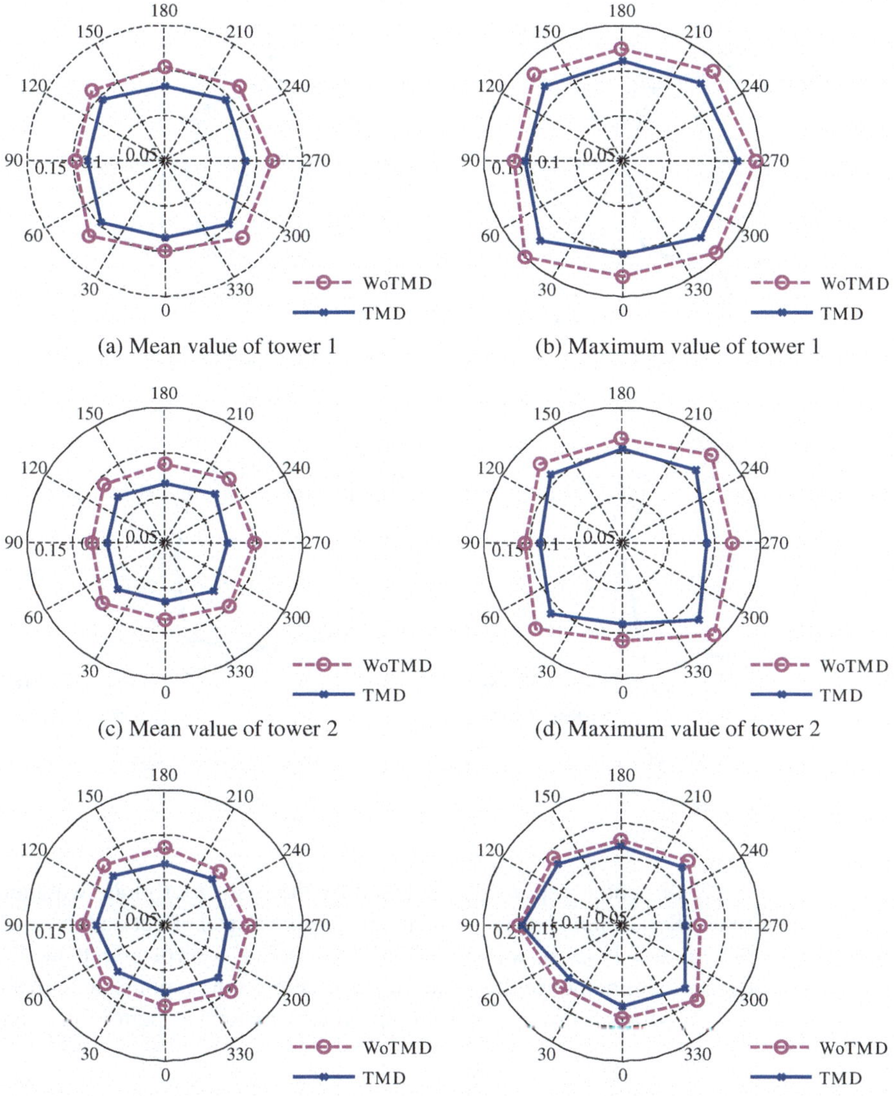

(a) Mean value of tower 1

(b) Maximum value of tower 1

(c) Mean value of tower 2

(d) Maximum value of tower 2

(e) Mean value of tower 3

(f) Maximum value of tower 3

Fig. 15.7 Acceleration response of the view hall of each tower

15.1.4.3 Time and Frequency Domain Analysis of Wind-Induced Vibration Response of the Structure

The time history and power spectral density of X-direction displacement and acceleration of the two points on viewing platform of tower 1 and tower 5 are shown in Figs. 15.8 and 15.9. It can be seen from the graph that: (1) TMD has better con-

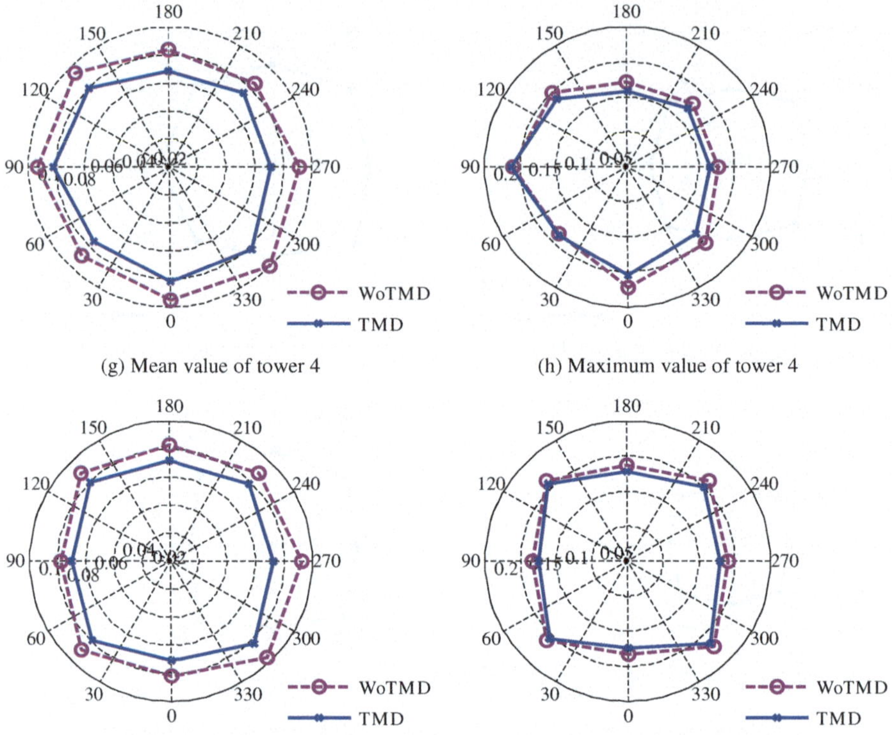

(g) Mean value of tower 4

(h) Maximum value of tower 4

(i) Mean value of tower 5

(j) Maximum value of tower 5

Fig. 15.7 (continued)

trol effect on the X-direction response contributed by the first-order mode whether displacement or acceleration response, while the response control effect on that contributed by other modes is close to 0; (2) TMD has better control effect on the X-direction displacement response than that on the acceleration response. Comparatively speaking, the vibration reduction of the points on tower 1 is better than that on tower 5.

15.1.5 Field Test and Analysis

15.1.5.1 Test Scheme and Equipment

In view of this project, the dynamic characteristics of the structure were measured by the random vibration method in the environment. In this method, the dynamic characteristics of buildings were measured by measuring the small vibration (i.e. "pulsation") generated by random excitation of the environment. The pulsation of

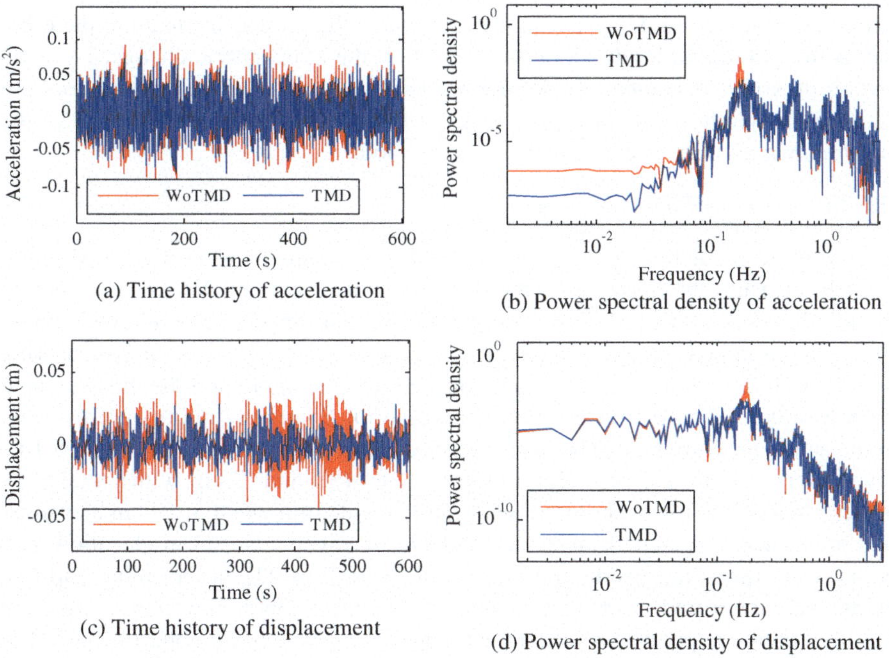

Fig. 15.8 X-directional responses of one point on the view platform of tower 1

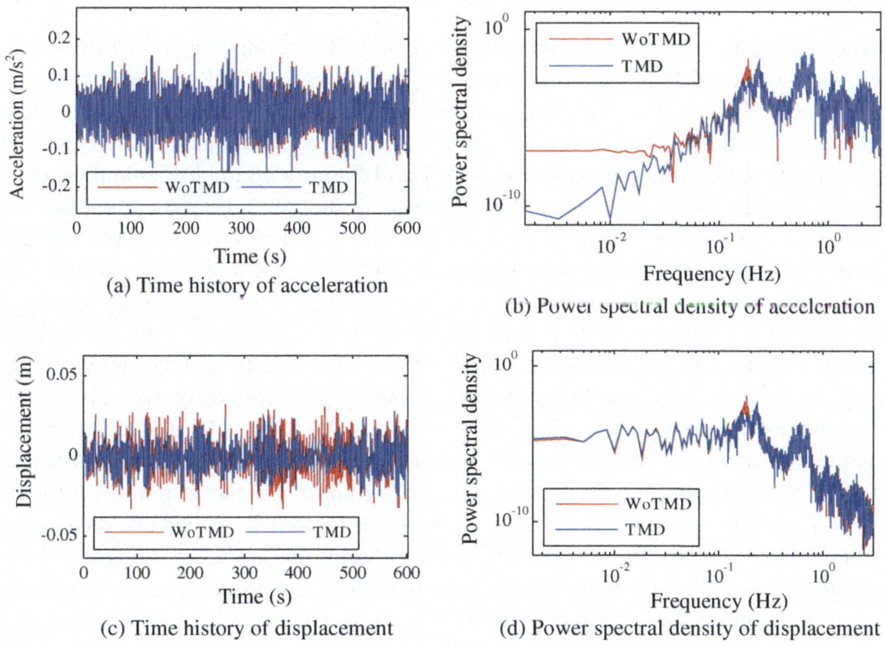

Fig. 15.9 X-directional responses of one point on the view platform of tower 5

buildings comes from two aspects, the first one is the ground fluctuation, the other one is the micro-amplitude vibration caused by the atmospheric change, i.e. wind and air pressure. This method does not need special excitation equipment, does not damage the building structure, and does not affect the normal interior work of the building, which is an effective and simple method.

(1) Structural modal testing scheme

According to the actual situation in the field, four channels were used to simultaneously measure the horizontal vibrations in X direction and Y direction, and an X-direction pick-up and a Y-direction pick-up were arranged at the reference point. An X-direction pick-up and a Y-direction pick-up were arranged at each measuring point, and four-channel measurements of the reference point and the measurement point were carried out at the same time. 63 measuring points on 8 floors were arranged in the actual measurement. The layout of measuring points is shown in Fig. 15.10a. Five towers were measured at elevations of 54, 96, 138, and 174 m, with a total of 11 points. Two points were measured inside and outside of tower 4 at 192 m elevation, two points were measured inside and outside of tower 3 at 203 m elevation, two points were measured inside and outside of tower 2 at 222 m elevation, and five points were measured inside and outside of tower 1 at 238 m elevation. During the test, the reference point was unchanged, which was the middle point of tower 1 at 174 m elevation.

(2) Structural wind vibration response testing scheme

In order to verify the wind-induced response and TMD control effect of the structure, the local wind load should be measured, on the other hand, the corresponding responses of the structure and TMD should be obtained. Considering that the horizontal acceleration values of towers 1 and 5 were larger in numerical analysis, the sensors were installed at the top of towers 1 and 5 to obtain the horizontal acceleration of the structure; considering the validation of the vibration reduction effect of TMD, the sensors were installed on the mass block of TMD and the floor surface where TMD was located to obtain the acceleration response at the same time.

As shown in Fig. 15.10b, the field tests were carried out at structural elevations of 186.00 m, 231.00 m, and 244.35 m, respectively. Because there is no monitoring system installed during the construction of the structure and the distance between the layers is relatively long, synchronous testing could not be realized on site. Therefore, a layered testing method was adopted, that is, only the wind velocity and structural acceleration at the same height were tested synchronously. The data at different altitudes were measured asynchronously. The layout of measuring points on top of towers 1 and 5 is shown in Fig. 15.10b, while the water tank layer had one measuring point on water tank and floor, respectively. Two accelerometers were arranged at each measuring point. Among them, the measuring points 1–2 were set in tower 1, and the measuring points 3–4 were set in tower 5. When tower 1 and tower 5 were tested, the anemometer was installed and fixed on the roof, and the corresponding wind speed

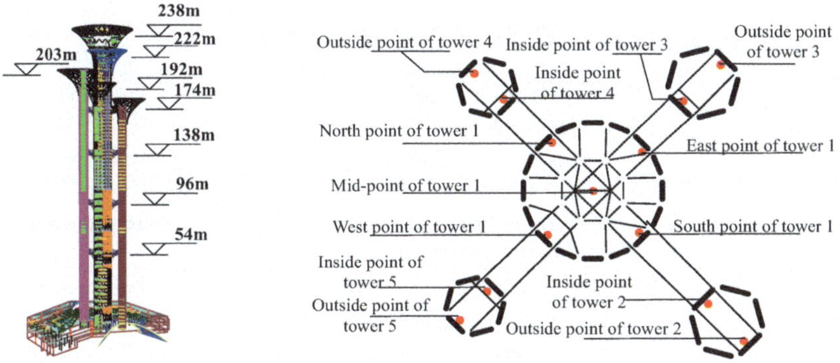

(a) Layout of measuring points for modal testing

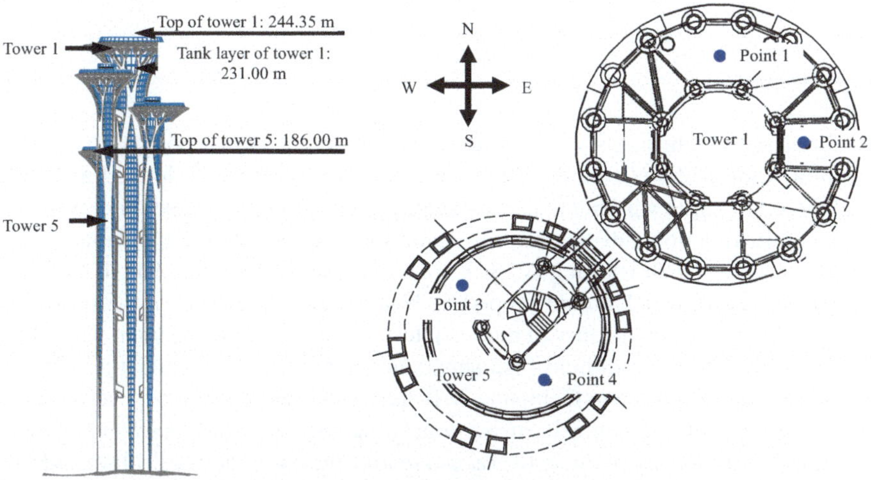

(b) Layout of test points for wind-induced vibration response

(c) Testing equipment

Fig. 15.10 Layout of measuring points

Table 15.4 Technical parameters of main test instruments

Test instrument	Number	Main technical indicators
941-B vibration pick-up	10	Passband 0.25–80 Hz
AZ808 signal filter	2	Passband 0.025–35 Hz
AZ308 signal acquisition box	1	16 pathways
Short cable	12	Length of 1.0 m
AZCRAS analysis System	1	
Cable	10 (100 m of each)	Fit the 941-B vibration pick-up
Two-way connector	5	

was measured synchronously. The north sign of anemometer was pointed to the north direction when installed. So the direction of the north sign was defined as 0°, the wind direction angle was calculated by analogy according to rotated clockwise.

The multi-function anti-aliasing filter amplifier was adopted in the dynamic test, as well as AZ316 acquisition box, AZCRAS analysis system, and 941-B ultra-low frequency pickup, which are developed by Nanjing Anzheng Software Engineering Co., Ltd. The type and quantity of the main test instruments are shown in Table 15.4. The 1590-PK-020 three-dimensional supersonic anemometer manufactured by Gill Instruments and Equipment Company in Britain was used for wind speed measurement in various environments. It can accurately measure the wind speed and direction at a sampling frequency of up to 40 Hz, and realize the automatic collection and storage of measured data. The working temperature of the anemometer is −40 to + 70e, the wind speed is 300–370 m/s, and the humidity is < (5–100%). Considering the long recording time and large amount of data, the sampling frequency was set to 10 Hz. In the test, the wind speed range was set to 0–45 m/s, the measurement accuracy was set to 0.01 m/s, and the wind vector range was set to 0°–359.9°.

15.1.5.2 Structural Modal Information

The dynamic characteristics of the structure were measured on December 19, 2013 and December 13, 2014, respectively. The first test was in the construction stage, most of the non-structural loads were not in place; the second test was in the decoration stage, and some of the non-structural loads were already in place. The first four natural frequencies of the structure obtained from two measured and numerical simulations are shown in Table 15.5. The design model and the modified model in numerical simulation represent the numerical model established in the design stage and the modified model based on the measured results, respectively. The errors were

Table 15.5 Comparison of natural frequencies of the structure

Mode shape	Measurement		Simulation		Error	
	1st test	2nd test	Design model	Modified model	Design model (%)	Modified model (%)
1	0.220	0.200	0.176	0.206	12.0	3.0
2	0.225	0.205	0.182	0.213	11.2	3.8
3	0.300	0.280	0.227	0.260	18.9	7.1

calculated according to the results of the two models relative to the second measured results. The comparisons show that: (1) the natural frequency of the structure decreases to a certain range due to the increase of the load on the structure during the second measurement, and the influence of the load on the natural frequency of the structure is obvious; (2) compared with the second measurement, the result of the design model has a larger error, which is mainly due to the large difference of the effective mass assumed in the design stage and the actual measurement, and partly because some non-structural members take part in the work, which provides additional stiffness which was not considered in the design stage for the structure. At the same time, the high complexity of the structure also has a certain impact on this result; (3) according to the actual load situation during the second construction, the effective quality of the model is adjusted appropriately to obtain the modified model. The analysis results are close to the measured results. It can be seen that the effective mass is indeed one of the important factors affecting the errors between the measured and analyzed results.

The structural damping ratio identified according to structural fluctuation is shown in Table 15.6. The results of the two tests have some errors, but they are still comparable with other similar high-rise or super-high-rise structures, which can provide reference for similar engineering design.

The measured vibration modes of the structure were obtained by processing the pulsation data of each measuring point, and compared with the results of finite element analysis, as shown in Fig. 15.11. The analysis shows that: (1) because of the high structure, the plane dimensions of the towers are relatively small, and the top mass is large, which has obvious long-period structural characteristics; (2) because of the large stiffness of the corridor, the towers are closely connected with tower 1, which makes the structural vibration show obvious overall vibration characteristics; (3) the first two vibration modes of the structure are translational in different directions, and the third one is torsional vibration, the 4th and 5th order are translational,

Table 15.6 Measured structural damping ratio

Mode shape	1st test/%	2nd test/%	Mean value
1	2.19	1.15	1.67
2	0.33	0.88	0.605
3	1.42	1.43	1.425

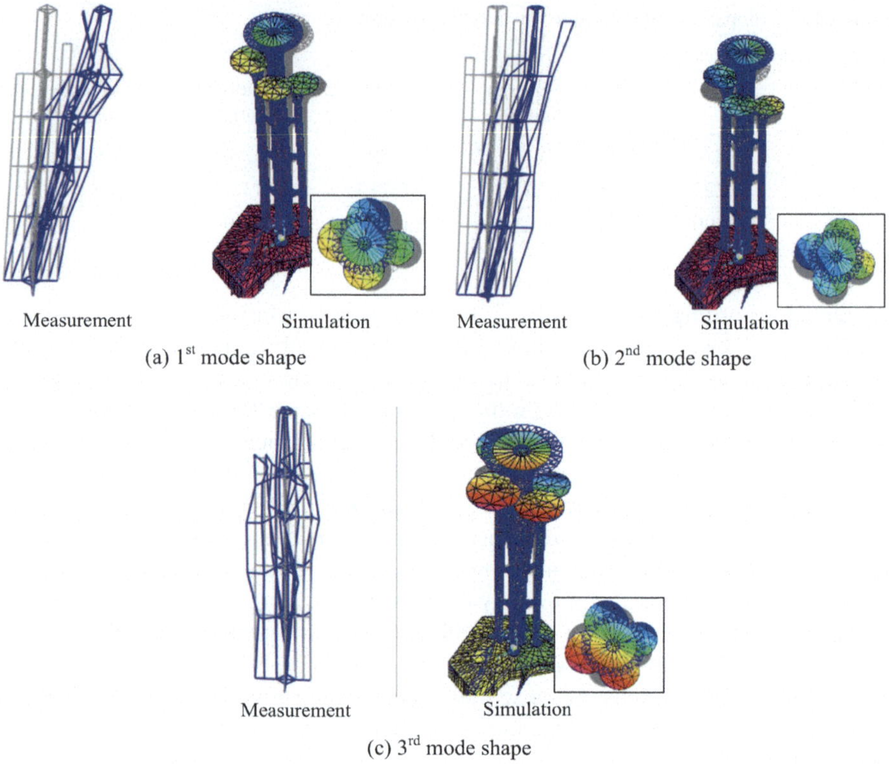

Fig. 15.11 Comparison of structural mode shapes

the 6th order are torsional, and the order of mode shapes is regular; (4) the numerical analysis shows that the modal mass participation coefficients in all directions of the structure can reach to more than 99% by the 300th order of mode shape, so it is necessary to select more mode shapes for dynamic response analysis using modal superposition method.

15.1.5.3 Structural Dynamic Response

(1) Structural response

The top acceleration responses of towers 1 and 5 were tested on December 15, 2014 and December 16, 2014, respectively. The two-way top acceleration responses were obtained after processing the test signals. The acceleration time history curves of the top of the tower at 17:10:00–17:23:20 (measuring points 1, 2) on December 15, 2014 and 17:12:40–17:26:00 (measuring points 3 and 4) on December 16, 2014 are shown in Fig. 15.12a, b. The statistics of the mean square root value and maximum

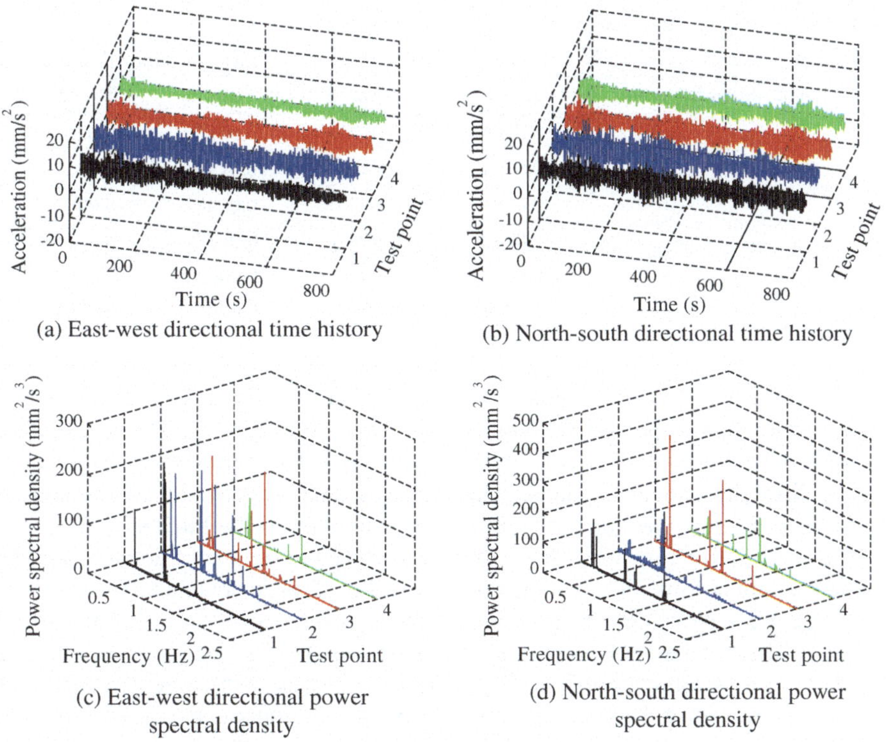

Fig. 15.12 Measured acceleration response of the structure

value of acceleration time history are shown in Table 15.7. The comparison shows that: (1) the maximum value of the two-way acceleration of the structure is smaller due to the small wind speed, and the acceleration response of the structure meets

Table 15.7 Statistical value of measured acceleration response of the structure

Position and direction of measuring points			Root mean square (mm/s^2)	Maximum (mm/s^2)	Test time
Tower 1	Point 1	North–south	2.36	9.53	2014/12/15 17:10:00–17:23:20
		East–west	1.78	8.83	
	Point 2	North–south	2.73	12.93	
		East–west	2.52	10.43	
Tower 5	Point 3	North–south	2.33	8.60	2014/12/16 17:12:40–17:26:0
		East–west	1.88	7.27	
	Point 4	North–south	1.74	7.00	
		East–west	1.16	4.60	

Table 15.8 Damping ratio after structural response identification

Direction	North–south direction		East–west direction	
Location of measuring points	Tower 1	Tower 5	Tower 1	Tower 5
1	0.0206	0.0191	0.0213	0.0201
2	0.0191	0.0160	0.0192	0.0121
3	0.0088	0.0080	0.0088	0.0083
4	0.0080	0.0087	0.0067	0.0075
5	0.0097	0.0040	0.0063	0.0042
6	0.0064	0.0047	0.0061	0.0054

the requirements of building comfort under the daily wind speed; (2) the north-south acceleration response is important due to the northerly wind. The acceleration response of the north-south acceleration is greater than that of the east-west direction, but the difference is not very large. The vertical wind vibration will also be caused by a one-way wind load due to the influence brought by the building shape.

The acceleration time history was further analyzed in frequency domain, and the power spectral density curves shown in Fig. 15.12c, d were obtained. By averaging the frequencies picked up by each curve, the first three frequencies of the structure after TMD installation were 0.200 Hz (east–west), 0.207 Hz (north–south) and 0.273 Hz (east–west and north–south, i.e. torsion). At the same time, the energy of the first two modes of the structure was reduced to some extent, and the contribution of higher order modes to acceleration response increased. The Frequency Domain Decomposition (FDD) method of modal identification was used to process the measured data. The data from different measuring points, at the same time, and in the same direction were used as the mutual power spectral density to identify the damping ratio of the structure, as shown in Table 15.8. The analysis shows that: (1) the average damping ratio of the first mode is 0.0207 at daily wind speed after TMD installation, the average damping ratio of the second mode is 0.0198, and the average damping ratio of the third mode is 0.0085; (2), the damping ratio of east–west direction is slightly larger than that of north-south direction at lower modes, while the damping ratio of north–south direction is slightly larger than that of east-west direction at higher modes; (3) in general, the damping ratio measured at tower 1 is slightly larger than that measured at tower 5. This is because that tower 1 is the main tower, the other towers are connected to tower 1 through corridors, and TMD was installed in tower 1. On the other hand, the average wind speed and structural response of tower 5 were smaller than that of tower 1.

(2) TMD response

The two-way sensors were installed on the water tank and floor of 231.00 m elevation, respectively. The acceleration response time histories are shown in Fig. 15.13a, b, and the time history data are counted in Table 15.9. It can be seen from the comparison that: (1) the acceleration response in the north–south direction is greater than that in the east–west direction in both the water tank and the floor, because the wind

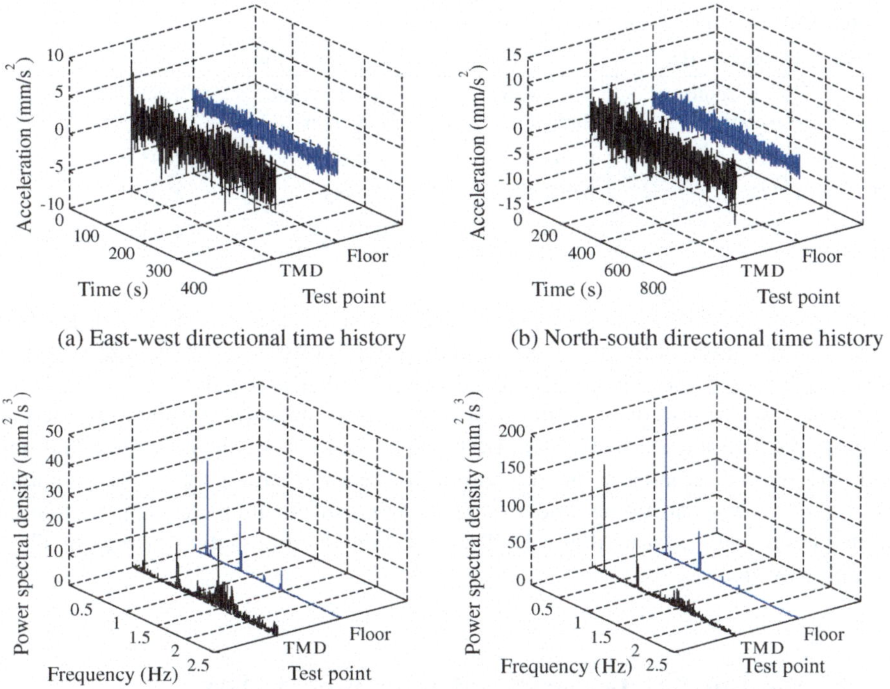

(a) East-west directional time history
(b) North-south directional time history
(c) East-west directional power spectral density
(d) North-south directional power spectral density

Fig. 15.13 Measured acceleration response of TMD

Table 15.9 Statistical value of measured acceleration response of TMD

Location of measuring points		North-south direction		East-west direction	
		Root mean square (mm/s^2)	Maximum (mm/s^2)	Root mean square (mm/s^2)	Maximum (mm/s^2)
Tower 1	TMD	2.06	8.77	1.62	7.23
	Floor	1.09	3.90	0.70	2.30

direction at that time was the north wind, but because of the influence of the building shape on the wind load, the difference between the two results is not very large; (2) under the condition of synchronous measurement, the acceleration of the water tank is greater than the that of the floor where it is located.

Similarly, the TMD acceleration time history curve was transformed in frequency domain to obtain the power spectral density shown in Fig. 15.13c, d. The frequency of floor pick-up is similar to that of tower 1. The frequency of TMD pick-up is shown in Table 15.9. The measured first-order frequency value is larger than the TMD design value of 0.180, and slightly smaller than the structure pick-up frequency of the corresponding direction. The power spectral density of the measured point in the

Table 15.10 Dynamic characteristic identification of TMD

Order	Frequency/Hz		Damping ratio	
	North–south direction	East–west direction	North–south direction	East–west direction
1	0.205	0.194	0.0221	0.0431
2	0.779	0.784	0.0083	0.0161
3	1.182	1.464	0.0431	0.0385

north–south direction is obviously larger than that in the east–west direction. The first order modal energy of the main vibration direction (north–south direction) accounts for a large proportion. The measured data were further processed by FDD, and the self-power spectral density of the data of the same measuring point was used to identify the damping ratio of TMD. As shown in Table 15.10, the measured damping ratio is obviously smaller compared with the designed damping ratio. This is mainly since the test was carried out under the daily wind speed and the energy dissipation of the damper played a smaller role. Therefore, the damping ratio reflected in TMD is relatively small.

15.2 Nanjing TV Tower (Wind Vibration, AMD)

15.2.1 Project Overview

Nanjing TV Tower is a high-rise structure with radio and television transmission as its main function, including sightseeing, entertainment and other functions, as shown in Fig. 15.14 [2, 3]. The total height of the tower is 310.1 m. The main body of the tower is composed of three pre-stressed concrete column limbs with 120° of angle between the limbs. The lower limb is big and the upper limb is small. The column limbs are characterized by thin-walled box-shaped sections, which are connected by pre-stressed concrete connecting beams every 25 m in the middle. The large tower at the height of 169.78–202.29 m is supported on these three columns, and there are revolving restaurants and other facilities on the large tower. The small tower with VIP hall is located at the height of 235.18–246.61 m. The upper part of the small tower is equipped with ventilator room, and the upper part are reinforced concrete cylinder, platform, and steel mast, respectively. The acceleration response of the small tower of Nanjing TV Tower is 0.239 m/s^2 under the condition of the wind load of grade 8 ($\overline{v}_{10} = 20.70\,\mathrm{m/s}$), which exceeds the human comfort limit of 0.15 m/s^2. In order to ensure the normal operation of the TV tower in the 8-grade wind and open to tourists, it is necessary to control the acceleration response of the small tower.

(a) Photo

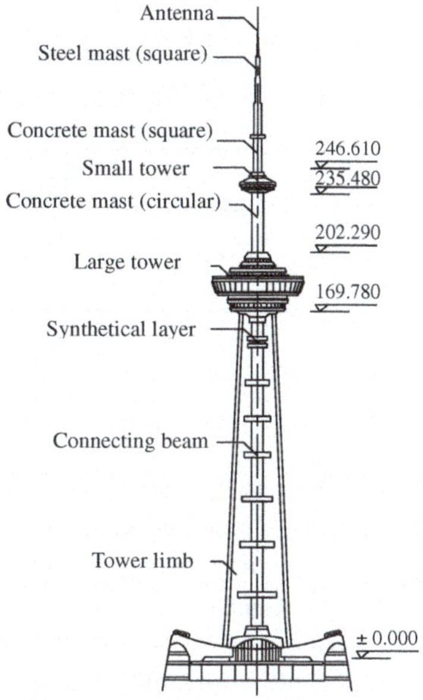

(b) Elevation diagram

Fig. 15.14 Nanjing TV Tower

15.2.2 Structural Vibration Reduction Design Using AMD

In the selection of control schemes, there were two main schemes: active control and passive control. The passive control is easy to achieve, but the additional mass and operating space required are relatively large. The passive control mainly plays a role in the first mode of vibration and has poor control effect on acceleration; while the active control is more flexible, the additional mass required is smaller than that of passive control, and it has control effect on both the first mode and the higher order mode of vibration, and its control effect on acceleration is good. Since the Nanjing TV Tower is a built structure, its bearing capacity and space capacity have been determined, it is decided to study the feasibility of the application of active mass damper (AMD) in the tower.

A typical closed-loop control system usually includes the controlled object, additional mass, energy, actuator, measurement system and control system. The active control system of wind-induced vibration for Nanjing TV Tower is a closed-loop quality control system. Its main components are shown in Fig. 15.15.

After the acceleration and displacement responses of the controlled object (the small tower part) were obtained by the measuring system, whether the control was needed was determined based on the calculation of controller. If the control was needed, the oil source and the control system were automatically activated, and the control instructions were applied to the actuator. The actuator exerted the specified

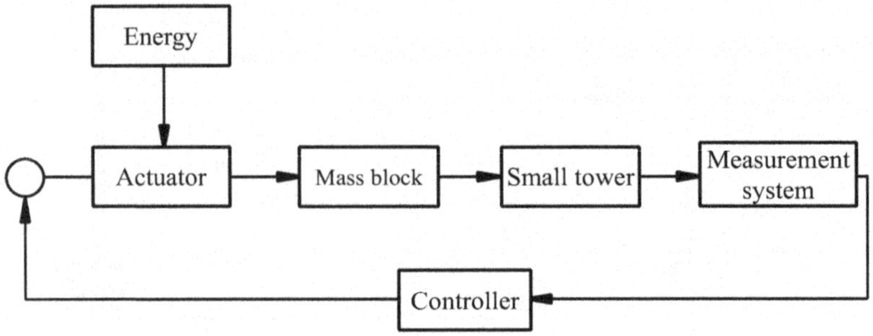

Fig. 15.15 Active control system of Nanjing TV Tower

control force to the mass loop. The dynamic response of the small tower was changed to a certain extent, and the new response value was transmitted to the control system at the same time, so as a new round of control was started, and so on, until the acceleration of the small tower reached the control goal. In order to make the whole system run smoothly as envisaged, it is necessary to design various links to meet the requirements of relevant performance.

15.2.2.1 Additional Mass System

Additional mass system is the source of active control force. It must have certain strength and stiffness to ensure that the control force can be effectively applied during the operation of the system. Secondly, it must have certain reliability to enable the additional mass body to move according to the set trajectory. As the Nanjing TV Tower is a structure that has been built and is in operation, the actual conditions of the site must be fully considered.

The ventilation room of the small tower of Nanjing TV Tower is a circular plane, in which a concrete core tube with a diameter of 6 m is in the middle, which has an outer diameter of 12 m. Its profile is shown in Fig. 15.16a.

All AMD facilities in this project were located in the ventilator room. In view of the circumstance that the plane is circular, the circular mass block was used to surround the core cylinder, and the force acting in any direction in the plane was generated through the movement of the center of mass (centroid) relative to the center of mass (centroid) of the core cylinder. Thus, the horizontal control force was exerted on the small tower. The working process is outlined as follows.

Because the actual tower tube is a spatial structure, there are three degrees of freedom in the horizontal plane, namely, two translational components and one rotational component. In order to uniquely determine the position of the mass ring and the acting force of the actuator in the control process, at least three actuators must be set in the same plane. Therefore, an actuator was set up every 120° along the circumference in engineering, as shown in Fig. 15.16b. In this way, the three driving forces of the actuator to the mass ring were in the same plane, but their directions were different and did not intersect at one point. For the mass ring, it was equivalent

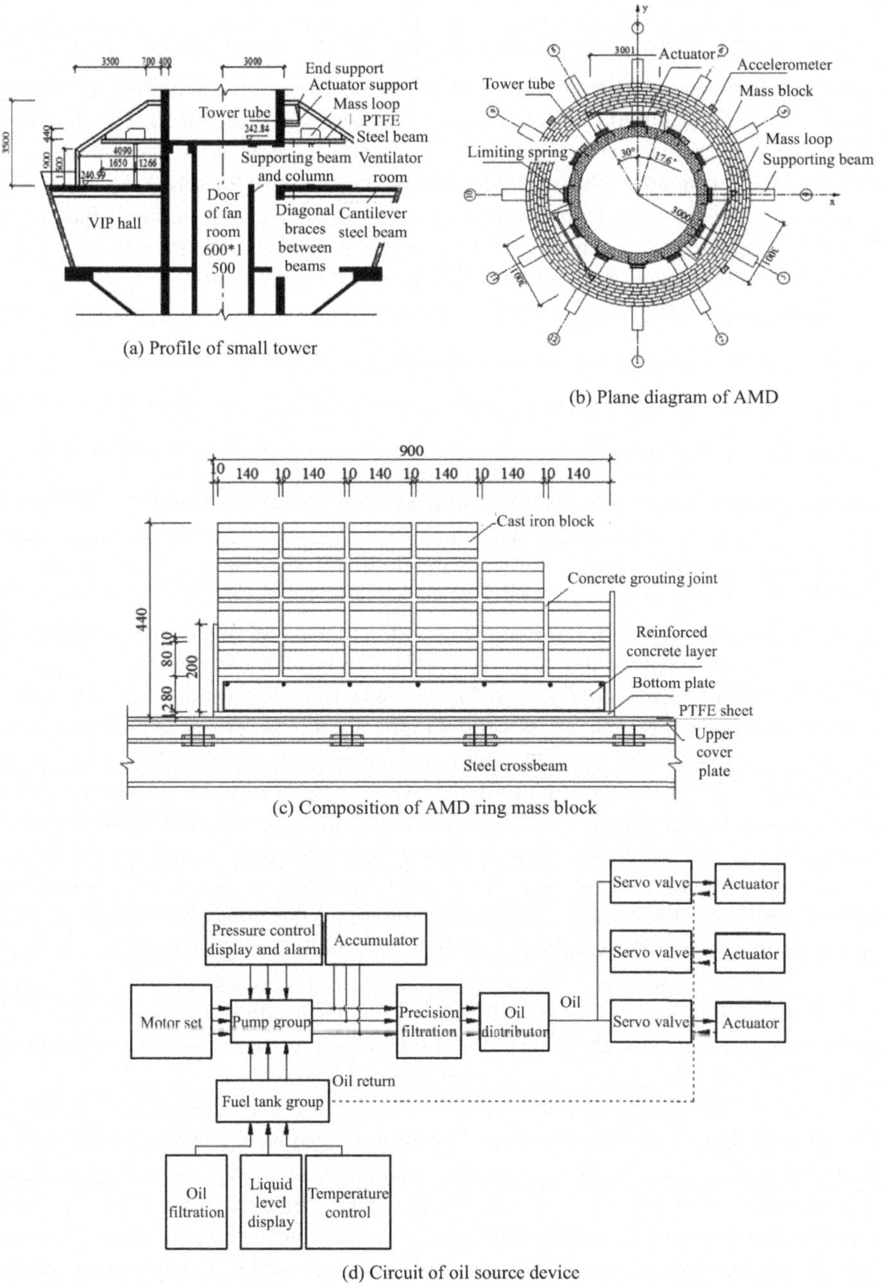

(a) Profile of small tower

(b) Plane diagram of AMD

(c) Composition of AMD ring mass block

(d) Circuit of oil source device

Fig. 15.16 Design of AMD system for Nanjing TV Tower

to adding three chain rod constraints in the plane so that its position was fixed relative to the tower barrel. According to the control force required by the tower barrel (two horizontal forces and one torque) and the initial position of the actuator, the driving force F_1, F_2, F_3 required by each actuator was uniquely determined. F_{1x}, F_{2x}, F_{3x} and F_{1y}, F_{2y}, F_{3y} are the components of F_1, F_2, F_3 on the X and Y axes of the XOY rectangular coordinate system originating from the center of the circle. x_1, x_2, x_3 and y_1, y_2, y_3 are the distances from the action points of F_1, F_2, F_3 to the coordinate axes Y and X, respectively. F_x and F_y are the components of the resultant forces of F_1, F_2, and F_3 in the X and Y directions, T is the resultant moments:

$$\begin{cases} F_{1x} + F_{2x} + F_{3x} = F_x \\ F_{1y} + F_{2y} + F_{3y} = F_y \\ F_{1x} \cdot y_1 + F_{1y} \cdot x_1 + F_{2x} \cdot y_2 + F_{2y} \cdot x_2 + F_{3x} \cdot y_3 + F_{3y} \cdot x_3 = T \end{cases} \tag{11.91}$$

where, x_1, x_2, x_3 and y_1, y_2, y_3 are determined by the initial position of the actuator, which are known values. At a certain moment, when the control force F_x, F_y, and T applied on the small tower are known values, the unknown force F_1, F_2, and F_3 given by the three actuators can be obtained by the three independent equations mentioned above.

In the vertical direction, the mass ring and actuator and other facilities must be placed in the air to ensure that there is a certain amount of clearance at the bottom, due to the need of normal operation in the ventilator room of the small tower building. In order to support the whole mass system in the air, it is necessary to provide a safe, smooth and rigid sliding plane to ensure that the mass ring can still operate in the air as planar operation according to the above scheme. Therefore, according to the actual structural characteristics of the ventilator room, 12 steel columns were erected on the floor of the ventilator room, and steel transverse beams were erected at the top of each steel column. One end of the steel transverse beams was anchored into the concrete core tube through expansion bolts, and the other end was suspended on the steel beams of the ventilator building roof with short steel bars, as shown in Fig. 15.16a. In this way, the rigid frame composed of 12 steel columns and beams constituted a rigid supporting platform. In order not to affect the normal force on the floor of the fan room, the steel column was supported at the position of the steel beam at the bottom of the floor, as shown in Fig. 15.16a.

Because of the limitation of the working elevator of TV tower and other field conditions, especially the size of the door of ventilator room is only 600 mm * 1500 mm, many equipment and accessories can only be assembled in zero and on-site. Therefore, the mass ring was made up of small cast iron blocks, each of which had a mass of 20 kg. In the course of stacking, it was also necessary to ensure that the mass ring itself had a certain stiffness, and there was no large deformation in the horizontal plane due to the actuator driving, so as to avoid actual unloading or displacement errors; in the vertical plane, it was necessary to be able to withstand the dead weight between two adjacent steel beams and control their deflection, so as to avoid additional resistance or deformation in the operation process, resulting in inaccurate control accuracy. Therefore, when stacking, the bottom of the mass ring was made of 12 mm thick steel plate, on which the reinforced concrete annular bottom plate was poured to form a rigid plane, and the side was wrapped with 10 mm thick steel plate. On the rigid plane, 140 mm * 300 mm * 80 mm cast iron blocks were piled, and fine stone concrete grouting was used between iron blocks and steel plates to form a circular whole. In order to make full use of the conical roof space of the fan room, the section of the mass ring was also stacked into trapezoids on the outside of the mass ring. In this way, on the one hand, the mass of the mass ring was increased, on the other hand, the space length that the mass ring can run was also increased. In this project, the radial length of the mass ring was ±450 mm. The whole mass ring had four layers, 440 mm high, 3856 mm inner diameter and 4746 mm outer diameter. The total design mass was 60 t, as shown in Fig. 15.16c.

15.2.2.2 Actuator Support

As can be seen from Fig. 15.16b, in order to make the three actuators not intersect at one point, and to ensure that the actuator and the mass ring have a certain operating space, an arm must be extended on the tower barrel as the support of the actuator. Similarly, in order to ensure that there is no large error caused by the deformation of the support during operation, the support itself should have a certain stiffness. In the process of implementation, a steel plate was anchored with expansion bolts on the concrete tower tube, and then the "[" shaped arm was welded on it. During construction, it was very difficult to anchor bolts because of the dense steel bars in the tower barrel. In order to minimize the deformation of the end of the arm, i.e. the support part of the actuator, under load, the angle steel was also used to diagonally support the steel beam on the roof vertically. By calculation, the displacement at the end of the arm is less than 2 mm under 160 kN load, which can meet the control accuracy requirements.

15.2.2.3 Hydraulic Servo System

The hydraulic servo system, namely electro-hydraulic servo system, is generally a closed-loop system. By applying circuit control to hydraulic equipment, this kind of system has both advantages, such as small size, light weight, large driving force, and fast response.

(1) **Oil source**

The oil source included hydraulic oil, oil tank, and oil pump. The hydraulic oil was stored in the tank and pumped to produce a certain pressure, which was then flowed into the actuator through the tubing, valves, etc. Because the site of the project is located in a small tower over 200 m high, and there are many important communication facilities on it, in the process of implementation, in line with the principle of safety first, it is necessary to ensure that no disasters, especially fires, occur. So when choosing hydraulic oil, according to this requirement and considering the economic factors, the HM46 hydraulic oil was selected.

In the process of hydraulic system implementation, the actual conditions of the site should also be taken into account. Because of the size limitation of the portal of the small tower, the fuel tank cannot be very large in the implementation process, and the power of the oil pump cannot be very large. In the process of implementation, six small oil tanks of 600 mm * 1200 mm * 1200 mm were used and connected by pipes to form an oil tank. Each tank was divided into two units, one main unit and one pair to supply three groups of 12 pumps. Each pump group consisted of four high-pressure vane pumps, which were connected in parallel with each other and could be combined flexibly according to the actual situation. The layout of the hydraulic system is shown in Fig. 15.16d.

In the system, the flow rate of a single pump was 15 L/min, and the total flow rate of 12 pumps in three groups was 180 L/min. The average flow rate was 180 L/min and the maximum instantaneous flow rate was 440 L/min. Therefore, in order to meet the requirement of instantaneous maximum flow, 9 nitrogen accumulators were added to the system, three in each group. In addition, the MTS oil distributor was installed between the high-pressure oil and the actuator, and the high-pressure oil was conveyed to the actuators through the distributor.

The oil distributor was directly controlled by the control part to ensure that the opening and closing of the hydraulic power part were directly controlled by the control part in the operation process.

The working oil pressure of the oil source system was 21 MPa (3000 psi), that is, the oil pressure under high pressure. In order to realize the monitoring of hydraulic system automatically, the pressure sensors, digital displays, and temperature controllers were installed in the system. When the temperature exceeded the control range, the system alarmed automatically and started the cooling system. The system can also alarm automatically when the pressure exceeds the control range. In order to protect the hydraulic equipment in the system, especially the hydraulic servo valve, it is necessary to filter the hydraulic oil, and the filter meets the requirement of 3 μs.

(2) **Actuator**

Actuator, also known as exciter, is the direct execution equipment of driving mass ring. In this project, considering that the tonnage required by the system was not particularly large and the displacement required was very large, the MTS247.21 hydraulic actuator was selected. Its performance parameters are shown in Table 15.11. One end of the actuator was connected with the mass ring, and the other end was fixed on the rigid arm of the tower barrel. A solenoid differential transformer displacement

Table 15.11 Parameters of actuator performance

Performance parameter	Calibration value of actuator load (kN)		Piston displacement (mm)	
Stress state	Tension	Compression	Dynamic	Static state
Design value	100	100	762	762
MTS247.21	50	126.3	1524	1557

sensor was also installed at the end of the piston rod of the actuator to transfer the position of the piston. In order to improve the dynamic performance of the actuator, a differential pressure sensor was also installed between the servo valve and the actuator. Through this sensor, the oil pressure on both sides was transmitted to the control loop, to adjust or monitor the oil pressure in the actuator, estimate the force exerted by the actuator, and transfer it to the digital control system by the customer self-designed system. Before installing the actuator, the displacement sensor in the piston rod was adjusted to ensure the accuracy of the feedback displacement signal in operation. In the process of implementation, the actuator was lifted by the manual hoist and separated from the connection of the mass ring in order to avoid driving the operation of the mass ring in debugging. Then the end of the actuator was pre-tightened to zero. After zero adjustment, the end spring washer was clamped and connected with the mass ring. Two MTS249.23 hinge supports were installed at both ends of the actuator. Obviously, the bearing must meet certain strength and stiffness requirements. The bearing capacity of the bearing was 160 kN.

(3) **Servo valve**

The hydraulic servo valve is the core part of the whole hydraulic servo system. All the control signals are realized by it. The basic principle of servo valve is to change the position of internal baffle using the magnetic field generated by the current in the coil, so as to change the direction and flow of hydraulic oil in the valve, and then change the direction of hydraulic flow in the actuator and the magnitude of the acting force. According to the magnification series of servo valve, there are valves of one-stage, two-stage and three-stage, the higher the level, the greater the oil flow. In general, the valves of two-stage are used widely in engineering. The three-stage valves are used only in the case of large flow at seismic stations.

Because of the long running distance of the mass ring and the long running distance (±75 cm) of the actuator in this project, the flow of hydraulic oil was also large, so the MTS256.09 three-stage magnifying servo valve was selected. Its performance parameters are shown in Table 15.12. The maximum working pressure of the servo valve is 21 MPa and the minimum working pressure is 1.3 MPa. The required filtration level should be up to 3 μm.

Table 15.12 Performance parameters of servo valve

Performance parameter	Flow calibration (L/min)	Piston flow rate (L/min)	Full Flow Frequency (Hz)	Opening time of spool (ms)	
				Open 90%	Open 10%
MTS256.09 (static force)	340	–	–	–	–
MTS256.09A01 (dynamic force)	340	3.8	30	9	3.5
MTS256.09A02 (dynamic force)	340	9.5	80	4.6	4.3

15.2.2.4 Control System Design

The control system of AMD control project of Nanjing TV Tower was roughly divided into three parts: analog control part, digital control part, and signal interface system. The control panel in the analog control part was directly connected with the servo valve of the actuator, and the actuator was controlled by the input of analog quantity, that is, voltage signal. The function of the digital control part was to convert the analog signal into digital information. After calculation, a new digital signal was obtained, and then the analog signal was transferred to the analog control part through digital-to-analog conversion. The customers designed their own signal interface system, which was mainly used to enhance the system's fail-safe function and the automatic start-up and shutdown function of active control system. When the acceleration of the tower exceeded a certain value, the signal interface system automatically opened the relevant circuit, and opened the oil distributor of the hydraulic system to start the whole active control system. Similarly, if the system was not working properly and made the movement of the small tower more disadvantageous, the system started automatically and closed the control circuit and hydraulic source if the acceleration exceeded a certain limit.

In addition, in order to detect the external conditions, the remote monitoring of structural response is needed in the operation of the whole control system. Therefore, the interface circuit generally included the following six subsystems: (1) judgment subsystem; (2) filtering subsystem; (3) monitoring subsystem; (4) fault-proof extremum detection subsystem; (5) signal exchange subsystem; and (6) remote start subsystem.

15.2.3 Structural Vibration Reduction Analysis

Under the active control of AMD, the wind-induced response equation of the simplified model of Nanjing TV Tower is as follows (Fig. 15.17):

$$[M]\{\ddot{x}(t)\} + [C]\{\dot{x}(t)\} + [K]\{x(t)\} = \{p(t)\} - [H]\{u(t) - c_d\dot{x}_d(t) - k_d x_d(t)\} \tag{15.1}$$

$$m_d\ddot{x}_d(t) + c_d\dot{x}_d(t) + k_d x_d(t) = -m_d[H]^T\{\ddot{x}(t)\} + u(t) \tag{15.2}$$

where,

$[M]$, $[C]$, $[K]$—Mass matrix, damping matrix, and stiffness matrix of TV tower model;
$\{x(t)\}$, $\{\dot{x}(t)\}$, $\{\ddot{x}(t)\}$—Displacement, velocity, and acceleration response vectors of simplified model of TV tower;
$\{p(t)\}$—Fluctuating wind load vector;
$[H]$—Position vector of control force, $[H] = [0, \ldots, 0, 1, 0, \ldots 0]^{\mathrm{T}}$;
$x_d, \dot{x}_d, \ddot{x}_d$—Displacement, velocity, and acceleration responses of AMD mass blocks relative to the structure layer (ventilator roof);

Fig. 15.17 Structure analysis model of Nanjing TV Tower

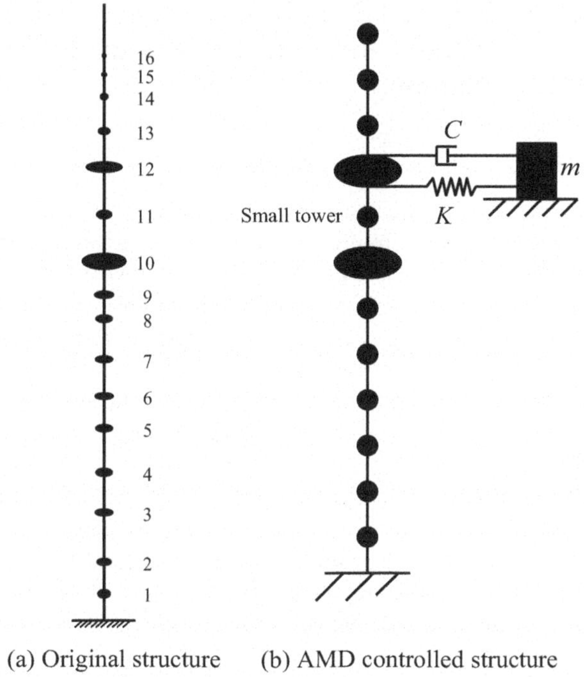

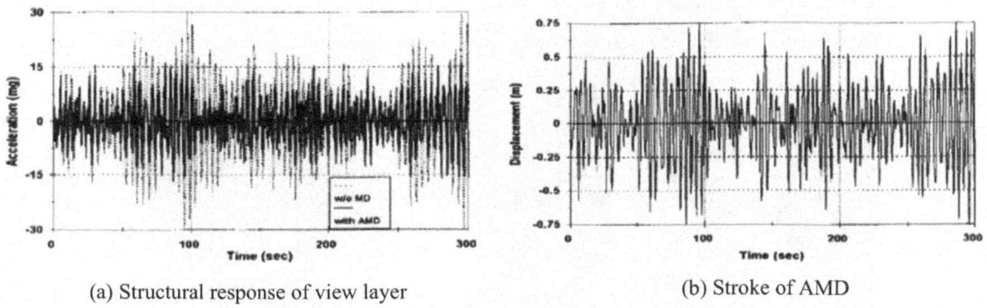

Fig. 15.18 Wind vibration response analysis of AMD controlled structure of Nanjing TV Tower

m_d, c_d, k_d—Mass, damping ratio, and stiffness of AMD;
$u(t)$—Active control force of AMD.

According to the simplified model of the TV tower structure and the active control algorithm, the control forces applied on the TV tower structure at different times can be obtained. Through parameter optimization, the acceleration response of the viewing layer after AMD installation is shown in Fig. 15.18. It can be seen that the acceleration response of the small tower after AMD active control is 0.148 m/s^2, which meets the requirement of human body comfort of 0.150 m/s^2 of the small tower in strong wind.

15.3 Beijing Olympic Multi-functional Broadcasting Tower (Wind Vibration, TMD+Variable Damping Viscous Damper)

15.3.1 Project Overview

Beijing Olympic multi-functional broadcasting tower is located in the central part of the Olympic Park [4]. It connects the west side with the central axis landscape avenue and the east side with the training ground of the National Stadium. It is a landmark building in the National Olympic Forest Park. As shown in Fig. 15.19, the total height of the studio tower is 135 m. There are seven studios along the height direction. The top floor is also the VIP sightseeing hall, and the first floor is the hall or studio. Each floor of tower is 13.5 m apart. Each tower has two floors. The upper floor is the studio platform, and the lower floor is the toilet and equipment room. The net height is 2.2 m. The first plane is an equilateral triangle with enlarged area. The underground floor is equipment floor equipped with a local interlayer.

The plane of the building is an equilateral triangle with a side length of 21 m. The three corners are vertical traffic space. There are two evacuation stairs and one fire elevator, and all the buildings and elevators are outdoor space. The three vertical

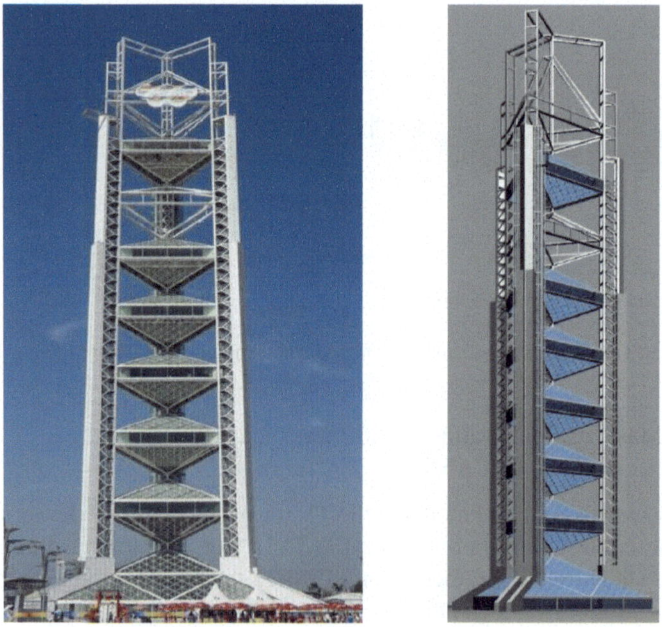

(a) Architectural scene (b) Architectural renderings

Fig. 15.19 Beijing Olympic multi-functional broadcasting tower traffic cores are at the bottom of the first floor, which are concrete bases and extend outward at the entrance. The structure system of the tower is mainly composed of seven repeated polyhedral structural units and three cylinders located at the corner.

15.3.2 Structural Vibration Reduction Design

The fire tank in seven-layer equipment interlayer (96.75 m elevation) was used as the quality of TMD system. The total mass was 27,000 kg. Four interlayer rubber cushions were installed at the bottom of the tank to completely separate the tank from the ground. The structure of TMD is shown in Fig. 15.20. The interlayer rubber cushion was used as the spring of TMD system. The horizontal shear stiffness of the interlayer rubber cushion was used to ensure that the inertia mass and the vibration mode resonance of the main structure had a definite natural vibration period. The viscous fluid damper was used to provide damping for the TMD system.

15.3.3 Structural Analysis Model

Beijing Olympic multi-functional broadcasting tower is a complex large-scale space truss structure. Firstly, a three-dimensional finite element model was established according to the structural design information as shown in Fig. 15.21a. The linear members were all spatial beam elements; the floor and wall were quadrilateral space shell elements, and the bottom joints were fixed. In this project, a variable damping viscous fluid damper was used, and there is no such element in the conventional analysis software. Therefore, a series multi-degree-of-freedom model was established based on the displacement method of structural mechanics as shown in Fig. 15.21b. The natural frequencies of the structures based on the modal analysis of the two models are shown in Table 15.13. The comparison shows that the maximum difference of natural frequencies between the two models is 4.4%, and that of the first-order modal error is only 1.1%. Therefore, the two models have similar structural dynamic characteristics.

15.3.4 Analysis of Vibration Absorption Performance of the Structure

The wind load of the structure was simulated based on the wind tunnel test results, and the wind-induced response analysis and the optimization design of the vibration reduction effect of the structure were carried out accordingly. The damping ratio was defined as the ratio of (structural response without setting TMD—structural

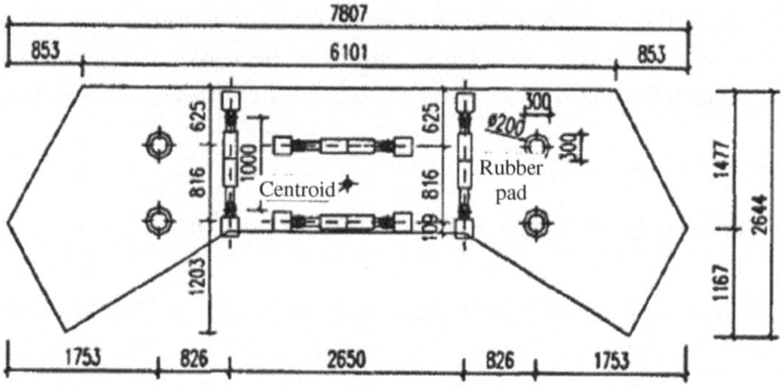

(a) Botom floor plan e layout

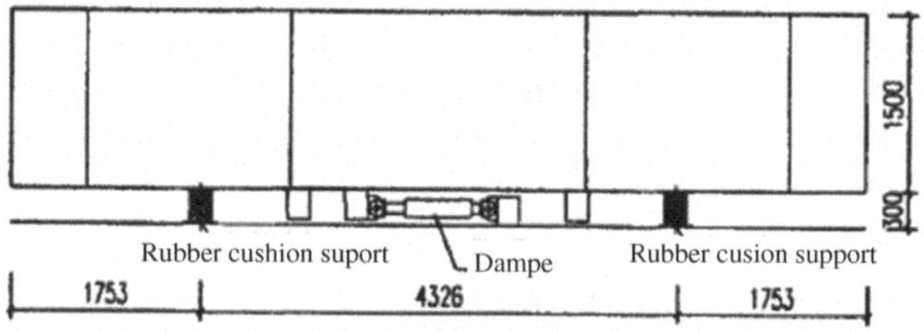

(b) Elevation layout

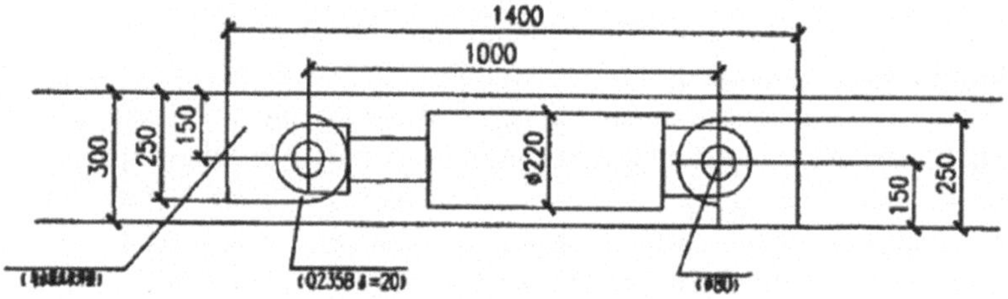

(c) Elevation layout of damper

(d) On-site photo

Fig. 15.20 Design of the supported TMD

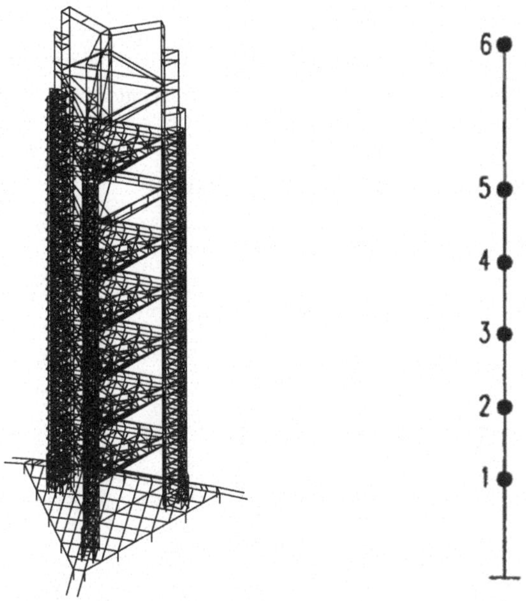

(a) 3-D finite element model (b) Simplified serial multi-degree -of-freedom model

Fig. 15.21 Structural analysis model

Table 15.13 Natural vibration circle frequency of the structure

Order	X-direction			Y-direction		
	3-D model	Simplified model	Error (%)	3-D model	Simplified model	Error (%)
1	4.163	4.208	1.1	4.236	4.254	0.4
2	12.310	12.442	1.1	12.517	12.779	2.1
3	22.577	22.479	0.4	21.950	22.922	4.4

response with setting TMD)/structural response without setting TMD. The results of parameter optimization when using TMD are shown in Table 15.14.

On this basis, the TMD with variable damping viscous fluid dampers was further optimized. Under 10-year and 50-year fluctuating wind loads, the variations of the acceleration responses of top floor with the damping coefficient ratio are shown in Fig. 15.22. Choosing the damping coefficient ratio C_1/C_2 to be 0.67, the wind-induced response of the structure with variable damper TMD was analyzed. As shown in Fig. 15.23, it can be seen that the response of the top floor of the structure with variable damper TMD is obviously attenuated, and the maximum stroke of TMD is only 15 mm under the action of fluctuating wind once in 50 years.

Table 15.14 Optimized results of TMD parameters

TMD parameter	Pulsatile wind once in 10 years		Pulsatile wind once in 50 years	
	Acceleration of vertex (m/s^2)	TMD displacement (mm)	Acceleration of vertex (m/s^2)	TMD displacement (mm)
$\zeta_{TMD} = 0.14, \lambda = 0.99$	0.1370	11.5	0.1530	13.0
$\zeta_{TMD} = 0.06, \lambda = 1.09$	0.1521	16.7	0.1471	20.5

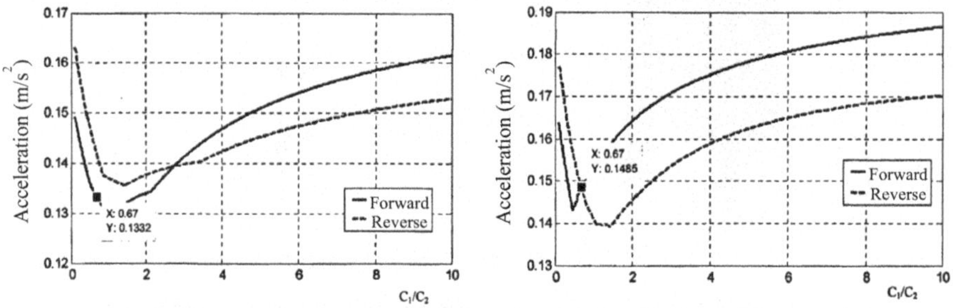

(a) Acceleration change of top layer under pulsating wind once in 10 years

(b) Acceleration change of top layer under pulsating wind once in 50 years

Fig. 15.22 Optimum curve of acceleration of structure top with variable damper viscous fluid damper

15.3.5 Field Test and Analysis

In order to verify the correctness of structural analysis and the effectiveness of variable damping TMD in vibration reduction, the field dynamic tests were carried out on Beijing Olympic multi-functional broadcasting tower on June 13, 2008 (without TMD) and July 14, 2008 (with TMD). There were five layers and ten test points in each test. The specific layout is shown in Fig. 15.24. The INV-6 multi-functional anti-aliasing filter amplifier developed by Beijing Oriental Institute of Vibration and Noise Technology and the AZ316 acquisition box and CRAS analysis system of Nanjing Anzheng Software Engineering Co., Ltd. were used for dynamic testing. The 941-B ultra-low frequency vibration pickup device produced by Institute of Engineering Mechanics, China Seismological Bureau was used as the sensor. The types and quantities of the main test instruments are shown in Table 15.15.

The natural frequencies and damping ratios of the structure were determined by identifying and processing the measured data and analyzing the vibration mode shapes, as shown in Table 15.16. It can be seen that, the change of structural natural frequency is small before and after the installation of TMD, the setting of TMD does not increase the quality of the structure and has little influence on the natural frequency of the structure; after the installation of TMD, the damping ratios of the

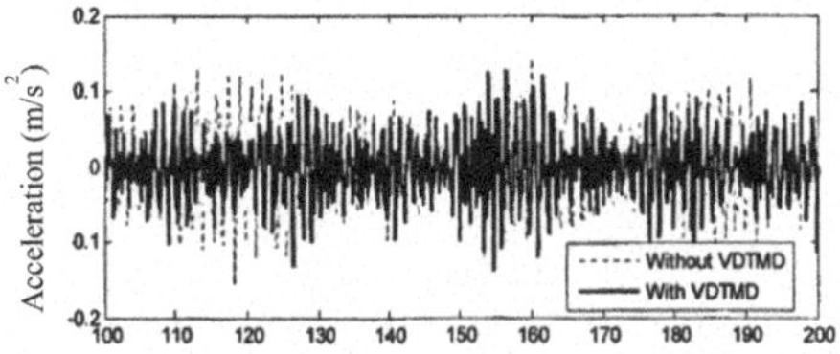

(a) Acceleration time history of top (Pulsating wind once in 10 years)

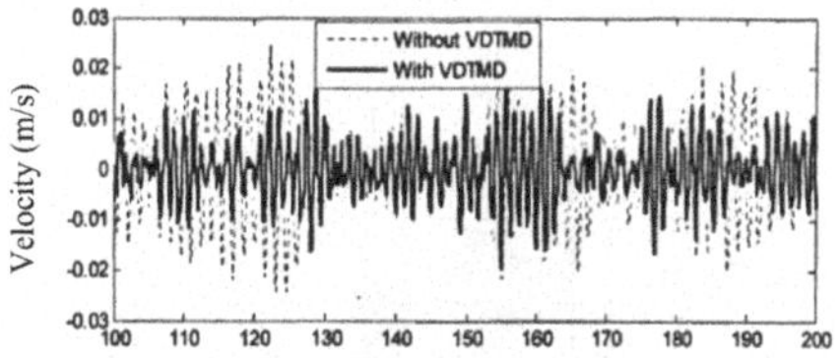

(b) Velocity time history of top (Pulsating wind once in 10 years)

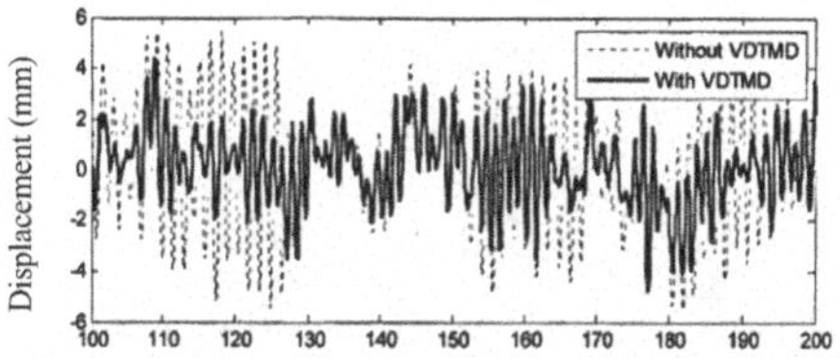

(c) Displacement time history of top (Pulsating wind once in 10 years)

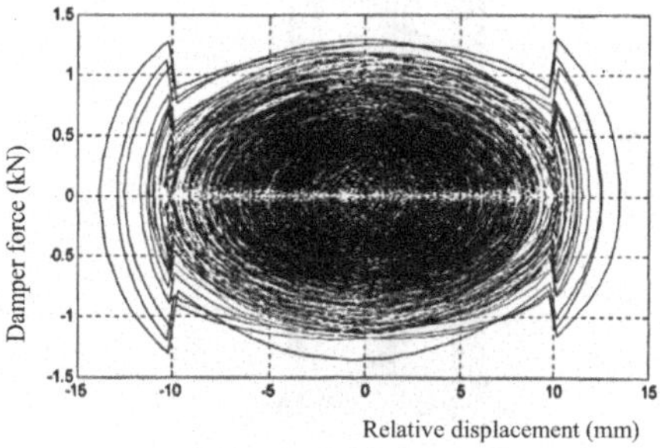

(d) Force-displacement curve of damper (Pulsating wind once in 10 years)

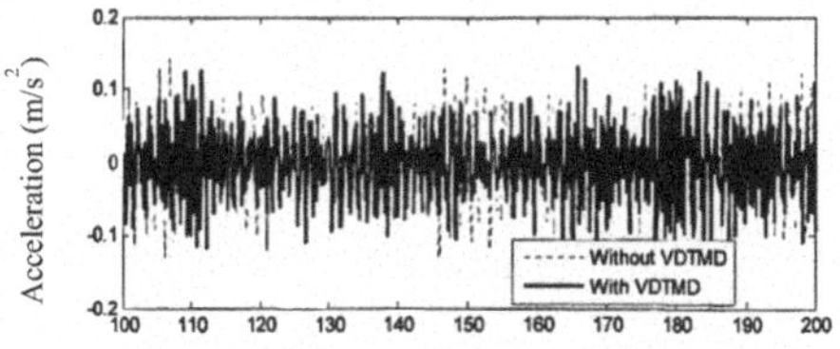

(e) Acceleration time history of top (Pulsating wind once in 50 years)

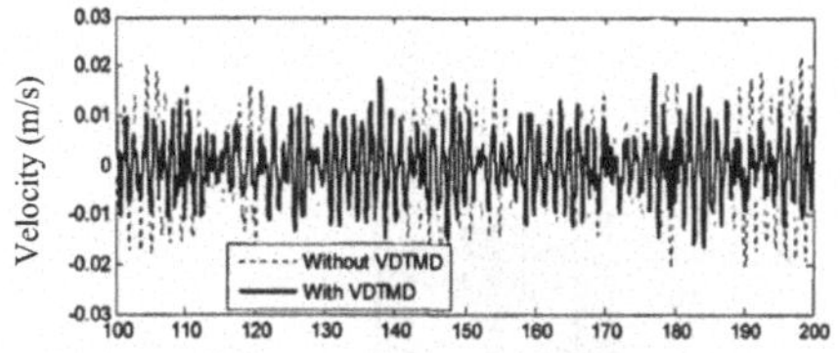

(f) Velocity time history of top (Pulsating wind once in 50 years)

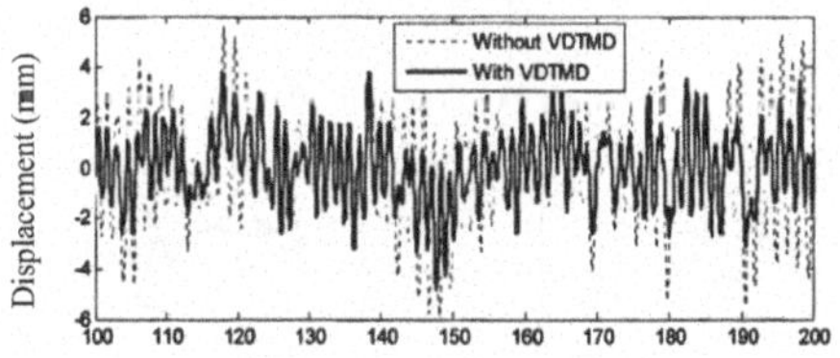

(g) Displacement time history of top (Pulsating wind once in 50 years)

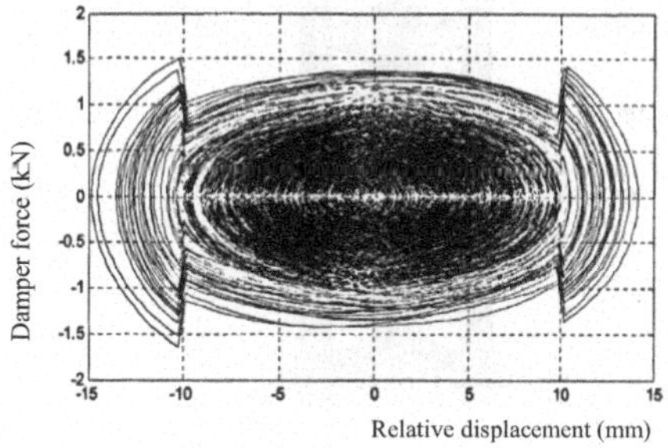

(h) Force-displacement curve of damper (Pulsating wind once in 50 years)

Fig. 15.23 Responses of structure and TMD under pulsating wind

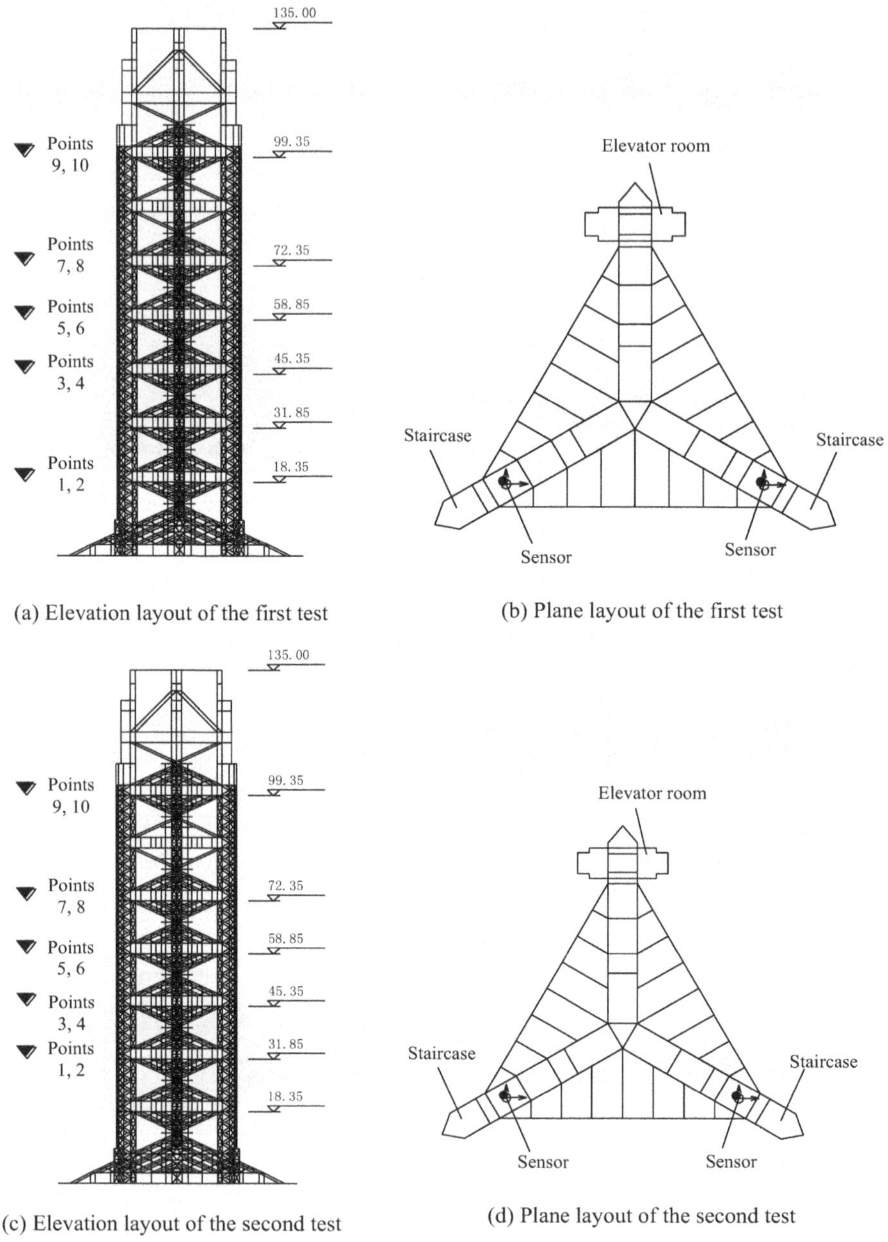

Fig. 15.24 Layout of measuring points

Table 15.15 Main test instruments and performance indicators

Instrument	Number	Main technical indicators
941-B vibration picker in horizontal direction	12	Passband 0.25–80 Hz
INV-6 intelligent signal conditioner	2	Passband 0.025–35 Hz
AZ316 signal acquisition instrument	1	16 passageways
Short cable	12	Length of 1.0 m
AZCRAS analysis system	1	–
Cable	12 (100 m)	Suitable for the 941-B Vibration Picker
Two-way connector	5	According to the actual situation on site

first three modal of the structure increase by 0.82%, 0.96%, and 0.87%, respectively, and the equivalent damping ratio of the structure system increases significantly, and TMD has played a good role in reducing the natural frequency. The test results of the natural frequencies are close to the results of finite element numerical simulation, and the analysis model has certain reliability. The comparison between the measured and calculated results of the first three mode shapes of the structure is shown in Fig. 15.25.

15.4 Proposed Hefei TV Tower (Earthquake, Wind Vibration, TMD)

15.4.1 Project Overview and Analysis Model

Hefei TV Tower was the highest steel structure tower to be built in China at that time [5]. It has comprehensive functions such as radio and television transmission, microwave communication, tourism, catering, entertainment, and accommodation (as shown in Fig. 15.26a) The tower is mainly composed of external spatial truss, upper and lower towers, bottom skirt house, mast and inner wellbore. The total height of the tower is 339, and 96 m high mast is installed at the top. The lower tower is located at the elevation of 60.50–69.50 m, and the upper tower is located at the elevation of 209.70–243.30 m. The inner shaft below 28 m is reinforced concrete structure, and the part form the elevation of 28 m to upper tower is composed of four vertical plane steel trusses.

As a complex large-scale spatial truss structure, the tower does not conform to the assumption of plane section in horizontal plane. Therefore, it should be analyzed according to three-dimensional spatial structure. However, considering that the vertical series multi-degree-of-freedom model is the simplest and feasible method for structural control analysis, the vertical series multi-degree-of-freedom system model

Table 15.16 Test results of the first three natural frequencies of the structure

Order	Without TMD			With TMD		
	Frequency (Hz)		Damping ratio (%)	Frequency (Hz)		Damping ratio (%)
	Measurement	Simulation	Measurement	Measurement	Simulation	Measurement
1	0.45 (Torsion)	0.40	1.17	0.45 (Torsion)	0.40	1.99
2	0.70 (X-direction translation)	0.66	1.64	0.73 (X-direction translation)	0.67	2.60
3	0.75 (Y-direction translation)	0.67	1.11	0.75 (Y-direction translation)	0.69	1.98

Fig. 15.25 Comparison of structural modal modes

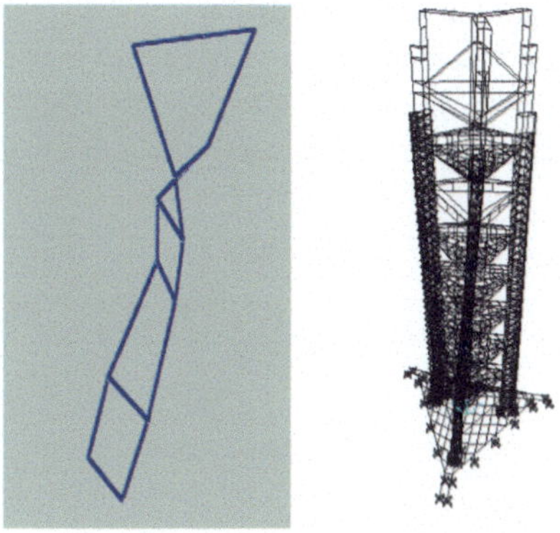

(a) 1st order mode shape (without TMD)

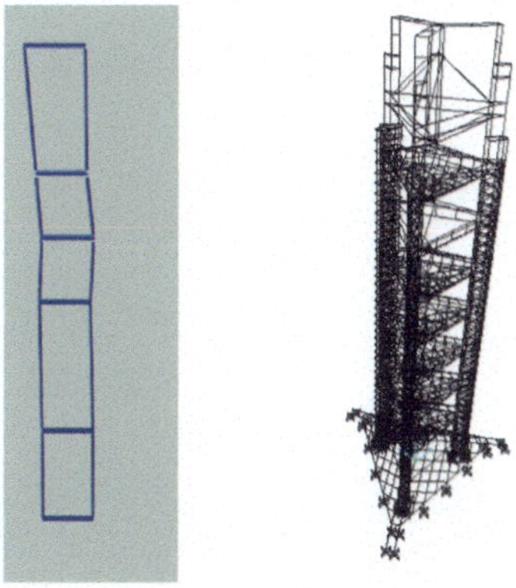

(b) 2nd order mode shape (without TMD)

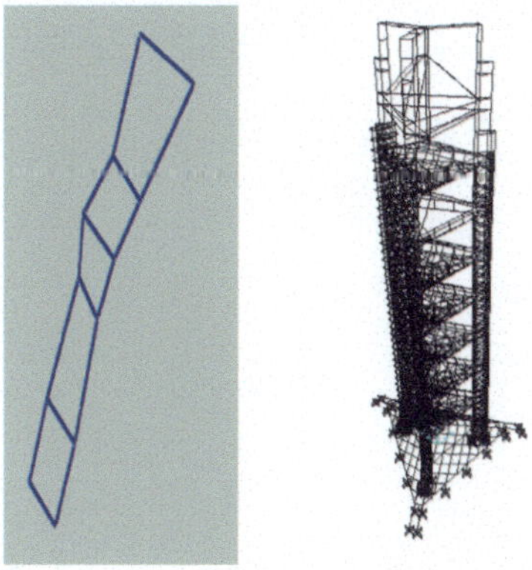

(c) 3rd order mode shape (without TMD)

Fig. 15.25 (continued)

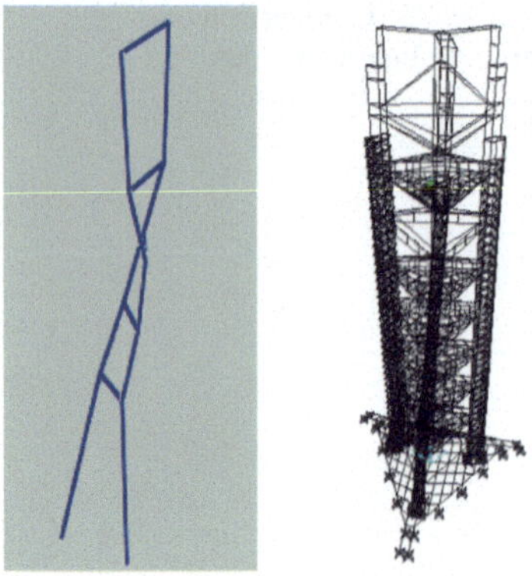

(d) 1st order mode shape (with TMD)

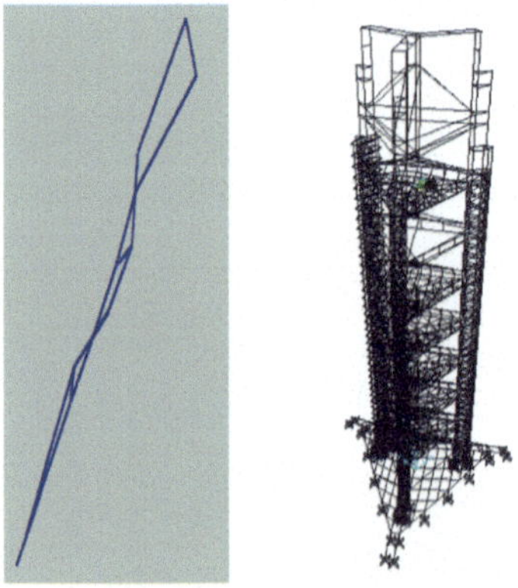

(e) 2nd order mode shape (with TMD)

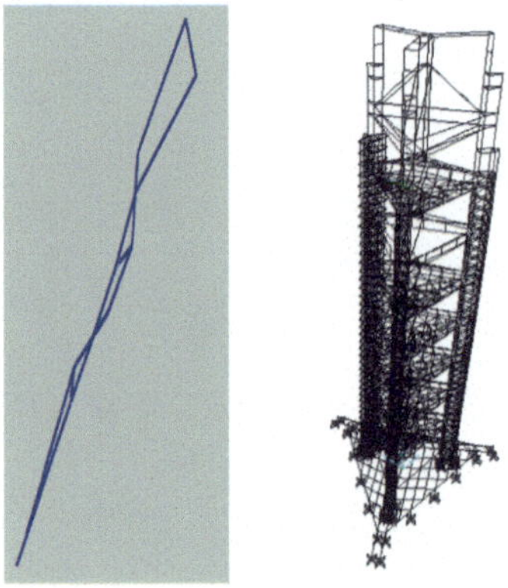

(f) 3rd order mode shape (with TMD)

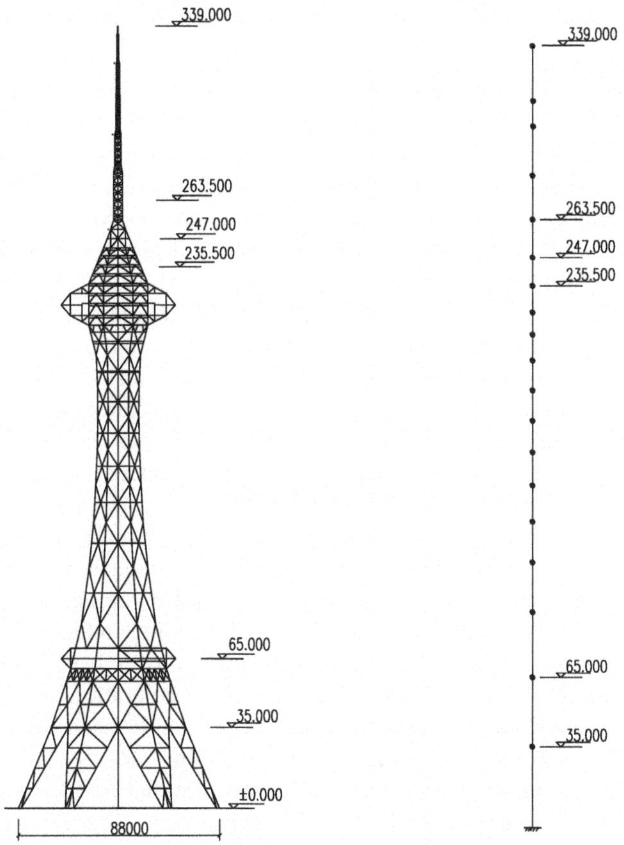

(a) Structural elevation layout

(b) Series multi-degree-of-freedom model

Fig. 15.26 Elevation layout and analysis model of Hefei TV Tower

($n = 19$) was established based on the actual structure of the TV tower, which is shown in Fig. 15.26b.

Because the stiffness and damping of Hefei TV Tower are relatively small, especially the antenna mast is more flexible, and the stiffness and mass distribution of the whole structure is extremely uneven. The dynamic response of Hefei TV Tower under strong wind and strong earthquake is larger, so it is necessary to control the dynamic response of the structure under strong vibration.

15.4.2 *Analysis of Wind-Induced Vibration Response Control*

15.4.2.1 Frequency Domain Analysis of Wind Vibration Control

In order to obtain the maximum damping rate of wind-induced vibration response of TV tower, it is necessary to optimize the TMD parameters (set TMD at the 13th particle) and determine the three important parameters of TMD, namely mass m_d, frequency ω_d, and damping ratio ζ_d. Because the wind-induced vibration response of TV tower is mainly the first mode, TMD should be tuned to the first frequency of the structure. The total mass of the water tank is 60,000 kg, so the mass of TMD is 60,000 kg. The ratio of TMD to the generalized mass of the first mode of the TV tower is calculated to be 0.0196. Under the condition of fixed mass ratio and wind load parameters, the relationship between structural vibration reduction coefficient β and frequency ratio λ_T and damping ratio ζ_T of FM mass damper was discussed, respectively. The vibration reduction coefficient β of wind-induced vibration response is defined as the ratio of the standard deviation response of the upper tower with damper to that of the upper tower without damper. The frequency ratio λ_T is defined as the ratio of the frequency of the damper to the first mode frequency of the TV tower. Therefore, the optimum value of frequency ratio and damping ratio is the minimum value of the vibration reduction coefficient of the wind-induced response of the upper tower structure. Figure 15.27a–c show the relationship between the vibration reduction coefficient β of the displacement, velocity, acceleration of the upper tower and the frequency ratio λ_T and damping ratio ζ_T. As can be seen from Fig. 15.27a, the optimal frequency ratio λ_T and the optimal damping ratio ζ_T are 0.99 and 0.07, respectively. As can be seen from Fig. 15.27c, the optimal frequency ratio λ_T and the optimal damping ratio ζ_T are 1.0 and 0.07, respectively. According to the human comfort requirement of the upper tower, the optimal frequency ratio and damping ratio of TMD are 1.0 and 0.07, respectively.

In the design of TMD, because the space of TMD motion is limited, it is necessary to calculate the displacement of TMD relative to the tower. The relationship between the vibration reduction coefficient β of displacement of the tower and the frequency ratio λ_T and damping ratio ζ_T of TMD is shown in Fig. 15.27a. As can be seen from the figure, the larger the vibration reduction coefficient of response of TV tower, the larger the relative displacement of TMD. The relationship between the standard deviation of TMD relative displacement response and frequency ratio λ_T and damping ratio ζ_T is shown in Fig. 15.28. It can be seen from the figure that the larger the damping ratio of TMD, the smaller the relative displacement response; when the optimal frequency ratio and the optimal damping ratio are 1.00 and 0.07, the standard deviation of the relative displacement of TMD is 2.5556 m.

The standard deviations of the displacement, velocity and acceleration of each particle in the TV tower with and without TMD are shown in Table 15.17. The maximum displacement response, velocity response and acceleration response of each particle in the TV tower with or without TMD are shown in Table 15.18. Compared with the mode response of TV tower without TMD, the displacement, velocity

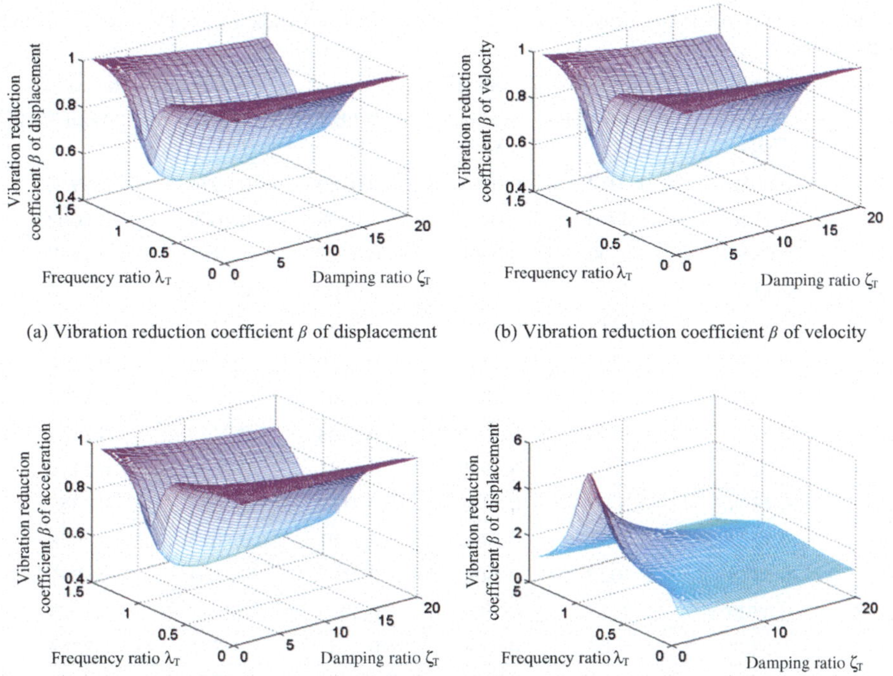

(a) Vibration reduction coefficient β of displacement (b) Vibration reduction coefficient β of velocity

(c) Vibration reduction coefficient β of acceleration (d) TMD displacement relative to tower displacement

Fig. 15.27 Variations of structural response and TMD stroke with parameters

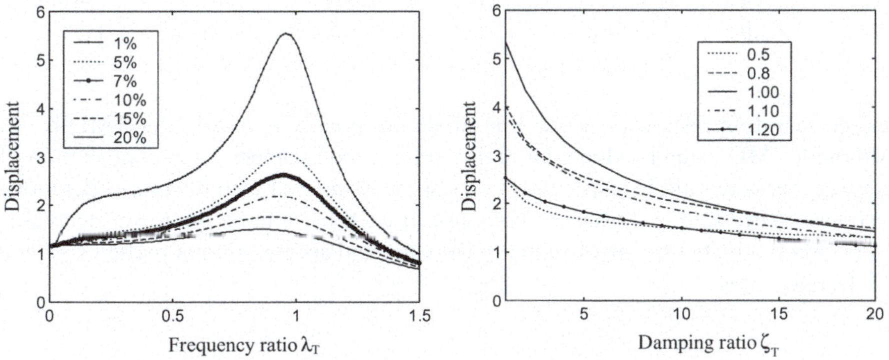

(a) Variation of TMD displacement response relative to frequency ratio (b) Variation of TMD displacement response relative to damping ratio

Fig. 15.28 Relation between displacement response and parameters of TMD

Table 15.17 Standard deviations of the displacement of each particle in the TV tower with and without TMD

Particle	Displacement (m)		Velocity (m/s)		Acceleration (m/s^2)	
	Without TMD	With TMD	Without TMD	With TMD	Without TMD	With TMD
1	0.0050	0.0032	0.0193	0.0189	0.0930	0.0928
2	0.0156	0.0090	0.0333	0.0308	0.1009	0.1009
3	0.0558	0.0300	0.0698	0.0523	0.1481	0.1467
4	0.1156	0.0615	0.1250	0.0800	0.1998	0.1960
5	0.1732	0.0918	0.1781	0.1045	0.2335	0.2265
6	0.2315	0.1225	0.2314	0.1280	0.2536	0.2421
7	0.2923	0.1545	0.2861	0.1501	0.2539	0.2353
8	0.3511	0.1854	0.3388	0.1708	0.2425	0.2134
9	0.4132	0.2181	0.3949	0.1931	0.2231	0.1774
10	0.4728	0.2495	0.4495	0.2160	0.2056	0.1355
11	0.5277	0.2784	0.5008	0.2391	0.2013	0.1038
12	0.5728	0.3022	0.5435	0.2595	0.2164	0.1086
13	0.6306	0.3327	0.5991	0.2873	0.2622	0.1620
14	0.6828	0.3603	0.6500	0.3137	0.3245	0.2354
15	0.7595	0.4009	0.7272	0.3579	0.4749	0.4055
16	0.8571	0.4534	0.8406	0.4441	0.8109	0.7635
17	0.9832	0.5246	1.0109	0.5978	1.2404	1.2031
18	1.0602	0.5715	1.1162	0.6932	1.4368	1.4011
19	1.2862	0.7382	1.5777	1.1980	1.6809	1.6407

and acceleration responses of the first mode are greatly reduced after TMD is set. Although TMD cannot reduce the higher-order mode response, the contribution of these higher-order mode responses is very small compared with the first-order mode response. After setting TMD, the response of each particle on the tower decreases correspondingly, but the displacement, velocity, and acceleration at particles 12 and 13 decrease most.

15.4.2.2 Time Domain Analysis of Wind Vibration Control

According to the direct dynamic method of structural dynamics, the governed vibration response equation of the structure and the motion equation of the control device were discretized, and the governed vibration response of the structure in each instantaneous state was obtained by step integration according to the time history record of the input load. Obviously, the time history record of the controlled wind-induced vibration response of the structure is a simulation of the actual controlled response of the

Table 15.18 Maximum displacement response of each particle in the TV tower with and without TMD

Particle	Displacement (m)		Velocity (m/s)		Acceleration (m/s^2)	
	Without TMD	With TMD	Without TMD	With TMD	Without TMD	With TMD
1	0.0125	0.0080	0.0483	0.0473	0.2325	0.2320
2	0.0390	0.0225	0.0833	0.0770	0.2523	0.2523
3	0.1395	0.0750	0.1745	0.1308	0.3703	0.3668
4	0.2890	0.1538	0.3125	0.2000	0.4995	0.4900
5	0.4330	0.2295	0.4453	0.2613	0.5838	0.5663
6	0.5788	0.3063	0.5785	0.3200	0.6340	0.6053
7	0.7308	0.3863	0.7153	0.3753	0.6348	0.5883
8	0.8778	0.4635	0.8470	0.4270	0.6063	0.5335
9	1.0330	0.5453	0.9873	0.4828	0.5578	0.4435
10	1.1820	0.6238	1.1238	0.5400	0.5140	0.3388
11	1.3193	0.6960	1.2520	0.5978	0.5033	0.2595
12	1.4320	0.7555	1.3588	0.6488	0.5410	0.2715
13	1.5765	0.8318	1.4978	0.7183	0.6555	0.4050
14	1.7070	0.9008	1.6250	0.7843	0.8113	0.5885
15	1.8988	1.0023	1.8180	0.8948	1.1873	1.0138
16	2.1428	1.1335	2.1015	1.1103	2.0273	1.9088
17	2.4580	1.3115	2.5273	1.4945	3.1010	3.0078
18	2.6505	1.4288	2.7905	1.7330	3.5920	3.5028
19	3.2155	1.8455	3.9443	2.9950	4.2023	4.1018

structure. From this, we can see the actual effect of the structure wind-induced vibration control, and judge whether the parameters of the TMD designed are reasonable or not.

The gust fluctuating wind loads on the structure are usually regarded as smooth random processes experienced by various states. The loads vector of fluctuating wind on the series multi-degree-of-freedom systems can be a cross-correlated, zero-mean Gauss stationary random process. The mathematical simulation of multidimensional stationary stochastic processes can be described by the trigonometric series model, which is a linear superposition of harmonic vibration with random amplitude and random phase. The M. Shinozuka method was used to simulate the downwind fluctuating wind load of Hefei TV Tower.

According to the central sampling theorem, when $N \rightarrow \infty$, the simulated stochastic process $f(t)$ is an asymptotic Gauss process, so $N = 4145$ was chosen in the simulation. In the simulation, the 12th order frequency of the TV tower was taken as the cut-off frequency ω_n, the sampling frequency was 20 Hz, which was twice the 12th order mode frequency of the TV tower, and the spectrum dividing $\Delta\omega$ was 0.0127 rad/s. The time step was 0.05 s.

The responses of the upper tower (12th particle), TMD, and mast (19th particle) under the fluctuating wind loads (standard values) with or without TMD are shown in Fig. 15.29. It can be seen from the figure that: (1) under the action of TMD, the energy input of wind load, the kinetic energy and potential energy of the structure are all greatly reduced; (2) the energy input of wind load is balanced by the damping energy of the structure, the energy consumption of TMD, the kinetic energy and the potential energy of the structure, and the energy consumption of TMD is the main factor. The TMD is a good energy-consuming device for the tower structure under the action of wind load.

The standard deviation of the response of each particle is shown in Table 15.19. The maximum peak responses of displacement, velocity and acceleration of the TV tower with or without TMD are shown in Table 15.20. Compared with the control analysis results in frequency domain shown in Tables 15.17 and 15.18, the standard deviation is close, but the control effects of peak responses are different. Because the length of the wind sample used in time domain analysis is only 10 min and there is only one wind sample curve, so the accuracy of the peak response of the structure-damper system calculated is slightly worse, and the cumulative error of the time domain analysis will also cause some errors in the accuracy of the peak response of the structure.

15.4.3 Analysis of Seismic Response Control

The El-Centro wave (N–S), Taft wave and synthetic Hefei wave were Used. The maximum acceleration of ground motion of the three waves was 2.2 m/s^2 without changing the spectral characteristics of the three waves. The displacement time history and energy distribution of the upper tower with and without TMD under El-Centro wave (N–S), Taft wave and synthetic Hefei wave (the optimum parameters were taken for TMD) are shown in Fig. 15.30. It can be seen from the graph that: (1) the seismic responses of the Hefei TV Tower under three kinds of seismic waves are decreased obviously by TMD; (2) under three kinds of seismic waves, the controlled displacement response of the upper tower in the first 10 s is basically the same as that without TMD, which indicates the lag of TMD behind the impulse seismic input; (3) after the seismic response of the tower is controlled by TMD, the input energy, kinetic energy and potential energy of the TV tower are all reduced; (4) the three seismic waves all reach the peak acceleration in the first 10 s, and the energy input is mainly concentrated in the first 10 s, but the energy consumption of TMD in the first 30 s is poor, TMD only plays a role after 30 s, which indicates that TMD has hysteresis phenomenon under earthquake action. (5) the input energy of earthquake during the first 20–30 s is mainly balanced by structural damping energy, structural kinetic energy, and structural potential energy; after 30 s, it is balanced by

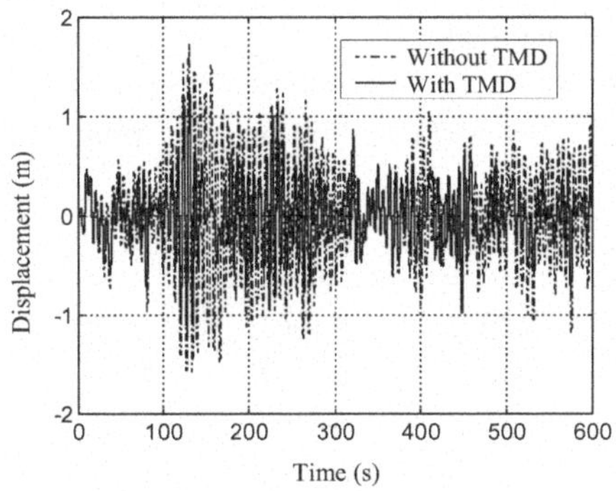

(a) Displacement time history of upper tower

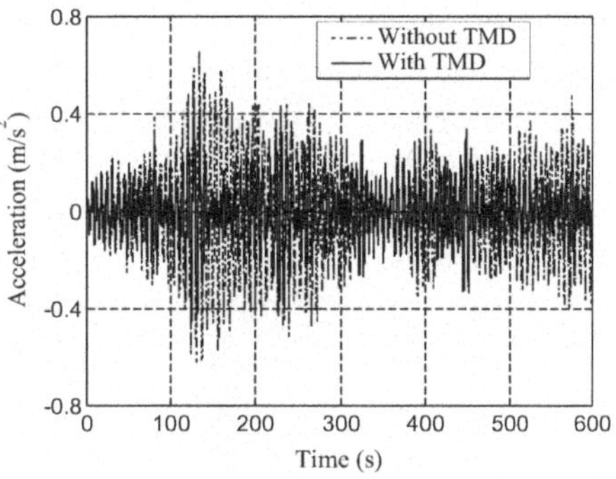

(b) Acceleration time history of upper tower

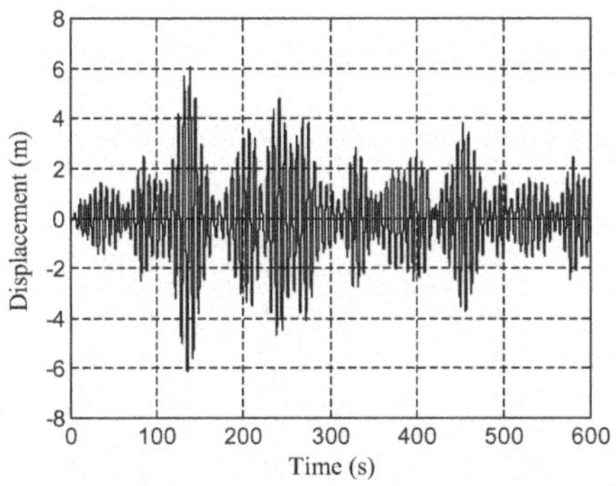

(c) Relative displacement time history of TMD

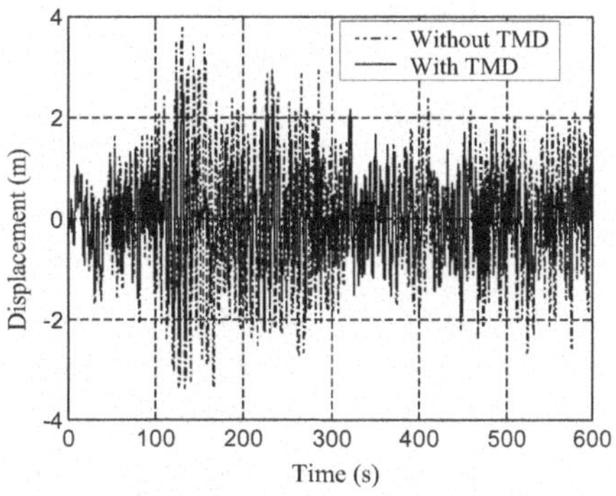

(b) Displacement time history of mast

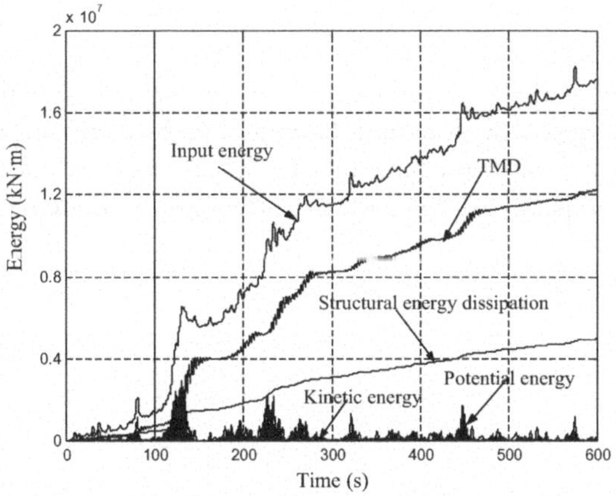

(e) Energy time-history curve of wind-induced vibration response of TV Tower under TMD

Fig. 15.29 Time domain responses of the structure and TMD under pulsating wind

Table 15.19 Standard deviation of the response of each particle

Particle	Displacement (m)		Velocity (m/s)		Acceleration (m/s^2)	
	Without TMD	With TMD	Without TMD	With TMD	Without TMD	With TMD
1	0.0045	0.0031	0.0116	0.0111	0.0428	0.0428
2	0.0148	0.0095	0.0246	0.0219	0.0646	0.0644
3	0.0534	0.0328	0.0593	0.0425	0.1003	0.0989
4	0.1106	0.0674	0.1118	0.0715	0.1357	0.1313
5	0.1656	0.1005	0.1624	0.0988	0.1597	0.1513
6	0.2211	0.1339	0.2135	0.1261	0.1769	0.1634
7	0.2788	0.1685	0.2663	0.1536	0.1835	0.1624
8	0.3345	0.2018	0.3173	0.1800	0.1824	0.1511
9	0.3933	0.2369	0.3715	0.2083	0.1779	0.1311
10	0.4496	0.2704	0.4238	0.2363	0.1775	0.1124
11	0.5014	0.3013	0.4725	0.2629	0.1865	0.1063
12	0.5439	0.3267	0.5128	0.2854	0.2023	0.1151
13	0.5985	0.3593	0.5649	0.3149	0.2320	0.1425
14	0.6479	0.3888	0.6122	0.3421	0.2672	0.1791
15	0.7205	0.4324	0.6831	0.3845	0.3386	0.2569
16	0.8128	0.4885	0.7819	0.4543	0.5328	0.4710
17	0.9326	0.5643	0.9327	0.5842	0.8360	0.7859
18	1.0062	0.6140	1.0405	0.6883	0.9343	0.8828
19	1.2226	0.7825	1.5273	1.2200	1.5282	1.4850

structural damping energy, structural kinetic energy, structural potential energy, and TMD damping energy, with structural damping energy as the main energy dissipation, and the energy dissipation capacity of TMD is small, which indicates that the energy dissipation capacity of TMD under earthquake is limited.

Table 15.20 Maximum of the response of each particle

Particle	Displacement (m)		Velocity (m/s)		Acceleration (m/s^2)	
	Without TMD	With TMD	Without TMD	Without TMD	Without TMD	With TMD
1	0.0190	0.0142	0.0457	0.0407	0.1497	0.1506
2	0.0493	0.0357	0.0934	0.0812	0.2473	0.2507
3	0.1652	0.1269	0.2229	0.1772	0.3706	0.3836
4	0.3414	0.2615	0.4141	0.3189	0.5063	0.5227
5	0.5129	0.3903	0.5864	0.4437	0.5962	0.6149
6	0.6886	0.5201	0.7524	0.5654	0.6701	0.6494
7	0.8734	0.6545	0.9144	0.6815	0.6986	0.5956
8	1.0520	0.7832	1.0622	0.7840	0.7453	0.5609
9	1.2406	0.9179	1.2098	0.8821	0.7279	0.5158
10	1.4212	1.0461	1.34518	0.9739	0.6793	0.4767
11	1.5878	1.1636	1.4713	1.0663	0.6245	0.4110
12	1.7247	1.2605	1.5852	1.1774	0.6526	0.4537
13	1.9006	1.3854	1.7567	1.3333	0.8407	0.6229
14	2.0599	1.4986	1.9142	1.4780	1.0035	0.7892
15	2.3031	1.6670	2.1619	1.6803	1.2564	0.9883
16	2.6251	1.8819	2.5817	1.9117	2.0539	1.7155
17	3.0366	2.1418	3.1218	2.1360	3.3081	2.8684
18	3.2689	2.2802	3.3886	2.5102	3.5790	3.0544
19	3.7934	2.6869	4.5801	4.0938	5.0762	5.3767

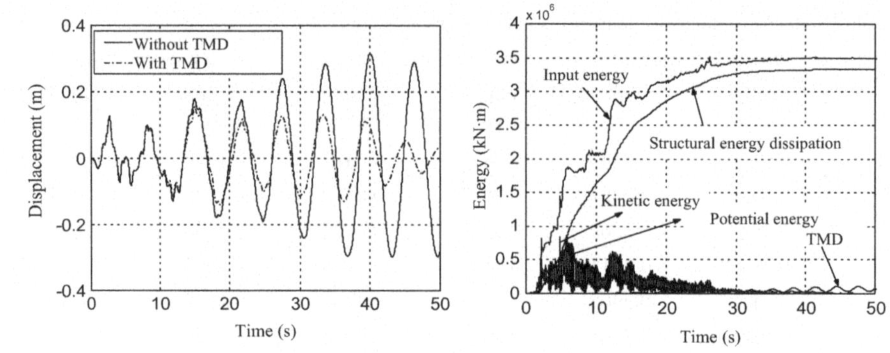

(a) El-Centro wave

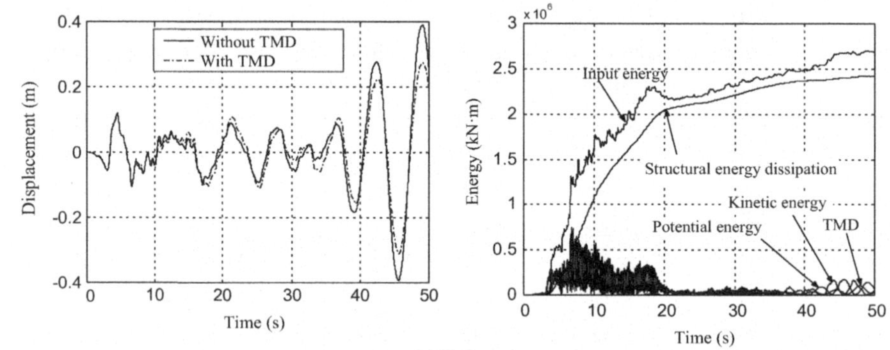

(b) Taft wave

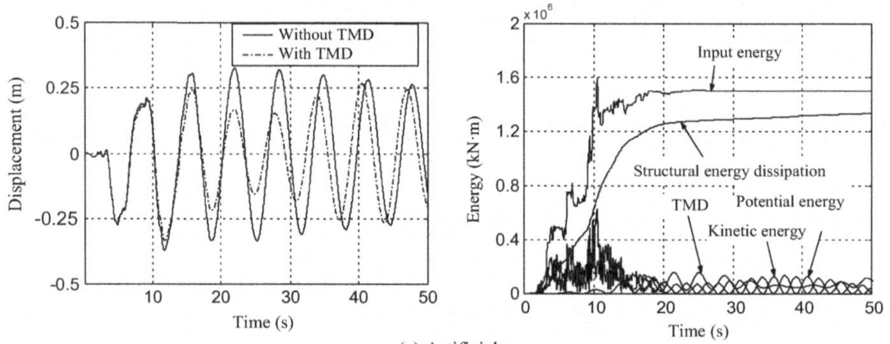

(c) Artificial wave

Fig. 15.30 Structural displacement response and energy dissipation distribution under earthquake

References

1. Zhang, Zhiqiang, Tong Guo, Kang Yang, et al. 2017. Simulation and measurement of human-induced vibrations of the Beijing Olympic watchtower with tuned mass dampers. *Journal of Performance of Constructed Facilities* 31(040170956).
2. Li, Aiqun. 2007. *Vibration control of engineering structure*. Beijing: China Mechine Press.

3. Fei, Lu. 2002. *Study on some problems of active control project of wind vibration control for Nanjing TV tower*. Nanjing: Southeast University. (in Chinese).
4. Huang, Ruixin. 2011. *Research on TMD control of vibration response of high-rise structure with variable damping*. Nanjing: Southeast University. (in Chinese).
5. Zhang, Zhiqiang. 2003. *Study on vibration control of wind vibration and seismic response of Hefei TV tower*. Nanjing: Southeast University. (in Chinese).

Chapter 16
Engineering Practice of Vibration Control for Long-Span Structures

Abstract Project cases of vibration control engineering practice for long-span structure are introduced. Case 1 is Beijing Olympic National Conference Center with a steel structure floor span of 60 m×81 m. Case 2 is the High-speed Railway Hub Station, including Changsha New Railway Station with the span 49 m, Xi'an North Railway Station with the span 43 m and Shenyang Railway Station with the span 31 m. In Case 1 and 2, the MTMD (Multi-TMD) are used for comfort control under pedestrian load. Case 3 is the Fuzhou Strait International Conference and Exhibition Center, the MTMD system was set up in the large span roof to control the wind-induced vibration response of the structure. In these cases, project overview, MTMD design, analysis model and calculation of structural control are introduced respectively.

16.1 Beijing Olympic National Conference Center (Pedestrian Load, TMD)

16.1.1 Project Overview

The National Conference Center of Beijing Olympic Park (Fig. 16.1) undertakes the main functions of the International Conference Hall and the News Center for the 2008 Olympic Games [1]. The project is divided into two major areas: conference and exhibition. On the fourth floor of the conference area, there is a steel structure floor with a span of 60 m * 81 m, which is the competition venue for the Olympic Games, and a great hall for 5500 people after the game. Owing to the complex function of this area, the owners require the live load of 7.5 kN/m^2 (10 kN/m^2 in local area of seat storage), which is much larger than the normal floor live load (2–3.5 kN/m^2), and belongs to the heavy load and long-span steel structure floor.

A. Li, *Vibration Control for Building Structures*, Springer Tracts in Civil Engineering, https://doi.org/10.1007/978-3-030-40790-2_16

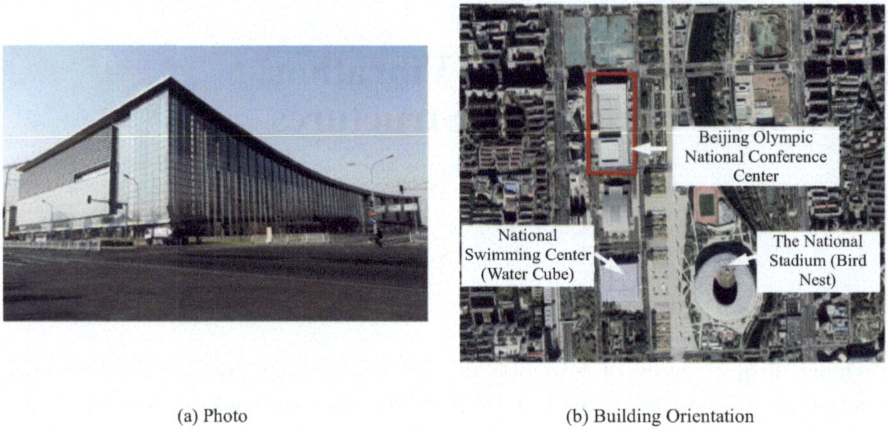

Fig. 16.1 Beijing Olympic National Conference Center

16.1.2 Structural Vibration Reduction Design

As a high-standard national conference center after the games, the Olympic Conference Center needs to meet various requirements for use. For this reason, walking, jumping, standing and other analysis conditions were used in the calculation. After many cycles of optimization calculation, 72 sets of vibration absorbers were arranged on the floor (Fig. 16.2, blue block is TMD1, red block is TMD2). The TMD parameters are shown in Table 16.1. Among them, the linear viscous dampers were used as dampers. The parameters are as follows: damping coefficient is 4000 N s/m, the maximum stroke is ±50 mm, the maximum output force is 10 kN.

16.1.3 Structural Analysis Model

16.1.3.1 Finite Element Model and Dynamic Characteristic Analysis

The finite element model of the fourth floor of the project was established as shown in Fig. 16.3a. Through modal analysis, the natural frequencies of the first vertical mode and the second vertical mode are 2.7866 Hz and 3.5039 Hz, respectively. The first two modes of the structure are shown in Fig. 16.3b, c. The first mode is a first-order vertical symmetrical bending mode with the largest amplitude in the middle span, and the second mode is a second-order vertical antisymmetric bending mode. The first natural frequency of the structure is close to the frequency of normal walking and jumping (1.8–2.7 Hz), so it is easy to produce resonance. It is unrealistic and unreasonable from the point of view of technology, economy and space utilization to rely on the method of enlarging section and changing structure type. Therefore, the MTMD system was adopted to control structure response.

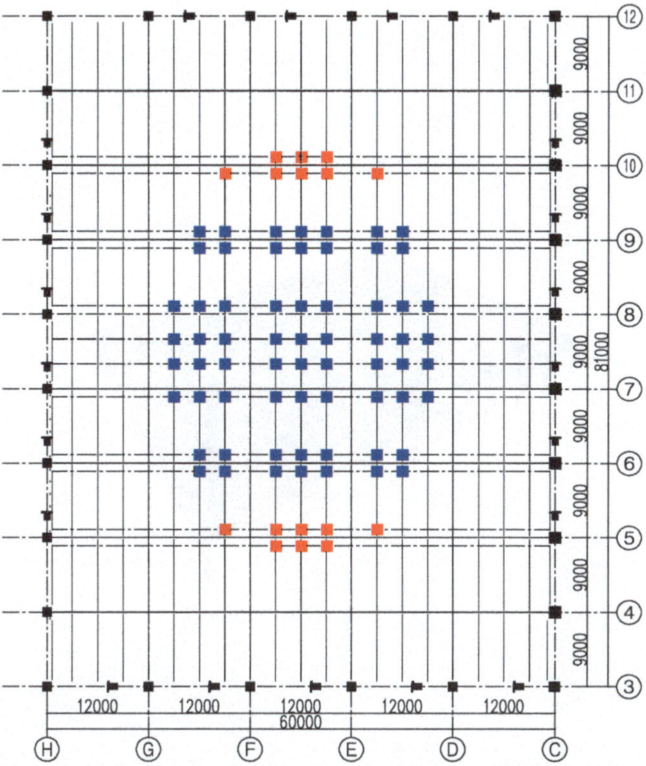

(a) TMD Layout sketch

(b) Scene of TMD layout and installation

Fig. 16.2 TMD layout of floor

Table 16.1 Damping system parameters

Damping system	Spring stiffness (N m^{-1})	Quality of mass block (kg)	Tuning frequency of TMD (Hz)
TMD1	47,016.5	580	2.9
TMD2	70,456.8	580	3.5

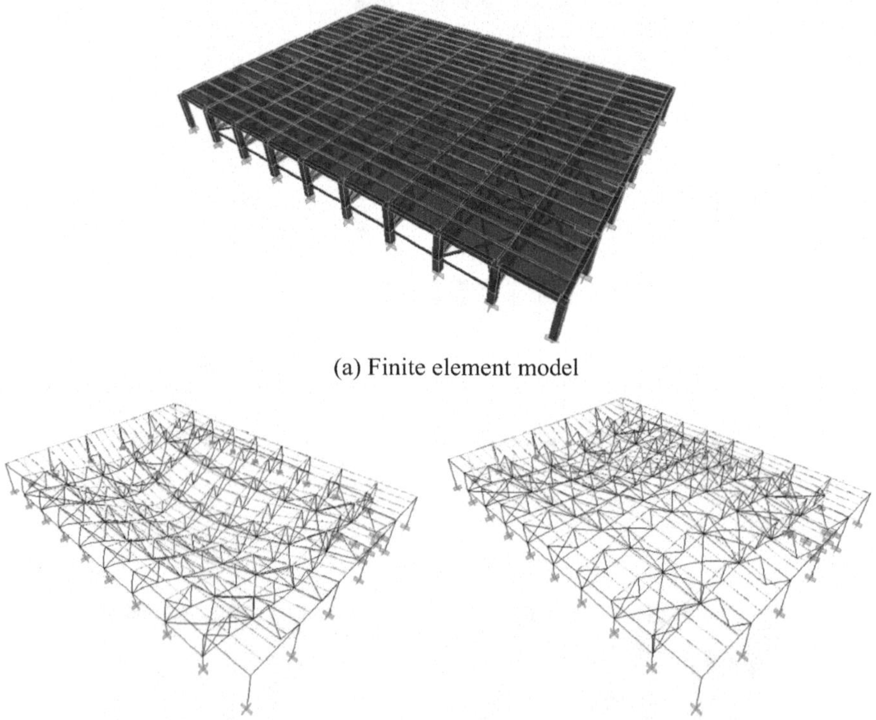

(a) Finite element model

(b) 1st mode shape (c) 2nd mode shape

Fig. 16.3 Finite element model and vibration modes of the structure

16.1.3.2 Human Induced Dynamic Load

(1) Erecting load

According to the characteristics of human erecting activities, it is assumed that: (1) the duration of human erecting is 1 s; (2) the impact load curve at the time of erecting is assumed to be a sinusoidal wave. The research shows that the expression of acceleration of human body gravity center movement is as follows:

$$a(t) = a_1 \sin \frac{2\pi}{T} t, \quad t \in [0, T] \tag{16.1}$$

The derivation of Eq. (16.1) shows that the velocity of the motion of the human body's center of gravity is as follows:

$$v(t) = -\frac{T}{2\pi}a_1 \cos\frac{2\pi}{T}t + C_1 \quad (16.2)$$

Continuous derivation of Eq. (16.2) shows that the displacement of the center of gravity of the human body is as follows:

$$s(t) = -\left(\frac{T}{2\pi}\right)^2 a_1 \sin\frac{2\pi}{T}t + C_1 t + C_2 \quad (16.3)$$

Then, according to the boundary conditions $s|_{t=0} = 0$, $s|_{t=T} = h_1$, $v|_{t=0} = 0$, $C_2 = 0$, $C_1 = Ta_1/2\pi$, the peak acceleration of the center of gravity of the human body is obtained as $a_1 = 2\pi C_1/T = 2\pi h_1/T^2$. Among them, T is the time spent during the whole erection process, and h_1 is the height of a person's weight gain.

Assuming the height difference between the center of gravity of the human body before and after standing as $h_1 = 0.4$ m, the peak acceleration is obtained as $a_1 = 2\pi h_1/T^2 = 2\pi \times 0.4/1^2 = 2.512$ m/s^2, so the dynamic coefficient is $\alpha = a_1/g = 0.256$. The weight of a person is taken as 70 kg/person according to the Sect. 2.2.1 of *AISC Steel Design Guide Series 11*. The equivalent uniform load is 0.7 kPa. Therefore, the impact force curve of the standing load is shown in Fig. 16.4a.

(2) Walking load

The curve of IABSE (*International Association for Bridge and Structural Engineering*) is selected as shown in Fig. 16.4b. The formula is as follows:

$$F(t) = G\left[1 + \sum_{i=1}^{3} \alpha_i \sin(2i\pi f_s t - \Phi_i)\right] \quad (16.4)$$

where F is pedestrian motivation, t is time, G is weight, f_s is walking frequency, and $\alpha_1 = 0.4 + 0.25(f_s - 2)$, $\alpha_2 = \alpha_3 = 0.1$, $\Phi_1 = 1$, $\Phi_2 = \Phi_3 = \pi/2$. The fast walking frequency of human is 2.3 Hz, the slow walking frequency of human is 1.7 Hz, and the weight of person is taken as 70 kg/person.

(3) Jumping load

The impact force curve of jumping activity on the floor can be simulated approximately by sinusoidal curve. The first natural frequency of the floor is taken as the frequency, the dynamic coefficient is taken as 1.5, and the weight of the person is taken as 70 kg/person. So the impact force curve of jumping load is shown in Fig. 16.4c.

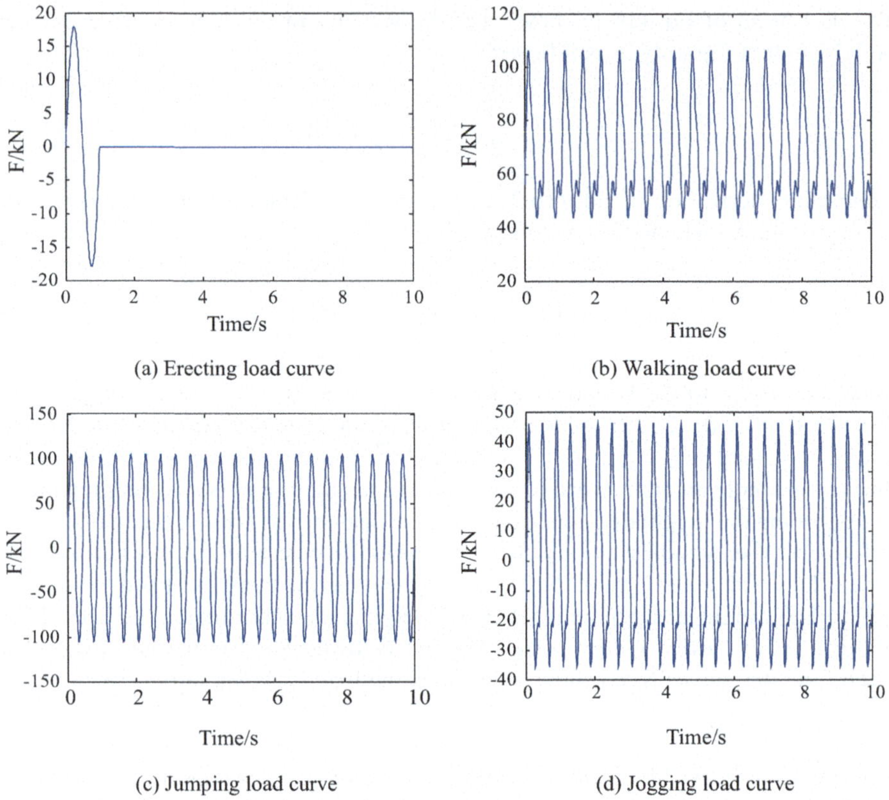

Fig. 16.4 Human actuated load curve

(4) Running load

It is approximated that the running load can be simulated according to the walking load formula (16.4). Only the excitation frequency is faster, and is selected as 2.5 Hz according to jogging. The human weight is 70 kg/person. The impact force curve of the human's running is shown in Fig. 16.4b.

16.1.4 Analysis of Structural Comfort Control

The results of dynamic analysis are shown in Table 16.2. It can be seen that the MTMD system greatly reduces the acceleration response of the floor, especially when the frequencies of fast walking and jumping are close to the natural frequencies of the structure. The acceleration response of the structure after vibration reduction basically meets the requirements of the code for human comfort.

Table 16.2 Peak acceleration of vertical vibration of the structure

Condition	Without damper (g)	With damper (g)	Vibration reduction rate (%)	Acceleration limit	Note
1	0.00969	0.00494	49.1	0.005 g ("Office")	Brisk walking
2	0.0185	0.0152	17.4	0.015 g ("Business")	Slow walking
3	0.0219	0.0117	46.7	0.015 g ("Business")	Beat
4	0.0106	0.0105	1.3	0.015 g ("Business")	Stand up slowly
5	0.0168	0.0151	10.3	0.015 g ("Business")	Stand up quickly
6	0.0339	0.0325	4.2	0.04 g ("Rhythmic activity only")	Concert

16.1.5 On-Site Dynamic Test

In order to investigate the effectiveness of the MTMD system installed and verify the correctness of the analysis results, the field dynamic tests were carried out for the structure. In order to verify and supplement the theoretical calculation, the natural vibration characteristics of floor and the vibration of floor before and after installation of MTMD system were measured.

The instrument used in the test mainly included the acceleration sensor and signal acquisition and analysis system. The ultra-low frequency 941-B sensor was selected in this test, with a passband range of 0.25–80 Hz, which meets the requirements of the engineering test. According to the dynamic characteristics of the floor before and after vibration reduction and its symmetry, the maximum points of the first three modes were selected as the monitoring objects. The measuring points were mainly arranged in the middle of span (E–F axis/7–8 axis), as shown in Fig. 16.5. D1, D2 and D3 are the main measuring points, D4 and D5 are the reference measuring points, M1–M18 are the measuring points needed for dynamic characteristics testing. The types and quantities of the main test instruments required are shown in Table 16.3.

According to the actual measurement, the dynamic characteristics of the structure were analyzed by SAP2000 software. The natural frequencies of the structure were determined according to the dynamic characteristics test, then the free attenuation signals of single mode were extracted by band-pass filtering, and the damping ratios were obtained by logarithmic reduction method as 0.021 and 0.026. The comparison of theoretical calculation and measured results of the first two vertical natural frequencies of floor is shown in Table 16.4. There is a big difference between

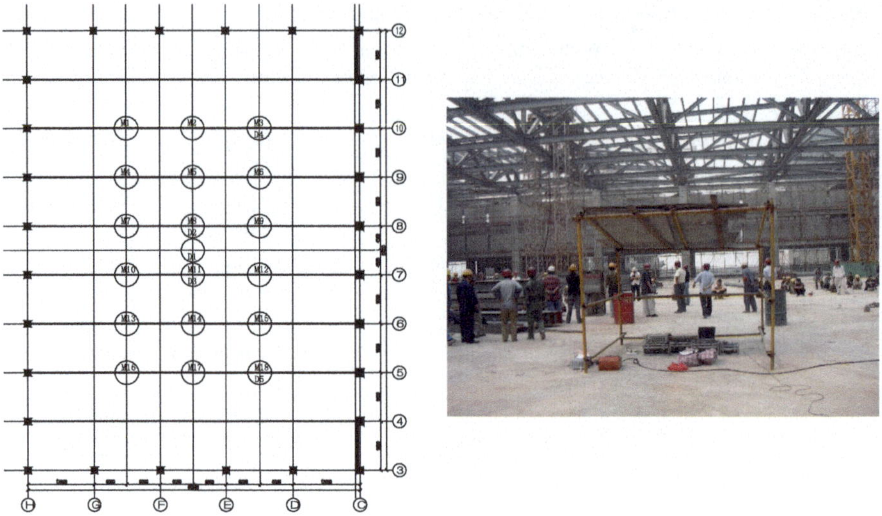

(a) Layout of vibration measuring points (b) Test site

Fig. 16.5 Field measurement

Table 16.3 Main experimental instruments and performance indicators

Instrument	Number	Main technical indicators
941-B vibration picker in vertical direction	6	Passband 0.25–80 Hz
941-B vibration picker in horizontal direction	4	Passband 0.25–80 Hz
991 six-wire amplifier	2	Passband 0.025–35 Hz
AZ_CRAS signal acquisition instrument	1	16 passageways
AZ_CRAS analysis system	1	6.2 version

Table 16.4 Comparison between theoretical calculation and experimental results of dynamic characteristics

Order	Theoretical calculation frequency of test state (Hz)	Measured frequency (Hz)
1	3.70	3.75
2	4.57	4.88

the structure in the test and that in actual use. The main reason is that due to the time limit, the ground method, hanging load and the service load of the hall were not applied during the test. Therefore, there is a big difference in the structure quality compared with the previous theoretical analysis, which leads to that the natural vibration frequency of the structure in the test and that in the normal use had a certain gap in vibration frequency.

Table 16.5 Comparison of theoretical calculation and measured results of test conditions

Condition	Without MTMD (mm·s^{-2})			With MTMD (mm·s^{-2})			
	Stability value	Maximum value	Theoretical value	Stability value	Vibration reduction rate (%)	Theoretical value	Maximum value
1	90–95	98.0	111.7	30–40	>55	50.0	83.3
2	40–45	68.1	91.4	25–30	>40	80.6	53.8

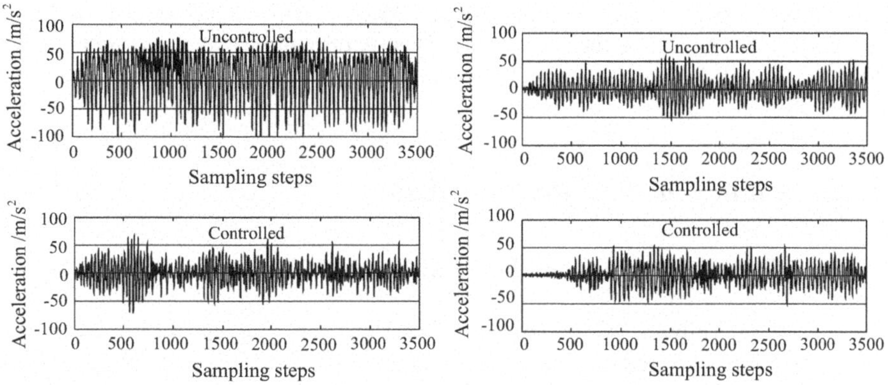

(a) Condition 1

(b) Condition 2

Fig. 16.6 Measured time-history response of acceleration

Based on the measured results, the responses of the structure under condition 1 (22 people in the center of the conference hall jumped together at 3 Hz) and condition 2 (90 people in the corridor walk quickly at 3 Hz) were analyzed by SAP2000 (the damping ratio of floor was 0.02). The variations of theoretical calculation results and measured results of vibration amplitude at a measuring point of the floor structure before and after TMD installation are shown in Table 16.5 and Fig. 16.6. The results show that the test results and calculation results of floor vibration response before and after MTMD installation are basically the same. The proposed analysis method can be used to analyze the dynamic response of large-span floor structures under human-induced loads. The floor vibration reduction effect after MTMD installation is obvious, and the most significant effect is for jumping vibration reduction. The comfort of the floor after the MTMD system installation has been significantly improved.

16.2 High-Speed Railway Hub Station (Pedestrian Load, TMD)

With the rapid development of China's high-speed railway industry, the demand for hub stations is growing. Such stations play an important role in the distribution of high-speed railway network. The trains enter and leave frequently and the traffic is large. The demands for buildings and functions make the floor structures have the characteristics of long-span, light weight, and low damping. Under the action of crowd loads, they are prone to produce large vibration, which leads to the problem of human comfort exceeding the limit [2–4]. This section mainly focuses on the theme of this book and gives some engineering practice cases.

16.2.1 Changsha New Railway Station

16.2.1.1 Project Overview and Finite Element Model

Changsha New Railway Station (Fig. 16.7a) is 266.25 m long and 177 m wide. The second floor is an elevated floor with the elevation of 10.50 m. Due to the requirements of construction and technology, the span between axis 10 and axis 13 is 49 m, and the steel truss of floor is supported on the columns on both sides. The floor is a composite floor with profiled steel plates, and the main beams and secondary beams are steel trusses. According to the design drawings and the finite element model, the finite element program SAP2000 was used to analyze the dynamic characteristics of the structure floor before and after vibration reduction. The calculation was carried out according to the three-dimensional spatial structure. The calculation model is shown in Fig. 16.7b. The spatial beam-column element was used to simulate the reinforced concrete beams and columns, and the influence of rigid zone of joints on dynamic characteristics was considered in reinforced concrete structures. The high-precision quadrilateral space shell element was used to simulate the reinforced concrete floors

(a) Photo

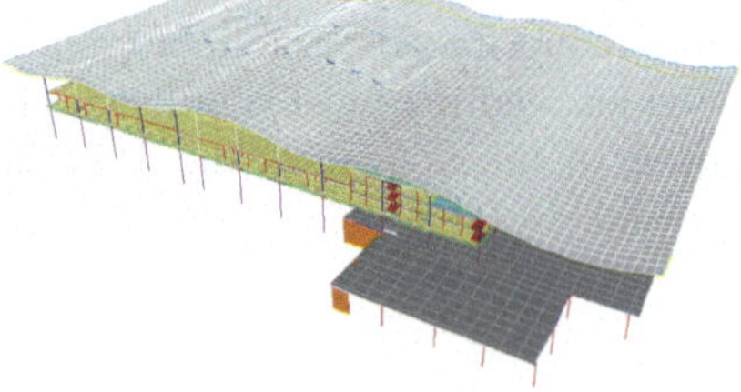

(b) Finite element model

Fig. 16.7 Changsha New Railway Station

and walls, which can accurately describe the mechanical characteristics of reinforced concrete floors and walls, and can also consider the in-plane and out-of-plane deformations of reinforced concrete floor and wall; the steel bundle element was used to simulate the flexible cables. The properties of the materials used were taken according to the specifications.

Because of the complex structure of the station building in Changsha New Railway Station, the natural vibration period of the structure is very intensive. The mode shapes of the floor at elevation of 10.5 m are mainly related to the project. Therefore, the modal analysis for this part was carried out. The first six vertical modes of the floor are shown in Fig. 16.8. It can be seen that the mode of vibration of the structure is local floor vibration, which is related to the function and structural characteristics of the station building. The vertical natural frequencies of each vibration block are close to each other, which are between 2.0 and 3.0 Hz, and close to the walking frequencies of human beings from 1.8 to 2.7 Hz. It is easy to produce resonance. The mode is obviously related to the distribution of the entrance.

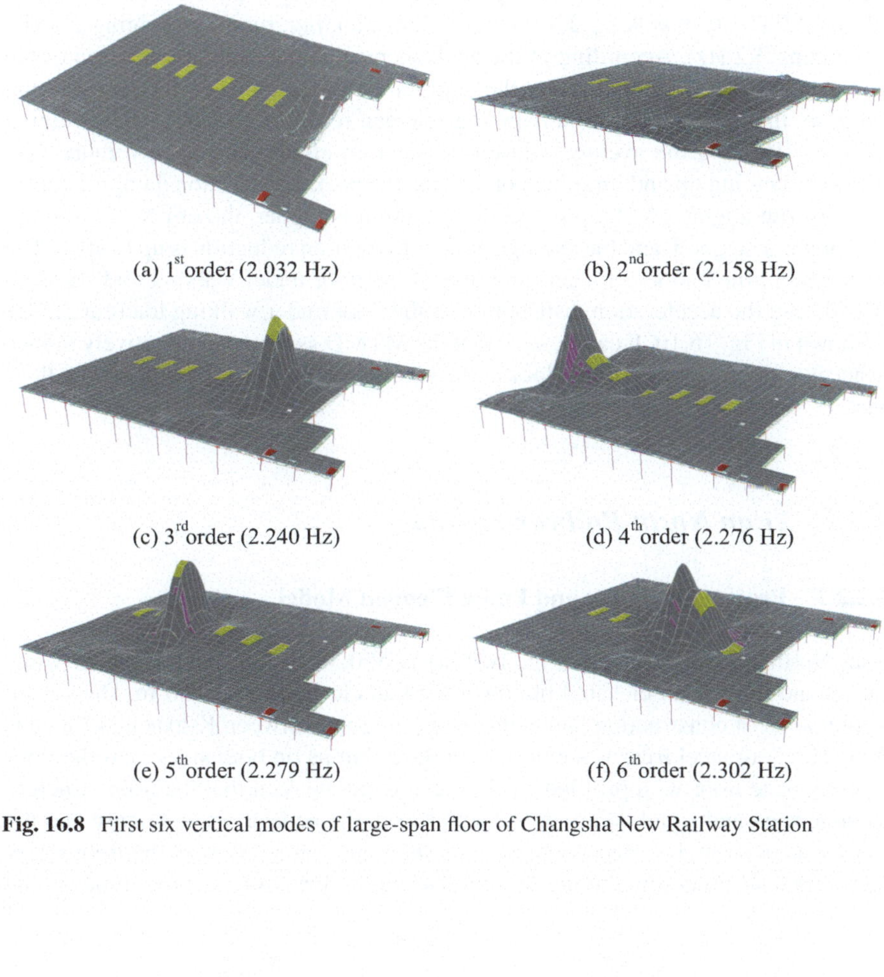

(a) 1storder (2.032 Hz)

(b) 2ndorder (2.158 Hz)

(c) 3rdorder (2.240 Hz)

(d) 4thorder (2.276 Hz)

(e) 5thorder (2.279 Hz)

(f) 6thorder (2.302 Hz)

Fig. 16.8 First six vertical modes of large-span floor of Changsha New Railway Station

16.2.1.2 Structural Vibration Reduction Design

The MTMD technology was used to control the floor vibration. After optimum design, 56 sets of TMD dampers were arranged on the floor (in Fig. 16.9, the black block is TMD1, the magenta block is TMD2, the blue block is TMD3, the red block is TMD4, the grey block is TMD5). The parameters of the device are shown in Table 16.6. The dampers are linear viscous dampers. The parameters are as follows: the damping coefficient is 1000 N s/m, the maximum stroke is ±50 mm, and the maximum output force is 8 kN.

16.2.1.3 Analysis of Structural Comfort Control

The crowd load simulation is the same as that in Sect. 16.1. According to the distribution of the entrance, the load distribution area was divided into nine parts. At the same time, different starting phases were also considered in the calculation. 49 kinds of analysis conditions were adopted, such as walking (jogging 1.7 Hz, ordinary walking 2.0 Hz, fast walking 2.3 Hz), jumping, standing, running (jogging 2.5 Hz, fast running 3.2 Hz). According to the analysis results, the peak acceleration of the floor is only 0.13 m/s^2, which meets the requirements of human comfort; for walking condition, the peak acceleration damping rate can reach to 40–70% when walking fast at 2.3 Hz, and the average vibration reduction effect can reach to more than 50%. For standing up and jumping conditions, the peak acceleration damping rate is 50%. For running at 2.5 Hz, the peak acceleration is higher, the effect of vibration reduction is also good, and the average effect of vibration reduction is up to 40%. The acceleration time history of a certain point of the floor under walking and standing up loads and the acceleration distribution of the floor under walking loads at 2.3 Hz are shown in Fig. 16.10. It can be seen that the MTMD system can effectively reduce the acceleration response of the floor in the whole load period and in the whole floor space.

16.2.2 Xi'an North Railway Station

16.2.2.1 Project Overview and Finite Element Model

Xi'an North Railway Station (Fig. 16.11a) is 550.38 m long and 450.5 m wide. The second floor is an elevated platform with an elevation of 10.00 m. Due to the requirements of construction and technology, the span between K-axis and J-axis is 43 m. The floor steel truss is supported on the columns on both sides, and the floor is a composite floor with profiled steel sheets. It can be seen that the floor structure has large span and small damping. As a platform, the pedestrians are dense and the crowd load is large. It is necessary to study the vibration comfort of the floor under pedestrian load. According to the structural design information, a three-dimensional

Fig. 16.9 TMD layout for the long-span floor of Changsha New Railway Station

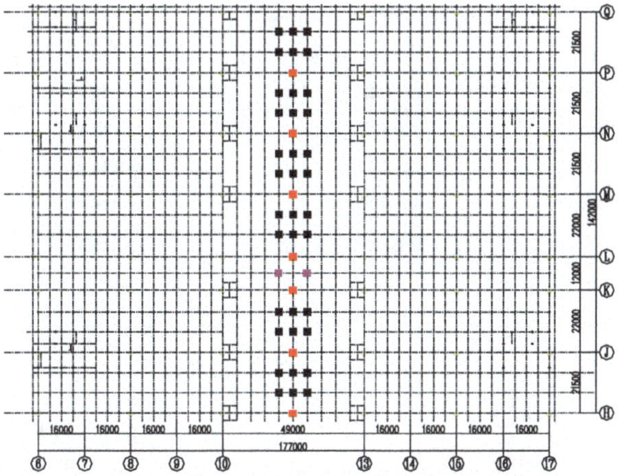

(a) TMD layout in area B of floor

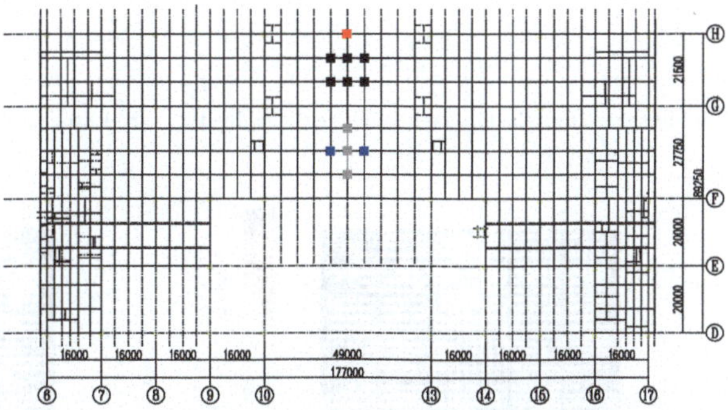

(b) TMD layout in area A of floor

(c) Scene of installation

Table 16.6 Damping system parameters

Serial number	Spring stiffness (N m^{-1})	Quality (kg)	Tuning frequency (Hz)
TMD1	40,644 ± 15%	800	2.27
TMD2	57,501 ± 15%	800	2.7
TMD3	34,784 ± 15%	800	2.1
TMD4	61,622 ± 15%	1000	2.5
TMD5	43,480 ± 15%	1000	2.1

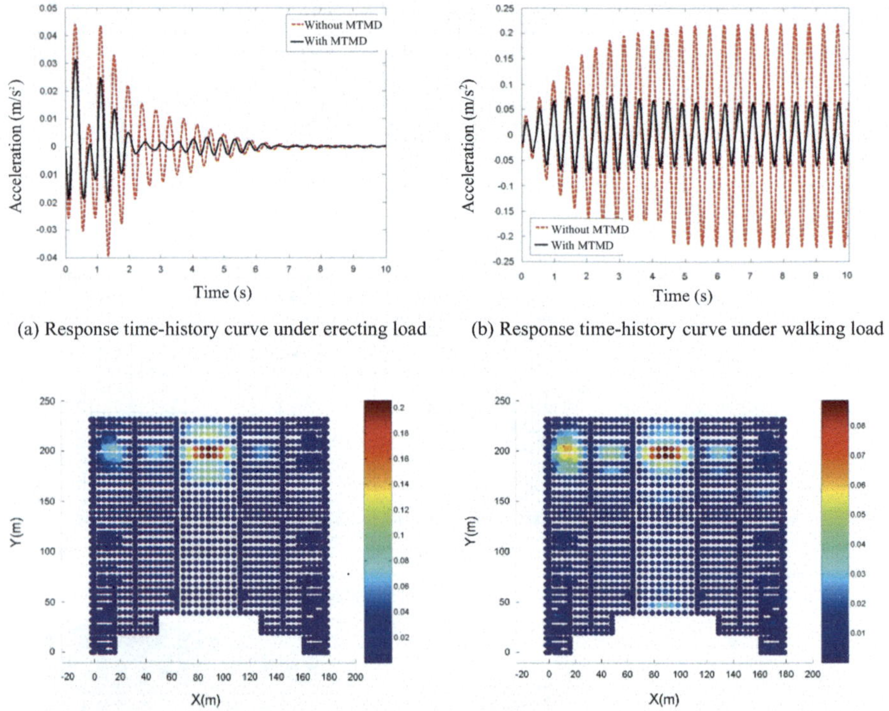

(a) Response time-history curve under erecting load

(b) Response time-history curve under walking load

(c) Acceleration distribution of floor under 2.3Hz walking load (before vibration reduction)

(d) Acceleration distribution of floor under 2.3Hz walking load (after vibration reduction)

Fig. 16.10 Acceleration response time-history curve of one point of long-span floor of Changsha New Railway Station

fine integral finite element model was established. As shown in Fig. 16.11b, the spatial beam-column element was used to simulate the reinforced concrete beams and columns, and the influence of rigid zone of joints on dynamic characteristics was considered in reinforced concrete structures. The high-precision quadrilateral space shell element was used to simulate the reinforced concrete floors and walls, which can accurately describe the mechanical characteristics of reinforced concrete floors and walls, and can also consider the in-plane and out-of-plane deformations

(a) Photo (b) Finite element model

Fig. 16.11 Xi'an North Railway Station

of reinforced concrete floor and wall; the steel bundle element was used to simulate the flexible cables.

In order to reasonably consider the constraints of other parts of the structure on the floor analyzed, the overall model shown in Fig. 16.11b was used for analysis. The modal analysis shows that the first two modes of the model are translational, and this project mainly studies the vertical vibration of the platform floor. The first four modes of the floor which needs vibration-absorbing are shown in Fig. 16.12. Among them, the first two vertical natural frequencies are 2.33 Hz and 2.65 6 Hz,

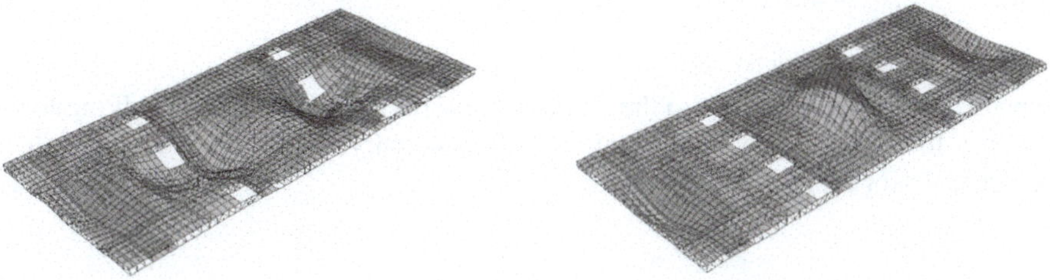

(a) 1st order (2.333 Hz) (b) 2nd order (2.656 Hz)

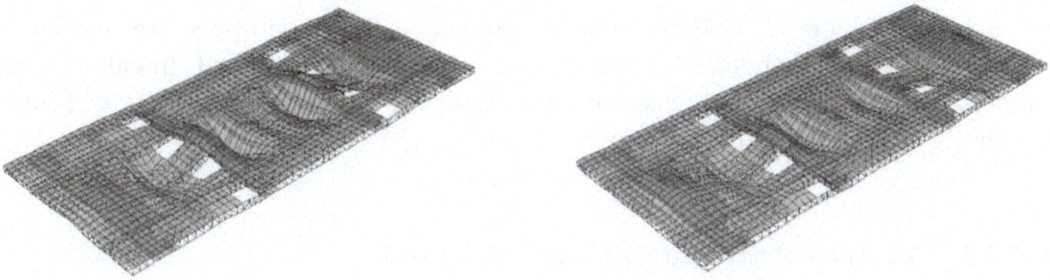

(c) 3rd order (3.076 Hz) (d) 4th order (3.108Hz)

Fig. 16.12 The first four vertical modes of large-span floor of Xi'an North Railway Station

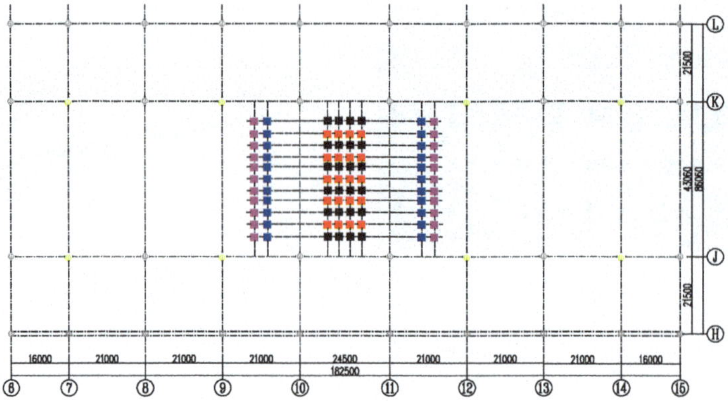

Fig. 16.13 TMD layout of large-span floor of Xi'an North Railway Station

Table 16.7 Damping system parameters

Serial number	Spring stiffness ($N{\cdot}m^{-1}$)	Quality of mass block (kg)	Tuning frequency (Hz)
TMD1	19,719 ± 10%	500	2.0
TMD2	35,938 ± 10%		2.7
TMD3	47,375 ± 10%		3.1
TMD4	26,078 ± 10%		2.3

respectively, which are close to the characteristic frequencies of daily walking loads and easy to cause resonance. Therefore, it is necessary to take measures to reduce the vertical vibration of the floor.

16.2.2.2 Structural Vibration Reduction Design

After optimum design, 88 sets of MTMD dampers were arranged on the floor (Fig. 16.13, black is block TMD1, red block is TMD2, blue block is TMD3, and magenta block is TMD4). The parameters of the dampers are shown in Table 16.7. Among them, the dampers were newly developed linear viscous dampers with variable dampers, and the damping coefficients are: $u < 1$ mm, $C = 0$; $u \geq 1$ mm, $C = 3000$ N s/m.

16.2.2.3 Analysis of Structural Comfort Control

The pedestrian frequency was defined as 1.7 and 2.0 Hz for walk slowly, 2.3 Hz for brisk walking, 2.5 Hz for jogging, and 2.7 Hz for fast run, and the load distribution area was divided into B_1–B_4 blocks (Fig. 16.14). Therefore, 20 kinds of analysis conditions were defined. From the calculation, it can be seen that when the loads are

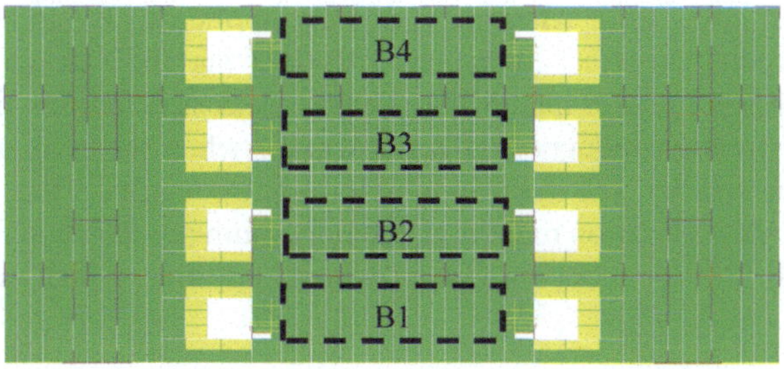

Fig. 16.14 Load layout of the long-span floor of Xi'an North Railway Station

applied to B_1 and B_4 plates, the peak acceleration of the structure is far less than the human comfort limit, which need not be considered; when the loads are applied to B_2 and B_3 plates, the distribution and value of the peak acceleration of the floor are similar, and exceed the prescribed human comfort limit. Limited to space, this section mainly gives the calculation results of five working conditions under which the load is applied on the B_3 plate, and only gives the peak acceleration distribution nephogram of the floor under the larger working conditions 3 and 5 (Fig. 16.15). The results show that the peak acceleration of the structure is far less than the human

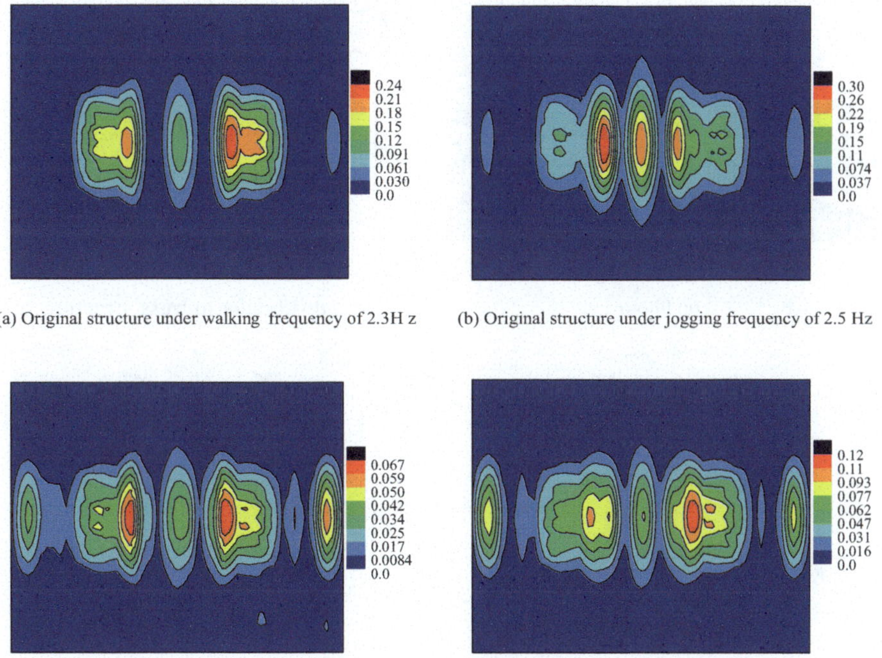

(a) Original structure under walking frequency of 2.3H z

(b) Original structure under jogging frequency of 2.5 Hz

(c) Damping structure under walking frequency of 2.3 Hz

(d) Damping structure under jogging frequency of 2.5 Hz

Fig. 16.15 Distribution cloud map of acceleration peak of long-span floor of Xi'an North Railway Station (m/s^2)

comfort limit when the loads are applied to B_1 and B_4 plates; the peak vibration reduction effect can reach to 30–70% and the average vibration reduction effect can reach to more than 50% when the peak acceleration of load condition is larger; the vibration reduction effect is weaker when the peak acceleration of load condition is smaller, but the average vibration reduction effect can reach to more than 15%. After vibration reduction, the maximum acceleration peak value of the floor is about 0.13 m/s^2 under various working conditions, which meets the requirements.

16.2.3 *Shenyang Railway Station*

16.2.3.1 Project Overview and Finite Element Model

This project is the large-span floor at the interlayer of building structure of Shenyang Railway Station (Fig. 16.16a), and is a part of the renovation project of Shenyang Station of Shenyang Junction of the newly-built Harbin-Dalian Passenger Dedicated

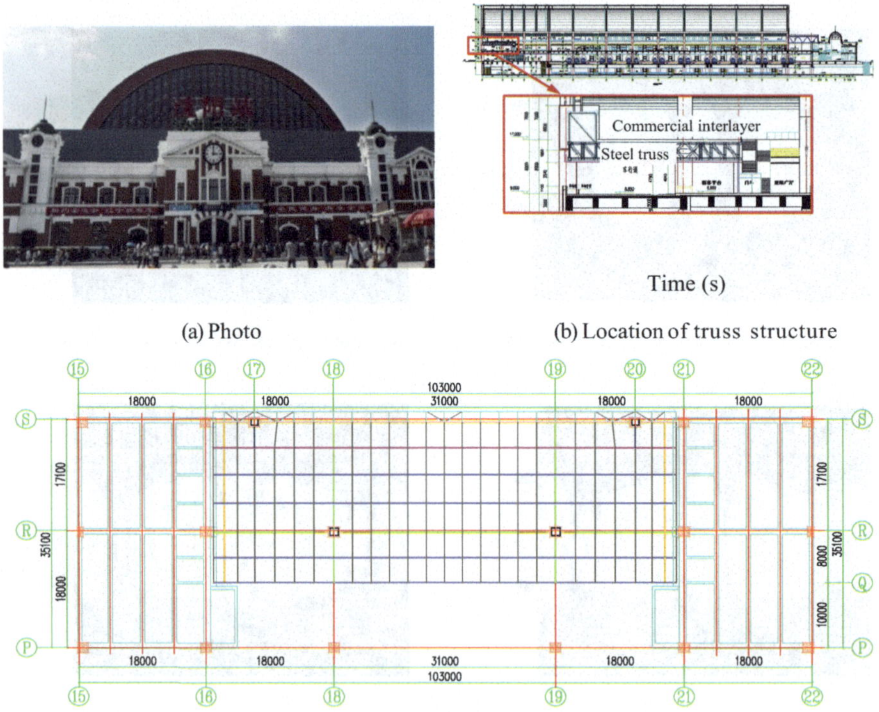

(a) Photo (b) Location of truss structure

(c) Plane layout of floor

Fig. 16.16 Shenyang Railway Station

Line. The roof of the station is partially steel structure, and the middle roof of the station is a 67 m span semi-circular arch, and both ends are supported on the reinforced concrete side beams. The arch is a rectangular three-dimensional truss consisting of pipe truss. The platform canopy is located on both sides of the main station building, and the platform canopy is separated from the station building by anti-seismic joints. Abdominal steel structure is used for the station canopy, and the light steel is used for the roofs of station and canopy. The columns are steel reinforced concrete and canopy columns are steel columns. Due to the requirements of construction and technology, the span of the interlayer at 17.0 m elevation between axis 18 and axis 19 (Fig. 16.16b) reaches to 31 m. The composite slabs of profiled steel sheet and concrete are used in this part of the floor. The H-shaped steel and box-shaped beams are used in the main and secondary beams. The plane layout of the structure is shown in Fig. 16.15c. The center distance of the upper and lower chords of the steel truss of interlayer floor is 1.1 m, the middle span is 31 m, the ratio of height to span of the truss is 1/28, and the vertical stiffness is small. The steel truss of interlayer floor is an entry platform according to the requirements of building function, so it is easy to gather dense crowds to form larger pedestrian loads.

According to the construction drawings and the finite element model, the interlayer floor at 17. 0 m elevation of Shenyang Station is calculated and analyzed by the finite element software SAP2000 according to the three-dimensional spatial model. The overall calculation model is shown in Fig. 16.17a. Among them, the frame element is used to simulate the reinforced concrete beams and columns, the upper and lower chords and web members of steel trusses; the high precision quadrilateral space thin shell element is used to simulate the reinforced concrete slabs, which can accurately describe the mechanical characteristics of reinforced concrete slabs and walls, and the deformation of reinforced concrete slabs and walls in and out of the plane can be considered simultaneously. The attributes of the materials used are taken in accordance with the specifications, and the specific values are shown in Table 16.8. In addition to the self-weight of the structure, the constant and live loads on the top chord surface of the floor are taken as 3.5 kN/m^2 according to the structural scheme of the design institute, in which the constant loads include the floor self-weight, the building surface, etc., and the live loads include the use loads; the decorative ceiling is installed on the bottom chord surface of the floor, and the constant load is 1.0 kN/m^2; and the mass source is 1.0*dead load + 0.5*live load for structural modal analysis. Through modal analysis, the first four modes of the structure are obtained as shown in Fig. 16.17b–e, and the first 10 frequencies are shown in Table 16.8. It can be seen that the first order vertical natural frequency of the structure is 2.83 Hz, which is close to the main frequency of walking load.

16.2.3.2 Structural Vibration Reduction Design

After optimum design, 16 sets of vibration absorbers were arranged on the floor, and the specific location is shown in Fig. 16.18. The shock absorber is divided into two types. The black block is TMD1 and the red block is TMD2. Each vibration absorber

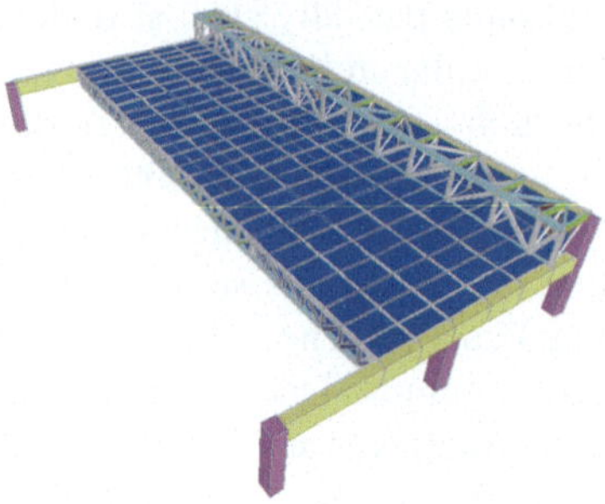

(a) Finite element model

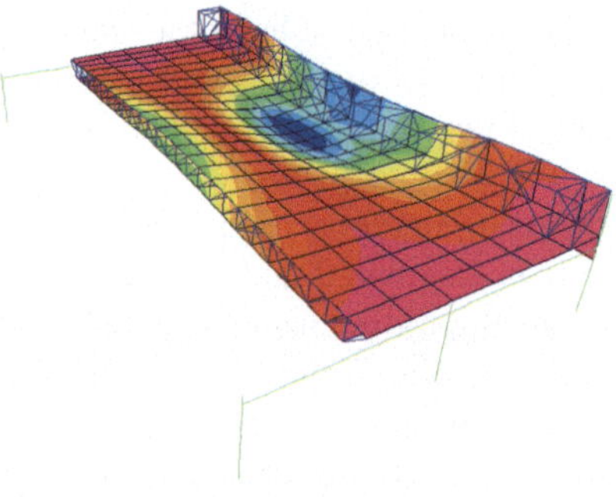

(b) 1st order mode shape

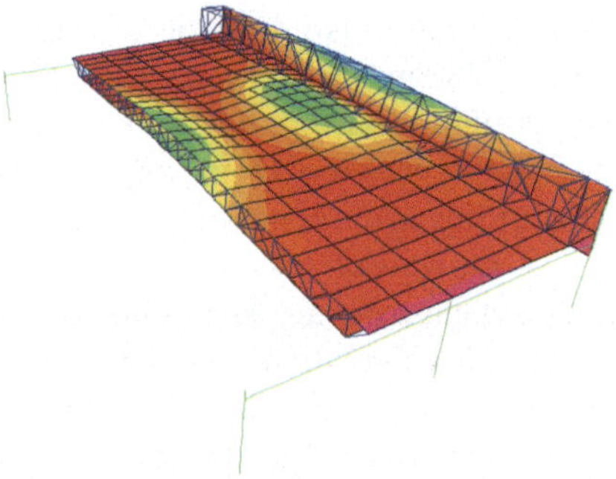

(c) 2nd order mode shape

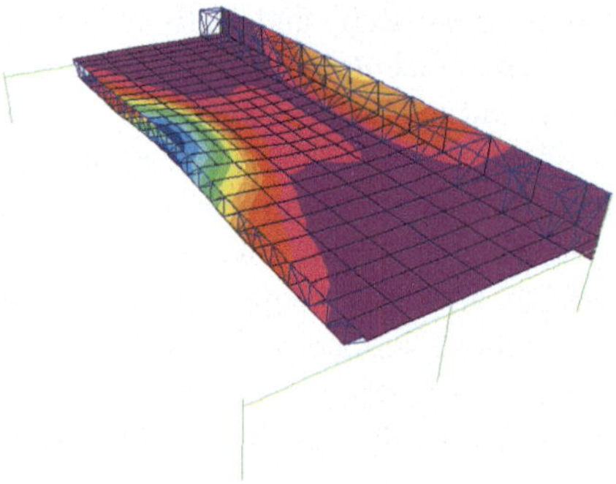

(d) 3rd order mode shape

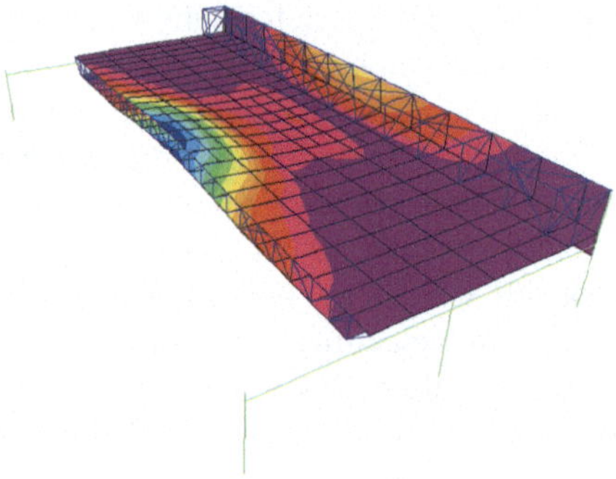

(e) 4th order mode shape

Fig. 16.17 Finite element model of Shenyang Railway Station and the first four vertical vibration modes

Table 16.8 First 10 natural frequencies of floor

Order	Frequency (Hz)	Main directions	Order	Frequency (Hz)	Main directions
1	2.83	Vertical	6	4.56	Torsion
2	3.23	Vertical	7	4.61	Vertical
3	3.57	Vertical	8	4.79	Vertical
4	3.66	Horizontal	9	5.43	Vertical
5	3.83	Vertical	10	5.6	Vertical

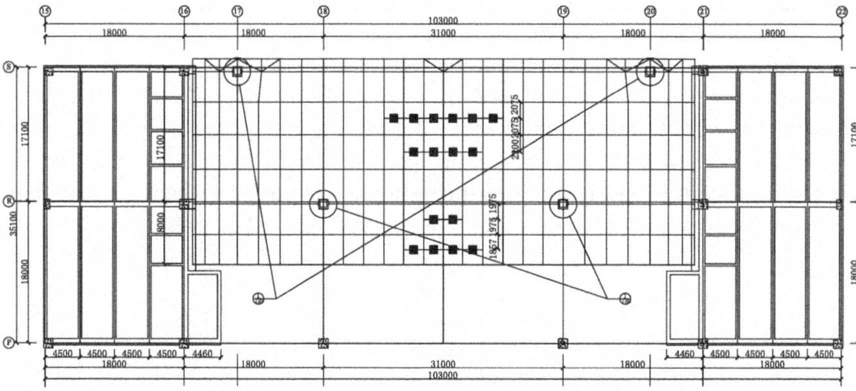

Fig. 16.18 TMD layout diagram of Shenyang Railway Station Floor

consists of a viscous damper and a frequency modulated mass damper, including four spring dampers, one viscous damper and several connectors, universal hinges, etc. The parameters of the shock absorber are shown in Table 16.9. The parameters in the table were derived from the repeated optimization results.

Table 16.9 Calculating parameters of vibration absorption system

Type of vibration absorber	Total spring stiffness (N/m)	Mass block (kg)	Tuning frequency (Hz)	Damper parameters			
				Damping index	Damping coefficient (N s/m)	Maximum stroke (mm)	Maximum output force (kN)
TMD1	416,568 ± 15%	1000	3.25	1	C = 1500	±30	0.91
TMD2	287,505 ± 15%	1000	2.7				

Note Considering the errors between the calculation model and the actual model, the spring stiffness in the table is adjustable (±15% of the calculated value)

16.2.3.3 Analysis of Structural Comfort Control

The frame columns on both sides and the steel trusses connected with them form the restraint to the vertical vibration of the floor, which is equivalent to the support. The floor near the support has a small vibration amplitude. From the modal analysis, it is also known that the range of vertical vibration of floor mainly concentrates on the middle part of the span, so the load arrangement is shown in Fig. 16.19, and the shadow part in the figure is the loading range of walking load. The walking load with random step frequency and starting phase distributed uniformly in the loading range is equivalent to the walking load with synchronous frequency modulation. The working condition definition is shown in Table 16.10.

According to the above definition of analysis condition, the dynamic response of the structure under load was analyzed using the simulated load curve. The peak acceleration of the structure is shown in Table 16.11. The plane distribution of the peak acceleration of the floor is shown in Fig. 16.20. It can be seen that the acceleration of the floor decreases significantly after vibration reduction. The plane distribution of the peak acceleration is close to the vertical vibration mode of the floor.

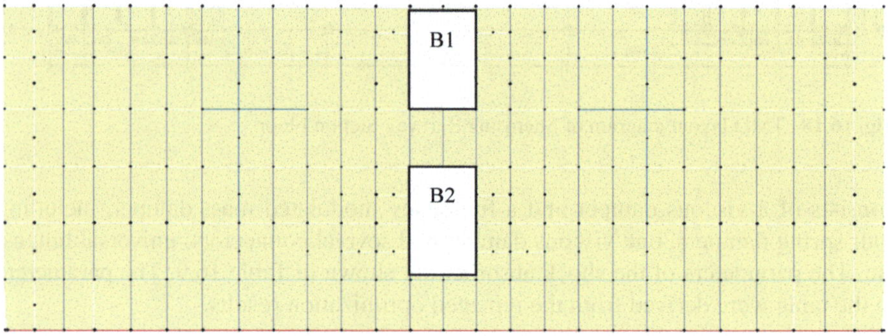

Fig. 16.19 Arrangement location of crowd load at Shenyang Railway Station

Table 16.10 Definition of analytical work

Layout position		B1	B2	B
Load condition	Frequency (Hz)			
Pedestrian load (A)	2.3 (Fast walking)			Condition A1
	2.8 (Fast running)			Condition A2
	3.0 (Near natural frequency)			Condition A3
	3.2 (Near natural frequency)			Condition A4
20 jumpers (B))		Condition B1	Condition B2	

Note The B in the table represents the whole floor, B1 and B2 are shown in Fig. 16.18

Table 16.11 Peak acceleration of floor under various working conditions (Unit: m/s^2)

Condition	Condition A1	Condition A2	Condition A3	Condition A4	Condition B1	Condition B2
Original structure	0.083	0.128	0.192	0.242	0.291	0.188
Damping structure	0.062	0.104	0.128	0.143	0.191	0.132
Vibration reduction rate (%)	25.30	18.75	33.33	40.91	34.36	29.79

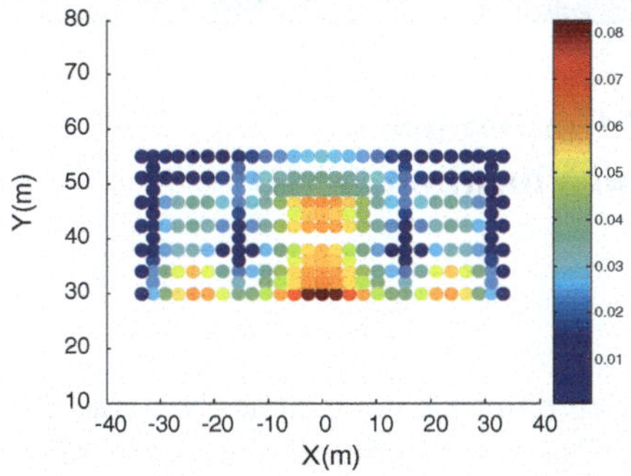

(a) Condition A1 (before vibration reduction)

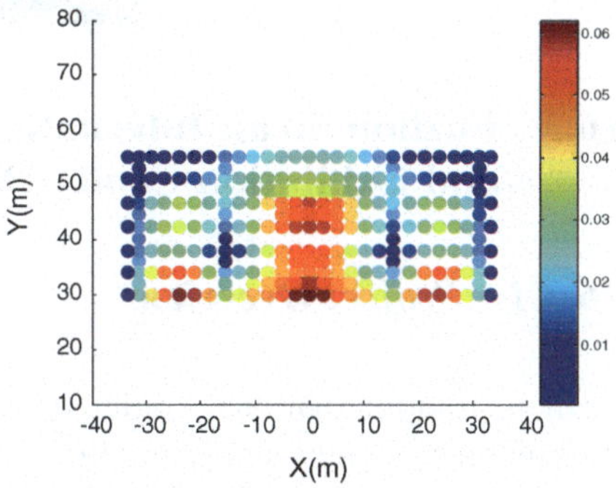

(b) Condition A1 (after vibration reduction)

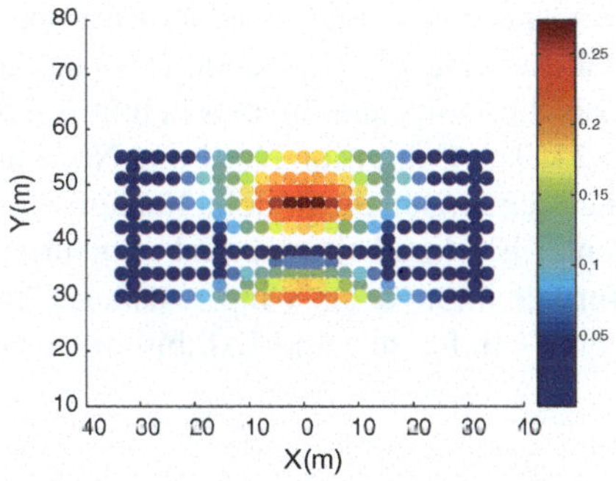

(c) Condition B2 (before vibration reduction)

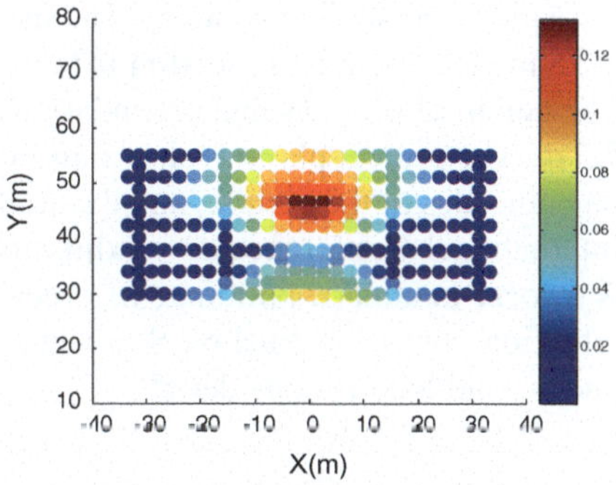

(d) Condition B2 (after vibration reduction)

Fig. 16.20 Planar distribution of peak acceleration of floor of Shenyang Railway Station

Fig. 16.21 Fuzhou Strait International Conference Center

16.3 Fuzhou Strait International Conference and Exhibition Center (Wind Vibration, TMD)

16.3.1 Project Overview

Fuzhou Strait International Conference and Exhibition Center includes No. 1 and No. 2 Conference Center and Exhibition Center. The two exhibition centers are 468 m long and 130 m wide, adopting primary and secondary steel pipe truss structure and truss pipe intersection welding. The main truss space is 12 m, secondary truss space is 18 m. The main truss section is trapezoidal, secondary truss section is triangular. The height of primary and secondary trusses is the same, and the axis height is 4.5–5.5 m. The project photo is shown in Fig. 16.21. The Exhibition No. 1 and No. 2 are on both sides of the picture, and the conference center is in the middle. Fuzhou Strait International Conference and Exhibition Center is a large-span spatial structure. It is a local landmark building. In order to further improve its wind resistance, the vibration control of Fuzhou Strait International Conference and Exhibition Center under wind load is considered.

16.3.2 Structural Vibration Reduction Design

Considering the characteristics of vibration mode and response of the structure, the MTMD system was set up in the large span roof to control the wind-induced vibration response of the structure. The location of TMD is shown in Fig. 16.22. There were 64 TMD1 and 60 TMD2, and the specific parameters of each TMD are shown in Table 16.12.

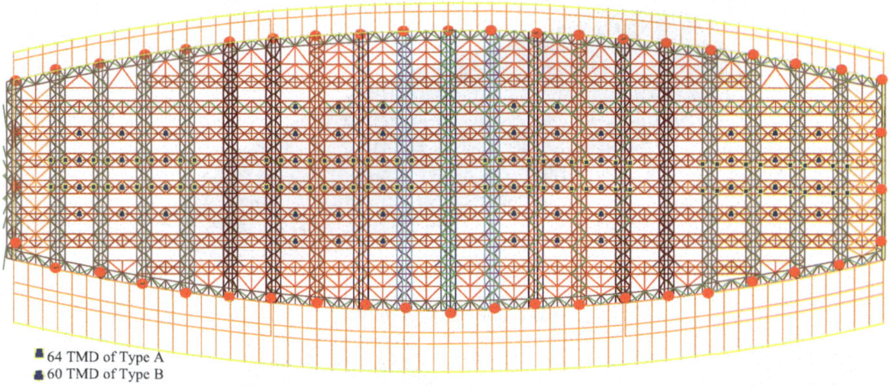

Fig. 16.22 Location of MTMD system in Fuzhou Strait International Conference Center

Table 16.12 MTMD system parameters

Serial number	Spring stiffness (kN/m)	Tuning quality (kg)	Tuning frequency (Hz)	Damping coefficient (Ns/m)	Damping index α
TMD1	60	480	1.78	5000	1.0
TMD2	120	440	2.63	5000	1.0

16.3.3 Structural Analysis Model

The professional finite element software Midas was used to carry out the finite element modeling as shown in Fig. 16.23a, and then the modal analysis was carried on. The typical four orders of modes and their periods are shown in Fig. 16.23b–e.

16.3.4 Comparative Analysis of Wind-Induced Vibration of the Structure

The wind tunnel test of rigid model pressure measurement was carried out for Fuzhou Strait International Conference and Exhibition Center by the Wind Engineering Research Center of Hunan University. The wind pressure time histories of 312 points were measured by wind tunnel tests. There are 203 measuring points on the roof surface of Exhibition Hall No. 1, which were numbered as 1–203. 109 measuring points on the 2-layers curved side wall of Exhibition Hall No. 1 (6 and 14 m elevation), which were numbered as 204–312. The time-domain signals with 6600 data of wind pressure were recorded at each wind direction angle at each measuring point, and a total of about 190 million data were recorded. Based on the wind load data provided by Hunan University, six wind load time history modes,

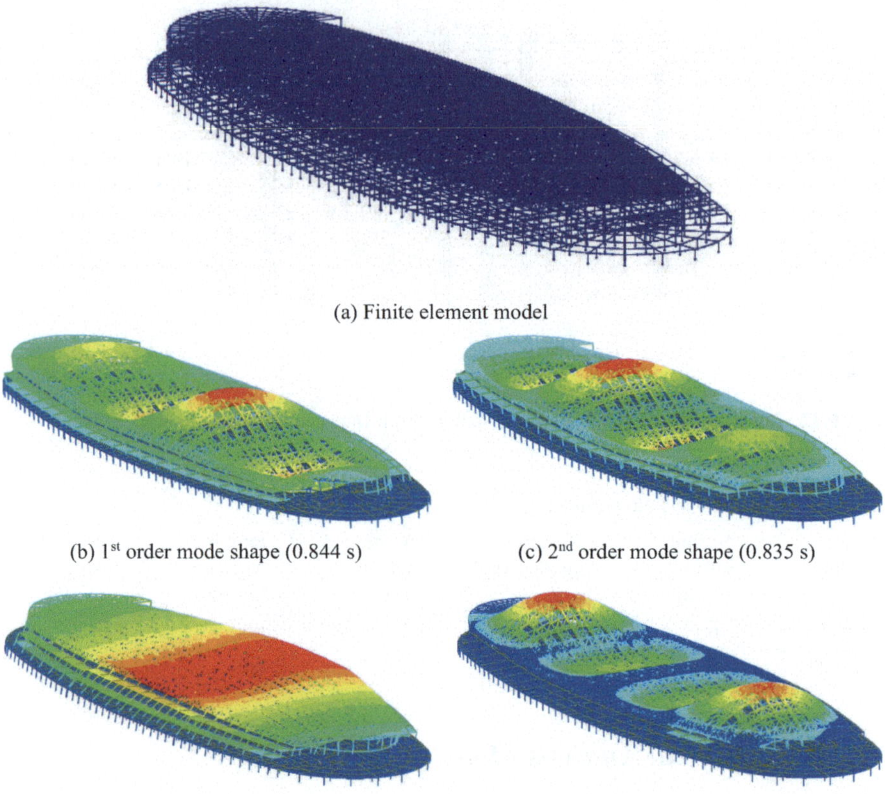

(a) Finite element model

(b) 1st order mode shape (0.844 s)

(c) 2nd order mode shape (0.835 s)

(d) 3rd order mode shape (0.783 s)

(e) 4th order mode shape (0.737 s)

Fig. 16.23 Finite element model and vibration mode of Fuzhou Strait International Conference Center

namely, W00 (0°), W45 (45°), W90 (90°), W135 (135°), W270 (270°) and W315 (315°), were selected as the analysis directions. The wind pressure time history data of each point were extracted. The time interval of wind pressure data was 0.1196 s, totaling 6600 data points and 789.36 s. Based on the test results, the wind load simulation and wind vibration response analysis of the structure were carried out. The wind-induced displacement responses of typical joints under different working conditions are shown in Table 16.13. It can be seen that the maximum vibration reduction rate is 29.33%, but the response of individual joints is slightly enlarged. Further, the time history and peak value distributions of nodal responses are given in Figs. 16.24 and 16.25, respectively. It can be seen that the magnitude of structural response is related to the magnitude of its vertical mode displacement. However, when MTMD system was installed, the overall wind-induced vibration response of the structure decreased significantly.

Table 16.13 Comparison of displacement responses of typical joints under different working conditions

Condition	Serial number of node	Before vibration reduction	After vibration reduction	Vibration reduction rate (%)
W00	3983	26.61	22.85	14.13
	3963	41.68	32.37	22.34
	10,029	20.1	17.43	13.28
	4278	31.36	22.38	28.64
	2459	27.45	19.40	29.33
	4258	29.40	22.26	24.29
W45	3983	18.21	14.94	17.96
	3963	25.66	20.24	21.12
	10,029	14.53	11.69	19.55
	4278	18.81	14.80	21.32
	2459	15.79	12.62	20.08
	4258	18.11	13.41	25.95
W90	3983	16.36	13.36	18.34
	3963	19.33	14.13	26.90
	10,029	13.00	10.72	17.54
	4278	15.37	13.20	14.12
	2459	13.10	11.21	14.43
	4258	17.34	12.67	26.93
W135	3983	10.91	11.77	−7.88
	3963	13.86	10.58	23.67
	10,029	9.672	9.954	−2.92
	4278	16.05	14.21	11.46
	2459	14.09	12.60	10.57
	4258	9.885	10.09	−2.07
W270	3983	14.6	11.38	22.05
	3963	17.91	14.54	18.82
	10,029	11.22	8.86	21.03
	4278	19.36	17.61	9.04
	2459	16.7	15.53	7.01
	4258	16.62	14.52	12.64
W315	3983	18.75	14.16	24.48
	3963	24.64	20.08	18.51
	10,029	14.52	10.93	24.72
	4278	23.18	19.2	17.17
	2459	19.45	16.7	14.14
	4258	20.6	16.95	17.72

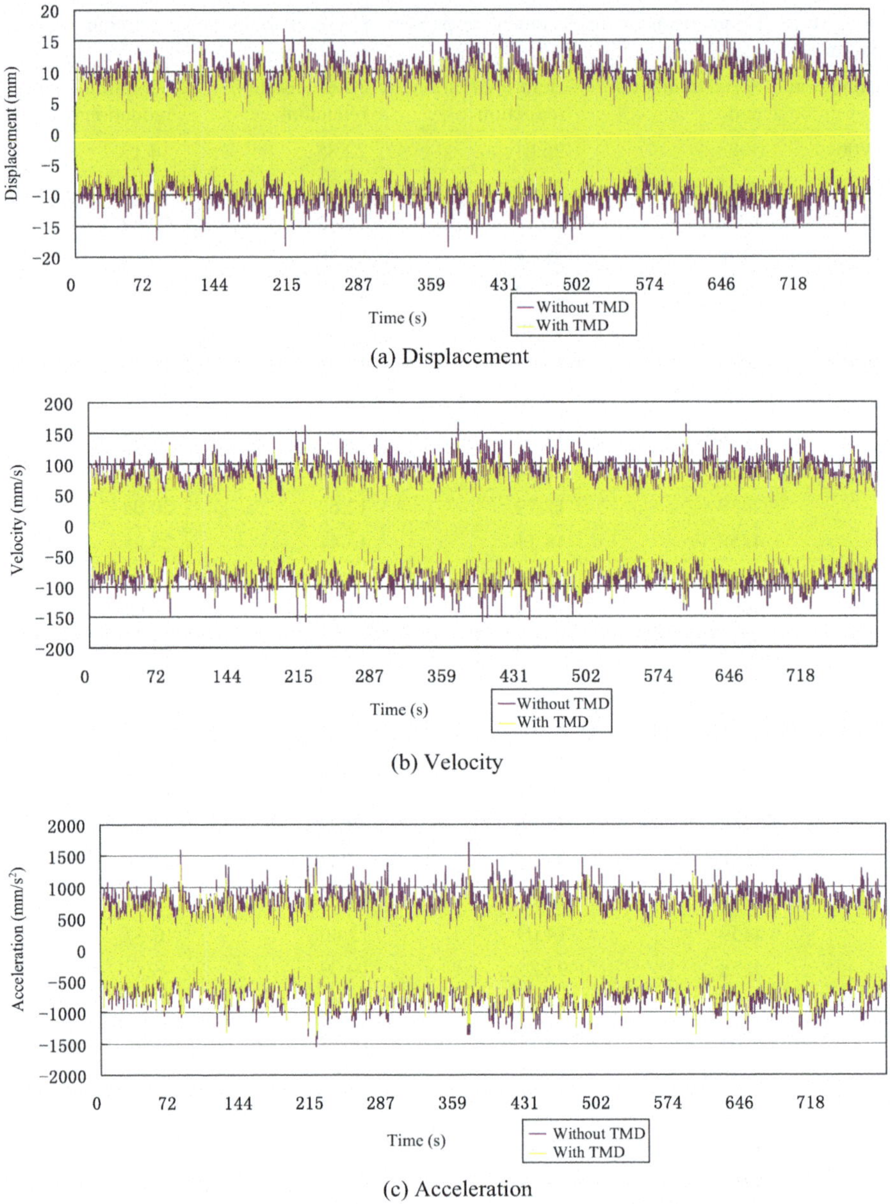

Fig. 16.24 Dynamic response time history of Node 3983 of Fuzhou Strait International Conference Center under condition W45

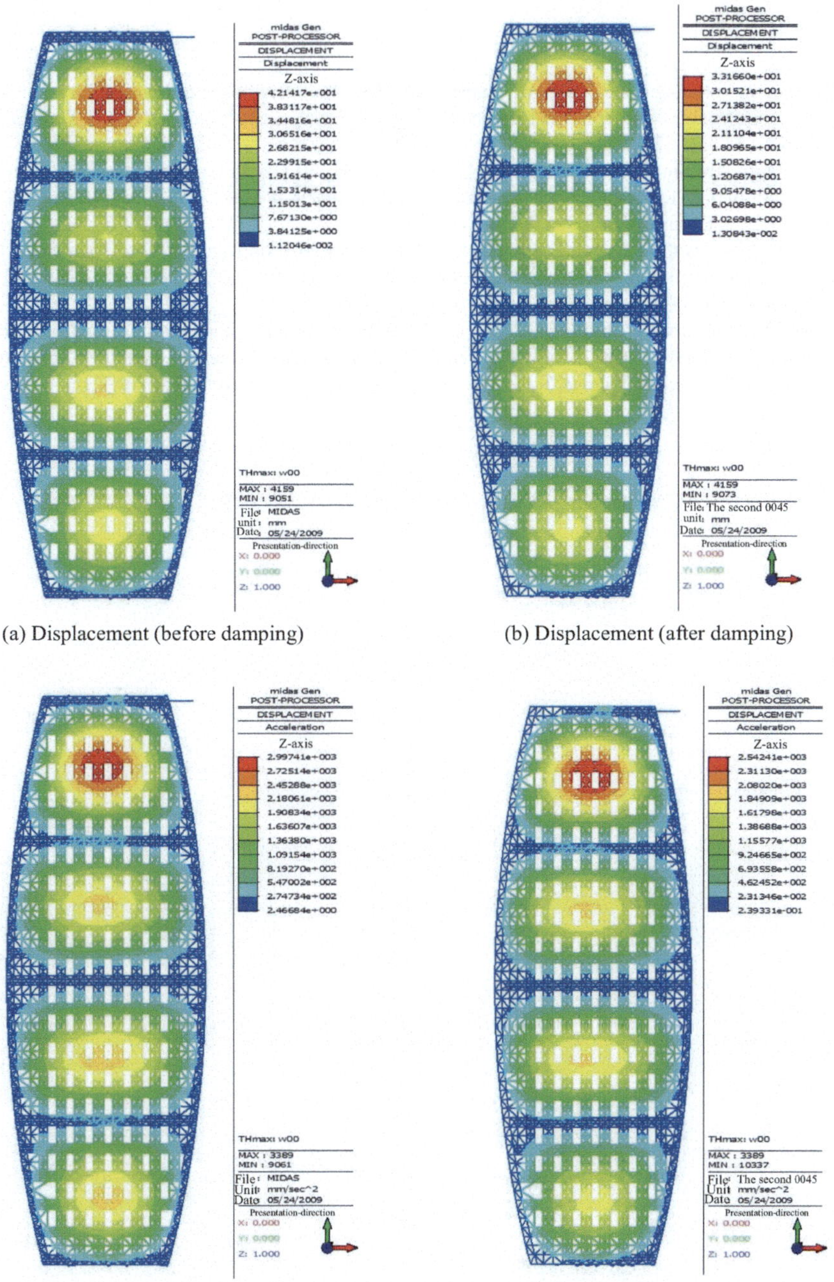

(a) Displacement (before damping)   (b) Displacement (after damping)

(c) Acceleration (before damping)   (d) Acceleration (after damping)

Fig. 16.25 Peak response distribution of Fuzhou Strait International Conference Center under condition W00

References

1. Chen, Xin, Youliang Ding, Zhiqiang Zhang, et al. 2012. Investigations on serviceability control of long-span structures under human-induced excitation. *Earthquake Engineering and Engineering Vibration* 11 (1): 57–71.
2. Chen, Xin, Ai-Qun Li, Zhi-Qiang Zhang, et al. 2017. Hybrid optimization of the multiple tuned mass dampers in long-span floor for human comfort. *Journal of Vibration Engineering* 30 (5): 827–836. (in Chinese).
3. Cao, Li-Lin, Ai-Qun Li, and Xin Chen. 2011. Vibration control of a long-span floor in a large station room considering human comfort. *Journal of Disaster Prevention and Mitigation Engineering* 31 (1): 75–79. (in Chinese).
4. Chen, Xin, Aiqun Li, Zhiqiang Zhang, et al. 2010. Design and analysis on vibration control of long-span floor in large station room under stochastic crowd-induced excitation. *Journal of Southeast University (Natural Science Edition)* 40 (3): 543–547. (in Chinese).

The manufacturer's authorised representative in the EU is Springer Nature Customer Service Centre GmbH, Europaplatz 3, 69115 Heidelberg, Germany. If you have any concerns regarding our products, please contact ProductSafety@springernature.com

Printed and bound by CPI Group (UK) Ltd, Croydon, CR0 4YY
30/09/2026
02225602-0018